Intermediate Dynamics

This advanced undergraduate physics textbook presents an accessible treatment of classical mechanics using plain language and clear examples. While comprehensive, the book can be tailored to a one-semester course. An early introduction of the Lagrangian and Hamiltonian formalisms gives students an opportunity to utilize these important techniques in the easily visualized context of classical mechanics. The inclusion of 321 simple in-chapter exercises, 82 worked examples, 550 more challenging end-of-chapter problems, and 65 computational projects reinforce students' understanding of key physical concepts and give instructors freedom to choose from a wide variety of assessment and support materials. This new edition has been reorganized. Numerous sections were rewritten. New problems, a chapter on fluid dynamics, and brief optional studies of advanced topics such as general relativity and orbital mechanics have been incorporated. Online resources include a solutions manual for instructors, lecture slides, and a set of student-oriented video lectures.

Patrick Hamill has taught physics at San José State University for over 30 years. During that time he was honored by student organizations for teaching excellence and was named a "President's Scholar" for his research activities in atmospheric science. He received the NASA Ames Julian Allen award for his studies of the role of polar stratospheric clouds in the formation of the ozone hole over the Antarctic. Professor Hamill has published over 100 peer-reviewed papers. He is the author of the Cambridge University Press text, *A Student's Guide to Lagrangians and Hamiltonians*.

Intermediate Dynamics

Second Edition

PATRICK HAMILL

San José State University, California

CAMBRIDGE
UNIVERSITY PRESS

CAMBRIDGE
UNIVERSITY PRESS

University Printing House, Cambridge CB2 8BS, United Kingdom

One Liberty Plaza, 20th Floor, New York, NY 10006, USA

477 Williamstown Road, Port Melbourne, VIC 3207, Australia

314-321, 3rd Floor, Plot 3, Splendor Forum, Jasola District Centre, New Delhi – 110025, India

103 Penang Road, #05–06/07, Visioncrest Commercial, Singapore 238467

Cambridge University Press is part of the University of Cambridge.

It furthers the University's mission by disseminating knowledge in the pursuit of education, learning, and research at the highest international levels of excellence.

www.cambridge.org
Information on this title: www.cambridge.org/highereducation/isbn/9781009098472
DOI: 10.1017/9781009089494

First published in 2008 by Jones and Bartlett Publishers, Inc.
Second edition 2022

Printed in the United Kingdom by TJ Books Limited, Padstow Cornwall

A catalogue record for this publication is available from the British Library.

ISBN 978-1-009-09847-2 Hardback

Additional resources for this publication at www.cambridge.org/hamill

Contents

Preface

Although this book begins at an introductory level, by the end of the book the student will have been exposed to all of the subject matter usually found in an intermediate mechanics course as well as a few advanced topics.

Organization

This book is divided into six parts. Part I is called "Kinematics and Dynamics." This part (Chapters 1–4) covers kinematics in various coordinate systems, the dynamical theory of Isaac Newton, the equations of motion given by Newton's second law, and the equations of Lagrange and Hamilton. Chapter 1 consists of a review of a few essential introductory concepts. This chapter can be skipped by well-prepared students, or assigned as reading for students who only need a quick refresher. The next chapter (Chapter 2) is called "Kinematics." This is the traditional starting point for courses in intermediate mechanics. Here, the student is exposed to relations between acceleration, velocity, and position in Cartesian, plane polar, cylindrical, and spherical coordinates. A few simple concepts from vector analysis are introduced. A number of reasonably difficult projectile problems are included in the problems. The next chapter (Chapter 3) considers Newton's laws. It includes a discussion on "Determining the Motion" in which the student learns techniques for integrating Newton's second law to obtain the position as a function of time. This is done for constant forces, and for forces that are functions of time, of velocity, and of position. A short section called "Numerical Solutions" gives a flavor for the use of computational techniques in physics. The role of computers in physics is not emphasized in this course. However, I realize that many instructors want to expose their students to computational methods, so I have included a few discussions of numerical techniques. Furthermore, nearly every chapter has a number of "Computational Projects." However, this text does not stress the role of computers in physics because I find that teaching the traditional material of intermediate mechanics takes most of two semesters and does not give enough time to delve into computational physics. Furthermore, many universities have included a computational physics course in the undergraduate physics curriculum. The next chapter (Chapter 4) is called "Lagrangians and Hamiltonians." I think it is important for physics students to be exposed to these concepts early on in their study of mechanics. The Lagrangian is presented, at first, as a simple technique for generating the equation of motion. Later in the chapter, I go through a derivation of the Lagrange equations using the calculus of variations. This section need not be covered if the instructor feels it is too advanced. The chapter ends with a discussion of the Hamiltonian and Hamilton's equations. There are several reasons why I chose to present the Lagrangian early in the course, perhaps the most important being that it gives the student a simple (almost "cookbook") technique for obtaining the equations of motion for a complicated dynamical system. For example, the Lagrangian technique allows one to determine the equations of motion for a double pendulum, for a spherical pendulum, or for coupled oscillators. More importantly, it allows one to introduce the concepts of generalized momentum and ignorable coordinates and leads to the relation between conservation laws and symmetries. Furthermore, it lets the student know that this course is not simply a rehash of concepts learned in introductory physics.

Part II (Chapters 5–8) is denoted "Conservation Laws" in which conservation principles are treated in depth. These chapters cover the conservations of energy, linear momentum, and angular momentum (Chapters 5, 6, and 7), followed by a short chapter (Chapter 8) on the relation between symmetry and conservation laws. Chapter 5, on the conservation of energy, discusses potential energy and the use of energy diagrams. Potential energy naturally leads to a discussion of the gradient of a scalar field. There is a section on the way the "Del"operator can be expressed in cylindrical and spherical coordinates; this allows one to discuss coordinate transformations in general. (I believe that introducing concepts from vector calculus as required by the physics is more effective than stuffing all of the vector concepts into a single introductory chapter.) Chapters 6 and 7, on the conservation of linear and angular momentum, cover the usual topics (rockets, collisions, etc.) as well as some less usual topics such as the fact that angular momentum is an axial vector.

Part III is called "Gravity" and consists of two chapters. The first one (Chapter 9) deals with Newtonian gravity and is an introduction to field theory. The study is limited to considerations of the gravitational field, because field theory is treated exhaustively in courses on electromagnetism. Additional vector concepts are introduced here and the student is exposed to Gauss's law and the equations of Poisson and Laplace. I felt that a chapter on gravity would be incomplete if Einstein's contributions were ignored. The topic is rather forbidding from a mathematical point of view, but I attempted to present it in a way that would make sense to students at this stage of their development. Nevertheless, I expect that many instructors will prefer not to discuss this admittedly superficial analysis, so I have labeled the section "Optional." The next chapter (Chapter 10) deals with central force motion in a gravitational field, as illustrated by the Kepler problem. Also considered is the stability of circular orbits, showing the student how to deal with small perturbations. Specifically, we imagine a comet striking a planet in a previously perfectly circular orbit and analyze the planet's subsequent motion to determine stability conditions and the frequency of radial oscillations.

Part IV (Chapters 11–14) is called "Oscillations and Waves." In Chapter 11 damped and driven harmonic oscillators are treated in depth. A rather thorough discussion on how to solve second order differential equations is included here. Coupled oscillators and normal modes are considered. Chapter 12 is on the motion of a pendulum. We begin with the motion of a simple pendulum of arbitrary amplitude and introduce elliptic integrals. Next we consider the physical pendulum, centers of oscillation and percussion, the spherical pendulum, and the conical pendulum. To spend a whole chapter on the pendulum may seem excessive, but it is a simple, easily visualized physical system that allows one to introduce many useful mathematical techniques without having to spend time explaining the motion. The next chapter in this part (Chapter 13) is an introduction to wave motion. This topic is not considered in great detail because it is treated extensively in the undergraduate electromagnetic theory class. Nevertheless, the student will receive a reasonably complete overview of mechanical waves. Sound waves require knowledge of fluid dynamics and are left to Chapter 19. The last chapter in Part IV is an analysis of small oscillations. It is rather advanced and the chapter is denoted as optional.

Part V (Chapters 15–17) is called "Rotation." Portions of the material in this part may be too advanced for some classes, but the instructor will probably want to cover Chapter 15 and some topics in rotational kinematics and dynamics. The first chapter in this part (Chapters 15) is called "Accelerated Reference Frames" in which we (mainly) consider motion on the surface of the rotating Earth. Coriolis forces and the Foucault pendulum are treated. Perturbation theory is used to solve these problems. Chapter 16 (on rotational kinematics) introduces orthogonal transformations. We obtain the Euler angles and consider Euler's theorem. The following chapter (Chapter 17) on rotational dynamics introduces the inertia tensor and some simple methods from tensor analysis.

The last part of the book is called "Special Topics" and consists of four chapters (18–21). The first chapter in this part (Chapter 18) is a fairly advanced study of statics, including a discussion of d'Alembert's principle and the concept of virtual work. The next chapter (Chapter 19) is called "Fluid Dynamics and Sound Waves." Logically, this material could be treated immediately after Chapter 13, but the mathematics gets fairly complicated so I placed it near the end of the book and marked it as "optional." The instructor may wish to cover the section on hydrostatics and skip the rest of the chapter. The next chapter (Chapter 20) "The Special Theory of Relativity" is an introduction to special relativity, and the final chapter (Chapter 21) "Classical Chaos" is a brief introduction to chaos. These two chapters are simply intended to give the student a flavor of these interesting subjects.

A One-Semester Course

Many instructors will find that the intermediate mechanics course in their department has been reduced to one semester. In such a situation it is impossible to cover all of the material in this book. From a personal perspective I feel that the essential material is covered in Parts I and II and Chapters 11, 15, and 16.

Exercises and Problems

Learning physics requires doing physics, so I have included a large number of "exercises." These are found at the end of nearly every section. Most of them are fairly easy. Some are merely "plug-ins" to get the student to look at a formula and (hopefully) to think about it. Others ask the student to fill in the missing steps in a derivation. A few require a bit of clever thinking. Nearly all have answers given. I hope that students studying this book will solve every one of these exercises. At the end of each chapter is a collection of problems that are of the degree of difficulty to be expected from a course at this level. Many of these will require significant effort on the part of the student. However, I believe that a student who has read the chapter and worked the exercises will be prepared to attack the problems.

Acknowledgments

I thank my colleagues at San Jose State University and NASA Ames Research Center, particularly Dr. Alejandro Garcia and Dr. Michael Kaufman. I am especially indebted to the many students in my mechanics courses whose influence on this book cannot be overestimated.

Part I

Kinematics and Dynamics

1 A Brief Review of Introductory Concepts

You already know a lot of physics and quite a bit of mathematics. You have been exposed to introductory courses in Mechanics, Electromagnetism, Thermodynamics, and Optics. You are now beginning your studies of these topics in a much more profound and rigorous manner. I hate to tell you this, but you are expected to know those concepts from the introductory courses! I know how easy it is to forget definitions and mathematical relationships if you are not using them all the time. So this chapter is a brief review of some of the concepts from your introductory mechanics course that you will be using in this intermediate level course. (I have included only those concepts that are absolutely necessary.) If the brief explanations in this chapter are not sufficient, please go back to your introductory physics textbook and review the material there. The standard introductory physics texts are well written and contain many instructive figures and diagrams. It is a good idea to refer to that text whenever you are exposed to the same material on a more advanced level.[1]

Just as in introductory physics, you will begin the upper-division physics sequence with this course in classical mechanics. This is the most beautiful of all physics courses (at least in my opinion!). It is also probably the most useful of all physics courses as it is the basis of essentially all advanced physics. The brilliant physicists who developed Quantum Mechanics, Relativity, Statistical Mechanics, etc., were experts in Classical Mechanics. The Lagrangian (which you will study in Chapter 4) is the basis of Elementary Particle Physics and the Hamiltonian (also in Chapter 4) is fundamental in Quantum Mechanics.

As you may recall, the mechanics section of your introductory physics book covered the following topics:

- Kinematics
- Newton's second law
- Work and energy
- Momentum
- Rotational motion.

We now very briefly review some concepts from each of these items.

1.1 Kinematics

Kinematics is the study of motion. Essentially, kinematics involves determining the relationships between position, velocity, acceleration, and time.

---

[1] If you are a particularly well-prepared student and feel that you know the material in this chapter, I suggest that you go to the end of each section where you will find a few exercises. If you can solve them, then skip to the next section, but if you feel uncertain or even somewhat confused, read the section. You might also try solving a few of the problems at the end of the chapter.

Position is denoted by the vector[2] **r** and the change in position (or displacement) can be written as $\Delta\mathbf{r}$.

Velocity is defined as the displacement with respect to time, so the average velocity is given by

$$\langle\mathbf{v}\rangle = \frac{\Delta\mathbf{r}}{\Delta t},$$

where Δt is the time during which the object had a displacement $\Delta\mathbf{r}$.

As the time interval becomes very small, we replace the *difference* (represented by Δ) with the *derivative* and write

$$\mathbf{v} = \frac{d\mathbf{r}}{dt}. \tag{1.1}$$

The change in velocity with respect to time is called the acceleration and is given by

$$\mathbf{a} = \frac{d\mathbf{v}}{dt}. \tag{1.2}$$

We can use the definitions of acceleration and velocity to write the inverse relation:

$$\int d\mathbf{v} = \int \mathbf{a}dt.$$

Integrating we get

$$\mathbf{v} = \int \mathbf{a}dt + \text{constant}.$$

Integrating again,

$$\mathbf{r} = \int \mathbf{v}dt + \text{constant}.$$

These are vector relationships and are valid in any coordinate system. In Chapter 2 you will find the relations between acceleration, velocity, and position in various coordinate systems.

1.1.1 Motion in a Straight Line at Constant Acceleration

For motion in a straight line,[3] we do not need to use vector notation. Let the position to be represented by x. If the acceleration is constant the integrals above lead to the familiar relations

$$v(t) = at + v_0, \tag{1.3}$$

and

$$x(t) = \frac{1}{2}at^2 + v_0t + x_0, \tag{1.4}$$

where v_0 and x_0 are the initial velocity and the initial position.

You probably memorized Equations (1.3) and (1.4) in your introductory physics course. But don't forget that they are only valid if the acceleration is constant. In this course, we shall frequently be concerned with nonconstant accelerations and these relations cannot be used. The appropriate relations are derived in Chapter 3.

Two other useful relations can be obtained from Equations (1.3) and (1.4). Solving one equation for t and substituting it in the other yields

$$2a(x - x_0) = v^2 - v_0^2.$$

---

[2] In this book vectors are represented in *bold face* (**r**), whereas scalars are represented by *italics* (*r*).

[3] Motion in a straight line is an example of one-dimensional motion.

Similarly, if we solve one equation for a and substitute it in the other equation we obtain

$$x - x_0 = \frac{1}{2}(v + v_0)t.$$

Again I emphasize that these relations are valid only if the acceleration is constant.

> **Comment** In this book you will find a large number of "exercises." They are not difficult. They are intended to give you a chance to review and understand the concepts in the preceding section. It would be a good idea to work out the solution for each exercise as you read through the text. At the end of each chapter you will find a set of problems that are significantly more difficult than the exercises. However, if you solve the exercises, you will be well prepared for solving the problems. Many students find working the exercises to be a good preparation for the examinations.

Exercise 1.1

You were driving your new Ferrari at 62 mph (= 100 km/h) when you spotted a police car. Naturally, you hit the brakes. You slowed to 31 mph, covering a distance of 50 m. (a) What is your constant acceleration? (b) How much time did it take to slow to 31 mph? Answers: (a) -5.79 m/s^2, (b) 2.4 s. ∎

1.2 Newton's Second Law

The study of the relation between the forces acting on a body and the motion of the body is called dynamics. In your introductory course you were exposed to dynamics in the form of Newton's second law. That law states that a body of mass m acted upon by a force **F** will accelerate at

$$\mathbf{a} = \mathbf{F}/m.$$

This relationship is usually remembered as

$$\mathbf{F} = m\mathbf{a}. \tag{1.5}$$

As we will discuss in Chapter 3, this form of Newton's second law is only valid if the mass is constant. The relation $\mathbf{F} = m\mathbf{a}$ is usually applied to a force that is acting on a *point mass* (often referred to as a "particle"). However, it can also be applied to an extended rigid body. Then **a** is defined as the acceleration of the center of mass.

> **Worked Example 1.1** Determine the acceleration of a block of mass m sliding down an inclined plane of angle θ. Assume the coefficient of sliding friction is μ. See Figure 1.1.
>
> **Solution** The forces acting on the block are gravity mg (downwards), the normal force N (perpendicular to the plane), and the frictional force μN (parallel to the plane). These forces are illustrated in Figure 1.1(a). The *free-body diagram* with all of the forces acting at the center of mass of the block is shown in Figure 1.1(b). In a problem such as this one, it is convenient to assume the axes are parallel to and perpendicular to the surface of the plane as indicated in Figure 1.1(c).

There is no acceleration perpendicular to the plane so the net force in that direction must be zero. Consequently, $N = mg \cos\theta$. The net force down the plane is $F_d = mg \sin\theta - \mu N$. The acceleration of the block is

$$a = \frac{F_d}{m} = g \sin\theta - \mu g \cos\theta.$$

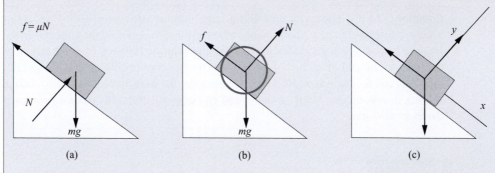

(a)   (b)   (c)

Figure 1.1 A block sliding down an inclined plane with friction. Sketch (a) shows the forces acting *on* the block. Sketch (b) shows the free-body diagram. Sketch (c) shows the *x*- and *y*-axes inclined so that they are parallel and perpendicular to the plane.

Exercise 1.2

Two blocks of mass M_1 and M_2 are tied together. They are sitting on a smooth frictionless surface as shown in Figure 1.2. A force F is applied to the free string attached to M_1. What is the tension in the string between the two blocks? Answer: $T = M_2 F/(M_1 + M_2)$. ■

Exercise 1.3

A block of mass 25 kg is held in place on an inclined plane of angle $30°$ as shown in Figure 1.3. The coefficient of static friction is 0.4. (a) Draw the free-body diagram. What forces act on the block? (b) What is the tension in the string? (c) If the string is cut, what is the acceleration of the block? Answers: (b) $T = 122.5$ N, (c) $a = 1.51$ m/s^2. ■

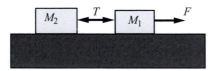

Figure 1.2 Two blocks on a smooth frictionless surface connected by a massless string.

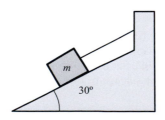

Figure 1.3 A block on an inclined plane.

1.3 Work and Energy

This section outlines some aspects of work and energy. An in-depth study is presented in Chapter 5.

Let us begin with the concept of work. Imagine pushing a box along a horizontal surface, such as a tabletop. The force you are applying can be denoted by $\mathbf{F}$. As you might expect, there will be opposing forces such as friction, air resistance, etc., but for now we are interested only in the force *you* are exerting. If you push the box from $\mathbf{r}_1$ to $\mathbf{r}_2$, the *work* you do on it is defined to be

$$W = \int_{\mathbf{r}_1}^{\mathbf{r}_2} \mathbf{F} \cdot d\mathbf{r}. \tag{1.6}$$

Power is defined as the work done per unit time. Suppose a force $\mathbf{F}$ acts on a particle for an infinitesimal time interval, from t to $t + dt$. The work done during this interval is $dW = \mathbf{F} \cdot d\mathbf{r}$. Therefore, the power is

$$P = \frac{dW}{dt} = \frac{\mathbf{F} \cdot d\mathbf{r}}{dt}.$$

Since $d\mathbf{r}/dt \equiv \mathbf{v}$, another expression for power is $P = \mathbf{F} \cdot \mathbf{v}$.

When we do work on an object, we usually change its *energy*. The mechanical energy of an object is either kinetic energy (energy of motion) or potential energy (energy of position). For a particle of mass m moving with speed v the kinetic energy (denoted by T) is given by

$$T = \frac{1}{2}mv^2. \tag{1.7}$$

The potential energy of an object of mass m raised a height h above the surface of the Earth is given by

$$V = mgh. \tag{1.8}$$

We shall often be interested in the potential energy of a stretched or compressed spring. The potential energy of such a system is given by

$$V = \frac{1}{2}kx^2, \tag{1.9}$$

where k is the spring constant and x represents the amount the spring is stretched or compressed.

There is a very important relationship between potential energy and force. Consider Equation (1.8), but express it as $V = mgz$, where z is the height above some reference surface and V is the potential energy of an object raised to height z in the gravitational field near the surface of the Earth. The force of gravity on the object is mg. Note that

$$F = \frac{dV}{dz} = \frac{d}{dz}(mgz) = mg.$$

Similarly, the force exerted by a spring is $F = kx$ and the potential energy stored in a spring is $V = \frac{1}{2}kx^2$. Consequently,

$$F = \frac{dV}{dx} = \frac{d}{dx}\frac{1}{2}kx^2 = kx.$$

Thus it would seem that force is the derivative of potential energy with respect to position. That is not quite correct. For one thing, the sign on F is wrong and for another thing, we obtained a scalar. That is, in our examples we just evaluated the magnitude of the force, but we know that force is a vector. The correct relationship between force and potential energy is

$$\mathbf{F} = -\nabla V. \tag{1.10}$$

This is discussed in detail in Chapter 5, but it is useful at this stage to remember that force and potential energy are related to one another. Forces that can be derived from a potential energy as given by Equation (1.10) are called conservative forces.

If the only work done on a system is due to conservative forces the sum of kinetic energy and potential energy is constant ($T + V =$ constant). This is called the law of conservation of energy.

Exercise 1.4

A rock is thrown upward from the top of a 30 m building with a velocity of 5 m/s. Determine its velocity (a) when it falls back past its original point, (b) when it is 15 m above the street, and (c) just before it hits the street. Answer: (a) -5 m/s, (c) 24.76 m/s. ∎

Exercise 1.5

A horse drags a 100 kg sled a distance of 4 km in 20 min. The horse exerts one horsepower, of course. What is the coefficient of sliding friction between the sled and the ground? Answer: $\mu_k = 0.23$. ∎

1.4 Momentum

In Chapter 6 you will encounter a detailed study of linear momentum. Here I just want to remind you of a few facts from introductory mechanics.

A moving particle is characterized by having a particular *momentum*. When we use the term "momentum" we usually are referring to the linear momentum, not to be confused with the angular momentum, which we will define in a little while.

Momentum is a vector defined as mass times velocity and is denoted by the letter **p**. Thus,

$$\mathbf{p} = m\mathbf{v}. \tag{1.11}$$

If the mass of a body is constant, the time derivative of the momentum is

$$\frac{d\mathbf{p}}{dt} = \frac{d(m\mathbf{v})}{dt} = m\frac{d\mathbf{v}}{dt} = m\mathbf{a} = \mathbf{F}.$$

In Equation (1.5) we wrote $\mathbf{F} = m\mathbf{a}$ and called it Newton's second law. But that is only valid if the mass is constant. We now appreciate that a more general expression of Newton's second law is

$$\mathbf{F} = \frac{d\mathbf{p}}{dt}. \tag{1.12}$$

This relationship is valid even if the mass is not constant. In fact, it is the most general statement of Newton's second law.

The force **F** appearing in Newton's second law is the net or total *vector sum* of all forces acting on the body. Consequently, we appreciate that if the net force is zero, the time derivative of the momentum is zero. That is, the momentum of the body is constant. This is called the law of conservation of momentum.

Exercise 1.6

A 1500 kg car traveling East at 40 km/h turns a corner and speeds up to a velocity of 50 km/h due North. What is the change in the car's momentum? Answer: 26 700 kg m/s at 38.7° West of North. ∎

1.5 Rotational Motion

1.5.1 Rotational Kinematics

The motion of a rigid body rotating about a *fixed axis* is mathematically identical to one-dimensional linear motion. Recall that kinematics is a study of the relationship between position, velocity, acceleration, and time. Rotational kinematics deals with angular position (θ), angular velocity (ω), and angular acceleration (α), where angular velocity is defined by

$$\omega \equiv \frac{d\theta}{dt},$$

and angular acceleration is

$$\alpha \equiv \frac{d\omega}{dt} = \frac{d^2\theta}{dt^2}.$$

You will discover in Chapters 7, 16, and 17 that rotational motion can be very complicated. To keep things simple for the moment, consider the special case of a *symmetrical* body rotating about a *fixed* axis, such as the wheel illustrated on the left side of Figure 1.4.

For a fixed, stationary axis, the center of the wheel is at rest. All other points are moving in circles around it. If you looked straight down the axis you would see the circle shown on the right side of Figure 1.4. Point P is on the rim of the wheel. The angular position of P is given by the angle between some fixed line and the radius vector to P. If the wheel is turning, after a time dt point P will have moved a distance ds to P'. Recall from geometry that $ds = r\,d\theta$, where $d\theta$ (in radians of course!) is the angle subtended by the arc $PP' = ds$. Point P moves with speed

$$v = \frac{ds}{dt} = \frac{r\,d\theta}{dt} = r\omega.$$

The speed of point P is called the "tangential speed" because instantaneously P is moving tangent to the rim of the wheel. It is sometimes convenient to write the tangential speed as v_T. Then

$$v_T = r\omega. \tag{1.13}$$

Taking the time derivative of v_T yields the tangential acceleration a_T,

$$a_T = \frac{dv}{dt} = \frac{d}{dt}(r\omega) = r\frac{d\omega}{dt},$$

where we used the fact that r is constant. But $\frac{d\omega}{dt} = \alpha$, so

$$a_T = r\alpha. \tag{1.14}$$

This is the relationship between the tangential acceleration a_T and the angular acceleration α.

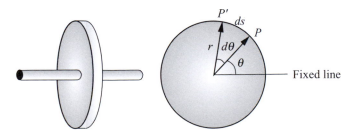

Figure 1.4 A wheel mounted on a fixed axis. The wheel is allowed to rotate but not to translate.

It is often convenient to express angular velocity and angular acceleration as vectors (ω) and (α) directed along the axis of rotation.

Exercise 1.7

A wheel initially spinning at $\omega_0 = 50.0$ rad/s comes to a halt in 20.0 seconds. Determine the constant angular acceleration and the number of revolutions it makes before stopping. Answers: -2.5 rad/s^2, 79.6 rev.   ■

1.5.2 Rotational Dynamics

Rotational dynamics is the analysis of the motion of a body subjected to external *torques*.

Consider a body constrained to rotate about a fixed axis, as shown in Figure 1.5. I drew the body in the shape of a plane lamina for simplicity. Let a force **F** act on the body at a point on its rim. Let us assume (again for simplicity) that **F** is perpendicular to the axis of rotation. The point of application of **F** is specified by the vector **r** whose origin is at the axis of rotation.

The body cannot accelerate linearly because the axis is fixed. The applied force causes the body to rotate about the axis. The tendency of a force to cause a rotation is called the *moment* of the force, or more commonly, the *torque*. Just as you can think of a force as a pull, you can think of a torque as a twist.

The ability of a force to produce a rotation depends not only on the *magnitude* of the force, but also on its *direction* and on the *location* of the point where the force is applied to the body.

To define torque draw a line having the direction of the force and passing through the point where the force is applied. This is called the "line of action of the force." (See Figure 1.6.) Next draw a line that starts at the axis of rotation and intersects the line of action at a 90° angle. This

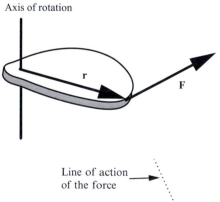

Figure 1.5 Illustration of torque. The laminar body is free to rotate around the fixed axis of rotation. The vector **r**, with origin at the axis, specifies the point of application of the force.

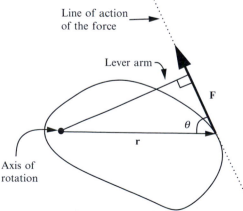

Figure 1.6 Definition of lever arm. The axis of rotation is perpendicular to the plane of the figure.

line is called the *lever arm.* The lever arm is the shortest distance from the axis of rotation to the line of action of the force. Its length is $r \sin\theta$ where θ is the angle between $\mathbf{r}$ and $\mathbf{F}$. The magnitude of the torque is given by the relation:

Torque = Force times Lever Arm.

This elementary definition of torque is complicated and cumbersome. Fortunately, vector concepts allow us to simplify the definition. Torque ($\mathbf{N}$) is the cross product of $\mathbf{r}$ and $\mathbf{F}$, thus:

$$\mathbf{N} \equiv \mathbf{r} \times \mathbf{F}. \tag{1.15}$$

In Equation (1.15) $\mathbf{r}$ is the vector from the axis of rotation to the point of application of the force. The direction of the torque is perpendicular to the plane defined by $\mathbf{r}$ and $\mathbf{F}$ and it is usually represented as a vector along the axis of rotation.

Just as a force produces a linear acceleration, a torque produces an angular acceleration. The rotational version of $\mathbf{F} = m\mathbf{a}$ is $\mathbf{N} = I\boldsymbol{\alpha}$, where I is a constant that depends on the mass and geometry of the body, as well as the location of the axis of rotation. It is called the "moment of inertia." For example, the moment of inertia of a particle of mass m a distance r from the axis of rotation is given by $I = mr^2$, and the moment of inertia of a sphere about a diameter is $I = \frac{2}{5}mr^2$.

Exercise 1.8

Show that a given force applied at any point along the line of action yields the same torque. ∎

Exercise 1.9

An athlete is holding a 2.5-meter pole by one end. The pole makes an angle of 60° with the horizontal. The mass of the pole is 4 kg. Determine the torque exerted by the pole on the athlete's hand. (The mass of the pole can be assumed to be concentrated at the center of mass.) Answer: Torque = 24.5 N m. ∎

1.6 Statics

This section is an overview of some of the basic concepts of statics. Chapter 18 is a more advanced analysis of the statics of a rigid body and Section 19.2 considers the statics of fluids (called "hydrostatics").

Consider a rigid body. Newton's second law tells us that if no net external force acts on the body, it will move at constant linear velocity. That is, zero force implies zero acceleration. Consequently, a body that is *not* accelerating must have *no* net external force acting on it. Similarly, a body that is rotating at a constant angular velocity has zero angular acceleration. Such a body has no net external *torque* acting on it.

A body that has no linear acceleration and no angular acceleration is said to be in *equilibrium.* The conditions for equilibrium are:

$$\sum \mathbf{F}_i = 0, \tag{1.16}$$

and

$$\sum \mathbf{N}_i = 0, \tag{1.17}$$

where $\mathbf{F}_i$ and $\mathbf{N}_i$ are the *external* forces and torques acting on the system.

A body in equilibrium has constant linear velocity and constant angular velocity. A particular but important special case occurs when both the linear and angular velocity are *zero*. This is called *static equilibrium*.[4]

The general conditions for equilibrium are given by Equations (1.16) and (1.17). I will state these conditions in words for a situation in which all of the forces applied to a body lie in the same plane (but it is not difficult to generalize). If a body is in equilibrium then:

(1) the algebraic sum of the components of the forces in each of two mutually perpendicular directions is zero, and,
(2) the algebraic sum of the torques about any point in the plane of the forces is zero.

The first condition guarantees the body is not accelerating. The second condition will be proved in Chapter 7, but we might note that if the body is *not rotating* the body can be considered to be not rotating about an axis through any arbitrary point. This fact is very useful for solving simple statics problems because it allows us to take the torques about an axis through any convenient point. This point does not even have to be in the body.

Worked Example 1.2 A steel beam of length 4 m and weight 300 N is bolted to two pillars that are 1.5 m apart. The leftmost bolt is a distance 0.2 m from the end of the beam. A weight $W = 600$ N is hanging from the free end of the beam. Determine the tension (or compression) acting on the bolts. See Figure 1.7.

Solution The beam is in equilibrium so

$$\text{forces up} = \text{forces down}$$

and

$$\text{torques clockwise (CW)} = \text{torques counterclockwise (CCW).}$$

The first condition leads to

$$300 + 600 = F_1 + F_2.$$

Placing the axis of rotation at the end of the beam, the second condition leads to

$$600 \times 4 + 300 \times 2 = F_1 \times 0.2 + F_2 \times 1.7.$$
$$2400 + 600 = (900 - F_2) \times 0.2 + F_2 \times 1.7.$$

Therefore, $F_2 = 1880$ N (up) and $F_1 = -980$ N (down).

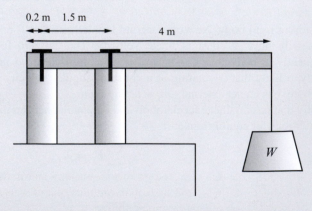

Figure 1.7 What forces are exerted on the bolts that hold the beam on the pillars?

---

[4] For fluids the term "equilibrium" implies static equilibrium.

1.7 Rotational Kinetic Energy

An extended rigid body can be considered to be a collection of particles. If the body is rotating about a fixed axis, then all the particles are moving in circles about the axis and the total rotational energy is the sum of the kinetic energies of all the particles due to the rotation. The linear velocity v of any particle is related to the angular velocity of the body, ω, according to $v = \omega r$, where r is the distance from the axis of rotation to the particle. The kinetic energy of the particle is $T = \frac{1}{2}mv^2 = \frac{1}{2}m(\omega r)^2$ and the total rotational kinetic energy is

$$T = \sum_i \frac{1}{2}m_i r_i^2 \omega^2.$$

The quantity $\sum_i m_i r_i^2$ is the moment of inertia of the body, I.[5] Therefore the rotational kinetic energy can be expressed as

$$T = \frac{1}{2}I\omega^2. \tag{1.18}$$

Note the similarity to the translational kinetic energy $T = \frac{1}{2}mv^2$.

Exercise 1.10

A disk of mass M and radius R is initially at rest. It is acted upon by a constant torque N for a time T. Determine the final angular velocity of the disk and the work required to spin it up to this final state. If the energy was supplied by a motor, what was the average power output of the motor? Answer: Average power $= N^2 T / M R^2$. ∎

1.8 Angular Momentum

The angular momentum of a particle with linear momentum $\mathbf{p}$ is defined to be

$$\mathbf{l} = \mathbf{r} \times \mathbf{p}.$$

The direction of $\mathbf{l}$ is perpendicular to both $\mathbf{r}$ and $\mathbf{p}$, as given by the right-hand rule.

The magnitude of the angular momentum, from the rule for cross products is

$$l = |\mathbf{l}| = |\mathbf{r} \times \mathbf{p}| = |\mathbf{r} \times m\mathbf{v}| = rmv \sin\theta,$$

where θ is the angle between the vectors $\mathbf{r}$ and $\mathbf{v}$.

For an extended rigid body, the total angular momentum is

$$\mathbf{L} = \sum \mathbf{l}_i = \sum (\mathbf{r}_i \times \mathbf{p}_i).$$

The relationship between the net external torque acting on a body and the rate of change of the angular momentum of the body is

$$\mathbf{N}_{\text{net}} = \frac{d\mathbf{L}}{dt}.$$

(I don't want to prove this now; the proof is a bit long and is found in Section 7.3.) If the net torque acting on a system is zero, the angular momentum of the system is constant. This is called the law of conservation of angular momentum.

---

[5] In Section 7.5 we will define the moment of inertia more generally and in Chapter 17 we introduce the inertia tensor.

1.9 Rotational Equivalents

It is interesting to note that one can obtain expressions for rotational quantities from the expressions for the corresponding linear quantities by replacing m with I and v with ω. We saw an example of this in a previous section where

$$\frac{1}{2}mv^2 \quad \Longrightarrow \quad \frac{1}{2}I\omega^2.$$

Similarly, the expression for linear momentum ($p = mv$) leads to the following expression for angular momentum (L).

$$\mathbf{p} = m\mathbf{v} \quad \Longrightarrow \quad \mathbf{L} = I\omega. \tag{1.19}$$

In like manner, Newton's second law for linear motion leads to the following rotational equivalent:

$$\mathbf{F} = m\mathbf{a} \quad \Longrightarrow \quad \mathbf{N} = I\alpha, \tag{1.20}$$

and, more generally,

$$\mathbf{F} = \frac{d\mathbf{p}}{dt} \quad \Longrightarrow \quad \mathbf{N} = \frac{d\mathbf{L}}{dt}.$$

Exercise 1.11

A spinning ice skater speeds up by pulling her hands to the side of her body. Approximate the ice skater by a cylinder of radius 10 cm and mass 30 kg. Her hands are two point masses of 0.25 kg each and her (massless) arms extend 90 cm from her torso (or 1 m from the axis of rotation). If she was initially rotating at 2 rev/s, what is her final angular velocity? Answer: 8.4 rev/s. ∎

1.10 Summary

The position of a particle is given by the vector $\mathbf{r}$. Velocity is defined by

$$\mathbf{v} = \frac{d\mathbf{r}}{dt},$$

and acceleration is defined by

$$\mathbf{a} = \frac{d\mathbf{v}}{dt}.$$

Newton's second law for a particle with constant mass is

$$\mathbf{F} = m\mathbf{a},$$

and the general form is

$$\mathbf{F} = \frac{d\mathbf{p}}{dt}.$$

Work is defined by the integral of the dot product of force and displacement,

$$W = \int \mathbf{F} \cdot d\mathbf{r}.$$

Kinetic energy is given by

$$T = \frac{1}{2}mv^2,$$

for linear motion, and

$$T = \frac{1}{2}I\omega^2,$$

for rotational motion.

The potential energy of an object raised a height h above the Earth's surface is

$$V = mgh.$$

The potential energy stored in a spring is

$$V = \frac{1}{2}kx^2.$$

Conservation of energy tells us that if all forces are conservative, then

$$T + V = \text{constant}.$$

The linear momentum of a particle of mass m moving at velocity $\mathbf{v}$ is defined as

$$\mathbf{p} = m\mathbf{v}.$$

The relation between angular velocity and linear velocity is given by the cross product:

$$\mathbf{v} = \omega \times \mathbf{r},$$

where ω is a vector pointing along the axis of rotation whose magnitude is equal to the angular velocity.

Torque is defined by

$$\mathbf{N} = \mathbf{r} \times \mathbf{F}.$$

The angular momentum of a particle can be written as

$$\mathbf{l} = \mathbf{r} \times \mathbf{p}.$$

The rotational version of Newton's second law is

$$\mathbf{N} = \frac{d\mathbf{l}}{dt}.$$

If the torque is zero, the angular momentum is constant.

1.11 Problems

Problem 1.1 Two ships are sailing in a thick fog. Initially, ship A is 10 miles North of ship B. Ship A sails directly East at 30 miles per hour. Ship B sails due East at constant speed v_B then turns and sails due North at the same speed. After two hours, the ships collide. Determine v_B.

Problem 1.2 You carefully observe an object moving along the x-axis and determine that its position as a function of time is given by

$$x(t) = 2t - 3t^2 + t^3.$$

(a) What is the position at time $t = 2$ s?
(b) What is the velocity at time $t = 2$ s?

(c) What is the acceleration at time $t = 2$ s?

(d) How far did it travel between times $t = 0$ and $t = 2$ s? (Note: Distance, not displacement! It might be helpful to plot x vs. t for $0 \leq t \leq 2$.)

Problem 1.3 The first train leaves the station and accelerates at a constant rate to its maximum speed of 100 km/h, reaching this speed at a distance of 2 km from the station. Five minutes later, a second train leaves the station and accelerates to 100 km/h in 4 km. What is the distance between the two trains when they both reach maximum speed?

Problem 1.4 A small helicopter is trying to land on a barge in the ocean. The propeller delivers an upward force of 36 000 N and the helicopter is observed to be descending at a constant safe speed of 3 m/s when it is 100 m above the barge. But suddenly there is a malfunction and the upward force is reduced to 30 000 N.

(a) What is the mass of the helicopter?

(b) What is the acceleration of the helicopter after the malfunction?

(c) Assume this acceleration is maintained constant during the final descent. What is the speed of the helicopter when it contacts the barge?

Problem 1.5 You travel a distance d in time t. (a) If you travel at speed v_1 for half the time and at v_2 for the other half of the time, what is your average speed? (b) If you travel at speed v_1 for half the distance and at speed v_2 for the other half of the distance, what is your average speed?

Problem 1.6 There is a long straight road out in the desert and it goes through a small town that has just one police car. The police car accelerates at 2 m/s^2 until it reaches a maximum speed of 200 km/h. A car full of escaped criminals speeds through the town at its top speed which is 150 km/h. The police car, starting from rest, gives chase. How far from the town do the police catch up to the criminals?

Problem 1.7 A police car is at rest on the side of the road when a wild teenager comes speeding by at 75 miles per hour. The police car starts immediately and accelerates at 8 miles per hour per second. At that same moment the teenager steps on the gas, but his car only accelerates at 2 miles per hour per second. (a) How far from the starting point does the police car overtake the speeder? (b) How fast are they going at that time? (c) Why is the speed you calculated for the police car unrealistic?

Problem 1.8 A brick is on a wooden plank that is resting on a table. One end of the plank is slowly raised so that it forms an angle θ with the horizontal table top. When $\theta = 60°$ the brick starts to slide down the plane. (a) Draw the free-body diagram. (b) Determine the coefficient of static friction between brick and plank. (c) If the coefficient of sliding friction is one half of the coefficient of static friction, determine the acceleration of the block.

Problem 1.9 A block of mass M is on an inclined plane of angle θ with a coefficient of sliding friction μ. At a given instant of time the block is at some point P on the plane and is moving up the plane with a speed v_0. (a) Obtain the time for the block to reach its highest point relative to P. (b) Obtain the time for the block to slide back down to point P. (c) Obtain an expression for the velocity of the block at the time it returns to P.

Problem 1.10 A railgun is a device that uses electromagnetic forces to accelerate a body along a set of conducting rails. Assume a railgun accelerates some object directly upward at 60 m/s^2 for 1.5 s. The object then coasts upwards to some maximum altitude before falling back down. Determine the maximum altitude reached. Ignore air resistance.

Problem 1.11 Atwood's machine consists of two weights (M_1 and M_2) suspended at the ends of a string that passes over a pulley. Assume massless, inextensible strings and a frictionless pulley. Let $M_1 = 6$ kg and $M_2 = 5.5$ kg. The masses are released from rest. Determine the distance descended by the 6 kg mass when its velocity reaches 0.5 m/s.

Problem 1.12 (a) Determine the rotational kinetic energy of a wheel of your bicycle when your linear speed is 20 km/h. You may assume the wheel is a hoop of mass 1.5 kg and radius 30 cm. (b) Compare your result

Figure 1.8 An object hanging from a pivot.

with the translational kinetic energy of the wheel. (c) Is the equality of parts (a) and (b) just a numerical coincidence or is it always true? (d) Would the energies be equal if the wheel were a disk rather than a hoop?

Problem 1.13 The frictional force between water and seabed in shallow seas causes an increase in the day by about 1 ms/century. Determine the torque that causes this change. Assume the Earth is a uniform sphere.

Problem 1.14 A meter stick has a pivot at one end. It is found to be in static equilibrium when acted upon by three forces that act at different points along the meter stick and act in different directions. We conclude that the net torque about the pivot is zero. Prove that the net torque about any other (arbitrary) point is also zero.

Problem 1.15 To build the pyramids it was necessary to pull heavy stones up inclined planes. Suppose a 2000 kg stone was dragged up a $20°$ incline at a speed of 0.25 m/s by a gang of 20 laborers. The coefficient of kinetic friction between the stone and the incline was 0.4. How much power was exerted by each laborer?

Problem 1.16 Two wooden blocks, both of mass 3 kg, are sliding in the same direction at 2 m/s on a frictionless horizontal surface. Let $M_1 = M_2 = 3$ kg and assume their speed is 2 m/s. A third block of mass 1 kg is also sliding in the same direction at a speed of 10 m/s, and it collides with the trailing 3 kg block. The third block is covered with a sticky, gooey substance, so it sticks to the trailing block. This combination block catches up with and collides *elastically* with the leading 3 kg block. Determine the final speed of the leading block. (Note that this is a one-dimensional problem. You may find it easier to solve the problem if you insert numerical values sooner rather than later.)

Problem 1.17 You are driving at 60 mph (=100 km/h). Your car's wheels have a radius of 35 cm. (a) Determine the angular velocity of the wheels. (b) What is the angular displacement of a wheel when you travel 1 km? (c) If you slow down and stop in 1 km, what is the angular acceleration of the wheel?

Problem 1.18 Aeronautical engineers have developed "tip jet" helicopters in which small jet engines are attached to the tips of the rotor. One such helicopter is powered by two ramjets. For a ramjet to develop thrust, it needs to be moving through the air quite rapidly, and is not efficient until it is moving at about 1000 km/h. Assume the rotor diameter is 10 m. Determine the angular speed of the rotor when the ramjet is moving through the air at 1000 km/h.

Problem 1.19 Figure 1.8 shows an object of mass M that is hanging from a pivot at point P. The three segments have equal length. (a) Show that the object is in equilibrium. (b) Determine whether or not this is a stable equilibrium orientation.

Problem 1.20 A solid ball of mass M and radius R rolls down an inclined plane. (a) What is its translational speed when it has descended a vertical distance h? (b) Determine its translational kinetic energy and its rotational kinetic energy.

Problem 1.21 A disk of mass 72 kg and radius 50 cm is rotating at 2000 rpm. (a) Determine its angular momentum. (b) If acted upon by a retarding force of 20 N acting tangent to the rim of the disk, determine the time required to stop the disk.

Computational Projects

The following problems can be solved easily if you know a computer language such as Matlab, Python, Fortran, or C++. A few can also be solved using a spreadsheet program such as Excel.

Although some of the computational projects in this book can be solved analytically, they will require a significant amount of brainless labor that can be done much more conveniently using a computer.

Computational Project 1.1 The position of a particle as a function of time is given by $x = 5t^3 - 2t$ (in meters). Plot the position as a function of time for the interval $t = -5$ s to $t = +10$ s. Using the relationship $\bar{v} = \Delta x / \Delta t$, obtain the average velocity at intervals of 1 second. Plot the average velocity as a function of time and, on the same graph, plot the analytical expression for the velocity.

Computational Project 1.2 The velocity of a certain particle is given by

$$v = 120 \left(1 - e^{-t/10}\right) + 0.5 \cos(t/2)$$

in meters per second. Plot the velocity as a function of time. Determine and plot the distance as a function of time by numerically adding up the area under the velocity vs. time curve.

Computational Project 1.3 This is a more realistic version of Problem 1.7. In that problem a teenager driving at 33.5 m/s (~75 mph) speeds past a parked policeman. The policeman and the teenager then accelerate at given constant rates and you are required to determine how far from the starting point the policeman catches up to the teenager. To make the problem somewhat more realistic, assume the teenager accelerates at $a_T = k_T e^{-b_T t}$ and the policeman accelerates at $a_P = k_P e^{-b_P t}$, where $k_T = 5.0$ mph/s and $k_P = 10$ mph/s. By plotting the positions as functions of time for various values of b_T and b_P obtain reasonable values for these constants. (Make sure your answers are reasonable. Eventually the policeman will catch the teenager, but your answer is not reasonable if the distance required is hundreds of miles!)

Computational Project 1.4 Electrons are randomly distributed in a small region of space. (The dimension is not important, but you can assume a cube with side 10 angstroms (Å).) Numerically determine the position of the center of mass, for 100, 1000, and 10 000 electrons. (You should use a random number generator to obtain the coordinates of each of the electrons.)

2 Kinematics

Recall that kinematics is the study of motion and, specifically, it is the study of the relationships between position, velocity, acceleration, and time.

Kinematics in one dimension is fairly simple because position, velocity, and acceleration can be treated as scalars. Motion in a straight line and rotation about a fixed axis are examples of one-dimensional motion.[1] Some two-dimensional problems, such as the motion of a projectile, can be resolved into two linked one-dimensional problems. However, in general, motion in two or three dimensions complicates the problem significantly because then you need to express the basic quantities as vectors.

In this chapter you will learn the relations between the position, velocity, and acceleration in the three main coordinate systems used in physics: Cartesian coordinates, cylindrical coordinates, and spherical coordinates.

Although some of the material presented in this chapter will be familiar to you, you will also find many new concepts. These concepts are used throughout the course, so please make sure you understand this chapter thoroughly. Be aware that many students find this material rather difficult.

I think it is important for you as a physicist to know something about the people upon whose shoulders you are standing, so in this book I have included a few sections marked "Historical Note." You will encounter very little physics in these sections, but you might find them interesting and helpful in placing a few famous physicists in historical context. The first historical note describes the life of the person who invented your profession.

2.1 Galileo Galilei (Historical Note)

Galileo Galilei (1564–1642) was a brilliant but difficult man whose studies of the physical world launched a scientific and cultural revolution. This cantankerous Italian genius was the first modern scientist. He rejected authority and based his conclusions on observation, experiment, and rational analysis. His ideas were opposed by the powers of the Church, and he ended his days under house arrest. Nevertheless, his view of the universe and his methods for discovering scientific truth eventually won out and today every physicist is an intellectual descendent of Galileo.[2]

Galileo did not invent the telescope, but he built one and was probably the first person to make a scientific study of what he saw in the sky. In short order, he discovered that the Moon has mountains and valleys, that the Milky Way is made up of individual stars, that Jupiter has satellites orbiting it, and that the Sun rotates and has spots on its surface. He became the first "celebrity scientist." He got into trouble, however, because he was outspoken and had little patience with people who put faith above reason. He was particularly offensive towards those who believed

---

[1] Rotational motion about an arbitrary (moving) axis is a more difficult problem and will be left to Chapters 16 and 17.
[2] An excellent biography of Galileo is *Galileo's Daughter* by Dava Sobel, Walker and Co., New York, 1999.

the Earth was the center of the universe and that the Sun, planets, and stars revolved around it. This "geocentric" universe was based on the philosophy of Aristotle which had been adapted to Christianity by Thomas Aquinas and was subsequently accepted as the official philosophy of the Church. Therefore, Galileo's outspoken attacks on the geocentric theory were considered by many to be heretical. Galileo was eventually called to Rome and ordered by Church authorities to neither teach nor defend the Copernican theory that the Sun is the center of the universe. Shortly thereafter, Galileo wrote a book in the form of a dialogue between three men. In this book he ridiculed the Aristotelians and even put some arguments used by the Pope in the mouth of a character named "Simplicio" who was pictured as a bit of a dunce. Galileo was called back to Rome by the Inquisition and was eventually condemned to life imprisonment, to take place in his own home. He was 70 years old at the time. He continued to work, publishing his last book, *Discourses and Mathematical Demonstrations Relating to Two New Sciences*, in 1638. This book describes much of the work in physics that he had carried out over the previous 30 years. He died at age seventy-eight.

Galileo's life is an instructive example of the conflict between a brilliant individual and the powers that rule society. There is no doubt that Galileo changed the world and our understanding of it. Some people condemn the Church authorities for trying to stop scientific progress; others are more understanding of the authorities who saw their entire world view threatened by a revolutionary iconoclast.

2.2 The Principle of Inertia

Among his many accomplishments, Galileo was arguably the first person to truly understand motion. He performed a series of experiments with pendulums of different lengths, as well as falling bodies and balls rolling down inclined planes. His experiments with these simple devices allowed him to grasp the essential aspects of motion, something that had eluded the greatest minds of previous ages. Prior to Galileo's work, many respected philosophers (including Aristotle) described motion in ways that modern scientists consider to be either meaningless or just plain silly. For example, they said, "Water flows downhill because it is trying to return to its natural place, which is the ocean." Or they said, "Smoke rises because smoke is air and the natural place for air is up in the sky." They explained the motion of the stars and planets with the preposterous statement that, "The celestial bodies are perfect, therefore they move about the Earth in perfect circles."

A crucial element in this erroneous theory of motion was the idea that an object will move to its "natural place" and then remain at rest in that place. This was called, of course, "natural motion." For something to move away from its natural place, it was necessary to give it a push or a pull. This kind of motion, requiring a force, they called "violent motion."

Galileo's most significant insight was that objects do *not* tend to come to rest, but rather they tend to *keep on moving*. In fact they tend to keep on moving at a constant speed in a straight line. (We call this *uniform motion*.) The property of material bodies to maintain their state of motion is called *inertia*. As you recall from your introductory course in physics, the law of inertia states that:

A body will remain in uniform motion as long as no net external force acts on it.

This basic principle has come to be known as Newton's first law. We will return to it later.

As we shall discuss in detail in Chapter 3, the law of inertia implies that the motion of the body is referred to a reference frame that is not accelerating. Such a reference frame is called an "inertial reference frame."

Although his equipment was crude, Galileo's experiments showed him that *time* is an essential component of motion. Mechanical clocks had not yet been invented, but Galileo devised a number of ingenious ways to estimate time intervals. He counted his pulse or watched the oscillations of a pendulum or weighed the amount of water dripping into a container.

Time is a very difficult concept to comprehend. Everyone has an intuitive, qualitative understanding of time. If you think about it, you will realize that your personal definition of time has some connection to the motion of material bodies. The length of a day is related to the rotation of the Earth, and a year is the Earth's orbital period. Most people describe an hour and a minute in terms of the motion of the hands on a clock. Scientists define the second in terms of the vibrations of certain molecules. Is it possible to define time without reference to motion? If all objects in the universe were perfectly still, would time exist? Newton avoided these philosophical questions by assuming time to be an absolute parameter that is continuously changing at a constant rate. Einstein gave an even simpler definition when he said, "Time is what you measure with a clock." We shall adopt Newton's point of view at present, then come back for a deeper look at the question of time when we consider Einstein's theory of relativity in Chapter 20.[3]

As mentioned above, kinematics is the study of the relationships between position, velocity, acceleration, and time. Galileo and his disciples grappled with these concepts. Fortunately, we can use many powerful mathematical tools that were not available to Galileo. For example, the calculus was invented by Newton and Leibniz after Galileo's death, and vector analysis was not developed until about 150 years ago. We will, of course, be taking full advantage of these techniques.

2.3 Basic Concepts in Kinematics

Let me remind you that the general relations between position ($\mathbf{r}$), velocity ($\mathbf{v}$), and acceleration ($\mathbf{a}$) are given by

$$\mathbf{v} = \frac{d\mathbf{r}}{dt} = \dot{\mathbf{r}},$$

$$\mathbf{a} = \frac{d\mathbf{v}}{dt} = \dot{\mathbf{v}} = \ddot{\mathbf{r}}.$$

One-dimensional motion refers to any kind of motion for which the position can be described in terms of a single parameter. For example, the position of a car driving down a curving road can be specified by its distance from the starting point. Similarly, the position of a bead sliding along a wire can be specified by the distance to the bead from some given point on the wire. A single number is enough to completely specify the position of the bead, no matter how the wire may be twisted or curled. However, the linear *distance* may not be the most convenient parameter to use. For example, the position of a bead sliding on a circular hoop might best be described in terms of an angle.

2.3.1 Motion in One Dimension with Constant Acceleration

As discussed in Section 1.1.1, if the acceleration is constant, we can integrate $dv = a\,dt$ to obtain

$$v(t) = v_0 + at, \tag{2.1}$$

---

[3] Some physicists believe that time, in the Newtonian sense of one event happening before another event, does not apply in the realm of quantum mechanics. For an interesting discussion of this problem see the article by Charles Seife, "Quantum Physics: Spooky Twins Survive Einsteinian Torture," *Science*, **294**, 1265 (2001).

and

$$x(t) = x_0 + v_0 t + \frac{1}{2} a t^2. \tag{2.2}$$

From these equations we can derive two additional relations:

$$x(t) - x_0 = \frac{1}{2}(v + v_0)t, \tag{2.3}$$

and

$$2a(x(t) - x_0) = v^2 - v_0^2. \tag{2.4}$$

I am sure you are very familiar with these equations from your introductory physics course. In fact, you probably memorized them. However, it is more important at this stage in your physics career to understand the process for *obtaining* these equations.

Exercise 2.1

Assuming a = constant, use the definitions of acceleration and velocity to obtain Equations (2.1) and (2.2), and from these obtain Equations (2.3) and (2.4). Are these expressions valid if the acceleration is not constant? ∎

I must emphasize that the formulas obtained above in Equations (2.1) through (2.4) are *only valid for constant acceleration.* If the acceleration is a function of time or velocity or position, then you *must* go through the entire procedure to obtain the correct expressions for $v(t)$ and $x(t)$.

2.3.2 Projectile Motion

A projectile is any object that is launched with some initial velocity and then moves under the action of the gravitational force which imparts to it a constant downward acceleration.

Ignoring air resistance, the horizontal motion is motion at constant velocity, and therefore

$$x = x_0 + v_{0x} t,$$

and the vertical motion is motion at constant acceleration, and therefore

$$y = y_0 + v_{0y} t - \frac{1}{2} g t^2,$$

where $g = 9.8$ m/s.

It is helpful to remember that at the top of the trajectory the vertical speed (v_y) is zero. This gives us a useful way to determine the time for the projectile to reach the top of its trajectory. Since $v_y = v_{0y} + a_y t$ we have

$$0 = v_{0y} - g t_{\text{top}}$$

so

$$t_{\text{top}} = v_{0y}/g.$$

Worked Example 2.1 Consider the projectile motion of a cannon ball. Assume the initial speed is v_0 and that the cannon is aimed at an angle θ above the horizontal. Determine the range.

Solution The *range* (R) is the horizontal distance traveled by the projectile along a flat horizontal plane; it is the value of x when $y = 0$. The initial velocity components are

$v_{0x} = v_0 \cos\theta$ and $v_{0y} = v_0 \sin\theta$. Because the total time of flight is twice the time required to reach the top of the trajectory, the range is given by $x(t = 2t_{\text{top}})$. It is easy to obtain a formula for the range in terms of the initial velocity as follows:

$$v_y = v_{0y} - gt.$$

But $v_y = 0$ when $t = t_{\text{top}}$ so $0 = v_{0y} - gt_{\text{top}} = v_0 \sin\theta - gt_{\text{top}}$. Therefore,

$$t_{\text{top}} = (v_0 \sin\theta)/g.$$

Consequently,

$$R = v_{0x}(2t_{\text{top}}) = 2v_0 \cos\theta \, (v_0 \sin\theta)/g,$$

$$R = 2\frac{v_0^2}{g} \sin\theta \cos\theta = \frac{v_0^2}{g} \sin(2\theta). \tag{2.5}$$

Our analysis of projectile motion made two assumptions: (1) the acceleration of gravity is constant, and (2) there are no horizontal forces acting on the projectile. These assumptions mean that we are assuming a flat, nonrotating Earth, and that there is no air resistance acting on the body. These seemingly preposterous assumptions (a flat, motionless, airless Earth!) illustrate an important technique in physics. When faced with a difficult problem, we solve a similar, simpler problem. Frequently, the solution to the simple problem is sufficiently accurate. The solution of the simple problem is often referred to as the "zeroth-order" approximation to the solution of the real problem. Later in this course you will learn techniques for including the factors that were ignored in the zeroth-order approximation to get results that are nearer and nearer to the exact solution. You will appreciate more fully how this technique works when you study projectile motion on a rotating Earth in Chapter 15. As you can imagine, a working physicist is continually faced with very complicated problems. It is part of a physicist's education to learn what is fundamentally important and what can be left until later as a refinement (or "higher-order approximation"). For example, you can estimate the trajectory of a thrown baseball with reasonable accuracy by assuming a flat, airless, nonrotating Earth, but if you want to calculate the trajectory of a rocket, those factors will have to be included.

Exercise 2.2

Determine the elevation angle for a projectile such that the range will be maximized. (Hint: The maximum value of a function is determined by setting its derivative to zero.) Answer: $45°$.

Exercise 2.3

An airplane traveling at 900 km/h at 5000 m altitude is directly over the target when it drops a bomb. How far from the target does the bomb hit? Answer: $\simeq 8$ km.

In working problems involving projectile motion in a uniform gravitational field, keep in mind that the horizontal motion has constant velocity (v_x = constant) and the vertical motion has constant acceleration (a_y = constant). Projectile motion is a combination of two independent one-dimensional motions. These are linked to one another through the time. If you eliminate the time from the equations, you obtain an equation for the *trajectory* of the projectile, as shown in the following example.

Worked Example 2.2 Prove that the path of a projectile in a uniform gravitational field is a parabola.

Solution The equation of a parabola whose axis is parallel to the y-axis in Cartesian coordinates is

$$y = ax^2 + bx + c.$$

In Section 2.3.2 we showed that for a projectile

$$y = -\frac{1}{2}gt^2 + v_{0y}t + y_0,$$

$$x = v_{0x}t + x_0.$$

Therefore,

$$t = \frac{x - x_0}{v_{0x}},$$

and consequently

$$y = -\frac{g}{2}\frac{(x - x_0)^2}{v_{0x}^2} + v_{0y}\frac{x - x_0}{v_{0x}} + y_0,$$

$$y = -\frac{g}{2v_{0x}^2}(x^2 - 2xx_0 + x_0^2) + \frac{v_{0y}}{v_{0x}}x - \frac{v_{0y}}{v_{0x}}x_0 + y_0,$$

$$y = -\frac{g}{2v_{0x}^2}x^2 + \left[\left(\frac{g}{2v_{0x}}2x_0 + \frac{v_{0y}}{v_{0x}}\right)\right]x + \left(-\frac{gx_0^2}{2v_{0x}^2} - \frac{v_{0y}}{v_{0x}}x_0 + y_0\right).$$

This has the required form as can more easily be appreciated if we set $x_0 = 0$ and $y_0 = 0$ to obtain

$$y = -\frac{g}{2v_{0x}}x^2 + \frac{v_{0y}}{v_{0x}}x.$$

Although projectile problems tend to be straightforward, sometimes it is necessary to use a more subtle approach, as illustrated in the following example.

Worked Example 2.3 A cannon that fires shells with a muzzle speed v_0 is mounted on the top of a cliff a height h above a level plain. Show that the angle at which the cannon should be aimed to give maximum range is $\theta = \sin^{-1}(v_0/\sqrt{2(v_0^2 + gh)})$.

Solution I am including this problem to show that a seemingly simple projectile problem can require considerable ingenuity to obtain a solution. The obvious approach would be to attempt to solve the problem by using the fact that if t is the time of flight (t_f), then the value of y is $-h$ (where we assume that the coordinate origin $(0,0)$ is at the cannon) and the value of x is R. That is,

$$R = (v_0 \cos\theta)(t_f),$$

and

$$-h = (v_0 \sin\theta)(t_f) - \frac{1}{2}g(t_f)^2.$$

You can solve the first equation for t_f, obtaining $t_f = R/(v_0 \cos \theta)$. Plugging into the second equation leads to

$$-h = R \tan \theta - \left(\frac{g R^2}{2 v_0^2} \right) \frac{1}{\cos^2 \theta}. \tag{2.6}$$

Now, you could solve this quadratic equation to obtain an expression for R as a function of θ. Then taking the derivative of R with respect to θ and setting it equal to zero gives an expression for θ that maximizes R. However, the resultant expression is so complicated that you will find it extremely difficult to solve for θ. (You might want to give it a try.)

However, the problem can be solved fairly simply by using a clever trick. Instead of solving for R and taking its derivative respect to θ, note that R depends on h, so

$$\frac{dR}{d\theta} = \frac{dR}{dh} \frac{dh}{d\theta}.$$

This expression will be zero if $\frac{dh}{d\theta} = 0$. Going back to the expression above for $-h$, Equation (2.6), we have

$$-\frac{dh}{d\theta} = \frac{d}{d\theta} \left[R \tan \theta - \left(\frac{g R^2}{2 v_0^2} \right) \frac{1}{\cos^2 \theta} \right] = 0,$$

$$0 = 1 - \frac{g R}{v_0^2} \tan \theta,$$

$$R = \frac{v_0^2}{g \tan \theta}.$$

Plugging this into Equation (2.6) yields

$$-h = \frac{v_0^2}{g} - \frac{g}{2 v_0^2 \cos^2 \theta} \left(\frac{v_0^2}{g \tan \theta} \right)^2 = \frac{v_0^2}{g} - \frac{v_0^2}{2g \sin^2 \theta}.$$

Therefore

$$\frac{v_0^2}{2g \sin^2 \theta} = \frac{v_0^2}{g} + h,$$

$$\sin \theta = \frac{v_0}{\sqrt{2(v_0^2 + gh)}}. \qquad \text{Q.E.D.}$$

2.3.3 Rotation About a Fixed Axis

Rotational motion about a fixed axis (as discussed in Section 1.5) is one-dimensional motion described by the variables θ, ω, and α, where

$$\omega = \frac{d\theta}{dt},$$

$$\alpha = \frac{d\omega}{dt} = \frac{d^2\theta}{dt^2}.$$

Worked Example 2.4 A physics student is carefully observing a wheel while it is being spin balanced. The wheel speeds up for 10 s and subsequently rotates at a constant angular speed of 50 rad/s. Going home, the student decides to write an expression for the acceleration of the wheel that would account for this behavior, and comes up with

$$\alpha = bte^{-ct} \ (\text{rad/s}^2).$$

Obtain an expression for the angular velocity (ω) as a function of time. What are reasonable values for b and c?

Solution Using the definition $\alpha = \frac{d\omega}{dt}$,

$$\int_{\omega_0}^{\omega(t)} d\omega = \int_{t=0}^{t} \alpha \, dt$$

$$\omega(t) - \omega_0 = \int_{0}^{t} bte^{-ct} \, dt.$$

Note that $\omega_0 = 0$. Integrating by parts,

$$\omega(t) = b \left[\frac{1}{c^2} e^{-ct} (-ct - 1) \right]_0^t$$

$$= -\frac{b}{c^2} \left[(e^{-ct})(ct + 1) - e^0(1) \right]$$

$$= \frac{b}{c^2} \left[1 - e^{-ct}(ct + 1) \right].$$

Recall that at time $t = 0$ the angular speed was zero. The speed increases rapidly for a time, and then asymptotically approaches a maximum value of b/c^2. If the time to spin up to this value is 10 seconds, then at $t = 10$ the term $e^{-ct} \simeq 0$, suggesting a value of c near unity. Let $c = 1$. The final angular speed of the wheel is 50 rad/s so $b/c^2 = 50$, and $b \simeq 50$. It is instructive to plot α vs. t and ω vs. t.

2.3.4 The Relation between Linear and Rotational Motion

Imagine a child has tied a string to a rock and is whirling it around in a circle. If the stone makes one revolution every T seconds, it has an angular speed given by

$$\omega = \frac{2\pi}{T}.$$

The stone also has a linear speed that is tangent to the circular path (as is observed if the child lets go of the string). If the angular speed is constant, the linear (or tangential) speed is given by

$$v_T = \frac{2\pi r}{T},$$

where r is the radius of the circular path. Comparing the last two equations we appreciate that the linear and angular speeds are related by

$$v_T = \omega r.$$

We can represent the angular *velocity* by a vector $\boldsymbol{\omega}$ whose magnitude is the angular speed ω and whose direction is perpendicular to the plane of the motion along the axis of rotation. (If you curl

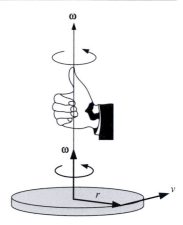

Figure 2.1 The direction of the vector angular velocity is given by a right-hand rule. The fingers of the right hand curl in the direction of the motion and the thumb points in the direction of the vector $\boldsymbol{\omega}$.

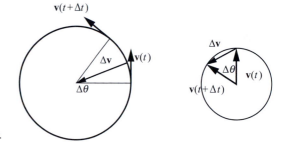

Figure 2.2 Illustration of centripetal acceleration. The sketch on the left represents the motion of a particle, and the sketch on the right is the *hodograph,* the path of the tip of the velocity vector.

the fingers of your right hand in the direction of the motion, your thumb will point in the direction of the vector $\boldsymbol{\omega}$. See Figure 2.1.)

The tangential velocity $\mathbf{v_T}$ is related to the angular velocity $\boldsymbol{\omega}$ by

$$\mathbf{v_T} = \boldsymbol{\omega} \times \mathbf{r}. \tag{2.7}$$

Rotational motion introduces two different accelerations. These are: (1) angular acceleration (α) and (2) centripetal acceleration (a_c).

The angular acceleration is due to any change in the angular speed of the body (as in a wheel that is speeding up or slowing down). It is given by $\alpha = d\omega/dt$ or, in vector notation,

$$\boldsymbol{\alpha} = \frac{d\boldsymbol{\omega}}{dt}.$$

Centripetal acceleration is associated with all rotational motion regardless of any change in the angular velocity. A car moving in a circle at constant speed is accelerating because acceleration is defined as a change in velocity. This acceleration is due to the change in the *direction* of the velocity vector. Similarly, a point on a rotating body is accelerating even if the body rotates at a constant angular speed. Figure 2.2 illustrates the change in the velocity vector of a point on the rim of a wheel that is rotating about a fixed axis with constant angular speed ω.

By definition,[4]

$$\mathbf{a} = \frac{d\mathbf{v}}{dt} = \lim_{\Delta t \to 0} \frac{\mathbf{v}(t + \Delta t) - \mathbf{v}(t)}{\Delta t} = \lim_{\Delta t \to 0} \frac{\Delta \mathbf{v}}{\Delta t}.$$

---

[4] In the rest of this section, the velocity $\mathbf{v}$ should perhaps be expressed as $\mathbf{v_T}$ but I left off the subscript to keep the notation simple.

Tip-to-tail addition of the vector $\mathbf{v}(t + \Delta t)$ and the vector $-\mathbf{v}(t)$ shows that $\Delta\mathbf{v}$ is a vector pointing towards the center of the circle. To help you see this, on the left side of Figure 2.2, I drew the vector $\Delta\mathbf{v}$ at a midpoint between the vectors $\mathbf{v}(t)$ and $\mathbf{v}(t + \Delta t)$. You can appreciate that the vector $\Delta\mathbf{v}$ is pointing approximately towards the center of the circle. As $\Delta t \to 0$, the vector $\Delta\mathbf{v}$ points more and more closely towards the center.

The same concept can be illustrated in a different way. If the angular velocity is constant then the speed (that is, the *magnitude* of the velocity vector) is a constant equal to v, but the *direction* of the vector changes with time. The tip of the velocity vector traces out a circle, as indicated on the right-hand side of Figure 2.2. (A plot showing the path of the tip of the velocity vector is called a hodograph.) For circular motion at constant speed, the hodograph is a circle of radius v. The vector $\Delta\mathbf{v}$ is a chord of this circle. Transferring the vector $\Delta\mathbf{v}$ back to the left-hand side of the figure shows that it points toward the center of the circle.

Consequently, a particle moving at a constant speed in a circular path is accelerating towards the center of the circle. We can determine the magnitude of this acceleration by considering the hodograph again. In the limit of infinitesimally small times, $\Delta\theta \to d\theta$, and $\Delta\mathbf{v} \to d\mathbf{v}$, and the chord approaches the subtended arc. From the basic relation $ds = rd\theta$, but applied to the hodograph, we see that in the limit, the magnitude of $\Delta\mathbf{v}$ is $dv = vd\theta$. Therefore,

$$a = \frac{dv}{dt} = \frac{vd\theta}{dt}.$$

But $d\theta = ds/r$, and $ds/dt = v$ so

$$a = \frac{v}{r}\frac{ds}{dt} = \frac{v^2}{r}.$$

That is, if a particle is moving in a circle at constant speed, it is accelerated towards the center with a "centripetal" acceleration given by

$$a_c = \frac{v^2}{r}.$$

Since $v = r\omega$, this can be written in the equivalent form

$$a_c = \omega^2 r.$$

Note that a point in a rotating body can actually experience three different types of acceleration:
(1) **linear acceleration** (if the body as a whole is accelerating),
(2) **tangential acceleration** ($a_T = r\alpha$; if the angular velocity of the body is changing),
(3) **centripetal acceleration** ($a_c = v^2/r = \omega^2 r$; due to the rotation of the body).

Exercise 2.4

(a) Determine the centripetal acceleration of the Earth due to its orbital motion about the Sun. Assume the orbit is circular. (b) Determine the centripetal acceleration of a point on the equator of the Earth due to the rotational motion of the Earth. Answers: (a) 5.95×10^{-3} m/s^2, (b) 3.37×10^{-2} m/s^2. ∎

Exercise 2.5

Show that the angles denoted $\Delta\theta$ on the right-hand and left-hand plots of Figure 2.2 are the same. ∎

2.4 The Position of a Particle on a Plane

The position of a point P on a flat surface can be specified in a variety of ways. One usually arbitrarily selects an origin and draws a pair of perpendicular reference lines through it. (These are denoted x and y in Figure 2.3.)

The most useful way to describe the position of P is by a vector $\mathbf{r}$ drawn from the origin to P, as in Figure 2.3. The magnitude of $\mathbf{r}$ is denoted by r and its direction is given by θ, the angle the vector makes with the x-axis.

To describe the position vector $\mathbf{r}$ we can either use its components (namely, x and y, the Cartesian coordinates), or its magnitude and direction (r and θ, the polar coordinates). These are equivalent descriptions of $\mathbf{r}$, so (x, y) and (r, θ) are related. A glance at Figure 2.3 shows that these relations are

$$x = r \cos \theta,$$
$$y = r \sin \theta.$$

The inverse relations (r and θ in terms of x and y) are

$$r = \sqrt{x^2 + y^2},$$
$$\theta = \tan^{-1}(y/x).$$

Equations such as these that relate two sets of coordinates are called "transformation equations."

In our study of mechanics, we will only be using three different coordinate systems, namely: Cartesian, cylindrical, and spherical coordinates. There are, however, many other coordinate systems, some being quite specialized and appropriate for only a few types of problems. In early editions of his book *Mathematical Methods for Physicists*, George Arfken included a long discussion describing fourteen different coordinate systems.[5] Just to give you a flavor of these, consider the "bipolar coordinates" η and ζ that are related to the Cartesian coordinates, x and y by the following transformation equations:

$$x = \frac{a \sinh \eta}{\cosh \eta - \cos \zeta} \quad \text{and} \quad y = \frac{a \sin \zeta}{\cosh \eta - \cos \zeta}.$$

You will be happy to know that we will not be using this particular set of coordinates! Fortunately, most coordinate transformations are much simpler than this one.[6]

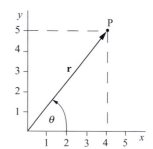

Figure 2.3 The position of the point P in rectangular coordinates (x, y) is $(4, 5)$ and in polar coordinates (r, θ) it is $(6.4, 0.9)$ where the angular measure is given in radians.

---

[5] *Mathematical Methods for Physicists*, 2nd ed., George Arfken, Academic Press, New York, 1970. The discussion was dropped in later editions of the book because computers allow one to solve most physics problems using simple coordinate systems. Many of the more exotic coordinate systems that were applicable only to one or two specific problems have fallen into disuse.

[6] It turns out that the capacitance of parallel charged cylinders can be easily evaluated using bipolar coordinates.

Exercise 2.6

A particle moves along the x axis at a constant speed of 3 m/s. Determine dr/dt and $d\theta/dt$. Answer: 3 m/s, zero. ■

Exercise 2.7

A particle travels in a circle of radius 3 m at a constant speed of 5 m/s. Determine dr/dt and $d\theta/dt$. Answer: zero, 5/3 rad/s. ■

Exercise 2.8

In bipolar coordinates a particle is at $\eta = 0.5$, $\zeta = 0.5$. Assuming $a = 1$ m, determine its position in plane polar coordinates r, θ. Answer: 2.832 m, 0.744 rad. ■

2.5 Unit Vectors

Unit vectors are the building blocks of kinematics. They are vectors that point in the increasing directions of the coordinates and have a length of one unit. (Unit vectors are unitless and have unit length.) Figure 2.4 illustrates the three unit vectors $\hat{\imath}, \hat{\jmath}, \hat{k}$ associated with the Cartesian coordinate system (x, y, z).

The coordinate system in Figure 2.4 is a right-handed coordinate system. This means that if you point the fingers of your right hand along the x-axis and bend them towards the y-axis, your thumb will be pointing in the direction of the z-axis.

Note that the Cartesian unit vectors are mutually perpendicular or *orthogonal*.

In your vector analysis course you learned the definition of the cross product (or vector product) of two vectors. Recall that if

$$\mathbf{c} = \mathbf{a} \times \mathbf{b}$$

then the magnitude of the vector $\mathbf{c}$ (denoted $|\mathbf{c}|$ or c) is

$$|\mathbf{c}| = |\mathbf{a}| |\mathbf{b}| \sin(\mathbf{a}, \mathbf{b}), \qquad (2.8)$$

where $\sin(\mathbf{a}, \mathbf{b})$ is the sine of the angle between $\mathbf{a}$ and $\mathbf{b}$, and the direction of $\mathbf{c}$ is given by the right-hand rule.

The definition of the dot product (or scalar product) of two vectors is:

$$\mathbf{a} \cdot \mathbf{b} = |\mathbf{a}| |\mathbf{b}| \cos(\mathbf{a}, \mathbf{b}). \qquad (2.9)$$

The cross (or vector) product and the dot (or scalar) product are the two most common ways of multiplying vectors. Later in this course you will learn another type of vector multiplication which involves mathematical objects called dyadics.

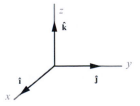

Figure 2.4 The unit vectors $\hat{\imath}, \hat{\jmath}, \hat{k}$ in the Cartesian coordinate system. The coordinate system is right-handed.

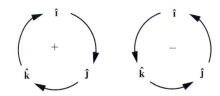

Figure 2.5 A mnemonic for the cross product of unit vectors. The sketches indicate, for example, that $\hat{i} \times \hat{j} = + \hat{k}$, and $\hat{j} \times \hat{i} = - \hat{k}$.

Consider the dot product of the unit vectors. From Equation (2.9) we can determine that

$$\hat{i} \cdot \hat{i} = 1, \qquad \hat{j} \cdot \hat{j} = 1, \qquad \hat{k} \cdot \hat{k} = 1,$$
$$\hat{i} \cdot \hat{j} = 0, \qquad \hat{i} \cdot \hat{k} = 0, \qquad \hat{j} \cdot \hat{k} = 0.$$

These relations describe the *orthogonality* of the unit vectors. Similarly, from the definition of the cross product, Equation (2.8), we find that

$$\hat{i} \times \hat{i} = 0, \qquad \hat{j} \times \hat{j} = 0, \qquad \hat{k} \times \hat{k} = 0,$$

and

$$\hat{i} \times \hat{j} = \hat{k}, \qquad \hat{j} \times \hat{i} = -\hat{k},$$
$$\hat{j} \times \hat{k} = \hat{i}, \qquad \hat{k} \times \hat{j} = -\hat{i},$$
$$\hat{k} \times \hat{i} = \hat{j}, \qquad \hat{i} \times \hat{k} = -\hat{j}.$$

The last six relations are easily remembered with the help of the mnemonic of Figure 2.5.

Exercise 2.9

Two vectors are given by $\mathbf{A} = 2\hat{i} + 3\hat{j} + 4\hat{k}$ and $\mathbf{B} = 3\hat{i} - 2\hat{j}$. Determine $\mathbf{A} + \mathbf{B}, \mathbf{A} - \mathbf{B}, \mathbf{A} \cdot \mathbf{B}$ and $\mathbf{A} \times \mathbf{B}$. ■

Exercise 2.10

Calculate the angle between the vector $\mathbf{V} = 2\hat{i} + 4\hat{j} + 6\hat{k}$ and the vector $\mathbf{W} = 2\hat{i} + 2\hat{j} + 2\hat{k}$. Answer: 22.21°. ■

Exercise 2.11

The direction angles of a vector are the angles between the vector and the x-, y-, and z-axes. Determine the direction angles of the vector $\mathbf{r} = 6\hat{i} + 5\hat{j} - 2\hat{k}$. Answer: $\alpha = 41.9°$, $\beta = 51.67°$, $\gamma = 104.3°$. ■

Exercise 2.12

A vector forms the diagonal of a cube of side a. Express it in terms of the unit vectors. ■

Exercise 2.13

Show that the cross product of two vectors $\mathbf{A}$ and $\mathbf{B}$ can be written as the following determinant:

$$\begin{vmatrix} \hat{i} & \hat{j} & \hat{k} \\ A_x & A_y & A_z \\ B_x & B_y & B_z \end{vmatrix}.$$

■

Exercise 2.14

Let **A** and **B** be the sides of a parallelogram. Show that the area of the parallelogram is **A** × **B**.

■

2.6 Kinematics in Two Dimensions

In two dimensions, the position of a point can be specified with two numbers. In Cartesian coordinates these are often the values of x and y and in plane polar coordinates they are r and θ. We now consider the kinematic relations between position, velocity, and acceleration in these two coordinate systems.

2.6.1 Cartesian Coordinates

When using Cartesian coordinates we usually set $z = 0$ and define the plane of motion to be the xy-plane. (As long as the motion is confined to a flat plane, we can let the plane be the $z = 0$ surface with no loss of generality.) In Cartesian coordinates the position vector is

$$\mathbf{r} = x\hat{\mathbf{i}} + y\hat{\mathbf{j}}.$$

The definition of velocity is $\mathbf{v} = \dot{\mathbf{r}} = d\mathbf{r}/dt$, so

$$\mathbf{v} = \frac{d}{dt}(x\hat{\mathbf{i}} + y\hat{\mathbf{j}}).$$

The Cartesian unit vectors $(\hat{\mathbf{i}}, \hat{\mathbf{j}}, \hat{\mathbf{k}})$ are constant in magnitude and direction so their time derivatives are zero. Consequently, writing $\dot{x}$ for dx/dt and $\dot{y}$ for dy/dt, the velocity is given by

$$\mathbf{v} = \dot{x}\hat{\mathbf{i}} + \dot{y}\hat{\mathbf{j}}.$$

Using the definition of acceleration, $\mathbf{a} = \dot{\mathbf{v}}$ we obtain

$$\mathbf{a} = \ddot{x}\hat{\mathbf{i}} + \ddot{y}\hat{\mathbf{j}}.$$

2.6.2 Plane Polar Coordinates

Let us now go through the same analysis as above but for the plane polar coordinates (r, θ). That is, let us derive expressions for position, velocity, and acceleration in plane polar coordinates. Although the analysis is complicated, it is very important for you to understand exactly what I am doing. The ideas presented in this section are the basis for the rest of this chapter.

The first thing we need to do is to define unit vectors for the polar coordinates. In Figure 2.6 the point P is located at (r, θ), where r is the length of the vector **r** and is equal to the distance from the origin to the point P. The angle between **r** and the x-axis is denoted θ. At the tip of vector **r** you can see the two unit vectors $\hat{\mathbf{r}}$ and $\hat{\boldsymbol{\theta}}$. These have unit length. The direction of $\hat{\mathbf{r}}$ is

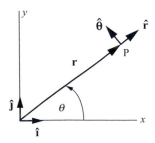

Figure 2.6 Definition of the plane polar unit vectors $\hat{\mathbf{r}}$ and $\hat{\boldsymbol{\theta}}$. The unit vector $\hat{\mathbf{r}}$ points in the direction of increasing **r**, and $\hat{\boldsymbol{\theta}}$ points in the direction of increasing θ.

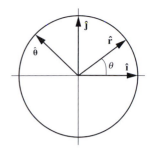

Figure 2.7 The unit vectors $\hat{\mathbf{r}}$ and $\hat{\boldsymbol{\theta}}$ and the unit circle. All the unit vectors originate at the center of the circle and all have magnitude unity, so they end on the circumference of the circle.

the direction of increasing $\mathbf{r}$. The direction of $\hat{\boldsymbol{\theta}}$ is the direction of increasing θ. This means that an increase in θ will cause the vector $\mathbf{r}$ to rotate in the direction of $\hat{\boldsymbol{\theta}}$.

As long as their magnitudes and directions are unchanged, vectors are allowed to "slide." We can slide $\hat{\mathbf{r}}$ and $\hat{\boldsymbol{\theta}}$ to the origin and represent them on a unit circle as shown in Figure 2.7.

The unit vectors $\hat{\mathbf{r}}$ and $\hat{\boldsymbol{\theta}}$ are *not* constant because they are not associated with a set of fixed axes. Rather, they are associated with a vector $\mathbf{r}$ that can change with time. If $\mathbf{r}$ changes in *magnitude,* this does not affect $\hat{\mathbf{r}}$ or $\hat{\boldsymbol{\theta}}$ but if $\mathbf{r}$ changes in *direction,* then the direction of the unit vectors $\hat{\mathbf{r}}$ and $\hat{\boldsymbol{\theta}}$ will change. This means that $\hat{\mathbf{r}}$ and $\hat{\boldsymbol{\theta}}$ are functions of θ. In mathematical form we express this as

$$\hat{\mathbf{r}} = \hat{\mathbf{r}}(\theta),$$

and

$$\hat{\boldsymbol{\theta}} = \hat{\boldsymbol{\theta}}(\theta).$$

The position of a point in plane polar coordinates is given quite simply by

$$\mathbf{r} = r\hat{\mathbf{r}}.$$

If you take the derivative of this equation with respect to time, you obtain an expression for the velocity in polar coordinates.

$$\mathbf{v} = \frac{d\mathbf{r}}{dt} = \frac{d}{dt}(r\hat{\mathbf{r}}) = \frac{dr}{dt}\hat{\mathbf{r}} + r\frac{d\hat{\mathbf{r}}}{dt}.$$

The last term comes from the product rule of differentiation and reflects the fact that the unit vector $\hat{\mathbf{r}}$ may be changing with time. Although the magnitude of $\hat{\mathbf{r}}$ is always unity, it may change in direction. Because $\hat{\mathbf{r}}$ is a function of θ, the differential of $\hat{\mathbf{r}}$ is

$$d\hat{\mathbf{r}} = \frac{d\hat{\mathbf{r}}}{d\theta}d\theta.$$

So,

$$\frac{d\hat{\mathbf{r}}}{dt} = \frac{d\hat{\mathbf{r}}}{d\theta}\frac{d\theta}{dt} = \frac{d\hat{\mathbf{r}}}{d\theta}\dot{\theta}.$$

But what is $d\hat{\mathbf{r}}/d\theta$? Looking at Figure 2.7 you see that $\hat{\mathbf{r}}$ and $\hat{\boldsymbol{\theta}}$ can be resolved into Cartesian components as follows:

$$\hat{\mathbf{r}} = 1\cos\theta\hat{\mathbf{i}} + 1\sin\theta\hat{\mathbf{j}},$$

where I explicitly included the number 1 to emphasize that the radius of the unit circle is 1. You obtain a similar expression for $\hat{\boldsymbol{\theta}}$ by inspection of Figure 2.7. Dropping the "1" we write:

$$\hat{\mathbf{r}} = \cos\theta\hat{\mathbf{i}} + \sin\theta\hat{\mathbf{j}}, \tag{2.10}$$

$$\hat{\boldsymbol{\theta}} = -\sin\theta\hat{\mathbf{i}} + \cos\theta\hat{\mathbf{j}}. \tag{2.11}$$

Therefore,

$$\frac{d\hat{\mathbf{r}}}{d\theta} = \frac{d}{d\theta}(\cos\theta\hat{\mathbf{i}} + \sin\theta\hat{\mathbf{j}}) = -\sin\theta\hat{\mathbf{i}} + \cos\theta\hat{\mathbf{j}} = \hat{\boldsymbol{\theta}}.$$

Furthermore,

$$\frac{d\hat{\boldsymbol{\theta}}}{d\theta} = -\cos\theta\hat{\mathbf{i}} - \sin\theta\hat{\mathbf{j}} = -\hat{\mathbf{r}}.$$

You may wish to memorize the following relations, as you will be using them frequently in this and other courses.

$$\frac{d\hat{\mathbf{r}}}{d\theta} = \hat{\boldsymbol{\theta}} \quad \text{and} \quad \frac{d\hat{\boldsymbol{\theta}}}{d\theta} = -\hat{\mathbf{r}}. \tag{2.12}$$

Having determined $d\hat{\mathbf{r}}/d\theta$ and $d\hat{\boldsymbol{\theta}}/d\theta$ we can obtain $d\hat{\mathbf{r}}/dt$ and $d\hat{\boldsymbol{\theta}}/dt$ as follows:

$$\frac{d\hat{\mathbf{r}}}{dt} = \frac{d\hat{\mathbf{r}}}{d\theta}\frac{d\theta}{dt} = \dot{\theta}\hat{\boldsymbol{\theta}},$$

and

$$\frac{d\hat{\boldsymbol{\theta}}}{dt} = \frac{d\hat{\boldsymbol{\theta}}}{d\theta}\frac{d\theta}{dt} = -\dot{\theta}\hat{\mathbf{r}}.$$

As noted previously, in polar coordinates the velocity is given by

$$\mathbf{v} = \frac{d}{dt}\mathbf{r} = \frac{d}{dt}(r\hat{\mathbf{r}}) = \frac{dr}{dt}\hat{\mathbf{r}} + r\frac{d\hat{\mathbf{r}}}{dt},$$

so

$$\mathbf{v} = \dot{r}\hat{\mathbf{r}} + r\dot{\theta}\hat{\boldsymbol{\theta}}. \tag{2.13}$$

This equation for the velocity can be differentiated to obtain the expression for acceleration in polar coordinates:

$$\mathbf{a} = (\ddot{r} - r\dot{\theta}^2)\hat{\mathbf{r}} + (r\ddot{\theta} + 2\dot{r}\dot{\theta})\hat{\boldsymbol{\theta}}. \tag{2.14}$$

We shall frequently refer to the expressions for velocity and acceleration in terms of polar coordinates, so you should either memorize Equations (2.12), (2.13), and (2.14) or mark this page so you can find them easily.

Worked Example 2.5 A turntable rotates at a constant angular speed ω. An ant crawls directly toward the rim along a radial line at a constant speed b. You observe the ant from above. From your point of view, the ant is moving in a spiral. Write an expression for the velocity and acceleration of the ant in polar coordinates.

Solution Given $\dot{r} = b$ and $\dot{\theta} = \omega$. Therefore, $r = bt + r_0$ and $\theta = \omega t + \theta_0$. Note that $\ddot{\theta} = 0$ and $\ddot{r} = 0$.

$$\mathbf{r} = r\hat{\mathbf{r}} = (bt + r_0)\hat{\mathbf{r}}$$

$$\mathbf{v} = \frac{d}{dt}[(bt + r_0)\hat{\mathbf{r}}] = b\hat{\mathbf{r}} + (bt + r_0)\frac{d\hat{\mathbf{r}}}{dt}$$

$$\mathbf{v} = b\hat{\mathbf{r}} + (bt + r_0)\omega\hat{\boldsymbol{\theta}},$$

and

$$\mathbf{a} = \frac{d}{dt}[b\hat{\mathbf{r}} + (bt + r_0)\omega\hat{\boldsymbol{\theta}}]$$

$$= b\frac{d\hat{\mathbf{r}}}{dt} + b\omega\hat{\boldsymbol{\theta}} + (bt + r_0)\omega\frac{d\hat{\boldsymbol{\theta}}}{dt}$$

$$= 2b\omega\hat{\boldsymbol{\theta}} - (bt + r_0)\omega^2\hat{\mathbf{r}}.$$

Exercise 2.15

Derive Equation (2.14) from Equation (2.13). ∎

Exercise 2.16

Given $\mathbf{r} = a(\hat{\mathbf{i}} \sin \omega t + \hat{\mathbf{j}} \cos \omega t)$, where ω is a constant. (a) Determine the magnitude of the velocity. (Recall that the magnitude of a vector $\mathbf{A}$ is $A = |\mathbf{A}| = \sqrt{\mathbf{A} \cdot \mathbf{A}}$.) (b) Prove the velocity is perpendicular to $\mathbf{r}$. (c) Describe the motion. Answer: (a) $a\omega$. ∎

Exercise 2.17

The position of a certain particle is described by $r = 4t$ (meters), $\theta = 0.2t$ (radians). Determine its velocity as a function of time in (a) polar coordinates and (b) Cartesian coordinates. Answers: (a) $\mathbf{v} = 4\hat{\mathbf{r}} + 0.8t\hat{\boldsymbol{\theta}}$, (b) $\mathbf{v} = (4\cos 0.2t - 0.8t \sin 0.2t)\hat{\mathbf{i}} + (4 \sin 0.2t + 0.8t \cos 0.2t)\hat{\mathbf{j}}$. ∎

Exercise 2.18

You are analyzing the motion of a red dot painted on the rim of a wheel of radius $r = 5.0$ cm. It returns to the same point every 2 s. At $t = 0$ the particle is at $(r, \theta) = (5, 0)$. Express the position and velocity vectors in Cartesian and polar coordinates. Answers: $\mathbf{r} = 5\hat{\mathbf{r}} = (5 \cos \pi t)\hat{\mathbf{i}} + (5 \sin \pi t)\hat{\mathbf{j}}$, $\mathbf{v} = 5\pi\hat{\boldsymbol{\theta}} = (-5\pi \sin \pi t)\hat{\mathbf{i}} + (5\pi \cos \pi t)\hat{\mathbf{j}}$. ∎

2.7 Kinematics in Three Dimensions

We now generalize to motion in three dimensions. We shall describe three different coordinate systems: Cartesian coordinates (x, y, z), cylindrical coordinates (ρ, ϕ, z), and spherical coordinates (r, θ, ϕ). These are the most commonly used three-dimensional coordinate systems in physics.

You will find that the cylindrical and spherical coordinates rely heavily on the plane polar coordinates described in the previous section.

2.7.1 Cartesian Coordinates

The position vector in a three-dimensional Cartesian coordinate system is given by

$$\mathbf{r} = x\hat{\mathbf{i}} + y\hat{\mathbf{j}} + z\hat{\mathbf{k}}.$$

In an inertial (nonaccelerating) coordinate system the three unit vectors $(\hat{\mathbf{i}}, \hat{\mathbf{j}}, \hat{\mathbf{k}})$ are constant in magnitude and direction so their time derivatives are zero. This makes it easy to obtain expressions for the velocity and acceleration in terms of Cartesian coordinates. The velocity is the time derivative of position:

$$\mathbf{v} = \frac{d\mathbf{r}}{dt} = \dot{x}\hat{\mathbf{i}} + \dot{y}\hat{\mathbf{j}} + \dot{z}\hat{\mathbf{k}}.$$

Similarly, the acceleration is obtained by taking the time derivative of the velocity:

$$\mathbf{a} = \frac{d\mathbf{v}}{dt} = \ddot{x}\hat{\mathbf{i}} + \ddot{y}\hat{\mathbf{j}} + \ddot{z}\hat{\mathbf{k}}.$$

2.7.2 Cylindrical Coordinates

Next we obtain expressions for position, velocity, and acceleration in cylindrical coordinates. In any coordinate system, the position vector is simply $\mathbf{r}$. Figure 2.8 shows $\mathbf{r}$ in a Cartesian coordinate system. We drop a perpendicular to the xy-plane from the tip of $\mathbf{r}$. This perpendicular has magnitude z and direction $\hat{\mathbf{k}}$ so we can denote it by the vector $z\hat{\mathbf{k}}$. The point at which the perpendicular intersects the $z = 0$ plane can be specified by its plane polar coordinates. (But when using cylindrical coordinates we generally use the symbols ρ and ϕ rather than r and θ.) The set of coordinates, ρ, ϕ, z are called *cylindrical coordinates*. Figure 2.8 gives a graphical representation of this coordinate system. Note that ϕ is measured from the x-axis to the projection of $\mathbf{r}$ onto the xy-plane, which is denoted ρ.

The next task is to determine unit vectors for the cylindrical coordinate system. This is quite simple because the new coordinate system is made up of old coordinates whose properties you have already learned. You can appreciate from Figure 2.9 that $\hat{\rho}$ and $\hat{\phi}$ are the same as the plane polar unit vectors $\hat{\mathbf{r}}$ and $\hat{\theta}$, and the unit vector in the z direction is just $\hat{\mathbf{k}}$ as in the Cartesian coordinate system.

Although $\hat{\mathbf{k}}$ is constant in magnitude and direction, $\hat{\rho}$ and $\hat{\phi}$ will change in direction as the angle ϕ varies. That is, $\hat{\rho}$ and $\hat{\phi}$ are functions of ϕ. We write:

$$\hat{\rho} = \hat{\rho}(\phi),$$

and

$$\hat{\phi} = \hat{\phi}(\phi).$$

The vectors $\hat{\rho}$, $\hat{\phi}$, $\hat{\mathbf{k}}$ form an *orthogonal* set. The orthogonality of these unit vectors leads to the following easily proved properties:

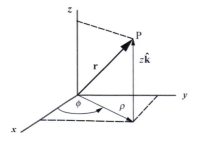

Figure 2.8 The position of point P can be specified either in terms of the Cartesian coordinates x, y, z or the cylindrical coordinates ρ, ϕ, z.

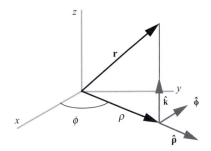

Figure 2.9 Definition of the cylindrical unit vectors $\hat{\boldsymbol{\rho}}$, $\hat{\boldsymbol{\phi}}$, and $\hat{\mathbf{k}}$.

$$\hat{\boldsymbol{\rho}} \cdot \hat{\boldsymbol{\rho}} = 1, \qquad \hat{\boldsymbol{\phi}} \cdot \hat{\boldsymbol{\phi}} = 1, \qquad \hat{\mathbf{k}} \cdot \hat{\mathbf{k}} = 1,$$
$$\hat{\boldsymbol{\rho}} \cdot \hat{\boldsymbol{\phi}} = 0, \qquad \hat{\boldsymbol{\rho}} \cdot \hat{\mathbf{k}} = 0, \qquad \hat{\boldsymbol{\phi}} \cdot \hat{\mathbf{k}} = 0,$$
$$\hat{\boldsymbol{\rho}} \times \hat{\boldsymbol{\phi}} = \hat{\mathbf{k}}, \qquad \hat{\boldsymbol{\phi}} \times \hat{\mathbf{k}} = \hat{\boldsymbol{\rho}}, \qquad \hat{\mathbf{k}} \times \hat{\boldsymbol{\rho}} = \hat{\boldsymbol{\phi}}.$$

The relationships between cylindrical coordinates and Cartesian coordinates (the transformation equations) are:

$$x = \rho \cos \phi \qquad y = \rho \sin \phi \qquad z = z.$$

The inverse relationships are:

$$\rho = \sqrt{x^2 + y^2} \qquad \phi = \tan^{-1}\left(\frac{y}{x}\right) \qquad z = z.$$

The unit vectors $\hat{\boldsymbol{\rho}}$ and $\hat{\boldsymbol{\phi}}$ can be expressed in terms of $\hat{\mathbf{i}}$ and $\hat{\mathbf{j}}$ as follows:

$$\hat{\boldsymbol{\rho}} = \cos \phi \hat{\mathbf{i}} + \sin \phi \hat{\mathbf{j}}, \tag{2.15}$$

$$\hat{\boldsymbol{\phi}} = -\sin \phi \hat{\mathbf{i}} + \cos \phi \hat{\mathbf{j}}. \tag{2.16}$$

The position vector in cylindrical coordinates is given by

$$\mathbf{r} = \rho \hat{\boldsymbol{\rho}} + z \hat{\mathbf{k}}. \tag{2.17}$$

The velocity is obtained by taking the time derivative of the position vector:

$$\mathbf{v} = \dot{\mathbf{r}} = \frac{d}{dt}(\rho \hat{\boldsymbol{\rho}} + z \hat{\mathbf{k}}) = \dot{\rho} \hat{\boldsymbol{\rho}} + \rho \frac{d \hat{\boldsymbol{\rho}}}{dt} + \dot{z} \hat{\mathbf{k}}.$$

From our study of polar coordinates we appreciate that $\frac{d \hat{\boldsymbol{\rho}}}{dt} = \dot{\phi} \hat{\boldsymbol{\phi}}$ and $\frac{d \hat{\boldsymbol{\phi}}}{dt} = -\dot{\phi} \hat{\boldsymbol{\rho}}$ so

$$\mathbf{v} = \dot{\rho} \hat{\boldsymbol{\rho}} + \rho \dot{\phi} \hat{\boldsymbol{\phi}} + \dot{z} \hat{\mathbf{k}}. \tag{2.18}$$

Similarly, the acceleration is given by

$$\mathbf{a} = \ddot{\mathbf{r}} = (\ddot{\rho} - \rho \dot{\phi}^2) \hat{\boldsymbol{\rho}} + (\rho \ddot{\phi} + 2\dot{\rho} \dot{\phi}) \hat{\boldsymbol{\phi}} + \ddot{z} \hat{\mathbf{k}}. \tag{2.19}$$

(You should be able to carry out the intermediate steps in obtaining this expression for **a**. Please don't just nod your head and think, "Of course I can do it." Pick up your pencil and work it out. The exercise will make the concepts stick in your mind.)

Worked Example 2.6 A bead slides on a wire bent into a helix. The position of the bead as a function of time is given by $\rho = c$, $\phi = \omega t$, $z = bt$, where c, ω, b are constants. Determine an expression for the velocity and acceleration of the bead as a function of time.

Solution In cylindrical coordinates the position of the bead is

$$\mathbf{r} = \rho\hat{\boldsymbol{\rho}} + z\hat{\mathbf{k}}, \ = c\hat{\boldsymbol{\rho}} + bt\hat{\mathbf{k}},$$

$$\mathbf{v} = \frac{d\mathbf{r}}{dt} = c\frac{d\hat{\boldsymbol{\rho}}}{dt} + b\hat{\mathbf{k}} = c\dot{\phi}\hat{\boldsymbol{\phi}} + b\hat{\mathbf{k}} = c\omega\hat{\boldsymbol{\phi}} + b\hat{\mathbf{k}}$$

$$\mathbf{a} = \frac{d\mathbf{v}}{dt} = c\omega\frac{d\hat{\boldsymbol{\phi}}}{dt} = c\omega(-\dot{\phi}\hat{\boldsymbol{\rho}}) = -c\omega^2\hat{\boldsymbol{\rho}}.$$

Note that $\mathbf{a}$ is the centripetal acceleration.

Exercise 2.19

Carry out the steps to obtain Equation (2.19).

Exercise 2.20

The position of a particle is given by $\rho = 3t^2$, $\phi = 2t$ and $z = 12t$ (distances in meters, angles in radians). What is the acceleration at time $t = 2$ seconds? Answer: $\mathbf{a} = (-42\hat{\boldsymbol{\rho}} + 48\hat{\boldsymbol{\phi}})$ m/s^2.

2.7.3 Spherical Coordinates

The last set of coordinates we will consider are called *spherical coordinates*. In these coordinates the position of a particle is specified by the vector $\mathbf{r}$ which has length r and direction (or *orientation*) specified by two angles, θ and ϕ, as shown Figure 2.10. These angles are defined as follows: The "polar" angle θ is measured from the z-axis to the vector $\mathbf{r}$. The "azimuthal" angle ϕ is measured from the x-axis to the projection of $\mathbf{r}$ onto the xy-plane. The component of $\mathbf{r}$ along the z-axis is $r\cos\theta$. The component of $\mathbf{r}$ in the xy-plane (denoted ρ) has a length of $r\sin\theta$. The x component of $\mathbf{r}$ is, therefore, $r\sin\theta\cos\phi$ and the y component is $r\sin\theta\sin\phi$.

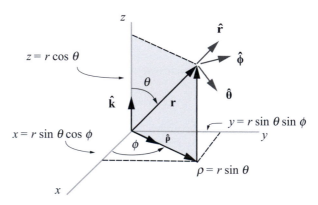

Figure 2.10 Spherical coordinates and the unit vectors $\hat{\mathbf{r}}, \hat{\boldsymbol{\theta}}, \hat{\boldsymbol{\phi}}$. Note that the Cartesian components of the vector $\mathbf{r}$ are $x = r\sin\theta\cos\phi$, $y = r\sin\theta\sin\phi$ and $z = r\cos\theta$. The unit vectors $\hat{\mathbf{r}}, \hat{\boldsymbol{\theta}}, \hat{\boldsymbol{\phi}}$ point in the direction of increasing r, θ, ϕ, respectively. Also shown are the cylindrical unit vectors $\hat{\boldsymbol{\rho}}$ and $\hat{\mathbf{k}}$.

That is, the relationships (transformation equations) between the spherical coordinates and the Cartesian coordinates are:

$$x = r \sin\theta \cos\phi, \tag{2.20}$$
$$y = r \sin\theta \sin\phi,$$
$$z = r \cos\theta.$$

The inverse transformation equations are:

$$r = (x^2 + y^2 + z^2)^{\frac{1}{2}},$$
$$\theta = \tan^{-1}\left(\frac{\sqrt{x^2 + y^2}}{z}\right),$$
$$\phi = \tan^{-1}\left(\frac{y}{x}\right).$$

What are the appropriate unit vectors? The unit vector $\hat{\mathbf{r}}$ is in the direction of increasing $\mathbf{r}$. The unit vector $\hat{\boldsymbol{\theta}}$ is in the direction that the tip of $\mathbf{r}$ will move if θ increases. In Figure 2.11 I sketched a quarter circle to illustrate the motion of $\mathbf{r}$ as θ changes. The unit vector $\hat{\boldsymbol{\theta}}$ is tangent to this circle and has the direction of increasing θ. The unit vector $\hat{\boldsymbol{\phi}}$ lies in the xy-plane and gives the direction that the tip of $\hat{\boldsymbol{\rho}}$ will move if ϕ increases. By sliding $\hat{\boldsymbol{\phi}}$ down to the xy-plane, you can see that $\hat{\boldsymbol{\phi}}$ in spherical coordinates is the same as the $\hat{\boldsymbol{\phi}}$ of the cylindrical coordinates.

Although it may not be immediately obvious, the spherical unit vectors $\hat{\mathbf{r}}, \hat{\boldsymbol{\theta}}, \hat{\boldsymbol{\phi}}$ form an orthogonal set. Figure 2.10 shows these three mutually perpendicular unit vectors at the tip of the vector $\mathbf{r}$. This is probably the easiest way to remember them. In that figure, both $\hat{\mathbf{r}}$ and $\hat{\boldsymbol{\theta}}$ lie in the shaded plane that contains $\mathbf{r}$ and the z-axis, and $\hat{\boldsymbol{\phi}}$ is perpendicular to that plane.

Equations (2.20) are extremely useful and you might want to memorize them. (However, they are easy to remember if you can visualize Figure 2.10.) You will use these equations throughout this course as well as in many upper-division and graduate physics courses.

Figure 2.12 shows a unit circle that contains both $\mathbf{r}$ and the z-axis. Its intersection with the xy-plane gives $\hat{\boldsymbol{\rho}}$, the unit vector for the cylindrical coordinates. From the figure,

$$\hat{\mathbf{r}} = \cos\theta\hat{\mathbf{k}} + \sin\theta\hat{\boldsymbol{\rho}}. \tag{2.21}$$

Recall that ρ and ϕ are the same as the plane polar coordinates (formerly denoted r and θ). Referring back to Equation (2.15), which is

$$\hat{\boldsymbol{\rho}} = \cos\phi\hat{\mathbf{i}} + \sin\phi\hat{\mathbf{j}},$$

we can express the unit vector $\hat{\mathbf{r}}$ in terms of the Cartesian unit vectors as

$$\hat{\mathbf{r}} = \sin\theta\cos\phi\hat{\mathbf{i}} + \sin\theta\sin\phi\hat{\mathbf{j}} + \cos\theta\hat{\mathbf{k}}. \tag{2.22}$$

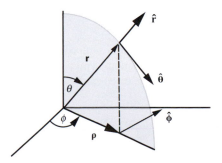

Figure 2.11 The unit vectors $\hat{\mathbf{r}}, \hat{\boldsymbol{\theta}}$, and $\hat{\boldsymbol{\phi}}$.

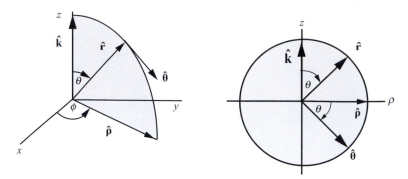

Figure 2.12 The unit vectors $\hat{\mathbf{r}}, \hat{\boldsymbol{\theta}}, \hat{\boldsymbol{\phi}}$ for spherical coordinates and the unit circle. Also shown are $\hat{\mathbf{k}}$ and $\hat{\boldsymbol{\rho}}$. In the sketch on the right, $\hat{\boldsymbol{\theta}}$ has been translated from the tip of $\hat{\mathbf{r}}$ to the origin.

We can also express $\hat{\boldsymbol{\theta}}$ and $\hat{\boldsymbol{\phi}}$ in terms of the Cartesian unit vectors. From the unit circle on the right in Figure 2.12, you can appreciate that

$$\hat{\boldsymbol{\theta}} = \cos\theta\,\hat{\boldsymbol{\rho}} - \sin\theta\,\hat{\mathbf{k}}.$$

Substituting for $\hat{\boldsymbol{\rho}}$ we obtain

$$\hat{\boldsymbol{\theta}} = \cos\theta\cos\phi\,\hat{\mathbf{i}} + \cos\theta\sin\phi\,\hat{\mathbf{j}} - \sin\theta\,\hat{\mathbf{k}}. \tag{2.23}$$

Finally, $\hat{\boldsymbol{\phi}}$ is the same as for the cylindrical coordinates, so

$$\hat{\boldsymbol{\phi}} = -\sin\phi\,\hat{\mathbf{i}} + \cos\phi\,\hat{\mathbf{j}}. \tag{2.24}$$

Now that we have determined $\hat{\mathbf{r}}, \hat{\boldsymbol{\theta}}, \hat{\boldsymbol{\phi}}$, we can evaluate their time derivatives. Expressions for the velocity and acceleration are somewhat more difficult to obtain in terms of spherical coordinates than in cylindrical coordinates because the unit vectors $\hat{\mathbf{r}}$ and $\hat{\boldsymbol{\theta}}$ are functions of *both* θ and ϕ. (Figure 2.10 indicates that a change in either θ or ϕ will cause a change in the directions of $\hat{\mathbf{r}}$ and $\hat{\boldsymbol{\theta}}$.) On the other hand, $\hat{\boldsymbol{\phi}}$ is a function only of ϕ. Thus we write, in the usual mathematical notation,

$$\hat{\mathbf{r}} = \hat{\mathbf{r}}(\theta, \phi),$$
$$\hat{\boldsymbol{\theta}} = \hat{\boldsymbol{\theta}}(\theta, \phi),$$
$$\hat{\boldsymbol{\phi}} = \hat{\boldsymbol{\phi}}(\phi).$$

Recall that if a quantity (say f) is a function of several variables such as x, y, and z, then $f = f(x, y, z)$, and by the rules of calculus, the differential quantity df is given by:

$$df = \frac{\partial f}{\partial x}dx + \frac{\partial f}{\partial y}dy + \frac{\partial f}{\partial z}dz.$$

(This simple relationship is used over and over again in physics. If you have already taken a course in thermodynamics, you are aware how important it is.)

Going back to the discussion of the unit vectors, you can appreciate that if $\hat{\mathbf{r}} = \hat{\mathbf{r}}(\theta, \phi)$, then

$$d\hat{\mathbf{r}} = \frac{\partial \hat{\mathbf{r}}}{\partial \theta}d\theta + \frac{\partial \hat{\mathbf{r}}}{\partial \phi}d\phi,$$

and consequently

$$\frac{d\hat{\mathbf{r}}}{dt} = \frac{\partial \hat{\mathbf{r}}}{\partial \theta}\frac{d\theta}{dt} + \frac{\partial \hat{\mathbf{r}}}{\partial \phi}\frac{d\phi}{dt},$$

or

$$\frac{d\hat{\mathbf{r}}}{dt} = \frac{\partial \hat{\mathbf{r}}}{\partial \theta}\dot{\theta} + \frac{\partial \hat{\mathbf{r}}}{\partial \phi}\dot{\phi}. \tag{2.25}$$

From Equation (2.22)

$$\frac{\partial \hat{\mathbf{r}}}{\partial \theta} = \cos\theta\cos\phi\hat{\mathbf{i}} + \cos\theta\sin\phi\hat{\mathbf{j}} - \sin\theta\hat{\mathbf{k}}. \tag{2.26}$$

But comparison with Equation (2.23) shows us that the right-hand side of this equation is simply $\hat{\theta}$.

Therefore,

$$\frac{\partial \hat{\mathbf{r}}}{\partial \theta} = \hat{\theta}.$$

We also need to determine $\partial \hat{\mathbf{r}}/\partial \phi$. Taking the derivative of Equation (2.22) with respect to ϕ we obtain

$$\frac{\partial \hat{\mathbf{r}}}{\partial \phi} = -\sin\theta\sin\phi\hat{\mathbf{i}} + \sin\theta\cos\phi\hat{\mathbf{j}}$$

$$= \sin\theta(-\sin\phi\hat{\mathbf{i}} + \cos\phi\hat{\mathbf{j}}),$$

or, comparing with Equation (2.24),

$$\frac{\partial \hat{\mathbf{r}}}{\partial \phi} = \sin\theta\,\hat{\phi}.$$

So finally, the expression for $d\hat{\mathbf{r}}/dt$, Equation (2.25) can be written

$$\frac{d\hat{\mathbf{r}}}{dt} = \dot{\theta}\hat{\theta} + \sin\theta\dot{\phi}\hat{\phi}.$$

Similarly, we obtain the following two relations (see Problem 2.36):

$$\frac{d\hat{\theta}}{dt} = \frac{\partial \hat{\theta}}{\partial \theta}\dot{\theta} + \frac{\partial \hat{\theta}}{d\phi}\dot{\phi} = -\dot{\theta}\hat{\mathbf{r}} + \cos\theta\dot{\phi}\hat{\phi}, \tag{2.27}$$

$$\frac{d\hat{\phi}}{dt} = \frac{\partial \hat{\phi}}{\partial \theta}\dot{\theta} + \frac{\partial \hat{\phi}}{\partial \phi}\dot{\phi} = 0 - \dot{\phi}\hat{\rho} = -\dot{\phi}\hat{\rho}. \tag{2.28}$$

Using Figure 2.12 you can show that,

$$\hat{\rho} = \sin\theta\hat{\mathbf{r}} + \cos\theta\hat{\theta},$$

so we can also write Equation (2.28) as

$$\frac{d\hat{\phi}}{dt} = -\dot{\phi}\sin\theta\hat{\mathbf{r}} - \dot{\phi}\cos\theta\hat{\theta}.$$

Having obtained the time derivatives of the three unit vectors $\hat{\mathbf{r}}, \hat{\theta}, \hat{\phi}$ it is not too difficult to determine expressions for the velocity and acceleration in spherical polar coordinates. Since

$$\mathbf{v} = \frac{d\mathbf{r}}{dt} = \frac{d}{dt}(r\hat{\mathbf{r}}) = \frac{dr}{dt}\hat{\mathbf{r}} + r\frac{d\hat{\mathbf{r}}}{dt}, \tag{2.29}$$

the velocity is

$$\mathbf{v} = \dot{r}\hat{\mathbf{r}} + r\dot{\theta}\hat{\boldsymbol{\theta}} + r\sin\theta\dot{\phi}\hat{\boldsymbol{\phi}}. \tag{2.30}$$

Taking the time derivative of both sides and using Equations (2.25) and (2.26) leads to the following complicated expression for the acceleration in spherical polar coordinates.

$$\begin{aligned} \mathbf{a} \;=&\; (\ddot{r} - r\dot{\theta}^2 - r\dot{\phi}^2\sin^2\theta)\hat{\mathbf{r}} + (r\ddot{\theta} + 2\dot{r}\dot{\theta} - r\dot{\phi}^2\sin\theta\cos\theta)\hat{\boldsymbol{\theta}} \\ &+ (r\ddot{\phi}\sin\theta + 2\dot{r}\dot{\phi}\sin\theta + 2r\dot{\theta}\dot{\phi}\cos\theta)\hat{\boldsymbol{\phi}}. \end{aligned} \tag{2.31}$$

Worked Example 2.7   Prove that the unit vectors in spherical coordinates are orthogonal. You may assume the orthogonality of the Cartesian unit vectors.

Solution   The unit vectors for the spherical coordinates can be expressed in terms of the Cartesian unit vectors by

$$\hat{\mathbf{r}} = \sin\theta\cos\phi\hat{\mathbf{i}} + \sin\theta\sin\phi\hat{\mathbf{j}} + \cos\theta\hat{\mathbf{k}}$$

$$\hat{\boldsymbol{\theta}} = \cos\theta\cos\phi\hat{\mathbf{i}} + \cos\theta\sin\phi\hat{\mathbf{j}} - \sin\theta\hat{\mathbf{k}}$$

$$\hat{\boldsymbol{\phi}} = -\sin\phi\hat{\mathbf{i}} + \cos\phi\hat{\mathbf{j}}.$$

If two vectors are orthogonal (mutually perpendicular) their dot product is zero. So

$$\begin{aligned} \hat{\mathbf{r}} \cdot \hat{\boldsymbol{\theta}} &= \left(\sin\theta\cos\phi\hat{\mathbf{i}} + \sin\theta\sin\phi\hat{\mathbf{j}} + \cos\theta\hat{\mathbf{k}}\right) \cdot \left(\cos\theta\cos\phi\hat{\mathbf{i}} + \cos\theta\sin\phi\hat{\mathbf{j}} - \sin\theta\hat{\mathbf{k}}\right) \\ &= \sin\theta\cos\phi\cos\theta\cos\phi + \sin\theta\sin\phi\cos\theta\sin\phi - \cos\theta\sin\theta \\ &= (\sin\theta\cos\theta)(\cos^2\phi + \sin^2\phi) - \cos\theta\sin\theta \\ &= \sin\theta\cos\theta - \cos\theta\sin\theta = 0. \end{aligned}$$

$$\therefore \hat{\mathbf{r}} \text{ and } \hat{\boldsymbol{\theta}} \text{ are perpendicular.}$$

Similarly

$$\begin{aligned} \hat{\mathbf{r}} \cdot \hat{\boldsymbol{\phi}} &= \left(\sin\theta\cos\phi\hat{\mathbf{i}} + \sin\theta\sin\phi\hat{\mathbf{j}} + \cos\theta\hat{\mathbf{k}}\right) \cdot \left(-\sin\phi\hat{\mathbf{i}} + \cos\phi\hat{\mathbf{j}}\right) \\ &= -\sin\theta\cos\phi\sin\phi + \sin\theta\sin\phi\cos\phi = 0. \end{aligned}$$

$$\therefore \hat{\mathbf{r}} \text{ and } \hat{\boldsymbol{\phi}} \text{ are perpendicular.}$$

Finally

$$\begin{aligned} \hat{\boldsymbol{\theta}} \cdot \hat{\boldsymbol{\phi}} &= \left(\cos\theta\cos\phi\hat{\mathbf{i}} + \cos\theta\sin\phi\hat{\mathbf{j}} - \sin\theta\hat{\mathbf{k}}\right) \cdot \left(-\sin\phi\hat{\mathbf{i}} + \cos\phi\hat{\mathbf{j}}\right) \\ &= -\cos\theta\cos\phi\sin\phi + \cos\theta\sin\phi\cos\phi = 0. \end{aligned}$$

$$\therefore \hat{\boldsymbol{\theta}} \text{ and } \hat{\boldsymbol{\phi}} \text{ are perpendicular.}$$

Exercise 2.21

A particle is at (3,4,5) meters in Cartesian coordinates. Determine r, θ, ϕ. Answer: $(7.07, \pi/4, 0.927)$. ∎

2.8 Summary

In this chapter you learned how to express the position, velocity, and acceleration of a particle in two dimensions in Cartesian coordinates and plane polar coordinates, and in three dimensions in Cartesian coordinates, cylindrical coordinates and spherical coordinates.

In three dimensions the position, velocity, and acceleration in Cartesian coordinates are given by the simple relations:

$$\mathbf{r} = x\hat{\mathbf{i}} + y\hat{\mathbf{j}} + z\hat{\mathbf{k}},$$

$$\mathbf{v} = \dot{x}\hat{\mathbf{i}} + \dot{y}\hat{\mathbf{j}} + \dot{z}\hat{\mathbf{k}},$$

$$\mathbf{a} = \ddot{x}\hat{\mathbf{i}} + \ddot{y}\hat{\mathbf{j}} + \ddot{z}\hat{\mathbf{k}}.$$

These relationships for two dimensions in plane polar coordinates are:

$$\mathbf{r} = r\hat{\mathbf{r}},$$

$$\mathbf{v} = \dot{r}\hat{\mathbf{r}} + r\dot{\theta}\hat{\boldsymbol{\theta}},$$

$$\mathbf{a} = (\ddot{r} - r\dot{\theta}^2)\hat{\mathbf{r}} + (r\ddot{\theta} + 2\dot{r}\dot{\theta})\hat{\boldsymbol{\theta}}.$$

In three dimensions using cylindrical coordinates, the relationships are:

$$\mathbf{r} = \rho\hat{\boldsymbol{\rho}} + z\hat{\mathbf{k}},$$

$$\mathbf{v} = \dot{\rho}\hat{\boldsymbol{\rho}} + \rho\dot{\phi}\hat{\boldsymbol{\phi}} + \dot{z}\hat{\mathbf{k}},$$

$$\mathbf{a} = (\ddot{\rho} - \rho\dot{\phi}^2)\hat{\boldsymbol{\rho}} + (\rho\ddot{\phi} + 2\dot{\rho}\dot{\phi})\hat{\boldsymbol{\phi}} + \ddot{z}\hat{\mathbf{k}}.$$

Finally, in spherical coordinates the position, velocity, and acceleration are given by:

$$\mathbf{r} = r\hat{\mathbf{r}},$$

$$\mathbf{v} = \dot{r}\hat{\mathbf{r}} + r\dot{\theta}\hat{\boldsymbol{\theta}} + r\sin\theta\dot{\phi}\hat{\boldsymbol{\phi}}.$$

$$\mathbf{a} = (\ddot{r} - r\dot{\theta}^2 - r\dot{\phi}^2\sin^2\theta)\hat{\mathbf{r}} + (r\ddot{\theta} + 2\dot{r}\dot{\theta} - r\dot{\phi}^2\sin\theta\cos\theta)\hat{\boldsymbol{\theta}}$$
$$+ (r\ddot{\phi}\sin\theta + 2\dot{r}\dot{\phi}\sin\theta + 2r\dot{\theta}\dot{\phi}\cos\theta)\hat{\boldsymbol{\phi}}.$$

2.9 Problems

Problem 2.1 A particle initially at rest undergoes an acceleration given by $a = 3e^{-0.5t}$ m/s^2. Determine the terminal velocity.

Problem 2.2 The nitrogen atom in an ammonia molecule can be assumed to oscillate in simple harmonic motion, so that its position at any time t is given by $z = A\cos\omega t$, where A and ω are constants. Obtain expressions for its velocity and acceleration as functions of time.

Problem 2.3 This is the story of Freddy Physics who got a ticket for going through a red light. He explained to the judge that if he slammed on the brakes when he saw the signal light turn yellow, he would not have been able to stop before the intersection, and if he continued at a constant speed the light would turn red before he reached the intersection. In this problem you will determine how far from the light Freddy was when the light turned yellow. Suppose he is traveling at the speed limit, v_0, that the light remains yellow for a time t_0 and that Freddy's reaction time is τ. His car decelerates at a rate a. (a) Determine the distance d_1 he will travel in time t_0 if he continues at the speed limit (b) Determine the distance d_2 required to stop. (c) Show that he is bound to go through the red light if $t_0 < \tau + v_0/2|a|$.

Problem 2.4 Assume a particle initially at the origin is given an acceleration $a = 10e^{-2t}$ m/s^2. The velocity at time $t = 0$ is zero. (a) Obtain expressions for the velocity and position of the particle as functions of time. (b) Determine the velocity as $t \to \infty$.

Problem 2.5 An object is passing through the origin at time $t = 0$ with a velocity of 1 m/s. It then undergoes an acceleration inversely proportional to the velocity: $a = \frac{k}{v}$. Determine the position of the object at time $t = 10$ s, if $k = 0.5$ m^2/s^3.

Problem 2.6 In some circumstances the acceleration is a function of the velocity. (An example is the motion of an object in a resistive medium such as air.) Assume $a = -kv^2$, where k is a constant. Determine the velocity and position as a function of time. What is the value of the velocity if the time is very large?

Problem 2.7 Mary is playing outfield and John is playing third base, 30.0 m away. Mary can throw a baseball at 20.0 m/s. Ignoring air resistance, determine the two launch angles she can use to get the ball to John.

Problem 2.8 A cannon launches a shell at 250 m/s at an elevation angle of 40°. Assuming level ground, determine the vertical and horizontal components of the velocity one second before the shell hits the ground.

Problem 2.9 When great athletes or ballet dancers leap into the air, they appear to "hang" at the top of the trajectory. (The reason for this is that the vertical speed of a projectile is smallest near the top of its path and greatest near the ground.) Show that half of the total time for the jump is spent in the top one quarter of the trajectory.

Problem 2.10 A long time ago, an imaginary country consisted of an island exactly 3 km wide with a 1.5 km high mountain running down its center, as shown in Figure 2.13. Owing to a serious misunderstanding, the army and navy went to war with each other. The navy gunboat was anchored offshore of the western side of the island, 2.5 km from the center. On the eastern shore a distance of 1.5 km from the center, the army had a cannon that fired shells at 200 m/s. The gunboat's cannon fired shells at 300 m/s. Find the angle at which the navy should aim its cannon. Show that the army cannot sink the navy ship.

Problem 2.11 A soldier at rifle practice is lying on the ground at the base of a hill that slopes upward at a constant angle of 10°. The soldier aims a rifle up the hill at an angle of 30° above the horizontal. If the initial speed of the bullet is 300 m/s, determine how far up the hill the bullet will hit.

Problem 2.12 A projectile is launched at an angle θ_0 with initial speed v_0. (a) Obtain an expression for y as a function of x. (b) Obtain an expression for the speed and direction of the projectile as a function of its position.

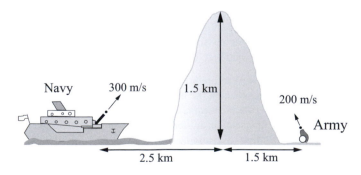

Figure 2.13 The navy gunboat and the army cannon of Problem 2.10. Do the shells of each one reach the other?

Problem 2.13 David is practicing with his sling before going out to meet Goliath. The sling is 1.3 m long and David whirls it in a horizontal circle over his head at 3 rev/s. The circular path of the sling is 2.0 m above the ground. How far does the stone fly horizontally before hitting the ground?

Problem 2.14 In projectile problems one usually assumes the gravitational acceleration is directed straight down. Assume the gravitational acceleration is a constant equal to 9.8 m/s^2 but pointing at an angle $30°$ below the horizontal (that is, at $330°$). Obtain an expression for the range of a projectile launched at $+20°$ to the horizontal with an initial velocity of 300 m/s.

Problem 2.15 (a) Determine the distance traveled along its parabolic path when a projectile moves from the origin to the top of its trajectory. It is launched with initial velocity v_0 at an angle θ. (b) Determine the displacement.

Problem 2.16 James Bond wants to impress Miss Moneypenny. They go out to an empty field and Bond fires his pistol at an elevation angle θ_1, then an instant later, he fires it again at a lower angle θ_2. The bullets collide in midair. Show that the time interval between the shots is

$$\Delta t = \frac{2v}{g} \frac{\sin(\theta_1 - \theta_2)}{\cos\theta_1 + \cos\theta_2}.$$

Problem 2.17 A projectile fired from $y = 0$ has reached the top of its trajectory. Its coordinates at this point are (x_t, y_t). Obtain a formula for the magnitude and direction of the initial velocity in terms of x_t, y_t.

Problem 2.18 A particle is moving in a circular path of radius 3 m with angular speed $\omega = 10$ rad/s. Place the origin of the Cartesian coordinate system at the center of the circle. Assume the initial value of θ is zero.

(a) Obtain expressions for its Cartesian coordinates (x, y) as functions of time.
(b) Express **r** and **v** in Cartesian coordinates.
(c) Determine the speed from your expression for **v**.

Problem 2.19 Assume the orbit of the Earth is perfectly circular. Using tabulated values for the radius of the Earth's orbit and the appropriate masses, determine the ratio of the centripetal accelerations of the Earth and the Sun. From this, determine the radius of the Sun's orbit around the center of mass of the system.

Problem 2.20 The position (x, y) of a particle (in meters) is given by

$$x = 3 \sin 5t$$
$$y = 3 \cos 5t.$$

(a) Plot the path of the particle. (b) Determine the tangential acceleration and (c) the centripetal acceleration of the particle.

Problem 2.21 Consider a "surface-skimming" satellite which is in a circular orbit whose radius is equal to the Earth's radius. (a) Obtain an expression for the period of such a satellite in terms of the acceleration and radius. (b) Look up the radius of the Earth and evaluate the acceleration, speed, and period of a surface-skimming satellite.

Problem 2.22 A spinning top is rotating at 20 rev/s about its axis. The axis itself is oriented at $15°$ from the vertical and is precessing at 0.5 rev/s about the vertical. What is the total vector angular velocity $\boldsymbol{\omega}$? (Express in Cartesian coordinates.)

Problem 2.23 A satellite is in a circular orbit around the Earth. The period, radius, and acceleration are related by $\tau = 2\pi \sqrt{r/a}$. Furthermore, it is known from Newton's law of universal gravitation that the acceleration decreases with distance from the center of the Earth according to

$$a \propto \frac{1}{r^2}.$$

If the orbital radius increases by a small amount Δr, the period will increase by $\Delta \tau$. Show that

$$\frac{\Delta \tau}{\tau} = \frac{3}{2} \frac{\Delta r}{r}.$$

Problem 2.24 Prove the triangle inequality which states that for two vectors **A** and **B**, $|\mathbf{A} + \mathbf{B}| \leq |\mathbf{A}| + |\mathbf{B}|$.

Problem 2.25 Show that $\mathbf{A} \cdot (\mathbf{B} \times \mathbf{C})$ is the volume of a parallelepiped whose edges can be represented by the vectors **A**, **B**, **C**.

Problem 2.26 Evaluate $(\mathbf{A} \times \mathbf{B}) \cdot \mathbf{C}$ if $\mathbf{A} = 3\hat{\imath} + 2\hat{\jmath} + 4\hat{k}, \mathbf{B} = 3\hat{\imath} + 2\hat{k}$ and $\mathbf{C} = 4\hat{k}$. Also show that you get the same result by evaluating $\mathbf{A} \cdot (\mathbf{B} \times \mathbf{C})$.

Problem 2.27 Show that

$$(\mathbf{A} \times \mathbf{B}) \cdot \mathbf{C} = \begin{vmatrix} A_x & A_y & A_z \\ B_x & B_y & B_z \\ C_x & C_y & C_z \end{vmatrix}.$$

Problem 2.28 Prove the "BAC minus CAB" rule:

$$\mathbf{A} \times (\mathbf{B} \times \mathbf{C}) = \mathbf{B}(\mathbf{A} \cdot \mathbf{C}) - \mathbf{C}(\mathbf{A} \cdot \mathbf{B}).$$

Problem 2.29 Derive the *law of cosines* for triangles, i.e., show that if the sides of a triangle have lengths a, b, c, and the angle between a and b is θ, then

$$c^2 = a^2 + b^2 - 2ab \cos \theta.$$

Problem 2.30 The position of a particle is given by $r = 2t$ (m), $\theta = 5t$ (rad). (a) Plot the path of the particle. (b) Obtain an expression for the speed as a function of time.

Problem 2.31 Suppose the position of a particle as a function of time is given by $r = 1 + \sin t$ (m) and $\theta = 1 - e^{-t}$ (rad). Obtain **v** and **a** in terms of $\hat{r}$ and $\hat{\theta}$.

Problem 2.32 A particle moves in a plane such that $r = 1 - \cos \theta$ m/s and $\dot{\theta} = 4$ rad/s. Determine **v** and **a**.

Problem 2.33 A particle moves in a plane in such a way that $r = 2 + \sin t$ (m) and the magnitude of the velocity is $v = \sqrt{2} \cos t$ (m/s). Find a formula for $\theta = \theta(t)$. Assume initial values of (r, θ) are (2,0).

Problem 2.34 For plane polar coordinates, determine

$$\frac{d^3 \mathbf{r}}{dt^3}.$$

Problem 2.35 Let $\hat{\mathbf{e}}_v$ be a unit vector parallel to the velocity vector **v**. Therefore, the acceleration is given by

$$\mathbf{a} = \left(\frac{dv}{dt}\right) \hat{\mathbf{e}}_v + v \left(\frac{d\hat{\mathbf{e}}_v}{dt}\right).$$

(a) Show that $(d\hat{\mathbf{e}}_v/dt)$ is a vector perpendicular to $\hat{\mathbf{e}}_v$. (b) Show that if r is the radius of curvature of the path of the motion, then $(d\hat{\mathbf{e}}_v/dt) = v/r$. (c) Finally, show that if $\hat{\mathbf{e}}_n$ is a unit vector directed towards the center of curvature on the concave side of the path, that we can write

$$\mathbf{a} = a_t \hat{\mathbf{e}}_v + a_n \hat{\mathbf{e}}_n = \dot{v} \hat{\mathbf{e}}_v + \frac{v^2}{r} \hat{\mathbf{e}}_n.$$

Problem 2.36 Derive Equations (2.27) and (2.28).

Problem 2.37 A bug is crawling on the curved surface of a cylinder of radius 25 cm. A scientist determines that the bug's coordinates vary according to $\phi = 4.0t$ and $z = 0.3t^2$. Write an expression for the bug's velocity in cylindrical coordinates and evaluate its speed at time $t = 2$ s.

Problem 2.38 A particle of mass m moves on the curved surface of a cylinder of radius R. Express the acceleration in cylindrical coordinates.

Problem 2.39 In cylindrical coordinates, the divergence of a vector **F** is given by

$$\nabla \cdot F = \frac{1}{\rho}\frac{\partial(\rho F_\rho)}{\partial \rho} + \frac{1}{\rho}\frac{\partial F_\phi}{\partial \phi} + \frac{\partial F_z}{\partial z}.$$

Evaluate $\nabla \cdot F$ for

$$\mathbf{F}(\rho,\phi,z) = \rho\hat{\rho} + z\sin\phi\,\hat{\phi} + \sqrt{\rho z}\hat{\mathbf{k}}.$$

Problem 2.40 An object initially at $(0,0,1)$ m in Cartesian coordinates, moves in the y-direction at a constant speed v. (a) Obtain an expression for its position as a function of time in spherical coordinates. (b) Determine $\dot{r}$ and $\dot{\theta}$ as functions of time. (c) Show that the speed, expressed in spherical coordinates, is constant.

Problem 2.41 A satellite moves with constant speed about the Earth in a circular orbit of radius a inclined at $90°$ to the equator, so that the satellite goes over the poles. A system of Cartesian coordinates fixed in the Earth has its origin at the center of the Earth, the z-axis passing through the North Pole, and the x-axis in the equatorial plane and passing through the Greenwich meridian. Obtain expressions for the position of the satellite (as functions of time) in Cartesian coordinates. The angular speed of the satellite is ω (rad/s) and the rotation rate of the Earth is Ω (rad/s).

Problem 2.42 Show that the element of arc length in spherical coordinates is

$$ds = \left(dr^2 + r^2 d\theta^2 + r^2 \sin^2\theta d\phi^2\right)^{1/2}.$$

Problem 2.43 Derive the equations of transformation between cylindrical and spherical coordinates.

Problem 2.44 Show that the unit vectors of the spherical coordinate system can be expressed in terms of Cartesian coordinates as follows:

$$\hat{\mathbf{r}} = \frac{x\hat{\mathbf{i}} + y\hat{\mathbf{j}} + z\hat{\mathbf{k}}}{\left(x^2 + y^2 + z^2\right)^{1/2}}$$

$$\hat{\theta} = \frac{z(x\hat{\mathbf{i}} + y\hat{\mathbf{j}}) - \left(x^2 + y^2\right)\hat{\mathbf{k}}}{\left(x^2 + y^2\right)^{1/2}\left(x^2 + y^2 + z^2\right)^{1/2}}$$

$$\hat{\phi} = \frac{-y\hat{\mathbf{i}} + x\hat{\mathbf{j}}}{\left(x^2 + y^2\right)^{1/2}}.$$

Problem 2.45 An object is moving in such a way that its position in cylindrical coordinates is given by $\rho = a - bt$, $\phi = \frac{1}{2}kt^2$, $z = 0$. (a) Describe the path. (b) Find the speed as a function of time.

Problem 2.46 An object moves at constant *speed*. Prove that any nonzero acceleration must be perpendicular to the velocity.

Problem 2.47 (a) Show that the component of the acceleration in the direction of the velocity can always be expressed as

$$a_{\parallel} = \frac{\mathbf{v} \cdot \mathbf{a}}{v}.$$

(b) Find the component of the acceleration perpendicular to the velocity.

Computational Projects

Computational Project 2.1 A certain (imaginary) comet is in orbit about the Sun. It follows an elliptical path that is given by

$$r = \frac{10}{1 + 0.8\cos\theta},$$

where r is measured in Astronomical Units (AU). Plot the orbit of the comet in polar coordinates.

Computational Project 2.2   A cannon fires a projectile at an initial speed of $v_1 = 250$ m/s at an elevation angle of $\theta_1 = 64.1°$. After a time interval Δt, it fires another projectile at $v_2 = 300$ m/s and $\theta_2 = 47.3°$. Determine the time interval Δt such that the two projectiles collide in midair. Check your answer against the analytical result

$$\Delta t = \frac{2v_1 v_2}{g} \frac{\sin(\theta_1 - \theta_2)}{v_1 \cos \theta_1 + v_2 \cos \theta_2}.$$

Computational Project 2.3   A volcanic eruption throws a boulder vertically into the air with an initial speed of 50 m/s. (a) Write a computer program to plot the position and velocity of the boulder as a function of time. (b) Now assume the boulder is ejected at an angle of ten degrees to the vertical. Plot the two components of position and velocity as functions of time and also plot the trajectory of the boulder (on an x, y plot). Ignore the effect of air resistance.

Computational Project 2.4 In Worked Example 2.3 we obtained an expression for the launch angle that would give the greatest range for a cannon mounted on the top of a hill. (a) Validate that expression by varying the launch angle θ. (b) Determine how the optimum launch varies as the height of the hill is increased from zero.

Computational Project 2.5 Worked Example 2.6 describes a bead sliding on a helical wire. Plot the trajectory of the bead.

Computational Project 2.6 In a shotput competition an athlete throws a heavy metal ball as far as possible. A champion athlete can throw the "put" a distance of 20 m. Assume the ball is thrown with an initial speed of 13.66 m/s at an angle of $45°$. If it is thrown from a height of one meter above the ground, the ball's initial position is $(0,1)$. Using the relations

$$x = x_0 + v_{0x}t,$$
$$y = y_0 + v_{0y}t - (1/2)gt^2,$$

obtain $x(t)$ and $y(t)$. Plot the trajectory for $y > 0$. Also plot trajectories for angles ranging from $20°$ to $70°$.

3 Newton's Laws: Determining the Motion

This chapter (and much of the rest of this book) deals with dynamics, that is, the relation between the forces acting on a body and its motion. A force is an interaction between a body (or particle) and its environment, often described as a push or a pull in a specified direction. In this book you will encounter a number of familiar forces, such as the gravitational force and forces exerted by springs, as well as a few less-familiar ones.

On a very basic level, all forces are manifestations of the "fundamental forces," namely, the gravitational force, the nuclear (or strong) force, and the electroweak force. The electroweak force is often thought of as two different forces, the electromagnetic force and the weak nuclear force. The electromagnetic force, in turn, is often thought of as the electric force and the magnetic force. All the known forces in nature are ultimately related to these fundamental forces.[1] For example, the force exerted by your muscles can be traced back to electrical forces.

Dynamics is neatly summarized by Newton's three laws. You studied Newton's laws in your introductory mechanics course. By this time, you surely know the laws by heart and you know how to use them to solve reasonably complicated physics problems. However, it is important for you as a physicist to have a thorough understanding of these fundamental statements about the nature of the physical universe. In this chapter I will discuss some aspects of Newton's laws that you may not have considered before. I hope this will give you a greater appreciation for the scope and significance of the laws.

3.1 Isaac Newton (Historical Note)

In some people's minds, Sir Isaac Newton was the greatest physicist who ever lived. Others might give that honor to Albert Einstein, but no one would deny the tremendous insight and genius of Isaac Newton.

Newton was born in 1642, the year Galileo died.[2] Newton's father died before Isaac was born, and his mother moved to her brother's farm where the boy grew up. When the time came, his uncle sent him to Cambridge. While he was studying at Cambridge, a plague swept through England, and the university authorities sent the students home for a year. Isaac went back to the farm where he spent his time thinking about the properties of the physical universe and doing simple but very clever experiments. After the year was up, the plague had run its course, and Newton returned to Cambridge. We can only imagine the scene when he and his major professor met.

---

[1] To learn more about the fundamental forces you might read the interesting article by Charles Seife, "Can the Laws of Physics be Unified?" *Science*, **309**, 82 (2005).

[2] Newton was born on December 25, 1642, according to the Julian calendar then in use in England. In 1753, England adopted the Gregorian calendar making Newton's birth date January 4, 1643.

Professor: Hello Isaac. Welcome back. Did you spend your time fruitfully while the University was closed?

Newton: Yes sir, I believe I did.

Professor: Very good, Isaac. Precisely what *did* you do?

Newton: Well, sir, I proved the binomial theorem, invented calculus, designed and built a reflecting telescope, derived the law of universal gravitation, developed a theory of optics, and determined three fundamental laws of nature governing the motion of any physical object.

According to legend, the professor quit his job and left the position to Newton.

Newton did not formulate his laws out of the blue. He was very familiar with the work of his predecessors, especially Galileo and Kepler. (As he stated, "If I have seen further than others, it is by standing on the shoulders of giants.") In fact, Newton's first law, the law of inertia, was formulated and demonstrated by Galileo. It is possible that Galileo got the idea from René Descartes. However, the second and third laws were first formulated by Isaac Newton.[3]

Although Newton deduced the law of universal gravitation when he was very young, he did not publish it until many years later because his early calculations did not give the correct value for the period of the Moon. Urged by his friend, the architect Sir Christopher Wren, Newton carried out the calculations more carefully, obtained the right answer, and convinced himself that his law was correct.

Newton invented the reflecting telescope and carried out a large number of experimental investigations, particularly in optics. He was also interested in alchemy and theology. But his greatest contribution was in formulating physics as an exact mathematical science. His genius was immediately recognized and he soon became one of the most famous men in Europe. After Newton, people realized that natural processes take place in a manner that can be analyzed and predicted. People began to think of the universe as a "clockwork" mechanism in which the future development of any system could (in principle) be determined from a knowledge of its present state. The poet Alexander Pope, eulogized Newton with the couplet,

Nature and Nature's Laws lay hid in Night,
God said, *Let Newton be!* and All was Light.

Personally, Newton was not easy to deal with. It was said of him that, "He suffered people poorly, and fools not at all." Because his position at Cambridge was a semiclerical post, he was not allowed to marry. It appears that the life of a bachelor suited him well. He died in 1727 at the ripe old age of eighty four.[4]

3.2 The Law of Inertia

The law of inertia, or Newton's first law,[5] can be stated as follows:

A body in motion will remain in uniform motion and a body at rest will remain at rest unless acted upon by a net external force.

---

[3] Newton expressed the laws of motion in his famous book, *Philosophia Naturalis Principia Mathematica*, published in 1687. A fairly recent English version is *The Principia: Mathematical Principles of Natural Philosophy* translated by I. Bernard Cohen and Anne Whitman, University of California Press, Berkeley, 1999.

[4] A recent very readable biography is *Isaac Newton* by James Gleick, Pantheon Books, New York, 2003.

[5] This law was first expounded by Galileo and Newton gave Galileo full credit for it, but over the years we have grown accustomed to calling it "Newton's first law."

   If you happen to be a purist, you might want to know Newton's precise words. He said, "Every body perseveres in its state of rest, or of uniform motion in a right line, unless it is compelled to change that state by forces impressed thereon." (*The Principia* by Isaac Newton, translated by Andrew Motte from the Latin. Published by D. Adee, 1848.)

   I urge you to express Newton's laws in your own words. In fact, sometimes I state them one way and sometimes another. The crucial thing is not the *words* we use but the *concept* expressed by those words.

You have certainly thought about this law and you realize that it says that in the absence of *external* forces a body will maintain its state of motion. This is a very profound concept. Like many profound concepts, after you have accepted it, it appears completely obvious. It is important to note that *internal forces* cannot change the motion of a system. If you sit on a chair and pull up on the seat, you remain at rest no matter how hard you pull. You cannot lift yourself into the air. The reason is, of course, that your pull is an internal force.

You might object that an exploding bomb is an example of internal forces causing a system to accelerate. After the explosion, pieces of the bomb are flying in all directions. This is true. But the center of mass of the bomb remains at rest. The position of an extended body is described by the position of its center of mass and if you trace the trajectories of all the pieces, you will find that the center of mass did not accelerate.

Perhaps the most amazing thing about the law of inertia is the idea that a moving body will continue to move even though no external force is applied to it. This was hard for people in Newton's time to understand. It is still hard for some people to understand. They say, "When I push a trunk across the floor, as soon as I quit pushing, the trunk quits moving." Of course, you and I know the trunk stops moving because there *is* a force acting on it – the force of friction. Furthermore, you are familiar with games such as air hockey in which the puck slides across the table at a (nearly) constant velocity because the frictional force has (very nearly) been eliminated. You have also been exposed to countless TV scenes of astronauts floating in space, so you do not have the conceptual difficulties your great-grandfather had in accepting that an object, not subjected to any force, will move forever in a straight line at a constant velocity.

Consider a moving body with no external forces acting on it. According to the first law, this body will move at a constant speed in a straight line. If you think about it for a moment, you will realize that this will only be true in certain specific reference frames. For example, if you imagine you are riding in a modern very smooth train and someone places a tennis ball in the aisle, it will remain at rest in the aisle as long as the train is moving at a constant velocity. But if the train accelerates the ball will roll towards the back. If the train slows down, the ball rolls forward. If the train turns left, the ball rolls to the right. If I were somebody who did not know very much physics, I might conclude that the ball does not obey Newton's first law. You realize, of course, that I am describing the motion of the ball *relative to the train* and the train is accelerating. So you would refute my statement by pointing out that *relative to the ground* the ball *did* obey the first law. It *did* maintain a state of uniform motion in the reference frame of the Earth. I might then point out that the Earth is also an accelerating reference frame because it is rotating. How would you respond?

If we follow such arguments to their logical conclusion we will decide that the first law is only valid in a nonaccelerating reference frame. (This is also true of Newton's other two laws.) A nonaccelerating frame is called an *inertial reference frame*. Does such a reference frame actually exist? Newton stated that his laws were valid in a reference frame that was "at rest with respect to the fixed stars." But we know the stars are all in motion! It is probably best to treat the concept of an inertial reference frame as a useful idealization. For many problems, the Earth can be treated as if it were at rest. This approximation breaks down when the rotation of the Earth must be considered. Then we usually assume that the inertial reference frame is a nonrotating reference frame with origin at the center of the Earth. That approximation breaks down if we need to include the orbital motion of the Earth around the Sun. If necessary, we can take the origin of coordinates to be at the center of the Sun, or even the center of the galaxy.

Some people think of the first law as a definition of an inertial reference frame and what it is really telling us is that the next two laws are only valid in such a reference frame.

The physical property called "inertia" is associated with the fact that a moving body tends to preserve its state of motion. The expression, "preserving its state of motion" means the body has constant velocity. A locomotive has more inertia than a ping-pong ball. The word inertia thus appears to be a synonym for *mass*. Because it requires a great force to change the motion of an object with great mass, one often hears the expression, "Mass is a measure of the inertia of a body." But the tendency of a body to maintain its state of motion also depends on its velocity. It is easier to deflect a slowly moving five gram ping-pong ball than a speeding five gram bullet. In this case, inertia appears to be a synonym for *momentum.*

It is very difficult to define fundamental quantities such as mass, distance, time, charge, and so on. Similarly, inertia is usually described by the somewhat vague expression that it is the tendency of a body to maintain its state of motion. (We do not have a formula for inertia!) On the other hand, the *law* of inertia (Newton's First Law) is perfectly well defined.

3.3 Newton's Second Law and the Equation of Motion

Newton's second law can be stated in the form:

**The rate of change of the momentum of an isolated body
is equal to the net external force applied to it.**

In equation form this is written:

$$\frac{d\mathbf{p}}{dt} = \mathbf{F}, \tag{3.1}$$

where $\mathbf{F}$ is the net or total force acting on the body. (You may prefer to write it as $\mathbf{F}_{net}$ or as $\Sigma\mathbf{F}$.) Now momentum is defined as $\mathbf{p} = m\mathbf{v}$, so

$$\mathbf{F} = \frac{d}{dt}(m\mathbf{v}) = \frac{dm}{dt}\mathbf{v} + m\frac{d\mathbf{v}}{dt}.$$

If the mass is constant, Newton's second law takes on the familiar form $\mathbf{F} = m\mathbf{a}$. Remember that Newton's second law is applicable only in an inertial reference frame.

Newton's second law is sometimes referred to as the "equation of motion." The reason is that if the forces are known, we can determine the acceleration of the body, and once the acceleration has been determined we can integrate to obtain the "motion," that is, the velocity and position as functions of time. According to classical mechanics, a knowledge of the position and velocity of all the particles in a system allows one to determine the forces and hence to predict the future development of the system.[6]

An expression for the acceleration in terms of the velocity, position, and time is called an equation of motion. Much of our work in this course will involve obtaining and solving the equation of motion for a variety of different physical systems. The equation of motion $a = a(x, v, t) = F/m$ can be obtained directly from an analytical expression for F. As you will learn in Chapter 4, the equation of motion can be obtained without explicitly using Newton's second law. However, for now, we will stick to the second law.

Although we shall not be particularly concerned with philosophical questions, it should be mentioned that since the publication of Newton's great work, *Principia*, there has been much debate and speculation on the meaning of the second law. Clearly it is a statement of how a body reacts when a force is applied to it. If the mass is constant, the second law tells us that the

---

[6] This is a wonderful fact of nature. As stated by Landau and Lifshitz on the very first page of their excellent advanced mechanics book, "If all the coordinates and velocities are simultaneously specified, *it is known from experience that the state of the system is completely determined and that its subsequent motion can, in principle, be calculated.*" (Emphasis added.) This suggests that there is no fundamental reason why the motion of a system depends only on position, velocity, and time; we must accept it as an experimental fact. (Reference: *Mechanics, Course in Theoretical Physics, Volume 1,* L. D. Landau and E. M. Lifshitz, Pergamon Press, New York, 1976, page 1.) Interestingly, in electrodynamics there is a quantity called the "Abraham–Lorentz force" that is usually expressed as a function of acceleration. This force leads to some strange behavior that has been referred to as "philosophically repugnant" by David Griffiths (*Introduction to Electrodynamics,* 3rd ed., Prentice Hall, 1999, page 467).

acceleration of the body is proportional to the force and inversely proportional to the mass ($a = F/m$). From this point of view, mass becomes the proportionality constant in the relationship $a \propto F$. That is, mass is a resistance to acceleration, the resistance of a body to changing its state of motion. Some thinkers have gone so far as to state that the second law is just a definition of mass. But if mass is nothing more than a measure of inertia, one cannot explain the Law of Universal Gravitation ($\mathbf{F} = -G\left(mM/r^2\right)\hat{\mathbf{r}}$), which states that the gravitational force between two bodies of masses m and M is proportional to the product of the masses of the bodies. The gravitational law implies that mass is the *source* of the gravitational force. Perhaps the "mass" in the gravitational force law is not the same as the "mass" in Newton's second law. Perhaps we are using the same word for two different things. In that case, the question could be resolved by using different symbols and different names for the two things. Thus, for example, we could define the quantities m_G and m_I as:

m_G = gravitational mass = the property by which a body exerts a gravitational force on other bodies, and Newton's gravitational law would be expressed as $\mathbf{F} = -G\left(m_G M_G/r^2\right)\hat{\mathbf{r}}$,

and

m_I = inertial mass = the property by which a body resists a change in its state of motion, so that Newton's second law would be expressed as $\mathbf{F} = m_I \mathbf{a}$.

However, many very careful experiments (some of the most famous were by Baron Lorand Eötvos in Hungary over 100 years ago), showed that m_G and m_I are equal to within one part in 20 million. This leads one to conclude that gravitational mass and inertial mass are indeed the same thing. The fact that inertial mass and gravitational mass are equal is called "The principle of equivalence." Albert Einstein used this equivalence as a basic postulate of his General Theory of Relativity (1915).[7]

Newton's second law is a cornerstone of classical physics because it can be used to calculate the acceleration of a body, given the force acting on it. Once the acceleration of a body is known, the laws of kinematics determine its velocity and position at any later time. This means that if you know the net force acting on a body, you can calculate its position at any future time. The ability to *predict* the motion is the power of the second law.

Recall that dynamics is the study of how a force affects the motion of a body. As you might suspect, dynamics usually involves accelerating bodies. However, in some situations there may be forces acting on a body but nevertheless the acceleration is zero. Consider, for example, a body acted upon by two equal and opposing forces. The effects of these forces cancel out and the body does not accelerate. Zero acceleration is an important special case in dynamics and is called *statics*. Statics is of particular interest to civil engineers who want to make sure the structures they design, such as bridges and skyscrapers, will have zero acceleration. Statics was treated briefly in Section 1.6 and will be dealt with in greater detail in Chapter 18.

The *principle of superposition* states that if two or more forces act on a particle, the net effect is the same as that of a single force equal to the *vector sum* of all the forces. You will be exposed to the principle of superposition in other areas of physics. For example, the net electric field at a point is the vector sum of all the electric fields acting at that point.[8]

---

[7] If you are interested in some of the philosophical implications of mass, force, inertia and Newton's laws, you might enjoy the three articles by Franck Wilczek entitled, "Whence the Force of $F = ma$?". These articles were published in *Physics Today* in the issues of December 2004, July 2005, and October 2005. The article "Drop Test" by Adrian Cho (*Science,* 6 March, 2015, p. 1096) describes three experiments to measure the equivalence of inertial mass and gravitational mass to better than one part in ten trillion.

[8] The principle of superposition is a property of linear systems in which the output (result) is proportional to the input (cause). Thus, $F = ma$ is linear because the result (an acceleration) is proportional to the input (a force). Except for a discussion of chaos in Chapter 21, we shall not consider nonlinear dynamics in this book.

Figure 3.1 What direction does the tricycle move when you pull on the bottom pedal as shown?

In our discussion of Newton's laws we considered interactions between particles. When we apply Newton's second law to an extended body the acceleration **a** refers to the acceleration of the center of mass of the body. A net external force **F** applied *at any point* of an extended rigid body will cause the center of mass of the body to accelerate according to $\mathbf{a} = \mathbf{F}/m$. This is often not at all intuitive. For example, consider a tricycle with the pedals in a vertical position, as shown in Figure 3.1. If you pull forward on the top pedal, the bicycle will move forward. But what happens if you pull forward on the bottom pedal? It might seem that this would propel the tricycle backward, but Newton's second law tells us that if we pull forward, the tricycle moves forward. (If you do not believe me, try it! You will be surprised by the motion of the pedal itself. This is not at all the same as what would happen if you were *riding* the tricycle. In that case, the force your foot exerts on the pedal is an *internal* force.)

Exercise 3.1

In a famous demonstration, while standing on the airless surface of the Moon, Astronaut David Scott dropped a hammer and a feather. The fact that they fell together has been cited as a proof of the equivalence of gravitational and inertial mass. Explain why this demonstration only proves that m_I and m_G are proportional to one another rather than equal to each other. ∎

Exercise 3.2

A box of mass 5 kg is placed on an inclined plane of angle 35°. A force of 10 N parallel to the plane in a direction up the plane is applied to the box. The coefficient of sliding friction is 0.07. Determine the acceleration of the box. (Hint: Is the net force up or down the plane?) Answer: -3.05 m/s^2 (down the plane). ∎

Exercise 3.3

A crate of mass 50 kg is sitting on the flat bed of a truck. The truck accelerates at 0.2 m/s^2. The coefficient of static friction is 0.15. Does the crate slide? Answer: No. ∎

Exercise 3.4

A spool of thread is lying on its side on a table and is free to roll. You pull on the free end of the string. Does the spool move towards you? Does it matter whether the string comes from the bottom side of the spool or the top side? (Try it!) ∎

Figure 3.2 Illustrating the strong form and the weak form of Newton's third law. The arrows represent the forces acting on the particles. In both cases the forces are equal and opposite, but in the strong form the forces act along the line joining the particles.

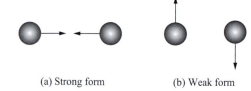

(a) Strong form (b) Weak form

3.4 Newton's Third Law: Action Equals Reaction

Using Isaac Newton's terminology of "action" and "reaction" we can formulate his third law as follows:

To every action there is always opposed an equal reaction.

In different words, if one body exerts a force on a second body, the second body exerts an equal but oppositely directed force of the same kind on the first body. In applying the third law, it is important to remember that the two forces involved act on *different* bodies.

The third law states that the force one body exerts on another body (the "action") is equal and opposite to the force the second body exerts on the first (the "reaction"). The law does not, however, state that these forces must lie along the same line. Thus, in Figure 3.2 we see that in both case (a) and case (b) the forces are equal and opposite, but in case (b) the forces do not act along the same line. When the action–reaction forces lie along the same line we say the third law is obeyed in its *strong* form. Otherwise, we say it is obeyed in the *weak* form.

Newton's third law is intimately related to the law of conservation of linear momentum. Consider, for example, an isolated system consisting of two particles that exert equal and opposite forces on each other:

$$\mathbf{F}_1 = -\mathbf{F}_2.$$

By the second law, the forces can be expressed as changes in the momentum of the particles:

$$\frac{d\mathbf{p}_1}{dt} = -\frac{d\mathbf{p}_2}{dt}.$$

Therefore,

$$\frac{d}{dt}(\mathbf{p}_1 + \mathbf{p}_2) = 0.$$

That is, the total momentum of an isolated system is constant.

Is the third law always obeyed? This is a question that physicists have debated for many years. We can state unequivocally that the third law is always obeyed in purely mechanical systems, but the situation is not quite as clear when we consider the electromagnetic interaction between charged particles, as illustrated by the following example.

Worked Example 3.1 Consider the action–reaction pair of forces for two charged particles interacting through the magnetic force. See Figure 3.3. Is the third law obeyed?

Solution Figure 3.3 shows two moving charged particles. The problem is to determine if the forces they exert on each other are equal and opposite.

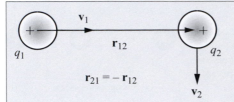

Figure 3.3 Two moving charged bodies give rise to an apparent counter example to Newton's third law. For this arrangement $\mathbf{v_1} \times \mathbf{r_{12}} = 0$ whereas $\mathbf{v_2} \times \mathbf{r_{21}} \neq 0$.

You remember from your introductory physics course in electricity and magnetism that the force on a charge Q moving at velocity $\mathbf{V}$ in a magnetic field $\mathbf{B}$ is

$$\mathbf{F} = Q\mathbf{V} \times \mathbf{B} .$$

The magnetic field $\mathbf{B}$ acting on Q is due to the motion of another charge Q' and is given by[a]

$$\mathbf{B} = \left(\frac{\mu_0}{4\pi r^3}\right)(Q'\mathbf{V}' \times \mathbf{r}),$$

where $\mathbf{r}$ is the vector from Q' to Q.

Applying to our problem, the $\mathbf{B}$ field at q_2 due to q_1 is

$$\mathbf{B_{12}} = \frac{\mu_0 q_1}{4\pi r_{12}^3}(\mathbf{v_1} \times \mathbf{r_{12}}) = 0$$

because $\mathbf{v_1} \parallel \mathbf{r_{12}}$. On the other hand, the $\mathbf{B}$ field at q_1 due to q_2 is

$$\mathbf{B_{21}} = \frac{\mu_0 q_2}{4\pi r_{21}^3}(\mathbf{v_2} \times \mathbf{r_{21}}) \neq 0$$

because $\mathbf{v_2} \perp \mathbf{r_{21}}$. Therefore, the force on q_1 is nonzero, but the force on q_2 is zero. We might therefore conclude that the third law has failed! This result is actually cited in many physics texts as evidence that the third law is not universal and does not apply to electromagnetic forces.

But wait! Our conclusion was reached too hastily! Upon further (and much deeper) analysis one finds that the two moving charged particles set up electromagnetic fields that have momentum and energy associated with them. When one takes into consideration the forces acting on the particles due to the rate of change of electromagnetic momentum, then the third law *is* found to hold. Furthermore, if the charges are part of electric currents in two nearby circuits, the forces exerted by the circuits on one another are equal and opposite. Consequently, it is safe to conclude that the third law has the same general range of validity as the first two.[b]

[a] This relation is based on the Biot–Savart law which, strictly speaking, is not valid for point charges. However, the correct expressions reduce to the given formula as long as the velocity of the particle is much less than the speed of light. (D. J. Griffiths, *Introduction to Electrodynamics*, 3rd ed., Prentice Hall, 1999, p. 439.)

[b] Roald K. Wangsness, *Electromagnetic Fields*, 2nd ed., Wiley and Sons, New York, 1986, pp. 219 and 359. For a different point of view, see *Classical Dynamics* by Jerry Marion and Stephen Thornton, 3rd ed., Harcourt, Brace, Jovanovich, 1988, p. 45.

The point of the preceding example is that you, as a physicist, must be very careful in applying the third law. As an illustration, let me give you a puzzle. You may have heard it before. It is the story of a donkey hitched to a cart. The donkey pulls on the cart. The cart pulls back on the donkey. By Newton's third law, action equals reaction, and these two forces are equal and opposite. Thus it would appear that the forces cancel out. How, then, is it possible for the cart to move?

Give up? Well, the answer is that the forces do not cancel because they are acting on different bodies! Suppose a (single) body is acted upon by two forces of equal magnitude but opposite directions. You can then say that these forces cancel. But when you are adding up the forces that are acting on a particular body, you certainly cannot include forces that are acting on a *different* body! You would never say the force the Sun exerts on the Earth is canceled by the force the Earth exerts on the Sun. These forces are indeed equal and opposite, but they act on different bodies. Similarly, for the cart and donkey problem, the donkey exerts a force *on the cart* and the cart exerts an equal and opposite force *on the donkey*. Considering only the cart, the forces on the cart are the force exerted by the donkey and frictional forces (the road on the wheels, etc.). If the force exerted by the donkey is greater than the frictional force, the cart accelerates forward. You should also think about the forces acting on the donkey and explain why the donkey is accelerating forward and not backward. The only forces that can make an object accelerate are *external* forces. You cannot pick yourself up by pulling on your shoestrings!

In determining action–reaction pairs, it is useful to remember that they are always the same *kind* of force. For example, a book on a tabletop is pulled downwards by the gravitational attraction of the Earth. It is easy to make the error that the reaction is the upward normal force of the table. But this is not the same *kind* of force. The reaction is the upward gravitational pull of the book on the Earth. What is the reaction force to the normal force exerted by the table?

3.5 Is Rotational Velocity Absolute or Relative?

Linear motion is relative. A person in a moving train is at rest relative to the train, but has nonzero velocity relative to the ground. The velocity of an object depends on (is *relative* to) the reference frame in which it is viewed.

Galileo suggested that a person in the hold of a ship sailing in perfectly smooth water would not be able to determine if the ship was moving. According to Newton (and later Einstein) it is impossible to distinguish between one inertial reference frame and another. If you were to place a TV camera inside a closed box, there would be no physical phenomenon you could observe and no experiment that you could carry out that would tell you whether the box was at rest, or moving at a constant velocity.

Is rotational motion also relative? Newton pondered this question and decided that rotation is absolute. He described an experiment in which he suspended a bucket full of water from a long rope. He turned the bucket, twisting the rope. Upon releasing the bucket, he observed the water level while the rope unwound and the bucket rotated. He noted that initially, when the bucket and water were at rest, the surface of the water was flat. Next, the bucket began to rotate, but the water was still at rest (for a while) and the surface of the water was still flat. Finally, the water took on the rotation of the bucket and the surface of the water became concave. (I am sure none of this surprises you.)

Now in the initial and final stages, the water was not moving relative to the bucket, but the surface was flat in the initial stage and concave in the final stage. Therefore, the behavior of the surface was not due to the motion of the water relative to the bucket. Also, when the bucket was rotating but the water was not rotating, the surface remained flat. Newton concluded that the concavity of the surface was due to the *absolute* rotation of the water and not its motion *relative* to the bucket.

If you placed the bucket of water in a closed box and set the box rotating, a TV camera inside the box would show the surface of the water to be concave. Thus there is a physical phenomenon that can be used to determine whether or not a reference frame is rotating. There is no other reference frame (Newton decided) in which the surface of the water is flat, so rotation is not relative.[9]

---

[9] Newton stated that centrifugal force is the feature that distinguishes absolute motion from relative motion.

Ernst Mach was fascinated by this simple experiment. However, he believed that all motion was relative, including rotational motion. Mach claimed that rotation was relative to all the mass in the universe. He reasoned that a bucket of water at rest would have a flat surface, just as Newton observed. But, he asked, would the surface be curved if the bucket were spun in a totally empty universe? In other words, what would the bucket be spinning relative to? From Mach's point of view, it is the rest of the universe that causes the surface to be curved, and the same effect would be obtained by spinning the entire universe around a stationary bucket. You cannot determine whether the water is rotating and the universe is at rest, or the water is at rest and the whole universe is rotating around it! Einstein's theory of general relativity does not include this "total relativity" of Mach, but later in his life Einstein tried to analyze the consequences of imposing it on his theory.

Although the question of whether or not rotation is relative or absolute is still being debated,[10] we shall go along with Newton and assume that rotation is absolute. This point of view is supported by an in-depth study of Mach's principle that shows there are many effects that cannot be explained by a rotating universe and a stationary bucket.[11] For example, Mach's principle cannot explain how two buckets, rotating in opposite directions, can both have concave surfaces. Finally, we might note that in quantum mechanics we find that rotational motion is quantized whereas linear motion is not. A body can have any linear velocity whatever, but a body can only have certain discrete values for rotational motion.

An important consequence of Newton's bucket experiment is that it demonstrates that acceleration is an absolute quantity and not relative because rotational motion is accelerated motion and rotation is absolute. In other words, the velocity of a particle depends on the reference frame but its acceleration is the same in all inertial reference frames.

3.6 Determining the Motion

Recall that "determining the motion" means obtaining an equation for the *position* of the body as a function of *time*. This is done by solving the *equation of motion*. Although we shall be considering other forms of the equation of motion, for now let us assume it is simply Newton's second law. Then the equation of motion is a second-order differential equation giving the acceleration as a function of position, velocity, and time, $a = \ddot{r} = \frac{d^2r}{dt^2} = F(r, \dot{r}, t)/m$. Consequently, solving for the motion involves carrying out two integrations.

In this section you will learn how to obtain expressions for the velocity and position of a particle subjected to several different kinds of force. Specifically, we will consider the following types of force.

Force is constant:	$F = \text{constant}$.
Force is a function of time:	$F = F(t)$.
Force is a function of velocity:	$F = F(v)$.
Force is a function of position:	$F = F(x)$.

Unless otherwise noted (and to keep things simple) the motion will be one dimensional and the object that is moving will be a particle of constant mass. The generalization to two or three dimensions is quite straightforward.

---

[10] An interesting article on this subject is "Total Relativity: Mach 2004" by Frank Wilczek, *Physics Today*, April 2004, page 10. Newton's bucket is also used as a starting point for the very readable book, *The Fabric of the Cosmos: Space, Time and the Texture of Reality* by Brian Greene, Knopf, New York, 2004.

[11] H. Hartman and C. Nissim-Sabat, "On Mach's Critique of Newton and Copernicus," *American Journal of Physics*, **71**, 1163 (2003).

Constant Force

If the force is constant, from $F = ma$ we appreciate that the acceleration is constant. The equation of motion is

$$\ddot{x} = F/m = \text{constant} = a,$$

and integrating $dv = adt$ yields

$$v(t) = v_0 + at. \tag{3.2}$$

Integrating again we obtain

$$x(t) = x_0 + v_0 t + \frac{1}{2}at^2. \tag{3.3}$$

Equation (3.3) giving $x = x(t)$ is the *solution* of the equation of motion. Thus, we have "determined the motion."

You learned Equations (3.2) and (3.3) in your introductory physics course. Perhaps you did not fully appreciate at that time that these equations *are only valid if the force is constant*. The rest of this chapter deals with forces that are *not* constant.

Exercise 3.5

It is observed that the position of an object of mass 3 kg is given by $x(t) = 3t + 6t^2$ m. Determine the force acting on it. Answer: 36 N. ∎

Force as a Function of Time

A somewhat more complicated situation arises when the force F can be expressed as a function of time, i.e., $F = F(t)$. Then the equation of motion is

$$a = \frac{F(t)}{m}.$$

or

$$\frac{dv}{dt} = \frac{1}{m}F(t).$$

Separating variables and integrating:

$$\int_{v_0}^{v(t)} dv = \frac{1}{m}\int_0^t F(t)dt,$$

or

$$v(t) = v_0 + \frac{1}{m}\int_0^t F(t)dt.$$

Integrating again,

$$\int_{x_0}^{x(t)} dx = \int_0^t v(t)dt,$$

so,[12]

$$x = x_0 + \int_0^t v(t)dt.$$

Worked Example 3.2 A particle of mass m is acted upon by a force $F(t) = Ae^{-bt}$. The particle is initially at rest at $x = 0$. Determine the velocity and position of the particle as a function of time.

Solution From $F = ma$

$$a = F/m = (A/m)e^{-bt} = \frac{dv}{dt},$$

so

$$v(t) - v_0 = \int_0^t (A/m)e^{-bt}dt = -\frac{A}{mb}e^{-bt}\Big|_0^t = -\frac{A}{mb}\left[e^{-bt} - 1\right],$$

where $v_0 = 0$.

To determine the position we integrate again.

$$x - x_0 = \int_0^t vdt = \int_0^t -\frac{A}{mb}\left[e^{-bt} - 1\right]dt$$

$$x(t) = \frac{A}{mb^2}\left(e^{-bt} - e^0\right) + \frac{A}{mb}t$$

$$= \frac{A}{mb^2}\left(e^{-bt} - 1 + bt\right),$$

where we used the fact that $x_0 = 0$.

Exercise 3.6

A force $F = 3t^2$ N acts on a particle for 3 s after which the particle moves freely. The particle is initially at the origin with zero velocity. Determine its position at time $t = 5$ s. Assume the mass of the particle is 0.1 kg. Answer: 742.5 m. ∎

Exercise 3.7

A certain electromagnetic wave has its electric field oriented along the z-axis. The magnitude of the field varies in time according to $E = E_0 \cos \omega t$. Initially, an electron is at rest at the origin. Determine the motion of the electron. (Recall that the force exerted on a charged body in an electric field is given by $\mathbf{F} = q\mathbf{E}$.) Answer: $x = -(qE_0/\omega^2 m)(\cos \omega t - 1)$. ∎

---

[12] In writing this last equation I was careless and used the same symbol (t) for a variable of integration and a limit of integration. (This is the kind of thing that drives mathematicians crazy!) You can usually get away with this sloppy notation, but sometimes it can get you into trouble. To avoid confusion, you might want to distinguish between variables and integration limits by using primed and unprimed symbols. You will end up with complicated expressions such as

$$x(t) = x_0 + v_0 t + \frac{1}{m}\int_0^t \left[\int_0^{t'} F(t'')dt''\right]dt'.$$

Force as a Function of Velocity

In nature, many forces depend on the velocity of the body. For example, a particle moving through a fluid, such as a marble falling through water or a baseball thrown through the air, will experience a *resistive* force. This force can usually be assumed to be proportional to the velocity for small speeds and proportional to the velocity squared for higher speeds. (Of course, this is just an approximation; the actual dependence of the resistive force on the velocity is more complicated. See, for example, the fluid mechanics books by Batchelor[13] or by Landau and Lifshitz.[14]) Another velocity-dependent force is the Lorentz force. This is the force on a charged particle moving with velocity **v** in a region of space where there is an electric field **E** and a magnetic field **B**. You probably remember from your introductory course in electricity and magnetism that the Lorentz force is given by $\mathbf{F} = q(\mathbf{E} + \mathbf{v} \times \mathbf{B})$. (The magnetic force was considered in Worked Example 3.1 in connection with Newton's third law.) This force presents us with a rather complicated situation because it depends on the *direction* of the velocity as well as on its *magnitude*. In this section we will only consider forces that act in a constant direction.

Consider the one-dimensional problem of determining the motion of a particle if the force F is a function of the speed (v). That is,

$$F = F(v).$$

The equation of motion is

$$\frac{dv}{dt} = \frac{F(v)}{m}.$$

Separating variables and integrating:

$$\int_{v_0}^{v(t)} \frac{dv}{F(v)} = \frac{1}{m} \int_0^t dt.$$

This yields

$$t = m \int_{v_0}^{v(t)} \frac{dv}{F(v)}.$$

Given an explicit expression for $F(v)$ you can carry out the integration and obtain an equation for t involving v and v_0 (as well as some other parameters such as the mass and any other constants that appear in the expression for the force). You have now obtained an expression of the form $t = t(v)$. This can usually be *inverted* to yield the desired form, $v = v(t)$. To be specific, after inverting you will have

$$v = v(t, v_0, \alpha),$$

where α is the set of constant parameters mentioned above. You then use the definition

$$v = \frac{dx}{dt},$$

and integrate again to obtain the position as a function of time. That is,

$$\int_{x_0}^{x(t)} dx = \int_0^t v(t, v_0, \alpha) dt,$$

[13] G. K. Batchelor, *An Introduction to Fluid Dynamics,* Cambridge University Press, 1970.
[14] L. D. Landau, and E. M. Lifshitz, *Fluid Mechanics,* Vol. 6 of *Course of Theoretical Physics,* Pergamon Press, Oxford, 1959.

or

$$x(t) = x_0 + \int_0^t v(t, v_0, \alpha) dt.$$

This procedure is a bit complicated so I will give you an example. Please go through the steps carefully.

Worked Example 3.3 Imagine you are paddling a canoe. When you approach the dock you quit paddling and let the resistance of the water bring you to a stop. Assume this force is proportional to the first power of the speed. Determine the motion.

Solution By Newton's second law, $F = m\frac{dv}{dt}$. According to the assumption, $F = -bv$, where b is the constant of proportionality and the negative sign indicates that it is a retarding force. Then, $-bv = m\frac{dv}{dt}$, or $\frac{dv}{v} = -\frac{b}{m}dt$. Hence,

$$\int_{v_0}^{v(t)} \frac{dv}{v} = -\frac{b}{m}\int_0^t dt.$$

Integrating we get

$$\ln v|_{v_0}^{v(t)} = -\frac{bt}{m}.$$

Because $\ln v(t) - \ln v_0 = \ln \frac{v(t)}{v_0}$, we obtain

$$v(t) = v_0 e^{-bt/m}.$$

Thus we have obtained an expression for the velocity at any given time; half of our job is done. Integrating again we will obtain the position as a function of time. Starting with the definition of velocity, $v = \frac{dx}{dt}$, we write

$$\frac{dx}{dt} = v(t) = v_0 e^{-bt/m}.$$

Consequently,

$$\int_{x_0}^{x(t)} dx = \int_0^t v_0 e^{-bt/m} dt.$$

Integrating,

$$x(t) - x_0 = \left[-\frac{v_0 m}{b} e^{-bt/m}\right]_0^t = -\frac{v_0 m}{b}(e^{-bt/m} - e^0),$$

or

$$x(t) = x_0 + \frac{v_0 m}{b}(1 - e^{-bt/m}).$$

It is interesting to consider the limiting value of our result by letting $t \rightarrow \infty$.[a] This leads to the rather unintuitive result for the velocity of the canoe because it indicates that the velocity of the canoe [$v_0 e^{-bt/m}$] does not go to zero until $t \rightarrow \infty$. The canoe is in motion for an infinite time! How far do you suppose it will travel in this infinite amount of time? To answer this question consider the expression for position and set $t = \infty$ to obtain $x(t = \infty) = x_0 + v_0 b/m$. Note that this is a *finite* distance. So, even though it takes an infinitely long time for the canoe to stop, it only goes a finite distance. Does this result make sense? Yes, it does. The retarding force gets smaller and smaller as the velocity decreases. Perhaps you might complain that the problem is poorly posed and it should ask for the time required for the velocity to fall below some particular value. For example, once the canoe is moving at some very small velocity, such as one inch per hour, then for all intents and purposes it is stopped.

^a It is always a good idea to subject your answers to a "sanity check" by seeing how they behave in limiting cases such as $t = 0$ and $t = \infty$.

Air Resistance If an object is moving in air at less than about 20 m/s ($\approx$ 45 mph), it is safe to assume the retarding force is proportional to the first power of the speed. For objects moving at higher speeds (but less than the speed of sound) it is more realistic to assume the resistive force is proportional to the speed *squared*. That is, $F = -Dv^2$. The equation of motion can then be written in the form

$$m\frac{dv}{dt} = -Dv^2.$$

The proportionality constant D depends on the size and shape of the body and the density of the fluid through which it is moving. A reasonable formula for calculating D is

$$D = \frac{1}{2}C_D A\rho,$$

where the "drag coefficient" C_D is a unitless parameter of the order unity which depends on the shape of the body. In practical applications the value $C_D = 0.2$ is often used. A is the cross-sectional area of the object and ρ is the air density and depends on both the altitude and the temperature. (You can appreciate that air resistance problems can be very complicated, but here we are only going to concentrate on the *method* for solving them and assume D is a known constant.)

If you apply this law to a body falling through air under the action of the force of gravity, the resistive force is upwards and the gravitational force is downwards, so the equation of motion becomes

$$m\frac{dv}{dt} = +Dv^2 - mg.$$

In applying this last equation you will have to be very careful with the signs. If the body is *rising*, then both gravity and the force of air resistance are acting downwards.

Worked Example 3.4 A body of mass m is acted upon by the gravitational force $(-mg)$ and the retarding force of air resistance given by Dv^2. It is dropped from rest at an initial height x_0. Obtain expressions for its velocity and position as a function of time.

Solution In problems such as this one, it is often convenient to express relations in terms of the "terminal velocity" which is the speed when the two forces are equal and opposite and the object is no longer accelerating. In this problem the net force is

$$F = Dv^2 - mg.$$

The terminal velocity is obtained by setting $F = 0$, so

$$v_T = \sqrt{mg/D}.$$

Now we solve for the velocity as a function of time using

$$a = \frac{dv}{dt} = \frac{F}{m} = \frac{D}{m}v^2 - g,$$

or

$$\frac{dv}{dt} = \frac{D}{m}\left(v^2 - \frac{gm}{D}\right) = \frac{D}{m}(v^2 - v_T^2).$$

Therefore,

$$\int_{v_0=0}^{v(t)} \frac{dv}{v^2 - v_T^2} = \frac{D}{m}\int_0^t dt,$$

which yields

$$-\frac{1}{v_T}\tanh^{-1}\frac{v}{v_T} = \frac{D}{m}t,$$

and consequently,

$$v(t) = -v_T \tanh\frac{v_T Dt}{m} = -v_T \tanh\sqrt{\frac{Dg}{m}}t.$$

The position is given by

$$x(t) - x_0 = \int_0^t v\,dt = -\int_0^t v_T \tanh\sqrt{\frac{Dg}{m}}t\,dt.$$

But $\int \tanh ax\,dx = \frac{1}{a}\log\cosh x$, so

$$x(t) = x_0 - \frac{m}{D}\log\cosh\sqrt{\frac{gD}{m}}t.$$

Exercise 3.8

A body is falling under the effects of gravity and a retarding force $F = -bv$. Determine its terminal velocity. Answer: $v_T = gm/b$. ∎

Exercise 3.9

A body is dropped from rest from a hot air balloon. Make the unrealistic assumption that the resistive force is proportional at all times to the velocity, i.e., $f_R = -bv$. (a) What is its terminal velocity v_T? (b) How much time is required for the body to reach a velocity of 0.9 v_T? Answer: (b) 2.3 m/b. ∎

Force as a Function of Position

The most important forces you will encounter in this course are forces that depend on position. These forces are most easily treated using conservation of energy, and we will do that later. However, for the sake of completeness, let us obtain the motion by a straightforward integration of the equation of motion.

If $F = F(x)$, the second law is

$$m\frac{d^2x}{dt^2} = F(x).$$

This differential equation is not hard to solve if you first separate the variables. Because the acceleration is a function of position the velocity will be a function of position. But if $v = v(x)$, we can use the chain rule to write

$$\frac{d^2x}{dt^2} = \frac{dv}{dt} = \frac{dv}{dx}\frac{dx}{dt} = \frac{dv}{dx}v. \tag{3.4}$$

Therefore,

$$v\frac{dv}{dx} = \frac{1}{m}F(x),$$

or

$$v\,dv = \frac{1}{m}F(x)dx.$$

Having separated the variables, you can now integrate to obtain

$$\tfrac{1}{2}v^2|_{v_0}^{v} = \frac{1}{m}\int_{x_0}^{x}F(x)dx. \tag{3.5}$$

You will need an explicit expression for $F(x)$ to carry out the integral on the right. A bit of algebra will then yield an expression for the velocity as a function of position

$$v = v(x). \tag{3.6}$$

But what you really want is an expression for the *position* as a function of *time,* $x = x(t)$. Writing dx/dt for v in Equation (3.6) and rearranging generates a differential equation involving x and t, namely,

$$\frac{dx}{v(x)} = dt.$$

Therefore,

$$\int_{x_0}^{x}\frac{dx}{v(x)} = \int_{0}^{t}dt.$$

That is,

$$t = \int_{x_0}^{x}\frac{dx}{v(x)}.$$

Assuming this integral can be carried out, you will obtain an expression of the form $t = t(x)$. Finally, this must be inverted to yield $x = x(t)$, a mathematical procedure that may involve a significant amount of algebra.

3.7 Simple Harmonic Motion

As an important application of a force that depends on position, we consider the one-dimensional force

$$F = -kx, \tag{3.7}$$

where x is the displacement from an equilibrium position. This is, for example, the force exerted by a spring on a block, as shown in Figure 3.4. Here x is the "stretch" or "compression" of the spring as the block is displaced from the equilibrium point. The minus sign in Equation (3.7) indicates that F is a restoring force that is always directed towards the equilibrium point. Consequently, the block will oscillate back and forth about this point. This is referred to as **simple harmonic motion** (SHM).

I can't overestimate the importance of simple harmonic motion in physics. It is an excellent approximation to any oscillatory motion, from the vibrations of a guitar string to the beating of

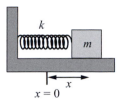

Figure 3.4 A mass m connected to a massless spring of constant k on a frictionless horizontal surface.

$x = 0$

x

k

m

your heart and the oscillations of the quartz crystal in your smartphone. In fact, systems involving repetitive motion can nearly always be analyzed in terms of simple harmonic motion.

In this section we will simply consider the spring and block system of Figure 3.4. However, you will encounter simple harmonic motion in detail in Chapter 11 as well as when we study the pendulum in Chapter 12, waves in Chapters 13 and 19, and small oscillations in Chapter 14.

You can consider Figure 3.4 to be a prime example of simple harmonic motion. It consists of a block of mass m attached to a spring of force constant k, and sliding on a frictionless horizontal surface.

The force exerted by the spring is $F = -kx$, where x is the amount the spring has been stretched. This is equal to the displacement of the block from the equilibrium position. The equation of motion is

$$m\ddot{x} = -kx. \tag{3.8}$$

Going through the steps from Equation (3.4) to Equation (3.5) you obtain

$$v^2 = v_0^2 + \frac{2}{m} \int_{x_0}^{x} (-kx)dx.$$

$$\therefore \quad v^2 = v_0^2 - \frac{k}{m}(x^2 - x_0^2)$$

or

$$v = \sqrt{v_0^2 - \frac{k}{m}(x^2 - x_0^2)}.$$

Note that $v_0^2 + \frac{k}{m}x_0^2$ is a constant whose value depends on the initial conditions. Denoting this constant by C you can write

$$v = \frac{dx}{dt} = \sqrt{C - \frac{k}{m}x^2}.$$

Separate variables again and integrate to obtain the motion:

$$\int_{x_0}^{x} \frac{dx}{\sqrt{C - \frac{k}{m}x^2}} = \int_0^t dt,$$

or

$$\int_{x_0}^{x} \frac{dx}{\sqrt{mC/k - x^2}} = \sqrt{\frac{k}{m}} \int_0^t dt.$$

The integral on the left can be found in any table of integrals, yielding

$$\sin^{-1} \frac{x}{\sqrt{mC/k}} \Big|_{x_0}^{x} = \sqrt{\frac{k}{m}} t \Big|_0^t.$$

Therefore

$$\sin^{-1}\frac{x}{\sqrt{mC/k}} - \sin^{-1}\frac{x_0}{\sqrt{mC/k}} = \sqrt{\frac{k}{m}}t.$$

The second term on the left is just a constant; call it β. Then

$$\sin^{-1}\frac{x}{\sqrt{mC/k}} = \sqrt{\frac{k}{m}}t + \beta,$$

or

$$x = \sqrt{\frac{mC}{k}}\sin\left(\sqrt{\frac{k}{m}}t + \beta\right).$$

This is usually written in the easily remembered form

$$x = A\sin(\omega t + \beta). \tag{3.9}$$

The constant

$$A = \sqrt{\frac{mC}{k}} = \sqrt{\frac{mv_0^2 + kx_0^2}{k}}$$

is the amplitude. The quantity

$$\omega = \sqrt{k/m}$$

is the (angular) frequency of the oscillation. The constant β is called the "phase constant" and is related to the position of the oscillator at time $t = 0$. By adding $\pi/2$ to β one can express $x(t)$ in terms of the cosine rather than the sine.

Exercise 3.10

Verify Equation (3.9) by direct substitution into the equation of motion.   ■

3.8 Closed-Form Solutions

It is usually fairly easy to determine the forces acting on a body and to write the equation of motion. However, it is not always possible to *solve* the equation of motion in *closed form*. That is, it is not always possible to write down an *analytical solution* in which the position is expressed as a function of time in terms of simple functions. For example, $x = 3t^2$, and $x = 4\cos 5t$ are analytical solutions. However, there are ways to describe the motion that do not involve analytical expressions. For example, you might measure the position of an object in the laboratory and draw up a table giving its position at various times. This would give you a collection of numbers that you could use to determine the position of the body at any given time, but you would not have a solution in *closed form*. Similarly, a graph of x vs. t is a visual representation of position as a function of time, but once again it is not in closed form. A closed-form (or analytical solution) is a relationship between x and t in terms of elementary mathematical functions, namely powers, roots, logarithms, exponentials, and trigonometric functions.

If you have the equation of motion for a system and you know the initial conditions, you can always write a computer program that will yield a "numerical solution" giving the position and velocity of the body at any desired time. Numerical solutions have become very common in physics. In fact, there is a whole branch of physics called "computational physics" in which you program a computer to solve physics problems by applying numerical methods.[15] Many important problems cannot be solved analytically; in recent years more and more work has gone into developing sophisticated techniques for obtaining numerical solutions using computers.

You may be wondering why we bother to study the techniques for obtaining analytical solutions instead of just writing a computer program to solve the problem. There are many reasons. For one thing, numerical techniques for solving physics problems are often based on the analytic solution of a similar (usually simpler) problem. Furthermore, numerical techniques are nearly always verified by determining whether or not they can reproduce a known analytic solution. Knowing analytical techniques for solving the equations of motion is extremely helpful in obtaining and interpreting numerical solutions. It might also be mentioned that numerical solutions have a number of disadvantages. A numerical solution, as the name implies, is just a number and not a formula. Numerical solutions yield a specific answer to the problem at hand but rarely lead to general formulas.

In this book, I emphasize obtaining analytical solutions for problems that have closed-form solutions. You will find that the techniques developed here will be used frequently in your other physics courses and in your professional career. If you end up working as a Computational Physicist you will often apply the material you are learning in this course.

You have seen that, in principle, the equation of motion can be solved analytically for a number of different types of forces, specifically for forces that are functions of time, velocity, or position.

All of the procedures described above basically boil down to integrating the equation of motion twice, leading to expressions of the form:

$$v = v(t, v_0, x_0),$$
$$x = x(t, v_0, x_0).$$

These give the position and velocity as functions of time and the initial conditions, v_0 and x_0. If you are lucky, these expressions are in closed form, that is, in terms of familiar functions such as logarithms, exponentials, trigonometric functions, etc. (You are comfortable with these so-called "elementary functions," but if you were confronted with an expression involving the gamma function or an elliptic integral, you might not feel so happy.)

If $v(t)$ and $x(t)$ are given in closed form, you have a great deal of knowledge about the motion. First of all, you can determine the velocity and position at any time by simply plugging the time into the equation. Furthermore, you can simply *look* at the equation and get a very good idea about the motion. For example, if you obtain the solution $x = A \cos \omega t$ you immediately know that the motion is oscillatory with amplitude A and angular frequency ω. Generally, it is easy to manipulate the solution to obtain more information about the system. For example, squaring the expression for the velocity to get v^2 immediately leads to an equation for the kinetic energy as a function of time. If the motion is two dimensional and you obtain solutions $x(t)$ and $y(t)$ you can easily combine them to get an equation for the path (or orbit) of the body.

3.9 Numerical Solutions (Optional)

A closed-form analytic solution of the equation of motion contains all the essential physical information about the system. (In this sense, it is similar to the quantum mechanical wave function from which one can extract all knowledge of the system.) However, it often happens that such a solution is not available. Perhaps we are not able to solve the equation of motion due to a lack of

---

[15] An excellent computational physics textbook is *Numerical Methods for Physics*, 2nd ed., by Alejandro L. Garcia, CreateSpace Publishing, 2015.

mathematical knowledge, or perhaps the equation of motion is actually insolvable. In such a case, we are forced to use a different technique, namely, we are forced to evaluate a *numerical solution.*

Because numerical solutions are seldom as useful as analytical solutions, you should make sure that you cannot find an analytical solution before you start writing a computer program. An analytical solution can usually be found in much less time than it takes to write, debug, and run a computer program. With this warning out of the way, I will now describe an elementary technique for obtaining a numerical solution to the equation of motion.[16]

A numerical solution of a differential equation essentially involves replacing the *differential equation* with a *difference equation.* For example, suppose you wanted to solve

$$\frac{dx}{dy} = f(x, y),$$

You would replace this expression, in which dx and dy are infinitesimal quantities, with the expression

$$\frac{\Delta x}{\Delta y} = f(x, y).$$

Here Δx and Δy are small quantities but are not infinitesimal. To generate a solution recall the definition of derivative:

$$\frac{dx}{dy} = \lim_{\Delta y \to 0} \frac{x(y + \Delta y) - x(y)}{\Delta y}.$$

Then replace $dx/dy = f(x, y)$ by the approximate relation

$$\frac{x(y + \Delta y) - x(y)}{\Delta y} \simeq f(x, y),$$

and rearrange to get

$$x(y + \Delta y) \simeq x(y) + f(x, y)\Delta y.$$

If you know x and y and can calculate $f(x, y)$ you can use this relationship to determine x for a slightly larger value of y, namely, $y + \Delta y$. Note that the relationship is not exact. However, the approximation becomes better and better as Δy becomes smaller and smaller.

Let us apply this technique to the equation of motion. We want to solve for $x(t)$ numerically, given an initial value x_0 and an explicit expression for the acceleration, $a = a(x, v, t)$. The procedure involves carrying out the following sequence of operations.

1. Select Δt to be a small time step.
2. Set $x = x_0$ and $t = t_0$.
3. Determine $a = a(x, v, t)$.
4. Obtain new time from $t = t + \Delta t$.
5. Obtain new velocity from $v = v + a \cdot \Delta t$.
6. Obtain new position from $x = x + v \cdot \Delta t$.
7. Go to Step 3.

---

[16] Physicists frequently use numerical techniques for solving differential equations which are difficult or impossible to solve analytically. This is especially true if they are analyzing fluid flow. Fluid dynamics is, of course, a branch of Newtonian mechanics and the flow of a fluid is controlled by Newton's laws. However, the forces involved are very complicated. Some are velocity-dependent quantities that depend on the pressure exerted by other parts of the fluid. Other forces include viscous drags and the force of gravity. When one writes Newton's second law for a fluid, it may have a large number of terms in it. Often, the only practical way to solve this equation and thus determine the motion of the fluid is to use numerical techniques and a computer.

You go through this procedure repeatedly. At each step you obtain new values for x, v, and t.

Consider a simple but important numerical technique called the Euler method. This is a technique for obtaining the numerical solution of an equation of the form

$$\frac{d^2x}{dt^2} = f(x,v,t).$$

If the function on the right-hand side is the force divided by the mass, this is just Newton's second law.

Write the second-order differential equation above as two first-order equations. These are

$$\frac{dx}{dt} = v(x,t),$$

$$\frac{dv}{dt} = f(x,v,t).$$

Solve as follows. Let x_1 and v_1 be the initial values of x and v. Let the initial time be t_1 and let τ be a small time step. Then

$$x_2 = x_1 + v_1\tau,$$

$$v_2 = v_1 + f(x_1,v_1,t_1)\tau.$$

The next step is to replace x_1 by x_2 and v_1 by v_2 and let t_1 be replaced by $t_2 = t_1 + \tau$. Then evaluate

$$x_3 = x_2 + v_2\tau,$$

$$v_3 = v_2 + f(x_2,v_2,t_2)\tau.$$

You can appreciate that, in general, the technique consists in replacing x_n by x_{n+1} and v_n by v_{n+1} and repeating as many times as desired. This is the essence of the Euler method.

Some years ago, Alan Cromer[17] noted that the Euler method, as expressed above, is unstable. That is, the answer diverges further and further from the correct solution. However, Cromer discovered that the equations could be made stable by a simple but crucial change, namely by calculating v first and using the new value of v to determine x. The so-called "Euler–Cromer" algorithm is

$$v_2 = v_1 + f(x_1,v_1,t_1) \cdot \tau,$$

$$x_2 = x_1 + v_2\tau.$$

This simple scheme allows you to numerically determine the velocity and position of a particle at any future time if you know its acceleration $f(x,v,t)$ at a prior time.

When carrying out a numerical integration, you should check frequently to determine if the solution is stable. For example, you might check at each time step to make sure the energy or angular momentum is conserved.

3.10 Summary

A number of important concepts were presented in this chapter. Basically, you learned (or reviewed) Newton's three laws and you learned how to determine the motion for a mechanical system (assuming you know the force).

---

[17] Alan Cromer, "Stable Solutions using the Euler Approximation," *American Journal of Physics*, **49**, 455 (1981). In this paper Cromer explains why one algorithm is stable and the other is not.

Newton's three laws are:

(1) The law of inertia. (A body tends to preserve its state of motion.)
(2) The second law. $\mathbf{F} = \frac{d\mathbf{p}}{dt}$ gives the relation between the *net external* force acting on a body and the rate of change of its momentum.
(3) Action equals reaction. (Two bodies always exert equal and opposite forces of the same kind on each other.)

If the mass is constant, Newton's second law can be expressed as

$$a = \frac{F}{m}.$$

We call such a relation an "equation of motion" because it can be integrated to determine the velocity v and the position x as functions of time. Obtaining an expression for $x = x(t)$ is called "determining the motion."

To determine the motion you need to integrate the equation of motion twice, thus:

$$v(t) = \int_0^t a\,dt$$

$$x(t) = \int_0^t v(t)\,dt.$$

We have described the techniques for determining the motion for four different kinds of forces. These forces are: (1) constant force, (2) force a function of time, (3) force a function of velocity, and (4) force a function of position. Make sure you understand the proper procedure for each of these cases.

Finally, you were exposed to the method of determining the motion using numerical techniques. The "Euler–Cromer" algorithm is a simple way to obtain $x = x(t)$.

3.11 Problems

Problem 3.1 A particle of mass 3 kg is acted upon by the two forces (in newtons)

$$\mathbf{F}_1 = 6\hat{\imath}$$

and

$$\mathbf{F}_2 = 3\hat{\imath} + 3\hat{\jmath}.$$

Determine the magnitude and direction of the acceleration.

Problem 3.2 A pail is filled with oil to a depth of 10 cm. A steel marble of mass 0.2 kg is released from rest 50 cm above the top surface. Assuming the oil exerts a constant resistive force of 2.4 N on the marble, determine the speed with which it reaches the bottom of the pail.

Problem 3.3 An automobile starts from rest and accelerates at a constant rate for 1 km, covering that distance in 20 s. Ignore air resistance. Determine the coefficient of static friction between the tires and the pavement.

Problem 3.4 A sailor on board a ship of mass 2×10^6 kg moving at 0.2 m/s throws a heavy rope to a man standing on the wharf who drops the noose around a bollard. After the rope is taut, the ship is brought to rest, stretching the rope 0.5 m. Find the average pull sustained by the rope.

Problem 3.5 An 80 kg physics student named George goes bungee jumping. The unstretched bungee cord has a length of 20 m, so after George has fallen 20 m, the cord slows his fall, stretching 5 m before bringing him to a stop. What was the average force exerted by the cord?

Problem 3.6 A particle of mass m is observed to have a velocity given by $v = A \cos \alpha x$, where A and α are constants. Obtain an expression for the force $F = F(x)$.

Problem 3.7 A particle of mass m is initially at rest at $x = 0$. It is acted upon by a force $F = A \cosh \beta t$.

(a) Determine the velocity and position of the particle at time t.
(b) Show that at very small values of t, the position is approximately given by $x(t) \doteq \frac{1}{2} \left(\frac{F_0}{m} \right) t^2$, where $F_0 = $ force at time $t = 0$.
(c) Show that for large values of t, the position is approximately given by $x(t) \doteq \frac{1}{2} \frac{F_0}{m} (e^{\beta t}/\beta^2)$.

Problem 3.8 (a) Two stars of masses M_1 and M_2 attract one another with the gravitational force $F = -G M_1 M_2 / r^2$. Determine the ratio of their accelerations. (b) Assume both stars move in circles about their common center of mass. Determine the radii of these circles in terms of M_1, M_2, and r. (Actually the stars will move in elliptical orbits about their center of mass.)

Problem 3.9 A jet boat is propelled by a jet of water that exerts a constant force on the boat. Call this force F_0. The boat is moving in still water that presents a retarding force of magnitude bv, where b is a constant and v is the velocity.

(a) Determine the terminal velocity v_T reached by the boat.
(b) Obtain an expression for the velocity as a function of time.
(c) Carefully plot $v(t)$ vs. t. Label the plot.

Problem 3.10 You are searching for an inertial reference frame. Determine the accelerations of the following reference frames. (a) A reference frame fixed to the surface of the Earth at latitude $37°$. (b) A reference frame with origin at the center of the Earth with one axis always pointing towards the Sun. (c) A reference frame with origin at the center of the Sun with one axis always pointing towards the center of the galaxy. (The Sun is about 3/5 of the distance from the center of our galaxy to the edge. It takes some 200 million years to "orbit" the galaxy. The diameter of the Milky Way is about 10^5 ly.)

Problem 3.11 Atwood's Machine consists of two masses m_1 and m_2 tied together by a light string that slides frictionlessly over a smooth pulley. (a) Obtain an equation for the acceleration of the masses and the tension in the string. (b) Obtain an expression for the acceleration if the pulley is perfectly rough, so it rotates with the string. Assume the pulley has a moment of inertia I and radius R.

Problem 3.12 One normally learns Newton's three laws as statements about the nature of the physical universe. However, some people prefer to consider only the third law as a law of nature and the first two as definitions. Assume that you subscribe to this point of view. Write a short essay (one or two paragraphs) explaining why you are interpreting Newton's laws in this manner.

Problem 3.13 The gravitational force between two masses is $F = -G \left(M_1 M_2 / r^2 \right)$ and the electrostatic force between two opposite charges is $F = -k \left(Q_1 Q_2 / r^2 \right)$. (a) Show that all massive objects attracted to the Earth will have the same acceleration independent of mass. (b) Show that two charged objects attracted to a point charge Q will *not* have the same acceleration unless both have the same charge to mass ratio.

Problem 3.14 A friend of yours, hearing that the universe is expanding at an increasing rate, theorizes that there is a universal repulsive force between any two masses, as well as the usual gravitational force. According to your friend, this hypothetical repulsive force decreases with the inverse of R rather than the inverse *square* of R and is given by

$$F = G' \frac{M_1 M_2}{R},$$

where $G' = 6.67 \times 10^{-31}$ Nm/kg^2. Determine the separation between two bodies when the repulsive force would become dominant. Could this explain the expansion of the universe? Would a galaxy the size of the Milky Way be stable under such a force? (The diameter of the Milky Way is about 100 000 ly.)

Problem 3.15 A certain satellite in a circular orbit around the Earth has a period T. Suppose the gravitational force were not exactly an inverse square force, but rather depended on distance as $1/r^\alpha$ with $\alpha = 2 + \epsilon$, where ϵ is a small number. The period of the satellite would then be T'. Determine the ratio T'/T. If the value of ϵ were 10^{-4}, would the difference in periods be measurable? (Assume the satellite is in a circular orbit of radius 7000 km.)

Problem 3.16 Suppose that gravitational mass and inertial mass are different but that they are proportional to one another. Would heavy objects and light objects then fall at the same rate? As an explicit example, assume that inertial mass is twice as great as gravitational mass. Determine the acceleration of a freely falling object near the surface of the Earth.

Problem 3.17 Consider the following three systems. (1) A book is sitting on a table which is at rest on the surface of the Earth. (2) A rocket is taking off from the surface of the Earth. A large cloud of burned fuel forms a contrail behind it. (3) A donkey that is hitched to a cart is trotting down a road. (a) Identify the action–reaction pairs in these three cases. (b) What is the force that is causing the rocket to accelerate? (c) What is the force on the donkey that keeps it in motion?

Problem 3.18 The Aristotelians noticed that if you push a box across the room, it moves with a constant speed, and if you push it harder the speed is greater, but as soon as you quit pushing, the box stops. This suggests that the equation of motion should be $F = mv$. Furthermore, the Aristotelians claimed that heavier objects fall to the ground faster than light objects. (Thus, an object four times heavier than a light object will fall to the ground in one fourth the time.) (a) Show that these assumptions lead to the absurd conclusion that the Earth exerts the same gravitational force on all bodies, regardless of their mass. (b) Describe a simple experiment to show that the Earth does not exert the same gravitational attraction on all bodies, regardless of their mass.

Problem 3.19 Two carts of masses M and $2M$ have springs attached to either end as shown in Figure 3.5. Cart 1 has mass M and cart 2 has mass $2M$. They are on a frictionless segment of track of length L with barriers at the ends. The two carts are brought together at the center of the track, the springs in contact are compressed, and the carts are released from rest. Cart 1 moves left, hits the barrier at $x = 0$ and bounces back. Cart 2 moves right, hits the barrier at $x = L$ and bounces back. The collisions are completely elastic. Where do the carts meet? The dimensions of the carts are much smaller than the length of track, so the carts can be treated as particles. (b) Where do the carts meet if cart 2 is four times more massive than cart 1?

Problem 3.20 The velocity of a particle of mass m varies with distance x as

$$v(x) = v_0 - bx,$$

where b and v_0 are constants. (a) Find the force acting on the particle, $F(x)$. (b) Determine $x(t)$. (c) Obtain an expression for the force as a function of time, $F(t)$.

Problem 3.21 A certain object is losing mass at a rate κ (kg/s). It is acted upon by a constant force F. (a) Determine its acceleration and its position as a function of time. Assume that at $t = 0$, the mass is m_0, the velocity is zero and the object is at $x = 0$. (b) Show that as $\kappa \to 0$, the expression for $x(t)$ reduces to $x = \frac{1}{2}\left(\frac{F}{m_0}\right)t^2$, that is, $x = \frac{1}{2}at^2$.

Problem 3.22 An object of mass m at rest is dropped from a high place. It falls under the action of gravity and a retarding force of Dv^2.

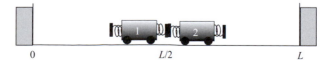

| 0 | | $L/2$ | | L |

Figure 3.5 Two carts with attached springs on a frictionless track. If the carts are squeezed together at the midpoint, where will they meet? The mass of cart 2 is twice the mass of cart 1. (Problem 3.19.)

(a) Write the equation of motion.
(b) Obtain an expression for the terminal velocity v_T.
(c) Obtain an expression for the velocity as a function of time in terms of the terminal velocity and other parameters.
(d) Obtain an expression for $x(t)$ the position as a function of time.
(e) Expand the hyperbolic functions to obtain approximations to $v(t)$ and $x(t)$ at small values of time ($t \ll v_T/g$).
(f) Similarly, obtain expressions for $v(t)$ and $x(t)$ at large values of time ($t \gg v_T/g$).
    (Hint: Taylor expansion, $\ln \cosh x = 0 + x(0) + \frac{1}{2}x^2 + \cdots$.)

Problem 3.23 A cannon is on a raised platform so its muzzle is 2 m above the level ground. It fires a 2 kg shell at 600 m/s horizontally. It does not acquire a large downward velocity so we can assume the upward force of air resistance is proportional to v (thus: $F_y(\text{air}) = K_y v$) but the horizontal force of air resistance is proportional to v^2 (thus: $F_x = K_x v^2$). Assume $K_y = 10^{-2}$ Ns/m and $K_x = 10^{-3}$ Ns^2/m^2.

(a) Write the two equations of motion and obtain $v_y(t), v_x(t), y(t), x(t)$.
(b) Show that the time of flight is essentially independent of air resistance. (Hint: Expand e^x and show that to order t^2 the time to fall is the same as in a vacuum.)
(c) Determine the range and compare to the value obtained if there is no air resistance. Plot the trajectories.

Problem 3.24 An object of mass 2 kg is thrown upwards with an initial velocity of 4 m/s. It is acted upon by the constant gravitational force of the Earth and a mysterious force of 2 N that always opposes the motion of the object. (This is like the force of air resistance, except we are assuming the force is constant.) Determine the position of the object at time $t = 0.5$ s. (Be careful; by this time the object is falling downwards and the mysterious force is acting upwards!)

Problem 3.25 A bullet of 5 g is fired vertically at $v_0 = 300$ m/s. If there is no air resistance it reaches a maximum height (where $v(t) = 0$) at a time given by $0 = v_0 - gt_{\text{top}}$. Assume air resistance acts on the bullet so that its time to reach the top of its trajectory will be reduced. Determine this time, assuming the force of air resistance is Dv^2, where $D = \frac{1}{2}C_v A\rho$. You may assume the density of air is 1.2 kg/m^3, that the bullet has a radius of 0.5 cm , and $C_v = 0.2$.

Problem 3.26 A builder plans to use a chain of length 1.5 m and linear density 7 kg/m to suspend a 220 kg mass from an overhead beam. Determine the forces at either end and at the middle of the chain. Obtain an equation for the force on a link at any point of the chain. If the maximum tension that can be supported by a link is 2200 N, where will the chain break?

Problem 3.27 A 200 kg gorilla walks into an elevator and steps onto a scale that just happens to be there. Conveniently, the scale is calibrated in kilograms. (This problem is somewhat unrealistic!)

(a) What does the scale read when the elevator is accelerating upward at 2 m/s^2?
(b) A while later, after pushing many buttons, the gorilla notices that the scale reads 150 kg. What is the acceleration of the elevator at that time?

Problem 3.28 A bathroom scale is placed in an elevator. A man weighing 180 lb stands on the scale and observes that as the elevator ascends, the scale reads 200 lb while the elevator is accelerating and 140 lb when it is slowing to a stop. What are the acceleration and the deceleration of the elevator ($g = 32.2$ ft/s^2)?

Problem 3.29 A bullet initially traveling at 400 m/s passes through a 2 cm thick slab of wood and emerges at speed v_f. When fired into a thick block of the same wood, the bullet is found to penetrate 10 cm. Determine v_f.

Problem 3.30 A physics student designed an apparatus for her experimental physics class. The apparatus consisted of a block $m = 0.25$ kg mass attached to a compressed spring of constant k mounted on a cart of mass $M = 2$ kg, as shown in Figure 3.6. The cart is pulled by a string that passes over a pulley to a 10 kg hanging weight. The student used photogate timers to determine the acceleration of the mass on the spring

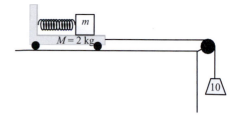

Figure 3.6 A block on an accelerated cart. (Problem 3.30.)

when it was 15 cm from its equilibrium position. She found this acceleration to be -2 m/s^2 relative to the lab bench. (a) What is the spring constant? (b) What is the acceleration of the block relative to the cart? (You may assume there is no friction in the system and the spring is massless.)

Problem 3.31 Imagine you are Isaac Newton. You know that Johannes Kepler determined that the period of a planet squared is proportional to the cube of the semimajor axis of the orbit. Assume planets move in circular orbits, in which case the semimajor axis is equal to the radius of the orbit. Using this fact, deduce that the gravitational force obeys an inverse square law. (Note: You cannot use Newton's law of universal gravitation!) Also show that the velocity of a planet decreases proportionally with the inverse square root of its distance from the sun ($v \propto 1/\sqrt{r}$).

Problem 3.32 A singly charged lithium ion is accelerated by an electric field $\mathbf{E} = 10\hat{\imath}$ volts/meter applied between two plates ten centimeters apart. After passing through a small hole in the negative plate, the ion enters a region of space where there is a magnetic field $\mathbf{B} = 5\hat{\jmath}$ millitesla. The ion then moves in a circular path. What is the radius of this circle? (You may assume the mass of the Li ion is 7 amu.)

Problem 3.33 A particle of mass 0.1 kg is acted upon by a force given by

$$F = A + B \left(\frac{t - C}{2} \right)^2,$$

where $A = 1$ N, $B = -1$ N/s^2 and $C = 2$ s. Determine the speed and position of the particle at time $t = 4$ s, if it is initially at the origin with a velocity $+6$ m/s.

Problem 3.34 A particle of mass m, initially at rest, is subjected to a force $F(t) = Ae^{-\alpha t}\cos(\beta t)$. Determine $v(t)$ and $x(t)$.

Problem 3.35 An object with initial velocity v_0 is slowed by a force given by $F = -Ae^{bv}$, where A and b are constants. Obtain an expression for the velocity as a function of time.

Problem 3.36 An airplane accelerates down the runway. Assume the jet engines exert a constant force on the airplane and assume air resistance is proportional to the velocity squared. Obtain an expression for the velocity as a function of time. What is the terminal velocity?

Problem 3.37 A falling body is subjected to the gravitational force (acting down) and the retarding force of air resistance (acting up). Suppose the retarding force is given by Dv^2, where D is a constant. Show that in the limit of small t ($t \ll \sqrt{m/Dg}$) the velocity and position are given by

$$v \simeq -gt, \qquad\qquad x \simeq -\frac{1}{2}gt^2,$$

and in the limit of large t ($t \gg \sqrt{m/Dg}$)

$$v \simeq -\sqrt{mg/D}, \qquad\qquad x \simeq \frac{m}{D}\ln 2 - \sqrt{mg/D}\, t.$$

(Hint: Expand the hyperbolic functions.)

Problem 3.38 You throw a ball vertically upward with an initial velocity v_0. Assume that air resistance generates a force given by Dv^2 opposing the motion. (a) Determine the terminal velocity (note that this involves the *falling* motion of the ball). (b) Determine the height attained by the ball. (c) Determine the speed of the ball when it returns to your hand (in general this will not be the terminal velocity).

Problem 3.39 A race car of mass 800 kg reaches a velocity of 200 km/h when suddenly the engine dies and the brakes go out! The only thing slowing the car down is air resistance (rolling friction is assumed negligible). Assume the drag force is proportional to the square of the velocity ($F = -Dv^2$). After the car coasts 500 m it is moving at 5 km/h. What is the drag coefficient D?

Problem 3.40 A skydiver has a mass of 100 kg (including the parachute). When the parachute is open, the retarding force is proportional to the velocity to the first power. Furthermore, it is known experimentally that the parachute exerts a retarding force per unit area of 90 N/m² when the speed is 3 m/s. (a) It is desired that the skydiver not exceed a safe speed of 1.5 m/s upon reaching the ground. What is the required area of the parachute? (b) Assuming an initial speed of zero, how much time does it take for the skydiver to reach a speed of 90 percent of the terminal velocity?

Problem 3.41 A bullet of mass m is fired vertically upward with an initial velocity v_0. Assume the resistive force of the air is given by mkv where m is the mass, v is the velocity and k is a constant. How much time is required for the bullet to reach the top of its path?

Problem 3.42 A physics student on the roof of a tall building drops a ball of diameter 10 cm and mass 0.3 kg. The air density is 1.2 kg/m³. Assume $C = 0.2$ and the drag force is proportional to the velocity squared. (a) What is the terminal speed? (b) Obtain an expression for the speed of the ball as a function of how far it has fallen. (c) If the roof is 100 m above the ground, what is the speed of the ball when it hits the sidewalk?

Problem 3.43 An object is dropped from an airplane that is flying at an altitude of 6000 m at a speed of 1000 km/h. Determine the horizontal distance traveled by the object by the time it hits the ground. For the sake of making the calculations somewhat simpler, you may assume that the resistive force of the air is proportional to the first power of the velocity and that the terminal velocity of the object is 98 m/s. (Justify neglecting small terms.)

Problem 3.44 An object of mass m in a uniform gravitational field is dropped from some initial height. Assume that the force of air resistance is proportional to the first power of the velocity ($\propto bv$). Show that the distance y it has fallen in time t is

$$y = \frac{mg}{b}\left(t + \frac{m}{b}\left(e^{-bt/m} - 1\right)\right).$$

Problem 3.45 Assume the force of air resistance is given by $F_{air} = Dv^2$, where v is the velocity and D is a constant equal to 0.01 kg/m. An object of mass 2 kg initially at rest, is dropped from a height of 1000 m. The gravitational force can be assumed constant. Determine how far the object has fallen in 20 s.

Problem 3.46 An object of mass 10 kg is dropped from a high place. Assume air resistance is proportional to the velocity squared ($= Dv^2$), where $D = 0.01$ Ns²/m². (a) Evaluate the terminal velocity. (b) Determine the time required for the ball to reach $0.9v_T$.

Problem 3.47 A bead on a long straight wire is repelled from the origin $x = 0$ by a force proportional to its distance to that point. The bead is initially at rest at position x_o. Determine its motion.

Problem 3.48 A particle of mass 2 kg is at rest at x_0. It is subjected to a force of magnitude $F = k/x$, where $k = 4$ Nm. The force is directed along the positive x-axis. Determine the speed of the particle when it passes through the point $x = 2x_0$.

Problem 3.49 A block of mass m on a frictionless horizontal surface is attached to two massless springs in series. The springs have force constants k_1 and k_2. Determine the angular frequency of oscillation of the block.

Problem 3.50 An object of mass m is initially at rest at $x = 2$ m. Determine the time for it to reach $x = 3$ m if it is acted upon by a force $F = kx$. Express your answer in terms of k and m.

Problem 3.51 Astronomers discover a comet that is a distance d from the Sun and is speeding directly toward the Sun with a speed v_0. What is its speed when it is at a distance $d/2$ from the Sun? (This problem can also be solved using energy methods, but do not use them.)

Problem 3.52 A mass m is attached to a spring of constant k. Determine the motion if it is initially at the unstretched position and is given an impulse $\mathbf{J}$. (Impulse is defined by $\mathbf{J} = \int \mathbf{F} dt$.)

Problem 3.53 A block of mass 0.75 kg is at rest on a frictionless horizontal surface. It is attached to a spring of constant $k = 30$ N/m. The block is struck with a hammer giving it an initial velocity of -3 m/s. By direct integration of the equation of motion obtain expressions for $v(x), v(t)$, and $x(t)$. Determine the frequency and amplitude of the subsequent motion.

Problem 3.54 (a) Obtain a general relation for the motion of a particle of mass m acted upon by a *repulsive* force $F = +kx$. Assume the particle is initially at the position $x = x_0$. Note that the solution will have the form $x = Ae^{\beta t} + Be^{-\beta t}$. (b) It may surprise you to know that for this force there is a possible solution in which the particle moves to the origin and remains at rest there. If this occurs, what was the initial value of the velocity?

Problem 3.55 An asteroid of mass m is initially very far from the Earth and has zero velocity relative to Earth. It falls under the action of the gravitational force, $F = -GmM/z^2$, where M is the mass of the Earth. Determine the speed of the asteroid as a function of z, its distance from the center of the Earth.

Problem 3.56 At the surface of the Earth the gravitational force on a mass m is

$$F = -G\frac{M_e}{R_e^2}m,$$

where M_e is the mass of the Earth and R_e is its radius. On the other hand, at the center of the Earth there is no net gravitational force. If you were to dig an extremely deep hole, the gravitational force would decrease as you get closer to the center. As Isaac Newton knew, the gravitational force at a distance r' from the center of the Earth is given by

$$F = -G\frac{M'}{r'^2}m,$$

where M' is the mass contained in a sphere of radius r'. You now have sufficient information to solve the following completely unrealistic problem.

   Superman has super powers, but he isn't very smart. He decides to stand on the surface of the Sun, believing it to be a solid sphere of radius 6.96×10^8 m and mass 1.99×10^{30} kg. Superman transports himself to the Sun and finds himself at rest at 6.96×10^8 m from the center. Of course, the Sun is not solid, so Superman sinks into the Sun and soon he finds himself bobbing back and forth through the center, like a harmonic oscillator. Ignore the drag force of the gases in the Sun and make the unrealistic assumption that the density of the Sun is constant. Determine the period (T) of Superman's oscillations.

Problem 3.57 An enterprising scientist drills a hole straight through the Earth, from North Pole to South Pole, and drops an object of mass m in the hole at the North Pole. How long does it take for the object to emerge from the hole at the South Pole? You may assume that there is no air in the hole. (Hint: The gravitational force inside a uniform sphere is given by $F = -GM'm/r^2$, where M' is the mass contained within a sphere of radius r and r is the distance from the center of the sphere to the object m.)

Computational Projects

Computational Project 3.1 Write a program to plot the velocity and position of the canoe of Worked Example 3.3. Assume the mass of the canoe plus person is 200 kg. Use $b = 150$ kg/s as a first estimate and get plots for various values of b. Use a reasonable choice for the initial velocity.

Computational Project 3.2 Demonstrate that the Euler–Cromer method is stable but the Euler method is not. Do this by assuming a particle moves in one dimension (x) under the influence of a force $F = -kx$. This will lead to simple harmonic motion. Compare the total energy as a function of time as obtained using the Euler method and the Euler–Cromer method. Also, on the same plot, present the total energy as a function of time as obtained from the exact (analytical) expression. You will appreciate that the Euler method does not conserve energy, the total energy obtained by the Euler–Cromer method oscillates around a constant value, whereas the analytical expression gives a constant value for the energy at all times.

Computational Project 3.3 A projectile is fired perpendicular to the Earth's surface with an initial velocity of 600 m/s. Assume a constant gravitational force acts on the projectile so its acceleration is 9.8 m/s^2 downwards. Write a program to determine the position of the particle as a function of time. Next, consider the same problem, but now include the effect of air resistance, assuming it is a retarding force that can be expressed as $f_{air} = 2.5 \times 10^{-4} v^2$, where v is the velocity of the projectile.

4 Lagrangians and Hamiltonians

In this chapter you will learn about two different approaches for solving physics problems. These approaches are associated with Lagrange and Hamilton, although other physicists and mathematicians contributed to them, including Leibniz and Euler.

In Chapter 3 we considered various methods for solving the equation of motion, but we assumed the equation of motion itself was a consequence of Newton's second law expressed as $a = \frac{F}{m}$. Recall that this equation is only valid for a particle of constant mass moving in an inertial reference frame. Furthermore, for complicated systems with several moving parts it is often quite difficult to obtain an expression for the acceleration. Therefore, it is important to develop a general method for obtaining the equation of motion.

In this chapter we introduce two methods for determining the equations of motion. These methods are based on functions called the Lagrangian (L) and the Hamiltonian (H).

The Lagrangian and the Hamiltonian are essential functions in advanced physics. If you study elementary particle physics you will find the theory is highly dependent on the Lagrangian. And when you take a course in Quantum Mechanics you will be working continually with Hamiltonians.

We will begin by introducing the Lagrangian. If you know the Lagrangian for a system, you can determine the equation of motion, the momentum, and all other relevant mechanical quantities. You will appreciate this eventually, but our main effort in this chapter is to learn how to obtain the Lagrangian and how to use it to determine the equations of motion (called "Lagrange's equations" in this context).

The derivation of Lagrange's equations is based on a branch of mathematics known as the "Calculus of Variations." After a brief consideration of this theory we introduce the Hamiltonian and "Hamilton's equations" which are yet another way to obtain the equations of motion.

Although the Lagrangian and Hamiltonian formulations are fairly advanced, to understand the material in this chapter it is sufficient to remember the formulas for kinetic energy (T) and potential energy (V) and the relationship $F = -\frac{dV}{dx}$ between force and potential energy for a one-dimensional conservative system.

4.1 Joseph Louis Lagrange (Historical Note)

Joseph Louis Lagrange lived a fairly uneventful life in a time and place where great events were taking place. He spent many years in the court of Frederich the Great, King of Prussia, and when Frederich died, Lagrange moved to Paris in time to experience the French Revolution.

Lagrange was born in Turin, Italy, in 1736. His grandfather was a Frenchman (hence the French surname) who moved to Turin and became quite rich. Unfortunately, Lagrange's father squandered all of the money. When Lagrange was a young man he attended the local university, planning to study law. However, he happened upon a paper by Edmund Halley, Newton's friend, and it changed his life. He taught himself mathematics and before long he wrote a series of articles about some mathematical results he had obtained. At age 19 he was appointed professor of mathematics at the Royal School of Artillery.

Lagrange's work came to the attention of Euler who was living in Berlin at the time. Euler convinced King Frederich to invite Lagrange to Berlin. But Lagrange turned down the invitation, commenting that because Berlin had Euler it did not need Lagrange. But when Euler left Berlin to live in St. Petersburg, the king repeated the invitation and this time Lagrange accepted and moved to Berlin. There he developed a number of important mathematical concepts, including work on the calculus of variations, and a theory of differential equations. His book on analytical mechanics is famous for not containing a single figure. Perhaps his greatest contribution to physics was a major work on Celestial Mechanics. Lagrange discovered that a two-body system (such as the Sun and the planet Jupiter) has five quasi-stable points, which are now called the "Lagrange points." Much later, it was discovered that at the two absolutely stable points in the Sun–Jupiter system there are clusters of asteroids that are called the "Trojan Asteroids."

After King Frederich died, there was an antiforeign sentiment in Berlin and Lagrange accepted an invitation from the King of France to come to Paris, where he spent the rest of his life. When the French Revolution broke out, many intellectuals were beheaded. Lagrange feared for his life, and was on the verge of escaping from Paris and returning to Turin. However, he was a quiet and self-effacing person and was never singled out by the revolutionaries. During the revolutionary period, he was ordered to teach classes at a university. It would seem that he was not very successful as a teacher. The students were unhappy with him, complaining that they could not understand him because he spoke quietly and had an Italian accent. However, other mathematicians attended his lectures and praised them highly. Later in life he was appointed to the Senate and received many awards and honors from Napoleon. He was named president of the committee for the reform of weights and measures and was instrumental in having France adopt the SI system of units. When he died in 1813, he was buried with great ceremony in the Pantheon in Paris.

4.2 The Equation of Motion by Inspection

Before discussing the Lagrangian, it is helpful to review how Newton's second law leads to an expression for the equation of motion.

We have considered the equation of motion to be an equation for the acceleration as a function of position, velocity, and time. That is,

$$a = a(x, v, t),$$

or

$$\ddot{x} = \ddot{x}(x, \dot{x}, t).$$

In Chapter 3, we considered how to *solve* the equation of motion for a variety of different force laws, but we did not emphasize how to *determine* the equations of motion. In fact, we assumed the equation of motion was given by Newton's second law in the form $F = ma$.

For example, the equation of motion for a simple pendulum is easily obtained by drawing the forces acting on the bob and resolving them into appropriate components, as illustrated in Figure 4.1. A glance at the figure will remind you that the gravitational force can be resolved into a component along the string ($mg \cos\theta$) and a component ($mg \sin\theta$) tangent to the circular path of the bob. The position of the bob can be specified by the arc length s, measured from the lowest point in the motion. Consequently, $F = ma$ becomes

$$-mg \sin\theta = m\ddot{s}.$$

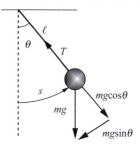

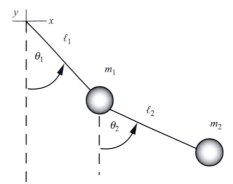

Figure 4.1 Forces acting on a simple pendulum. The length of the string is l.

Figure 4.2 A double planar pendulum.

This equation involves two variables, s and θ. But $s = l\theta$, where l is the length of the string. So the equation of motion can be written in terms of the single parameter θ as:

$$\ddot{\theta} = -\frac{g}{l}\sin\theta. \tag{4.1}$$

I went through this example to remind you that it is easy to determine the equation of motion for a simple system, such as a pendulum.

But now consider the problem of determining the equation of motion for the double pendulum shown in Figure 4.2. For simplicity, assume the double pendulum is constrained to oscillate in the plane of the page so the positions of the bobs can be described by the two angles θ_1 and θ_2. Observe that the motion of m_2 depends on the motion of m_1. The point of support of the lower pendulum is accelerating; it would be very difficult to obtain the accelerations of the two masses by applying Newton's second law. A different approach is needed.

In a little while I will describe a very powerful but fairly simple technique to obtain the equations of motion of complicated systems (like the double pendulum) by writing an equation called the Lagrange equation. First, however, I need to tell you about a function called the Lagrangian.

Exercise 4.1

Draw a force diagram for each of the masses of the double planar pendulum. Consider the motion of mass m_2. What is wrong with writing $m_2\ddot{s}_2 = -m_2 g \sin\theta_2$? ∎

4.3 The Lagrangian

Studying Lagrangian dynamics helps to develop an understanding of the framework of mechanics and an appreciation of how the physical universe behaves. Lagrangian dynamics goes to the very core of the science of mechanics. In fact, as you will see in Section 4.8.2, the Lagrangian approach can be used to derive Newton's laws.

If you take a graduate courses in mechanics, you will find yourself deeply immersed in Lagrangian methods. Here I will only use it as a technique – an easy way to obtain the equations of motion for a complicated system. Nevertheless, later in this chapter I will present an elementary derivation of the technique, showing how Lagrange's equations are obtained using the calculus of variations. You will learn that Lagrangian dynamics is based on a profound statement about the nature of the physical world called Hamilton's principle. Later (in Chapter 8) you will see how the Lagrangian is related to symmetries in physical systems, and to the conservation laws of physics.

The Lagrangian method, as described here, is applicable to systems in which all the forces are conservative. Conservative forces are forces that can be obtained from the potential energy by taking its derivative. For example, the gravitational force is conservative, but the force of friction is not. There are ways of incorporating dissipative forces into the Lagrangian formulation but they involve advanced methods and we do not consider them in this book, except for a short discussion of nonconservative forces in Section 4.7. Generally speaking, if friction is present, you will need to go back to Newton's second law.

The study of Lagrangian dynamics begins with the definition of a physical quantity called the *Lagrangian.* It is denoted by L and can be expressed as the difference between the kinetic energy (T) and the potential energy (V):

$$L = T - V. \tag{4.2}$$

As an application, I will evaluate the Lagrangian for the simple pendulum of Figure 4.1. Assume the origin of coordinates is at the point of support. In Cartesian coordinates the potential energy is $V = mgy$. The kinetic energy is $T = \frac{1}{2}m(\dot{x}^2 + \dot{y}^2)$. Therefore the Lagrangian for the simple pendulum is

$$L = T - V = \frac{1}{2}m(\dot{x}^2 + \dot{y}^2) - mgy. \tag{4.3}$$

There are *two* coordinates in this expression for L, namely x and y. As you will see shortly, this leads to *two* equations of motion. Normally it is desirable to have as few equations of motion as possible. Therefore, it is desirable to use a coordinate system (such as r, θ) that will minimize the number of equations of motion. Physicists frequently need to transform from one set of coordinates to another. This is done by using *transformation equations.* Consider transforming from the Cartesian coordinates, x, y, z, to a different set of coordinates, say q_1, q_2, q_3. The transformation equations are a set of equations having the general form

$$x = x(q_1, q_2, q_3, t)$$
$$y = y(q_1, q_2, q_3, t)$$
$$z = z(q_1, q_2, q_3, t).$$

The qs are called "generalized coordinates." In practice they will often be Cartesian, cylindrical, or spherical coordinates, but they are not limited to these familiar coordinates.[1]

In the example of the pendulum, the transformation equations from x and y to θ are

$$x = l \sin \theta,$$
$$y = -l \cos \theta,$$

---

[1] The generalized coordinates are normally a "minimal set" of coordinates. For example, in Cartesian coordinates the simple pendulum requires two coordinates (x and y), but in polar coordinates only one coordinate (θ) is required. So θ is the appropriate generalized coordinate for the pendulum problem.

where the coordinate origin is at the point of suspension and θ is measured from the negative y-axis (this is *not* the way we usually define θ!)

Differentiating these equations with respect to time,

$$\dot{x} = l\dot{\theta}\cos\theta,$$

$$\dot{y} = l\dot{\theta}\sin\theta.$$

Plugging these expressions into the Lagrangian for the pendulum (Equation (4.3)) leads to

$$L = \frac{1}{2}m(l\dot{\theta}\cos\theta)^2 + \frac{1}{2}m(l\dot{\theta}\sin\theta)^2 + mgl\cos\theta$$

$$= \frac{1}{2}ml^2\dot{\theta}^2(\cos^2\theta + \sin^2\theta) + mgl\cos\theta$$

$$= \frac{1}{2}ml^2\dot{\theta}^2 + mgl\cos\theta. \tag{4.4}$$

Equation (4.4) gives the same information as Equation (4.3) that was previously obtained for the Lagrangian in terms of x and y, except that now L depends only on θ.

All I have done so far is to write the Lagrangian in Cartesian coordinates and then transform to polar coordinates. In doing so, the Lagrangian went from being a function of two variables to being a function of one variable. This procedure illustrates a very important point: the Lagrangian should be expressed in terms of the least possible number of coordinates.

Let me give you a simple "cookbook" procedure for obtaining the Lagrangian in terms of a set of generalized coordinates q_1, q_2, q_3.

(1) If possible, write the Lagrangian $L = T - V$ in terms of *Cartesian coordinates*. In Cartesian coordinates the translational kinetic energy is

$$T = \frac{1}{2}m\left(\dot{x}^2 + \dot{y}^2 + \dot{z}^2\right).$$

(2) Write the transformation equations:

$$x = x(q_1, q_2, q_3, t),$$

$$y = y(q_1, q_2, q_3, t),$$

$$z = z(q_1, q_2, q_3, t).$$

(3) Take time derivatives to obtain expressions for $\dot{x}$, $\dot{y}$, $\dot{z}$ in terms of qs, $\dot{q}$s, and t.
(4) Write $L = T - V$ in terms of the qs, $\dot{q}$s, and t.

You might wonder why I am emphasizing that you should first write the Lagrangian in Cartesian coordinates. The answer is simple. The translational kinetic energy in Cartesian coordinates always has the simple form $T = \frac{1}{2}m\left(\dot{x}^2 + \dot{y}^2 + \dot{z}^2\right)$, whereas it can be very complicated in other coordinate systems. Furthermore, the potential energy is often (but not always) easier to express in Cartesian coordinates.

A very simple (but important) system is a block of mass m on a frictionless surface and connected to a spring of constant k. See Figure 4.3. The Lagrangian for this system is

$$L = \frac{1}{2}m\dot{x}^2 - \frac{1}{2}kx^2. \tag{4.5}$$

Note that x is measured from the equilibrium position of the block. (In this example, the generalized coordinate is simply x.)

Exercise 4.2

A rock of mass m is dropped from a location near the surface of the Earth. Write the Lagrangian. Answer: $L = \frac{1}{2}m\dot{z}^2 - mgz$. ∎

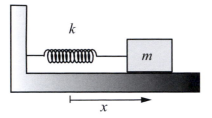

Figure 4.3 A mass m on a frictionless surface acted upon by a spring of constant k.

Worked Example 4.1   Determine the Lagrangian of the double planar pendulum.

Solution See Figure 4.2. In Cartesian coordinates the kinetic and potential energies are:

$$T = \frac{1}{2}m_1(\dot{x}_1^2 + \dot{y}_1^2) + \frac{1}{2}m_2(\dot{x}_2^2 + \dot{y}_2^2),$$

$$V = m_1 g y_1 + m_2 g y_2,$$

where (x_1, y_1) are the coordinates of the bob of mass m_1 and (x_2, y_2) are the coordinates of the other bob. The origin of coordinates is at the point of suspension. The transformation equations from the Cartesian coordinates to the angles θ_1 and θ_2 are:

$$x_1 = l_1 \sin \theta_1$$
$$x_2 = l_1 \sin \theta_1 + l_2 \sin \theta_2$$
$$y_1 = -l_1 \cos \theta_1$$
$$y_2 = -l_1 \cos \theta_1 - l_2 \cos \theta_2.$$

The velocity components are obtained by taking derivatives of position with respect to time:

$$\dot{x}_1 = l_1 \dot{\theta}_1 \cos \theta_1$$
$$\dot{x}_2 = l_1 \dot{\theta}_1 \cos \theta_1 + l_2 \dot{\theta}_2 \cos \theta_2$$
$$\dot{y}_1 = l_1 \dot{\theta}_1 \sin \theta_1$$
$$\dot{y}_2 = l_1 \dot{\theta}_1 \sin \theta_1 + l_2 \dot{\theta}_2 \sin \theta_2.$$

Therefore,

$$\dot{x}_1^2 + \dot{y}_1^2 = l_1^2 \dot{\theta}_1^2 (\cos^2 \theta_1 + \sin^2 \theta_1) = l_1^2 \dot{\theta}_1^2,$$

and

$$\dot{x}_2^2 + \dot{y}_2^2 = (l_1 \dot{\theta}_1 \cos \theta_1 + l_2 \dot{\theta}_2 \cos \theta_2)^2 + (l_1 \dot{\theta}_1 \sin \theta_1 + l_2 \dot{\theta}_2 \sin \theta_2)^2$$
$$= l_1^2 \dot{\theta}_1^2 + l_2^2 \dot{\theta}_2^2 + 2 l_1 l_2 \cos(\theta_1 - \theta_2).$$

Consequently

$$L = T - V = \frac{1}{2}(m_1 + m_2) l_1^2 \dot{\theta}_1^2 + \frac{1}{2} m_2 [l_2^2 \dot{\theta}_2^2 + 2 l_1 l_2 \dot{\theta}_1 \dot{\theta}_2 \cos(\theta_1 - \theta_2)]$$
$$+ m_1 g l_1 \cos \theta_1 + m_2 g (l_1 \cos \theta_1 + l_2 \cos \theta_2). \tag{4.6}$$

I hope you appreciate that obtaining the Lagrangian is a straightforward procedure although it may be somewhat tedious. This example also illustrates the value of using Cartesian coordinates as your starting point.

Worked Example 4.2   Determine the Lagrangian for a disk of mass m and radius R rolling down an inclined plane of angle α. See Figure 4.4. The plane is perfectly rough so the disk rolls without slipping.

Solution   Because the disk is rolling, its total kinetic energy is rotational kinetic energy plus translational kinetic energy. That is, $T = T_{\text{rot}} + T_{\text{trans}}$. The rotational kinetic energy is not, of course, expressed in Cartesian coordinates; it is $T_{\text{rot}} = \frac{1}{2}I\dot{\theta}^2$. (The moment of inertia is I and the angle through which the disk has rotated is θ.) The moment of inertia of a disk is $\frac{1}{2}mR^2$. Therefore,

$$T_{\text{rot}} = \frac{1}{2}\left(\frac{1}{2}mR^2\right)\dot{\theta}^2 = \frac{1}{4}mR^2\dot{\theta}^2.$$

As always, $T_{\text{trans}} = \frac{1}{2}m(\dot{x}^2 + \dot{y}^2)$. Let s be the distance measured down the plane. It is convenient to use s as the generalized coordinate. The transformation equations are

$$x = s\cos\alpha,$$
$$y = -s\sin\alpha.$$

The translational velocity components are

$$\dot{x} = \dot{s}\cos\alpha,$$
$$\dot{y} = -\dot{s}\sin\alpha.$$

Consequently the translational kinetic energy is

$$T_{\text{trans}} = \frac{1}{2}m(\dot{x}^2 + \dot{y}^2) = \frac{1}{2}m\dot{s}^2.$$

I expressed the translational kinetic energy in terms of $\dot{s}$ and the rotational kinetic energy in terms of $\dot{\theta}$. But these two quantities are related. As the disk rolls down the plane, the relationship between the angle θ through which it has rotated, and the distance s it has moved down the plane is

$$s = R\theta.$$

(A relationship between coordinates, such as this one, is called a "constraint." A constraint allows us to express one coordinate in terms of the others and to describe the problem with one less coordinate.) Therefore, $\dot{s} = R\dot{\theta}$. Consequently, $T_{\text{rot}} = (1/4)m\dot{s}^2$, and the total kinetic energy can be written in terms of the single parameter $\dot{s}$:

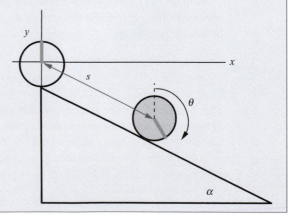

Figure 4.4   A disk of mass m and radius R rolls down a perfectly rough plane of angle α.

$$T = T_{\text{trans}} + T_{\text{rot}} = \frac{1}{2}m\dot{s}^2 + \frac{1}{4}m\dot{s}^2 = \frac{3}{4}m\dot{s}^2.$$

The potential energy is

$$V = mgy = -mgs\sin\alpha.$$

The Lagrangian is defined as $L = T - V$, so,

$$L = \frac{3}{4}m\dot{s}^2 + mgs\sin\alpha. \tag{4.7}$$

Exercise 4.3

A pendulum is made of a bob of mass m but the string is replaced by a spring of constant k. Write the Lagrangian in terms of the length of the spring and the angle it makes with the vertical. Let the unstretched length of the spring be l_0. Answer:

$$L = \frac{1}{2}m\dot{l}^2 + \frac{1}{2}ml^2\dot{\theta}^2 + mgl\cos\theta - \frac{1}{2}k(l - l_0)^2.$$

■

4.4 Lagrange's Equations

As you have seen, the Lagrangian for a system can be obtained by using the definition $L = T - V$. The algebra may get a bit complicated at times, but basically all you need to do is express the translational kinetic energy and the potential energy in terms of Cartesian coordinates (if possible) and then transform to the appropriate set of "generalized" coordinates. Perhaps the most difficult thing is deciding what to use as a generalized coordinate.

By this time you are probably wondering what the Lagrangian is used for. What good is it? The simplest answer is that the Lagrangian allows us to determine the equations of motion for a mechanical system. In fact, the Lagrangian technique is usually the *easiest* way to find the equations of motion for a complicated system. (As I mentioned earlier, the Lagrangian method is much more that just a useful technique, but for the present let's take a purely utilitarian approach and consider the Lagrangian simply as a tool for generating the equations of motion.)

Suppose the Lagrangian is expressed as a function of a single generalized coordinate q, the corresponding generalized velocity $\dot{q}$, and the time t. That is, $L = L(q, \dot{q}, t)$. Then the equation of motion is given by:

$$\frac{d}{dt}\left(\frac{\partial L}{\partial \dot{q}}\right) - \frac{\partial L}{\partial q} = 0. \tag{4.8}$$

This is called *Lagrange's equation.* You should memorize it. Note particularly that the first partial derivative is taken with respect to the generalized velocity ($\dot{q}$) and the second partial derivative is taken with respect to the generalized position (q).

If the Lagrangian is a function of n coordinates, $q_1, q_2, \ldots, q_n$, there are n equations of motion, each having the form of Equation (4.8). That is, the equations of motion are

$$\frac{d}{dt}\left(\frac{\partial L}{\partial \dot{q}_i}\right) - \frac{\partial L}{\partial q_i} = 0 ; \quad i = 1, 2, \ldots, n. \tag{4.9}$$

Consider, for example, the mass on the spring of Figure 4.3. As noted in Equation (4.5), the Lagrangian is $L = \frac{1}{2}m\dot{x}^2 - \frac{1}{2}kx^2$. For this system the generalized coordinate q is just the Cartesian coordinate x. There are no other coordinates. Therefore, there is only one Lagrange equation, namely,

$$\frac{d}{dt}\left(\frac{\partial L}{\partial \dot{x}}\right) - \frac{\partial L}{\partial x} = 0.$$

But

$$\frac{\partial L}{\partial \dot{x}} = m\dot{x},$$

and

$$\frac{\partial L}{\partial x} = -kx.$$

Consequently, Lagrange's equation is

$$\frac{d}{dt}(m\dot{x}) + kx = 0,$$

or

$$m\ddot{x} + kx = 0.$$

Of course, you get the same result immediately by applying Newton's second law.

To illustrate the process for a slightly more complicated system, the following worked example asks for the equation of motion for a simple pendulum and the next one asks for the equations of motion for a double pendulum.

Worked Example 4.3 Determine the equation of motion for a simple pendulum.

Solution The Lagrangian for a pendulum is $L = \frac{1}{2}ml^2\dot{\theta}^2 + mgl\cos\theta$. This is shown in Equation (4.4). In this expression, the single coordinate is θ and Lagrange's equation (4.8) becomes

$$\frac{d}{dt}\left(\frac{\partial L}{\partial \dot{\theta}}\right) - \frac{\partial L}{\partial \theta} = 0.$$

The partial of L with respect to the angular velocity $\dot{\theta}$ is

$$\frac{\partial L}{\partial \dot{\theta}} = \frac{\partial}{\partial \dot{\theta}}\left(\frac{1}{2}ml^2\dot{\theta}^2 + mgl\cos\theta\right) = ml^2\dot{\theta},$$

and the partial of L with respect to the angular position θ is

$$\frac{\partial L}{\partial \theta} = \frac{\partial}{\partial \theta}\left(\frac{1}{2}ml^2\dot{\theta}^2 + mgl\cos\theta\right) = -mgl\sin\theta.$$

Therefore, Lagrange's equation is

$$\frac{d}{dt}\left(ml^2\dot{\theta}^2\right) + mgl\sin\theta = 0,$$

or,

$$\ddot{\theta} = -\frac{g}{l}\sin\theta.$$

This is, of course, exactly the same equation of motion obtained earlier by elementary methods, for example in Equation (4.1).

Worked Example 4.4 Determine the equations of motion for the double planar pendulum illustrated in Figure 4.2.

Solution Recall that the double pendulum was too complicated to analyze using Newton's second law. The Lagrangian for the double pendulum is given by Equation (4.6), which is repeated here:

$$L = \frac{1}{2}(m_1 + m_2)l_1^2\dot{\theta}_1^2 + \frac{1}{2}m_2[l_2^2\dot{\theta}_2^2 + 2l_1l_2\dot{\theta}_1\dot{\theta}_2\cos(\theta_1 - \theta_2)]$$
$$+ m_1gl_1\cos\theta_1 + m_2g(l_1\cos\theta_1 + l_2\cos\theta_2).$$

This problem involves two coordinates, θ_1 and θ_2, and therefore there are two Lagrange equations, namely:

$$\frac{d}{dt}\left(\frac{\partial L}{\partial \dot{\theta}_1}\right) - \frac{\partial L}{\partial \theta_1} = 0, \tag{4.10}$$
$$\frac{d}{dt}\left(\frac{\partial L}{\partial \dot{\theta}_2}\right) - \frac{\partial L}{\partial \theta_2} = 0.$$

To facilitate writing them, first evaluate the derivatives:

$$\frac{\partial L}{\partial \theta_1} = -m_2l_1l_2\dot{\theta}_1\dot{\theta}_2\sin(\theta_1 - \theta_2) - m_1gl_1\sin\theta_1 - m_2gl_1\sin\theta_1,$$

$$\frac{\partial L}{\partial \theta_2} = m_2l_1l_2\dot{\theta}_1\dot{\theta}_2\sin(\theta_1 - \theta_2) - m_2gl_2\sin\theta_2,$$

$$\frac{\partial L}{\partial \dot{\theta}_1} = (m_1 + m_2)l_1^2\dot{\theta}_1 + m_2l_1l_2\dot{\theta}_2\cos(\theta_1 - \theta_2),$$

$$\frac{\partial L}{\partial \dot{\theta}_2} = m_2l_2^2\dot{\theta}_2 + m_2l_1l_2\dot{\theta}_1\cos(\theta_1 - \theta_2).$$

Plugging into Equations (4.10) yields two coupled equations of motion:

$$\frac{d}{dt}\left[m_1l_1^2\dot{\theta}_1 + m_2l_1^2\dot{\theta}_1 + m_2l_1l_2\dot{\theta}_2\cos(\theta_1 - \theta_2)\right]$$
$$+ m_2l_1l_2\dot{\theta}_1\dot{\theta}_2\sin(\theta_1 - \theta_2) + (m_1 + m_2)gl_1\sin\theta_1 = 0,$$

and

$$\frac{d}{dt}\left(m_2l_2^2\dot{\theta}_2 + m_2l_1l_2\dot{\theta}_1\cos(\theta_1 - \theta_2)\right)$$
$$- m_2l_1l_2\dot{\theta}_1\dot{\theta}_2\sin(\theta_1 - \theta_2) + m_2gl_2\sin\theta_2 = 0.$$

These equations can be further simplified and written in the form

$$(m_1 + m_2)l_1^2\ddot{\theta}_1 + m_2l_1l_2\left[\ddot{\theta}_2\cos(\theta_1 - \theta_2) + \dot{\theta}_2^2\sin(\theta_1 - \theta_2)\right]$$
$$+ (m_1 + m_2)gl_1\sin\theta_1 = 0,$$

and

$$m_2l_2^2\ddot{\theta}_2 + m_2l_1l_2\left[\ddot{\theta}_1\cos(\theta_1 - \theta_2) - \dot{\theta}_1^2\sin(\theta_1 - \theta_2)\right] + m_2gl_2\sin\theta_2 = 0.$$

(I'm sure you now appreciate why I did not attempt to solve the double pendulum problem by using $F = ma$!)

Exercise 4.4

Find the equation of motion for a disk rolling down a perfectly rough inclined plane of angle α. (The Lagrangian was obtained in Worked Example 4.2.) Answer: $\ddot{s} = \frac{2}{3}g\sin\alpha$. ∎

Exercise 4.5

Find the equation of motion for a sphere rolling down a perfectly rough inclined plane of angle α. (Use the Lagrangian technique.) The moment of inertia of a sphere is $\frac{2}{5}mR^2$. Answer: $\ddot{s} = \frac{5}{7}g\sin\alpha$. ∎

Exercise 4.6

Using the Lagrangian technique, determine the equation of motion for a body of mass m falling in a constant gravitational field. ∎

Exercise 4.7

Find the Lagrangian and the equation of motion for two astronomical bodies on a collision course that are attracting one another according to Newton's Law of Universal Gravitation. (Place the origin of coordinates at the point where the bodies will collide.) Answer: $m_1\ddot{r}_1 = -G\frac{m_1 m_2}{(1+m_1/m_2)^2 r_1^2}$ (where r_1 is the distance from the center of mass to body m_1). ∎

4.5 Degrees of Freedom

The Lagrangian technique often involves changing from one set of coordinates (usually x, y, z) to another (say, r, θ, ϕ). (It turns out that coordinate transformations are central to the Lagrangian method; they are considered in excruciating detail in graduate courses in mechanics. If you want to know more about this subject, I recommend the excellent, but advanced, books by Goldstein[2] and by Fetter and Walecka.[3])

In discussing coordinates it is important to realize that every physical system has a particular number of *degrees of freedom*. The number of degrees of freedom is the number of independent coordinates needed to completely specify the position of every part of the system. To describe the position of a free particle you must specify the values of three coordinates (say, x, y, and z). Thus a free particle has three degrees of freedom. For a system of two free particles you need to specify the positions of both particles. Each particle has three degrees of freedom, so the system as a whole has six degrees of freedom. In general a mechanical system consisting of N free particles will have $3N$ degrees of freedom.

For many systems, however, the number of *independent* coordinates is much less than $3N$. For example, the position of a particle on a flat surface such as a tabletop can be described

---

[2] H. Goldstein, *Classical Mechanics*, Addison-Wesley Publishing Co., Reading, MA, 1950. Note that there are three editions. The first edition is the shortest and probably the best. The second edition included a significant amount of new material but unfortunately it had many typographical errors which made it rather difficult to read. The third edition (by H. Goldstein, C. Poole, and J. Safko) has corrected the errors from the second edition and contains some interesting new material.

[3] A. L. Fetter, and J. D. Walecka, *Theoretical Mechanics of Particles and Continua*, McGraw-Hill, NY, 1980.

with two coordinates, such as x and y. A condition that specifies the value of a coordinate or gives a relation between coordinates is called a *constraint*.[4] For the particle on the tabletop, the constraint is

$$z = \text{constant}.$$

A pendulum bob is constrained by the string to remain a constant distance from the point of support. That is, the bob is constrained to the surface of a sphere whose radius equals the length of the string. A bar of soap slipping in a sink is constrained to the surface of the sink. If the sink is a hemispherical bowl of radius a, the equation of constraint is

$$x^2 + y^2 + z^2 = a^2.$$

This relationship indicates that there are only two *independent* coordinates because the third coordinate (say z) is related to x and y through

$$z = \sqrt{a^2 - x^2 - y^2}.$$

Constraints are important to us because **each constraint reduces the number of degrees of freedom by one.**

Going back to the Lagrangian procedure, recall that there is one Lagrange equation of motion for each coordinate; see Equation (4.9). It is obviously beneficial to reduce the number of equations. You can do this by using constraints to get rid of as many nonindependent coordinates as possible. When you describe the system in terms of independent coordinates (say $q_1, q_2, \ldots, q_n$), you have minimized the number of Lagrange equations. These *independent coordinates* are referred to as generalized coordinates and are usually denoted by q_i. A system with n degrees of freedom can be described in terms of the n generalized coordinates $q_1, \ldots, q_n$. For such a system, the Lagrangian will be a function of the generalized coordinates, the generalized velocities, and possibly the time. That is,

$$L = L(q_1, q_2, \ldots q_n; \dot{q}_1, \dot{q}_2, \ldots \dot{q}_n; t).$$

In this case, there will be n Lagrange equations of motion.[5]

4.6 Generalized Momentum

Consider again the problem of a mass m connected to a spring of constant k as illustrated in Figure 4.3. If the mass is moving in the x direction with speed $\dot{x}$ it has momentum $p_x = m\dot{x}$. We have seen that the Lagrangian for this system is

$$L = \tfrac{1}{2}m\dot{x}^2 - \tfrac{1}{2}kx^2.$$

Taking the derivative of the Lagrangian with respect to $\dot{x}$ yields

$$\frac{\partial L}{\partial \dot{x}} = \frac{\partial}{\partial \dot{x}}\left(\tfrac{1}{2}m\dot{x}^2 - \tfrac{1}{2}kx^2\right) = m\dot{x}.$$

---

[4] An equation giving the value of one coordinate in terms of the other coordinates and possibly the time is called a "holonomic" constraint. The relationship between coordinates allows us to reduce by one the number of degrees of freedom for each holonomic constraint. If the equation contains other quantities, such as velocities, it is "nonholonomic." We will be assuming all constraints are holonomic.

[5] I must warn you that physicists are somewhat careless in their usage of the term "generalized coordinate." You will often hear it applied in situations in which the q_i are not all independent.

But $m\dot{x}$ is just the linear momentum! Therefore, for this system, the linear momentum is related to the Lagrangian by

$$p_x = \frac{\partial L}{\partial \dot{x}}.$$

Next, consider the problem of a simple pendulum consisting of a mass m hanging from a string of length l. According to Equation (4.4), the Lagrangian for this system is

$$L = \tfrac{1}{2}ml^2\dot{\theta}^2 + mgl\cos\theta.$$

Taking the derivative of the Lagrangian with respect to $\dot{\theta}$ yields

$$\frac{\partial L}{\partial \dot{\theta}} = ml^2\dot{\theta}.$$

But $ml^2\dot{\theta}$ is the angular momentum of the pendulum! In this case the derivative of the Lagrangian with respect to the angular velocity $\dot{\theta}$ is the *angular* momentum.

In the first case, the Lagrangian was expressed in terms of x and $\dot{x}$. That is, $L = L(x,\dot{x})$. The generalized coordinate was x and the generalized velocity was $\dot{x}$. In the second case, the Lagrangian was a function of θ and $\dot{\theta}$, that is, $L = L(\theta,\dot{\theta})$. Here the generalized coordinate was θ and the generalized velocity was $\dot{\theta}$. These two cases illustrate the general functional form of the Lagrangian for a one variable system, $L = L(q,\dot{q},t)$. That is, the Lagrangian depends on the generalized position (q), the generalized velocity ($\dot{q}$), and possibly the time (t).

The linear momentum of a mass on a spring was given by $\partial L/\partial \dot{x}$ and the angular momentum for the pendulum was given by $\partial L/\partial \dot{\theta}$. It is reasonable to define a quantity called the "generalized momentum" by

$$p_i = \frac{\partial L}{\partial \dot{q}_i}. \tag{4.11}$$

The generalized momentum p_i is associated with the generalized coordinate q_i. These two quantities are called *conjugates*. Thus, for the mass on a spring, the generalized momentum conjugate to x is $m\dot{x}$ and for the pendulum, the generalized momentum conjugate to θ is $ml^2\dot{\theta}$. In the first case the generalized momentum is the linear momentum and in the second case the generalized momentum is the angular momentum.

Although the quantity $\partial L/\partial \dot{q}_i$ is called the generalized momentum, and is denoted by the symbol p_i, be aware that it is frequently not anything you would normally think of as a momentum. For example, when we discussed a disk rolling down an inclined plane (see Worked Example 4.2) we found the Lagrangian to be $L = \tfrac{3}{4}m\dot{s}^2 + mgs\sin\alpha$. Consequently the generalized momentum is

$$p_s = \frac{\partial L}{\partial \dot{s}} = \frac{6}{4}m\dot{s},$$

which is not the same as the linear momentum.[6]

4.6.1 Ignorable Coordinates

We now come to a very interesting point. For a mechanical system the Lagrangian might not depend *explicitly* on a particular coordinate, say q_i. Although q_i may not appear in the

---

[6] It is interesting to note that $\frac{d}{dt}\left(\frac{\partial L}{\partial \dot{x}}\right) = \frac{dp}{dt}$ and $\frac{\partial L}{\partial x} = -\frac{\partial V}{\partial x} = F$ so Lagrange's equation reduces to $\frac{dp}{dt} - F = 0$, the original form of Newton's second law.

Lagrangian, the generalized velocity $\dot{q}_i$ will be present in the expression for L. For example, the Lagrangian for a spherical pendulum is $L = \frac{1}{2}ml^2(\dot{\theta}^2 + \sin^2\theta\dot{\phi}^2) - mgl\cos\theta$. This expression contains $\dot{\phi}$ but does not contain ϕ. If q_i does not appear explicitly in the Lagrangian, then q_i is called an *ignorable coordinate*.[7] The partial derivative of the Lagrangian with respect to an ignorable coordinate is zero: $\partial L/\partial q_i = 0$. Recall that Lagrange's equation states that

$$\frac{d}{dt}\left(\frac{\partial L}{\partial \dot{q}_i}\right) - \frac{\partial L}{\partial q_i} = 0.$$

But if $\partial L/\partial q_i = 0$, then

$$\frac{d}{dt}\left(\frac{\partial L}{\partial \dot{q}_i}\right) = 0.$$

The term in parenthesis is the generalized momentum p_i, so this equation states that $dp_i/dt = 0$. If the time derivative of a quantity is zero, the quantity is a *constant*. Therefore, if the coordinate q_i does not appear explicitly in the Lagrangian, then the conjugate generalized momentum p_i is constant.

I will summarize all of this in a single phrase which is easy to remember, but perhaps not so easy to understand:

If the coordinate q_i is ignorable, the conjugate momentum p_i is constant.

In studying a physical system, whenever you encounter a quantity that remains constant during the motion, you have a very useful tool for solving problems. Such constant quantities are called "constants of the motion" or "first integrals" for reasons we shall discuss later.

Worked Example 4.5 The Lagrangian for a planet in orbit about a star has the form $L = (1/2)m\dot{r}^2 + (1/2)mr^2\dot{\phi}^2 - \frac{b}{r}$, where r and ϕ are generalized coordinates and m and b are constants. Determine the generalized momenta p_r and p_ϕ. Is either of these a constant of the motion?

Solution The generalized coordinates are r and ϕ, so the conjugate generalized momenta are

$$p_r = \frac{\partial L}{\partial \dot{r}} \quad \text{and} \quad p_\phi = \frac{\partial L}{\partial \dot{\phi}}.$$

Therefore,

$$p_r = \frac{\partial L}{\partial \dot{r}} = \frac{\partial}{\partial \dot{r}}\left(\frac{1}{2}m\dot{r}^2 + \frac{1}{2}mr^2\dot{\phi}^2 - \frac{b}{r}\right) = m\dot{r}$$

and

$$p_\phi = \frac{\partial L}{\partial \dot{\phi}} = \frac{\partial}{\partial \dot{\phi}}\left(\frac{1}{2}m\dot{r}^2 + \frac{1}{2}mr^2\dot{\phi}^2 - \frac{b}{r}\right) = mr^2\dot{\phi}.$$

Since the Lagrangian does not depend on ϕ (it is ignorable), we conclude that the angular momentum of a planet is constant (p_ϕ = constant).

---

[7] When we say that q_i does not appear explicitly in the Lagrangian we mean that if the Lagrangian is expressed in terms of the minimal set of generalized coordinates, the quantity q_i is not present. (Nevertheless, the generalized velocity $\dot{q}_i$ does appear in the Lagrangian.) Just because a coordinate is called "ignorable" does not mean we can ignore it!

Exercise 4.8

Determine the generalized momenta p_{θ_1} and p_{θ_2} for the double planar pendulum. Is either one of them a constant? Answer: $p_{\theta_2} = m_2 l_2^2 \dot{\theta}_2 + m_2 l_1 l_2 \dot{\theta}_1 \cos(\theta_1 - \theta_2)$. No. ∎

Exercise 4.9

(a) Determine the generalized momentum for a sphere of mass m and radius R rolling down an inclined plane of angle α. Let the distance down the plane be given by s. (b) Is the generalized momentum a constant? Answers: (a) $p_s = (5/2)m\dot{s}$, (b) No. ∎

4.7 Generalized Force

Lagrangian mechanics introduces the concepts of generalized coordinates and generalized momenta. So it should not surprise you that there is another quantity called the *generalized force*. The generalized force is denoted by Q_j and is defined in terms of a quantity called the "virtual work." However, before we can consider these concepts further, we need to introduce yet another new concept, the "virtual displacement" δx. The virtual displacement is an imaginary infinitesimal displacement that is similar to an actual infinitesimal displacement denoted dx, except that time is frozen for the virtual displacement. That is, δx is *imaginary, infinitesimal*, and *instantaneous*. Furthermore, it satisfies any constraints on the system.[8]

We now consider virtual work. Actual, real, physical work is defined in terms of forces and displacements and we can write

$$dW = \mathbf{F} \cdot d\mathbf{s} = F_x dx + F_y dy + F_z dz = \sum_i F_i dx_i.$$

Similarly, as you might suspect, the virtual work is given by

$$\delta W = \sum_i F_i \delta x_i.$$

Let us transform from Cartesian coordinates to generalized coordinates. We can express x_i in terms of the generalized coordinates, thus, $x_i = x_i(q_1, q_2, \ldots q_n, t)$. By the rules of calculus, the differential dx_i is given by

$$dx_i = \frac{\partial x_i}{\partial q_1} dq_1 + \frac{\partial x_i}{\partial q_2} dq_2 + \cdots \frac{\partial x_i}{\partial q_n} dq_n + \frac{\partial x_i}{\partial t} dt.$$

Since the virtual displacement is instantaneous we write

$$\delta x_i = \frac{\partial x_i}{\partial q_1} \delta q_1 + \frac{\partial x_i}{\partial q_2} \delta q_2 + \cdots \frac{\partial x_i}{\partial q_n} \delta q_n = \sum_j \frac{\partial x_i}{\partial q_j} \delta q_j.$$

Note that the dt has disappeared because time is frozen during a virtual displacement. Consequently, the virtual work can be expressed as

$$\delta W = \sum_i F_i \sum_j \frac{\partial x_i}{\partial q_j} \delta q_j = \sum_j \left(\sum_i F_i \frac{\partial x_i}{\partial q_j} \right) \delta q_j.$$

---

[8] For example, if a particle is constrained to a tabletop, the virtual displacement will take place in that plane.

If we define the generalized force as

$$Q_j = \sum_i F_i \frac{\partial x_i}{\partial q_j},$$

we can write the virtual work in the form

$$\delta W = \sum_j Q_j \delta q_j,$$

and it has the same form as our usual expression for work, that is, as a force times a displacement.

We will primarily be interested in generalized forces that can be derived from a potential energy, thus: $Q_i = -\frac{\partial V}{\partial q_i}$. Such forces are called "conservative" and can be denoted Q_i^c. Forces (such as friction or air resistance) not associated with a potential energy are called nonconservative (Q_i^{nc}).

To include nonconservative forces in our analysis we write the Lagrange equation in the form

$$\frac{d}{dt} \frac{\partial L}{\partial \dot{q}_i} - \frac{\partial L}{\partial q_i} = Q_i^{nc}. \tag{4.12}$$

This expression is often written in terms of the kinetic energy by replacing L with $T - V$ and noting that $\frac{\partial L}{\partial q_i} = -\frac{\partial V}{\partial q_i} = Q_i^c$. This yields

$$\frac{d}{dt} \frac{\partial T}{\partial \dot{q}_i} - \frac{\partial T}{\partial q_i} = Q_i^c + Q_i^{nc}. \tag{4.13}$$

Worked Example 4.6 Show that the Lagrange equation can be expressed in the "Nielsen form"

$$\frac{\partial \dot{T}}{\partial \dot{q}_i} - 2 \frac{\partial T}{\partial q_i} = Q_i^c + Q_i^{nc}.$$

Solution Start with Equation (4.13), i.e.,

$$\frac{d}{dt} \frac{\partial T}{\partial \dot{q}_i} - \frac{\partial T}{\partial q_i} = Q_i^c + Q_i^{nc}.$$

Note that $T = T(q_i, \dot{q}_i, t)$, $i = 1, 2, \ldots, n$. Consequently, by the rules of calculus

$$\dot{T} = \frac{dT}{dt} = \sum_i \left(\frac{\partial T}{\partial q_i} \frac{dq_i}{dt} + \frac{\partial T}{\partial \dot{q}_i} \frac{d\dot{q}_i}{dt} \right) + \frac{\partial T}{\partial t},$$

$$\dot{T} = \sum_i \left(\frac{\partial T}{\partial q_i} \dot{q}_i + \frac{\partial T}{\partial \dot{q}_i} \ddot{q}_i \right) + \frac{\partial T}{\partial t}.$$

$$\therefore \frac{\partial \dot{T}}{\partial \dot{q}_j} = \frac{\partial}{\partial \dot{q}_j} \sum_i \left(\frac{\partial T}{\partial q_i} \dot{q}_i + \frac{\partial T}{\partial \dot{q}_i} \ddot{q}_i \right) + \frac{\partial}{\partial \dot{q}_j} \frac{\partial T}{\partial t}.$$

Note: Be aware of the different subscripts on the $\dot{q}$s. Continuing the derivation,

$$\frac{\partial \dot{T}}{\partial \dot{q}_j} = \sum_i \left(\frac{\partial^2 T}{\partial \dot{q}_j \partial q_i} \dot{q}_i + \frac{\partial T}{\partial q_i} \frac{\partial \dot{q}_i}{\partial \dot{q}_j} \right) + \sum_i \left(\frac{\partial^2 T}{\partial \dot{q}_j \partial \dot{q}_i} \ddot{q}_i + \frac{\partial T}{\partial \dot{q}_i} \frac{\partial \ddot{q}_i}{\partial \dot{q}_j} \right)$$

$$+ \frac{\partial^2 T}{\partial \dot{q}_j \partial t}.$$

Observe that $\sum_i \frac{\partial \dot{q}_i}{\partial \dot{q}_j} = \delta_{ij}$. That is, the quantity is zero unless $i = j$. Since we are summing over i eventually the value of i is equal to j and the derivative is equal to unity.

Consequently, the term $\sum_i \frac{\partial T}{\partial q_i} \frac{\partial \dot{q}_i}{\partial \dot{q}_j} = \sum_i \frac{\partial T}{\partial q_i} \delta_{ij} = \frac{\partial T}{\partial q_j}$.

Furthermore,

$$\sum_i \frac{\partial T}{\partial \dot{q}_i} \frac{\partial \ddot{q}_i}{\partial \dot{q}_j} = \sum_i \frac{\partial T}{\partial \dot{q}_i} \frac{d}{dt}\left(\frac{\partial \dot{q}_i}{\partial \dot{q}_j}\right) = \sum_i \frac{\partial T}{\partial \dot{q}_i} \frac{d}{dt} \delta_{ij} = 0,$$

leaving us with

$$\frac{\partial \dot{T}}{\partial \dot{q}_j} = \sum_i \frac{\partial^2 T}{\partial \dot{q}_j \partial q_i} \dot{q}_i + \frac{\partial T}{\partial q_j} + \sum_i \frac{\partial^2 T}{\partial \dot{q}_j \partial \dot{q}_i} \ddot{q}_i + \frac{\partial^2 T}{\partial \dot{q}_j \partial t}$$

$$= \frac{\partial}{\partial \dot{q}_j}\left(\sum_i \left(\frac{\partial T}{\partial \dot{q}_i} + \frac{\partial T}{\partial \dot{q}_i}\ddot{q}_i\right) + \frac{\partial T}{\partial t}\right) + \frac{\partial T}{\partial q_j}.$$

But the term in parentheses is $\frac{dT}{dt}$ so

$$= \frac{\partial}{\partial \dot{q}_j} \frac{dT}{dt} + \frac{\partial T}{\partial q_j}$$

$$= \frac{d}{dt} \frac{\partial T}{\partial \dot{q}_j} + \frac{\partial T}{\partial q_j}.$$

$$\therefore \frac{d}{dt} \frac{\partial T}{\partial \dot{q}_j} = \frac{\partial \dot{T}}{\partial \dot{q}_j} - \frac{\partial T}{\partial q_j}.$$

Plug into Equation (4.13) to get

$$\frac{\partial \dot{T}}{\partial \dot{q}_j} - 2\frac{\partial T}{\partial q_i} = Q_i^c + Q_i^{nc}.$$

Q.E.D.

Exercise 4.10

Derive Equation (4.13) from Equation (4.12).   ■

4.8 The Calculus of Variations

I am now going to discuss a branch of mathematics known as *the Calculus of Variations*. I am only doing this to derive the Lagrange equations, so you should consider this section to be just a brief outline of the theory.

The calculus of variations considers problems involving maximums and minimums for certain definite integrals. In doing so, it arrives at answers to various interesting questions, such as: (1) What is the shape of the curve of given length that encloses the largest area? (A circle.) (2) What curve gives the shortest distance between two points on a plane? (A straight line.) (3) What curve gives the shortest distance between two points on a sphere? (A segment of a great circle.) (4) What is the shape of a curve between two points, one higher than the other, such that a bead slides from one point to the other in minimum time? (A cycloid.)

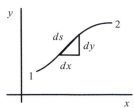

Figure 4.5 The quantity ds is an element of the path from 1 to 2. Note that $ds^2 = dx^2 + dy^2$.

In such problems you are asked to maximize or minimize some quantity. The quantity itself will be expressed as a definite integral. You may have to exert considerable cleverness to figure out this integral, as it is usually not obvious.

Consider a specific example. Suppose you want to determine the equation of the curve between two points in a plane such that the distance along the curve is a minimum. (You already know the answer: It is a straight line, and its equation is $y = mx + b$.) Now the distance between two points along a curve s is $\int_1^2 ds$. But $ds^2 = dx^2 + dy^2$ (Figure 4.5) so

$$I = \int_1^2 ds = \int_1^2 \sqrt{dx^2 + dy^2} = \int_{x_1}^{x_2} \sqrt{1 + \left(\frac{dy}{dx}\right)^2}\, dx$$

$$= \int_{x_1}^{x_2} \sqrt{1 + (y')^2}\, dx. \tag{4.14}$$

Here $y' \equiv dy/dx$. (In the rest of this chapter, a prime will represent differentiation with respect to x.) The integral I is the distance between points 1 and 2. The problem is to find the curve $y = y(x)$ that minimizes I.

I will come back to this problem in a moment, but first I want to consider the calculus of variations in a general way. Problems in the calculus of variations always lead to definite integrals of the form

$$I = \int_{x_1}^{x_2} \Phi\left(x, y, \frac{dy}{dx}\right) dx = \int_{x_1}^{x_2} \Phi(x, y, y')dx.$$

The solution of the problem is the function $y = y(x)$ that makes the integral an extremum.[9] Note that Φ is a *function of a function* because Φ is a function of y, and y itself is a function of x. Φ is called a "functional" to distinguish it from an ordinary function. For the problem of determining the shortest path, the functional is $\Phi(x, y, y') = \sqrt{1 + (y')^2}$.

Keep in mind that the integral I is a line integral between limits x_1 and x_2. It is the distance from point 1 to point 2 along a particular path $y = y(x)$. Naturally, if you use a different function $y_1 = y_1(x)$, you will integrate along a different path and get a different value for the integral.

Figure 4.6 illustrates the problem of the shortest distance between two points in a plane. The three paths $y_1(x)$, $y_2(x)$, and $y_3(x)$ have different path lengths. The problem asks us to find the shortest of all the *possible* paths between the *fixed end points*. Let us assume that $y_0(x)$ is the shortest path, and let us further assume that $y_1(x)$ is a path that only differs infinitesimally from $y_0(x)$. That is, at each point on the path (except at the end points) $y_1(x)$ is slightly different from $y_0(x)$. We can write the relationship between $y_1(x)$ and $y_0(x)$ in the form

$$y_1(x) = y_0(x) + \epsilon_1 \eta(x),$$

---

[9] By "extremum" we mean a maximum or a minimum, that is, values of a function where the derivative is zero. The derivative of a function is also zero at an inflection point, but in the calculus of variations we are always interested in either the maximum or the minimum of the integral.

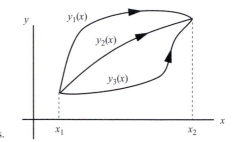

Figure 4.6 Three possible paths between two fixed end points.

where ϵ_1 is a small quantity and $\eta(x)$ is a function of x. Note that $\eta(x)$ is completely arbitrary except that it must be zero at the end points. That is,

$$\eta(x_1) = \eta(x_2) = 0,$$

because all the paths between x_1 and x_2 must meet at the end points.

Remember that the problem is to find the minimal path $y_0(x)$. I will denote this minimal path simply by $y(x)$ from now on. In general, a nearby path is a member of a family of paths that are described by the relationship

$$Y = Y(x, \epsilon) = y(x) + \epsilon \eta(x), \tag{4.15}$$

where $y(x)$ is the minimal path. Here ϵ is taken as a continuous variable. Different paths correspond to different values of ϵ. The minimal path is obtained when $\epsilon = 0$.

The path length along any particular member of this family of curves is

$$I = I(\epsilon) = \int_{x_1}^{x_2} \Phi\left(x, Y, \frac{dY}{dx}\right) dx = \int_{x_1}^{x_2} \Phi\left(x, Y, Y'\right) dx.$$

Here the integral I is explicitly expressed as a function of ϵ. The desired curve has $\epsilon = 0$; it is the curve that minimizes I. Therefore, the condition on I is

$$\left[\frac{dI}{d\epsilon}\right]_{\epsilon=0} = 0.$$

Observe that

$$\frac{dI}{d\epsilon} = \frac{d}{d\epsilon} \int_{x_1}^{x_2} \Phi(x, Y, Y') dx = \int_{x_1}^{x_2} \left(\frac{\partial \Phi}{\partial Y} \frac{\partial Y}{\partial \epsilon} + \frac{\partial \Phi}{\partial Y'} \frac{\partial Y'}{\partial \epsilon}\right) dx.$$

From Equation (4.15)

$$\frac{\partial Y}{\partial \epsilon} = \eta(x).$$

Also,

$$\frac{\partial Y'}{\partial \epsilon} = \frac{\partial}{\partial \epsilon}\left(\frac{dY}{dx}\right) = \frac{\partial}{\partial \epsilon}\left(y' + \epsilon \eta'\right) = \eta'.$$

If $\epsilon = 0$ then $Y = y$ and $Y' = y'$, so

$$\left[\frac{dI}{d\epsilon}\right]_{\epsilon=0} = \int_{x_1}^{x_2}\left(\frac{\partial \Phi}{\partial y}\eta + \frac{\partial \Phi}{\partial y'}\eta'\right) dx = 0. \tag{4.16}$$

The formula for integration by parts is

$$\int_a^b u\, dv = uv\Big|_a^b - \int_a^b v\, du.$$

Applying this to the second term in Equation (4.16) leads to

$$\int_{x_1}^{x_2} \frac{\partial \Phi}{\partial y'} \frac{d\eta}{dx} dx = \left[\frac{\partial \Phi}{\partial y'}\eta\right]_{x_1}^{x_2} - \int_{x_1}^{x_2} \eta \frac{d}{dx}\frac{\partial \Phi}{\partial y'} dx.$$

But $\eta = 0$ at the limits x_1 and x_2 so the first term drops out, and Equation (4.16) becomes

$$0 = \int_{x_1}^{x_2} \left(\frac{\partial \Phi}{\partial y} - \frac{d}{dx}\frac{\partial \Phi}{\partial y'}\right)\eta dx.$$

Because $\eta \neq 0$ (except at the end points) this integral can be zero if and only if the term in parenthesis is zero.[10] That is,

$$\frac{\partial \Phi}{\partial y} - \frac{d}{dx}\frac{\partial \Phi}{\partial y'} = 0. \qquad (4.17)$$

This is called the *Euler–Lagrange equation*. It gives the condition on Φ that minimizes the integral I.

To appreciate how to use the Euler–Lagrange equation, let us return to the problem of the shortest distance between two points. From Equation (4.14) we found that for this problem the functional Φ is

$$\Phi = \sqrt{1 + \left(\frac{dy}{dx}\right)^2} = \left(1 + y'^2\right)^{\frac{1}{2}}.$$

Plugging this functional into the Euler–Lagrange equation (4.17) gives

$$\frac{\partial}{\partial y}\left(1 + y'^2\right)^{\frac{1}{2}} - \frac{d}{dx}\frac{\partial}{\partial y'}\left(1 + y'^2\right)^{\frac{1}{2}} = 0,$$

$$0 - \frac{d}{dx}\left[\frac{1}{2}\left(1 + y'^2\right)^{-\frac{1}{2}}\left(2y'\right)\right] = 0,$$

$$\frac{d}{dx}\left[\frac{y'}{\sqrt{1 + y'^2}}\right] = 0.$$

If the derivative of a function is zero, the function is constant, so

$$\frac{y'}{\sqrt{1 + y'^2}} = \text{constant} = C_1.$$

Squaring both sides and rearranging,

$$y'^2 = (C_1)^2(1 + y'^2). \qquad (4.18)$$

Solving for y'^2,

$$y'^2(1 - C_1^2) = C_1^2,$$

---

[10] You might argue that the equation has the form

$$\int_a^b g(x)\eta(x)dx = 0,$$

and this can be satisfied even if $g(x)$ is not zero everywhere for *some* function $\eta(x)$. But that argument does not hold for an *arbitrary* function $\eta(x)$.

$$y' = \sqrt{\frac{C_1^2}{1 - C_1^2}} = \text{constant} = m.$$

But $y' = dy/dx$, so

$$\frac{dy}{dx} = m.$$

A final integration yields

$$y = mx + b,$$

the equation for a straight line! Thus we have proved that a straight line is the shortest distance between two points. (Whew!)

4.8.1 Hamilton's Principle

By this time you are surely wondering what any of this has to do with physics. The answer lies in *Hamilton's principle.* Hamilton's principle states that **the behavior of any physical system will minimize the time integral of the Lagrangian.** For example, suppose some physical system is described in terms of the single generalized coordinate q. As time goes on, the value of q will change. The quantity we want to determine is $q = q(t)$. Hamilton's principle tells us that $q(t)$ is the function that will minimize the integral

$$I = \int_{t_1}^{t_2} L(q, \dot{q}, t) dt.$$

The condition for this integral to be minimized is given by the Euler–Lagrange equation of the calculus of variations. By replacing q for y and t for x and changing the order of the terms, the Euler–Lagrange equation (4.17) becomes

$$\frac{d}{dt} \frac{\partial L}{\partial \dot{q}} - \frac{\partial L}{\partial q} = 0,$$

which you recognize as Lagrange's equation!

The integral $\int L dt$ is called the "action." Using that terminology, Hamilton's principle is:

The time development of a dynamical system always minimizes the action.

Hamilton's principle is sometimes referred to as, "the fundamental principle of mechanics."[11]

To make this a bit more explicit, consider a simple physical process, such as the motion of a dropped rock. This physical system can be considered to evolve from some initial situation at time t_1 to a different final situation at time t_2. Initially the rock is at height h and has zero velocity. Therefore the initial conditions are $(z, \dot{z}) = (h, 0)$. The final conditions (just before it hits the ground) are $(0, \sqrt{2gh})$. Hamilton's principle states that the behavior of the rock minimizes the quantity $\int_{t_1}^{t_2} L(z, \dot{z}, t) dt$, where the Lagrangian is $(1/2)m\dot{z}^2 - mgz$. Applying the methods of the calculus of variations we obtain the Lagrange equation and the Lagrange equation generates the equation of motion, which for the falling rock, is $d^2z/dt^2 = -g$.

---

[11] You may have studied Fermat's principle in optics which states that the path of a ray of light from one point to another is such as to minimize the time of flight. This is a special case of Hamilton's principle.

4.8.2 Relation to Newton's Second Law

As mentioned previously, Newton's second law can be derived from the Lagrange equation. As an illustration, consider the one-dimensional motion of a particle of mass m. Assume the particle is acted upon by a conservative force F. As we have pointed out, for a conservative force in one dimension, the force can be obtained from the potential energy $V = V(x)$ by $F = -dV/dx$. The Lagrangian is

$$L = T - V = \frac{1}{2}m\dot{x}^2 - V(x),$$

and Lagrange's equation is

$$0 = \frac{d}{dt}\frac{\partial L}{\partial \dot{x}} - \frac{\partial L}{\partial x} = \frac{d}{dt}\left[\frac{\partial}{\partial \dot{x}}\left(\frac{1}{2}m\dot{x}^2 - V(x)\right)\right] - \frac{\partial}{\partial x}\left(\frac{1}{2}m\dot{x}^2 - V(x)\right)$$

$$= \frac{d}{dt}(m\dot{x}) + \frac{\partial V(x)}{\partial x} = m\ddot{x} - F.$$

So

$$F = m\ddot{x}$$

as expected.

Thus, Newton's second law is a consequence of Lagrange's equations and Lagrange's equations are a consequence of Hamilton's principle.[12] I suppose you might conclude that Hamilton's principle is more basic and important than Newton's laws. Nevertheless, it is true that Newton's laws are the basic relations from which all of physics was derived. Hamilton's principle is a beautiful, sophisticated, and elegant way of describing the way nature behaves, but it is not very easily understood. Newton's laws, on the other hand, can be grasped by anyone with an elementary knowledge of mathematics.

4.9 The Hamiltonian and Hamilton's Equations

Another formulation of classical mechanics was developed by the Irish mathematician and physicist W. R. Hamilton.[13] Hamiltonian mechanics is of fundamental importance in quantum mechanics as well as in modern studies in field theory.

It is interesting to note that there is no general technique for solving a set of n coupled Lagrange equations of motion. Now Hamiltonian mechanics is based on different equations of motion (as you will see in a moment) and the German mathematician Jacobi developed a general technique for solving Hamilton's equations of motion. Unfortunately, this "Hamilton–Jacobi equation" is so difficult to solve that it is not a practical tool for determining the motion. We will not consider the Hamilton–Jacobi equation.[14]

Generally speaking, we will not be using Hamiltonian mechanics for the topics considered in this book, the exception being when we study chaos in Chapter 21. However, in your quantum mechanics course the Hamiltonian will be used over and over again.

---

[12] You might ask: "Why does nature always minimize the time integral of the Lagrangian?" I do not know the answer to this question. I know that Hamilton's principle is an expression of the way the physical universe is put together. I know that is *how* nature behaves, but I cannot tell you *why* it behaves that way. (Physics tells us *how*, not *why*.)

[13] Hamilton lived 1805–1865.

[14] If you are interested, it is discussed in Chapter 6 of *A Student's Guide to Lagrangians and Hamiltonians* by Patrick Hamill, Cambridge University Press, Cambridge, 2014.

In this section I will briefly discuss Hamilton's formulation of the laws of mechanics. First, however, I have to define a quantity called the *Hamiltonian*. The Hamiltonian is a function of position and momentum. For a single particle the one-dimensional Hamiltonian is defined as

$$H(p,q,t) = p\dot{q} - L(q,\dot{q},t), \tag{4.19}$$

where p stands for the generalized momentum and q stands for the generalized coordinate. Note that the Lagrangian is a function of *velocity,* position, and time, $L = L(q,\dot{q},t)$. The Hamiltonian is a function of *momentum*, position, and time, $H = H(p,q,t)$. If you are asked to determine a Hamiltonian, make sure your final expression does not explicitly contain any velocities!

For more than one dimension and/or more than one particle the Hamiltonian is:

$$H = \sum_i p_i \dot{q}_i - L. \tag{4.20}$$

More explicitly, for a system of N particles in three dimensions,

$$H(p_1, p_2, \ldots, p_{3N}; q_1, q_2, \ldots q_{3N}; t) = \sum_{i=1}^{3N} p_i \dot{q}_i - L(q_1, \ldots q_{3N}; \dot{q}_1, \ldots \dot{q}_{3N}; t).$$

Having obtained the Hamiltonian you can use it to obtain the equations of motion of the system. Recall that the equation of motion as expressed by Newton's second law as well as by Lagrange's equation is a second-order differential equation. On the other hand the Hamiltonian formulation yields two *first*-order equations for the momentum and position. They are:

$$\dot{p}_i = -\frac{\partial H}{\partial q_i} \quad \text{and} \quad \dot{q}_i = +\frac{\partial H}{\partial p_i}. \tag{4.21}$$

These are *Hamilton's equations of motion* and they are completely equivalent to Newton's second law and to the Lagrange equations of motion.

To derive Hamilton's equations of motion we obtain expressions for dH two different ways and then equate the two expressions. The first way is to use the definition of H so that

$$dH = d\left(\sum_i p_i \dot{q}_i - L(q_i, \dot{q}_i, t) \right)$$

$$= \sum_i \left(p_i d\dot{q}_i + \dot{q}_i dp_i - \frac{\partial L}{\partial q_i} dq_i - \frac{\partial L}{\partial \dot{q}_i} d\dot{q}_i \right) - \frac{\partial L}{\partial t}$$

$$= \sum_i \left(p_i d\dot{q}_i + \dot{q}_i dp_i - \frac{d}{dt}\frac{\partial L}{\partial \dot{q}_i} dq_i - p_i d\dot{q}_i \right) - \frac{\partial L}{\partial t}$$

$$\therefore dH = \sum_i (\dot{q}_i dp_i - \dot{p}_i dq_i) - \frac{\partial L}{\partial t},$$

where we used $p_i = \frac{\partial L}{\partial \dot{q}_i}$ and Lagrange's equation in the form $\frac{d}{dt}\frac{\partial L}{\partial \dot{q}_i} = \frac{\partial L}{\partial q_i}$.

The second way to obtain dH is to use the fact that $H = H(p_i, q_i, t)$, so that

$$dH = \sum_i \left(\frac{\partial H}{\partial p_i} dp_i + \frac{\partial H}{\partial q_i} \right) + \frac{\partial H}{\partial t}.$$

A term-by-term comparison of the two expressions for dH yields Equations (4.21) as well as

$$\frac{\partial H}{\partial t} = -\frac{\partial L}{\partial t}.$$

For the sake of a specific example, consider the Hamiltonian of a particle of mass m moving vertically in a uniform gravitational field g. (For example, you might want to determine the Hamiltonian and Hamilton's equations for a ball of mass m thrown vertically upward.) Begin with the Lagrangian. It is

$$L = T - V = \frac{1}{2}m\dot{z}^2 - mgz.$$

By the definition of generalized momentum,

$$p = \frac{\partial L}{\partial \dot{z}} = m\dot{z}.$$

Therefore,

$$\dot{z} = \frac{p}{m}.$$

Equation (4.19) then gives

$$H = p\dot{z} - L$$
$$= p\left(\frac{p}{m}\right) - \frac{p^2}{2m} + mgz$$
$$= +\frac{p^2}{2m} + mgz.$$

Note that I was careful to replace $\dot{z}$ by p/m. Inserting this Hamiltonian into Hamilton's equations yields the following equations of motion:

$$\dot{p} = -\frac{\partial H}{\partial z} = -\frac{\partial}{\partial z}\left[\frac{p^2}{2m} + mgz\right] = -mg$$
$$\dot{q} = \dot{z} = +\frac{\partial H}{\partial p} = \frac{\partial}{\partial p}\left[\frac{p^2}{2m} + mgz\right] = \frac{p}{m}.$$

The second equation reads $\dot{z} = p/m$ or $p = m\dot{z}$, which is just the definition of momentum. The first equation states that $\dot{p} = -mg$ or

$$\frac{d}{dt}p = \frac{d}{dt}m\dot{z} = m\ddot{z} = -mg.$$

That is, $\ddot{z} = -g$, as expected.

Now

$$\frac{p^2}{2m} = \frac{m^2v^2}{2m} = \frac{1}{2}mv^2 = T,$$

and

$$V = mgz,$$

so

$$H = T + V = E = \text{total energy.}$$

Therefore, in this case, $H = E$. In fact, the Hamiltonian is nearly always the total energy expressed in terms of generalized momentum and position. Students frequently assume that H is *always* equal to E. This is not true. The following conditions determine when H is constant and when it is equal to E.

(1) If the Lagrangian does not depend explicitly on time, the Hamiltonian is constant but *not* necessarily equal to the energy.
(2) If the transformation equations do not depend on time and the potential energy does not depend on velocity, then the Hamiltonian is equal to the total energy but it may not be constant.
(3) If the constraints and the transformation equations and the potential energy are all time independent, then the Hamiltonian is equal to the total energy and is constant.

The third condition is satisfied by many systems, and frequently when requested to write the Hamiltonian (especially in quantum mechanical problems) a physics student will simply write $H = p^2/2m + V$. However, this can be dangerous because there are problems in which the Hamiltonian is *not* equal to the total energy. Fortunately, you probably will not encounter any of these situations in your undergraduate courses in physics.

Students are often asked to determine the Hamiltonian for a particular system. A common error is to write an expression for H that involves generalized velocities. This is *not* the way a Hamiltonian should be written. The final expression for H must contain *only* momenta and positions (and possibly the time) because $H = H(p,q,t)$.

Finally, I would like to mention that if you transform to a new coordinate system, you will have a new set of generalized momenta and generalized coordinates, say P_i and Q_i. If these new coordinates maintain the form of Hamilton's equations unchanged, then they are called "canonical conjugates." Although we will not consider this subject any further, I can assure you that you will be hearing much more about canonical conjugates in your advanced courses.

Worked Example 4.7 (a) Write the Hamiltonian in polar coordinates for a planet (mass m) in the gravitational field of a star (mass M). (Assume the planet is a particle moving in a planar orbit. The star may be assumed to remain at rest. The potential energy of this system is $V(r) = -\frac{GmM}{r}$.)

Solution The Lagrangian for this system in polar coordinates (r,θ) can be obtained from the fact that the velocity is

$$\mathbf{v} = \dot{r}\hat{\mathbf{r}} + r\dot{\theta}\hat{\boldsymbol{\theta}}$$

so

$$T = \frac{1}{2}mv^2 = \frac{1}{2}m\left(\dot{r}^2 + r^2\dot{\theta}^2\right).$$

Therefore,

$$L = T - V = \frac{1}{2}m\dot{r}^2 + \frac{1}{2}mr^2\dot{\theta}^2 + \frac{GMm}{r}.$$

The generalized momenta are

$$p_r = \frac{\partial L}{\partial \dot{r}} = m\dot{r},$$

$$p_\theta = \frac{\partial L}{\partial \dot{\theta}} = mr^2\dot{\theta}.$$

The Hamiltonian is

$$H = \Sigma p_i \dot{q}_i - L$$
$$= p_r \dot{r} + p_\theta \dot{\theta} - L$$
$$= p_r \dot{r} + p_\theta \dot{\theta} - \left(\frac{1}{2}m\dot{r}^2 + \frac{1}{2}mr^2\dot{\theta}^2 + \frac{GMm}{r}\right).$$

Replacing $\dot{r}$ by p_r/m and $\dot{\theta}$ by p_θ/mr^2 we obtain

$$H = p_r \frac{p_r}{m} + p_\theta \frac{p_\theta}{mr^2} - \frac{1}{2}m \left(\frac{p_r}{m}\right)^2 - \frac{1}{2}mr^2 \left(\frac{p_\theta}{mr^2}\right)^2 - \frac{GMm}{r}$$

$$= \frac{p_r^2}{m} + \frac{p_\theta^2}{mr^2} - \frac{1}{2}\frac{p_r^2}{m} - \frac{1}{2}\frac{p_\theta^2}{mr^2} - \frac{GMm}{r}$$

$$= \frac{1}{2}\frac{p_r^2}{m} + \frac{1}{2}\frac{p_\theta^2}{mr^2} - \frac{GMm}{r}.$$

Exercise 4.11

Write the Hamiltonian for a free particle. Answer: $p^2/2m$. ∎

Exercise 4.12

A student writes the Hamiltonian for a particle falling in a uniform gravitational field as $H = \frac{1}{2}m\dot{z}^2 + mgz$. What is wrong with this? ∎

Exercise 4.13

Write the Hamiltonian for a simple pendulum. ∎

4.10 Summary

To determine the equation(s) of motion do one of the following.

(1) Draw the free body diagram and apply $F = ma$. (That is, determine the equation of motion by inspection.)
(2) Write the Lagrangian and apply Lagrange's equations.
(3) Write the Hamiltonian and apply Hamilton's equations.

The Lagrangian is given by

$$L = L(q_i, \dot{q}_i, t) = T - V$$

and the Lagrange equations are

$$\frac{d}{dt}\left(\frac{\partial L}{\partial \dot{q}_i}\right) - \frac{\partial L}{\partial q_i} = 0 \; ; \quad i = 1, 2, \ldots, n.$$

The q_i and $\dot{q}_i$ are called generalized coordinates and generalized velocities. The generalized momentum is defined by

$$p_i = \frac{\partial L}{\partial \dot{q}_i}.$$

If the generalized coordinate q_i does not appear in the Lagrangian, it is called an ignorable coordinate. The generalized momentum conjugate to an ignorable coordinate is constant.

The number of generalized coordinates is equal to the number of degrees of freedom. Each constraint reduces by one the number of degrees of freedom (and hence, the number of generalized coordinates).

Lagrange's equations are obtained from the calculus of variations. The calculus of variations allows one to determine the extremum conditions for a given parameter (such as time, distance, area, etc.). To determine the function $y = y(x)$ that will minimize or maximize the quantity

$$\int \Phi(x, y, y')dx,$$

one applies the Euler–Lagrange equation

$$\frac{\partial \Phi}{\partial y} - \frac{d}{dx}\frac{\partial \Phi}{\partial y'} = 0.$$

Similarly, Lagrange's equations are obtained from Hamilton's principle. This principle states that the time evolution of a mechanical system will minimize the action, defined by

$$\text{Action} \equiv \int L dt.$$

Applying the technique of the calculus of variations to minimize the action leads to Lagrange's equations.

The Hamiltonian is defined by

$$H \equiv \sum p_i \dot{q}_i - L.$$

The equations of motion (Hamilton's equations) are

$$\dot{p}_i = -\frac{\partial H}{\partial q_i} \qquad \text{and} \qquad \dot{q}_i = +\frac{\partial H}{\partial p_i}.$$

4.11 Problems

Problem 4.1 The gravitational potential energy of a particle of mass m a distance r from a star is given by $V = -\frac{K}{r}$, where K is a constant. Determine the Lagrangian. (Express your answer in plane polar coordinates.)

Problem 4.2 Determine the Lagrangian for a particle sliding on the inner surface of a hemisphere of radius a (the "bar of soap in a sink" problem). See Figure 4.7.

Problem 4.3 A bead of mass m, acted upon by a uniform gravitational force, can slide freely on a frictionless circular hoop of radius R. The hoop is oriented vertically and is rotating about a vertical axis with constant angular velocity ω. Assume the hoop is massless. Determine the Lagrangian for the bead.

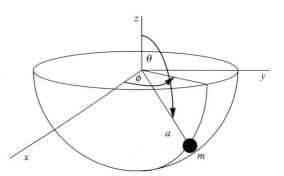

Figure 4.7 A particle of mass m slides on the inside of a perfectly smooth hemispherical surface. (Problem 4.2.)

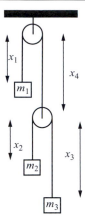

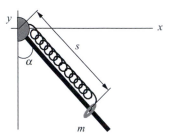

Figure 4.8 A double Atwood's machine. (Problem 4.4.)

Figure 4.9 The ring connected to the spring is free to slide on the inclined rod. (Problem 4.6.)

Problem 4.4 The double Atwood's machine is illustrated in Figure 4.8. Both strings have the same length (l) and we will assume the pulleys are frictionless and massless. The strings are inextensible. What are the constraints? Write the Lagrangian for this system.

Problem 4.5 A certain molecule is made up of four atoms. Three of them have mass M and are located at the vertices of an equilateral triangle of side a. The fourth atom has mass m and is free to move along a line perpendicular to the triangle and passing through its center. Assume this atom is attracted to the other atoms by a force proportional to the distance between them. (You can imagine the atom of mass m is attached by three springs of force constant k to the atoms of mass M.) Determine the Lagrangian for the system.

Problem 4.6 A heavy ring of mass m can slide on a frictionless rod. The rod is attached to a wall at one end with a bracket and the other end hangs downward, so the constant angle between the rod and the wall is α. One end of a massless spring of constant k is attached to the bracket and the other end is connected to the ring. Determine the Lagrangian for the system and the equation of motion. See Figure 4.9.

Problem 4.7 A binary star system consists of two stars of masses m_1 and m_2 in circular orbits about the center of mass. The potential energy of the system is $V = -G\frac{M_1 M_2}{r}$, where r is the distance between the stars. Obtain the Lagrangian in polar coordinates, and from it show that the angular momentum is constant.

Problem 4.8 Write the Lagrangian for a simple Atwood's machine and obtain the equations of motion. Assume the string is inextensible. The pulley is a disk of mass M and radius R and is perfectly rough so the string does not slip.

Problem 4.9 A cart of mass M slides freely (no friction) on elevated horizontal tracks. Hanging from the cart is a pendulum with string of length l and bob of mass m. Determine the Lagrangian for this system. Obtain the equations of motion. By considering whether or not any coordinates are ignorable, determine any constants of the motion. See Figure 4.10.

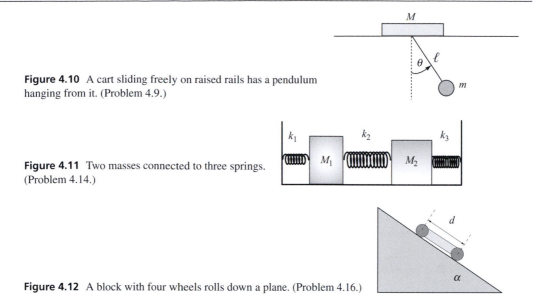

Figure 4.10 A cart sliding freely on raised rails has a pendulum hanging from it. (Problem 4.9.)

Figure 4.11 Two masses connected to three springs. (Problem 4.14.)

Figure 4.12 A block with four wheels rolls down a plane. (Problem 4.16.)

Problem 4.10 Obtain the equations of motion for a spherical pendulum, that is, a pendulum that is not constrained to oscillate in a plane.

Problem 4.11 A wedge of mass M and angle α is resting on a frictionless horizontal plane. A rectangular box of mass m is on the wedge. Because all surfaces are frictionless, the box will slide down the wedge and the wedge will slide in the opposite direction on the plane. Determine the accelerations of the two bodies.

Problem 4.12 A ball of radius a and mass m rolls down a perfectly rough wedge of angle α and mass M. The wedge sits on a frictionless surface, so as the ball rolls down the plane, the wedge will move in the opposite direction. Obtain the Lagrangian and the equations of motion.

Problem 4.13 A disk of mass M and radius R has a light string wound around its circumference. The end of the string is held steady and the disk is allowed to fall under gravity (like an unwinding yoyo). Determine the equation of motion.

Problem 4.14 A system of two masses and three springs is illustrated in Figure 4.11. Write the Lagrangian. Determine the generalized momenta for this system. (Hint: Measure the positions of the masses from their equilibrium points.)

Problem 4.15 Obtain all the components of generalized momentum for the following situations.

(a) A block is sliding down a smooth inclined plane of angle α. Use the distance down the plane as the generalized coordinate.
(b) A bar of soap slides in a frictionless hemispherical sink of radius a. Use spherical coordinates and the constraint $r = a$.
(c) A bead of mass m is sliding on a stationary vertical hoop of radius R. (Use θ as the generalized coordinate.)

Problem 4.16 A toy cart is made up of a block of mass M and four cylindrical wheels of radius R and mass m. It is rolling down a fixed plane of angle α. Let the cart have a length d and let the wheels be at positions $\pm d/2$ relative to the center of mass of the cart. Assume the center of mass lies on a line joining the centers of the wheels. See Figure 4.12. Determine the Lagrangian for the system.

Problem 4.17 A rope passes over a frictionless pulley. A monkey of mass M is hanging from one end of the rope and a bunch of bananas of mass m is attached to the other end. Because $M > m$, the monkey is

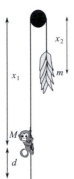

Figure 4.13 The monkey and the bananas. (Problem 4.17.)

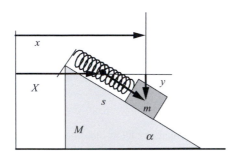

Figure 4.14 A block and spring on a sliding wedge. The dot in the middle of the spring represents the equilibrium position of the block, and is a distance X from some fixed point. The stretch of the spring is measured from the equilibrium position. The position of the block is given by x and y, where $x = X + s\sin\alpha$. (Problem 4.18.)

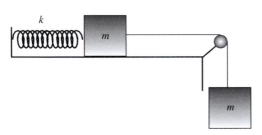

Figure 4.15 Two masses, a frictionless surface, a spring, and an ideal pulley. (Problem 4.20.)

descending and the bananas are ascending and getting further and further away from the monkey. Naturally, the monkey begins to climb the rope. See Figure 4.13. You are asked to obtain the motion ($x = x(t)$) for the monkey under two different assumptions. (a) Assume the monkey climbs at a constant velocity relative to the rope. (b) Assume the monkey exerts a constant force on the rope.

Problem 4.18 A wedge of mass M and angle α is free to slide on a frictionless horizontal surface. On the wedge is mounted a mass and spring arrangement that allows the mass m to slide up and down on the wedge. All surfaces are frictionless and the spring is massless. (a) Write the Lagrangian for the system. (b) Obtain expressions for the generalized momenta. (c) Obtain the equations of motion. See Figure 4.14.

Problem 4.19 A pendulum is made of a bob of mass M but, instead of a string, the bob is suspended from a massless spring of constant k. The system can swing back and forth and also move up and down, but it is restricted to motion in a plane. Determine the equations of motion.

Problem 4.20 Figure 4.15 shows two equal masses m. One is hanging and the other is on a frictionless surface. They are connected by an ideal horizontal string that passes over an ideal massless pulley. The mass on the surface is connected to a fixed point by a massless spring of constant k. Place the origin of coordinates at the pulley. (a) Write the Lagrangian for the system in terms of a single variable. (b) Obtain the equation of motion. (c) Determine the frequency of oscillation of the system.

Problem 4.21 A marble of mass M and radius R rolls back and forth on the bottom of a semicircular bowl of radius a. (a) Write the Lagrangian for the system. (b) Obtain the equation of motion. (c) Determine the period with which the marble rolls back and forth. You may assume the amplitude of the oscillations is small.

Problem 4.22 It is desired to find the equation for the shortest distance between two points on a sphere. Determine the functional for this problem. (Use spherical coordinates.)

Problem 4.23 A particle is constrained to the surface of a cylinder of radius a. We want to determine the shortest distance between two points on the surface of a cylinder.

(a) Determine the functional.
(b) Write the appropriate Euler–Lagrange equation.
(c) Show that $\phi = k_1 z + k_2$ (a helix) satisfies the Euler–Lagrange equation. (Note that k_1 and k_2 are constants.)

Problem 4.24 The speed of light in a medium of index of refraction n is $v = c/n = ds/dt$. The time for light to travel from point A to point B is

$$\int_A^B \frac{ds}{v}.$$

Obtain the law of reflection and the law of refraction (Snell's law) by using Fermat's principle of least time.

Problem 4.25 Determine the relation $y = y(x)$ such that the following integral is an extremum. What is the shape of the curve?

$$\int_{x_1}^{x_2} \left[(x)(1 + y'^2)\right]^{1/2} dx.$$

Problem 4.26 A bead slides down a wire. You want to determine the shape the wire must have for the bead to slide from its initial position to its final position in the least amount of time. (This is called the "brachistochrone" problem.) Note that the quantity to be minimized is the time and that $dt = ds/v$, where the velocity v is given by conservation of energy. I do not expect you to actually solve this problem; it will be sufficient if you write the functional that describes the system.[15]

Problem 4.27 Two pegs are at the same height above the ground. A rope is draped over them, with both ends of the rope resting on the ground. We wish to know the shape of the rope between the pegs. The quantity that is minimized in this case is the potential energy. Determine the appropriate functional.

Problem 4.28 A vase is designed by drawing a curve and then rotating it about the z-axis, forming a "surface of revolution." It is desired that the vase have the minimum possible surface area. Obtain the functional for this problem. See Figure 4.16 to get an idea of the geometry.

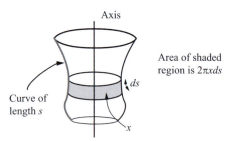

Figure 4.16 A curve of length s is revolved about an axis to generate a surface of revolution. (Problem 4.28.)

---

[15] This problem was apparently first devised by one the Bernoulli brothers who sent it out as a "challenge problem" to the scientists of that day. It is said that Newton solved the problem in a few hours. (The calculus of variations had not been invented yet!)

Problem 4.29 In quantum mechanics, one replaces the momentum p by the "momentum operator" $i \frac{h}{2\pi} \frac{\partial}{\partial x}$. Using this replacement, write the quantum mechanical Hamiltonian operator for a free particle. (The quantity $h/2\pi$ is usually written as $\hbar$.)

Problem 4.30 Obtain the Schrödinger equation as follows: Write the quantum mechanical version of the Hamiltonian for a particle whose potential energy is $V(x)$. Let $H = E$, and allow the operators to act on the function $\Psi(x)$.

Problem 4.31 Write the Hamiltonian for a planar pendulum and obtain the Hamilton's equations of motion for the system.

Problem 4.32 Write the Hamiltonian for a mass on a spring (as in Figure 4.3) and obtain the Hamilton's equations of motion for the system.

Problem 4.33 A spherical pendulum consists of a bob of mass m connected to a massless rigid rod of length a. It is free to move anywhere on the surface of a sphere of radius a so its position at any time is given by the values of r, θ, ϕ. The bob is acted upon by gravity. Using spherical coordinates, write the Lagrangian and the Hamiltonian for the spherical pendulum.

Computational Projects

Computational Techniques To solve the first computational problem below you will have to use a Computer Algebra System (CAS) such as Mathematica or Maple. We do not have the space nor the time to get into the details of these excellent programs, but you should learn some of the basics of one such system. They are easily available. For example, Matlab incorporates symbolic computation using the Maple kernel. (Maple is a CAS that was developed primarily at the University of Waterloo in Canada and more recently at ETH in Zurich – Einstein's Alma Mater.) In Matlab's version of Maple, if you want to obtain, say, an expression for $y = \cos^2 x + \sin^2 x$ you simply declare $x, y,$ and z as symbolic variables, using the command

$$syms \; x \; y \; z,$$

then write

$$y = \cos(x)^\wedge 2 + \sin(x)^\wedge 2.$$

Then write

$$simplify(y).$$

You will obtain (of course) y = 1.
    Another example, is to write

$$z = \sin(x)/\cos(x),$$

then write

$$simple(z),$$

to get

$$z = \tan(x).$$

Of course, you will have to consult the "help" files to get further information, but let me assure you that using Maple is not difficult.

Computational Project 4.1 Use a symbolic manipulator such as Maple to obtain the Hamiltonian for the double planar pendulum.

Computational Project 4.2 Solve for the motion of the double pendulum. Assume the two bobs have the same mass and that the two strings are of length 0.1 m. Compute examples of the motion for several initial conditions. Show that the motion is regular for small angle oscillations but is chaotic for large angles.

Computational Project 4.3 A 2 kg object is launched horizontally from a 5 m cliff with an initial velocity of 20 m/s. As you know, it will follow a parabolic path. It hits the ground a distance 20.2 m horizontally from the launch point and will have been in the air for 1.01 s. The purpose of this project is to evaluate the *action* for this process. The action is defined by the integral $\int L dt$, where $L = T - V$. After you have written a computer program to evaluate the action for the actual path of the projectile, evaluate the action for the following hypothetical path between the same two end points and taking the same amount of time. The hypothetical path is a straight line from (0,5) to (20.2,0) at a constant speed of 20.6 m/s. You will find that the action for the hypothetical path is greater than the action for the actual path since the actual path for any dynamical system always minimizes the action. You may wish to construct other paths between the end points and show that their action is greater than that for the actual path. (Suggestion: Use small time steps (about 1/1000 of a second); evaluate L at each time step and estimate the integral by summing the quantities $L(t)\Delta t$.)

Part II

Conservation Laws

5 Energy

This chapter analyzes the conservation of energy in a much more rigorous manner than the review in Section 1.3. You will be exposed to a number of useful concepts such as energy diagrams, metrics, and a review of the del operator. The techniques presented here are used quite frequently, not only in this course but also in other advanced physics courses.

5.1 The Work–Energy Theorem

When you push an object from one place to another, you do work on it. To be precise, the work done on a particle by a force **F** during a displacement from point 1 to point 2 along some given path C, is defined as

$$W = \left[\int_1^2 \mathbf{F(s)} \cdot d\mathbf{s} \right]_C , \tag{5.1}$$

where $\mathbf{F(s)}$ is the force acting on the particle at location **s**, C is the path from 1 to 2, and $d\mathbf{s}$ is an infinitesimal displacement. If there is no displacement, no work is done. By definition, the displacement $d\mathbf{s}$ is the displacement of the point of application of the force. For a particle, this is equal to the displacement of the particle. For an extended rigid body $d\mathbf{s}$ refers to the displacement of the center of mass.[1]

It is reasonable to assume that doing work on a body will increase its kinetic energy. For an extended body, this would include the kinetic energy of rotation as well as the kinetic energy of translation. Because a particle cannot rotate, it is easiest to consider, for the time being, the work done on a particle.

The relation between work and kinetic energy is expressed by the **work–energy theorem:**

Work–Energy Theorem. The increase in kinetic energy of an object is equal to the *total* work done on it.

This theorem refers to the total or *net* work done by *all* the forces acting on the particle.

---

[1] The definition of work is often written as $W = \int \mathbf{F} \cdot d\mathbf{r}$. This definition may cause some confusion because in this formula $d\mathbf{r}$ represents the displacement of the body, which is not necessarily in the $\hat{\mathbf{r}}$ direction. It is *wrong* to write $d\mathbf{r} = dr\hat{\mathbf{r}}$ unless the displacement happens to be in the $\hat{\mathbf{r}}$ direction. For example, in spherical coordinates, you would have to write

$$d\mathbf{r} = dr\hat{\mathbf{r}} + rd\theta\hat{\boldsymbol{\theta}} + r\sin\theta d\phi\hat{\boldsymbol{\phi}}.$$

To avoid confusion, in this chapter I will use $d\mathbf{s}$ for the infinitesimal vector displacement.

The proof of this theorem is very simple. Let $\mathbf{F}$ be the net force acting on a particle of mass m. Then using the definition $\mathbf{v} = d\mathbf{s}/dt$ and Newton's second law in the form $\mathbf{F} = m\frac{d\mathbf{v}}{dt}$, we can write

$$W = \int_{\mathbf{r}_1}^{\mathbf{r}_2} \mathbf{F} \cdot d\mathbf{s} = \int_{t_1}^{t_2} \mathbf{F} \cdot \mathbf{v} \, dt = \int_{t_1}^{t_2} m\frac{d\mathbf{v}}{dt} \cdot \mathbf{v} \, dt.$$

Now

$$\frac{d}{dt}(v^2) = \frac{d}{dt}(\mathbf{v} \cdot \mathbf{v}) = \frac{d\mathbf{v}}{dt} \cdot \mathbf{v} + \mathbf{v} \cdot \frac{d\mathbf{v}}{dt} = 2\frac{d\mathbf{v}}{dt} \cdot \mathbf{v}.$$

Therefore, for constant mass,

$$m\frac{d\mathbf{v}}{dt} \cdot \mathbf{v} = \frac{1}{2}m\frac{d}{dt}(v^2) = \frac{d}{dt}\left(\frac{1}{2}mv^2\right).$$

So

$$W = \int_{\mathbf{r}_1}^{\mathbf{r}_2} \mathbf{F} \cdot d\mathbf{s} = \int_{t_1}^{t_2} \frac{d}{dt}\left(\frac{1}{2}mv^2\right) dt = \int_{T_1}^{T_2} d(T) = T_2 - T_1 = \Delta T,$$

and the theorem is proved.

There are two things in this proof that I would like you to note. One is the fact that

$$\frac{d}{dt}v^2 = 2\frac{d\mathbf{v}}{dt} \cdot \mathbf{v}.$$

This relationship will help you in solving a number of problems. The other thing I would like you to be aware of is the way the limits of integration change when the variable of integration changes. Be sure to change the limits whenever you make a change of variables!

5.2 Work Along a Path: The Line Integral

Equation (5.1) states that the work done by a force on a particle as it is displaced from one point to another depends on the path between the two points. For some forces, the work does not depend on the path; those forces are called "conservative" and have certain special properties that we will study shortly. In general, however, the work done along one path is different from the work done along some other path. Therefore, to evaluate work, you will often have to evaluate a *path* or *line integral*.

An example of a line integral is the work done by a force $\mathbf{F}$ as a particle moves along a path described by the curve C:

$$W = \int_C \mathbf{F} \cdot d\mathbf{s}.$$

I will describe two ways to evaluate line integrals. The first is to express the force and the displacement in Cartesian coordinates. Thus:

$$\mathbf{F} = F_x\hat{\mathbf{i}} + F_y\hat{\mathbf{j}} + F_z\hat{\mathbf{k}}$$

and

$$d\mathbf{s} = dx\hat{\mathbf{i}} + dy\hat{\mathbf{j}} + dz\hat{\mathbf{k}}.$$

So

$$W = \int_C \mathbf{F} \cdot d\mathbf{s} = \int_C F_x dx + \int_C F_y dy + \int_C F_z dz. \tag{5.2}$$

Here the three integrals are evaluated *along the curve C*. Note that in the first integral the force component F_x must be expressed in terms of x, and likewise F_y is expressed in terms of y and F_z in terms of z, where the relations between $x, y,$ and z are determined by the path.

Worked Example 5.1 A particle is acted upon by the force

$$\mathbf{F} = a(x + y)\hat{\mathbf{i}} + b(y - x)\hat{\mathbf{j}}.$$

Determine the work done by this force as the particle moves along a straight line from the origin to the point $(2, 1)$.

Solution Using Equation (5.2),

$$W = \int_C \mathbf{F} \cdot d\mathbf{s} = \int_C F_x dx + \int_C F_y dy$$

$$= \int_C a(x + y)dx + \int_C b(y - x)dy.$$

Along the straight line, the relation between x and y is $y = mx$, where $m = 1/2$. In the first integral, replace y by $x/2$ (thus expressing y as a function of x) and in the second integral write $x = 2y$ (thus expressing x as a function of y). Then,

$$W = \int_{x=0}^{2} a\left(x + \frac{1}{2}x\right)dx + \int_{y=0}^{1} b(y - 2y)dy$$

$$= \int_{x=0}^{2} \frac{3}{2}ax\,dx - \int_{y=0}^{1} by\,dy$$

$$= 3a - \frac{1}{2}b.$$

The technique used in this example required having an equation for the curve in Cartesian coordinates, but the procedure can easily be generalized to other coordinate systems.

It is interesting to note that if the path is first along the x-axis to $(2,0)$ then up along the y-axis to $(2,1)$ we have the same initial and end points, but the work done is quite different:

$$W = \int_0^2 a(x + 0)dx + \int_0^1 b(y - 2)dy = 2a - \frac{3}{2}b.$$

The second approach is, in principle, the same as the first approach, but it may be easier to apply in certain situations. This approach is based on the fact that for motion along a smooth curve, the position can be specified by a single variable, say λ. This single independent variable can be the distance along the curve from some starting point, or the time, or some other parameter (as the angle θ in the example below). Then writing both $\mathbf{F}$ and $d\mathbf{s}$ in terms of λ, the integral $\int \mathbf{F} \cdot d\mathbf{s}$ reduces to a *single* integral over *one* variable. Thus, if $\mathbf{F} = \mathbf{F}(\lambda)$ and $\mathbf{s} = \mathbf{s}(\lambda)$, then

$$W = \int_C \mathbf{F} \cdot d\mathbf{s} = \int_C \mathbf{F}(\lambda) \cdot \frac{d\mathbf{s}}{d\lambda}d\lambda.$$

The last integral can be expressed in a variety of ways. For example, if $\mathbf{F}$ is given in Cartesian coordinates, we write

$$W = \int_{\lambda_1}^{\lambda_2} \left(F_x\frac{dx}{d\lambda} + F_y\frac{dy}{d\lambda} + F_z\frac{dz}{d\lambda}\right)d\lambda. \tag{5.3}$$

This expression shows the close relationship to the first method. Nevertheless, the second method is often more convenient, as illustrated by the following example.

Worked Example 5.2 Consider the force

$$\mathbf{F} = \frac{-y}{x^2 + y^2}\hat{\mathbf{i}} + \frac{x}{x^2 + y^2}\hat{\mathbf{j}}.$$

Evaluate $\int_C \mathbf{F} \cdot d\mathbf{s}$ along a semicircular path from $(-1,0)$ to $(+1,0)$, as shown in Figure 5.1.

Solution The shape of the path suggests using polar coordinates and the angle θ as a parameter. Because the radius of the semicircle is unity, we have $x = \cos\theta$, and $y = \sin\theta$ and $x^2 + y^2 = 1$. Note that in terms of θ, $F_x = -\sin\theta/(\sin^2\theta + \cos^2\theta) = -\sin\theta$ and similarly $F_y = \cos\theta/(\sin^2\theta + \cos^2\theta) = \cos\theta$. Therefore,

$$W = \int_{\theta=\pi}^{\theta=0} \left(F_x \frac{dx}{d\theta} + F_y \frac{dy}{d\theta} \right) d\theta = \int_{\pi}^{0} \left(\sin^2\theta + \cos^2\theta \right) d\theta$$

$$= \int_{\pi}^{0} d\theta = -\pi.$$

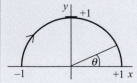

Figure 5.1 A semicircular path from $(-1,0)$ to $(1,0)$.

Worked Example 5.3 A particle is in a region of space where the force on it is given by

$$\mathbf{F} = xy\hat{\mathbf{i}} - x^2\hat{\mathbf{j}}.$$

The particle is dragged along a path that is described by

$$\mathbf{s} = \mathbf{s}(t) = 3t\hat{\mathbf{i}} + 2t^2\hat{\mathbf{j}},$$

for $0 \le t \le 1$. Here t is just a parameter describing the position as the particle moves from $(0,0)$ to $(3,2)$. The force is in newtons and the displacement is in meters. Determine the work done by the force during the displacement. Do this two ways, first by treating t as a parameter (as in the second method described above), and then by evaluating the work with Equation (5.2). The force field and the trajectory are shown in Figure 5.2. Note that as the particle moves along the trajectory the force on it due to the force field is changing. We are not considering the work done by the force that is dragging the particle through this field.

Solution Treating t as a parameter, we write $\mathbf{F}$ and $d\mathbf{s}$ in terms of t. Since $\mathbf{s} = x\hat{\mathbf{i}} + y\hat{\mathbf{j}}$ we appreciate that $x = 3t$ and $y = 2t^2$. Furthermore,

$$d\mathbf{s} = \frac{d\mathbf{s}}{dt}dt = (3\hat{\mathbf{i}} + 4t\hat{\mathbf{j}})dt$$

Then the force can be expressed in terms of t as

$$\mathbf{F} = (3t)(2t^2)\hat{\mathbf{i}} - (3t)^2\hat{\mathbf{j}} = 6t^3\hat{\mathbf{i}} - 9t^2\hat{\mathbf{j}}.$$

Consequently,

$$W = \int \mathbf{F} \cdot d\mathbf{s} = \int_{t=0}^{t=1} (6t^3\hat{\mathbf{i}} - 9t^2\hat{\mathbf{j}}) \cdot (3\hat{\mathbf{i}} + 4t\hat{\mathbf{j}})dt$$

$$= \int_{t=0}^{t=1} (18t^3 - 36t^3)dt = -\frac{18}{4}\left[t^4\right]_0^1 = -\frac{9}{2}J.$$

Now let us solve the problem using Equation (5.2). Note that we need to have expressions for dx and for dy. Since $x = 3t$ and $y = 2t^2$ we can solve for t and obtain y as a function of x; that is, $t = x/3$, so $y = 2(x/3)^2 = (2/9)x^2$. Then

$$W = \int_{x=0}^{3} F_x dx + \int_{y=0}^{2} F_y dy = \int_0^3 xy dx - \int_0^2 x^2 dy$$

$$= \int_0^3 x\left(\frac{2}{9}x^2\right)dx - \int_0^2 \left(\frac{9}{2}y\right)dy$$

$$= \frac{2}{9}\left[\frac{x^4}{4}\right]_0^3 - \frac{9}{2}\left[\frac{y^2}{2}\right]_0^2 = \frac{81}{18} - \frac{36}{4} = -\frac{9}{2}J.$$

As expected, the two methods yield the same answer.

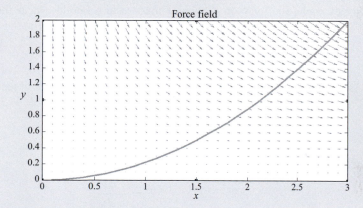

Figure 5.2 The force field (arrows) as a function of position, and the trajectory (solid line) for the problem considered in Worked Example 5.3.

Exercise 5.1

A charged particle in an electric field $\mathbf{E}$ is acted upon by a force $q\mathbf{E}$. Assume the field is constant and evaluate the work done by this force for the closed path illustrated in Figure 5.3(a). Does the result depend on whether you traverse the path clockwise or counterclockwise? Answers: zero, no.

Exercise 5.2

Evaluate the work done by the force $\mathbf{F} = y\hat{\mathbf{i}} + x\hat{\mathbf{j}}$ N along the parabolic path illustrated in Figure 5.3(b) as the particle moves from (0,0) to (3,18) m. Answer: 54 J.

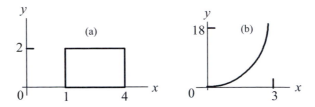

Figure 5.3 (a) A closed path. The corners are at (1,0), (1,2), (4,2) and (4,0). (b) A parabolic path going through the origin and the point (3,18).

5.3 Potential Energy

Forces are vectors. So are many other physical quantities and it is hard to imagine trying to do physics without them. Nevertheless, it is usually much easier to deal with scalar quantities. As mentioned in Section 1.3 some forces can be represented in terms of a *scalar* quantity called the potential energy. Examples of forces that can be represented in terms of the potential energy are the electrostatic force, the gravitational force, and the force exerted by a spring. Scalars will usually make your life much simpler and it is a good idea to use them whenever possible.

For example, the force on a body of mass m at a height z near the surface of the Earth is $\mathbf{F} = -m\mathbf{g}$. If we define the scalar function $V = -mgz$ we can write the magnitude of the force as

$$F = -\frac{dV}{dz}.$$

As a second example, consider a stretched spring acting on an object of mass m as illustrated in Figure 3.4. The force exerted by the spring is $\mathbf{F} = -kx\hat{\mathbf{i}}$. If $V = \frac{1}{2}kx^2$, then

$$\frac{dV}{dx} = kx = -F,$$

and

$$F = -\frac{dV}{dx}.$$

As a third example, consider the electrostatic force:

$$\mathbf{F} = \frac{Q_1 Q_2}{4\pi\epsilon_0}\frac{\hat{\mathbf{r}}}{r^2}.$$

We can define $V = \frac{Q_1 Q_2}{4\pi\epsilon_0 r}$ such that the electrostatic force is given by

$$F = -\frac{dV}{dr}.$$

You appreciate that in each case the derivative of V is the negative of the magnitude of the force. (As mentioned above, the quantity V is the potential energy. But don't worry about the physical meaning of V just yet. I'll get to that pretty soon. Just consider V to be a function whose derivative is related to the force.)

Replacing F by $-dV/dr$ or $-dV/dx$ in the examples above only gave the *magnitude* of the force but not its direction. Is there some way to obtain both the magnitude and the direction of the force from the potential energy? The answer is yes. However, before continuing with the discussion, I need to make a small digression to consider the properties of the mathematical object called "del."

5.3.1 The Del Operator

The del operator[2] is defined as

$$\nabla \equiv \hat{\mathbf{i}}\frac{\partial}{\partial x} + \hat{\mathbf{j}}\frac{\partial}{\partial y} + \hat{\mathbf{k}}\frac{\partial}{\partial z}. \tag{5.4}$$

Del is an *operator,* that is, it is a mathematical object that acts on a function to produce a different function.

We shall be interested in three different kinds of operations involving del, and these are:

$$\begin{array}{ll}
\nabla f & \text{the \textbf{gradient} of a scalar function } f \\
\nabla \cdot \mathbf{F} & \text{the \textbf{divergence} of a vector function } \mathbf{F} \\
\nabla \times \mathbf{F} & \text{the \textbf{curl} of a vector function } \mathbf{F}.
\end{array}$$

Note that the *gradient* will generate a *vector*, the *divergence* (being a dot product) will generate a *scalar*, and the *curl* (being a cross product) will generate a *vector*.

5.3.2 The Gradient

First consider the gradient, ∇f, where $f = f(x, y, z)$ is a scalar function. Using the definition of del, we have

$$\text{gradient of } f = \mathbf{grad}\, f = \nabla f = \hat{\mathbf{i}}\frac{\partial f}{\partial x} + \hat{\mathbf{j}}\frac{\partial f}{\partial y} + \hat{\mathbf{k}}\frac{\partial f}{\partial z}. \tag{5.5}$$

For example, if $f = f(x, y, z) = 3x + 2y^2$ then

$$\mathbf{grad}\, f = \left(\hat{\mathbf{i}}\frac{\partial}{\partial x} + \hat{\mathbf{j}}\frac{\partial}{\partial y} + \hat{\mathbf{k}}\frac{\partial}{\partial z} \right)\left(3x + 2y^2 \right) = 3\hat{\mathbf{i}} + 4y\hat{\mathbf{j}}.$$

The gradient of the scalar function $f = 3x + 2y^2$ produced the vector $3\hat{\mathbf{i}} + 4y\hat{\mathbf{j}}$. The fact that taking the gradient of a scalar function generates a vector is one of the most important properties of the gradient.

The geometrical interpretation of the gradient is quite interesting. To best appreciate it consider a specific example. Figure 5.4 is a contour map, as on a regular geographic map, and it represents the shape of a region containing two mountains. The curves on the figure are contour lines or altitude "isopleths," that is, lines drawn through points having the same value of height or altitude.[3] The map in our figure shows two peaks, one higher than the other. You could (in principle at least) determine an equation of the form $h = h(x, y)$ that would give the height above sea level at any point (x, y) on the map.

If you were at some spot on the mountain, say at the point marked A, you could climb towards the peak directly (a very steep climb) or you could walk along the isopleth passing through A. That would be a level walk. If you walk along a contour line your altitude above sea level does not change ($dh = 0$). By looking at the sketch, you can figure out which of the three arrows points along the path of steepest ascent. It is, of course, the arrow pointing in the direction in which the altitude isopleths are closest together so that in moving a given distance in that direction, you cross the largest number of isopleths. This is the direction of the gradient, ∇h. The gradient does not necessarily point toward the peak of the mountain. (What is the direction of ∇h at point B?)

It is not difficult to prove that the gradient ∇h is a vector pointing in the direction of steepest ascent and whose magnitude gives the "steepness" along this path of steepest ascent.[4]

---

[2] The symbol ∇ is also called "nabla."
[3] In general an isopleth is a line drawn on a map through all points having the same numerical value. You are familiar with the temperature isopleths ("isotherms") and pressure isopleths ("isobars") drawn on weather maps.
[4] By "steepness" I mean the change of h for a given displacement.

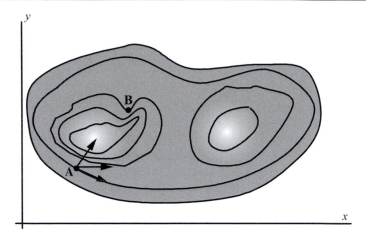

Figure 5.4 Contour plot of two mountains (altitude isopleths).

Let us begin by considering the differential of an arbitrary scalar function. In your calculus class you learned that the differential of a function of several variables, such as $u = u(x, y, z)$, is given by

$$du = \frac{\partial u}{\partial x}dx + \frac{\partial u}{\partial y}dy + \frac{\partial u}{\partial z}dz.$$

Keep this expression for du in mind for a moment. Now consider the dot product of ∇u with the infinitesimal displacement vector $d\mathbf{s}$. Since

$$d\mathbf{s} = dx\hat{\mathbf{i}} + dy\hat{\mathbf{j}} + dz\hat{\mathbf{k}},$$

the dot product is

$$\nabla u \cdot d\mathbf{s} = \left(\hat{\mathbf{i}}\frac{\partial u}{\partial x} + \hat{\mathbf{j}}\frac{\partial u}{\partial y} + \hat{\mathbf{k}}\frac{\partial u}{\partial z} \right) \cdot \left(dx\hat{\mathbf{i}} + dy\hat{\mathbf{j}} + dz\hat{\mathbf{k}} \right),$$

$$= \frac{\partial u}{\partial x}dx + \frac{\partial u}{\partial y}dy + \frac{\partial u}{\partial z}dz.$$

But this is precisely the expression for du. Therefore,

$$du = \nabla u \cdot d\mathbf{s}. \tag{5.6}$$

The relationship given by Equation (5.6) is so important that some people consider it to be the *definition* of du. (I guess it just depends on whether you think of du or ∇u as the more fundamental quantity.) I will refer to relationship (5.6) as "the vector definition of the differential."

By Equation (5.6)

$$\int_1^2 \nabla u \cdot d\mathbf{s} = \int_1^2 du = u_2 - u_1,$$

which is the change in u due to the displacement from point 1 to point 2.

∇u and $d\mathbf{s}$ are both vectors. Their dot product, du, depends on the angle between the vectors ∇u and $d\mathbf{s}$. From the definition of dot product, if $d\mathbf{s}$ is perpendicular to ∇u, then $\nabla u \cdot d\mathbf{s} = 0$, and so $du = 0$. Because $d\mathbf{s}$ is a displacement, this tells you that for a displacement perpendicular to ∇u there is no change in u. Therefore, the lines of constant u (contour lines) must be perpendicular to ∇u because a displacement along a contour line results in no change in the value of the function. (Actually, I should say *surfaces* of constant u are perpendicular to ∇u because we have defined $u = u(x, y, z)$ to be a three-dimensional function.)

Similarly, the largest change in $du = \nabla u \cdot d\mathbf{s}$ comes when the displacement $d\mathbf{s}$ is *parallel* to ∇u, that is, in the direction of ∇u. The largest change in du for a given displacement is along the path of steepest ascent, so ∇u points in the direction of steepest ascent. Furthermore, $du = |\nabla u||d\mathbf{s}|\cos\theta$, and the maximum value of du occurs when $\cos\theta = 1$, so

$$\left.\frac{du}{ds}\right|_{\text{max}} = |\nabla u|.$$

That is, the magnitude of ∇u gives the value of the greatest increase in u per unit displacement.

To summarize the preceding discussion you could state that the *direction* of the gradient indicates the direction of greatest change of a function and the *magnitude* of the gradient gives the maximum change in the function per unit displacement.

Exercise 5.3

The surface of a sphere is described by $x^2 + y^2 + z^2 = $ constant. Show that the gradient of this function points in the radial direction. ∎

Exercise 5.4

Evaluate the gradient of $f = 5x^3 + 6y^2 + 2z$ at (1,2,3). Answer: $15\hat{\mathbf{i}} + 24\hat{\mathbf{j}} + 2\hat{\mathbf{k}}$. ∎

Exercise 5.5

A family of coaxial cylinders is described by

$$f(x, y, z) = x^2 + y^2 = \text{constant}.$$

Show that ∇f points away from the axis. ∎

Exercise 5.6

The pressure in a body of still water is given by $P = \rho g z + P_0$, where P_0 is atmospheric pressure and z is the depth below the surface. Show that the direction of maximum increase in pressure is straight down. (If you got the wrong sign it is because you did not consider that z is measured positive downward!) ∎

5.3.3 The Relationship between Force and Potential Energy

As described above, the potential energy is a function whose derivative with respect to position is related to the force. The expressions $F = -dV/dr$ and $F = -dV/dx$ only gave the magnitude of the forces, not their direction. The description is not complete. It turns out (as mentioned in Chapter 1) that the *gradient* of the potential energy is related to the vector force. Specifically, *the force is equal to the negative gradient of the potential energy*. In symbols,

$$\mathbf{F} = -\nabla V. \tag{5.7}$$

This equivalence means you can work with the scalar function V instead of the vector function **F**. This is a great advantage in solving problems because it simplifies the calculations significantly. (You probably recollect from your introductory electricity class that determining the electric field **E** by direct application of Coulomb's law required resolving **E** into components and then integrating over the charge distribution. You soon discovered it was much easier to first evaluate the scalar function V, the *electric potential,* and then determine **E** by differentiation.)

Not all forces can be represented by a potential energy function. Forces that do not have a potential energy associated with them are called *nonconservative forces.* Fortunately, most of the important forces, such as the gravitational force and the electrical force, are conservative. All forces acting on the molecular and atomic level are believed to be conservative. On the other hand, forces such as friction and air resistance are nonconservative and do not allow one to take advantage of the simplifications afforded by the potential energy.

Worked Example 5.4   A particle has a potential energy given by $V = V(x) = kx^2 e^{x/a}$ (in joules), where $k = 3$ J/m^2 and $a = 1$ m. What is the force on the particle at $x = 0$? What is the force on the particle at $x = 2.0$ m? What is the work done by the conservative force during a displacement from $x = 0$ to $x = 2$ m?

Solution Because this is a one-dimensional problem, the relation $\mathbf{F} = -\nabla V$ reduces to $F = -\frac{dV}{dx}$. Consequently, writing $V(x) = 3x^2 e^x$,

$$F = -\frac{dV}{dx} = -\frac{d}{dx}(3x^2 e^x) = -\left[6xe^x + 3x^2 e^x\right] = -e^x(6x + 3x^2).$$

Therefore, for $x = 0, F = 0$, and for $x = 2$,

$$F = -e^2(12 + 12) = -24e^2 = -177 \text{ N}.$$

The work is

$$W = -\Delta V = -[V(x = 2) - V(x = 0)] = -\left[3(2^2)e^2 - 0\right] = -88.7 \text{ J}.$$

How can you determine whether or not a force is conservative? The answer is quite simple. A force is conservative *if and only if* its curl is zero:

F is conservative iff $\nabla \times F = 0$.

This statement is easy to prove. If **F** is conservative, then it can be represented by the negative gradient of the potential energy, i.e., $\mathbf{F} = -\nabla V$. Consequently,

$$\nabla \times \boldsymbol{F} = \nabla \times (-\nabla V) = -\nabla \times (\nabla V) = -\mathbf{curl}\ (\mathbf{grad}\ V).$$

But the curl of the gradient of *any* scalar function is zero. Therefore, if we can write **F** as the gradient of V, then $\nabla \times \boldsymbol{F} = 0$.

For the sake of completeness and because the steps in the proof are important to understand, I will now prove that **curl grad** $V = 0$ for any function V. (Here $V = V(x, y, z)$ can be any arbitrary scalar function and not necessarily the potential energy.) Using Cartesian coordinates,

$$\nabla \times \nabla V = \left(\hat{\mathbf{i}}\frac{\partial}{\partial x} + \hat{\mathbf{j}}\frac{\partial}{\partial y} + \hat{\mathbf{k}}\frac{\partial}{\partial z}\right) \times \left(\hat{\mathbf{i}}\frac{\partial V}{\partial x} + \hat{\mathbf{j}}\frac{\partial V}{\partial y} + \hat{\mathbf{k}}\frac{\partial V}{\partial z}\right)$$

$$= \begin{vmatrix} \hat{\mathbf{i}} & \hat{\mathbf{j}} & \hat{\mathbf{k}} \\ \partial/\partial x & \partial/\partial y & \partial/\partial z \\ \partial V/\partial x & \partial V/\partial y & \partial V/\partial z \end{vmatrix}$$

$$= \hat{\mathbf{i}}\left(\frac{\partial^2 V}{\partial y \partial z} - \frac{\partial^2 V}{\partial z \partial y}\right) - \hat{\mathbf{j}}\left(\frac{\partial^2 V}{\partial x \partial z} - \frac{\partial^2 V}{\partial z \partial x}\right) + \hat{\mathbf{k}}\left(\frac{\partial^2 V}{\partial x \partial y} - \frac{\partial^2 V}{\partial y \partial x}\right).$$

But

$$\frac{\partial^2 V}{\partial y \partial z} = \frac{\partial^2 V}{\partial z \partial y},$$

because the order of taking partial derivatives does not affect the result. Therefore, each term in parenthesis is zero and

$$\nabla \times \nabla V = 0.$$

Thus, we have proved that

curl grad (any scalar function) = 0.

5.3.4 Del in Other Representations

So far we have been using Cartesian coordinates, but it is often useful to work in some other set of coordinates, such as cylindrical coordinates or spherical coordinates. I am going to show you how to obtain representations for ∇ in these coordinate systems. However, before doing so we must discuss coordinate transformations.[5]

Coordinate Transformations

Suppose you want to transform from the Cartesian coordinates x, y, z to a different coordinate system in which the coordinates are q_1, q_2, q_3. In the Cartesian coordinate system the location of a point is specified by the intersection of three planes, $x = $ constant, $y = $ constant, and $z = $ constant, as shown in Figure 5.5(a). For example, the point (1,2,0) is located at the intersection of the planes $x = 1$, $y = 2$, $z = 0$.

In Figure 5.5(b) I generalized the situation and specified the location of the point by the intersection of three *surfaces* ($q_1 = $ constant, $q_2 = $ constant, $q_3 = $ constant). Although it is not necessary, it is usually convenient to consider only *orthogonal* coordinate systems. For an orthogonal coordinate system the surfaces are always perpendicular at the lines of intersection. (All of the coordinate systems used in this book are orthogonal.)

The relationship between the Cartesian coordinates (x, y, z) and the "new" coordinates (q_1, q_2, q_3) is given by the transformation equations:

$$x = x(q_1, q_2, q_3),$$
$$y = y(q_1, q_2, q_3),$$
$$z = z(q_1, q_2, q_3).$$

[5] If you would like to have more information on this subject, see Chapter 2 of *Mathematical Methods for Physicists*, 5th ed., by G. Arfken and H. Weber, Academic Press, New York, 2001, or Section 10.9 of *Mathematical Methods in the Physical Sciences*, 3rd ed., by Mary L. Boas, John Wiley & Sons, New Jersey, 2006.

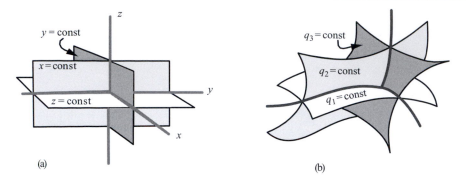

Figure 5.5 (a) A Cartesian coordinate system. (b) A general coordinate system.

The *inverse* transformations are:

$$q_1 = q_1(x, y, z),$$
$$q_2 = q_2(x, y, z),$$
$$q_3 = q_3(x, y, z).$$

Just as $\hat{\imath}$, $\hat{\jmath}$, $\hat{k}$ are defined as the unit vectors of the Cartesian system, so too you can define the quantities $\hat{e}_1, \hat{e}_2, \hat{e}_3$ as the unit vectors of the new coordinate system. These are defined so that $\hat{e}_1$ points in the direction of increasing q_1 and is perpendicular to the surface $q_1 = $ constant, and similarly for $\hat{e}_2$ and $\hat{e}_3$.

Our analysis of coordinate transformations is based on the behavior of the element of infinitesimal displacement, ds. In the Cartesian system ds is related to dx, dy, dz by the generalized Pythagorean relation

$$ds^2 = dx^2 + dy^2 + dz^2.$$

As I will show in a moment, when we express ds in terms of the new coordinates we get a fairly complicated expression that reduces to the form

$$ds^2 = \sum_{ij} h_{ij}^2 dq_i dq_j \quad i, j = 1, 2, 3. \tag{5.8}$$

To see how this comes about, note that because x is a function of $q_1, q_2,$ and q_3, you can write

$$dx = \frac{\partial x}{\partial q_1} dq_1 + \frac{\partial x}{\partial q_2} dq_2 + \frac{\partial x}{\partial q_3} dq_3.$$

Squaring this expression leads to

$$dx^2 = \left(\frac{\partial x}{\partial q_1}\right)^2 dq_1^2 + \left(\frac{\partial x}{\partial q_1}\right)\left(\frac{\partial x}{\partial q_2}\right) dq_1 dq_2 + \left(\frac{\partial x}{\partial q_1}\right)\left(\frac{\partial x}{\partial q_3}\right) dq_1 dq_3 + \ldots.$$

Since Problem (5.4) asks you to go through this process, I did not write out the whole expression, which has nine terms (some terms are the same, so there are only six *different* terms). You can obtain similar expressions for dy^2 and dz^2. Each term involves a product of the form $dq_i dq_j$. If you add and combine the three terms $dx^2 + dy^2 + dz^2$ to get ds^2 you will obtain a sum of nine terms involving $dq_i dq_j$ and it will have the form of Equation (5.8) above. The coefficients of $dq_i dq_j$ (the h_{ij}^2 in Equation (5.8)) are called *scale factors*. The set of nine quantities $h_{ij}^2 (i, j = 1, 2, 3)$ is called the *metric* or *metric tensor* for the coordinate system. It is easy to appreciate that

$$h_{ij}^2 = \frac{\partial x}{\partial q_i}\frac{\partial x}{\partial q_j} + \frac{\partial y}{\partial q_i}\frac{\partial y}{\partial q_j} + \frac{\partial z}{\partial q_i}\frac{\partial z}{\partial q_j}.$$

A tensor can be expressed in matrix form, so the metric tensor can be written as

$$\begin{pmatrix} h_{11}^2 & h_{12}^2 & h_{13}^2 \\ h_{21}^2 & h_{22}^2 & h_{23}^2 \\ h_{31}^2 & h_{32}^2 & h_{33}^2 \end{pmatrix}.$$

For an orthogonal coordinate system, the scale factors have the property

$$h_{ij}^2 = 0 \text{ for } i \neq j,$$

and the expression for ds^2 in Equation (5.8) reduces to

$$ds^2 = h_1^2 dq_1^2 + h_2^2 dq_2^2 + h_3^2 dq_3^2, \tag{5.9}$$

where I made a slight change in notation and replaced h_{ii} by h_i.

With this expression for ds^2 it is clear that for an orthogonal coordinate system, the element of displacement along a given coordinate is

$$ds_i = h_i dq_i.$$

In vector form an infinitesimal displacement is

$$d\mathbf{s} = h_1 dq_1 \hat{\mathbf{e}}_1 + h_2 dq_2 \hat{\mathbf{e}}_2 + h_3 dq_3 \hat{\mathbf{e}}_3.$$

The element of area in the new system of coordinates is

$$d\sigma_{ij} = ds_i ds_j = h_i h_j dq_i dq_j.$$

For example, an element of area perpendicular to $\hat{\mathbf{e}}_3$ would be

$$d\sigma_{12} = ds_1 ds_2 = h_1 h_2 dq_1 dq_2.$$

The element of volume is

$$d\tau = ds_1 ds_2 ds_3 = h_1 h_2 h_3 dq_1 dq_2 dq_3. \tag{5.10}$$

We can now obtain an expression for del in generalized coordinates. In Cartesian coordinates, $\partial u / \partial x$ describes how the function $u = u(x)$ varies for a given change in x. A small displacement along the x-axis leads to a small change in u given by

$$du = \left(\frac{\partial u}{\partial x}\right) dx.$$

Similarly, a small displacement along the q_1-axis (denoted by ds_1) leads to a change in u given by

$$du = \left(\frac{\partial u}{\partial s_1}\right) ds_1.$$

But $ds_1 = h_1 dq_1$, so

$$\frac{\partial u}{\partial s_1} = \frac{\partial u}{h_1 \partial q_1}.$$

Thus the definition of del in Cartesian coordinates

$$\nabla \equiv \hat{\mathbf{i}}\frac{\partial}{\partial x} + \hat{\mathbf{j}}\frac{\partial}{\partial y} + \hat{\mathbf{k}}\frac{\partial}{\partial z},$$

leads to the following general definition of del for *any* coordinate system

$$\nabla = \hat{\mathbf{e}}_1\frac{1}{h_1}\frac{\partial}{\partial q_1} + \hat{\mathbf{e}}_2\frac{1}{h_2}\frac{\partial}{\partial q_2} + \hat{\mathbf{e}}_3\frac{1}{h_3}\frac{\partial}{\partial q_3}. \qquad (5.11)$$

Exercise 5.7

For cylindrical and spherical coordinates, sketch the three planes that define the location of point (1,2,0). Label the axes and the planes clearly. ■

Exercise 5.8

Sketch the element of area $d\sigma_{12}$ for Cartesian coordinates. ■

Exercise 5.9

Spherical coordinates are orthogonal. Show that $h_{13}^2 = 0$. The transformation equations are given by Equation (2.20). ■

5.3.5 Cylindrical Coordinates

Let us apply these general concepts to obtain an expression for del in cylindrical coordinates. Begin with the transformation equations for cylindrical coordinates:

$$x = \rho \cos \phi,$$
$$y = \rho \sin \phi,$$
$$z = z.$$

Then,

$$dx = d\rho \cos \phi - \rho \sin \phi d\phi,$$
$$dy = d\rho \sin \phi + \rho \cos \phi d\phi,$$
$$dz = dz,$$

and

$$ds^2 = dx^2 + dy^2 + dz^2, \qquad (5.12)$$
$$= d\rho^2 + \rho^2 d\phi^2 + dz^2.$$

By inspection of this last equation and comparing it to Equation (5.9), the scale factors for cylindrical coordinates are

$$h_1^2 = 1, \quad h_2^2 = \rho^2, \quad h_3^2 = 1.$$

The unit vectors are $\hat{\rho}, \hat{\phi}, \hat{\mathbf{k}}$, so using Equation (5.11) yields the following expression for ∇:

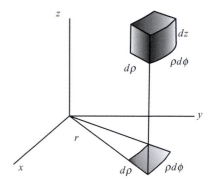

Figure 5.6 Volume element in cylindrical coordinates. The projected shaded area in the xy-plane is the base of the volume element. Note that the edge marked $\rho d\phi$ is an arc of a circle of radius ρ. The sides of the volume element become flatter as they shrink to zero.

$$\nabla = \hat{\rho}\frac{\partial}{\partial\rho} + \hat{\phi}\frac{1}{\rho}\frac{\partial}{\partial\phi} + \hat{\mathbf{k}}\frac{\partial}{\partial z}.$$

The element of volume in cylindrical coordinates is obtained immediately from Equation (5.10) as

$$d\tau = \rho d\rho d\phi dz.$$

Geometrically this is the infinitesimal parallelepiped obtained from a displacement of $d\rho$ along $\hat{\rho}$, a displacement of $\rho d\phi$ along $\hat{\phi}$, and a displacement of dz along $\hat{\mathbf{k}}$. For the purposes of remembering this expression, it is probably best to visualize the volume element as illustrated in Figure 5.6.

Exercise 5.10

Fill in the missing steps in the derivation of Equation (5.12).   ∎

5.3.6 Spherical Coordinates

The relationships between Cartesian coordinates and spherical coordinates are

$$x = r\sin\theta\cos\phi,$$
$$y = r\sin\theta\sin\phi,$$
$$z = r\cos\theta.$$

Consequently,

$$dx = dr\sin\theta\cos\phi + d\theta r\cos\theta\cos\phi - d\phi r\sin\theta\sin\phi,$$
$$dy = dr\sin\theta\sin\phi + d\theta r\cos\theta\sin\phi + d\phi r\sin\theta\cos\phi,$$
$$dz = dr\cos\theta - d\theta r\sin\theta.$$

In Exercise 5.11 you will show that these relations lead to:

$$ds^2 = dx^2 + dy^2 + dz^2$$
$$= dr^2 + r^2 d\theta^2 + r^2\sin^2\theta d\phi^2. \tag{5.13}$$

By inspection, the elements of the metric (the scale factors) for spherical coordinates are

$$h_1^2 = 1, \quad h_2^2 = r^2, \quad h_3^2 = r^2\sin^2\theta.$$

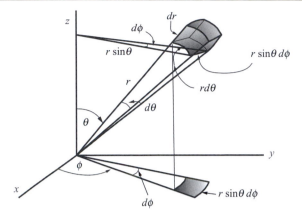

Figure 5.7 Volume element in spherical coordinates. The volume element has curved edges but in the limit of infinitesimal quantities the volume is given by (base × width × height).

The unit vectors are $\hat{\mathbf{r}}, \hat{\boldsymbol{\theta}}, \hat{\boldsymbol{\phi}}$, so using Equation (5.11) the gradient is expressed as:

$$\nabla = \hat{\mathbf{r}} \frac{\partial}{\partial r} + \hat{\boldsymbol{\theta}} \frac{1}{r} \frac{\partial}{\partial \theta} + \hat{\boldsymbol{\phi}} \frac{1}{r \sin \theta} \frac{\partial}{\partial \phi}.$$

From Equation (5.10), the volume element (illustrated in Figure 5.7) is

$$d\tau = h_1 h_2 h_3 dq_1 dq_2 dq_3 = r^2 \sin \theta \, dr \, d\theta \, d\phi.$$

Exercise 5.11

Fill in the missing steps between the two equations in (5.13). ∎

5.4 Force, Work, and Potential Energy

Consider the difference between conservative forces and nonconservative forces. Work done by a nonconservative force (such as friction) is at least partially converted into a nonrecoverable form of energy (such as thermal energy which is quickly dissipated to the environment as heat). You cannot define a potential energy function for a nonconservative force.

Think about lifting a rock. Imagine we raise it very slowly so the kinetic energy ($T = \frac{1}{2}mv^2$) is essentially zero at all times. Call the force that our muscles exert on the rock F_{us}. Then the work done by us is

$$\int_{z_1}^{z_2} F_{us} dz.$$

But F_{us} involves the tension in our muscles, the friction in our joints, and so on. I'm pretty sure nobody has come up with a mathematical expression for this force. Then how in the world can we evaluate the integral? (Now comes the clever part.) To evaluate the integral we use the fact that to lift the rock we exerted a force equal and opposite to the force of gravity.[6]

---

[6] To be precise, we need to exert a force slightly greater than the force of gravity for an instant to get the rock moving, and when it reaches the top we exert a force slightly less than the force of gravity so the rock will slow down and stop. Overall, the average force we exert on the rock is equal to the gravitational force.

$$F_{us} = -F_{grav}.$$

Therefore,

$$\int_{z_1}^{z_2} F_{us} dz = -\int_{z_1}^{z_2} F_{grav} dz,$$

where the gravitational force is mg, directed downward.

When we raised the rock, the change in its potential energy was

$$V_2 - V_1 = -\int_{z_1}^{z_2} F_{grav} dz$$

$$= \text{negative of work done by the gravitational force}$$

$$= \text{(positive) work done by us.}$$

Thus, the increase in potential energy is equal to the work done by us, just as you would expect.

The work done by the nonconservative force (us) plus the work done by the conservative force (gravity) is zero and since there is no *net* work done, the work–energy theorem tells us there is no increase in kinetic energy.

Now suppose we drop the rock. Ignoring air resistance, there are no nonconservative forces acting. The work done by the conservative gravitational force equals the decrease in potential energy, i.e.,

$$W_g = \int F_{grav} dz = \int_{z=h}^{z=0} -mg \, dz = mgh = -\Delta V.$$

According to the work–energy theorem, the work done by the gravitational force is equal to the increase in the kinetic energy. Therefore, the decrease in potential energy is equal to the increase in kinetic energy.

Generalizing to any conservative force ($\mathbf{F} = -\nabla V$), the work done by the force during a displacement from $\mathbf{r}_1$ to $\mathbf{r}_2$ is

$$W = \int_{\mathbf{r}_1}^{\mathbf{r}_2} \mathbf{F} \cdot d\mathbf{s} = \int_{\mathbf{r}_1}^{\mathbf{r}_2} -\nabla V \cdot d\mathbf{s} = -\int_{\mathbf{r}_1}^{\mathbf{r}_2} dV = V(\mathbf{r}_1) - V(\mathbf{r}_2).$$

Note the use of the vector definition of the differential (see Equation (5.6)). Furthermore, $\Delta V \equiv V(\mathbf{r}_2) - V(\mathbf{r}_1)$, so

$$W = -\Delta V.$$

That is, the work done by the conservative force will cause the potential energy to decrease by the same amount. This might be the opposite of what you expected. You have known for a long time that doing work on an object will *increase* its potential energy. But please note carefully that I said the work done by the *conservative* force causes an equal decrease in the potential energy. When you raise a body from the ground to a height h the conservative force (gravity) does negative work and the potential energy increases. When the body falls back to the ground, the conservative force does positive work and the potential energy decreases.

You might wonder where the zero point of potential energy should be situated. Recall that the *change* in potential energy is equal to the negative of the work done by the conservative forces. Thus, you are dealing only with the *changes* in V and the actual numerical values of V at the end points of the process make no difference. The choice of the zero level of V is completely arbitrary (but remember that once you have made the choice you must stick with it).

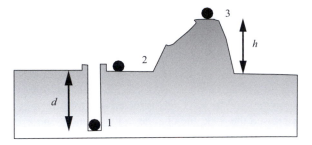

Figure 5.8 Illustration that potential energy is always relative to an arbitrary value.

For example, Figure 5.8 shows a mass m at three different locations: at the bottom of a well of depth d, on level ground, and on top of a hill of height h.

If you chose the zero level of potential energy to be the level ground, then $V_2 = 0$; the particle in the well has negative potential energy $V_1 = -mgd$, and the particle on the hill has positive potential energy, $V_3 = +mgh$. The work done by the nonconservative force to take the particle from the bottom of the well to the top of the hill is

$$W = V_3 - V_1 = mg(h + d).$$

If you select the zero of potential energy at the bottom of the well, the *values* of V_1, V_2, V_3 all change, but the work required to take the particle to the top of the hill is still $W = mg(h + d)$.

The quantity $W = mg(h + d)$ is the work *you* have to do to take the particle from the bottom of the well to the top of the hill. This is *not* the work done by the conservative gravitational force. *That* force was directed opposite to the displacement. The work done *by* the gravitational field as the particle went from $z_1 = -d$ to $z_3 = +h$ is $W = -mg(h + d)$.

When dealing with potential energy, it is often convenient to define some arbitrary reference position $\mathbf{r}_0$ as the position of *zero* potential energy. In general,

$$V(\mathbf{r}) - V(\mathbf{r}_0) = -\int_{\mathbf{r}_0}^{\mathbf{r}} \mathbf{F} \cdot d\mathbf{s}.$$

If $V(\mathbf{r}_0) = 0$, then

$$V(\mathbf{r}) = -\int_{\mathbf{r}_0}^{\mathbf{r}} \mathbf{F} \cdot d\mathbf{s}.$$

(The reference position is often taken to be at infinity.)

Worked Example 5.5 A rope of mass 2 kg and length 3 m is partially hanging off the edge of a platform, as shown in Figure 5.9. Initially, one meter of the rope is on the platform and two meters are hanging down. Determine the work required to slowly pull the rope onto the platform. (Assume that the platform is frictionless.)

Solution The force required to pull the rope onto the platform is equal to the weight of the portion of the rope hanging off the edge. Initially this is two meters and finally it is zero meters. The mass per unit length of the rope is $\lambda = (2/3)$ kg/m and the weight of the hanging portion is $\lambda g x$, where x is the length hanging off the edge. That is $F = -\lambda g x$. The work done to pull it up is

$$W = \int F dx = \int_2^0 (-\lambda g x) dx = \lambda g \left[\frac{x^2}{2} \right]_0^2 = \frac{2}{3}(9.8)\frac{4}{2} = 13.1 \text{ J}.$$

This problem can also be solved by determining the change in potential energy of the rope. Taking the platform as the zero of potential energy we have $V_{final} = 0$ because the entire rope is on the platform. The initial value of the potential energy is

$$V_{initial} = \int_{x=-2}^{0} dmgx = g\lambda \int_{-2}^{0} x\,dx = -2\lambda g.$$

So the work done by the conservative force (gravity) is

$$W_g = -\Delta V = -(V_{final} - V_{initial}) = -2g(2/3) = -13.1 \text{ J}, \qquad (5.14)$$

and the work done by the applied force is $W = -W_g = +13.1$ J.

A third approach is to simply determine the change in potential energy of the center of mass of the hanging portion of the rope. Once again, $V_{final} = 0$ whereas

$$V_{cm(initial)} = mgh_{cm} = (2/3)(2)(9.8)(-1) = -13.1 \text{ J}$$

and once again $W_g = -\Delta V = -13.1$ J, and the work done by the applied force is 13.1 J.

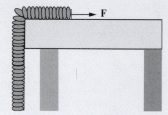

Figure 5.9 A rope hanging off of a frictionless platform.

Exercise 5.12

A particle of mass m is in a uniform gravitational field where the acceleration is g. Is this field conservative? ■

Exercise 5.13

Show that the gravitational force, as given by Newton's law of universal gravitation, $\mathbf{F} = -\left(GM_1M_2/R^2\right)\hat{\mathbf{r}}$, is conservative. ■

Exercise 5.14

A particular force can be expressed as $\mathbf{F} = 3x^2\hat{\mathbf{i}} + 2y^2\hat{\mathbf{j}}$. Is this force conservative? (Yes.) ■

Exercise 5.15

Two positive charges (Q_1 and Q_2) are separated by a distance r_1. They are pushed toward one another until their separation is r_2. What is the change in potential energy? How much work was done by external forces in pushing the charges together? Recall that the electrostatic force is given by $\mathbf{F} = (Q_1Q_2)/(4\pi\varepsilon_0 r^2)\hat{\mathbf{r}}$. Answer: $Q_1Q_2(r_1 - r_2)/4\pi\varepsilon_0 r_1 r_2$. ■

Exercise 5.16

The force exerted by a spring is $F = -kx$, where k is a constant of proportionality and x is the displacement from equilibrium. Obtain an expression for the potential energy of a spring. Answer: If $x_0 = 0$, $V = (1/2)kx^2$. ■

The gravitational potential energy of an object of mass m at a distance r from the center of the Earth is $V = -GmM_E/r$, where M_E is the mass of the Earth. From this obtain Newton's law of universal gravitation.   ■

What is the change in potential energy of a 1 kg mass when it is taken from the surface of the Earth to infinity? (Ignore the effect of any other bodies in the universe.) Answer: 6.24×10^7 J. ■

5.5 The Conservation of Energy

The work–energy theorem states that the net work done on a particle is equal to the increase in its kinetic energy. If W_{net} is the algebraic sum of *all* the work, then

$$\Delta T = T_f - T_i = W_{\text{net}}.$$

This principle is never violated. (However, it is easy to think of situations in which it does not *appear* to be true.)

Suppose the *only* forces acting on the particle are conservative. In that case, the work done can be written as $W = -(V_f - V_i) = V_i - V_f$. Combining this with the work–energy theorem leads to

$$T_f - T_i = V_i - V_f$$

or

$$T_f + V_f = T_i + V_i.$$

This equation is the simplest expression of the law of *conservation of energy*. It states that if the only work performed on a system is done by conservative forces, then the sum of the initial kinetic energy plus the initial potential energy (before the work was performed) is equal to the sum of the final kinetic energy plus the final potential energy (after the work was performed).

The sum of kinetic energy and potential energy is called the total energy. It is denoted by E:

$$E = T + V.$$

The conservation law simply states that the total energy is constant. This statement holds **as long as the only forces acting on the system are conservative forces.**

In high-school physics courses one often hears the energy conservation principle expressed as, "Energy can neither be created nor destroyed but it can be converted from one form to another." A more sophisticated version of this statement might be, "In mechanical systems the effect of a conservative force is to transform potential energy into kinetic energy, or kinetic energy into potential energy."

The situation gets a little more complicated when nonconservative forces are involved. Suppose you pick up a book from the floor and place it on a table, at height h above the floor. The initial energy of the book was zero because it had zero kinetic energy to begin with (and the floor is the zero point of potential energy). When the book is on the table, once again it has zero kinetic energy and its total energy is $E = V = mgh$. Where did this energy come from? Clearly *you* are the ultimate source of the energy of the book. The *net* work done on the book (by you and gravity) was zero, so the book's kinetic energy did not increase. The increase in potential energy was equal to the negative of the work done by the gravitational force, but this energy can be recovered (if the book falls off the table, for example). The work done by *you* is not recoverable because the force you exerted was nonconservative. If you could evaluate the chemical energy supplied to your muscles, do you think it would be equal to mgh? (Actually it would not, because energy is also internally converted into heat in a working person's body.)

Mechanics primarily deals with problems involving conservative forces, so the conservation of energy principle is nearly always written in the simple form $E = T + V = \text{constant}$, or equivalently, $T_i + V_i = T_f + V_f$, where subscript i represents some initial state of the system and subscript f represents some final state. If you wish to use the energy principle when *nonconservative* forces are acting you can include them as a work term. That is, if W_{nc} is the work done *on* the system by nonconservative forces as the system evolves from initial state i to final state f, then

$$T_f + V_f = T_i + V_i + W_{nc}$$

$$= T_i + V_i + \int_i^f \mathbf{F}_{nc} \cdot d\mathbf{s}.$$

In solving problems, make sure you have the right sign on W_{nc}, depending on whether the nonconservative work *increases* or *decreases* the total energy of the system. This is usually clear from the statement of the problem or from a consideration of the behavior of the system.

Exercise 5.19

A 40 kg child climbs 5 m up a flagpole then slides to the bottom. The child is moving at 3 m/s upon reaching the bottom. What was the work done by friction? Answer: 1780 J. ∎

Exercise 5.20

Evaluate the *escape velocity* of an object from the Earth. (This means that the object is initially on the surface of the Earth and ends up with zero velocity an infinite distance away.) Ignore any other bodies in the universe. Answer: 11.2 km/s. ∎

Exercise 5.21

Evaluate the velocity required for a body at 1 AU from the Sun to escape from the solar system. Assume the solar system consists only of the Sun (a good approximation). ∎

5.5.1 A Reflection on the Conservation of Energy

Conservation laws are extraordinarily useful for solving physics problems. To determine the speed of a falling object it is easier to apply the law of conservation of energy than to integrate the equation of motion. However, conservation laws are much more than mere computational aids; they are fundamental laws of nature that tell us how the physical universe behaves.[7] Consider, for example, the universe defined as the total collection of all material objects. There are no external forces doing work on the universe (because there is nothing outside of it!). Therefore, the total energy of the universe must be a constant quantity. Similarly, the total linear momentum and the total angular momentum of the universe are also constant quantities.

---

[7] A physicist who finds that a conservation law has been violated knows that it is either a mistake or a very important discovery. For example, it was believed for many years that parity was always conserved. The discovery of a reaction in which parity did not remain constant caused a furor among physicists and eventually led to a much deeper insight into the nature of the physical world. (We will consider the conservation of parity briefly in Chapter 8.)

It has been suggested that conservation laws are not laws of nature, but just mental constructs designed to help us comprehend and deal with the physical world. The law of conservation of energy is often cited as an example of this point of view. In your introductory physics course you learned the law of conservation of energy in the form:

$$\text{kinetic energy} + \text{potential energy} = \text{constant.}$$

When you studied thermodynamics you learned that mechanical energy could be converted into heat and you modified the law to:

$$\text{kinetic energy} + \text{potential energy} + \text{thermal energy} = \text{constant.}$$

Then you studied electricity and magnetism and learned of other forms of energy associated with the electromagnetic fields. These also had to be included in the conservation law. And so on...

This procedure gives the impression that every time we find a new form of energy we simply add it to the left-hand side of the equation and *define* energy as that quantity whose total value is constant. However, such an approach tends to trivialize the conservation laws and we shall not adopt it. We shall consider the conservation laws to be fundamental to the structure of the physical universe. As shown in Chapter 8, the conservation laws are related, on a very profound level, to *symmetries* found in nature. Thus, energy conservation is related to a symmetry in time and momentum conservation is related to a symmetry in space. (I really don't expect you to understand what I am talking about here – but don't worry, it will all become clear as we go along!)

5.6 Energy Diagrams

Imagine a particle that is initially at the top of a perfectly smooth hill, as illustrated in Figure 5.10 which shows the particle perched on a peak at point A. There are peaks at A, C, and E and valleys at B and D. The only force acting on the particle is gravity.

The particle is given an infinitesimal push to the right. It slides down the hill towards B speeding up as it goes. It climbs the hill at C, slowing down somewhat, then speeding up as it descends into the valley at D. Finally it climbs another big hill and stops on the tip of the hill at point E.

In other words, the motion is what you would expect from an ordinary amusement park roller coaster ride, if you could find a completely frictionless roller coaster.

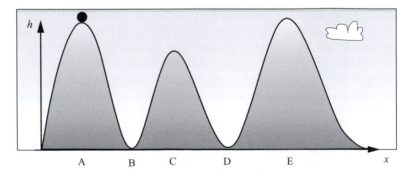

Figure 5.10 A particle on the top of a frictionless hill. If given an infinitesimal shove, the particle will slide past B, C, D, and end up on the tip of E.

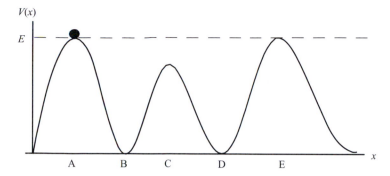

Figure 5.11 An energy diagram. The vertical axis is potential energy and the horizontal axis is position. Note that the horizontal line at energy E indicating the total energy is constant and equal to the same value for all x.

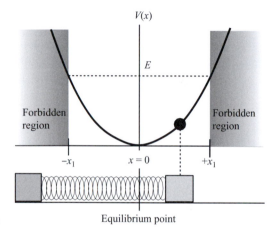

Figure 5.12 Energy diagram for a mass on a spring.

You could consider the sketch in Figure 5.10 to be a diagram of potential energy versus position because gravitational potential energy (mgh) is proportional to h, the height of the particle above zero level. Figure 5.11 shows the same situation, but now I labeled the vertical axis $V(x)$ and I drew a line intersecting the vertical axis at a value labeled E. This line gives the total energy. We call such a plot an *energy diagram*.

This plot is a very simple energy diagram, but it illustrates a number of important points. First of all, note that the total energy is constant (the horizontal line indicates the particle has the same total energy whatever the value of x, the position of the particle). Secondly, I knew where to draw the total energy line because I knew that at point A the particle had no kinetic energy $(T_A = 0)$ and the total energy at that location was equal to the potential energy.

Because $E = T + V$ at any point, the kinetic energy is given by the difference between the total energy and the potential energy at that point. The energy diagram allows you to determine the kinetic energy for any value of x by simply noting the "distance" between the line for the total energy and the potential energy curve. The velocity is proportional to the square root of T so this also gives an estimate of how fast the particle is moving.

It is very important for you to know how to interpret energy diagrams. Consider, for example, the energy diagram of Figure 5.12 in which I plotted the potential energy of a mass m connected to a spring of constant k. (Recall that for this system the potential energy is $V = \frac{1}{2}kx^2$.) The total energy is E. The mass will therefore oscillate between the points labelled $\pm x_1$. These two points are called "turning points." Even though the actual physical system is a block of mass m sliding back and forth horizontally, you can represent it by the round dot sliding up and down in a "potential well," as shown in the upper part of the figure. If the dot starts from rest at a turning point (say $-x_1$) it will slide down to $x = 0$, picking up speed and reaching maximum velocity at the bottom, then slowing down as it climbs the potential hill and reaching zero velocity at the other turning point (at $+x_1$). The physical motion of the block is given by the projection of the dot onto the horizontal (x) axis. At the turning points the velocity of the block is zero. The regions beyond the turning points are labeled "forbidden region." A particle in the forbidden region would have $V(x) > E$ (because the turning point is the location where $V(x) = E$). The allowed region is where $V(x) < E$. Note that $V(x) > E$ is impossible because $T = E - V$, and T would be negative if $V > E$. Since $T = \frac{1}{2}mv^2$, a negative value of the kinetic energy requires either a negative mass or an imaginary value for the velocity. There is no such thing as a negative mass, and an imaginary value for velocity is not allowed because velocity is an observable quantity and all observable physical quantities are real.[8]

5.7 The Energy Integral: Solving for the Motion

Recall that "solving for the motion" means obtaining an expression for the position as a function of time. In Chapter 3 we solved for the motion by integrating the acceleration twice. In this section we solve for the motion using energy techniques.

Consider the motion of a particle of mass m and total energy E in a region where the potential energy has a form such as illustrated in Figures 5.11 or 5.12 and all the forces are conservative. (Let's limit ourselves to one-dimensional motion for simplicity.) To obtain an expression for the position as a function of time, we can start with the fact that the total energy is constant:

$$T + V = E = \text{constant.}$$

Using the definition of kinetic energy write

$$\frac{1}{2}mv^2 + V(x) = E.$$

Then,

$$v^2 = \frac{2}{m}(E - V(x)),$$

or

$$\frac{dx}{dt} = \sqrt{\frac{2}{m}}\sqrt{E - V(x)}.$$

This fairly simple differential equation can be solved by separation of variables. Since E is a constant and V is a function only of x,

$$\frac{dx}{\sqrt{E - V(x)}} = \sqrt{\frac{2}{m}}dt.$$

---

[8] Physicists frequently use complex quantities to make the mathematics of a problem easier, but only the real part has physical significance. Remembering this simple fact will help you to avoid confusion, especially in your study of electromagnetic waves where the electric and magnetic fields are expressed as complex functions whose real parts correspond to the actual measurable quantities. This concept is also important in quantum mechanics where measurable physical quantities are expressed as real numbers even though the wave functions are complex.

The next step is to integrate both sides. The resulting expression,

$$\int_{x_0}^{x} \frac{dx}{\sqrt{E - V(x)}} = \sqrt{\frac{2}{m}} t, \tag{5.15}$$

is called the energy integral.

You can go no further unless you have an explicit expression for $V(x)$. If such an expression is known, you can carry out the integration over x and obtain a mathematical expression involving x and x_0 and E. You will obtain

$$t = t(x, x_0, E).$$

Finally, you need to invert this equation and solve for x in terms of $x_0, E,$ and t to get an expression of the form

$$x = x(t) = x(x_0, E, t).$$

The result is an expression for x as a function of time and various constants. The constants are often the initial position and the initial velocity. Here $x(t)$ is given in terms of initial position and total energy, which is not quite the same thing, but is equivalent.

Note that solving for the motion using the energy requires only *one* integration, whereas starting with the equation of motion requires *two* integrations. The reason is that the conservation of energy equation can be expressed as

$$\frac{1}{2} mv^2 = E - V(x),$$

which is essentially an expression for the velocity. If we start with an expression for the acceleration, the first integration gives the velocity. For this reason, the conservation of energy equation is often called a "first integral."

Because the total energy E remains constant (as long as the forces are conservative), E is called a "constant of the motion." In general, knowing the constants of the motion makes problem solving much easier.

Worked Example 5.6 A particle of mass $m = 0.5$ kg moves in a one-dimensional potential field given by $V = \frac{1}{2} kx^2$ with $k = 2$ N/m. Determine the motion if the total energy is 2 J. Assume that at time $t = 0$ the particle is at $x = 0$.

Solution Begin with Equation (5.15):

$$\sqrt{\frac{2}{m}} t = \int_{x_0}^{x} \frac{dx}{\sqrt{E - V(x)}} = \int_{0}^{x(t)} \frac{dx}{\sqrt{2 - x^2}} = \sin^{-1} \frac{x}{2} \Big|_0^{x(t)}.$$

Consequently

$$\sqrt{4} t = \sin^{-1} \frac{x}{2} - \sin^{-1} 0$$

$$x(t) = 2 \sin(2t).$$

The particle oscillates sinusoidally in a potential well between the turning points at $x = -2$ and $x = +2$. You probably recognized this as a simple harmonic motion problem.

Exercise 5.22

A particle moves in a constant potential energy field, $V = V_0 =$ constant. Assume E is known. Determine the position, velocity, and acceleration as a function of time, that is, $x = x(t), v = v(t),$ and $a = a(t)$. Answer: $x(t) = \sqrt{2(E - V_0)/m} \; t + x_0$. ∎

Exercise 5.23

Show that a particle in a region of space where the potential energy is constant will move with zero acceleration.   ■

5.8 The Kinetic Energy of a System of Particles

Nearly everything that has been said in this chapter has referred to a single particle or to a translating rigid body that can be treated as a particle located at the center of mass of the body. We now generalize to a system of particles that may either be moving independently or bound to one another as in a rigid body. We will show that the kinetic energy of a system of particles can be expressed as the energy of translation of the center of mass, plus the energy of the particles relative to the center of mass. That is,

$$T_{\text{Total}} = T_{\text{CM}} + T_{\text{wrt CM}},$$

where T_{CM} is the translational kinetic energy of a particle whose mass is the total mass of the body and which is moving with the center of mass of the body, and $T_{\text{wrt CM}}$ is the kinetic energy of all the particles with respect to the center of mass. This is essentially what we assumed when we considered a disk rolling down an inclined plane and wrote

$$T = T_{\text{trans}} + T_{\text{rot}} = \frac{1}{2}Mv^2 + \frac{1}{2}I\omega^2.$$

Before going any further, let us recall the definition of center of mass. For a collection of N particles located at $\mathbf{r}_i$, the position of the center of mass, denoted $\mathbf{r}_c$, is

$$\mathbf{r}_c = \frac{\sum_{i=1}^{N} m_i \mathbf{r}_i}{\sum_{i=1}^{N} m_i}. \tag{5.16}$$

So

$$M\mathbf{r}_c = \sum_{i=1}^{N} m_i \mathbf{r}_i, \tag{5.17}$$

where $M = \sum m_i$ is the total mass.

To prove that $T_{\text{tot}} = T_{\text{CM}} + T_{\text{wrt CM}}$ we begin with the kinetic energy of particle i:

$$T_i = \frac{1}{2}m_i v_i^2 = \frac{1}{2}m_i (\dot{\mathbf{r}}_i \cdot \dot{\mathbf{r}}_i).$$

The vector $\mathbf{r}_i$ is measured from the origin. As shown in Figure 5.13 the position vector of particle i relative to the origin is

$$\mathbf{r}_i = \mathbf{r}_c + \mathbf{r}'_i,$$

where $\mathbf{r}'_i$ is the position of particle i relative to the center of mass. Therefore, the kinetic energy of particle i can be expressed as

$$T_i = \frac{1}{2}m_i \left(\dot{\mathbf{r}}_c + \dot{\mathbf{r}}'_i\right) \cdot \left(\dot{\mathbf{r}}_c + \dot{\mathbf{r}}'_i\right) = \frac{1}{2}m_i \left(\dot{\mathbf{r}}_c^2 + 2\dot{\mathbf{r}}_c \cdot \dot{\mathbf{r}}'_i + \dot{\mathbf{r}}'^2_i\right).$$

The total kinetic energy is obtained by adding the kinetic energies of all the particles in the system.

$$T = \sum_{i}^{N} T_i = \frac{1}{2}\dot{\mathbf{r}}_c^2 \sum m_i + \dot{\mathbf{r}}_c \cdot \sum m_i \dot{\mathbf{r}}'_i + \sum \frac{1}{2}m_i \dot{\mathbf{r}}'^2_i.$$

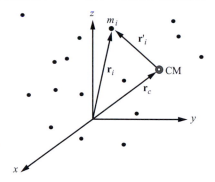

Figure 5.13 A system of particles. Particle i with mass m_i is located at position $\mathbf{r}_i$. The center of mass (CM) is at $\mathbf{r}_c$. The arrow marked $\mathbf{r}'_i$ is the position of particle i with respect to the center of mass.

The middle term is zero because

$$\sum m_i \dot{\mathbf{r}}'_i = \frac{d}{dt} \sum m_i \mathbf{r}'_i = \frac{d}{dt} \sum m_i (\mathbf{r}_i - \mathbf{r}_c),$$

and

$$\sum m_i (\mathbf{r}_i - \mathbf{r}_c) = \sum m_i \mathbf{r}_i - \sum m_i \mathbf{r}_c = M\mathbf{r}_c - \mathbf{r}_c M = 0.$$

where we used Equation (5.17).

Consequently, the total kinetic energy of a system of particles is given by

$$T = \tfrac{1}{2} \dot{r}_c^2 \sum m_i + \sum \tfrac{1}{2} m_i \dot{r}_i'^2,$$

or

$$T = \tfrac{1}{2} M v_c^2 + \sum \tfrac{1}{2} m_i v_i'^2. \tag{5.18}$$

If you imagine a particle of mass M at the center of mass, then the total kinetic energy of a system of particles is equal to the sum of the translational kinetic energy of M plus the kinetic energy of all of the particles *relative* to the center of mass, and the proposition is proved.

Exercise 5.24

Two particles are on a collision course along the x-axis. In the laboratory coordinate system a particle of mass 2 kg moves towards the right with a speed of 2 m/s, and a particle of mass 1 kg moves to the left with a speed of 1 m/s. (a) Determine the total kinetic energy of the system. (b) Determine the velocity of the center of mass. (c) Determine the kinetic energy of each particle relative to the center of mass. (d) Verify Equation (5.18). Answers: (a) 4.5 J, (b) 1 m/s. ∎

5.9 Work on an Extended Body: Pseudowork

The work done on a *particle* was defined by Equation (5.1),

$$W = \int_{\mathbf{r}_1}^{\mathbf{r}_2} \mathbf{F} \cdot d\mathbf{s},$$

where $\mathbf{F}$ was the force acting on the particle as it moved from $\mathbf{r}_1$ to $\mathbf{r}_2$. The displacement $d\mathbf{s}$ can be interpreted either as the displacement of the particle or as the displacement of the point of application of the force, since the two are equivalent. Furthermore, the source of the increase in the particle's kinetic energy is the agent that is producing the force. However, the situation gets more complicated when you consider the work done when a force acts on an *extended body*.

A simple example of an external force acting on an extended body is an ice skater who puts her hand on a wall and pushes herself away from it. The external force is the reaction force of the wall on her hand. (She pushes on the wall and the wall pushes back on her with an equal and opposite force.) The force acting on the skater stops as soon as her hand loses contact with the wall. The point of application of the force did not move at all. Did the wall do work on the skater? No. The energy of the wall did not change. The source of the kinetic energy imparted to the skater was chemical energy from her own body. This chemical energy was converted into mechanical energy through the action of her muscles. Nevertheless, the only unbalanced external force acting on the skater was the reaction force the wall exerted on her hand.

To determine the appropriate form of the equation for work in the case of an extended body, we begin with Newton's second law in the form

$$F_e = ma_c,$$

where a_c is the acceleration of the center of mass of the extended body and F_e is the external force. Multiplying both sides of this equation by an infinitesimal displacement of the center of mass, ds_c leads to

$$F_e ds_c = ma_c ds_c = m\frac{dv_c}{dt}ds_c = mdv_c\frac{ds_c}{dt} = mv_c dv_c.$$

Integrating from initial point i to final point f, gives

$$\int_i^f F_e ds_c = \int_i^f mv_c dv_c = \left(\frac{1}{2}mv_c^2\right)_f - \left(\frac{1}{2}mv_c^2\right)_i = \Delta T.$$

Therefore, the change in kinetic energy of the extended body is equal to the expression $\int_i^f F_e ds_c$. This looks suspiciously like the work. But it isn't *exactly* like the work because the point of application of the force did not move. Furthermore, the body exerting the force was not the source of the energy increase. For this reason, some authors prefer to call this expression the "pseudowork" and they call the equation

$$\int_i^f \mathbf{F}_e \cdot d\mathbf{s}_c = \Delta T$$

the "energy equation" to distinguish it from the work–energy theorem in which the external force is the source of the energy.[9]

5.10 Summary

In this chapter you have been exposed to a number of *mathematical concepts* and a number of *physical concepts*. In this summary the two sets of concepts are listed separately, first the math, then the physics.

5.10.1 Mathematical Concepts

Line Integral By definition, work $= \int_C \mathbf{F} \cdot d\mathbf{s}$. The integral is a *line integral*, and it must be evaluated along the path taken by the particle. You have seen two ways to evaluate a line integral:

---

[9] A very interesting article on this subject is the paper by Bruce Sherwood, "Pseudowork and Real Work," *American Journal of Physics*, **51**, 597–602 (1983).

(1) express the force and the differential of displacement in component form, thus

$$W = \int_C \mathbf{F} \cdot d\mathbf{s} = \int_C F_x dx + \int_C F_y dy + \int_C F_z dz,$$

or (2) express the force and displacement in terms of some parameter λ, and evaluate

$$W = \int_C \mathbf{F}(\lambda) \cdot d\mathbf{s} = \int_C \mathbf{F}(\lambda) \cdot \frac{d\mathbf{s}}{d\lambda} d\lambda.$$

Del Operator The del operator in Cartesian coordinates is defined as

$$\nabla = \hat{\mathbf{i}} \frac{\partial}{\partial x} + \hat{\mathbf{j}} \frac{\partial}{\partial y} + \hat{\mathbf{k}} \frac{\partial}{\partial z}.$$

Operations involving del include

∇f	the **gradient** of the scalar function f
$\nabla \cdot \mathbf{F}$	the **divergence** of the vector function $\mathbf{F}$
$\nabla \times \mathbf{F}$	the **curl** of the vector function $\mathbf{F}$.

This chapter concentrated on the gradient (∇f). Geometrically, the gradient is a vector pointing in the direction of the greatest increase in f, and its magnitude gives the rate of increase of f in that direction.

Metric The differential displacement vector $d\mathbf{s}$ has different forms in different coordinate systems. In terms of the generalized coordinates q_1, q_2, q_3,

$$ds^2 = \sum_{ij} h_{ij}^2 dq_i dq_j, \quad i, j = 1, 2, 3.$$

The scale factors h_{ij} depend on the geometrical properties of the coordinate space and collectively are called the *metric*.

Del and Volume Element in Other Representations Using the transformation properties of coordinates we derived expressions for the volume element and del in terms of generalized coordinates as

$$d\tau = ds_1 ds_2 ds_3 = h_1 h_2 h_3 dq_1 dq_2 dq_3,$$

and

$$\nabla = \hat{\mathbf{e}}_1 \frac{1}{h_1} \frac{\partial}{\partial q_1} + \hat{\mathbf{e}}_2 \frac{1}{h_2} \frac{\partial}{\partial q_2} + \hat{\mathbf{e}}_3 \frac{1}{h_3} \frac{\partial}{\partial q_3}.$$

In cylindrical coordinates these yield

$$d\tau = \rho d\rho d\phi dz,$$

$$\nabla = \hat{\rho} \frac{\partial}{\partial \rho} + \hat{\phi} \frac{1}{\rho} \frac{\partial}{\partial \phi} + \hat{\mathbf{k}} \frac{\partial}{\partial z},$$

and in spherical coordinates they yield

$$d\tau = r^2 \sin\theta dr d\theta d\phi,$$

$$\nabla = \hat{\mathbf{r}} \frac{\partial}{\partial r} + \hat{\theta} \frac{1}{r} \frac{\partial}{\partial \theta} + \hat{\phi} \frac{1}{r \sin\theta} \frac{\partial}{\partial \phi}.$$

5.10.2 Physical Concepts

I will now summarize the many important physical concepts introduced in this chapter.

Work The work done on a particle by a force $\mathbf{F}$ during a displacement from $\mathbf{r}_1$ to $\mathbf{r}_2$ is

$$W = \int_{\mathbf{r}_1}^{\mathbf{r}_2} \mathbf{F} \cdot d\mathbf{s}.$$

As noted above, the integral is a line integral and must be evaluated along the path followed by the particle. However, if the force is conservative, then the value of the integral depends only on the end points.

Work–Energy Theorem The total work done by *all* of the external forces equals the increase in kinetic energy of the particle.

$$W = \Delta T = T_f - T_i.$$

Potential Energy If a force is conservative ($\nabla \times \mathbf{F} = 0$) we can associate with it a potential energy V. The relation between force and potential energy is

$$\mathbf{F} = -\nabla V.$$

The work done by a conservative force can be expressed in terms of the change of potential energy, thus

$$W = -\Delta V = -V_f + V_i.$$

Conservation of Energy For conservative forces, equating the two expressions for work leads to the conservation law

$$T_f + V_f = T_i + V_i.$$

Energy Diagrams An energy diagram is a plot of the potential energy (V) as a function of position (x). The total energy is represented by a horizontal line on such a diagram. The "distance" between the total energy line and the potential energy curve ($E - V$) is equal to the kinetic energy and is proportional to the speed of the particle squared.

Solving for the Motion using Energy When the force is a function of position, energy methods are the easiest way to determine the motion. Basically, this involves evaluating an integral of the form

$$\int_{x_0}^{x} \frac{dx}{\sqrt{E - V(x)}} = \sqrt{\frac{2}{m}} t.$$

Energy of a System of Particles The kinetic energy of a system of particles can be expressed as the sum of the kinetic energy of the center of mass and the kinetic energy with respect to the center of mass:

$$T = \tfrac{1}{2} M v_c^2 + \sum \tfrac{1}{2} m_i v_i'^2 = T_c + T_{\text{wrt CM}}.$$

Work on an Extended Body Although the equations for the work done on an extended body are the same as the equations for the work done on a particle, the two concepts are not the same. In particular, the source of energy may be different in the two cases. Nevertheless, the correct answer is obtained by simply replacing the displacement of the particle by the displacement of the center of mass of the extended body.

5.11 Problems

Problem 5.1 The force on a particle is given by $\mathbf{F} = 3x\hat{\mathbf{i}} + 2y\hat{\mathbf{j}}$ N. Determine the line integral $\int \mathbf{F} \cdot d\mathbf{s}$ for the straight-line path that starts at the origin and ends at the point (3,6) m.

Problem 5.2 In this problem we assume that somehow an electron is dragged past a proton in a straight-line path. The electron is subject to the electrostatic force and to whatever external force keeps it moving in a straight line at constant speed. (a) Evaluate the work done by the electrostatic force as the electron is taken from x_1 to x_2 along the trajectory of Figure 5.14. (b) What is the work done by the external force required to keep the electron on the straight-line path? (c) Evaluate the work done by the electrostatic force if the electron is taken around a semicircular path with the proton at the center.

Problem 5.3 The magnitude of the force on a particle is $F = -kr$, where r is the distance from the origin to the particle. The force is directed towards the origin at all times. (This is an example of a central force.) Determine the work done when the particle is moved along a semicircular path of radius R from the origin to point $(2R,0)$. Determine the work done when the particle is moved from the origin to the point $(2R,0)$ along a straight-line path. (Note: The origin is not at the center of the semicircular path.)

Problem 5.4 Derive relations (5.8) and (5.10) from the transformation equations and the expression for ds^2 in Cartesian coordinates.

Problem 5.5 (a) Determine the metric for plane polar coordinates. (b) Obtain an expression for $d\mathbf{s}$ in polar coordinates, as well as the element of area.

Problem 5.6 Determine ds^2, the scale factors, the vector $d\mathbf{s}$, the volume element, and the $\hat{\mathbf{e}}$ vectors for the paraboloidal coordinates u, v, ϕ:

$$x = uv \cos \phi,$$
$$y = uv \sin \phi,$$
$$z = \frac{1}{2}(u^2 - v^2).$$

Problem 5.7 The "elliptic cylindrical" coordinates, u, v, z, are defined by

$$x = a \cosh u \cos v,$$
$$y = a \sinh u \sin v,$$
$$z = z.$$

Determine the metric. Write an expression for $d\mathbf{s}$.

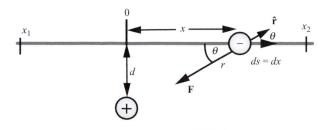

Figure 5.14 An electron is dragged past a proton from x_1 to x_2 along the straight-line path.

Problem 5.8 Consider the u, v, z coordinates defined by

$$x = \frac{1}{2}(u^2 - v^2),$$

$$y = uv,$$

$$z = z.$$

(a) Determine the metric for these coordinates. (b) Evaluate ∇ in these coordinates.

Problem 5.9 The prolate spheroidal coordinates η, θ, ϕ are related to the Cartesian coordinates by the following transformation equations

$$x = a \sinh \eta \sin \theta \cos \phi,$$

$$y = a \sinh \eta \sin \theta \sin \phi,$$

$$z = a \cosh \eta \cos \theta.$$

Determine the expression for ∇ using these coordinates.

Problem 5.10 (a) Using the expression for ∇ given by Equation (5.11), obtain $\nabla \cdot V$ in cylindrical coordinates. Note that the derivatives also act on the unit vectors, as in going from Equation (2.17) to Equation (2.18). (b) Show that the same result is obtained from

$$\nabla \cdot \mathbf{V} = \frac{1}{h_1 h_2 h_3} \left[\frac{\partial}{\partial q_1} (h_2 h_3 V_1) + \frac{\partial}{\partial q_2} (h_1 h_3 V_2) + \frac{\partial}{\partial q_3} (h_1 h_2 V_3) \right].$$

Problem 5.11 Using spherical coordinates, determine the volume of a sphere. Pay particular attention to the limits on the integrals over θ and ϕ.

Problem 5.12 Near the surface of the Earth the potential energy is mgh. Use this fact to obtain an equation for the position of a falling particle as a function of time. Show that the acceleration of the particle is g.

Problem 5.13 Upon finding that a certain mountain is a perfect cone, a surveyor describes it mathematically in cylindrical coordinates by the formula $z = h_0 - \rho$. (a) Sketch the contour lines. (b) Show that the direction of steepest ascent everywhere points toward the summit.

Problem 5.14 A mountain rises above a flat plain. The height of the mountain above the plane is described by the relation

$$z(x, y) = 2000 \exp\left[-\left(x^2 + 2y^2\right)/8000\right] \text{m}.$$

(a)How high is the mountain? (b) If you are standing at $x = 20$ m, $y = 10$ m, how high above the plain are you? (c) At that location, what is the direction of quickest descent? (d) What is the maximum rate of descent at this point? (That is, how many meters do you descend for every meter of horizontal displacement?)

Problem 5.15 The temperature in a certain location is given by $T = T_0 - A(2x^2 + y^2 + z^2)$ (in kelvins). Distances are in meters and the constant A has the value 0.5 K/m^2. The value of T_0 is 300 K. (a) At point (1,2,2) what is the direction and rate (in K/m) of maximum temperature increase? (b) If you move 1 m in that direction what is the temperature at the new position? (c) The temperature at the new position as calculated by the formula is not equal to the value obtained by multiplying the rate of temperature decrease by the displacement. Explain this discrepancy.

Problem 5.16 (a) Determine the center of mass of the system composed of three particles of masses 1 kg, 2 kg, and 3 kg located at (0,1), (1,0), and (1,1) respectively. (b) Assume the 1 kg particle has a velocity of $1\hat{\jmath}$ m/s, the 2 kg particle has a velocity of $2\hat{\imath}$ m/s, and the 3 kg particle has a velocity of $3(\hat{\imath} + \hat{\jmath})$ m/s. Determine the total kinetic energy of the system. (c) What is the velocity of the center of mass? (d) Determine the kinetic energy of the particles relative to the center of mass.

Problem 5.17 Two positive charges (Q_1 and Q_2) are located on the x-axis at positions $x = \pm a$. A negative charge ($-q$) of mass m is constrained to move along the y-axis. The electrostatic potential energy of the system is

$$V = -\frac{qQ_1}{4\pi\epsilon_0 r_1} - \frac{qQ_2}{4\pi\epsilon_0 r_2},$$

where r_1 and r_2 are the distances from the negative charge to the positive charges. The negative charge is released from rest at $y = b$. Obtain an expression for the maximum speed of the negative charge. Where does this occur?

Problem 5.18 A football player of mass M decides to go bungee jumping. He suits up in a halter with a long elastic cord of unstretched length b. After tying the free end of the rope to a high bridge he leaps out into space. A few moments later he finds himself suspended far above the ground. He climbs the bungee cord back up to the bridge. Determine the ratio of the work done climbing the elastic rope to the work done in climbing an inelastic rope of length b. You may assume the elastic rope obeys Hooke's law and has a force constant k.

Problem 5.19 A girl on a swing is pushed harder and harder by her older brother. Eventually, she is swinging in an arc whose highest point is 1.5 m above its lowest point. Assume the swing has massless ropes 2.8 m long and that the mass of the girl (plus swing seat) is 30 kg. What is the tension in the ropes when the girl passes through the lowest point?

Problem 5.20 A block of mass 5 kg is on a horizontal surface. The coefficient of sliding friction is 0.5. The mass has a speed of 10 m/s and is moving towards a stationary spring that is 3 m away. The spring constant is 4000 N/m. The block strikes the spring, compresses it, and bounces back. How far from the spring does the block come to rest?

Problem 5.21 A quarter circle is cut out of a wooden slab as shown in Figure 5.15. The sides of the slab are d and the radius of the circular cut is R. The slab is placed on a table. A marble of radius a is placed at the top of the cutout and rolls without slipping to the bottom where it is launched outwards. Determine the distance (x) from the base of the slab where the marble hits the tabletop.

Problem 5.22 A radioactive nucleus decays by emitting an alpha particle with energy 6.0 MeV. The total energy released in the disintegration is 6.2 MeV. Determine the mass of the recoiling nucleus (ignore relativistic effects).

Problem 5.23 A particle of mass m and total energy E is moving in a one-dimensional potential given by $V(x) = -bx$. Determine the motion.

Problem 5.24 A rail gun is a device that accelerates a projectile to extremely high speeds using magnetic forces. Suppose a particle of mass m is fired vertically from the surface of the Earth. Ignoring air resistance, show that the maximum height reached by the particle is given by

$$\frac{GMm}{|E|} - R,$$

where R is the radius of the Earth and M is its mass. (Assume that its total energy E is negative.)

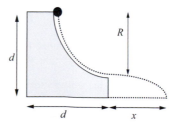

Figure 5.15 Illustration for Problem 5.21.

Problem 5.25 Prove that for a particle the rate of change of kinetic energy $\frac{dT}{dt}$ is equal to the dot product of the force acting on it and the instantaneous velocity of the particle, that is, $\mathbf{F} \cdot \mathbf{v}$.

Problem 5.26 Let $h = r - R$ be the position of a particle above the surface of the Earth where R is the radius of the Earth. Prove that in the limit

$$h \ll R$$

the gravitational potential reduces to mgh.

Problem 5.27 A particle of mass m is in a one-dimensional potential energy field given by

$$V(x) = -Ae^{-ax^2},$$

where A and a are constants. (a) Plot the energy diagram. (b) Determine the turning points if the particle has a total energy $E = -0.5A$. (c) Determine the turning points if the particle has a total energy $E = -Ae^{-1}$. (d) Assume the particle has zero energy and is located at $x = -\infty$. It is given an infinitesimal shove towards the origin. What is its velocity when it passes through the origin?

Problem 5.28 In a certain region of space the potential energy can be expressed as

$$V = -\frac{A}{\sqrt{x^2 + y^2 + z^2}},$$

where A is a constant, and the origin is excluded. (a) Obtain an expression for the force. (b) Determine the work required to take a particle from (x_1, y_1, z_1) to ∞. (c) Express the force in spherical coordinates.

Problem 5.29 The potential energy of a vibrating diatomic molecule as a function of the separation (s) between the two atoms is approximately given by the "Morse function,"

$$V(s) = V_0 \left(1 - e^{-(s-s_0)/\delta}\right)^2 - V_0,$$

where s_0, δ, V_0 are constant parameters. (a) Obtain an expression for the force on the atoms. (b) Determine the separation between the atoms when the potential energy is a minimum. (c) What is the minimum value of the potential energy?

Problem 5.30 Coulomb's law tells us that the electrostatic force between two charged particles is

$$\mathbf{F} = \frac{Q_1 Q_2}{4\pi \varepsilon_0 r^2} \hat{\mathbf{r}},$$

where Q_1 and Q_2 are the charges on the particles and r is the separation between them. The quantity $4\pi \varepsilon_0$ is constant. Show that the electrostatic force is conservative. Obtain an expression for the potential energy.

Problem 5.31 A particle is subjected to a force given by

$$\mathbf{F} = K[(2x + y)\hat{\mathbf{i}} + (x + 2y)\hat{\mathbf{j}}].$$

(a) Prove that this force is conservative. (b) Obtain an expression for the potential energy.

Problem 5.32 Newton's law of universal gravitation states that the force between two particles of masses m_1 and m_2 is

$$\mathbf{F} = -\frac{Gm_1 m_2}{r^2} \hat{\mathbf{r}},$$

where G is a constant and r is the distance between the particles. Use this force expression to obtain the gravitational potential energy of the system. Select an appropriate point for the zero of potential energy.

Problem 5.33 By evaluating the work required to bring three masses m_1, m_2, and m_3 from infinity to their final positions at $\mathbf{r}_1, \mathbf{r}_2$, and $\mathbf{r}_3$ determine the potential energy of the system. (The only force acting on the particles is their mutual gravitational attraction.)

Problem 5.34 A particle of mass 2 kg moves along the x-axis. Its potential energy as a function of position is $V(x) = -3x + x^2$ J. (Here x is measured in meters.) The particle passes through the origin with a speed of 4 m/s. (a) Sketch the potential energy as a function of x. (b) Determine the speed of the particle when it is at $x = 1$ m. (c) Where are the turning points?

Problem 5.35 Consider the potential

$$V = -\frac{a}{r} + \frac{b}{r^2}.$$

(a) Plot the potential as a function of r. (b) Obtain an expression for the force. (c) Obtain an expression for the turning points as a function of a, b, and the total energy E.

Problem 5.36 A particle moving in a potential given by $V(x, y) = ax^2 + by^2$ is a two-dimensional oscillator with different force constants in the two directions. (This is called a nonisotropic oscillator.) Determine the motion, assuming that at time $t = 0$ the particle is passing through the origin with velocity $\mathbf{v} = v_{0x}\hat{\mathbf{i}} + v_{0y}\hat{\mathbf{j}}$.

Computational Projects

Computational Project 5.1 Assume that the potential energy is given by $V(x) = -(1/x^6) + (1/x^{12})$. Integrate Equation (5.15) numerically to determine the position of a particle of unit mass and total energy 10^{-4} energy units. Plot $x = x(t)$ and $v = v(t)$ for a particle coming in from infinity. (Infinity is approximately at $x = 5$.)

Computational Project 5.2 An asteroid of mass 5000 kg is at 40 AU from the Sun (beyond the orbit of Pluto) when it undergoes a collision with another asteroid, leaving it with zero velocity. The asteroid then slowly (at first) falls toward the Sun. Write a program to evaluate the velocity of the asteroid as it reaches the surface of the Sun. Use the Euler–Cromer algorithm. Check your result by comparing with the value obtained analytically. The purpose of this project is for you use variable time steps. Assume the radius of the Sun is 5.6×10^7 m. (1 AU $= 1.5 \times 10^{11}$ m).

6 Linear Momentum

In this chapter we turn our attention to problems that can be solved using the law of conservation of linear momentum. Examples of such problems include the motion of a rocket, collisions in one and two dimensions, and the behavior of a system when an impulsive force acts on it.

6.1 The Law of Conservation of Momentum

The law of conservation of linear momentum can be obtained from Newton's second law: The rate of change of momentum of a system is equal to the net external force acting on the system:

$$\frac{d\mathbf{P}}{dt} = \mathbf{F},$$

where $\mathbf{F}$ is the net external force and $\mathbf{P}$ is the total linear momentum.[1] Therefore, if $\mathbf{F} = 0$, then $\mathbf{P} = \sum_i m_i \mathbf{v_i} = $ constant. In words, the law of conservation of linear momentum is:

If no net external force acts on a system, the total momentum of the system is constant.

For example, in the collision of two bodies, if external forces are negligible, the conservation of momentum principle can be expressed as

$$\mathbf{P}_{\text{final}} = \mathbf{P}_{\text{initial}},$$

where $\mathbf{P}_{\text{initial}}$ is the total momentum of the two bodies before the collision and $\mathbf{P}_{\text{final}}$ is the total momentum after the collision. Many problems can be solved using this simple relation.

If a net external force *does* act on a particle, and if the mass of a body changes with time, then Newton's second law must be expressed in the form

$$\mathbf{F} = \frac{d\mathbf{P}}{dt} = m\frac{d\mathbf{v}}{dt} + \mathbf{v}\frac{dm}{dt}. \tag{6.1}$$

Note that $d\mathbf{v}/dt$ is the acceleration of the body; in this case, one cannot write Newton's second law in the form $\mathbf{F} = m\mathbf{a}$.

Exercise 6.1

A bullet of mass m and speed v is fired at a wooden block of mass M at rest on a frictionless surface. The bullet embeds itself in the block. (a) Determine the speed of bullet plus block after the impact. (b) Is kinetic energy conserved? Explain. (c) Evaluate the loss of kinetic energy. Answers: (a) $mv/(M + m)$, (c) $\Delta T = -\frac{1}{2}\frac{mM}{m+M}v^2$. ■

---

[1] In Chapter 8 you will be exposed to a more general derivation of the law of conservation of linear momentum.

6.2 The Motion of a Rocket

As an application of the law of conservation of linear momentum to a system of variable mass, let us consider the motion of a rocket. Imagine the rocket to be in empty space, far from any stars or other material bodies that might exert an external force on it. For simplicity let us assume the motion is one dimensional. The rocket has (initial) mass M_R and moves with an initial speed V in an inertial coordinate system.

As illustrated in Figure 6.1, the pilot fires the rocket engine for a short time, say dt. The engine burns an amount of fuel that we will denote dM_F. This burned fuel is expelled from the rocket exhaust tubes at a speed U *with respect to the rocket*. In inertial space, the burned fuel has speed $V - U$ (because the fuel is expelled backward to propel the rocket forward).

In this problem, the mass of the burned fuel and the change in mass of the rocket are related to each other through $dM_F = -dM_R$.

The momentum of the rocket before the fuel burn is

$$P_i = M_R V.$$

After the burn, the total momentum (rocket plus fuel) is

$$P_f = (M_R - dM_F)(V + dV) + dM_F(V - U).$$

Since there are no external forces acting on the system, the initial momentum is equal to the final momentum, so

$$P_i = P_f$$
$$M_R V = (M_R - dM_F)(V + dV) + dM_F(V - U)$$
$$M_R V = M_R V + M_R dV - V dM_F - dM_F dV + V dM_F - U dM_F.$$

The term $dM dV$ is a second-order differential, that is, the product of two infinitesimal quantities, so it is an extremely small quantity. Discarding this second-order differential and cancelling several terms we obtain

$$0 = M_R dV - U dM_F,$$

or

$$M_R dV = U dM_F. \tag{6.2}$$

If we divide both sides by dt we obtain

$$M_R \frac{dV}{dt} = U \frac{dM_F}{dt}.$$

Here dM_F/dt is the rate of fuel burn. The left-hand side of this equation *looks like* the ma in Newton's second law (but M_R is not constant), so the right-hand side *looks like* a force. It is

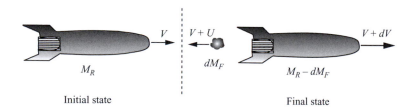

Figure 6.1 A rocket before and after it fires an amount of fuel dM_F.

called the "thrust" and is indeed the force that a rocket engine exerts on a test stand during a stationary test. Note that the thrust depends on the rate at which fuel is burned and the speed with which it is ejected through the exhaust tubes. To generate a large thrust, a rocket motor is designed to burn fuel very rapidly (large dM_F/dt) and to expel it at as high a speed as possible (large U). The whole purpose of *burning* the fuel is to convert it to a gas, thus greatly increasing its volume and causing it to be expelled through the rocket tubes at the highest possible velocity.

Going back to $M_R dV = U dM_F$ we can obtain a very useful relation by using the fact that $dM_F = -dM_R$, leading to

$$dV = -U \frac{dM_R}{M_R}.$$

Integrating this equation yields the change in speed of the rocket as a function of the change in mass of the rocket (equal to the mass of fuel burned). Note that we could not integrate the expression $\frac{dM_F}{M_R}$ because the numerator and denominator refer to different quantities. Thus, we had to replace dM_F with $-dM_R$. Ignoring this simple fact leads to many errors in solving rocket problems.

Similarly, the equation following Equation (6.2) can be expressed as

$$M_R \frac{dV}{dt} = -U \frac{dM_R}{dt}.$$

Worked Example 6.1   A rocket of total mass m_0 in interplanetary space is coasting at speed V_0. It approaches an interesting asteroid and must slow down to the speed of the asteroid which happens to be $V_0/2$. Determine the amount of fuel burned to achieve this reduction in speed.

Solution   Since the rocket is slowing down, the speed of the burned fuel is $V + U$ rather than $V - U$. Inserting $V + U$ into the relation for $P_i = P_f$ we obtain

$$dV = U \frac{dM}{M}.$$

Integrating

$$\frac{1}{U} \int_{V_0}^{V_0/2} dV = \int_{m_0}^{m_f} \frac{dM}{M},$$

$$\frac{1}{U}(V_0/2 - V_0) = \ln \frac{m_f}{m_0},$$

$$-\frac{V_0}{2U} = \ln \frac{m_f}{m_0},$$

$$m_f = m_0 e^{-V_0/2U}.$$

The amount of fuel that must be burned is

$$m_0 - m_f = m_0(1 - e^{-V_0/2U}).$$

Worked Example 6.2   The rocket equation was obtained by considering the difference in momentum between the initial and final states, assuming a small amount of fuel was burned. Obtain the same result by showing that the total momentum is constant. Assume the rate of fuel burn is constant and equal to $\dot{m}$.

Solution At any given moment the total momentum of the system is the momentum of the rocket (including unburned fuel) and the momentum of the expelled (burned) fuel. Let the mass of the rocket at any instant be M_R and the mass of burned fuel be m_f. The ejection speed of the fuel in inertial space is $V_R - U$, where U is a positive quantity. The rate of change of mass of the rocket is $dM_R/dt = -\dot{m}$ and the rate of change of the mass of burned fuel is $dm_f/dt = +\dot{m}$. We assume $\dot{m}$ is constant. To determine the momentum of the burned fuel at time τ we need to note that while it is being burned, the speed of the rocket is changing. Therefore, the momentum of the burned fuel at time τ is

$$P_f(\tau) = \int_0^\tau \dot{m}(V_R(t) - U)dt = \dot{m}\int_0^\tau (V_R - U)dt.$$

The momentum of the rocket at time τ is

$$P_R(\tau) = M_R(\tau)V_R(\tau).$$

The total momentum is $P_{\text{tot}} = P_R + P_f$ so the rate of change of momentum is

$$\frac{dP_{\text{tot}}}{dt} = \frac{d}{dt}\left[M_R(\tau)V_R(\tau) + \dot{m}\int_0^\tau (V_R - U)dt\right]$$

$$= \frac{d}{dt}(M_R V_R) + \dot{m}(V_R - U)$$

$$= \frac{dM_R}{dt}V_R + M_R\frac{dV_R}{dt} + \dot{m}V_R - \dot{m}U = 0.$$

But $\dot{m} = -\frac{dM_R}{dt}$, so

$$M_R\frac{dV_R}{dt} = -U\frac{dM_R}{dt}.$$

This is the same expression we found in our simpler derivation.

Exercise 6.2

A rocket of initial mass 500 kg burns fuel at a rate of 5 kg/s. The exhaust speed of the gases is 300 m/s. What is the initial acceleration of the rocket? (Ignore gravity.) What is its acceleration after one minute? Answers: 3 m/s^2, 7.5 m/s^2. ∎

Exercise 6.3

Joe and Bill are railroad men who are riding side by side on flatcars rolling on parallel tracks. The tracks are straight and perfectly horizontal. There are absolutely no frictional forces or air resistance. It starts to snow. Joe sweeps the snow off his flatcar as soon as it lands, sweeping it off the side, perpendicular to the direction of motion of the flatcar. Bill, the lazy one, simply lets the snow accumulate on his flatcar. Who travels further in the same interval of time? Answer this question conceptually and also mathematically. (The snow falls perfectly vertically.) ∎

Exercise 6.4

Beginning with Newton's second law in the form $\mathbf{F} = d\mathbf{P}/dt$ and the definition of derivative, generalize Equation (6.2) to include the effect of an external force. Answer: $M\frac{d\mathbf{V}}{dt} + \mathbf{u}\frac{dM}{dt} = \mathbf{F}$. ∎

Exercise 6.5

Consider a rocket that is rising into the air. The burning exhaust gases stream out through the nozzles and push down on the air below. But what pushes *upward* on the rocket? (An equivalent question: When you inflate a balloon and then release it, it flies wildly about the room. What is pushing the balloon?) ∎

6.3 Collisions

Another application of the principle of conservation of linear momentum is the analysis of collisions. Generating and studying collisions between elementary particles are a principal way physicists explore the underlying properties of nature. A collision between two bodies often involves a strong, short-range interactive force between the bodies. When two extended bodies (such as automobiles or billiard balls) come in contact, we can idealize the collision by assuming no forces act except during the instant of contact. Taking the two bodies as the entire system, these forces are *internal* forces. During an ideal collision, *no external forces* are acting. Consequently, the total momentum of the system is constant.

In a contact collision, there is a very short-range repulsive force that acts while the surfaces of the two bodies are touching.[2] A *glancing* collision, such as illustrated in Figure 6.2, occurs when the velocity vectors of the two bodies are not aligned along the *line of centers*. The distance *b* between the initial velocity vectors is called the *impact parameter*.

A collision may not involve the actual physical contact of the two bodies. The force exerted by one body on the other may be a *long-range force*, such as the force of electric repulsion between an alpha particle and the nucleus as in Rutherford's experiment (considered below), or the force of gravity when a comet in a hyperbolic orbit approaches from infinity, swings about the Sun, and travels back out to infinity. The same physics applies regardless of the range of the forces.[3]

As a basic collision problem, consider the situation illustrated in Figure 6.3, where body M_1 with velocity $\mathbf{V}_1$ makes a glancing collision with body M_2 that is initially at rest. (We can always find an inertial coordinate system in which one body is initially at rest.) The two bodies move off with velocities $\mathbf{V}'_1$ and $\mathbf{V}'_2$ at angles θ_1 and θ_2 relative to $\mathbf{V}_1$, as shown in the figure. To keep things simple, assume the bodies are not rotating.

There are no external forces acting on the system so the law of momentum conservation states that

$$\mathbf{P}_i = \mathbf{P}_f. \tag{6.3}$$

That is, the initial momentum and the final momentum are equal. This is a vector equation so it is equivalent to the three scalar equations

$$P_{xi} = P_{xf}, \quad P_{yi} = P_{yf}, \quad P_{zi} = P_{zf}.$$

(If two vectors are equal, their components must be equal.) It is convenient to place the origin of coordinates at the original position of body M_2 and to let the x-axis be defined by the direction

---

[2] The origin of these forces is the repulsion between the electrons in one body and the electrons in the other. When the two "electron clouds" start to overlap, there is a repulsive Coulomb force between them. Fortunately, we do not need detailed information about the force between the bodies because we can solve the problem using the law of conservation of momentum in which internal forces play no part.

[3] We often refer to colliding bodies as particles, even though they may be astronomical objects. It is appropriate to use the term particle when we are dealing with long-range forces. In a glancing collision, the surfaces of two extended bodies come in contact and they should not be called particles. Nevertheless, physicists are a bit careless in the usage of this term, and in dealing with collisions the word particle is often used when, strictly speaking, it should not be.

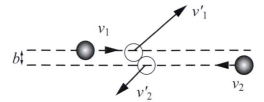

Figure 6.2 A glancing collision.

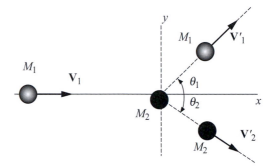

Figure 6.3 Illustration of the parameters involved when two bodies undergo a glancing collision.

of $\mathbf{V}_1$. Select the z-axis perpendicular to the plane of the motion, i.e., perpendicular to the plane containing $\mathbf{V}'_1$ and $\mathbf{V}'_2$. Then $P_{zf} = 0$. By momentum conservation, $P_{zi} = 0$. Therefore the problem is *two dimensional;* the motion takes place entirely in the xy-plane.

The momentum conservation equations in the xy-plane can be written

$$P_{xi} = P_{xf}, \quad \text{or} \quad M_1 V_{1x} = M_1 V'_{1x} + M_2 V'_{2x},$$
$$P_{yi} = P_{yf}, \quad \text{or} \quad 0 = M_1 V'_{1y} - M_2 V'_{2y}.$$

In terms of the angles θ_1 and θ_2 these two equations are

$$M_1 V_1 = M_1 V'_1 \cos\theta_1 + M_2 V'_2 \cos\theta_2, \tag{6.4}$$
$$0 = M_1 V'_1 \sin\theta_1 - M_2 V'_2 \sin\theta_2. \tag{6.5}$$

Note the minus sign on the last term.

Conservation of momentum led to two equations. These equations involve the seven parameters $M_1, M_2, V_1, V'_1, V'_2, \theta_1, \theta_2$. Five of these parameters must be "known" quantities so that you can solve for two unknowns. In most collision problems you will know the speed of the incoming particle, V_1. You will also probably know the masses of the particles, or at least their ratio (which is sufficient). This leaves you with four unknowns, namely the final speeds, V'_1 and V'_2, and the final directions, θ_1 and θ_2. Two of these must be determined (perhaps experimentally) before you can solve the problem for the other two.

If, however, the collision is *elastic,* then there is one more equation you can use, namely the equation expressing the conservation of kinetic energy. By definition, an *elastic collision* is one in which the kinetic energy is conserved.[4] That is, $T_i = T_f$, or in terms of our problem,

$$\frac{1}{2} M_1 V_1^2 = \frac{1}{2} M_1 V_1'^2 + \frac{1}{2} M_2 V_2'^2. \tag{6.6}$$

---

[4] A collision is usually elastic if the two bodies do not come in contact, or if neither body is deformed by the collision. Thus the interactions between celestial bodies and between elementary particles are often elastic. Collisions of billiard balls are frequently assumed to be elastic because the deformation is negligible.

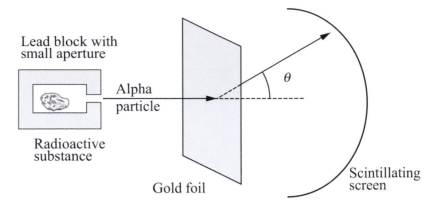

Figure 6.4 Rutherford's experiment. An alpha particle is scattered through the angle θ by a gold nucleus. This collision experiment was crucial in determining the structure of atoms.

This condition gives you a third equation, allowing you to solve for *three* unknowns. Frequently, the unknown quantities will be the final speeds and the direction of one particle, for example, V_1', V_2', and θ_2. The quantities you will need to know are (usually) the speed of the incoming particle and its deflection angle, as well as the masses.

To help you visualize a typical problem, Figure 6.4 illustrates Rutherford's famous experiment in which he bombarded the nuclei of gold atoms with alpha particles emitted by a radioactive substance. The gold nuclei were essentially at rest and the alpha particles approached with a known velocity. After interacting with a gold nucleus, the alpha particles hit a screen painted with a scintillating material. This caused a tiny flash of light to appear at the point where an alpha particle hit the screen. The pinpoints of light were observed by eye by Rutherford's graduate students who tabulated the final positions of all the alpha particles. In this problem the known quantities were M_1, M_2, V_1, and θ_1. The unknown quantities were V_1', V_2', and θ_2. (You will solve this problem after studying central force problems; see Problem 10.24.)

Going back to the general case, I will now show you how to manipulate Equations (6.4), (6.5), and (6.6) for momentum conservation and kinetic energy conservation to solve for the unknown quantities in terms of the known quantities. Let me warn you that the algebra is a bit tedious so you might want to get up and get a cup of coffee before we start. Also, I am not going to give you all the intermediate steps, so while you are up you had better get your pencil because you will not understand the final result unless you can work out all the steps.

To begin, rewrite Equations (6.4) and (6.5), placing all the terms with subscript 1 on one side. Thus:

$$M_1 V_1 - M_1 V_1' \cos \theta_1 = M_2 V_2' \cos \theta_2, \tag{6.7}$$
$$M_1 V_1' \sin \theta_1 = M_2 V_2' \sin \theta_2. \tag{6.8}$$

Next square both equations and add them together to eliminate θ_2. You obtain

$$M_1^2 V_1^2 - 2M_1^2 V_1 V_1' \cos \theta_1 + M_1^2 V_1'^2 = M_2^2 V_2'^2. \tag{6.9}$$

This equation, together with the kinetic energy equation (6.6), give you two equations in the two unknowns V_1' and V_2'. (The initial velocity V_1 and the angle θ_1 are assumed known.)

Now eliminate V_2' from Equations (6.6) and (6.9). If you take the expression for $M_2^2 V_2'^2$ given by Equation (6.9) and plug it into Equation (6.6), you obtain the following equation for V_1':

$$[(M_1 + M_2)] V_1'^2 - [2M_1 V_1 \cos \theta_1] V_1' + \left[(M_1 - M_2) V_1^2\right] = 0. \tag{6.10}$$

All the terms in square brackets are known, so this is a quadratic equation for V_1' whose solution is

$$V_1' = \frac{2M_1 V_1 \cos\theta_1 \pm \sqrt{(2M_1 V_1 \cos\theta_1)^2 - 4(M_1 + M_2)(M_1 - M_2) V_1^2}}{2(M_1 + M_2)},$$

or

$$V_1' = \frac{M_1}{M_1 + M_2} V_1 \cos\theta_1 \pm \sqrt{\left(\frac{M_1 V_1}{M_1 + M_2}\right)^2 \cos^2\theta_1 - \frac{M_1 - M_2}{M_1 + M_2} V_1^2}.$$

Dividing through by V_1 you obtain a nicer-looking expression:

$$\frac{V_1'}{V_1} = \frac{M_1}{M_1 + M_2} \left[\cos\theta_1 \pm \sqrt{\cos^2\theta_1 - 1 + \left(\frac{M_2}{M_1}\right)^2} \right]. \tag{6.11}$$

Let us interpret this result by discussing several special cases. Use your experience with colliding bodies to visualize each situation.[5]

Case 1: Head-on Collisions

If two objects hit "head-on" then the motion is one-dimensional and $\theta_1 = \theta_2 = 0$. The velocity vector of the incoming particle, V_1, is pointed directly at the center of body M_2. Assume M_1 is moving from left to right. Recall from experience that if the masses are equal, M_1 stops and M_2 moves off to the right. If $M_1 > M_2$ both masses move off to the right with M_2 moving faster than M_1. If $M_1 < M_2$, then M_1 bounces back (to the left) and M_2 moves off to the right. All of this information is incorporated in Equation (6.11). We now consider how to "read" Equation (6.11) to extract this information.

Recall that we are imagining a head-on collision so the angle θ_1 is zero. Setting $\cos\theta_1 = 1$ in Equation (6.11), and doing a little bit of algebra, leads to

$$V_1' = V_1 \frac{M_1 \pm M_2}{M_1 + M_2}. \tag{6.12}$$

Subcase 1.1: $M_1 = M_2$ Consider first the situation in which $M_1 = M_2$. Equation (6.12) then reduces to

$$\frac{V_1'}{V_1} = \frac{1}{2}[1 \pm 1] = \left\{ \begin{matrix} 1 \\ 0 \end{matrix} \right\}.$$

That is, V_1' is either equal to V_1 or it is equal to zero. From experience we expect V_1' to be zero (the incoming object stops). What does $V_1' = V_1$ correspond to? It corresponds to a *miss*, that is, no collision at all. It tells us that the final velocity of M_1 is equal to its initial velocity. Furthermore, Equations (6.6) and (6.9) both tell us that for a miss, $V_2' = 0$. The condition $\theta_1 = \theta_2 = 0$ is equally satisfied by a head-on collision or a miss. (Note how much physical information is coded into the equations!)

We are not interested in the situation where one particle misses the other, so let us set $V_1' = 0$. What is V_2', the final velocity of the particle that was initially at rest? Again, from experience we expect it to be equal to V_1, the initial velocity of the incoming object. And, indeed, Equation (6.7) for $M_1 = M_2$ and $\theta_1 = 0$ yields

---

[5] You can make a rough check of the results we obtain by carrying out experiments on a smooth surface with coins of the same or different masses.

$$V_1 - V_1' = V_2'$$

But $V_1' = 0$, so,

$$V_2' = V_1.$$

This equality corresponds to our experience that the struck object flies off with the velocity the incoming particle had before the collision.

Subcase 1.2 $M_1 \neq M_2$ Now allow the two colliding bodies to have different masses, but still assume a head-on collision so $\theta_1 = \theta_2 = 0$. Again, Equation (6.11) reduces to Equation (6.12). Selecting the plus sign gives $V_1' = V_1$, implying no collision at all. This case is of no interest, so select the minus sign. That is,

$$V_1' = V_1 \frac{M_1 - M_2}{M_1 + M_2}. \tag{6.13}$$

Therefore, if $M_1 > M_2$, the velocity V_1' is positive. Plugging Equation (6.13) into (6.7) and keeping in mind that $\theta_1 = \theta_2 = 0$, you get the velocity acquired by M_2 :

$$V_2' = 2V_1 \frac{M_1}{M_1 + M_2}. \tag{6.14}$$

Equations (6.13) and (6.14) agree with your experience and intuition. If a heavy (massive) object collides with a light (less massive) object, the heavy object will keep on moving in its original direction. If a billiard ball strikes a ping-pong ball, $M_1 \gg M_2$, and

$$V_1' \doteq V_1 \frac{M_1}{M_1} = V_1.$$

That is, the billiard ball keeps on moving at essentially its original velocity. What happens to the ping-pong ball? According to Equation (6.14), if $M_1 \gg M_2$, then $V_2' = 2V_1$. That is, the light object moves off at twice the speed of the incoming heavy object.

It is easy to appreciate that if $M_1 < M_2$, the situation is simply reversed. If a light object hits a heavy object then $M_1 < M_2$ and V_1' is negative. That is, the light object bounces back. For example, if you throw a ball against a wall, then M_1 is the mass of the ball and M_2 is the mass of the Earth (assuming the wall is attached to the Earth). So $M_1 \ll M_2$ and according to Equation (6.13), V_1', the final velocity of the ball, will be $V_1' = -V_1$. The ball bounces back with its initial speed. Similarly, $V_2' = 0$. (The Earth stands still.)

Worked Example 6.3 A spacecraft flyby of a planet can be used to give the spacecraft a "boost" and increase its speed by the process known as the "slingshot effect." Let us consider an unrealistic situation in which a spacecraft approaches Mars, swings around the planet, and heads off in the opposite direction. This would be (essentially) a head-on collision. As I said, this is unrealistic because a 180° deflection cannot be achieved. In a realistic problem the spacecraft approaches the planet at some angle to the planet's velocity vector and is "scattered" at another angle. However, the unrealistic example gives us a better appreciation of the physical principles involved.

In this problem we assume the spacecraft has a mass of 2000 kg and that Mars has a mass of 6.4×10^{23} kg. The initial speed of the spacecraft is -12 km/s and the initial speed of Mars is $+23.36$ km/s. (The speeds are relative to a coordinate system at rest with respect to the Sun.) (a) Determine the final speed of the spacecraft. (b) Evaluate the ratio of the final to initial kinetic energy of the spacecraft.

Solution As mentioned, this is (essentially) a two-body head-on collision. To analyze it we transform to a reference frame in which one particle (Mars, in this case) is at rest by subtracting 23.36 km/s from both velocities. Then the initial velocities are:

$$V_{\text{mars}} = V_2 = 0,$$
$$V_{\text{sc}} = V_1 = -23.36 - 12 = -35.36 \text{ km/s}.$$

Equations (6.13) and (6.14) give the final velocities.

$$V_1' = V_1 \frac{M_1 - M_2}{M_1 + M_2} = -35.36 \left(\frac{2000 - 6.4 \times 10^{23}}{2000 + 6.4 \times 10^{23}} \right) \doteq +35.36 \text{ km/s}.$$

$$V_2' = 2V_1 \frac{M_1}{M_1 + M_2} = (2)(35.36) \left(\frac{2000}{2000 + 6.4 \times 10^{23}} \right) \doteq 0.$$

These are the speeds in a reference frame in which Mars is at rest. Transforming back to the Sun-based reference frame by adding 23.36 km/s to both velocities, we find the speed of the spacecraft is

$$v_{\text{sc}}' = V_1' + 23.36 = 35.36 + 23.36 = +58.72 \text{ km/s}.$$

The ratio of final to initial kinetic energy of the spacecraft is

$$\frac{T_f}{T_i} = \frac{\frac{1}{2}m_s v_s^2}{\frac{1}{2}m_s v_s'^2} = \frac{(58.72)^2}{(12)^2} = 23.9.$$

Exercise 6.6

Fill in the missing steps to obtain Equation (6.12) from Equation (6.11).   ∎

Exercise 6.7

(a) A ball of mass m traveling at speed v hits a ball of mass $3m$ at rest. Determine the final velocities of the two balls. (b) A ball of mass $3m$ traveling at speed v hits a ball of mass m at rest. Determine the final velocities of the two balls. Assume head-on elastic collisions. Answers: (a) $-v/2$ and $v/2$, (b) $v/2$ and $3v/2$.   ∎

Case 2: Glancing Collisions

Let us now analyze a *glancing* collision (or "oblique collision"). The final directions of the two objects are given by the nonzero angles θ_1 and θ_2, shown in Figure 6.3. As before, there are three possibilities, $M_1 = M_2$, $M_1 > M_2$, and $M_1 < M_2$. Once again the discussion is based on Equation (6.11). Note that $\cos^2 \theta$ ranges from 1 to 0 so the term under the radical varies from a minimum of $(M_2/M_1)^2 - 1$ to a maximum of $(M_2/M_1)^2$.

Subcase 2.1: $M_1 = M_2$ Let us begin by assuming $M_1 = M_2$. Equation (6.11) then reduces to

$$\frac{V_1'}{V_1} = \frac{1}{2} \left[\cos \theta_1 \pm \sqrt{\cos^2 \theta_1} \right] = \begin{cases} 0 \\ \cos \theta_1 \end{cases}. \tag{6.15}$$

Thus, the final velocity of M_1 is either zero or $V_1 \cos \theta_1$. The solution $V_1' = 0$ can be discarded because it implies a head-on collision, as considered previously. Recall that V_1 and θ_1 are assumed known, so the lower solution of Equation (6.15) yields the first of our unknown quantities, the velocity of M_1 after the collision:

$$V_1' = V_1 \cos \theta_1.$$

Plugging this into Equation (6.6) yields

$$V_1^2 = (V_1 \cos \theta_1)^2 + V_2'^2,$$

or

$$(V_2')^2 = V_1^2(1 - \cos^2 \theta_1) = V_1^2 \sin^2 \theta_1, \qquad (6.16)$$

so

$$V_2' = V_1 \sin \theta_1.$$

This gives us the second unknown quantity, V_2'. The third "unknown" is θ_2 and it can be obtained immediately from the conservation of momentum along the y-axis, Equation (6.5), as

$$\theta_2 = \sin^{-1} \frac{V_1'}{V_1}.$$

Worked Example 6.4   Prove that in any glancing elastic collision between bodies of equal mass, the sum of the deflection angles is $\theta_1 + \theta_2 = \pi/2$.

Solution   Given $V_1' = V_1 \cos \theta_1$ and $V_2' = V_1 \sin \theta_1$. Plug into the conservation of momentum equation $V_1' \sin \theta_1 = V_2' \sin \theta_2$ to get

$$V_1 \cos \theta_1 \sin \theta_1 = V_1 \sin \theta_1 \sin \theta_2$$

$$\cos \theta_1 = \sin \theta_2.$$

But $\cos \theta_1 = \sin \left(\frac{\pi}{2} - \theta_1 \right)$, so $\theta_2 = \frac{\pi}{2} - \theta_1$, or

$$\theta_1 + \theta_2 = \frac{\pi}{2}.$$

Exercise 6.8

Show that the null solution for Equation (6.15) implies a head-on collision.   ∎

Exercise 6.9

Equation (6.16) yields two solutions: $V_2' = \pm V_1 \sin \theta_1$. Using the facts that if $M_1 = M_2$ then $\theta_1 + \theta_2 = \pi/2$ and $V_1' = V_1 \cos \theta_1$, show that only the positive solution is obtained. (The negative solution corresponds to $\theta_1 = 0$ and $V_1' = V_1$, that is, a miss.)   ∎

Subcase 2.2: $M_1 > M_2$ If the mass of body 1 is greater than the mass of body 2, you expect the heavier object (M_1) to continue moving in the forward direction. That is, θ_1 will be less than $\pi/2$. To appreciate that this is true, note that the quantity under the radical in Equation (6.11) must be positive. (Otherwise the velocity V_1' would be a complex number and this is not possible because physically measurable quantities must be real.) The quantity under the radical is nonnegative for $M_1 > M_2$ only if

$$\cos^2 \theta_1 \geq 1 - \frac{M_2^2}{M_1^2}.$$

This means that the angle θ_1 can range from zero to some maximum value θ_{max} given by

$$\theta_{max} = \cos^{-1} \left(1 - \frac{M_2^2}{M_1^2} \right)^{\frac{1}{2}}.$$

In the limit $M_2/M_1 \rightarrow 1$ you obtain $\theta_{max} = \cos^{-1}(0) = \pi/2$. In the limit $M_2/M_1 \rightarrow 0$ you obtain $\theta_{max} = \cos^{-1}(1) = 0$. Therefore,

$$0 < \theta_1 < \frac{\pi}{2}.$$

This tells you that if the incoming object is the more massive body, it will be scattered through an angle θ_1 smaller than $\pi/2$.

Subcase 2.3: $M_1 < M_2$ The case of a glancing collision in which $M_1 < M_2$ can also be analyzed with Equation (6.11). It is left as an exercise to show that if the target M_2 is much more massive than M_1, then the incoming body will bounce off with its original speed (but with a change in direction). Note that this case reduces to Subcase 2.2 if you interchange the two bodies.

In conclusion, you have seen that an elastic collision between two bodies can be analyzed using the conservation of linear momentum and the conservation of kinetic energy. The equations obtained are amazingly complex for such a simple problem. An important benefit you should get from the analysis of collisions is an appreciation for how to extract physical meaning from mathematical relations. Be aware that in a collision the conservation of momentum always holds, but conservation of kinetic energy may not.

Exercise 6.10

Show that if $M_1 \ll M_2$ the speed of M_1 after the collision is (essentially) the same as its speed before the collision.   ■

Exercise 6.11

A billiard ball with velocity v strikes a second billiard ball at rest with a glancing collision. The first ball is observed to emerge from the collision at an angle of $20°$. Determine the speed and direction of the second ball.   ■

6.4 Inelastic Collisions: The Coefficient of Restitution

In the previous section the collision between the two bodies was *elastic:* There was no loss of kinetic energy. Of course, this is not always the case. It is customary to represent the gain or loss of kinetic energy by a quantity called the "Q value," defined by

$$Q = T_f - T_i,$$

where T_f is the total final kinetic energy and T_i is the total initial kinetic energy.

For an elastic collision, $Q = 0$. This condition is generally not met; most collisions are either *endoergic* in which kinetic energy is lost, or *exoergic* in which kinetic energy is gained. For example, in the collision of two putty balls having equal but opposite momenta, all of the kinetic energy is lost. This collision is completely inelastic and Q is negative. On the other hand, a collision between two molecules may involve an exothermic chemical reaction in which chemical energy is transformed into mechanical energy. For such a collision, Q is positive.

A closely related concept is the *coefficient of restitution,* denoted e. This was originally described by Isaac Newton who observed that for any head-on collision of two nonrotating bodies the ratio of relative final velocities to relative initial velocities is a constant. That is, if V_1 and V_2 are the initial velocities and V_1' and V_2' are the final velocities, then *Newton's rule* is

$$e = \frac{|V_2' - V_1'|}{|V_2 - V_1|}.$$

For an elastic collision, $e = 1$. This is easily demonstrated from Equations (6.13) and (6.14). For a completely inelastic collision in which all energy is lost, $e = 0$.

If the collision is a glancing collision, then the velocities to be used in Newton's formula are the velocity components along the line joining the two bodies.[6]

Exercise 6.12

Prove that $e = 0$ for a completely inelastic collision and $e = 1$ for an elastic collision. You may assume a head-on collision. ∎

Exercise 6.13

Using Equations (6.13) and (6.14) show that $Q = 0$ for a head-on elastic collision. (Of course, this is true from the definition of elastic collision; the purpose of the exercise is to give you experience in manipulating the relations.) ∎

6.5 Impulse

When a bat hits a baseball, a large force acts for a short period of time. Such a blow gives rise to an *impulse.* By definition, an impulse is the time integral of a force. Denoting the impulse by $\mathbf{J}$, we can write

$$\mathbf{J} = \int_0^\tau \mathbf{F} dt,$$

where τ is the time during which the force acts. In general, it would be difficult to evaluate the integral on the right because the force is usually an unknown, and probably a complicated, function of time. Nevertheless, it is easy to evaluate $\mathbf{J}$ because from Newton's second law, $\mathbf{F} = \frac{d\mathbf{p}}{dt}$, so

$$\mathbf{J} = \int_0^\tau \mathbf{F} dt = \int_0^\tau \frac{d\mathbf{p}}{dt} dt = \int_{p_i}^{p_f} d\mathbf{p} = \mathbf{p}_f - \mathbf{p}_i = \Delta\mathbf{p}.$$

That is, the impulse is simply equal to the change in momentum.

Exercise 6.14

A force $F = 3 \sin 5t$ N acts on a particle of mass 2 kg that was initially at rest. The force acts during the time interval from $t = 0$ to $t = \pi/10$ s. What is the final velocity of the particle? Answer: 0.3 m/s. ∎

---

[6] The coefficient of restitution actually depends on various other factors, such as the medium in which the collision occurs, but Newton's formula is a good approximation.

Exercise 6.15

A stationary block sitting on a frictionless surface is acted upon by a force (in newtons) given by

$$F = 2t \quad \text{for } 0 \leq t \leq 2,$$
$$F = 4 \quad \text{for } 2 \leq t \leq 5,$$
$$F = -t \quad \text{for } 5 \leq t \leq 7.$$

(Times in seconds.) Determine the final momentum of the block. Answer: 4 kg m/s. ∎

6.6 Momentum of a System of Particles

In this book we have claimed more than once that internal forces do not affect the total linear momentum of a mechanical system. For example, if a bomb explodes, pieces fly off in all directions, but the total momentum is unchanged. We now prove the claim by considering a system of N particles and determining the effect of internal and external forces on the momentum. The particles have masses $m_1, m_2, \ldots, m_N$ and are located at positions $\mathbf{r}_1, \mathbf{r}_2, \ldots, \mathbf{r}_N$.

All of the particles exert forces on each other. These forces are *internal* forces. Denote by $\mathbf{F}_{ij}$ the force exerted *on* particle i *by* particle j:

$$\mathbf{F}_{ij} = \text{force acting } \mathbf{on} \text{ particle } i, \mathbf{due\ to} \text{ particle } j.$$

Newton's second law, applied to particle i is, then,

$$\frac{d\mathbf{p}_i}{dt} = \mathbf{F}_i^{(e)} + \sum_{\substack{j=1, N \\ j \neq i}} \mathbf{F}_{ij},$$

where $\mathbf{F}_i^{(e)}$ is the *external* force acting on i. Note that the summation over the internal forces is subject to the condition $j \neq i$ because a particle cannot exert a force on itself.

There is one such equation for each particle. Adding all N equations yields

$$\sum_i^N \frac{d\mathbf{p}_i}{dt} = \frac{d\mathbf{P}_{\text{tot}}}{dt} = \sum_i^N \mathbf{F}_i^{(e)} + \sum_i^N \sum_{j \neq i}^N \mathbf{F}_{ij},$$

where $\mathbf{P}_{\text{tot}} = \sum \mathbf{p}_i$ is the total momentum of the system. The last term (the double sum) is zero because by Newton's third law, $\mathbf{F}_{ij} = -\mathbf{F}_{ji}$. By writing out a few terms you will see that the double sum consists of pairs of terms that cancel each other out. Therefore,

$$\frac{d\mathbf{P}_{\text{tot}}}{dt} = \sum_i \mathbf{F}_i^{(e)} = \mathbf{F}_{\text{tot}}^{(e)}, \tag{6.17}$$

where $\mathbf{F}_{\text{tot}}^{(e)}$ is the total (or net, or resultant) external force acting on the system. Thus, we have shown that the internal forces have no effect on the total momentum.

This result has another important consequence. If the masses of the particles are constant, we can write

$$\frac{d\mathbf{P}_{\text{tot}}}{dt} = \sum_i m_i \ddot{\mathbf{r}}_i,$$

and Equation (6.17) can be written

$$\sum_i m_i \ddot{\mathbf{r}}_i = \mathbf{F}_{tot}^{(e)}.$$

The definition of center of mass ($\mathbf{r}_c$) for a system of particles having total mass M, see Equation (5.17), is

$$M\mathbf{r}_c = \sum_i m_i \mathbf{r}_i.$$

Differentiating twice with respect to time gives

$$M\ddot{\mathbf{r}}_c = \sum_i m_i \ddot{\mathbf{r}}_i. \tag{6.18}$$

But $\sum_i m_i \ddot{\mathbf{r}}_i = \mathbf{F}_{tot}^{(e)}$ so

$$M\ddot{\mathbf{r}}_c = \mathbf{F}_{tot}^{(e)}. \tag{6.19}$$

This important result states that the center of mass of a system of particles moves like a particle of mass M acted upon by the resultant of the sum of all the external forces, regardless of their point of application. A corollary is that if no net external force acts on a system, its center of mass will move with constant velocity.

6.7 Relative Motion and the Reduced Mass

When we study the motion of two interacting bodies such as a star and a planet, or an electron and a nucleus, we are often interested in their *relative motion* and do not care about their motion with respect to an inertial reference frame. For such problems it is convenient to introduce the concepts of *relative coordinate* and *reduced mass*. The introduction of these quantities allows us to replace the two-body problem (with two equations of motion) by a single one-body problem (and only one equation of motion).[7]

Consider a system consisting of two particles (m_1 and m_2) that are exerting equal and opposite forces on each other. Assume no external forces are acting. Let $\mathbf{F}$ be the force on m_2 due to m_1. Then the force on m_1 is $-\mathbf{F}$. Consequently, the equations of motion of the two particles are:

$$m_1 \ddot{\mathbf{r}}_1 = -\mathbf{F},$$

and

$$m_2 \ddot{\mathbf{r}}_2 = +\mathbf{F}.$$

Figure 6.5 shows two particles, m_1 and m_2, at positions $\mathbf{r}_1$ and $\mathbf{r}_2$ with respect to the inertial origin O. The "relative" vector $\mathbf{r}$ gives the position of m_2 with respect to m_1 and (by tip-to-tail addition) is given by

$$\mathbf{r} = \mathbf{r}_2 - \mathbf{r}_1.$$

Differentiating the relative coordinate twice with respect to time yields

$$\ddot{\mathbf{r}} = \ddot{\mathbf{r}}_2 - \ddot{\mathbf{r}}_1.$$

---

[7] This is a very important simplification allowing us to obtain (for example) the motion of a planet relative to the Sun. There is no such simplification for a system of three bodies, although physicists have been trying to solve the "three-body problem" for hundreds of years. Some mathematicians have stated that the problem is unsolvable.

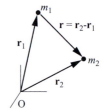

Figure 6.5 Relative positions of two particles. Particles m_1 and m_2 are located at $\mathbf{r}_1$ and $\mathbf{r}_2$ with respect to the origin. The relative coordinate $\mathbf{r}$ gives the position of m_2 with respect to m_1.

Substituting for $\ddot{\mathbf{r}}_2$ and $\ddot{\mathbf{r}}_1$ from the equations of motion gives

$$\ddot{\mathbf{r}} = \frac{\mathbf{F}}{m_2} + \frac{\mathbf{F}}{m_1} = \left(\frac{m_1 + m_2}{m_1 m_2}\right) \mathbf{F}$$

or

$$\left(\frac{m_1 m_2}{m_1 + m_2}\right) \ddot{\mathbf{r}} = F.$$

The quantity $\frac{m_1 m_2}{m_1 + m_2}$ is called the "reduced mass." It is usually denoted by μ. So the equation of motion for the relative coordinate is

$$\mu \ddot{\mathbf{r}} = F. \tag{6.20}$$

As an example, consider a system composed of a star and a planet. The star is much more massive than the planet, and for all intents and purposes the star remains at rest and the planet orbits around it. If the mass of the star is m_1 and the mass of the planet is m_2 and if $m_1 \gg m_2$, then $\mu \cong m_2$. On the other hand, for a binary star system both masses may be approximately equal and the two stars orbit around their common center of mass. If both stars have the same mass, say m, the reduced mass is $\mu = \frac{1}{2}m$.

Consider the Sun–Earth system. The relative coordinate gives the distance from the Sun to the Earth. $\mathbf{F}$ is the force the Sun exerts on the Earth. In an analysis of this system you should write $\mu \ddot{\mathbf{r}} = F$ and *not* $m_2 \ddot{\mathbf{r}} = F$. The reason is that the Sun is accelerating, so a coordinate system with origin at the Sun is not an inertial coordinate system and Newton's second law does not hold. (Actually for the Sun and Earth, m_2 and μ are so nearly equal that the error in using $m_2 \ddot{\mathbf{r}} = F$ is negligible.)

Exercise 6.16

Where is the center of mass of the Sun–Earth system? (Look up the necessary values.) Answer: 4.5×10^5 m from the center of the Sun. ∎

Exercise 6.17

Determine the reduced mass of the Sun–Jupiter system and the reduced mass of the Earth–Moon system. Answer: For Earth–Moon, $\mu = 7.26 \times 10^{22}$ kg. ∎

6.8 Collisions in Center of Mass Coordinates (Optional)

When studying two-body collisions, physicists frequently use a coordinate system that is moving with the center of mass because it is often an easier and quicker way to solve the problem. This approach is particularly useful when studying collisions between elementary particles. If you specialize in high-energy physics, you will become very familiar with center-of-mass coordinates.

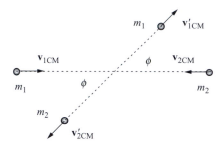

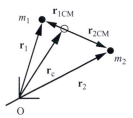

Figure 6.6 A collision as seen in the center of mass coordinate system.

Figure 6.7 The center of mass coordinates; $\mathbf{r}_{1CM}$ and $\mathbf{r}_{2CM}$ give the positions of m_1 and m_2 with respect to the center of mass. The open circle specifies the position of the center of mass.

We shall be using two coordinate systems that we can call the "laboratory frame" and the "center of mass frame." If there are no external forces acting on the system, the center of mass moves with constant velocity, so a coordinate system moving with the center of mass is an inertial (nonaccelerating) coordinate system.

From the center of mass point of view, in a two-body collision both bodies are moving, approaching one another and the center of mass. They collide at the center of mass. After the collision, they recede from the center of mass.

In the center of mass frame, the velocities of the two particles before the collision can be denoted $\mathbf{v}_{1CM}$ and $\mathbf{v}_{2CM}$. After the collision the particles recede in opposite directions with velocities $\mathbf{v}'_{1CM}$ and $\mathbf{v}'_{2CM}$. The angle between the incoming and outgoing directions is denoted ϕ, as shown in Figure 6.6. Note that there is now a single angle, so we have already achieved some simplification of the problem.

Figure 6.7 shows the positions of the two particles and their center of mass relative to the origin O of the laboratory frame (or any arbitrary inertial frame). The center of mass (indicated by the small open circle) lies on the line joining the two particles and is at $\mathbf{r}_c$ relative to O. The coordinates $\mathbf{r}_{1CM}$ and $\mathbf{r}_{2CM}$ give the positions of the particles relative to the center of mass. Note that

$$\mathbf{r}_{1CM} = \mathbf{r}_1 - \mathbf{r}_c$$
$$\mathbf{r}_{2CM} = \mathbf{r}_2 - \mathbf{r}_c$$

The total initial momentum in the center of mass system is zero. To prove this, take the time derivative of the definition of the center of mass, $\mathbf{r}_c = (m_1\mathbf{r}_1 + m_2\mathbf{r}_2)/M$, where $M = m_1 + m_2$:

$$M\dot{\mathbf{r}}_c = (m_1\dot{\mathbf{r}}_1 + m_2\dot{\mathbf{r}}_2) = (m_1\dot{\mathbf{r}}_{1CM} + m_1\dot{\mathbf{r}}_c + m_2\dot{\mathbf{r}}_{2CM} + m_2\dot{\mathbf{r}}_c),$$

so,

$$M\dot{\mathbf{r}}_c - (m_1 + m_2)\dot{\mathbf{r}}_c = m_1\dot{\mathbf{r}}_{1CM} + m_2\dot{\mathbf{r}}_{2CM},$$
$$0 = m_1\mathbf{v}_{1CM} + m_2\mathbf{v}_{2CM}.$$

Therefore, the total initial momentum in the center of mass system is

$$\mathbf{p}_{1CM} + \mathbf{p}_{2CM} = 0. \tag{6.21}$$

The total final momentum must also be zero,

$$\mathbf{p}'_{1CM} + \mathbf{p}'_{2CM} = 0. \tag{6.22}$$

The energy equation in the center of mass system is

$$\frac{p^2_{1CM}}{2m_1} + \frac{p^2_{2CM}}{2m_2} = \frac{p'^2_{1CM}}{2m_1} + \frac{p'^2_{2CM}}{2m_2} + Q, \tag{6.23}$$

where the factor Q represents any energy gained or lost during the collision. For the rest of this section I will assume elastic collisions so $Q = 0$.

Because the linear momentum in the center of mass system is zero, Equations (6.21) and (6.22) yield

$$m_1 \mathbf{v}_{1CM} + m_2 \mathbf{v}_{2CM} = 0,$$

$$\therefore \mathbf{v}_{2CM} = -\frac{m_1}{m_2} \mathbf{v}_{1CM}, \tag{6.24}$$

and similarly

$$\mathbf{v}'_{2CM} = -\frac{m_1}{m_2} \mathbf{v}'_{1CM}. \tag{6.25}$$

The conservation of kinetic energy – Equation (6.23) – leads to

$$m_1 v^2_{1CM} + m_2 v^2_{2CM} = m_1 v'^2_{1CM} + m_2 v'^2_{2CM}.$$

Inserting Equations (6.24) and (6.25) yields

$$v'_{1CM} = v_{1CM} \quad \text{and} \quad v'_{2CM} = v_{2CM}.$$

That is, the *speeds* of the particles are unchanged by the collision. Consequently, you can show that the relative velocity $\mathbf{v}_{rel} = \mathbf{v}_{2CM} - \mathbf{v}_{1CM}$ is a constant and the same in both coordinate systems (see Problem 6.31).

Let us go back to the basic collision of Section 6.3 in which a particle of mass m_1 and speed V_1 strikes a particle of mass m_2 at rest, but now we analyze the problem in the center of mass coordinate system. I will use capital letters (such as V_{1lab}) for speeds in the laboratory frame, and small letters (such as v_{1CM}) for speeds in the center of mass frame. (Note that particle m_2 is initially at rest at the origin of the laboratory frame.)

The speed of the center of mass in the laboratory frame is denoted V_{CM} and is obtained by taking the time derivative of the definition of center of mass,

$$\mathbf{r}_c = \frac{m_1 \mathbf{r}_1 + m_2 \mathbf{r}_2}{M},$$

$$\therefore \mathbf{V}_{CM} = \frac{m_1 \dot{\mathbf{r}}_1 + m_2 \dot{\mathbf{r}}_2}{M}.$$

Because initially $\dot{\mathbf{r}}_1 = \mathbf{V}_{1lab}$ and $\dot{\mathbf{r}}_2 = \mathbf{V}_{2lab} = 0$ the velocity of the center of mass is

$$\mathbf{V}_{CM} = \frac{m_1}{M} \mathbf{V}_{1lab}. \tag{6.26}$$

In the laboratory system the origin of coordinates is located at the initial position of the particle at rest. After the collision the two particles move off at angles θ_1 and θ_2. In other words, the laboratory frame is described by Figure 6.3.

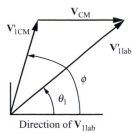

Direction of $\mathbf{V}'_{1lab}$

Figure 6.8 Relationship between velocities in the center of mass system and the laboratory system for the final velocities of particle number 1.

The transformation between laboratory frame velocities and center of mass velocities is illustrated in Figure 6.8, which shows the relationship between $\mathbf{v}'_{1CM}$ and $\mathbf{V}'_{1lab}$, as seen in the laboratory frame of reference. In constructing the figure I used the vector relationship

$$\mathbf{V}'_{1lab} = \mathbf{v}'_{1CM} + \mathbf{V}_{CM}. \tag{6.27}$$

This equation tells you that to convert velocities from the center of mass frame to the laboratory frame you simply add the velocity of the center of mass.

We now obtain some useful relationships between the velocities in the two coordinate systems. The velocity of the center of mass in the laboratory system is given by Equation (6.26). Then, according to Equation (6.27),

$$\mathbf{v}_{1CM} = \mathbf{V}_{1lab} - \mathbf{V}_{CM}, \tag{6.28}$$
$$= \mathbf{V}_{1lab}\left(1 - \frac{m_1}{M}\right),$$
$$= \mathbf{V}_{1lab}(m_2/M),$$

and similarly for the other velocities.

Looking at Figure 6.8 you can see that the velocity components are given by

$$V'_{1lab} \sin\theta_1 = v'_{1CM} \sin\phi,$$

and

$$V'_{1lab} \cos\theta_1 = v'_{1CM} \cos\phi + V_{CM}.$$

These two equations allow you to determine the relationship between θ_1 and ϕ. Dividing one equation by the other yields

$$\tan\theta_1 = \frac{v'_{1CM} \sin\phi}{v'_{1CM} \cos\phi + V_{CM}} = \frac{\sin\phi}{\cos\phi + (V_{CM}/v'_{1CM})}. \tag{6.29}$$

Although this gives an expression for θ_1 in terms of ϕ, it is not in a very convenient form. As you will see in a moment, it is possible to express θ_1 in terms of ϕ and the masses of the particles.

To do so, begin by expressing both V_c and v'_{1CM} in terms of the relative velocity,

$$\mathbf{v}_{rel} = \mathbf{V}_{2lab} - \mathbf{V}_{1lab} = \mathbf{V}'_{2lab} - \mathbf{V}'_{1lab}.$$

Because $\mathbf{V}_{2lab} = 0$, you can write

$$\mathbf{v}_{rel} = -\mathbf{V}_{1lab}.$$

Use Equation (6.26) to form the ratio

$$\frac{\mathbf{V}_c}{\mathbf{v}_{rel}} = \frac{\frac{m_1}{M}\mathbf{V}_{1lab}}{-\mathbf{V}_{1lab}} = -\frac{m_1}{M},$$

so

$$V_c = |\mathbf{V}_c| = \left| -\frac{m_1}{M} \mathbf{v}_{\text{rel}} \right| = \frac{m_1}{M} v_{\text{rel}}.$$

Similarly, you obtain an expression for $v'_{1\text{CM}}$ as follows:

$$\mathbf{v}'_{1\text{CM}} = \mathbf{V}'_{1\text{lab}} - \mathbf{V}_c = \mathbf{V}'_{1\text{lab}} - \left(\frac{m_1}{M} \mathbf{V}'_{1\text{lab}} + \frac{m_2}{M} \mathbf{V}'_{2\text{lab}} \right),$$

$$= \mathbf{V}'_{1\text{lab}} \left(1 - \frac{m_1}{M} \right) - \frac{m_2}{M} \mathbf{V}'_{2\text{lab}} = \frac{m_2}{M} \left(\mathbf{V}'_{1\text{lab}} - \mathbf{V}'_{2\text{lab}} \right),$$

$$= \frac{m_2}{M} \left(-\mathbf{v}_{\text{rel}} \right),$$

and consequently,

$$v'_{1\text{CM}} = \frac{m_2}{M} v_{\text{rel}}.$$

Having obtained expressions for V_c and $v'_{1\text{CM}}$ you can express Equation (6.29) as

$$\tan \theta_1 = \frac{\sin \phi}{\cos \phi + \left(\frac{(m_1/M)v_{\text{rel}}}{(m_2/M)v_{\text{rel}}} \right)} = \frac{\sin \phi}{\cos \phi + (m_1/m_2)}. \tag{6.30}$$

It is left as a problem to show that

$$\tan \theta_2 = \frac{\sin \phi}{1 - \cos \phi}. \tag{6.31}$$

If $V_{2\text{lab}} = 0$ and $V_{1\text{lab}}$ and θ_1 are known variables in the lab frame, you can determine the center of mass coordinate ϕ from Equation (6.30). Then Equation (6.31) yields θ_2. The value of $v_{1\text{CM}}$ is obtained from Equation (6.28). Because $\mathbf{v}_{\text{rel}} = -\mathbf{V}_{1\text{lab}}$ you can determine $v'_{1\text{CM}}$ from $v'_{1\text{CM}} = (m_2/M)v_{\text{rel}}$. Finally, $v'_{2\text{CM}} = -(m_1/m_2)v'_{1\text{CM}}$. If desired, you can then transform back to laboratory coordinates.

Worked Example 6.5 Using the center of mass coordinate system, show that in an elastic collision between two particles of equal mass, the scattering angles in the laboratory system add up to $\pi/2$. (That is, show that $\theta_2 + \theta_1 = \pi/2$.)

Solution If $m_1 = m_2$, Equation (6.30) can be written as

$$\tan \theta_1 = \frac{\sin \phi}{\cos \phi + 1} = \tan \frac{\phi}{2}.$$

(The last step requires the "half angle" trigonometric identity.) This result indicates that $\phi/2 = \theta_1$. Similarly, Equation (6.31) leads to

$$\tan \theta_2 = \frac{\sin \phi}{1 - \cos \phi} = \cot \frac{\phi}{2}.$$

Combining these relations gives

$$\tan \theta_2 = \cot \theta_1.$$

Another trigonometric identity is

$$\tan(\theta_1 + \theta_2) = \frac{\tan \theta_1 + \tan \theta_2}{1 - \tan \theta_1 \tan \theta_2}$$

so

$$\tan(\theta_1 + \theta_2) = \frac{\tan \theta_1 + \tan \theta_2}{1 - \tan \theta_1 \cot \theta_1} = \frac{\tan \theta_1 + \tan \theta_2}{0} \to \infty.$$

Therefore,

$$\theta_2 + \theta_1 = \pi/2.$$

This result indicates that if the two masses are equal, the angle between the two outgoing particles in the laboratory system is a right angle.

Exercise 6.18

Use the center of mass system to show that if $m_1 \ll m_2$ the scattering angle in the CM system is (almost) equal to the scattering angle in the lab system. That is, show that $\theta_1 \approx \phi$. ∎

6.9 Summary

The law of conservation of linear momentum is applicable to numerous problems. In this chapter we have been particularly interested in two problems: the motion of a rocket and the collision of two masses.

Conservation of momentum is based on Newton's second law. Since $\mathbf{F} = d\mathbf{p}/dt$, if the net external force is zero, the momentum is constant.

A system such as a rocket whose mass is changing with time but which is not acted upon by external forces, is analyzed by requiring that the initial and final momenta be equal. For the rocket, this leads to

$$M_R \frac{dV}{dt} = U \frac{dM_F}{dt},$$

and

$$dV = -U \frac{dM_R}{M_R},$$

where U is the speed of the ejected gases relative to the rocket.

The analysis of a glancing collision of two objects is also based on the conservation of linear momentum. If the masses are given, the problem can be expressed in terms of the parameters $V_1, V_1', V_2', \theta_1, \theta_2$. Momentum conservation yields two equations. Thus you need to be given three of these parameters. In the case of *elastic collisions,* conservation of kinetic energy gives you an additional equation and you only need to be given two of the parameters.

Assuming an elastic collision and that V_1 and θ_1 are given, the conservation laws lead to an equation for V_1', the final velocity of M_1:

$$\frac{V_1'}{V_1} = \frac{M_1}{(M_1 + M_2)} \left[\cos \theta_1 \pm \sqrt{\cos^2 \theta_1 - 1 + \left(\frac{M_2}{M_1}\right)^2} \right].$$

The final velocity of M_2 is given by

$$V_2'^2 = [M_1^2 V_1^2 - 2M_1^2 V_1 V_1' \cos \theta_1 + M_1^2 V_1'^2]/M_2^2.$$

Finally, θ_2 can be determined from Equation (6.8),

$$\sin \theta_1 = \frac{M_2 V_2'}{M_1 V_1'} \sin \theta_2.$$

A sharp blow involves a force F that acts during a short time interval τ. The *impulse* $\mathbf{J}$ of such a force is the time integral of the force and is equal to the change in momentum. Thus,

$$\mathbf{J} = \int_0^\tau \mathbf{F} dt = \Delta \mathbf{p} = \mathbf{p}_f - \mathbf{p}_i.$$

A system of particles is acted upon by internal and external forces. The center of mass moves like a particle of mass $M = \sum m_i$ acted upon by the total *external* force:

$$M \ddot{\mathbf{r}}_c = \mathbf{F}_{\text{tot}}^{(e)}.$$

When studying the *relative motion* of two particles, it is convenient to introduce the *reduced mass* μ, given by

$$\mu = \frac{m_1 m_2}{m_1 + m_2}.$$

Then, the equation of motion for the *relative coordinate* $\mathbf{r}$ is $\mathbf{F} = \mu \ddot{\mathbf{r}}$.

Collisions are frequently studied in the center of mass coordinate system in which the total initial and final momenta are zero. For elastic collisions, the scattering angles in the lab and CM systems are related by

$$\tan \theta_1 = \frac{\sin \phi}{\cos \phi + (V_c/v_{1CM}')},$$

$$\tan \theta_2 = \frac{\sin \phi}{1 - \cos \phi},$$

where θ_1 and θ_2 are the scattering angles in the laboratory coordinate system and ϕ is the (single) scattering angle in the center of mass system.

6.10 Problems

Problem 6.1 A 90 kg railroad worker is on a handcar of mass 200 kg. The handcar is moving at 5 m/s when it passes under a tree. (a) The railroad worker leaps upwards, grabs a limb, and hangs on. Does the speed of the handcar change? If so, determine its final velocity. (b) Now consider the converse problem. The empty handcar is moving at 5 m/s when it passes under a tree and a 90 kg railroad worker drops out of the tree onto the handcar. In this case, does the speed of the handcar change? If so, determine its final velocity.

Problem 6.2 Prove that in a one-dimensional elastic collision between two bodies, the relative velocity between the bodies has the same magnitude before and after the collision, but the opposite sign.

Problem 6.3 A ballistic pendulum can be used to determine the speed of the bullet fired from a rifle by determining its effect when it hits and is embedded in a pendulum. Consider a ballistic pendulum that consists of a suspended block of wood of mass M. A bullet of mass m and initial velocity v is fired into and becomes embedded in the block. To block swings upward a height h. (See Figure 6.9.) Derive an equation for v in terms of the given quantities.

Problem 6.4 A raindrop is falling through fog and is picking up tiny water droplets as it falls. (a) Justify that the rate of change of mass of the raindrop is proportional to $r^2 v$, where r is the radius of the raindrop and v is its downward velocity. (b) Prove that the acceleration of the drop is $g/7$. (Ignore air resistance.)

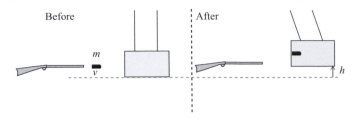

Figure 6.9 The ballistic pendulum.

Problem 6.5 A box filled with sand is placed on a sled and slides down an ice-covered hill (so that friction is negligible). Sand is leaking out of a hole in the box at a constant rate. The slope of the hill is α. (a) Show that the equation of motion of the box (plus sled) is just

$$m\frac{dv}{dt} = mg\sin\alpha.$$

(b) Now assume that the sand is somehow thrown out of the box with a velocity $-v$. (This is a velocity in the direction opposite to the motion of the box but with the same speed as the box.) Show that the equation of motion in this case is

$$m\frac{dv}{dt} = v\frac{dm}{dt} + mg\sin\alpha.$$

Problem 6.6 A jet boat (or jet ski) operates on the following principle: Water is drawn through an inlet into a turbine and pumped out at high speed through a smaller opening. The manufacturer of a jet boat states that the turbine draws 50 gallons of water per second and expels it at a pressure of 80 psi. The manufacturer states that the thrust developed is over 2000 lb. Determine whether or not this is a realistic value. (Hint: Look up Bernoulli's equation.)

Problem 6.7 A spherical asteroid of mass m_0 is moving freely in interstellar space with velocity v_0. It runs into a dust cloud whose uniform density is ρ_d. Assume that every particle of dust that hits the asteroid sticks to it. (a) Obtain an expression for the velocity of the asteroid as a function of time. (b) Obtain an expression for the force exerted on the asteroid by the dust as a function of time.

Problem 6.8 During a war being waged in Antarctica, an armored car of mass 2000 kg with a machine gun mounted on its roof drives onto a frozen lake at 30 km/h. The ice is, of course, perfectly frictionless, and the truck will continue to slide in a straight line at a constant velocity directly into the enemy camp unless it can make a 90° turn and slide to safety. G. I. Joe (who studied physics in college) jumps to the roof, swivels the machine gun, and begins firing in a direction perpendicular to the motion. The bullets have a mass of 500 g each and leave the gun with a velocity of 800 m/s. The gun fires at a rate of 200 bullets per minute. How long must G. I. Joe fire the machine gun for the car's motion to be deviated by 90°? Since the bullets are being fired so rapidly, you can assume the mass decrease is continuous.

Problem 6.9 The man on the flying trapeze is hanging by his knees from the cross bar. The woman trapeze artist stands on the circus floor. The trapeze starts at an angle of 60° from the vertical. At the bottom of the swing, the man grabs the woman and they both swing upward. To what angle will the trapeze swing with both artists on it? For simplicity, assume the man is a point mass m_M and the woman is a point mass m_W. The rope has length l and negligible mass.

Problem 6.10 A rocket motor is undergoing a "bench test." It is attached to a fixed support by four large springs of constant 10^6 N/m. The motor burns fuel at a rate of 50 kg/s. When the motor is running the springs are observed to stretch 1.5 cm. Determine the exhaust speed of the burned fuel.

Problem 6.11 Most airports have a conveyer belt behind the check-in counter for luggage to be transported to the airplane. Bags of mass m are dropped onto the belt a rate of k per second. (You can assume the mass increase is constant.) What is the increased power load during the time the bags are being placed on the belt?

Show that the extra power required is twice the rate of increase of kinetic energy. Explain what is happening to the "missing" power.

Problem 6.12 The "people mover" at the San Francisco airport is essentially a horizontal, very long conveyor belt running at speed v. Suppose that initially there are no people on the belt. The motor driving the belt is drawing W watts of electrical power. A flight arrives and passengers (all having the same mass) step onto the belt, one after another at one second intervals, so that the mass being carried increases at a constant rate k kg/s. Assume the people stepping onto the belt had an initial speed of $v/2$. If the belt is to continue running at the same speed, how much extra electrical power must be supplied to the motor?

Problem 6.13 A physics student is holding a vertical hanging chain by its top link. The bottom link is just touching the top surface of a scale. The student lets go of the chain and observes the reading on the scale while the chain is dropping. The student claims that the reading on the scale is three times the weight of the length of the chain on the scale. The laboratory instructor doubts the result obtained by the student. Show that the student's observation is correct. (The reading of the scale is the force exerted by the scale to stop the downward motion of the chain.)

Problem 6.14 Many years ago on a cold winter morning in Chicago, Bonnie and Clyde stole an armored truck full of money. (The mass of the truck was 2000 kg and its top speed was 240 km/h or 66.6 m/s.) Officer Dick Tracy and his driver spotted them and gave chase. (The police car had a mass of 1500 kg and its top speed was also 240 km/h.) As luck would have it, the two vehicles ran off the bank onto Lake Michigan, which was covered with perfectly frictionless ice, so the two vehicles continued to move at constant speed and maintained a constant separation. Dick Tracy grew impatient, so he opened the moon roof, stood up, and started to shoot at the armored truck with his machine gun. The machine gun fired 120 bullets per minute with a muzzle speed of 1000 m/s. Each bullet had a mass of 0.05 kg. All of the bullets hit and were embedded into the armored truck. After one minute of this, what was the speed of the police car and what was the speed of the truck?

Problem 6.15 An Atwood's machine uses two containers filled with water on either side of the ideal, frictionless pulley. Initially, both buckets contain the same amount of water and have the same mass m. However, one of the containers has a small hole in it, and water is leaking out at a rate k (kg/s). The leaking container will, therefore, move upward. Obtain an expression for the velocity of this container.

Problem 6.16 A rocket of mass 40 000 kg is in empty space. Determine its velocity increase after burning all its fuel if the mass of the fuel is 90 percent of the mass of the rocket. The rate of fuel burn is constant. The speed of the exhaust gas relative to the rocket is 3000 m/s.

Problem 6.17 A rocket of mass 1500 kg is designed to expel exhaust gas at 1000 m/s. Determine the minimum required burn rate if the rocket is to rise from the surface of the Earth. What burn rate is required for it to have an initial acceleration of 1 m/s^2?

Problem 6.18 A small rocket is launched from the surface of the Earth. Because it does not rise very high, we are justified in assuming $\mathbf{g}$ = constant. Obtain an equation for the height reached when all of the fuel is burned. The mass of the fuel is m and the initial mass of rocket plus fuel is M_0. The exhaust speed of the gases is u. Assume the exhaust speed and the burn rate are constant. Ignore air resistance.

Problem 6.19 Johnny Whizz, the inventor, designs an automobile that will just hover at the surface of the Earth. It has four rocket motors, one in each wheel well. The rocket motors expel burned fuel with an exhaust velocity of 4000 m/s. Assume the mass of the fuel is 80 percent of the total mass of the car. Evaluate the maximum time the car can hover above the ground.

Problem 6.20 Rockets are not always launched straight up. Consider a rocket launcher consisting of a ramp inclined at 30° above the horizontal. The mass of the rocket is 4000 kg of which 3000 kg are fuel. The exhaust speed is 1000 m/s and the fuel is burned at a constant rate of 200 kg/s. You may assume a flat, airless, nonrotating Earth. The rocket does not rotate about its center of mass. (a) Determine the velocity and direction of the rocket at the time all the fuel is burned. (b) Determine its position at this time.

Problem 6.21 A billiard ball is placed in contact with the upper surface of a bowling ball and they are dropped from a height h onto a cement floor. (Because they fall at the same rate, they are essentially in contact but you might like to think of the billiard ball as lagging the bowling ball by an infinitesimal amount.) The bowling ball hits the ground and bounces back up, immediately striking the billiard ball. How high does the billiard ball rise? You may assume the bowling ball has a mass 20 times greater than a billiard ball. All collisions are elastic.

Problem 6.22 A particle of mass 5 kg and initial speed 5 m/s undergoes a head-on elastic collision with a particle of mass 3 kg with initial speed -3 m/s where the negative sign indicates that it is approaching the first particle. Determine the final speeds of the two particles.

Problem 6.23 An air hockey puck of mass $2M$ collides in a glancing collision with a puck of mass M. The heavier puck had an initial velocity of 3 m/s and comes off at an angle of 30° after the collision. Determine the velocity and direction of the lighter puck. Assume an elastic collision.

Problem 6.24 Consider a glancing collision between two objects of nearly equal mass ($M_2 = M_1 + \delta$, where δ is a small quantity). Obtain an expression for V_1'/V_1 and show that it reduces to 1 or $\cos\theta$ as $\delta \to 0$, as in Subcase 2.1. What is the value of V_1'/V_1 in the extreme of δ being a large value (that is, $M_2 \gg M_1$)?

Problem 6.25 When a spacecraft carries out a flyby maneuver near a planet, the speed of the spacecraft can be dramatically increased. The Cassini spacecraft approached Jupiter at a speed of 9.36 km/s as measured relative to the Sun. Jupiter's orbital speed is 13.02 km/s. As it approached the planet, the spacecraft velocity formed an angle of about 37° to Jupiter's velocity vector. Cassini was deflected through an angle of 52.8°. Determine the increase in speed of the spacecraft due to this slingshot maneuver.

Problem 6.26 At time $t = 0$ three particles of masses $m_1 = 1$ g, $m_2 = 2$ g, and $m_3 = 3$ g are at rest and lined up along the x-axis at positions $x_1 = -1$ cm, $x_2 = 0$ cm, and $x_3 = +2$ cm. The forces acting on the particles are $\mathbf{F}_1 = -3\hat{\jmath}$ dynes, $\mathbf{F}_2 = 0$, and $\mathbf{F}_3 = +12\hat{\jmath}$ dynes. (a) Determine the position of the center of mass at $t = 0$. (b) Determine the positions of the three particles at $t = 10$ s. (c) Determine the position of the center of mass at $t = 10$ s from Equation (5.17). (d) Determine the position of the center of mass at $t = 10$ s by using Equation (6.19).

Problem 6.27 Two masses, m_1 and m_2, undergo a completely inelastic collision. Show that the loss of kinetic energy is equal to $\frac{1}{2}\mu v^2$, where μ is the reduced mass and v is the relative velocity of the two masses.

Problem 6.28 Consider a planet–star system with masses M_S and M_P. The planet is located at a distance r from the star. As seen from an inertial coordinate system, the star is located at $\mathbf{R}_S$. (a) Determine the acceleration (in inertial space) of the star and the planet due to their mutual gravitational attraction. (b) Determine the acceleration of the planet with respect to the star. (Note that it is not equal to $-(GM_S/r^2)\hat{\mathbf{r}}$ but it reduces to this value if $M_S \gg M_P$.)

Problem 6.29 A ball is dropped from a height h. The coefficient of restitution is e. Determine that the time required for the ball to come to rest is

$$t = \sqrt{\frac{2h}{g}} \left(\frac{1+e}{1-e} \right),$$

and that the total distance the ball travels is

$$d = h \left(\frac{1+e^2}{1-e^2} \right).$$

(Hint: Express the time as an infinite series. Note that the sum of a geometric series of the form $a, ar, ar^2, ar^3, \ldots$ is $S = a/(1-r)$.)

Problems 6.30–6.32 are based on Optional Section 6.8.

Problem 6.30 Show that in the center of mass system, the conservation of kinetic energy from Equation (6.23) reduces to

$$\frac{P_{1CM}^2}{2\mu} = \frac{P_{1CM}'^2}{2\mu}$$

if $Q = 0$.

Problem 6.31 Show that in an elastic collision between two particles the relative velocity does not change and that it is the same in both the laboratory reference frame and the center of mass frame.

Problem 6.32 Show that the scattering angle ϕ in center of mass coordinates is related to the angle θ_2 in the laboratory coordinates by

$$\tan \theta_2 = \frac{\sin \phi}{1 - \cos \phi}.$$

In other words, derive Equation (6.31).

Computational Projects

Computational Techniques: The Runge–Kutta Method

In Section 3.7 I described the Euler–Cromer algorithm for solving ordinary differential equations (ODEs). Here I will describe the Runge–Kutta method, which is more accurate and perhaps more elegant, but less transparent. It is based on the truncated Taylor series expansion for a function $g(t)$:

$$g(t + \tau) = g(t) + \tau \left. \frac{dg}{dt} \right|_\xi .$$

The second term is usually evaluated at t, but to make the solution more accurate, the Runge–Kutta technique evaluates it half way through the time step. That is $\xi = t + \tau/2$.

An economical way of expressing the ODE is to define two vectors, $\mathbf{x}$ and $\mathbf{f}$ in the following way. Assume a two-dimensional system so the position is given by x and y and the velocity by v_x and v_y. Then

$$\mathbf{x}(t) = [x(t) \;\; y(t) \;\; v_x(t) \;\; v_y(t)]$$
$$\mathbf{f}(\mathbf{x}, t) = [v_x(t) \;\; v_y(t) \;\; a_x(t) \;\; a_y(t)]$$

and the ODE can be written

$$\frac{d\mathbf{x}}{dt} = \mathbf{f}(\mathbf{x}(t), t).$$

The second-order Runge–Kutta algorithm is obtained by first defining a new vector $\mathbf{x}^*$ by:

$$\mathbf{x}^* = \mathbf{x}^*(t + \tau/2) = \mathbf{x}(t) + \frac{1}{2}\tau \mathbf{f}(\mathbf{x}(t), t).$$

Then

$$\mathbf{x}(t + \tau) = \tau \mathbf{f}(\mathbf{x}^*, t + \tau/2).$$

It is not too difficult to appreciate that this is equivalent to the truncated Taylor series. However, the most common ODE solver is not the second-order Runge–Kutta, but the fourth-order Runge–Kutta, which is also based on the Taylor series but in a much less transparent way. In vector form, the value of $\mathbf{x}$ at time $t + \tau$ is given by

$$\mathbf{x}(t + \tau) = \mathbf{x}(t) + \frac{1}{6}\tau \left(\mathbf{F}_1 + 2\mathbf{F}_2 + 2\mathbf{F}_3 + \mathbf{F}_4\right),$$

where

$$\mathbf{F}_1 = \mathbf{f}(\mathbf{x}, t),$$

$$\mathbf{F}_2 = \mathbf{f}\left(\mathbf{x} + \frac{1}{2}\tau \mathbf{F}_1, t + \frac{1}{2}\tau\right),$$

$$\mathbf{F}_3 = \mathbf{f}\left(\mathbf{x} + \frac{1}{2}\tau \mathbf{F}_2, t + \frac{1}{2}\tau\right),$$

$$\mathbf{F}_4 = \mathbf{f}\left(\mathbf{x} + \frac{1}{2}\tau \mathbf{F}_3, t + \tau\right).$$

Computational Project 6.1 A man is standing on a stationary railroad flatcar loaded with large rocks, all having the same mass of 10 kg. He is able to throw the rocks at a speed of 2 m/s, relative to himself. There are 50 rocks on the flatcar. Determine the speed of the flatcar after the man has thrown all of them straight back off the end of the car. The mass of the man plus empty flatcar is 1000 kg. Naturally, this advanced propulsion system depends on the development of completely frictionless railroad cars. Solve this problem numerically, and also solve it by using Equation (6.2). Compare your answers. Why do they not agree? Which answer do you think is correct?

Computational Project 6.2 Write a program to determine the altitude reached by a rocket launched from the surface of the Earth. Assume the rate at which fuel is burned is proportional to the amount of fuel left and the exhaust speed of the ejected gases is constant. Plot the position of the rocket as a function of time for $dM/dt = -0.01M$ if the initial mass of the fuel is 50 000 kg and $u = 2500$ m/s. The mass of the empty rocket is 10 000 kg. Ignore air resistance, but keep in mind that the gravitational force decreases with the distance from the center of the Earth.

Computational Project 6.3 A rocket is launched from the surface of the Earth. The initial mass of the rocket is 1000 kg, and it is 90 percent fuel. The exhaust gases have a speed (relative to the rocket) of 250 m/s, and the burn rate is 50 kg/s. The resistance of the air can be expressed as a retarding force given by $F = -0.5C_d \rho A v^2$, where the drag coefficient C_d is 0.35 and the density of the air can be assumed to be constant and equal to 1.20 kg/m^3. The cross-sectional area of the rocket, A, is 0.8 m^2. Do not assume $\mathbf{g}$ is constant, but you may set it equal to 9.8 m/s^2 at the ground. Determine the altitude reached by the rocket when all of its fuel is burned. Plot altitude as a function of time. Explain the shape of the curve.

Computational Project 6.4 Solve Computational Project 6.3 using a realistic profile for air density. (You can obtain tables of air density as a function of altitude using the United States Standard Atmosphere or the Smithsonian Meteorological Tables. These can be found in your library or on the Internet.)

Computational Project 6.5 A 5 MeV alpha particle is approaching a gold nucleus. The impact parameter is 1 Å. You may assume the gold nucleus is initially at rest. Plot the trajectory of the two particles and determine the angles at which both are scattered.

Computational Project 6.6 You are required to design a two-stage rocket that will accelerate a 5000 kg payload to Earth's escape velocity. Assume that 95 percent of the mass of the rockets is fuel. Assume the exhaust velocity of the rocket motors is 2000 m/s. Investigate the possible ranges of masses for the two stages and determine the configuration that will minimize the take-off weight. Determine why a single-stage rocket, burning the same amount of fuel, cannot accomplish the same objective.

7 Angular Momentum

In this chapter we consider the conservation of angular momentum. Although some of the concepts discussed here were mentioned in Chapter 1, you will now be exposed to them in much greater detail.

We begin by applying the conservation of angular momentum to a *particle,* and then generalize to a rigid body rotating about a fixed axis. To keep the analysis simple, it will be limited to *symmetrical* bodies with *fixed* rotational axes. In Chapters 16 and 17 you will study the general rotational motion of a rigid body.

7.1 Definition of Angular Momentum

Suppose a particle of mass m is located at position $\mathbf{r}$ and is moving with velocity $\mathbf{v}$. Its linear momentum is $\mathbf{p} = m\mathbf{v}$. Its angular momentum is defined by

$$\mathbf{l} \equiv \mathbf{r} \times \mathbf{p}. \tag{7.1}$$

See Figure 7.1. According to the definition of the cross product, the angular momentum is perpendicular to the plane defined by $\mathbf{r}$ and $\mathbf{p}$. By the right-hand rule, the angular momentum of the particle illustrated in Figure 7.1 points into the page.

The *magnitude* of the angular momentum of a particle is given by

$$l = mvr \sin\theta.$$

The angular momentum of an extended rigid body is given by

$$\mathbf{L} = \int_{\text{body}} \mathbf{r} \times \mathbf{v}\, dm.$$

Note that the angular momentum depends on the choice of the origin of the coordinate system (because the vector $\mathbf{r}$ is part of the definition). When working with angular momentum, remember that once you have chosen an origin, you must not change it.

An important special case is the angular momentum of a particle moving in a circular path. If the center of the circle is chosen as the origin, the angle between $\mathbf{r}$ and $\mathbf{v}$ is $\pi/2$ and

$$l = mvr \sin(\pi/2) = mvr.$$

But for circular motion, $v = \omega r$, where ω is the angular velocity – see Equation (1.13). Consequently,

$$l = mr^2\omega.$$

This is probably the formula you memorized in your introductory mechanics course.

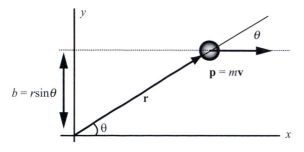

Figure 7.1 Illustrating angular momentum for a particle. A particle of mass m moving with velocity $\mathbf{v}$ is located at $\mathbf{r}$. The linear momentum of the particle is $\mathbf{p} = m\mathbf{v}$ and its angular momentum is $\mathbf{l} = \mathbf{r} \times \mathbf{p}$, a vector pointing into the page.

Exercise 7.1

Evaluate the magnitude and direction of the angular momentum vector for a 1000 kg automobile driving down a straight road at 100 km/h, with respect to a point 20 m to the side of the road. What is the angular momentum with respect to a point 20 m on the other side of the road? Answer: $l = 5.56 \times 10^5$ kg m^2/s. Directions = down and up.   ∎

7.2 Conservation of Angular Momentum

7.2.1 Torque

Before we express the law of conservation of angular momentum, let us recall the definition of torque. Suppose that an extended rigid body is mounted on a fixed axis and is free to rotate about that axis. (See Figures 1.5 and 1.6.) Let a force $\mathbf{F}$ be applied at some point. The torque exerted by the force is

$$\mathbf{N} = \mathbf{r} \times \mathbf{F},$$

where $\mathbf{r}$ is the vector from the origin (usually the axis of rotation) to the point of application of the force. This definition tells us that the torque depends on the location of the axis or rotation as well as the point of application of the force and, of course, the magnitude and direction of the force. However, if the net torque is *zero* then the torque does not depend on the location of the axis of rotation. This useful fact is explored in the following worked example.

Worked Example 7.1   Prove that for a body in rotational equilibrium (no angular acceleration) the net torque about any point is zero.

Solution   Consider the relation between the torque about some point (call it O) and the torque about some other point (O′) displaced by a distance $\mathbf{d}$ from O. The total torque about O is

$$\mathbf{N}_O = \sum \mathbf{N}_{iO} = \sum \mathbf{r}_{iO} \times \mathbf{F}_i,$$

where $\mathbf{r}_{iO}$ is the vector from O to the point of application of force $\mathbf{F}_i$. See Figure 7.2.
   Similarly, the torque about O′ is

$$\mathbf{N}_{O'} = \sum \mathbf{N}_{iO'} = \sum \mathbf{r}_{iO'} \times \mathbf{F}_i,$$

where $\mathbf{r}_{iO'}$ is the vector from O′ to the point of application of force $\mathbf{F}_i$. But as shown in Figure 7.2

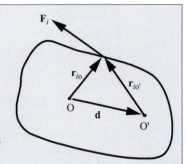

Figure 7.2 A body acted upon by various forces (only one of them is shown). If the sum of the forces is zero and the torque about point O is zero, then the torque about any other point O' is also zero.

$$\mathbf{r}_{iO} = \mathbf{d} + \mathbf{r}_{iO'}$$

so

$$\sum \mathbf{N}_{iO'} = \sum \mathbf{r}_{iO'} \times \mathbf{F}_i = \sum (\mathbf{r}_{iO} - \mathbf{d}) \times \mathbf{F}_i \tag{7.2}$$

or

$$\sum \mathbf{N}_{iO'} = \sum \mathbf{N}_{iO} - \mathbf{d} \times \sum \mathbf{F}_i.$$

For a body in equilibrium, the net force is zero ($\sum \mathbf{F}_i = 0$). Therefore, if the torque about a particular point is zero, then the torque about any other point will also be zero. This is a useful fact. For a body in rotational equilibrium, the torque on the body *with respect to any point* is zero.

7.2.2 Torque and Angular Momentum

In Section 1.8, we noted that the rate of change of the angular momentum is equal to the applied torque. We now prove this assertion, first for a particle, and then, in Section 7.3, for a collection of particles.

For a particle, $\mathbf{l} = \mathbf{r} \times \mathbf{p}$ and so

$$\frac{d\mathbf{l}}{dt} = \frac{d}{dt}(\mathbf{r} \times \mathbf{p}) = \frac{d\mathbf{r}}{dt} \times \mathbf{p} + \mathbf{r} \times \frac{d\mathbf{p}}{dt}.$$

The first term is zero because

$$\frac{d\mathbf{r}}{dt} \times \mathbf{p} = \mathbf{v} \times \mathbf{p} = m(\mathbf{v} \times \mathbf{v}) = 0.$$

Because $d\mathbf{p}/dt = \mathbf{F}$, and because the torque is defined by $\mathbf{N} = \mathbf{r} \times \mathbf{F}$, the equation for $d\mathbf{l}/dt$ reduces to

$$\frac{d\mathbf{l}}{dt} = \mathbf{r} \times \frac{d\mathbf{p}}{dt} = \mathbf{r} \times \mathbf{F} = \mathbf{N}. \tag{7.3}$$

That is, the rate of change of the angular momentum of a particle is equal to the net torque exerted on it by external forces. If there is no net external torque acting on a particle, the time rate of change of its angular momentum is zero. That is,

$$\boxed{\text{if } \mathbf{N} = 0 \text{ then } \tfrac{d\mathbf{l}}{dt} = 0 \text{ and } \mathbf{l} = \text{constant.}}$$

If the torque acting on the particle is zero, its angular momentum is conserved. This is one way of stating the law of conservation of angular momentum.

Exercise 7.2

A particle of mass m has an initial velocity $\mathbf{v} = v_0\hat{\mathbf{i}}$. It is subjected to a constant force $\mathbf{F} = f_0\hat{\mathbf{j}}$. Evaluate $\frac{d^2\mathbf{l}}{dt^2}$ at the initial time. Answer: $\frac{d^2\mathbf{l}}{dt^2} = v_0 f_0\hat{\mathbf{k}}$. ∎

7.3 Angular Momentum of a System of Particles

Having defined the angular momentum of a single particle, let us consider the angular momentum of a collection of particles. Since the angular momentum of the ith particle is $\mathbf{l_i} = \mathbf{r_i} \times \mathbf{p_i}$, the total angular momentum of N particles is the vector sum of the angular momenta of all the particles. That is,

$$\mathbf{L} = \sum_{i=1}^{N} \mathbf{l_i} = \sum_{i=1}^{N}(\mathbf{r_i} \times \mathbf{p_i}).$$

You will not be surprised to find that the internal torques these particles exert on each other do not affect the total angular momentum. That is, just as internal *forces* do not affect the total *linear* momentum, so too, internal *torques* do not affect the total *angular* momentum.

Suppose the N particles have masses $m_1, m_2, \ldots, m_N$ and at some instant of time these particles are located at positions $\mathbf{r}_1, \mathbf{r}_2, \ldots, \mathbf{r}_N$. All of the particles exert (internal) forces on each other. The force *on* particle i, *due to* particle j, will be denoted $\mathbf{F}_{ij}$. Particle i may also be subjected to an external force (denoted $\mathbf{F}_i^{(e)}$). Consequently, the equation of motion for particle i is,

$$m_i\ddot{\mathbf{r}}_i = \mathbf{F}_i^{(e)} + \sum_{\substack{j=1 \\ j\neq i}}^{N} \mathbf{F}_{ij}.$$

Crossing $\mathbf{r}_i$ into this equation of motion yields

$$\mathbf{r}_i \times (m_i\ddot{\mathbf{r}}_i) = \mathbf{r}_i \times \mathbf{F}_i^{(e)} + \sum_{j\neq i} \mathbf{r}_i \times \mathbf{F}_{ij}. \tag{7.4}$$

Now note that the rate of change of the angular momentum vector of particle i is

$$\begin{aligned}
\frac{d\mathbf{l}_i}{dt} &= \frac{d}{dt}(\mathbf{r}_i \times m_i\dot{\mathbf{r}}_i) \\
&= m_i(\dot{\mathbf{r}}_i \times \dot{\mathbf{r}}_i) + m_i(\mathbf{r}_i \times \ddot{\mathbf{r}}_i) \\
&= \mathbf{r}_i \times m_i\ddot{\mathbf{r}}_i.
\end{aligned}$$

Therefore, the left-hand side of Equation (7.4) is just the rate of change of the angular momentum, so

$$\frac{d\mathbf{l}_i}{dt} = \mathbf{r}_i \times \mathbf{F}_i^{(e)} + \sum_{j\neq i} \mathbf{r}_i \times \mathbf{F}_{ij}.$$

There is one such equation for each particle. Adding all these equations gives

$$\sum_i \frac{d\mathbf{l}_i}{dt} = \sum_{i=1}^{N}\left(\mathbf{r}_i \times \mathbf{F}_i^{(e)}\right) + \sum_i\left(\sum_{j\neq i} \mathbf{r}_i \times \mathbf{F}_{ij}\right). \tag{7.5}$$

But

$$\sum_i \frac{d\mathbf{l}_i}{dt} = \frac{d}{dt} \sum_i \mathbf{l}_i = \frac{d}{dt}\mathbf{L},$$

where $\mathbf{L}$ is the total angular momentum of the system, defined as the vector sum of the angular momenta of all the particles. Furthermore, $\sum_i \left(\mathbf{r}_i \times \mathbf{F}_i^{(e)} \right)$ is the total torque $\left(\mathbf{N}_{\text{tot}}^{(e)} \right)$ on the system due to external forces. Consequently, Equation (7.5) can be written as

$$\frac{d\mathbf{L}}{dt} = \mathbf{N}_{\text{tot}}^{(e)} + \sum_i \sum_{j \neq i} \mathbf{r}_i \times \mathbf{F}_{ij}.$$

The last term, containing the double summation, is the total torque due to internal forces. We now prove that this term is equal to zero. Consider two particles, i and j. The corresponding terms in the double sum are:

$$\mathbf{r}_i \times \mathbf{F}_{ij} + \mathbf{r}_j \times \mathbf{F}_{ji}.$$

By Newton's third law, $\mathbf{F}_{ij} = -\mathbf{F}_{ji}$, so we can write this pair of terms as

$$\mathbf{r}_i \times \mathbf{F}_{ij} - \mathbf{r}_j \times \mathbf{F}_{ij} = (\mathbf{r}_i - \mathbf{r}_j) \times \mathbf{F}_{ij} = \mathbf{r}_{ij} \times \mathbf{F}_{ij},$$

where $\mathbf{r}_{ij}$ is the relative coordinate and points from particle j to particle i. If Newton's law is obeyed in the strong form, then the force between the particles is directed along $\mathbf{r}_{ij}$ and so $\mathbf{r}_{ij}$ and $\mathbf{F}_{ij}$ are parallel. Therefore, the cross product $\mathbf{r}_{ij} \times \mathbf{F}_{ij}$ is zero,[1] and all the terms in the double sum cancel out pairwise, leaving

$$\frac{d\mathbf{L}}{dt} = \mathbf{N}_{\text{tot}}^{(e)}. \tag{7.6}$$

That is, the rate of change of the total angular momentum is equal to the net *external* torque on the system. In particular, the law of conservation of angular momentum can be expressed as follows:

If the net external torque acting on a system of particles (or an extended body) is zero, the total angular momentum of the system will remain constant.

This is the *law of conservation of angular momentum.* It is one of the basic physical principles governing material objects and, as far as we know, it is never violated. When you study quantum mechanics and are introduced to the concepts of the orbital and the spin angular momentum of electrons, you will begin to appreciate the power of this conservation law.

Furthermore, the law of conservation of angular momentum is a very useful tool for solving physics problems. There is a wide variety of problems in which no external torques act on a particle or system of particles. Then the angular momentum is constant and the angular momentum before some event is equal to the angular momentum after the event.

For example, suppose a comet is attracted to some star, approaches it, swings about the star in a hyperbolic path, and then travels back out to "infinity." The force exerted by the star on the comet is along the line joining them, so the torque is zero. Therefore, the initial angular momentum is equal to the final angular momentum, where "initial" and "final" can refer to any two points along the trajectory.

---

[1] A Lagrangian analysis of the constancy of angular momentum does not invoke the strong form of Newton's third law. See, for example, *A Student's Guide to Lagrangians and Hamiltonians* by Patrick Hamill, Cambridge University Press, 2014, pages 33–36.

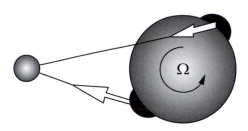

Figure 7.3 The Earth–Moon system, looking down on the North Pole. The Earth rotates counterclockwise and the Moon orbits the Earth counterclockwise. The Moon exerts a force on the tidal bulges of the Earth. The force on the further bulge is somewhat smaller than the force on the nearer bulge so there is a net torque on the Earth acting clockwise in the figure, causing the Earth to slow down. The figure is extremely out of scale. The Earth–Moon distance is roughly 60 Earth radii.

A common problem found in introductory physics textbooks has a running child leap onto a playground merry-go-round. This type of problem can be solved by setting the total initial angular momentum (the value for child plus merry-go-round before the child jumps aboard) equal to the total final angular momentum (the value after the child has jumped on).

Exercise 7.3

A little girl (mass = 20 kg) runs at 3 m/s and jumps onto the rim of a playground merry-go-round that consists of a 50 kg disk with radius 1 m. The frictionless merry-go-round was initially at rest. What is the final angular velocity of the system (child plus disk)? Answer: 4/3 rad/s. ∎

Another type of angular momentum problem involves bodies upon which a net external torque *is* acting. Consider, for example, a spool with thread wrapped around it. A tension in the thread exerts a torque on the cylinder causing its angular momentum to increase.

An interesting example in which a torque causes a change in angular momentum is the effect of the Moon on the rotation rate of the Earth. As you probably know, the Moon raises two tides a day in the oceans of the Earth.[2] If the Earth were not rotating, one tidal bulge would be on the surface of the Earth facing the Moon and the other on the opposite side. But the rotation of the Earth (and various other factors) causes the nearer bulge to be dragged "ahead" of the Moon whereas the other bulge lags the Moon, as illustrated schematically in Figure 7.3. The force on the nearer bulge is greater than the force on the further bulge so there is a net torque on the Earth. The effect of this torque is to slow down the rotation rate of the Earth. As the Earth rotates more and more slowly, the day grows longer. In several billion years the day will be so long that it will be equal to the orbital period of the Moon. That is, the day will be as long as the month.

If we consider Earth–Moon to be an isolated system, there are no external torques on it, and the total angular momentum must be constant. We have seen that the angular momentum of the Earth is decreasing, so how can the total angular momentum remain constant? The answer is that the angular momentum of the Moon is increasing. This requires an increase in the distance between the Earth and the Moon. That is, the Earth is slowing down, losing rotational angular momentum, and the Moon is receding from the Earth, gaining orbital angular momentum in such a way as to conserve the total angular momentum of the system.[3]

---

[2] The reason there are *two* tidal bulges is discussed in Section 15.7.

[3] The tidal interaction between Earth and Moon has many fascinating aspects. It has been used to explain the eccentricity of the Moon's orbit, but recent studies seem to indicate that the gravitational force of Jupiter may be more important. If you are interested, you might look up the paper by Matija Cuk, "Excitation of Lunar Eccentricity by Planetary Resonances," *Science*, **318**, 244 (2007).

Worked Example 7.2 Imagine a perfectly frictionless puck on a perfectly smooth table. An ideal massless string is attached to the puck. The string passes through a small hole in the tabletop and the other end is tied to a nail on the floor. When the puck is given an impulse, it moves at constant angular speed ω in a circle of radius r centered on the hole. (The setup is similar to that of Figure 10.15.)

Now a mischievous child crawls under the table, unties the string and pulls it down, reducing the radius of the puck's circular path to half of its original value. (a) Evaluate the change in angular momentum. (b) Compare the final angular velocity to the initial angular velocity. (c) Evaluate the change in kinetic energy. (d) How much work was done in pulling the mass closer to the center?

Solution (a) The initial angular momentum is $l = md_0^2\omega_0$. The force is along the string so the torque $= \mathbf{N} = \mathbf{r} \times \mathbf{F} = 0$ and angular momentum is conserved. Therefore $\Delta l = 0$.

(b) By conservation of angular momentum

$$l_f = l_i \implies md_f^2\omega_f = md_0^2\omega_0.$$

But $d_f = 0.5d_0$, so

$$\omega_f = \frac{d_0^2}{(d_0/2)^2}\omega_0 = 4\omega_0.$$

(c) The change in kinetic energy is

$$\Delta T = T_f - T_0.$$

Since $v = \omega r$ and $T = (1/2)mv^2$, the kinetic energy can be expressed as $T = (1/2)m\,(\omega r)^2$. Therefore, $T_0 = \frac{1}{2}md_0^2\omega_0^2$ and $T_f = \frac{1}{2}m(d_0/2)^2\omega_f^2$. Consequently,

$$\Delta T = \frac{1}{2}m(d_0/2)^2\,(4\omega_0)^2 - \frac{1}{2}md_0^2\omega_0^2$$

$$= \frac{1}{2}md_0^2\omega_0^2\left[\frac{1}{4}(16) - 1\right] = \frac{3}{2}md_0^2\omega_0^2 = 3T_0.$$

(d) By the work energy theorem, the work done is $3T_0 = \frac{3}{2}md_0^2\omega_0^2$.

Exercise 7.4

An Atwood's machine has two weights tied to the ends of a string that passes over a pulley of radius 2 cm. The two masses are 500 g and 700 g. Assuming the string does not slip, determine the rate of change of angular momentum of the pulley. The moment of inertia of the pulley is 100 g cm^2. (Note: Because the masses are accelerating, you cannot assume tension equals mg.) Answer: 8000 dyne cm. ∎

Exercise 7.5

A child's top of mass M is spinning at an angular velocity ω. When it slows down, it leans over a bit so that the axis is at an angle θ with the vertical. The center of mass is a distance R from the tip. Determine the gravitational torque with respect to the point of contact with the floor. What is the direction of this torque? Draw a diagram indicating the direction of the rate of change of angular momentum. (This is why the top precesses.) Answer: $N = RMg\sin\theta$ perpendicular to $\boldsymbol{\omega}$ and $\mathbf{g}$. ∎

> **Exercise 7.6**
>
> A dumbbell consists of two point masses M connected by a massless rod of length d. It is rotating about an axis through its center of mass at a rate ω_0. One of the masses comes loose and flies off. Determine the angular velocity of the remaining mass. ∎

7.3.1 Angular Momentum Relative to the Center of Mass

Recall that the kinetic energy of a system of particles can be expressed as the sum of the kinetic energy of the center of mass plus the kinetic energy of all the particles with respect to the center of mass. (See Section 5.8.) A similar relation holds for the linear momentum. (See Section 6.6.) As we now demonstrate, the angular momentum of a system of particles can also be expressed as the angular momentum *of* the center of mass, plus the angular momentum *about* the center of mass.

The total angular momentum of a system of particles is

$$\mathbf{L} = \sum_i \mathbf{r}_i \times m_i \dot{\mathbf{r}}_i.$$

The position vector of particle i can be expressed as

$$\mathbf{r}_i = \mathbf{r}_c + \mathbf{r}'_i,$$

where $\mathbf{r}_c$ is the vector to the center of mass and $\mathbf{r}'_i$ is the position of particle i with respect to the center of mass (see Figure 5.13). Using this relationship, the angular momentum is

$$\begin{aligned}
\mathbf{L} &= \sum_i m_i \left(\mathbf{r}_c + \mathbf{r}'_i \right) \times \left(\dot{\mathbf{r}}_c + \dot{\mathbf{r}}'_i \right) \\
&= \sum_i m_i \mathbf{r}_c \times \left(\dot{\mathbf{r}}_c + \dot{\mathbf{r}}'_i \right) + \sum_i m_i \mathbf{r}'_i \times \left(\dot{\mathbf{r}}_c + \dot{\mathbf{r}}'_i \right) \\
&= \sum_i m_i \mathbf{r}_c \times \dot{\mathbf{r}}_c + \sum_i m_i \mathbf{r}_c \times \dot{\mathbf{r}}'_i + \sum_i m_i \mathbf{r}'_i \times \dot{\mathbf{r}}_c + \sum_i m_i \mathbf{r}'_i \times \dot{\mathbf{r}}'_i.
\end{aligned}$$

The quantities $\mathbf{r}_c$ and $\dot{\mathbf{r}}_c$ do not depend on i and can be taken out of the summations. Also note that $\sum_i m_i = M$ (the total mass). Consequently,

$$\mathbf{L} = M \left(\mathbf{r}_c \times \dot{\mathbf{r}}_c \right) + \mathbf{r}_c \times \left(\sum_i m_i \dot{\mathbf{r}}'_i \right) - \dot{\mathbf{r}}_c \times \left(\sum_i m_i \mathbf{r}'_i \right) + \sum_i m_i \left(\mathbf{r}'_i \times \dot{\mathbf{r}}'_i \right).$$

But the two middle terms are zero because $\sum_i m_i \mathbf{r}'_i = 0$ and $\sum_i m_i \dot{\mathbf{r}}'_i = 0$ by the definition of the center of mass. (The first of these is proportional to the position of the center of mass with respect to itself and the second is proportional to the velocity of the center of mass with respect to itself.) The expression for $\mathbf{L}$ then reduces to

$$\mathbf{L} = M \left(\mathbf{r}_c \times \dot{\mathbf{r}}_c \right) + \sum_i m_i \left(\mathbf{r}'_i \times \dot{\mathbf{r}}'_i \right),$$

which can be written

$$\mathbf{L} = \mathbf{L}_c + \mathbf{L}'.$$

Here $\mathbf{L}_c = M \left(\mathbf{r}_c \times \dot{\mathbf{r}}_c \right)$ is the angular momentum of a particle of mass M located at the center of mass and moving with the center of mass, and $\mathbf{L}' = \sum_i m_i \mathbf{r}'_i \times \dot{\mathbf{r}}'_i$ is the sum of the angular momenta of all the particles relative to the center of mass. That is what we set out to prove.

Exercise 7.7

Prove that $\sum_i m_i \mathbf{r}'_i = 0$.   ■

7.4 Rotation of a Rigid Body about a Fixed Axis

A rigid body is a system of particles in which all the particles are constrained to remain at constant distances from one another. If the body is rotating about a fixed axis of rotation, the particles move in circular paths centered on the axis. These paths lie in planes perpendicular to the axis of rotation. All the particles have the same angular velocity. The angular velocity $\boldsymbol{\omega}$ is a vector directed along the rotation axis whose magnitude is equal to the rotation rate. The sense of $\boldsymbol{\omega}$ is given by the right-hand rule, as illustrated in Figure 7.4. The magnitude of $\boldsymbol{\omega}$ is

$$|\boldsymbol{\omega}| = \omega = \frac{2\pi}{T} = 2\pi f,$$

where T is the period of revolution and f is the frequency.

As noted in Equation (2.7), the linear velocity $\mathbf{v}_i$ of particle i is related to the angular velocity $\boldsymbol{\omega}$ by

$$\mathbf{v}_i = \boldsymbol{\omega} \times \boldsymbol{r}_{i\perp},$$

where $\boldsymbol{r}_{i\perp}$ is a vector in the plane of the circular path from the axis of rotation to the particle. Because the angle between $\boldsymbol{\omega}$ and $\mathbf{r}_{i\perp}$ is 90°, the magnitude of $\mathbf{v}_i$ is $r_{i\perp}\omega$, as expected, and the angular momentum of particle m_i is $l_i = m_i r_{i\perp}^2 \boldsymbol{\omega}$. I wrote the distance from the axis of rotation as $\mathbf{r}_{i\perp}$ to emphasize that it is the *perpendicular* distance from the axis and for future convenience.

Now recall that angular momentum depends on the choice of origin. If we place the origin at the center of the circular path of some particle, say m_i, then it will not be at the center of the path of another particle, say m_j, that is displaced somewhat along the axis of rotation. We need an expression for the angular momentum of a particle relative to an arbitrary origin. However, to keep our analysis simple, let us restrict the location of the origin to a point on the axis of rotation. This is not the most general position, but it is sufficient for our purposes at this time.

Our generalization of the origin to an arbitrary location on the axis of rotation is illustrated in Figure 7.5.

The angular momentum of particle i can be expressed in terms of the angular velocity as follows:

$$\mathbf{l}_i = \mathbf{r}_i \times \mathbf{p}_i = m_i (\mathbf{r}_i \times \mathbf{v}_i) = m_i \mathbf{r}_i \times (\boldsymbol{\omega} \times \mathbf{r}_i). \tag{7.7}$$

Figure 7.4 The direction of the vector angular velocity $\boldsymbol{\omega}$ is given by the right-hand rule. If the fingers of your right hand curl in the direction of the motion, your thumb points in the direction of $\boldsymbol{\omega}$.

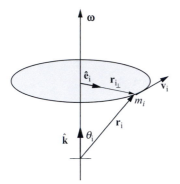

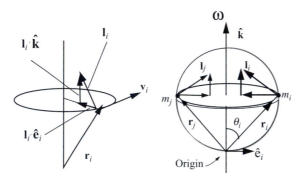

Figure 7.5 The origin is located on the axis of rotation. The vector $\mathbf{r}_i$ specifies the location of particle m_i relative to the origin. $\mathbf{r}_i$ is resolved into a component along the rotation axis ($r_i \cos\theta_i \hat{\mathbf{k}}$), and a component perpendicular to the rotation axis ($r_i \sin\theta_i \hat{\mathbf{e}}_i = \mathbf{r}_{i\perp}$). The unit vector $\hat{\mathbf{e}}_i$ points towards the instantaneous position of the particle.

Figure 7.6 The angular momentum of a particle (relative to the given origin) has components in the direction of the axis of rotation and perpendicular to it, as shown in the panel on the left. The angular momentum of a symmetric particle has an equal but opposite component perpendicular to the rotation axis.

Using the "BAC – CAB" rule for the vector triple product,[4] this becomes

$$\mathbf{l}_i = m_i \left[r_i^2 \boldsymbol{\omega} - (\mathbf{r}_i \cdot \boldsymbol{\omega})\, \mathbf{r}_i \right].$$

From Figure 7.5, we appreciate that if the z-axis is along $\boldsymbol{\omega}$, then

$$\mathbf{r}_i = r_i \cos\theta_i \hat{\mathbf{k}} + r_i \sin\theta_i \hat{\mathbf{e}}_i,$$

where θ_i is the angle between $\mathbf{r}_i$ and $\boldsymbol{\omega}$, and $\hat{\mathbf{e}}_i$ is a unit vector in the plane of the circular path and pointing (instantaneously) towards m_i, that is, along $\mathbf{r}_{i\perp}$. Using this expression for $\mathbf{r}_i$ and doing a little bit of algebra, you can show that

$$\mathbf{l}_i = m_i r_i^2 \omega \sin^2\theta_i \hat{\mathbf{k}} - m_i r_i^2 \omega \sin\theta_i \cos\theta_i \hat{\mathbf{e}}_i. \tag{7.8}$$

This expression shows us that when taken about an origin displaced along the axis, the angular momentum of particle i does not point along the axis of rotation.

Now I am going to place an important restriction on the analysis. I will only consider bodies that are symmetrical about the rotation axis. (When this restriction is lifted in Chapter 17 you will need to use tensors to describe the motion. I don't want to talk about tensors just yet, so let's assume the rotating body is symmetrical about the axis of rotation.) In a symmetrical body, for every particle i there is an identical particle j diametrically opposite to it at the same distance from the axis of symmetry. Consider the ith particle which has mass m_i and angular momentum $\mathbf{l}_i$, as shown in Figure 7.6. The jth particle is identical, except for the direction of its angular momentum, $\mathbf{l}_j$. The components of $\mathbf{l}_i$ and $\mathbf{l}_j$ parallel to the rotation axis add together, but the components perpendicular to the rotation axis cancel. So adding equations of the form (7.8) to get the total angular momentum of a symmetrical body yields

---

[4] The "BAC – CAB" rule is $\mathbf{A} \times (\mathbf{B} \times \mathbf{C}) = \mathbf{B}(\mathbf{A} \cdot \mathbf{C}) - \mathbf{C}(\mathbf{A} \cdot \mathbf{B})$.

$$\mathbf{L} = \sum_i m_i r_i^2 \omega \sin^2 \theta_i \hat{\mathbf{k}}.$$

The quantity $r_i \sin \theta_i$ is the perpendicular distance from the axis of rotation to the ith particle. If this distance is denoted by $r_{i\perp}$ we can write

$$\mathbf{L} = \left(\sum_i m_i r_{i\perp}^2 \right) \omega \hat{\mathbf{k}}. \tag{7.9}$$

Exercise 7.8

Fill in the missing steps between Equations (7.7) and (7.8). ∎

7.5 The Moment of Inertia

The quantity $\sum_i m_i r_{i\perp}^2$ in Equation (7.9) depends only on the mass of the body and how that mass is distributed with respect to the axis of rotation. For any given body rotating about some specific axis, this quantity is a characteristic constant. We call it the *moment of inertia* of the body,[5] and denote it by the letter I:

$$I = \sum_i m_i r_{i\perp}^2. \tag{7.10}$$

We can generalize Equation (7.10) to a continuous body of density ρ, and write

$$I = \int \int \int_{\text{body}} r_\perp^2 dm = \int \int \int_{\text{body}} \rho r_\perp^2 d\tau, \tag{7.11}$$

where $dm = \rho d\tau$ is a mass element of the body and $r_\perp$ is the perpendicular distance from mass element dm to the axis of rotation.

Using the definition for moment of inertia from Equation (7.10), we can write the angular momentum from Equation (7.9) in the form[6]

$$\mathbf{L} = I\boldsymbol{\omega}.$$

We now return to the relationship between the torque and the rate of change of angular momentum. If a torque $\mathbf{N}$ acts on a body there will be a change in its angular momentum given by

$$\mathbf{N} = \frac{d\mathbf{L}}{dt}.$$

Replacing $\mathbf{L}$ by $I\boldsymbol{\omega}$,

$$\mathbf{N} = \frac{d}{dt}(I\boldsymbol{\omega}).$$

---

[5] Although you studied the moment of inertia in your introductory mechanics course, you will find that this topic is more complicated than you might expect. In discussing the general motion of a rigid body we need to introduce a quantity called the *inertia tensor.* For the moment, however, we will restrict ourselves to the simple situation of a symmetrical rigid body rotating about a fixed axis and the scalar quantity I is all we need.

[6] The relationship $\mathbf{L} = I\boldsymbol{\omega}$ is based on the moment of inertia being a scalar. Then the angular momentum is parallel to the angular velocity. If the body is not symmetric about the axis of rotation, the angular momentum vector will not be parallel to the angular velocity vector. (Such a body will tend to "wobble" about the axis.)

For a body with constant moment of inertia, this becomes

$$\mathbf{N} = I\frac{d\omega}{dt}$$

or

$$\mathbf{N} = I\alpha. \tag{7.12}$$

This has the same form as $\mathbf{F} = m\mathbf{a}$ with I replacing m, α replacing $\mathbf{a}$, and $\mathbf{N}$ replacing $\mathbf{F}$.

The conservation of angular momentum for a rigid body when no external torques are acting on it is just $\mathbf{L}_i = \mathbf{L}_f$. For a symmetrical body this can be written as

$$I_i\omega_i = I_f\omega_f.$$

Exercise 7.9

A spinning disk of radius 10 cm and mass 15 g is rotating at 50 rad/s. A sticky piece of chewing gum (mass = 3 g) falls on the rim of the disk. Determine the final angular velocity of the system. Answer: 35.7 rad/s. ∎

7.5.1 Two Theorems

Two useful theorems for calculating the moment of inertia are the *parallel axis theorem* and the *perpendicular axis theorem*.

The *parallel axis theorem* states that the moment of inertia of a body about a given axis is equal to the moment of inertia about a parallel axis through the center of mass plus Md^2, where M is the mass of the body and d is the distance between the axes. The parallel axis theorem is a very valuable labor saving device. If we write $I_\parallel$ for the moment of inertia about the parallel axis and I_c for the moment of inertia about an axis that passes through the center of mass, we have

$$I_\parallel = I_c + Md^2.$$

The *perpendicular axis theorem* can be applied to plane laminar bodies. Consider two mutually perpendicular axes in the plane of the lamina. Let the moments of inertia about these two axes be I_1 and I_2. Then the moment of inertia about an axis perpendicular to the lamina and passing through the point of intersection of the first two axes is

$$I_3 = I_1 + I_2.$$

Worked Example 7.3 (a) Determine the moment of inertia of a cylindrical rod of mass M and length l about an axis passing through the center of mass of the rod (line AA), as illustrated in Figure 7.7. (b) Determine the moment of inertia about line BB at the end of the rod.

Solution (a) The mass per unit length (linear mass density) is $\lambda = M/l$. The moment of inertia about the axis (AA) is

$$I = \int\int\int r_\perp^2 dm = \int r_\perp^2 \lambda dx = \lambda \int_{-l/2}^{l/2} x^2 dx = \lambda \left[\frac{x^3}{3}\right]_{-l/2}^{l/2}$$

$$= \frac{\lambda}{3}\left[2l^3/8\right] = \frac{1}{12}Ml^2.$$

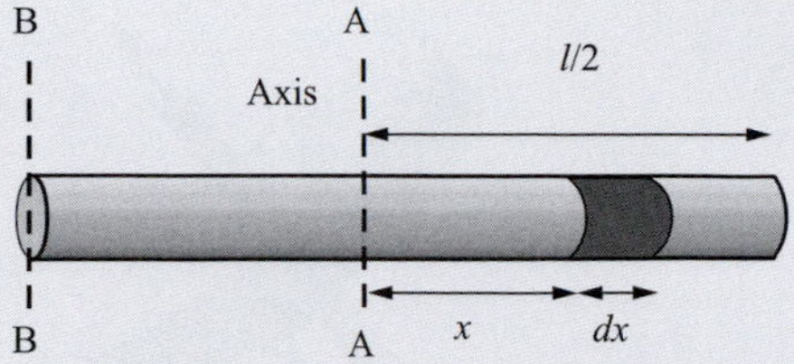

Figure 7.7 A rod of length l with axis passing through the center of mass. The mass element $dm = \lambda dx$ is a distance x from the axis.

(b) The moment of inertia about BB can be obtained by using the parallel axis theorem

$$I_\| = I_c + Md^2$$

$$= \frac{1}{12}Ml^2 + M\left(\frac{l}{2}\right)^2 = \frac{1}{3}Ml^2.$$

Exercise 7.10

Determine the moment of inertia of a rod of length l and mass M about an axis perpendicular to the axis of the rod and a distance a beyond one end of the rod. Answer: $Ma^2 + Mal + \frac{1}{3}Ml^2$. ∎

Exercise 7.11

A force of 2 N is applied tangentially on the rim of a disk of mass 15 kg and radius 20 cm. The disk is mounted on a frictionless fixed axis perpendicular to the plane of the disk and passing through its center. Determine the angular acceleration of the disk. Answer: 4/3 rad/s^2. ∎

Exercise 7.12

The moment of inertia of a uniform square plate of side a about a perpendicular axis through its center is $(1/6)Ma^2$. Determine the moment of inertia about an axis in the plane of the plate. Assume this axis passes through the center of the plate. Answer: $(1/12)Ma^2$. ∎

Exercise 7.13

From the definition $I = \int \int \int r_\perp^2 dm$, obtain the moment of inertia for a ring or hoop of mass M and radius R, relative to an axis through its center and perpendicular to the plane of the ring. Answer: MR^2. ∎

Exercise 7.14

A meter stick of mass 0.25 kg is placed on a frictionless table. It is constrained to rotate about a vertical axis at the 20 cm mark. A horizontal force of 3 N is applied perpendicular to the ruler at the 80 cm mark. Determine the angular acceleration of the ruler. Answer: 41.5 rad/s^2. ∎

7.6 The Gyroscope

Figure 7.8 illustrates a simple gyroscope. The disk has mass M and the axis has negligible mass. For convenience we place the origin of coordinates at the base (point Q in the figure).

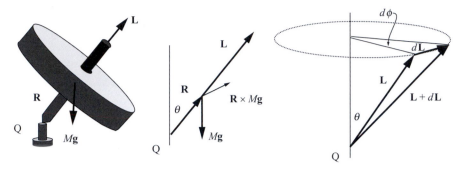

Figure 7.8 A gyroscope. The middle sketch shows the vectors involved. Note that the torque ($\mathbf{R} \times M\mathbf{g}$) is perpendicular to both $\mathbf{R}$ and $M\mathbf{g}$. The sketch to the right illustrates that the tip of the angular momentum vector traces out a circle.

The gyroscope is symmetrical, so the angular momentum vector and the angular velocity vector are parallel and point along the axis of rotation. The gravitational force acts at the center of mass, which is also on the axis of rotation, a distance R from Q. The torque about Q due to this force is

$$\mathbf{N}_Q = \mathbf{R} \times M\mathbf{g}.$$

Since $\frac{d\mathbf{L}}{dt} = \mathbf{N}$ we have

$$\frac{d\mathbf{L}}{dt} = \mathbf{R} \times M\mathbf{g} = RMg \sin\theta\, \hat{\boldsymbol{\phi}}.$$

The torque points in a direction perpendicular to both $\mathbf{R}$ and $M\mathbf{g}$ (as well as $\mathbf{L}$). In spherical coordinates this torque points in the direction of the unit vector $\hat{\boldsymbol{\phi}}$. Hence, $\mathbf{N} = N\hat{\boldsymbol{\phi}}$. Further, note that $\mathbf{L} = I\boldsymbol{\omega}$ and I is a constant, so

$$\mathbf{N} = \frac{d\mathbf{L}}{dt} = I\frac{d\boldsymbol{\omega}}{dt} = I\frac{d\omega}{dt}\, \hat{\boldsymbol{\phi}}.$$

This tells us that the angular momentum vector (as well as the angular velocity vector) is changing. It is not changing in *magnitude* because the direction of $\mathbf{N}$ is perpendicular to the direction of the angular momentum. Consequently, the angular momentum vector must be changing in *direction,* as indicated in the rightmost sketch in the figure, and, according to the formula above, $\frac{d\mathbf{L}}{dt}$ is directed along $\hat{\boldsymbol{\phi}}$.

The gyroscope *precesses,* that is, the axis of rotation moves around the vertical in such a way that the tip of the angular momentum vector traces out a horizontal circle centered on the vertical, as indicated in the figure. The precession rate Ω_p is the speed with which the tip of the angular momentum vector goes around this circle. That is,

$$\Omega_p = \frac{d\phi}{dt},$$

where ϕ is the azimuthal angle of the spherical coordinates. Figure 7.9 gives three different views of the precession of the angular momentum vector around the vertical. The leftmost figure shows the angular momentum vector and the circle traced out by the tip of this vector. Note that as time passes, the angular momentum changes from $\mathbf{L}$ to $\mathbf{L} + \Delta\mathbf{L}$. The magnitude of the angular momentum does not change, only its direction, so $\Delta\boldsymbol{L}$ lies in the plane of the circle (and is perpendicular to both $\mathbf{L}$ and $\mathbf{L} + \Delta\mathbf{L}$). The middle figure is a side view, indicating that θ is the

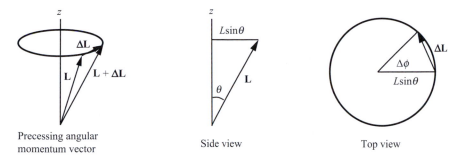

Precessing angular momentum vector

Side view

Top view

Figure 7.9 Left figure: The tip of the precessing angular momentum vector traces out a circle. Middle figure: Side view showing the angle θ between the angular momentum vector and the vertical. The horizontal component of angular momentum is $L \sin \theta$. Right figure: Top view showing that $\Delta L = L \sin \theta \Delta \phi$.

angle between $\mathbf{L}$ and the vertical, and that the horizontal component of $\mathbf{L}$ is $L \sin \theta$. The rightmost figure is a top view, looking down on the circular path, indicating that $\mathbf{\Delta L}$ is the chord of the arc subtended by the angle $\Delta \phi$. Setting the chord equal to the arc we have

$$\Delta L = L \sin \theta \Delta \phi.$$

Divide by Δt and let $\Delta t \to dt$, to obtain

$$\frac{dL}{dt} = L \sin \theta \frac{d\phi}{dt}.$$

Now recall that $\frac{dL}{dt} = RMg \sin \theta$, and $\frac{d\phi}{dt} = \Omega_p$ so

$$RMg \sin \theta = L \sin \theta \Omega_p,$$

and consequently

$$\Omega_p = \frac{RMg}{L} = \frac{RMg}{I\omega}.$$

It is interesting to note that the precession rate increases as the top slows down.

Because the angular momentum vector and the angular velocity vector ($\mathbf{\omega}$) lie along the same line, they vary together. The angular velocity vector is precessing at the same rate as the angular momentum vector.

A gyroscope also "nods" as it precesses. This is called *nutation*. We shall consider it in Chapter 17.

Exercise 7.15

A gyroscope is made of a very light axle and a heavy ring. The ring (of mass M and radius a) is connected to the axle by nearly massless spokes. The ring is a distance d from the point of the axis in contact with the floor. Determine the precession rate of this gyroscope. Answer: $\Omega_p = gd/a^2\omega$. ∎

7.7 Angular Momentum is an Axial Vector

Angular momentum is a vector, but it is a somewhat different type of vector than familiar "ordinary" vectors such as the velocity vector and the position vector. The difference between

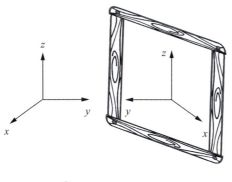

Figure 7.10 The reflection of a right-handed Cartesian coordinate system. The reflection is in the xz-plane. Note that the axis perpendicular to the reflecting plane changes direction. The reflected axes form a left-handed coordinate system.

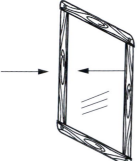

Figure 7.11 The reflection of the position vector **r** in a mirror.

the angular momentum vector and an ordinary vector is related to its behavior under a coordinate transformation. But before describing this difference, I have to tell you that there are two kinds of coordinate transformations. They are called "proper" transformations and (of course) "improper" transformations, depending on whether or not the coordinate axes retain their cyclic order. We usually draw a Cartesian coordinate system so that $\hat{\mathbf{i}} \times \hat{\mathbf{j}} = \hat{\mathbf{k}}$. Such a coordinate system is called "right-handed." Using your right hand, if you point your fingers along the x-axis and bend them into the y-axis, your thumb will be pointing in the direction of the z-axis. If you rotate a right-handed coordinate system, you still get a right-handed coordinate system, even though the axes will now be pointing in different directions. However, if you *reflect* a right-handed coordinate system, you end up with a left-handed coordinate system. This is illustrated in Figure 7.10 where the right-handed coordinate system reflected in the xz-plane is transformed into a left-handed coordinate system. This is, of course, an example of an *improper* coordinate transformation.

Consider the behavior of a vector under a transformation of the coordinate system. To begin, consider the behavior of the position vector **r**. The components of **r** may or may not change sign during a coordinate transformation. That detail is not important. However, what is important is whether or not the components of *other* vectors have the *same* sign changes as the components of **r** under the same transformation.

We call **r** a "polar" or "true" vector. Feynman called it an "honest" vector.[7] Many other vectors, such as velocity, momentum, and force, are polar vectors. They behave the same way as **r** under a coordinate transformation such as a rotation or a reflection.

For example, if you reflect the vector **r** it changes direction. (More precisely, the component of **r** perpendicular to the plane of reflection, changes sign.) See Figure 7.11. Under a similar reflection, velocity and force behave the same way.

[7] Richard P. Feynman, *The Feynman Lectures on Physics,* Addison-Wesley, Reading, MA, 1963.

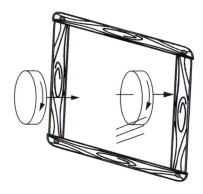

Figure 7.12 The angular velocity vector does not change directions under reflection.

There are other vectors, particularly those related to rotations, that do not transform like **r**. These are called "axial" or "pseudo" vectors. Recall that angular velocity $\boldsymbol{\omega}$ was defined to be a vector lying along the axis of rotation. The *sense* or direction of $\boldsymbol{\omega}$ was given by the right-hand rule (see Figure 7.4). If you observe a rotating wheel in a mirror, you see that the angular velocity vector does *not* change direction. This is illustrated in Figure 7.12. Angular momentum is another axial vector. The component perpendicular to the plane of reflection does not change direction. Therefore, angular velocity and angular momentum do not transform like the "honest" vector **r**. This means that $\boldsymbol{\omega}$ and **L** are axial vectors. Axial vectors behave differently than **r** under improper transformations.

You will study the transformation properties of vectors in much more detail in a course in mathematical methods. At this stage, it is sufficient for you to realize that there are different kinds of vectors and that they are distinguished by their transformation properties.

7.8 Summary

The angular momentum of a particle is defined as

$$\mathbf{l} = \mathbf{r} \times \mathbf{p}$$

and, consequently, it depends on the choice of origin. Since

$$\frac{d\mathbf{l}}{dt} = \frac{d}{dt}(\mathbf{r} \times \mathbf{p}) = \mathbf{r} \times \mathbf{F} = \mathbf{N},$$

we appreciate that if $\mathbf{N} = \mathbf{0}$ then $\mathbf{l} = $ constant. This relation summarizes the law of conservation of angular momentum.

The angular momentum of a system of particles (or an extended body) is

$$\mathbf{L} = \Sigma \mathbf{l}_i,$$

and the law of conservation of angular momentum can be written

$$\frac{d\mathbf{L}}{dt} = \mathbf{N}_{\text{tot}}^{(e)},$$

where $\mathbf{N}_{\text{tot}}^{(e)}$ is the net torque due to all external forces. Internal forces do not contribute to the rate of change of angular momentum.

The angular momentum of a system of particles can be written as the sum of the angular momentum of the center of mass and the angular momentum relative to the center of mass, that is,

$$\mathbf{L} = \mathbf{L}_c + \mathbf{L'},$$

where

$$\mathbf{L}_c = M(\mathbf{r}_c \times \dot{\mathbf{r}}_c),$$

and

$$\mathbf{L'} = \Sigma m_i(\mathbf{r}'_i \times \dot{\mathbf{r}}'_i).$$

The angular momentum of a symmetrical body rotating about an axis of symmetry is

$$\mathbf{L} = \left(\sum_i m_i r_{i\perp}^2\right)\boldsymbol{\omega}.$$

The quantity in parenthesis is the moment of inertia of the body and is denoted I. Therefore,

$$\mathbf{L} = I\boldsymbol{\omega},$$

and we can write

$$\mathbf{N}_{\text{tot}}^{(e)} = \frac{d\mathbf{L}}{dt} = \frac{d}{dt}I\boldsymbol{\omega} = \frac{d}{dt}I\boldsymbol{\omega} = I\boldsymbol{\alpha}.$$

If $\mathbf{N}_{\text{tot}}^{(e)} = 0$, the angular momentum is constant and consequently

$$\mathbf{L}_f = \mathbf{L}_i \Rightarrow I_f\boldsymbol{\omega}_f = I_i\boldsymbol{\omega}_i.$$

The motion of a gyroscope illustrates the concepts presented in this chapter. The torque due to gravity causes the angular momentum vector to precess at a rate

$$\Omega_p = \frac{RMg}{I\omega}.$$

7.9 Problems

Problem 7.1 Two particles of equal charge and equal mass travel at the same speed v but in opposite directions. Their initial trajectories are straight lines separated by a distance b. Owing to their mutual repulsion, they will reach a point of closest approach and then move away from each other. The electrical force between the charges is given by $\mathbf{F} = (q^2/4\pi\epsilon_0 r^2)\hat{\mathbf{r}}$. The vector $r\hat{\mathbf{r}}$ is drawn from one particle to the other. Although it is unrealistic to do so, ignore the magnetic force between the moving charges. (a) Draw the trajectories of the two particles, indicating the electric forces acting on them. (b) Explain whether or not the angular momentum of the system is constant. (c) Do your answers depend on your choice of coordinate origin?

Problem 7.2 A uniform rod of length d and mass M is moving at a constant velocity v in the direction of its length. (a) Obtain its angular momentum relative to an arbitrary origin and show that this is the same as the angular momentum of a particle of mass M moving along the trajectory of the rod. (b) If the rod is inclined at an angle α to its velocity vector, what is the position of the equivalent particle?

Problem 7.3 Prove that the total angular momentum of a system of two particles is independent of the displacement of the origin only if the total linear momentum is zero or if the displacement of the origin is parallel to the total linear momentum.

Problem 7.4 In proving the law of conservation of angular momentum for an extended body (or system of particles) we assumed that Newton's third law is obeyed in the strong form. In this problem you will investigate this assumption for the case of particles interacting through magnetic forces. (The magnetic force between moving charged particles is not directed along the line joining the particles.) The orbital motion of an electron around the nucleus is equivalent to a small current loop, which we usually call a "magnetic

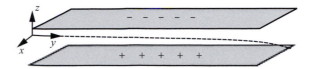

Figure 7.13 An electron between charged plates.

moment." A magnetic moment generates a magnetic field. Another electron orbiting around another nucleus will feel a force due to this magnetic field. Let us generalize the problem to two current loops of arbitrary shape that carry currents I_1 and I_2. Considering small elements of the current loops, $I_1 d\mathbf{l}_1$ and $I_2 d\mathbf{l}_2$, the force on $I_1 d\mathbf{l}_1$ due to $I_2 d\mathbf{l}_2$ is

$$d\mathbf{F}_{12} = I_1 d\mathbf{l}_1 \times \mathbf{B},$$

where

$$\mathbf{B} = \frac{\mu_0}{4\pi} \oint_2 \frac{I_2 d\mathbf{l}_2 \times \mathbf{r}_{12}}{|\mathbf{r}_{12}|^3}.$$

Show that the force between the two current loops obeys Newton's third law in the strong form. (Hint: This involves showing that $d\mathbf{l}_1 \cdot \mathbf{r}_{12}/|\mathbf{r}_{12}|^3$ is an exact differential.)

Problem 7.5 An electron is fired horizontally with speed v_{oy} between parallel charged electrical plates. Assume the plates are oriented horizontally and that the top plate is negatively charged and the bottom plate is positively charged, so the electron feels a uniform force downward. Denote this force by F_e. Using Cartesian coordinates, calculate the angular momentum as a function of time and show that the time rate of change of angular momentum is equal to the torque exerted on the electron by the electric field. Evaluate the angular momentum (and torque) relative to an origin located at the position of the electron when it first enters the electric field. See Figure 7.13.

Problem 7.6 In analyzing the collision of two particles in Chapter 6 we stated that the conservation of linear momentum ($\mathbf{p} = $ constant) led to three equations describing the motion. Because there are no external torques acting on the system, it might seem that conservation of angular momentum ($\mathbf{l} = $ constant) would lead to three additional equations. Show that this is not true, that is, in the collision problem, the conservation of angular momentum does not lead to any new relations.

Problem 7.7 A roll of rather heavy gift wrapping paper is placed on a frictionless roller and suspended by a bracket of length b as shown in Figure 7.14. The roll is in contact with the wall. The coefficient of static friction between wall and paper is μ. A portion d of the paper is hanging down. Assume the total length of paper on the roll is D and that it has a linear mass density λ. Determine the length d such that the paper will unroll on its own. At the instant the roll begins to unwind on its own, it has a radius R.

Problem 7.8 Professor Ptolemy of the Astronomy Department claims that there is a planet with the same mass as Earth that is directly on the other side of the Sun in an orbit of the same radius as Earth's. This planet, he says, has the same mass as Earth, but it is cylindrical in shape. The axis of the cylinder lies in the orbit plane and its length is one Earth diameter. Determine the torque the Sun exerts on this planet, as a function of the angle between the axis of the cylinder and a line from the planet to the Sun. You may assume the angle between the axis of the cylinder and the line to the Sun is small. (Hint 1: Determine the gravitational force on the planet, then find the center of gravity and finally calculate the torque. By the way, if there were a planet on the other side of the Sun from us, we would be able to see it at times, because of the eccentricity of Earth's orbit. But in this problem you may assume circular orbits. Hint 2: The center of gravity is the point in the body where the gravitational force acts. In this problem it is not at the same point as the center of mass.)

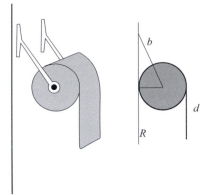

Figure 7.14 A roll of paper that will unwind under its own weight.

Problem 7.9 A neutron star whose mass is 1.5 times the mass of the Sun has collapsed to a sphere of radius 10 km. If the original angular velocity was one revolution per month (before the collapse), what is its final angular velocity? Assume the initial density of the star was equal to the mean density of the Sun.

Problem 7.10 Prove the parallel axis theorem.

Problem 7.11 Prove the perpendicular axis theorem.

Problem 7.12 (a) Derive an expression for the moment of inertia of a disk of mass M and radius R about an axis perpendicular to its plane and passing through its center of mass. (b) Obtain an expression for the moment of inertia about a parallel axis that is tangent to the edge of the disk.

Problem 7.13 Determine the moment of inertia of an annular cylinder (or ring) about an axis perpendicular to the plane of the ring and passing through its center. The ring has uniform density, mass M, inner radius R_1, and outer radius R_2.

Problem 7.14 Derive an expression for the moment of inertia of a uniform, solid sphere about a diameter.

Problem 7.15 Determine the moment of inertia of a flat disk of mass M and radius R about an axis tangent to the edge of the disk and lying in the plane of the disk.

Problem 7.16 Assume the Earth is a sphere of mass M and radius R. Remove the Southern Hemisphere. By direct integration determine the moment of inertia of the remaining hemisphere about an axis perpendicular to the equatorial plane and passing through the North Pole.

Problem 7.17 A disk of mass M and moment of inertia I has a hole of radius a in its center. An axle, which is slightly smaller in diameter than the hole, passes through the hole. The axle is horizontal and the plane of the disk is vertical. The disk spins smoothly on the axle. There is friction between the disk and the axle, and the coefficient of friction is μ. Assume the disk has an initial angular speed ω_0. Determine the number of turns and the time for the disk to stop. (Hint: The point of contact is not over the center of mass.)

Problem 7.18 A dumbbell is made up of a rod of length l and mass m connected on either end to spheres of mass M and radius R. (a) Determine the moment of inertia of the dumbbell about an axis perpendicular to the rod and going through its center. (b) You suspend the dumbbell from its center of mass and apply a force of 1.5 N at the center of mass of one of the spheres. The force is in the horizontal plane and perpendicular to the rod. Determine the angular acceleration of the system. Assume $M = 1$ kg, $R = 10$ cm, $l = 20$ cm, and $m = 0.1$ kg.

Problem 7.19 The gyroscope of Figure 7.8 is made up of a disk of mass M and radius a and an axis of negligible mass that is free to swivel in any manner about the bottom end. It is inclined at an angle $\theta = 90°$ so the axis is horizontal and the gyroscope is precessing. The disk spins about the axis with angular speed ω.

Write an expression for the total angular velocity vector in Cartesian coordinates in terms of M, a, R, ω, the time, and any other appropriate parameters.

Problem 7.20 A gyroscope that is free to move about the support point is in the upright position and is spinning at an angular velocity $\omega \hat{\mathbf{k}}$ (see Figure 7.8). Assume the axis has a length d and negligible mass. The disk of mass M and radius a is mounted at $d/2$. A frictionless ball-bearing device allows you to grasp the top of the axis and pull it to a horizontal orientation. After pulling the gyroscope down, you hold it stationary with the axis horizontal. (a) How much work did you do to rotate the gyroscope axis? (b) What is the final angular velocity of the gyroscope? (c) What force do you need to exert on the tip of the axis to keep it stationary? You may assume that ω is large so Ω_p is small enough that the energy associated with precession can be neglected.

Problem 7.21 A string is partially wound around a cylindrical rod of radius a, as shown in Figure 7.15. A particle of mass m is tied to the end of the string. The mass m is given an initial velocity v_0 and the string winds itself onto the rod. Make the unrealistic assumption that there is no gravitational force. Show that the linear velocity of the mass is constant. Do this is in the following two ways: (a) using the conservation of energy, and (b) using the fact that the rate of change of angular momentum is equal to the torque. (Hint: There is a torque on the particle because the rod does not have zero radius and the instantaneous velocity of the particle is perpendicular to the string at all times.)

Problem 7.22 A pendulum consists of a bob of mass M and a massless string of length l. The pendulum is initially at rest. It is struck by a bullet of mass m and velocity v. Let $m = 0.1M$. Assume the pendulum hangs vertically and the bullet travels horizontally. (a) Determine the initial angular velocity of the pendulum if the bullet becomes embedded in the bob. (b) Determine the initial angular velocity of the pendulum bob if the bullet bounces back off the bob elastically.

Problem 7.23 Generalize the derivation of $\frac{d\mathbf{L}}{dt} = \mathbf{N}$ to the case of a moving origin. If the origin (Q) is at $\mathbf{r}_Q$, and $\mathbf{L}_Q$ and $\mathbf{N}_Q$ are the total angular momentum and torque with respect to Q show that

$$\frac{d\mathbf{L}_Q}{dt} = \mathbf{N}_Q$$

as long as the point Q has zero acceleration or is accelerating along the line from Q to the center of mass.

Problem 7.24 Consider a symmetrical rigid body to be composed of N particles of masses m_i ($i = 1, 2, \ldots, N$). Assume the body is rotating about a fixed axis with angular speed ω. Prove that the kinetic energy is $\frac{1}{2}I\omega^2$, where I is the moment of inertia.

Problem 7.25 A particle moves in the field of an attractive central force whose potential is given by $V = -k/r$. Show that the angular momentum does not change.

Problem 7.26 Figure 7.16 is a view from above, showing two identical disks of mass M and radius R on a frictionless surface. One disk is at rest, the other is rotating counterclockwise with angular velocity ω and is moving with a linear velocity $v = \frac{1}{2}\omega R$. It makes a grazing collision with the second disk at point P. After the collision the two disks stick together. Determine the final angular momentum of the system with respect to P.

Figure 7.15 The speed of the particle m does not increase as the string winds up on the cylindrical rod.

m

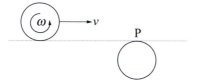

Figure 7.16 Two disks collide and stick together. See Problem 7.26.

Computational Project

Computational Project 7.1 A playground merry-go-round has been fitted with frictionless bearings. If air resistance can be ignored, this "ideal merry-go-round" will rotate indefinitely. Suppose that it is spinning at 3 rad/s when it starts to snow. The rate at which snow accumulates on the disk is 1 g/s. The disk has a radius of 1.5 m and a mass of 40 kg. Determine the angular speed of the merry-go-round after 30 min of steady snowfall. Plot ω vs. t.

8 Conservation Laws and Symmetries

8.1 Emmy Noether (Historical Note)

Emmy Noether is recognized as one of the greatest mathematicians of the twentieth century. She was born in Germany in 1882 to an intellectual Jewish family and died in the United States in 1935. Emmy trained as a language teacher, but after passing the qualifying exams to teach, she decided to study mathematics at the University of Erlangen. At that time in Germany a university education was limited to men, although women were allowed to attend classes if given permission by the professor. (She was half of the total female student body at that university.) She spent a semester at the University of Gottingen, at that time a world leader in mathematics and physics. There she attended lectures from a number of leading mathematicians, including Hermann Minkowski (who you will run into in Chapter 20) and Karl Schwarzschild (whose theory of black holes you will encounter in Chapter 9).

Then the rules changed and women were actually allowed to register as students at German universities. She returned to Erlangen and was awarded a doctorate in mathematics in 1907. She taught mathematics courses at Erlangen for several years, publishing a number of significant papers and gaining recognition as a first-rate mathematician. The mathematics department at Gottingen invited her back to work on problems involving Einstein's theory of relativity. In short order she resolved a problem with the relativistic conservation of energy. Einstein wrote to a friend, "The old guard at Gottingen should take some lessons from Miss Noether! She seems to know her stuff." But at Gottingen she found herself in an extremely unpleasant situation because the professors of history and philosophy objected strongly to a female professor. She found herself teaching classes which were officially listed as being taught by a male member of the department, and she was not paid. Eventually, however, the mathematics department prevailed and she was given a faculty position.

She remained at Gottingen until April 1933 when she was fired because the Nazi-dominated government passed a law forbidding Jews to teach at German universities. She continued to meet with students, holding classes in her home, but she realized that she would have to leave Germany. International efforts by mathematicians and physicists eventually paid off and she was offered a position at Bryn Mawr, a women's college in Pennsylvania near Philadelphia and only 50 miles from Princeton University, where she collaborated with several mathematicians and presented lectures at the Institute for Advanced Studies.[1] In April 1935 she underwent an operation for a pelvic tumor and died a few days later. Einstein wrote her obituary for the *New York Times* stating that "She found in America grateful pupils whose enthusiasm made her last years the happiest and perhaps the most fruitful of her entire career."

---

[1] At that time, Einstein was a member of the Institute for Advanced Studies.

Emmy Noether was not interested in the trivial concerns of everyday life. Her only interest was mathematics and she would converse happily with mathematicians for hours. She was respected for her generosity and empathy and often gave her students credit for work which she had mainly done herself. Although life had treated her harshly, she never complained and seemed unable to see evil in other people.[2]

Her main contributions to mathematics were in the field of abstract algebra. She is recognized in mathematical circles for subjects such as "Noetherian rings" and "Noetherian modules." A lunar crater and a minor planet are named after her, but physicists remember her for "Noether's theorem" that relates conserved physical quantities with symmetries; it is the main concept considered in this chapter.

8.2 Symmetry

Nature exhibits many symmetries, from the structure of atoms to the patterns of snowflakes. As physicists, we exploit the concept of symmetry to understand the nature of physical reality. On a more mundane level, we find symmetry concepts to be a valuable aid in solving physics problems.

All of us have an intuitive concept of what is meant by symmetry. We recognize spatial symmetry in flowers and snowflakes and in the bodies of most living creatures. We say these things "look the same" when viewed from different points of view. For example, a snowflake looks the same when rotated through 60° (known as hexagonal symmetry). A human body looks the same when viewed in a mirror. A flower vase looks the same when rotated about a vertical axis. Although its mirror image may be indistinguishable from the vase itself, it will look different when it is turned upside down. A physicist would say the vase has rotational symmetry about a vertical axis and symmetry under reflection, but it does not have symmetry under a rotation about a horizontal axis.

From these examples you can appreciate that a symmetry is related to an invariance under some particular operation. In simpler words, a symmetry means that a system is not changed when something is done to it. If there is no way we can tell whether or not a vase has been rotated through some angle, then it has rotational symmetry. If a pendulum oscillates in exactly the same way on one side of the room as on the other, then it has translational symmetry. A mechanical system that does not change in time has temporal symmetry.

Among the simpler symmetry operations are spatial translations, rotations, reflections, and translations in time. Other symmetry operations are not as familiar to you. For example, "charge conjugation" symmetry refers to interchanging positive and negative charges. Does an "antihydrogen atom" with a negative proton and a positive electron behave the same as an ordinary hydrogen atom? If so, it has charge conjugation symmetry. Another way of expressing invariance under charge conjugation is to call it matter–antimatter symmetry. This means that matter and antimatter particles are attracted to one another by the gravitational and electromagnetic forces and appear to be very much the same except for the fact that when they come into contact they annihilate one another!

An interesting symmetry is called "chirality."[3] This is the symmetry between your left hand and the mirror image of your right hand. There is no combination of translations or rotations that will transform one hand into the other, but a mirror reflection does produce the change.

---

[2] You can find more information about her in *Emmy Noether's Wonderful Theorem* by Dwight E. Neuenschwander, Johns Hopkins University Press, 2017.

[3] The name comes from the Greek word for "hand." The word "chiral" is pronounced "kai-ral" and rhymes with "spiral."

This is important in chemistry because some molecules are chiral, that is, they exist in both "left-hand" and "right-hand" forms, which may have different properties, such as their effect on polarized light. For example, proteins in sugars contain the "right-handed" form of the alanine molecule. This molecule also comes in a mirror imaged or "left-handed" form. As a result of some evolutionary quirk, biological systems (such as you and me) can only digest the right-handed alanine. However, it is possible to manufacture "left-handed" sugar that tastes the same as "ordinary" sugar, but it is not digested. This product is useful for diabetics and people who are avoiding sugar for medical reasons. It has the drawback of being expensive to produce.

For a simple example, consider a translation operation. Suppose a particle is lying on a smooth, infinite, featureless table. If we slide the particle to some other point on the table, all of its physical properties are unchanged. The system has symmetry under horizontal translations. On the other hand, if we move the particle vertically, its potential energy changes. The state of the system is different. The system exhibits symmetry for translations along the two horizontal axes but not for translations along the vertical axis.

If a system behaves the same at a later time as at an earlier time, we say it is symmetric with respect to a translation in time (temporal symmetry). Invariance with respect to time is an important property of a clock. There is a more complex symmetry operation involving time called "time reversal." Time reversal is analogous to spatial reflection. Imagine a box full of Mexican jumping beans. Suppose you used your smartphone and made a video of the beans while they were jumping wildly all over the place. There are apps that allow you to run the video backwards. If you show the video to your friends, they will not be able to tell whether it is running forward or backward. A box full of Mexican jumping beans is symmetric under time reversal. But if you make a video of your friend jumping off a diving board into a pool, you will certainly know if it is running backwards.

8.3 Symmetry and the Laws of Physics

The *laws of physics* are a set of relations between physical quantities (usually expressed in mathematical form) that describe how the universe behaves. For example, $\mathbf{F} = \frac{d\mathbf{p}}{dt}$ and the conservation of energy are laws of physics.

Do the laws of physics exhibit symmetry? That is, do the laws of physics remain constant under symmetry operations? Suppose I use some equipment and verify that for a constant mass system the relation $\mathbf{F} = m\mathbf{a}$ is true in my laboratory. If I perform a spatial translation by taking my equipment to your laboratory, will $\mathbf{F} = m\mathbf{a}$ still be true? I am sure you will agree that Newton's second law is invariant under a space translation. But is $\mathbf{F} = m\mathbf{a}$ invariant under a reflection? If you look at a physical system in a mirror, will Newton's law be $\mathbf{F} = m\mathbf{a}$ or will it be $\mathbf{F} = -m\mathbf{a}$? If you recall that both $\mathbf{F}$ and $\mathbf{a}$ are polar ("honest") vectors, you can easily convince yourself that in the reflected system the law is still $\mathbf{F} = m\mathbf{a}$. Would $\mathbf{F} = m\mathbf{a}$ hold in a time reversed system? (The answer is yes.) You see that this question of the symmetry of physical laws is rapidly becoming complicated!

Perhaps the most surprising thing about symmetry and the laws of physics is the relationship between symmetries and conservation laws. This relationship is expressed in Noether's theorem. The theorem tells us that:

For every symmetry there is a corresponding constant of the motion.

In the next section we shall explore this relationship using Lagrange's equations.

8.4 Symmetries and Conserved Physical Quantities

Consider a physical system. If it is symmetrical with respect to a rotation, then it is unchanged by the rotation. The system is the same after it was rotated as it was before.

A physical system can be described in many ways; for example, you could generate a table giving all of the physical properties of the system (velocity, position, etc.) as functions of time. However, a much better and simpler way to describe a physical system is to give its Lagrangian. If you can write down the Lagrangian, you know all the essential mechanical properties of a physical system.[4] You can use Lagrange's equations to determine the equations of motion and you can solve them to evaluate how the system will evolve in time.

Suppose the Lagrangian of some system does not contain a particular coordinate. Specifically, suppose that the angle ϕ does not appear in the Lagrangian. Remember that we call such a coordinate "ignorable" and as we showed in Chapter 4, the generalized momentum conjugate to an ignorable coordinate is constant. The generalized momentum conjugate to the angle ϕ is the angular momentum, so for this system, the angular momentum is constant.

That is, if a system is symmetrical with respect to a rotation, then the Lagrangian of the system does not depend in any way on ϕ and ϕ will not appear in the Lagrangian. Consequently, the angular momentum associated with rotations through ϕ will be conserved. Similarly, if a system is symmetrical with respect to a translation, then its *linear* momentum will be conserved.

Let us consider these concepts in more detail.

8.4.1 Conservation of Linear Momentum

Imagine a particle in a homogeneous region of space. By spatial homogeneity we mean that the space is the same at all points. A particle in this space will exhibit translational symmetry. We now show analytically that its momentum must remain constant.

The Lagrangian for such a particle cannot depend on its position, so instead of $L = L(q,\dot{q},t)$ the Lagrangian is just $L = L(\dot{q},t)$. (I am considering a single generalized coordinate for simplicity.) Lagrange's equation is

$$\frac{d}{dt}\frac{\partial L}{\partial \dot{q}} - \frac{\partial L}{\partial q} = 0.$$

Since L does not depend on q, this relation reduces to

$$\frac{d}{dt}\frac{\partial L}{\partial \dot{q}} = 0.$$

Therefore,

$$\frac{\partial L}{\partial \dot{q}} = \text{constant}.$$

But recall that

$$\frac{\partial L}{\partial \dot{q}} = p = \text{generalized momentum}.$$

Therefore, $p = $ constant and the proposition is proved.

Exercise 8.1

Show that if a system is symmetric under rotations through some angle ϕ then the angular momentum conjugate to ϕ is constant.   ∎

---

[4] To predict the future evolution of the system you would also like to have a set of initial conditions.

8.4.2 Conservation of Energy

Perhaps the most surprising of the symmetry/conservation relationships involves temporal symmetry. A system that is symmetrical with respect to a translation in *time* exhibits conservation of *energy*. If the system does not depend on time in any way, the Lagrangian is not a function of time. Therefore, instead of $L(q,\dot{q},t)$ the Lagrangian will be $L = L(q,\dot{q})$. Consequently, $\partial L/\partial t = 0$. Recall that if the constraints and the transformation equations and the potential energy are all time independent, the Hamiltonian is constant and equal to the total energy (see Section 4.9). If a system does not depend explicitly on time, we expect these conditions to be met. Therefore, it is sufficient to prove that if the Lagrangian is not a function of time, then the Hamiltonian is constant.

If L does not depend explicitly on t, then $\partial L/\partial t = 0$, but it could have an implicit time dependence if q or $\dot{q}$ depend on time. That is, the total time derivative of L will be

$$\frac{dL}{dt} = \frac{\partial L}{\partial \dot{q}}\frac{d\dot{q}}{dt} + \frac{\partial L}{\partial q}\frac{dq}{dt}.$$

By Lagrange's equation,

$$\frac{\partial L}{\partial q} = \frac{d}{dt}\frac{\partial L}{\partial \dot{q}},$$

so

$$\frac{dL}{dt} = \frac{\partial L}{\partial \dot{q}}\frac{d\dot{q}}{dt} + \frac{dq}{dt}\frac{d}{dt}\frac{\partial L}{\partial \dot{q}}.$$

Now note that

$$\frac{d}{dt}\left(\dot{q}\frac{\partial L}{\partial \dot{q}}\right) = \frac{\partial L}{\partial \dot{q}}\frac{d\dot{q}}{dt} + \frac{dq}{dt}\frac{d}{dt}\left(\frac{\partial L}{\partial \dot{q}}\right),$$

or

$$\dot{q}\frac{d}{dt}\frac{\partial L}{\partial \dot{q}} = \frac{d}{dt}\left(\dot{q}\frac{\partial L}{\partial \dot{q}}\right) - \frac{d\dot{q}}{dt}\frac{\partial L}{\partial \dot{q}},$$

so

$$\frac{dL}{dt} = \frac{\partial L}{\partial \dot{q}}\frac{d\dot{q}}{dt} - \frac{\partial L}{\partial \dot{q}}\frac{d\dot{q}}{dt} + \frac{d}{dt}\left(\dot{q}\frac{\partial L}{\partial \dot{q}}\right).$$

The first two terms on the right cancel, leaving us with

$$\frac{dL}{dt} = \frac{d}{dt}\left(\dot{q}\frac{\partial L}{\partial \dot{q}}\right),$$

or

$$\frac{d}{dt}\left[L - \dot{q}\frac{\partial L}{\partial \dot{q}}\right] = 0.$$

The definition of the Hamiltonian is

$$H = p\dot{q} - L,$$

where $p = \frac{\partial L}{\partial \dot{q}}$. Therefore the quantity in brackets is $-H$ and

$$\frac{d}{dt}[-H] = 0.$$

Consequently,

$$H = \text{constant}.$$

That is, for a system in which L is independent of the time, the Hamiltonian (and hence the energy) is constant.

8.5 Are the Laws of Physics Symmetrical?

So far we have seen that if a physical system exhibits some sort of symmetry, it will be characterized by having a corresponding conserved quantity. (Thus, we showed that if the system is symmetrical with respect to translations, then its linear momentum is constant, and if it is symmetrical with respect to time, its energy is constant.)

However, we have not shown that the *laws of physics* are invariant under symmetry operations, such as translations, reflections, inversions, and so on.

The discussion so far could lead you to believe that the laws of physics are invariant under any kind of transformation, but this is not true. For example, Galileo noted that the laws of physics are not invariant under a change of scale. To use his simple example, the bones of a dog would be shaped differently if dogs were as big as horses. Feynman pointed out that a model cathedral built out of toothpicks has very different properties from a full-sized cathedral. There are transformations (such as change in scale) under which physical relationships do not remain invariant. Aeronautical engineers are well aware that a scale model airplane in a wind tunnel does not behave the same as the full-scale airplane. However, you probably expect the laws of physics to be invariant under a reflection or a translation. In fact, the force of gravity, the electromagnetic force, and the strong nuclear force all exhibit symmetry under reflections and translations. On the other hand, as we saw in Section 3.5, a rotating water-filled bucket exhibits different behavior than a bucket full of water at rest. In Chapter 15 we will see that $F = ma$ takes on a more complicated form in a rotating coordinate frame. A less well-known lack of symmetry is related to the parity operation.

8.5.1 Nonconservation of Parity

The parity operation is essentially a quantum mechanical operation, but in classical terms we can think of it as a reflection followed by a 180° rotation about the reflected axis. The effect of this inversion operation is to reverse each coordinate axis; in a parity operation all three coordinates change sign:

$$(x, y, z) \rightarrow (-x, -y, -z).$$

A physical system can be assigned a value for parity, either even or odd, depending on how it behaves under a parity operation.[5]

This can be more easily understood in terms of a specific example. Consider a classical system described by a position vector $\mathbf{r}$ and an angular momentum $\mathbf{L}$. The position vector has odd parity because under the parity operation $\mathbf{r} \rightarrow -\mathbf{r}$. Likewise, the linear momentum vector $\mathbf{p} = m\mathbf{v}$ has odd parity. The angular momentum vector, however, has even parity because it is given by $\mathbf{L} = \mathbf{r} \times \mathbf{p}$. (The spin vector $\mathbf{S}$ also has even parity.)

---

[5] In quantum mechanics the wave function describing a system is characterized by its parity; for example, the parity of a system with orbital angular momentum number l is $(-1)^l$. If the wave function is unchanged under an inversion, we say the system has "even" parity, but if it is transformed into the negative of the original wave function, we say it has "odd" parity.

It seems reasonable that the parity of a physical system should remain constant. The invariance of the parity of a physical system is referred to as "conservation of parity." For example, in the reaction

$$\pi^- + d \rightarrow n + n,$$

the total parity of the products (the two neutrons) is the same as the total parity of the pion and deuteron. (When we speak of the parity of a particle, we mean the parity of the quantum-mechanical wave function that describes the particle.)

In the 1950s there was a great deal of interest in interactions involving elementary particles. Physicists were surprised to discover that parity is not conserved for particles undergoing a particular decay process called beta decay. This discovery was related to a problem that high-energy physicists called the tau–theta puzzle. The particle known as the τ meson was observed to decay into three pions, and the final state had odd parity. The particle known as the θ meson decayed into two pions and the final state had even parity. But the tau and theta mesons were identical in all respects except for their decay products. They had to be the same particle! The tau–theta puzzle was solved in 1956 by T. D. Lee and C. N. Yang who suggested that the tau and theta mesons were indeed the same particle but that parity was not conserved in the decay process. Their idea was verified by a series of clever experiments carried out by C. S. Wu.

The nonconservation of parity was important because it shed light on the weak force which is responsible for beta decay. One can think of the weak force as analogous to the electric force between charged particles. In fact, the elementary particles have a "weak charge" which is similar to electric charge in some respects. (Electric forces are transferred by photons, whereas weak forces are transferred by particles called the W and Z bosons.) The weak charge of the proton was measured in 2018.[6]

High energy physicists questioned some other conservation laws. They denoted the parity operation by P. The operation called charge conjugation, in which a particle is transformed into its antiparticle was denoted by C. It was found that in nature there are certain reactions involving elementary particles in which CP is found to be violated. (The CP operation involves changing all the particles into antiparticles, and then reflecting all the axes.)[7]

8.6 Strangeness (Optional)

Conservation laws are used in an unusual but interesting way by high energy physicists. In studying the elementary particles, it is found that some reactions never occur. There is no apparent reason why a reaction such as

$$\pi^- + p \rightarrow \pi^\circ + \Lambda$$

is never observed. It does not violate any of the everyday conservation laws. But it does not happen. Using the principle that "what is not forbidden is required," high-energy physicists decided that there must be a conservation principle at work. They called it "conservation of strangeness." The way it works is this: Each particle is given a "strangeness quantum number." For

---

[6] "Precision measurement of the weak charge of the proton," The Jefferson Lab Q_{weak} Collaboration, *Nature*, **557**, 207, (2018).

[7] However, relativistic quantum mechanics shows that CPT must be invariant in any physical process. Here the letter T stands for the time reversal process. It is interesting to note that CP violation was confirmed experimentally at the Stanford Linear Accelerator Center (SLAC) in 2001 based on the observation of 32 million decay events by a 1200 ton detector. The violation of CP explains why the universe is predominantly made of ordinary matter rather than being 50 percent antimatter. (See Colin Macilwain, "Physicists Show What Really Matters," *Nature*, **412**, 105 (2001).)

example, the strangeness of a π particle is 0, that of a proton is also 0, and the strangeness of a Λ particle is -1. Note that the total strangeness on the right-hand side of the reaction is not equal to the total strangeness on the left-hand side. Therefore, in this reaction, strangeness is not conserved and the reaction is "forbidden." This is not any weirder than requiring that linear momentum be conserved during a collision. However, we feel at home with momentum conservation because we have an analytical expression for momentum and we know that conservation of momentum implies that no net external forces are acting on the system. We do not have an analytical expression for strangeness nor do we know the symmetry implied by strangeness conservation. We can be sure, however, that there is some symmetry in nature that requires strangeness to be conserved.[8]

8.7 Symmetry Breaking

You may have heard the expression "symmetry breaking." This is used to describe a situation in which a symmetry ceases to exist. Consider, for example, the freezing of a liquid. An atom in a liquid is as likely to move in one direction as in another. There are no preferred directions for the coordinate axes. There is isotropy and spatial homogeneity so the system exhibits symmetry under translation. If the liquid freezes, the situation changes drastically because now the crystal *does* have preferred directions. The space is no longer isotropic. The symmetry between different directions has been "broken." Theoreticians interested in the evolution of the early universe believe that at the high temperatures shortly after the Big Bang, the electromagnetic interaction and the weak interaction were indistinguishable. But as the universe cooled down there was some sort of "phase transition" causing these two forces to have vastly different magnitudes and to act on vastly different spatial scales. This symmetry breaking is sometimes referred to as a "freezing out" of the different kinds of forces.[9]

8.8 Problems

Problem 8.1 Consider a double pendulum (as in Chapter 4). What physical quantities are conserved?

Problem 8.2 (a) Prove that if a certain coordinate q_i does not appear in the Lagrangian, it is also ignorable in the Hamiltonian.
(b) Show that

$$\frac{\partial H}{\partial q_i} = -\frac{\partial L}{\partial q_i}.$$

Problem 8.3 Does the law

$$\mathbf{N} = \frac{d\mathbf{L}}{dt}$$

exhibit symmetry under reflections? Explain.

Problem 8.4 Are your footprints chiral? Does the pattern formed by your footprints when you are walking exhibit chirality? Would chirality be exhibited if the pattern were of infinite length? (Hint: What is a "glide reflection"?)

Problem 8.5 Sketch the effect of the parity operation on a right-handed Cartesian coordinate system.

---

[8] If you wish to delve further into these questions, a good source is Chapter 16 of J. Brehm and W. J. Mullin, *Introduction to the Structure of Matter*, John Wiley and Sons, New York, 1989.
[9] For a good discussion see E. Whitten, "When Symmetry Breaks Down," *Nature*, **429**, 507–508 (2004).

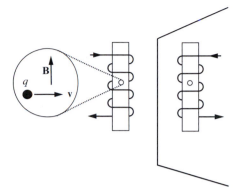

Figure 8.1 Illustration for Problem 8.7.

Problem 8.6 The position vector **r** has components (1,2,3). Express the components of this vector after (a) an inversion and (b) a reflection in the xy-plane. (c) Let **r**′ be the vector obtained by reflecting **r** in a plane perpendicular to the unit vector $\hat{\mathbf{n}}$. Show that $\mathbf{r}' = \mathbf{r} - 2(\mathbf{r} \cdot \hat{\mathbf{n}})\hat{\mathbf{n}}$.

Problem 8.7 An electromagnet consists of a current-carrying coil wrapped around an empty cylinder, as illustrated in Figure 8.1. (a) What is the direction of **B** in the mirror image? (b) If a positive charge inside the cylinder is moving towards the mirror, it will feel a force $\mathbf{F} = q\mathbf{v} \times \mathbf{B}$ directed out of the page. What is the direction of the force in the reflected system?

Problem 8.8 Protons repel according to Coulomb's law:

$$\mathbf{F}_{21} = \frac{q_1 q_2}{4\pi r^2} \hat{\mathbf{r}} = \frac{q_1 q_2 (\mathbf{r}_2 - \mathbf{r}_1)}{4\pi \epsilon_0 |\mathbf{r}_2 - \mathbf{r}_1|^3}.$$

(a) Is the force affected by the parity operation?
(b) Is the force affected by charge conjugation?

Part III

Gravity

9 The Gravitational Field

Classical field theory is primarily a study of electromagnetic and gravitational fields. This chapter is an *introduction* to field theory and is limited to a few aspects of the gravitational field.

Perhaps the most thoroughly studied field is the electromagnetic field. You will learn about it in your E&M course. However, field theory is used to study many other physical phenomena. For example, field theory is used in fluid dynamics and in studies of elasticity, not to mention such important research areas in physics as the unified field theory (which hopes to unite gravity and electromagnetism) and quantum field theory. An interesting aspect of quantum field theory is that it associates particles with the fields. Relativistic quantum field theory has become an important tool in high energy physics. As you might imagine, field theory is quite complicated, both conceptually and mathematically. You may be relieved to know that in this chapter we will limit ourselves to a few fairly straightforward concepts.[1]

An interesting and important aspect of field theory is the formation and propagation of waves. When you study the electromagnetic field you will spend a significant amount of time and mental energy learning about electromagnetic waves. These waves are of great practical importance, particularly because X-rays, visible light, radio waves, and infrared radiation are all electromagnetic waves. The gravitational field also gives rise to waves. Einstein's general theory of relativity predicts that gravity waves are generated by the acceleration of massive bodies. Thus, gravity waves are expected to be produced by supernova explosions, black hole mergers, and orbiting pulsars. The first direct detection of gravitational waves was made with a very large (and complex) system called LIGO.[2]

9.1 Newton's Law of Universal Gravitation

The concept of gravitation was put on a mathematical basis by Isaac Newton when he formulated the law of universal gravitation. He postulated that every particle in the universe attracts every other particle with a force that is proportional to the product of their masses and inversely

---

[1] To learn more about field theory and the closely related subject called potential theory, see L. D. Landau and E. M. Lifshitz, *The Classical Theory of Fields: Course of Theoretical Physics*, Vol. 2. Pergamon Press, 1975. A very old but very good book that was first published in 1929 is Oliver Dimon Kellogg, *Foundations of Potential Theory*, J. Springer, 1929 (available as a republication from Dover Publications).

[2] LIGO stands for Laser Interferometer Gravitational-Wave Observatory. On September 14, 2015, LIGO observed gravitational waves from the merger of two black holes. More details are found in the article by B. P. Abbott and many co-authors in *Physical Review Letters* (DOI:10.1103/PhysRevLett.116.061102), and in the news article by Adrian Cho in *Science*, February 11, 2016.

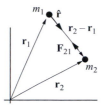

Figure 9.1 A body of mass m_1 is located at $\mathbf{r}_1$ and a body of mass m_2 is located at $\mathbf{r}_2$. The relative position of m_2 with respect to m_1 is $\mathbf{r}_2 - \mathbf{r}_1$.

proportional to the square of the distance between them. In equation form, Newton's law of universal gravitation states that the force on a particle of mass m_2 due to a particle of mass m_1 is

$$\mathbf{F}_{21} = -G\frac{m_1 m_2}{|\mathbf{r}_2 - \mathbf{r}_1|^2}\hat{\mathbf{r}}, \tag{9.1}$$

where G is a constant of proportionality, known as the universal gravitational constant and found experimentally[3] to be approximately equal to 6.67×10^{-11} Nm^2/kg^2.

As shown in Figure 9.1, the vectors $\mathbf{r}_1$ and $\mathbf{r}_2$ are the position vectors of the two particles, and $\hat{\mathbf{r}}$ is a unit vector pointing from m_1 to m_2. This unit vector is given by

$$\hat{\mathbf{r}} = \frac{\mathbf{r}_2 - \mathbf{r}_1}{|\mathbf{r}_2 - \mathbf{r}_1|}.$$

Therefore, a slightly different way to express the law of gravitation is

$$\mathbf{F}_{21} = -Gm_1 m_2 \frac{\mathbf{r}_2 - \mathbf{r}_1}{|\mathbf{r}_2 - \mathbf{r}_1|^3}. \tag{9.2}$$

Although this is more complicated, it avoids confusion as to the direction of the force. Recall that $\mathbf{F}_{21}$ is the force *on* m_2 due to m_1 and that the negative sign in the equation indicates that the force is attractive. By Newton's third law, the force on particle m_1 due to particle m_2 is

$$\mathbf{F}_{12} = -Gm_2 m_1 \frac{\mathbf{r}_1 - \mathbf{r}_2}{|\mathbf{r}_2 - \mathbf{r}_1|^3}.$$

What is the force law for extended bodies? If the distance between two bodies is much larger than the dimensions of the bodies (as for astronomical objects), it is usually safe to treat the bodies as particles. The quantity $|\mathbf{r}_2 - \mathbf{r}_1|$ is the distance between the centers of the two bodies. If an extended body cannot be treated as a particle, you will need to determine the gravitational field of the extended body, as discussed below in Section 9.3. (For example, if a satellite is in a near-Earth orbit, the Earth cannot be considered a point mass.)

9.1.1 Universality of the Law of Gravitation

According to Newton's law of gravitation, every object (every atom) in universe is exerting a force on every other object in the universe. It is true that the force decreases as $1/r^2$ (inverse square force law) so the force we exert on the distant stars is entirely negligible. Nevertheless, a person on the Earth is exerting a tiny force on all the planets and stars and galaxies.[4]

---

[3] A recent measurement of G by Guglielmo Tino and collaborators at the University of Florence, Italy, yielded a value of $G = 6.671\,91(99) \times 10^{-11}$ m^3 kg^{-1} s^{-2}. See the article "Precision Measurement of Newtonian Gravitational Constant Using Cold Atoms," by G. Rosi *et al.*, *Nature*, **510**, 518–521 (2015).

[4] The British physicist P. A. M. Dirac is reputed to have expressed this aspect of the law of gravitation with the poetic sentiment, "When you pluck a flower, you move a distant star."

9.1.2 Action at a Distance

Newton's law of universal gravitation is a prime example of the concept of "action at a distance" which means that body A exerts a force on B, no matter how far apart the bodies happen to be. Naturally, the question arises, "How can bodies that are separated by great distances exert forces on one another?". Newton stated, "Hypothesis non fingo," which means, "I make no hypothesis." This is a perfect example of physics telling us *how* nature works but making no assumptions (or hypotheses) as to *why* it works that way.

The biggest problem with the concept of action at a distance is that it implies that the force between two bodies is propagated instantaneously. That is, if body A is moved to another position, body B will instantaneously notice a change in the force acting on it, even if B is halfway across the universe. Clearly, this is impossible. We believe (but cannot yet prove) that the gravitational force is transmitted at the speed of light. The conceptual difficulties inherent in Newton's law of universal gravitation expressed in terms of action at a distance are effectively dealt with when one considers the gravitational *field*, rather than the gravitational *force*.

Exercise 9.1

Determine the force exerted by Mars on the Sun. What is the acceleration of Mars? What is the acceleration of the Sun? ∎

Exercise 9.2

A body falls in the gravitational field of the Earth. Show that its acceleration is independent of its mass. ∎

Exercise 9.3

A particle of mass m is located at $(1,2)$, a particle of mass $2m$ is located at $(3,4)$, and a particle of mass $3m$ is located at $(-2, -2)$. Determine the gravitational force acting on the particle of mass $2m$. Answer: $F = -Gm^2(0.24\hat{\mathbf{i}} + 0.25\hat{\mathbf{j}})$. ∎

Exercise 9.4

Prove that the gravitational force is conservative by evaluating the curl of Newton's law of universal gravitation. ∎

9.2 The Gravitational Field

By definition, *a field is a physical quantity that is defined at all points in some region of space.* For example, the temperature is a physical quantity and using a thermometer you could determine the temperature at every point in this room. This scalar field might be described by an equation $T = T(x, y, z)$, or by a table of values, or by a drawing showing the temperature isopleths of the room. Similarly, the water in a river has a velocity at every point. The region of space is the river and the physical quantity is velocity. This vector field might be represented graphically or perhaps by an equation for the velocity as a function of position. Suppose that somehow you could determine the gravitational force on a mass at all points near a planet. This would define the gravitational force field of the planet. Note that in all of these examples, the value of the field depends on position.

To define the gravitational field more explicitly, consider a region of space that is empty except for two particles. One particle has mass M and it is located at a fixed point in space. At some other point you place a particle of much smaller mass m. Assume you have a way to determine the force acting on the small mass, wherever it may happen to be. The small mass will be the "test body."

Now imagine you move this test body from one point to another and you measure the magnitude and the direction of the gravitational force acting on it at every location. If you represent the force by a vector, you will end up with a lot of force vectors all pointing towards M. Unfortunately, as indicated by Equation (9.1), these force vectors depend on the mass of the test body so they are not a good measure of the gravitational field of M. To avoid any dependence on the test body, you can define the gravitational field of M as the force *per unit test mass* at all points in space around M. Furthermore, because you don't want m to have any influence on the field, you take the limit as $m \to 0$.

These concepts are expressed more simply using mathematics. Let $\mathbf{F}$ be the *force* on m due to M and let $\mathbf{g}$ be the gravitational *field* of M. Then, by definition, if m is at $\mathbf{r}$ and M is at $\mathbf{r}'$,

$$\mathbf{g}(\mathbf{r}) = \lim_{m \to 0} \left(\frac{\mathbf{F}}{m} \right) = -GM \frac{\mathbf{r} - \mathbf{r}'}{|\mathbf{r} - \mathbf{r}'|^3}. \tag{9.3}$$

The notation $\mathbf{g} = \mathbf{g}(\mathbf{r})$ reminds us that $\mathbf{g}$ is a function of position.

The location of body M is called the "source point," and its position is denoted by $\mathbf{r}'$. The point specified by $\mathbf{r}$ is called the "field point." It is the place where the field is evaluated. See Figure 9.2. Note that the test mass does not enter into Equation (9.3), the expression for $\mathbf{g}(\mathbf{r})$.

Introductory physics books often place the origin of coordinates at the mass M. This gives a simpler but less descriptive expression for the field:

$$\mathbf{g}(\mathbf{r}) = -GM \frac{\mathbf{r}}{|\mathbf{r}|^3} = -\frac{GM}{r^2} \hat{\mathbf{r}}, \tag{9.4}$$

where $\mathbf{r}$ is the location of the field point.

Let me be very clear about the difference between the force and the field. The force $\mathbf{F}$ depends on the masses of both of the particles and their positions at some instant of time. The *field* $\mathbf{g}$ depends only on the mass of the source particle M and the relative position of the field point. The force $\mathbf{F}$ describes an interaction between two objects, but the field $\mathbf{g}$ is a property of a point in space.[5]

The point mass M is the *source* of the gravitational field. The field extends throughout all space. It is not easy to grasp the field concept; it is particularly difficult to describe exactly *what* is filling all of space. We know that if we place a test mass at any point in the field it will feel a

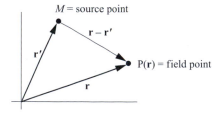

Figure 9.2 Particle M at the source point $\mathbf{r}'$ generates a gravitational field everywhere. The gravitational field at point $P(\mathbf{r})$ (the "field point") is given by Equation (9.3).

---

[5] Sometimes, I like to change the definition of field slightly and state, "A field is a region of space in which some physical quantity is defined at every point." This focuses the mind on the *space* rather than on the physical quantity. It is not, however, the standard definition of field.

force, so we think of the mass M as producing *something* which exerts a force on any other mass in this space. The field is everywhere, but it exerts a force only when a material body is placed in the field.

At the surface of the Earth the gravitational field is approximately given by $\mathbf{g} = -9.8 \text{ m/s}^2\,\hat{\mathbf{k}}$. Here $\hat{\mathbf{k}}$ is a unit vector pointing upward at the surface. For points far above the surface, the field is approximately $\mathbf{g}(\mathbf{r}) = -(GM_E/r^2)\hat{\mathbf{r}}$, where the origin is at the center of the Earth and M_E is the mass of the Earth. (Do not confuse the field vector $\mathbf{g}(\mathbf{r})$ with the quantity g which is an abbreviation for the numerical value 9.8 m/s^2.)

The field concept is quite different from action at a distance. In action at a distance we consider the force one body exerts directly on another. When using the field concept, we think of the interaction of bodies M and m as a two-step process in which mass M generates a field and then mass m interacts with the field, rather than interacting with M directly. Thus the field approach *decouples* the sources from the test body used to determine the field. If two different source arrangements generate the same field at some point, a test body at that point will feel the same force.

It is tempting to consider the field to be nothing but a convenient way of expressing the force, but it turns out that the field has an existence of its own, independent of the force. A force is just a force, but a field has energy, momentum, and angular momentum as well as the ability to exert a force. A field can propagate through space as a wave and can exist independent of the source. (The field can persist even after the source has ceased to exist.)

Although the field approach seems more complicated because it introduces a new concept ("the field"), it does resolve a number of difficulties with the action at a distance concept. As mentioned, in the scenario of action at a distance, forces are transmitted instantaneously. In the field concept, the response of the field to changes in the source propagate at a finite speed. Because the source (M) generates the field, if the source is moved, the field will change. However, this change does not occur instantaneously at all points in space. For example, we believe that changes in the gravitational field are propagated at the speed of light.

Furthermore, the field approach leads to a conceptual framework for understanding how the force is transmitted. In quantum field theory the interaction between particles can be described as an interchange of "virtual" particles. The electromagnetic force, for example, is believed to be transported by virtual photons. This is often represented graphically by "Feynman diagrams." For example, the electromagnetic interaction between two electrons ("electron–electron scattering") is represented by the Feynman diagram of Figure 9.3. In the figure, the two electrons are represented by the straight lines and the wavy line represents a virtual photon where the word "virtual" is used because these photons are not (and cannot be) observed. The photon transmits the electromagnetic force. It is emitted by one of the electrons and absorbed by the other. Both particles are deflected from their original paths. Similarly, the gravitational force is assumed to be transmitted by undetected particles called "gravitons." Because we have our hands full with just Newton's theory of gravitation, we will not consider these advanced concepts in this book.

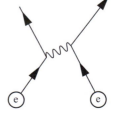

Figure 9.3 A Feynman diagram for electron–electron scattering. The wavy line represents the virtual photon.

However, if you are interested in delving deeper into quantum field theory, I recommend you read a very short but very profound book by Richard Feynman called *QED*.[6]

Exercise 9.5

Two particles of mass M are separated by a distance a. Determine the gravitational field at a distance z along the perpendicular bisector of the line joining the masses. (Remember that in Equation (9.3) the field point is at $\mathbf{r}$ and the source point is at $\mathbf{r}'$.) Answer: $\mathbf{g} = -2GMz\hat{\mathbf{z}} / \left[(a/2)^2 + z^2 \right]^{3/2}$. ■

Exercise 9.6

A point mass M is located at the origin. Another point mass M is located at (0,4). Determine the gravitational field at (0,3) and at (3,0). Answer: Field at (3,0) is $\mathbf{g} = GM \left[-0.14\hat{\mathbf{i}} + 0.03\hat{\mathbf{j}} \right]$. ■

Exercise 9.7

Determine the gravitational field due to the Earth and the Sun at the position of the Moon: (a) at the time of a lunar eclipse, and (b) at the time of a solar eclipse. (c) Determine the force on the Moon at these two times. ■

9.3 The Gravitational Field of an Extended Body

The gravitational field produced by an extended body can be determined by treating the body as a collection of point masses and summing the fields due to each individual particle. (Gravitational fields obey the principle of superposition and add vectorially.) Although this is a perfectly *reasonable* thing to do, it is not a *practical* thing to do. So we assume that the material of the extended body is *continuous*. This assumption breaks down on the atomic level where there are huge differences in mass density between the small, heavy nuclei and the empty space that comprises most of the volume of atoms. However, a volume element that contains billions of atoms is still very small on the macroscopic scale and we are justified in thinking of it as containing a continuous distribution of matter with a well-defined density.

Figure 9.4 shows an infinitesimal volume element ($d\tau'$) in a continuous body. The mass contained in this volume element is $dm' = \rho d\tau'$, where $\rho = \rho(\mathbf{r}')$ is the density. (For a surface, the element of mass is σdA, for a line it is λds and for an extended body it is $\rho d\tau$, where σ, λ, ρ are the mass per unit area, per unit length, and per unit volume.) The density may vary from one

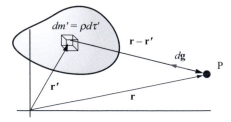

Figure 9.4 An extended body. The mass element $dm' = \rho d\tau'$ is located at $\mathbf{r}'$ (the source point) and it produces an infinitesimal gravitational field $d\mathbf{g}$ at point P located at $\mathbf{r}$ (the field point).

---

[6] Richard Feynman, *QED: The Strange Theory of Light and Matter*, Princeton University Press, Princeton, NJ, 1985.

part of the body to another so it is written as $\rho = \rho(\mathbf{r}')$ to remind us that ρ is a function of position.

In Figure 9.4 the point P is the field point. Note that the source point is at $\mathbf{r}'$ and the field point is at $\mathbf{r}$. The infinitesimal portion of the gravitational field at $\mathbf{r}$ due to an infinitesimal mass element dm' at $\mathbf{r}'$ is

$$d\mathbf{g}(\mathbf{r}) = -Gdm'\frac{\mathbf{r} - \mathbf{r}'}{|\mathbf{r} - \mathbf{r}'|^3}$$

or

$$d\mathbf{g}(\mathbf{r}) = -G\rho(\mathbf{r}')d\tau'\frac{\mathbf{r} - \mathbf{r}'}{|\mathbf{r} - \mathbf{r}'|^3},$$

where $d\tau'$ is so small that $\rho(\mathbf{r}')d\tau'$ can be considered a particle, yet large enough to treat the mass as continuous. To obtain the field at P due to the whole body, we integrate over the body. That is,

$$\mathbf{g}(\mathbf{r}) = -G\int_{\text{body}} \rho(\mathbf{r}')\frac{\mathbf{r} - \mathbf{r}'}{|\mathbf{r} - \mathbf{r}'|^3}d\tau'. \tag{9.5}$$

For an extended body with sufficient symmetry, it is usually not too difficult to evaluate this integral. To find the field due to an arbitrarily shaped body, it is usually best to first determine the *potential* and then obtain the field, using the technique described in Section 9.4 below. Therefore, I will not emphasize the direct integration of Equation (9.5). Nevertheless, you should have some experience with this technique so I will work through two examples (Worked Examples 9.1 and 9.2). You can get some additional experience by working out the exercises below. It might help you to review the section on the electric field of a continuous charge distribution in your introductory physics textbook, as the techniques are exactly the same. It is usually easier to solve such problems by resolving the vector $d\mathbf{g}$ into its components before carrying out the integration.

Worked Example 9.1 A long thin rod of length L lies along the x-axis. The end points of the rod are at $x' = 0$ and $x' = L$. Assume the rod can be represented as a continuous mass distribution with linear density λ mass per unit length. Determine the gravitational field at the point $x = 4L$.

Solution In this problem, the mass element is $dm = \lambda dx'$, where the variable x' represents the distance from the origin to the mass element. The infinitesimal field at P due to this portion of the rod is

$$d\mathbf{g} = -Gdm\frac{\mathbf{r} - \mathbf{r}'}{|\mathbf{r} - \mathbf{r}'|^3} = Gdm\frac{\mathbf{x} - \mathbf{x}'}{|\mathbf{x} - \mathbf{x}'|^3}$$

$$= -G\lambda dx'\frac{(x - x')\hat{\mathbf{i}}}{|x - x'|^3} = -G\lambda\hat{\mathbf{i}}\frac{dx'}{(x - x')^2}.$$

The field point is at $x = 4L$, so integrating we have

$$\mathbf{g} = -G\lambda\hat{\mathbf{i}}\int_{x'=0}^{x'=L}\frac{dx'}{(4L - x')^2} = -G\lambda\left[\frac{1}{4L - x'}\right]_0^L\hat{\mathbf{i}} = -\frac{1}{12L}\lambda G\hat{\mathbf{i}}.$$

Worked Example 9.2 Determine the gravitational field at a point P a distance $+z$ along the axis of a very thin disk of mass M, uniform density, and radius R. See Figure 9.5.

Solution Treat the disk as if it had zero thickness. Then the appropriate mass element is $\sigma\,dA$, where $\sigma = M/\pi R^2$. As indicated in Figure 9.5 the field at point P due to mass element dm' is the vector $d\mathbf{g}$ pointing towards dm'. The distance from the source point dm' to the field point is $|\mathbf{r} - \mathbf{r}'|$, denoted D for simplicity. Consider all the mass elements lying on a ring a distance r' from the axis. The $d\mathbf{g}$ vectors from P to these mass elements form a cone of vectors with vertex at P. The horizontal components of these vectors will cancel. The components along the axis will sum. The resultant field is given by

$$\mathbf{g} = \int (dg)\cos\theta\,\hat{\mathbf{k}} = -\int \frac{G\,dm'}{D^2}\cos\theta\,\hat{\mathbf{k}} = -G\int \frac{\sigma\,dA'}{D^2}\cos\theta\,\hat{\mathbf{k}}.$$

But $\cos\theta = z/D = z/(r'^2 + z^2)^{1/2}$, and $dA' = r'\,d\phi'\,dr'$, so

$$\mathbf{g} = -\sigma G\hat{\mathbf{k}} \int_{r'=0}^{R}\int_{\phi'=0}^{2\pi} z\frac{(dr')(r'\,d\phi')}{(r'^2 + z^2)^{3/2}} = -\sigma G\hat{\mathbf{k}}(2\pi)z \int_0^R \frac{r'\,dr'}{(r'^2 + z^2)^{3/2}}$$

$$= -\sigma G\hat{\mathbf{k}}(\pi)z \int_{u=z^2}^{u=R^2+z^2} \frac{du}{u^{3/2}} = -2\pi\sigma G\hat{\mathbf{k}}z\left[\frac{1}{z} - \frac{1}{\sqrt{z^2 + R^2}}\right]$$

$$= -\frac{2GM}{R^2}\left(1 - \frac{z}{(z^2 + R^2)^{1/2}}\right)\hat{\mathbf{k}}.$$

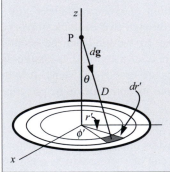

Figure 9.5 The gravitational field of a thin disk is determined by integration of the elementary field vector $d\mathbf{g}$.

Exercise 9.8

Determine the gravitational field at a point z on the axis of a ring of mass M and radius R. Answer: $-GMz/(z^2 + R^2)^{3/2}$. ■

Exercise 9.9

Show that a disk can be treated as a collection of concentric rings. Write an integral expression for $\mathbf{g}$ using the result of Exercise 9.8 and show that it yields the expression obtained in Worked Example 9.2. ■

9.4 The Gravitational Potential

The gravitational *potential* is a scalar quantity related to the gravitational *field*, which is a vector quantity. It is usually easier to work with the potential than the field. If you know the potential you can always determine the field.

Recall the discussion in Section 5.3.3 proving that if $\mathbf{F}$ is a conservative force, then $\nabla \times \boldsymbol{F} = 0$. It is easy to show that the gravitational force is conservative. (See Exercise 9.4.) Using the fact that the curl of the gradient of any scalar function is zero, we conclude that since $\nabla \times \boldsymbol{F} = 0$, then $\mathbf{F}$ can be expressed as the gradient of some scalar function. That is, we can define the potential energy $V = V(\mathbf{r})$ such that

$$\mathbf{F} = -\nabla V.$$

Then

$$\int_{\mathbf{r}_1}^{\mathbf{r}_2} \mathbf{F} \cdot d\mathbf{s} = -\int_{\mathbf{r}_1}^{\mathbf{r}_2} \nabla V \cdot d\mathbf{s} = -\int_1^2 dV = V_1 - V_2.$$

Thus, the difference in potential energy between two points is defined by

$$V_2 - V_1 = -\int_{\mathbf{r}_1}^{\mathbf{r}_2} \mathbf{F} \cdot d\mathbf{s}.$$

In working with the gravitational potential energy, it is often convenient to define the zero of potential energy at infinity, so let $\mathbf{r}_1 = \infty$ and $V_1 = 0$. Writing $\mathbf{r}$ for $\mathbf{r}_2$ we have

$$V = -\int_\infty^{\mathbf{r}} \mathbf{F} \cdot d\mathbf{s}.$$

If $\mathbf{F}$ is the gravitational force acting on test body m located a distance r from point mass M, you can write

$$V(\mathbf{r}) = -\int_\infty^{\mathbf{r}} -\frac{GMm}{r^2} \hat{\mathbf{r}} \cdot d\mathbf{s} = -G\frac{Mm}{r}, \tag{9.6}$$

where $d\mathbf{s} = -\hat{\mathbf{r}} dr$. This is the gravitational potential energy of a system of two point masses.

When using the field approach, it is convenient to define a quantity called the *"gravitational potential"* which we will denote by the symbol Φ. The gravitational potential is obtained from the gravitational potential *energy* by dividing by m, the mass of the test body. The potential energy (V) of two interacting masses is an example of action at a distance and depends on the presence of two bodies. The potential (Φ) does not depend on the presence of a test body and is defined at every point in space. Thus, $\Phi = \Phi(\mathbf{r})$ is a field. The potential is related to the potential energy by

$$\Phi(\mathbf{r}) = \lim_{m \to 0} \frac{V(\mathbf{r})}{m},$$

where m refers to an infinitesimal test mass located at $\mathbf{r}$.

Dividing Equation (9.6) by m shows that the gravitational potential at $\mathbf{r}$ due to a particle of mass M located at the origin is

$$\Phi(\mathbf{r}) = -\frac{GM}{r}. \tag{9.7}$$

The potential of an extended body is obtained from

$$\Phi(\mathbf{r}) = -G \int_{\text{body}} \frac{\rho(\mathbf{r}')d\tau'}{|\mathbf{r} - \mathbf{r}'|}. \tag{9.8}$$

where $\mathbf{r}$ is the field point and $\mathbf{r}'$ is the location of the source point $dm = \rho(\mathbf{r}')d\tau'$.

Worked Example 9.3 Determine the gravitational potential at a point outside of a spherical shell of mass M and radius a. See Figure 9.6.

Solution The mass per unit area (σ) of the (uniform) shell is the total mass of the shell divided by its surface area, or

$$\sigma = \frac{M}{4\pi a^2}.$$

Selecting the mass element is more or less an "art" and you will become proficient at making a good choice after you have worked out a number of problems. For a mass shell, pick a small element of area on the surface whose mass is

$$dm = \sigma dA = \sigma a^2 \sin\theta d\theta d\phi.$$

Integrating over ϕ generates a ring on the surface of the shell, as indicated in the figure. (Integrating over ϕ is equivalent to letting the line $a\sin\theta$ rotate around $\mathbf{r}$. Note that all points on this ring are equidistant from the field point.) The width of the ring is $ad\theta$. The potential at P due to the entire shell is given by the double integral

$$\Phi = -G \int_{\text{surface}} \frac{\sigma dA}{|\mathbf{r} - \mathbf{r}'|} = -G \int_{\phi=0}^{2\pi} \int_{\theta=0}^{\pi} \frac{\sigma a^2 \sin\theta d\theta d\phi}{|\mathbf{r} - \mathbf{r}'|},$$

$$\Phi = -2\pi G \int_{\theta=0}^{\pi} \frac{\sigma a^2 \sin\theta d\theta}{|\mathbf{r} - \mathbf{r}'|},$$

where the factor 2π came from integrating over ϕ. Using the law of cosines, the figure shows that because $|\mathbf{r}'| = a$,

$$|\mathbf{r} - \mathbf{r}'| = R = \sqrt{a^2 + r^2 - 2ar\cos\theta},$$

and $R^2 = a^2 + r^2 - 2ar\cos\theta$. Therefore $2RdR = 2ar\sin\theta d\theta$ so $\sin\theta d\theta = \frac{R}{ar}dR$ and the integral takes the form

$$\Phi = -2\pi G \int_0^{\pi} \frac{\sigma a^2 \sin\theta d\theta}{R} = -2\pi G\sigma a^2 \int_{R=r-a}^{R=r+a} \frac{(R/ar)dR}{R}$$

$$= -G\sigma a^2 (2\pi/ar) \int_{r-a}^{r+a} dR = \frac{-2\pi G\sigma a}{r}[(r+a) - (r-a)]$$

$$= -G\frac{4\pi a^2 \sigma}{r} = -\frac{GM}{r},$$

where we used the fact that as θ goes from 0 to π, R ranges from $r - a$ to $r + a$.

Our result shows that for points *outside* the shell, the potential of a shell of mass M is the same as the potential of a point mass M at the origin. As a problem you can show that the *field* inside the shell is zero. (See Problem 9.15.)

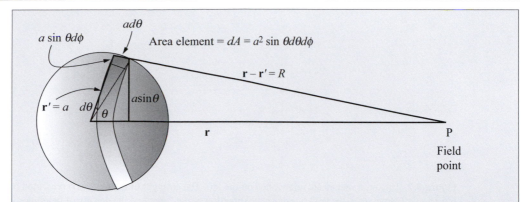

Figure 9.6 The potential at P due to a shell of radius a. The mass element dm is the mass of the element of area $dA = a^2 \sin\theta d\theta d\phi$. If σ is the surface mass density, $dm = \sigma dA$.

Exercise 9.10

Determine the potential on the axis of a disk of radius R, density ρ, and thickness t. Assume that $t \ll z$ so the thin disk approximation is still valid. Answer: $\Phi = -2\pi\rho t G[(z^2 + R^2)^{1/2} - z]$. ∎

9.5 Field Lines and Equipotential Surfaces

We now illustrate how to represent the field and potential graphically.

Consider the gravitational field ($\mathbf{g}$) due to a point mass M, recall that $\mathbf{g}$ is the force per unit mass on an infinitesimal test body; see Equation (9.3). You could, in principle, transport the test body to every point in space around M and measure the force on it at each location. Then you could *represent* the field $\mathbf{g}$ in a number of different ways. For example, you could make up a large table giving the magnitude and direction of $\mathbf{g}$ at every point. Or you could represent $\mathbf{g}$ by an equation, such as Equation (9.4). Or you could represent $\mathbf{g}$ graphically. The graphical representation would show many arrows pointing toward M. The field at some given point might be represented by drawing a vector of appropriate length at that point, as illustrated by the sketch on the left in Figure 9.7. Note that the vectors are longer at points close to the mass and shorter at points further away. But this is not a very good way to represent the field. The sketch on the right in Figure 9.7 illustrates a much more convenient way. In this representation the lines give the direction of the force and the *areal density* of the lines at a point gives the magnitude of the force at that point. Imagine the lines are drawn in such a way that the number of lines crossing a unit area perpendicular to the field is numerically equal to the field at that location. If the field at some point has a magnitude of 15 N/kg, it could be represented by drawing 15 lines (per unit area) at that point.

The total number of lines passing through spherical surfaces centered on M will be the same regardless of the radius of the sphere. Because the area of a sphere increases as r^2, the density of lines per unit area falls off as $1/r^2$, just like the gravitational field.

The lines representing the field are traditionally called "lines of force," although most modern books refer to them with the better terminology of "field lines."

Another graphical representation of the field consists in sketching the surfaces on which the potential is constant. For a point mass, the equipotential surfaces are concentric spheres centered on the mass. As we will show in a moment, the field and the potential are related by $\mathbf{g} = -\nabla\Phi$,

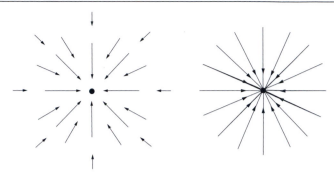

Figure 9.7 The force vectors are represented on the left. The strength of the force is represented by the length of the arrows. On the right, the field lines give the direction of the force and the density of field lines at any point is proportional to the strength of the field.

and you know from Chapter 5 that the gradient of Φ is perpendicular to the surfaces of constant Φ. Therefore, the two representations (field lines and equipotential surfaces) are equivalent.

9.6 The Newtonian Gravitational Field Equations

The relationship between potential energy and force is $\mathbf{F} = -\nabla V$, as shown in Equation (5.7). Dividing both sides of this equation by m yields the relation between the gravitational field $\mathbf{g}(\mathbf{r})$ and the gravitational potential $\Phi(\mathbf{r})$:

$$\mathbf{g} = -\nabla\Phi. \tag{9.9}$$

(In many books – particularly older books – the gravitational potential is defined as the *negative* of Φ and the relation equivalent to Equation (9.9) does not have the minus sign.)

The curl of a gradient is zero. Therefore,

$$\nabla \times \mathbf{g} = -\nabla \times (\nabla\Phi) = -\mathbf{curl\ grad}\Phi = 0. \tag{9.10}$$

(I am sure you expected this result because you know that $\mathbf{g}$ is a conservative field.)

9.6.1 Gauss's Law

Having found the curl of $\mathbf{g}$, it is reasonable to go on and determine the divergence of $\mathbf{g}$. The process is a bit involved, so please follow carefully. Consider a mass point M. Draw a surface S around it, as illustrated in Figure 9.8. Let dS be an infinitesimal patch of the surface. Let $\hat{\mathbf{n}}$ be a unit vector perpendicular to dS and let $\mathbf{r}$ be the vector from M to dS.

Place the origin of a coordinate system at the point mass M. Then the gravitational field of M at the position of the patch of surface dS can be written as

$$\mathbf{g} = -\frac{GM}{r^2}\hat{\mathbf{r}},$$

so

$$\mathbf{g} \cdot \hat{\mathbf{n}} = -\frac{GM}{r^2}\hat{\mathbf{r}} \cdot \hat{\mathbf{n}} = -\frac{GM}{r^2}\cos\theta,$$

where θ is the angle between $\hat{\mathbf{r}}$ and $\hat{\mathbf{n}}$ (see Figure 9.8).

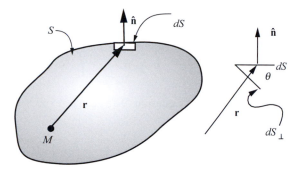

Figure 9.8 A closed surface S encloses a point mass M. The unit vector $\hat{\mathbf{n}}$ is perpendicular to the surface at the location of dS, an infinitesimal portion of the surface. Let $dS_\perp$ be the projection of dS perpendicular to $\mathbf{r}$. Note that $d\mathbf{S} \cdot \hat{\mathbf{r}} = dS\hat{\mathbf{n}} \cdot \hat{\mathbf{r}} = dS \cos\theta = dS_\perp$. The sketch on the right views the areas dS and $dS_\perp$ edge-on. The angle θ is the angle between the radial vector and $\hat{\mathbf{n}}$, or equivalently, between dS and $dS_\perp$.

Note that the solid angle[7] $d\Omega$ subtended by dS at M is

$$d\Omega = \frac{dS_\perp}{r^2} = \frac{dS \cos\theta}{r^2}, \tag{9.11}$$

where $dS_\perp$ is the perpendicular projection of dS on $\mathbf{r}$. Next consider the surface integral

$$I = \oint_S \mathbf{g} \cdot \hat{\mathbf{n}} dS = -\oint_S \frac{GM}{r^2} \cos\theta dS$$
$$= -GM \oint_S \frac{dS \cos\theta}{r^2} = -GM \oint_S d\Omega$$
$$= -4\pi GM.$$

This result is very interesting. It tells us that the surface integral I is proportional to the mass M inside the surface. This fact, in itself, is not too surprising. However, the result also tells us that *it does not matter where the mass is*, as long as it is inside the closed surface. If there are several masses ($m_1, m_2, \ldots$) inside S, then, by the principle of superposition,

$$\oint_S \mathbf{g} \cdot \hat{\mathbf{n}} dS = -\sum 4\pi G m_i = -4\pi G \sum m_i = -4\pi G M_{\text{enc}}, \tag{9.12}$$

where M_{enc} is the total mass enclosed by S. The equation

$$\oint_S \mathbf{g} \cdot \hat{\mathbf{n}} dS = -4\pi G M_{\text{enc}}$$

is called Gauss's law. It is particularly useful in the study of electrostatics.

Assume mass is continuously distributed in a volume V bounded by a closed surface S. The element of mass is $\rho d\tau$, so the mass enclosed is $M_{\text{enc}} = \int_V \rho d\tau$. Therefore,

$$\oint_S \mathbf{g} \cdot \hat{\mathbf{n}} dS = -\int_V 4\pi G\rho d\tau. \tag{9.13}$$

---

[7] The solid angle is the three-dimensional angle formed at the vertex of a cone. Let point O be at the origin and a surface of area dA be a distance r from the origin. Then, if dA is perpendicular to r, the solid angle subtended by dA at O is $d\Omega = dA/r^2$. If dA is not perpendicular to r, $d\Omega = (dA \cos\theta)/r^2 = dA_\perp/r^2$. The units of solid angle are *steradians*. The solid angle subtended by a spherical shell at a point inside the shell is 4π steradians.

You may recall from your vector analysis course that Gauss's divergence theorem states that for any vector function **A**,

$$\oint_S \mathbf{A} \cdot \hat{\mathbf{n}}\, dS = \int_V \nabla \cdot A\, d\tau. \tag{9.14}$$

Therefore, the left-hand side of Equation (9.13) can be expressed as $\int_V \nabla \cdot \mathbf{g}\, d\tau$, and consequently

$$\int_V \nabla \cdot \mathbf{g}\, d\tau = -\int_V 4\pi G\rho d\tau,$$

or

$$\nabla \cdot \mathbf{g} = -4\pi G\rho. \tag{9.15}$$

Thus, the divergence of **g** is proportional to ρ, the mass density. An important theorem called the Helmholtz theorem states that for any vector field, if one knows the divergence and the curl of the field, then one can determine the field itself. For this reason, $\nabla \cdot \mathbf{g}$ and $\nabla \times \mathbf{g}$ are called the "sources" of **g**. Equations (9.10) and (9.15) indicate that the source of the gravitational field is the mass density ρ. In other words, masses generate gravitational fields.

Worked Example 9.4 Use Gauss's law to determine the gravitational field outside of an infinitely long cylinder of radius a with constant linear mass density λ.

Solution To solve, we construct a *Gaussian surface* which is a closed surface that has the same symmetry as the mass distribution. We select this surface so that on the surface $\mathbf{g} \cdot \hat{\mathbf{n}}$ is either constant or zero. The field point lies at an arbitrary point on this surface. For this problem the appropriate Gaussian surface is a cylinder of length L and radius r, where $r > a$. See Figure 9.9.

On the curved face of this "tin can" surface, the field is directed towards the axis of the cylinder, and **g** is constant. On the two end faces, **g** and $\hat{\mathbf{n}}$ are perpendicular. Therefore, Gauss's law gives

$$\oint_S \mathbf{g} \cdot \hat{\mathbf{n}}\, dS = \int_{\text{side}} \mathbf{g} \cdot \hat{\mathbf{n}}\, dS + \int_{\text{ends}} \mathbf{g} \cdot \hat{\mathbf{n}}\, dS = -4\pi G M_{\text{enc}}.$$

Now

$$\int_{\text{side}} \mathbf{g} \cdot \hat{\mathbf{n}}\, dS = -g \int dS = -g(2\pi r)L,$$

P = field point

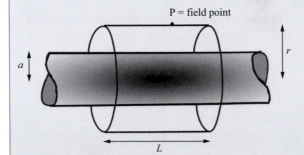

Figure 9.9 A Gaussian surface. The outer cylinder is the Gaussian surface appropriate for determining the field of the cylindrical mass. Note that the field point is on the Gaussian surface.

and

$$\int_{\text{ends}} \mathbf{g} \cdot \hat{\mathbf{n}} \, dS = 0.$$

The mass enclosed by the Gaussian surface is

$$M_{\text{enc}} = \lambda L.$$

Therefore,

$$g = -\frac{4\pi G \lambda L}{2\pi r L} = -\frac{2G\lambda}{r}.$$

By symmetry, this is pointing toward the axis of the cylinder so

$$\mathbf{g} = -\frac{2G\lambda}{r} \hat{\rho},$$

where $\hat{\rho}$ is the unit vector in cylindrical coordinates.

Exercise 9.11

Show that the solid angle subtended by a spherical shell at any interior point is 4π steradians.

■

Exercise 9.12

Use Gauss's law to determine the gravitational field of a point mass.

■

Exercise 9.13

Maxwell's equations for the electric field ($\mathbf{E}$) and the magnetic field ($\mathbf{B}$) can be written

$$\nabla \cdot \mathbf{E} = \rho/\epsilon_0 \quad \nabla \times \mathbf{E} = -\frac{\partial \mathbf{B}}{\partial t}$$

$$\nabla \cdot \mathbf{B} = 0 \quad \nabla \times \mathbf{B} = \mu_0 \mathbf{J} + \epsilon_0 \mu_0 \frac{\partial \mathbf{E}}{\partial t},$$

where ρ is the charge density and J is the current density. The quantities ϵ_0 and μ_0 are constants. What are the sources of the electric field? What are the sources of the magnetic field?

■

9.7 The Equations of Poisson and Laplace

Recall that $\mathbf{g} = -\nabla\Phi$. Taking the divergence of both sides,

$$\nabla \cdot \mathbf{g} = -\nabla \cdot \nabla \Phi = -\nabla^2 \Phi.$$

Using the relation $\nabla \cdot \mathbf{g} = -4\pi G\rho$ leads to

$$\nabla^2 \Phi = 4\pi G\rho. \tag{9.16}$$

This is called "Poisson's equation."

Poisson's equation is a second-order partial differential equation whose solution depends on the boundary conditions. In a region of space where there is no mass, the density ρ is zero, and Poisson's equation reduces to

$$\nabla^2 \Phi = 0. \tag{9.17}$$

This very important special case is called "Laplace's equation."

You will use Laplace's equation to determine the field in a region of space where $\rho = 0$, for example in empty space outside of a star. If, however, you want to determine Φ *inside* the star then you solve Poisson's equation.

Since Poisson's equation is an inhomogeneous differential equation, its solution can be expressed as the sum of two parts.

- The general solution of the homogeneous differential equation (Laplace's equation).
- A particular solution of the inhomogeneous differential equation (Poisson's equation).

In your undergraduate course in electromagnetic theory you will learn techniques for solving the Laplace equation under a variety of different conditions. In your graduate course in electromagnetic theory you will learn a number of methods for solving the Poisson equation.

Poisson's equation, $\nabla^2 \Phi = 4\pi G\rho$ is the "field equation" of the Newtonian gravitational field. If you solve for Φ you know the potential everywhere and you can determine the gravitational field **g** everywhere.

9.8 Einstein's Theory of Gravitation (Optional)

We cannot leave this chapter on the gravitational field without a few words about Einstein's theory of gravitation, which is referred to as "The General Theory of Relativity." This theory supersedes Newton's theory of gravitation and is one of the greatest achievements in physics.

An example that illustrates the great accuracy of Einstein's theory is the precession of the orbit of Mercury. Astronomers had known for many years that the semimajor axis of the orbit of Mercury precesses at the extremely precisely measured rate of 5601 seconds of arc per century. Newton's theory, including the effects of the other planets which perturb the motion of Mercury, yielded a precession rate of 5558 seconds of arc per century. The discrepancy (43 seconds per century) was explained by Einstein's theory. Note that we are talking about a difference of less than one degree per century! There are numerous other examples of situations in which Einstein's theory gives the correct value whereas Newton's theory is slightly off.

In that case, you might ask, why do we bother studying Newton's theory? The answer is that calculations in Einstein's theory are extremely complicated, and that Newton's theory gives results that are extremely close to the correct values.

This section will give you a rough idea of Einstein's theory.[8] You will probably not be using the theory of general relativity unless you decide to specialize in cosmology – the study of the structure of the universe. Nevertheless, as a physicist, you should have some idea about the theory.

You may be aware that in 1905 Einstein developed the theory called the "Special Theory of Relativity." (We will consider this theory in Chapter 20.) The special theory deals with the relationship between two inertial reference frames. That is, it considers two reference frames that are moving at *constant* velocities. The laws of physics are expected to be the same in the two reference frames. In other words, a transformation from one reference frame to the other should

---

[8] A very readable book on the topic is *Relativity, Gravitation and Cosmology* by Robert J. A. Lambourne, Cambridge University Press, 2010.

leave physical equations and relationships unchanged. The theory was extremely successful and was particularly applicable to electromagnetism. However, Einstein was unsatisfied with one aspect of the special theory, and that was that Newton's law of universal gravitation did not transform as expected. Furthermore, Einstein was interested in developing a theory that would be applicable to *accelerated* (that is, noninertial) reference frames. In 1915 he published the "general" theory of relativity that solved both of these problems. However, to do so, he had to delve deeply into very advanced concepts in geometry.

Einstein's theories required understanding how to transform from one reference frame to another, a procedure that has much in common with transforming from one coordinate system to another. Recall that in Section 5.3.4 we introduced the concept of the "metric tensor" that allowed us to transform from Cartesian coordinates to spherical or cylindrical coordinates. Letting h_{ij}^2 represent the elements (or components) of the metric tensor, we wrote the differential (infinitesimal) displacement – see Equation (5.8) – as

$$ds^2 = \sum_{i,j} h_{ij}^2 dq_i dq_j.$$

In Sections 5.3.5 and 5.3.6 we used this relation to obtain expressions for the volume element and for del in cylindrical and spherical coordinates.

In special relativity, Einstein used a four-dimensional coordinate system with x, y, z being the usual Cartesian space coordinates and the fourth coordinate being ct, where c is the constant speed of light. Note that ct has the dimensions of distance, but it depends only on t. Thus, time is the fourth dimension. Including an extra dimension requires a somewhat more complicated expression for the differential displacement, namely:

$$ds^2 = \sum_{\mu, \nu=0}^{3} \eta_{\mu\nu} dx^{\mu} dx^{\nu}.$$

There are two things to note about this expression. First of all it is customary to use upper and lower indices which are called "contravariant" and "covariant." The difference need not concern us at present, but it is important to appreciate that dx^{μ} does *not* mean dx raised to the power μ! It is just a way to denote the component. Secondly, note that the indices μ and ν range from 0 to 3. Here 1, 2, 3 represent the spatial coordinates, dx, dy, dz and zero denotes the "time" coordinate cdt. For example, dx^3 represents dz and dx^0 represents cdt. The quantity ds is the infinitesimal "line element" in four-dimensional "space-time." For constant-velocity reference frames, the metric $\eta_{\mu\nu}$ is quite simple. It is called the "Minkowski metric"[9] and, in matrix form, it can be written as

$$\eta_{\mu\nu} = \begin{pmatrix} 1 & 0 & 0 & 0 \\ 0 & -1 & 0 & 0 \\ 0 & 0 & -1 & 0 \\ 0 & 0 & 0 & -1 \end{pmatrix}.$$

Minkowski space-time is described as "flat" because Euclidean geometry is valid in this space. This means, for example, that parallel lines never meet and that the angles in a triangle add up to 180°. The geometry of the surface of a sphere is non-Euclidean. Parallel lines that start out perpendicular to the equator meet at the pole. The angles of a triangle do not sum to 180°. In three-dimensional space, the line element ds on a flat surface is

---

[9] This metric was developed by Hermann Minkowski, who was one of Einstein's mathematics professors. Minkowski is credited for having developed the concept of space-time.

$$ds^2 = dx^2 + dy^2 + dz^2,$$

but on the surface of a sphere of radius R,

$$ds^2 = R^2 d\theta^2 + R^2 \sin^2 \theta d\phi^2.$$

Not all surfaces that we might think of as curved are curved in the mathematical sense. For example, the curved side of a cylinder is mathematically flat because (as you can easily show) parallel lines never meet, triangles have 180°, etc.

Bernhard Riemann[10] used the concept of line element to generalize geometry. In Riemann's formulation, the line element in four-dimensional space-time is

$$ds^2 = \sum_{\mu, \nu=0}^{3} g_{\mu\nu} dx^\mu dx^\nu$$

In a flat region of space-time, the metric reduces to the Minkowski metric.

Einstein formulated the laws of physics in terms of tensors that reduce to the well-known forms of classical physics in the appropriate limit.

The basic relationship in Newtonian gravity is the Poisson equation:

$$\nabla^2 \Phi = 4\pi G\rho.$$

Einstein wanted to replace this expression with an equivalent tensor equation. In Newtonian mechanics, mass is a conserved quantity. But according to relativity theory, mass and energy are interchangeable quantities. (Recall the famous equation $E = mc^2$.) Einstein decided to replace the mass density ρ with the energy–momentum tensor $T^{\mu\nu}$.[11] This is a 4×4 symmetric tensor which has 16 components, although only 10 of them are independent. These components are:

T^{00} = local energy density (including mass energy),
T^{0i} = momentum density in i direction times the speed of light,
T^{ij} = rate of flow per unit area of ith component of momentum perpendicular to the j
      direction ($i, j > 0$).

In Einstein's theory, the energy–momentum tensor $T^{\mu\nu}$ multiplied by a constant of proportionality replaces $4\pi G\rho$ on the right-hand side of Poisson's equation.

Next Einstein needed to replace $\nabla^2 \Phi$, the left-hand side of Poisson's equation, with a tensor. Note that $\nabla^2 \Phi$ is a quantity that depends on the geometry of the space being considered. From Riemannian geometry the curvature of space-time can be expressed as a tensor which is denoted $G^{\mu\nu}$. Thus Einstein expressed the Poisson equation as $G^{\mu\nu} \propto T^{\mu\nu}$. It turns out that the proportionality constant is $8\pi G/c^2$ where G is the universal gravitational constant in Newton's law of gravity.[12] Consequently,

$$G^{\mu\nu} = -\frac{8\pi G}{c^2} T^{\mu\nu}. \tag{9.18}$$

This is, essentially, the Einstein expression for the gravitational field. It is a set of 10 coupled second-order differential equations for the metric $g^{\mu\nu}$ and is referred to as "Einstein's field

---

[10] Bernhard Riemann (1826–1866), who was a student of Carl Friedrich Gauss, is considered one of the world's greatest mathematicians. Among other achievements, he developed the geometry of curved surfaces.

[11] As you will learn in your E&M course, the basic laws of electricity and magnetism (Maxwell's equations) can be expressed in tensor form. (See, for example, Chapter 12 of *Introduction to Electrodynamics* by David. J. Griffiths, 4th ed., Cambridge University Press, 2017).

[12] G and $G^{\mu\nu}$ are two completely different things. Unfortunately, traditionally they are represented by the same letter. G is a scalar constant and $G^{\mu\nu}$ is a tensor describing the geometry of space-time and is a function of the $g^{\mu\nu}$.

equations." An interesting sidelight is the fact that when Einstein studied the implications of this set of equations, he discovered that they implied that space-time was increasing. That is, the equations predicted that the universe is expanding. At that time, astronomers were convinced that the universe was in a steady state, neither expanding nor collapsing. Bothered by this "fact," Einstein added a term to the equation to cancel out the expansion, thus

$$G^{\mu\nu} = -\frac{8\pi G}{c^2} T^{\mu\nu} - \Lambda g^{\mu\nu}.$$

The additional term is known as the "cosmological constant."

Then, in 1929, the astronomer Edwin Hubble discovered that distant galaxies are receding from us, and determined that the universe is, indeed, expanding.

Einstein called the introduction of the cosmological constant the "biggest blunder" of his life.

For many years, the cosmological constant was not included in Einstein's equations, but then in 1998 astronomers suggested that the universe is not only expanding, but this expansion is actually accelerating! Consequently, the cosmological constant has been reintroduced into Einstein's field equations to account for this increase in the rate of expansion. The present value of Λ is about $10^{-52}/m^2$.

Now that we have expressed Einstein's equations as an alternative to the Poisson equation, let us consider some of the consequences of this alternative approach to gravitation.

Perhaps the most fundamental question to be answered is what the theory predicts as to the effect on a particle (planet?) of mass m. How will a mass particle move in curved space-time? Recall our discussion of the calculus of variations in Chapter 4 in which we showed that the shortest distance between two points in a plane is a straight line and the shortest distance between two points on the surface of a sphere is a section of a great circle (or geodesic). It is an interesting consequence of general relativity that a particle in a gravitational field will move along a geodesic.

Furthermore, the theory predicts that a massive body will distort space-time, changing the curvature of space in its vicinity and thus affecting the geodesics.

The motion of a particle in curved space-time is often described in terms of the analogy of a bowling ball placed on a soft mattress, thus deforming the surface of the mattress. A marble rolling along the mattress will roll into the depression, as if attracted to the bowling ball. See Figure 9.10.

John Wheeler summarized the rules of general relativity in the following way:

Matter tells space how to curve.
Space tells matter how to move.

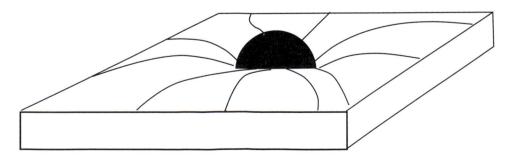

Figure 9.10 A bowling ball on a mattress as an analogue to curved space-time.

The theory of general relativity is basically a four-dimensional field theory of gravitation. It has been submitted to numerous experimental tests and has been proved correct over and over. It is unfortunate that it is so mathematically complex that it is only accessible to a small community of physicists, astronomers, and mathematicians.

9.9 Summary

A *field* is defined as a physical quantity whose value can be determined at every point in some given region of space. The gravitational field $\mathbf{g}(\mathbf{r})$ due to a particle of mass M located at point $\mathbf{r}'$ can be defined in terms of the force it exerts on an infinitesimal point mass m located at $\mathbf{r}$,

$$\mathbf{g}(\mathbf{r}) = \lim_{m \to 0} \left(\frac{\mathbf{F}}{m} \right) = -GM \frac{\mathbf{r} - \mathbf{r}'}{|\mathbf{r} - \mathbf{r}'|^3}.$$

The field produced by an extended body can be determined by evaluating the expression

$$\mathbf{g}(\mathbf{r}) = -G \int_{\text{body}} \rho(\mathbf{r}') \frac{\mathbf{r} - \mathbf{r}'}{|\mathbf{r} - \mathbf{r}'|^3} d\tau'.$$

To evaluate the integral it is usually necessary to resolve the vectors into components, using the symmetry of the problem as a guide.

The gravitational *potential* $\Phi(\mathbf{r})$ is given by

$$\Phi(\mathbf{r}) = -G \int_{\text{body}} \frac{\rho(\mathbf{r}') d\tau'}{|\mathbf{r} - \mathbf{r}'|}.$$

A field can be represented graphically by drawing the field lines (as illustrated in Figure 9.7) or by drawing equipotential surfaces.

The Helmholtz theorem tells us that we can determine a vector field if we know its "sources" defined as the gradient and curl of the field. For the gravitational field these are

$$\nabla \times \mathbf{g} = 0,$$
$$\nabla \cdot \mathbf{g} = -4\pi G \rho.$$

The first of these relations indicates that the gravitational field is conservative, thus allowing us to define a potential. The second equation is Gauss's law and in integral form it is expressed as

$$\oint_S \mathbf{g} \cdot \hat{\mathbf{n}} \, dS = -4\pi G M_{\text{enc}},$$

where M_{enc} is the mass enclosed by the surface S. In terms of the potential, this leads to

$$\nabla^2 \Phi = 4\pi G \rho,$$

which is called *Poisson's equation*. In regions of space where $\rho = 0$ this reduces to *Laplace's equation*

$$\nabla^2 \Phi = 0.$$

Finally, although Newton's theory of gravitation has been superseded by Einstein's theory of general relativity, we still use Newtonian mechanics for most practical problems because it is much easier to apply and for most situations the differences are miniscule.

9.10 Problems

Problem 9.1 Three particles of masses 1 kg, 2 kg, and 3 kg are located at (1,1), (2,0) and (3,3), respectively. They attract one another gravitationally. Determine the vector force acting on the 1 kg mass.

Problem 9.2 Assume the Earth's orbit is a circle of radius 1.496×10^{11} m. It orbits the Sun once a year. Using these facts, determine the mass of the Sun.

Problem 9.3 Suppose a person stood on a scale at the South Pole and the scale read 60 kg. What would it read at the equator? (Assume a perfectly spherical, homogeneous Earth.)

Problem 9.4 The Earth is suddenly brought to a standstill. Evaluate the time required for it to collide with the Sun. You can assume the Sun does not move and that the center of mass of the system is at the center of the Sun. How good is this approximation?

Problem 9.5 A neutron star has a mass of 10^{30} kg and a radius of 5 km. A body is dropped from a height of 20 cm above the surface. Determine the speed of the body when it hits the surface.

Problem 9.6 A neutron star is very small and very dense and rotates very rapidly. Assume the density of a particular neutron star is 5×10^{17} kg/m^3 and its mass is 5×10^{30} kg.
(a) Determine the gravitational force on a particle of mass 1 kg on the surface of the star.
(b) Assume the particle is on the equator of the star. Determine the maximum rotational velocity of the star if the gravitational force is sufficient to keep the particle on the surface.

Problem 9.7 Determine the period of a surface-skimming satellite in a circular orbit about a uniform, perfectly spherical planet of radius R and density ρ. Determine the period of such a satellite about a different planet which has radius $2R$ but the same density as the first. Explain your result.

Problem 9.8 (a) Determine the gravitational field at an arbitrary point above an infinite, thin, flat sheet having mass per unit area σ. (Hint: Use cylindrical coordinates.)
(b) A spherical shell of mass M and radius a is placed a distance b above the infinite flat plane ($b > a$). Determine the force on the sheet due to the shell.

Problem 9.9 Two concentric spherical shells of radii a and b have masses m_a and m_b. Assume $b > a$. What is the pressure (force per unit area) on the outer sphere?

Problem 9.10 Consider an infinitely long, straight string whose linear mass density is λ (mass per unit length). By direct integration, determine the gravitational field a distance r from the string.

Problem 9.11 By direct integration, find the field at a point on the axis of symmetry of a cylinder of length L, radius R and uniform density ρ. (Hint: Let the mass element be an infinitesimally thin disk and use the result of Worked Example 9.2.)

Problem 9.12 Consider an infinite string with linear mass density λ. A particle of mass m is a distance d from the string. Determine the force on the particle.

Problem 9.13 Obtain an expression for the potential at a distance z above the center of a circular disk of radius a and mass M.

Problem 9.14 Find the potential at an external point on the axis of a homogeneous right-circular cylinder which has mass M, length L, and radius a. The field point is a distance $z > L/2$ from the center of the cylinder. (Use the fact that a cylinder can be considered a collection of coaxial disks. The potential of a disk of mass dm is $d\Phi_{disk} = \frac{2Gdm}{a^2}[t - \sqrt{a^2 + t^2}]$, where t is the distance to the field point.)

Problem 9.15 Determine the potential *inside* a spherical shell of mass M. Use the technique of our analysis for the potential at an external point. Explain why your result predicts that there is no force on an object on the inside of a mass shell.

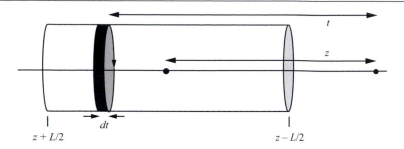

$z + L/2$ $\quad$ dt $\qquad\qquad\qquad\qquad\qquad$ $z - L/2$

Figure 9.11 A homogeneous cylinder. Determine potential on axis. (Problem 9.14.)

Problem 9.16 Out in interstellar space there is a large spherical gas cloud of constant density ρ and radius a. Determine the gravitational potential at points inside the cloud.

Problem 9.17 Consider a planet of mass M and radius R. Assume the planet is spherical and has a constant density. By direct integration determine the gravitational field and the gravitational potential at all points inside and outside the planet. (Assume the potential is zero at infinity and there are no other bodies in the universe.)

Problem 9.18 (a) Using Gauss's law, determine the gravitational field a distance a from a flat infinite sheet of uniform mass density σ.

(b) Using Gauss's law, determine the field outside a spherical shell of radius a having the same mass density as the infinite plane.

(c) Show that just outside the shell, the field is twice that of the infinite plane. Interpret your result.

Problem 9.19 Suppose an intergalactic gas cloud were found to have a density given by

$$\rho = \frac{Mb^2}{2\pi r(r^2 + b^2)^2}.$$

Here M and b are constants related to the size and mass of the cloud. Note that the density extends out to infinity, but the density is infinitely small for very large values of r. Determine the gravitational field and the gravitational potential as functions of r.

Problem 9.20 Using Gauss's law, determine the gravitational potential inside and outside of a constant density sphere of radius R and mass M. Plot Φ vs. r.

Problem 9.21 Use Gauss's law to determine the field inside and outside of: (a) a sphere of uniform mass density, (b) a hollow spherical shell. Let the mass and radius be M and R in both cases.

Problem 9.22 Use Gauss's law to determine the field inside and outside an infinite cylinder of radius R and uniform mass density ρ. Express your answer in terms of r, the distance from the axis.

Problem 9.23 A sphere of radius R and constant mass density ρ has a spherical cavity of radius r, where $r = R/2$. The center of the cavity is a distance $R/4$ from the center of the sphere. A particle of mass m is located at an outside point, a distance z from the center of the larger sphere and along the line going through the centers of the sphere and the cavity. Determine the force on the particle. (Note: $z > R$.)

Problem 9.24 (a) Determine the gravitational potential at a point on the axis of a ring of mass of radius a and mass m. (b) Determine the potential for this ring at an off-axis point located a distance r from the center of the ring and making an angle θ with the axis. You may assume $r \gg a$ and you can disregard any terms of order $(a/r)^3$ or smaller. (Hint: The angle γ between two lines whose directions are specified by θ_1, ϕ_1 and θ_2, ϕ_2 is given by $\cos\gamma = \cos\theta_1 \cos\theta_2 + \sin\theta_1 \sin\theta_2 \cos(\phi_1 - \phi_2)$.)

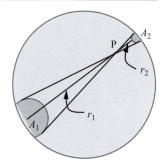

Figure 9.12 A spherical mass shell and two cones with common vertex at P.

Problem 9.25 Suppose a tunnel were dug through the Earth along a diameter. Show that an object dropped into this tunnel would undergo simple harmonic motion. Determine the period of the motion. (Air resistance is, of course, neglected.)

Problem 9.26 A particle is at an arbitrary point P inside a mass shell. (a) Draw the line through P so that it passes through the center of the shell (that is, the line is a diameter). Construct a "double cone" with vertices at P as shown in Figure 9.12. Show that the resultant gravitational force on the particle due to the two surface elements A_1 and A_2 is zero. (b) Generalize to any line through P and show that there is no net force acting on the particle at any interior point. (This is the way Newton showed that there was no force on a particle inside a shell.)

Problem 9.27 An infinite mass plane gives rise to a constant gravitational field, and a sphere of mass M gives rise to the gravitational field given by Newton's law of universal gravitation. Consider an object that looks like Saturn, but instead of rings, the equatorial plane of the sphere consists of a uniform mass plane of infinite extent and mass density σ (per unit area). Let the radius of the sphere be R. Determine the work done as a particle of mass m is moved from the "north pole" of the sphere to a distance h above it ($h > R$). (The principle of superposition tells us the field of two mass distributions is just the vector sum of the field due to each.)

Problem 9.28 The gravitational field in some region of space is given by $\mathbf{g} = -kr^3\hat{\mathbf{r}}$, where k is a constant. What is the mass density ρ? What is ρ if the field is given by $\mathbf{g} = -(k/r^2)\hat{\mathbf{r}}$?

Problem 9.29 Consider a point P in the plane of a ring of mass M and radius a. Assume P lies a distance $r > a$ from the center of the ring. Determine the potential at r to an accuracy of order a^3/r^3. (Hint: To evaluate the integral use the binomial expansion and discard terms of order a^3/r^3 and higher.)

Problem 9.30 A collision between Earth and an asteroid of diameter 200 m would cause widespread damage and loss of life. Space scientists are interested in devising ways to deflect an approaching asteroid. Not too long ago, two NASA engineers[13] suggested placing a spacecraft near such an asteroid and using the gravitational attraction between spacecraft and asteroid to tow it away. This "gravitational tractor" would hover a distance d from the asteroid. Its engines would be directed as shown in Figure 9.13 so that they would not blast the surface of the asteroid. The angle ϕ is the half angle of the rocket jet; you may assume a value of $20°$ for this angle. (a) Show that the minimum thrust required to just balance the gravitational attraction of the asteroid is

$$T = \frac{GMm/d^2}{\cos[\sin^{-1}(r/d) + \phi]},$$

where r is the radius of the asteroid. (b) Evaluate T for a 200 m diameter asteroid ($r = 100$ m), having a density of 2×10^3 kg/m^3, if a 20 tonne spacecraft hovers at $d = 2r$. (c) Show that the velocity change imparted to the asteroid per second is $\Delta v = Gm/d^2$.

---

[13] E. T. Lu and S. G. Love, "Gravitational Tractor for Towing Asteroids," *Nature*, **438**, 177–178 (2005).

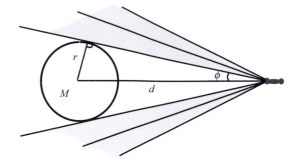

Figure 9.13 A "gravitational tractor." The shaded areas are the exhaust plumes from the rocket motors which are tilted away from the asteroid so as not to push on it. The angle ϕ is the half angle of the plume.

Problem 9.31 Show that the Minkowski metric yields $ds^2 = c^2 dt^2 - dx^2 - dy^2 - dz^2$.

Computational Projects

Computational Technique: Numerical Integration

Two simple numerical integration schemes are called the "trapezoidal rule" and "Simpson's rule." These are two applications of a more general method called the "Newton–Cotes technique." Both schemes assume you know the value of the integrand at the end points and at a number of intermediate points.

The trapezoidal rule assumes the integrand $f(x)$ varies linearly between the "data" points. Thus, if we know $f(x)$ at x_i and at x_{i+1}, then

$$\int_{x_i}^{x_{i+1}} f(x)dx = \frac{1}{2}(x_{i+1} - x_i)(f_{i+1} + f_i).$$

The total integral from initial to final values of x will be the sum of such terms. (Note that the area under the $f(x)$ curve is broken up into trapezoids. The results are often not very accurate unless the x values are very closely spaced.)

Simpson's rule uses a quadratic fit to a series of three equally spaced points, separated by distances d. Then

$$\int_{x_i}^{x_{i+1}} f(x)dx = \frac{d}{3}(f_i + 4f_{i+1} + f_{i+2}).$$

Note that the number of values of x must be odd.

Computational Project 9.1 A sphere of radius 1 m has a density that varies as $\rho(r) = 5r^3$ for $0 \leq r \leq 1$. Obtain the mass of the sphere analytically and numerically with the trapezoidal rule and Simpson's rule for nine equally spaced values of r. Compare results of the three calculations.

Computational Project 9.2 Write a computer program to determine the position and velocity of a projectile fired vertically from the Earth's surface with an initial speed of 3000 m/s. Plot the position and velocity as a function of time. The Earth is assumed to be a sphere. You may neglect air resistance, the rotation of the Earth, and the effect of any other astronomical bodies, but note that the acceleration due to gravity varies with distance from the center of the Earth.

Computational Project 9.3 Plot the potential for two equal point masses. (To make things easier, assume you are using a system of units in which the masses are unity, they are separated by unit distance and the gravitational constant is unity.) You can let the two particles lie on the x-axis.

Computational Project 9.4 Imagine a planet in the shape of a cube with side 10^5 m and density 1000 kg/m^3. Draw a Cartesian coordinate system with origin at the center of the cube and axes passing through the centers of the faces. Obtain an integral expression for the potential at a point 10^6 m from the origin along one of the axes. Write a computer program to carry out the integration and obtain a numerical answer.

10 Central Force Motion: The Kepler Problem

The orbital motion of a planet around the Sun was one of the first important problems to be analyzed in terms of Newton's laws of motion. The gravitational force attracting a planet to the Sun is a *central force,* that is, a force directed towards a fixed point. The motion of a planet is a prime example of the more general problem of the behavior of a particle acted upon by a central force.

Although we shall be primarily concerned with the motion of planets and satellites, the techniques you will learn in this chapter are applicable to any kind of central force. In this chapter, as well as learning the laws governing the motion of celestial bodies, you will be exposed to the concept of *effective potential energy,* and how constants of the motion are used in solving physics problems.

Historically, the quantitative analysis of orbital motion began with Kepler's realization that the motion of planets can be described by three empirical laws. An important point made in this chapter is that Kepler's laws of planetary motion can be derived theoretically from Newton's laws of motion. Additionally, Newton's laws give us a much deeper understanding of Kepler's laws. This application of Newton's ideas amazed and fascinated the "natural philosophers" of his era and was one of the most important events in the history of science.[1]

10.1 Johannes Kepler (Historical Note)

Johannes Kepler lived from 1571 to 1630. He was born into a poor family in a small town in Germany. His father was a professional soldier who spent much of his time away from home and his mother was a quarrelsome woman who was accused of witchcraft in her old age. It would be fair to assume that Kepler had an unhappy childhood. Nevertheless, he was very intelligent and an excellent student. The prince who ruled the region sent him to study at a Lutheran seminary. Eventually, Kepler graduated from the University of Tubingen. Originally he planned to study divinity, but before he was ordained, the authorities in the seminary convinced him that he was not cut out for the clergy. With the help of his advisers, he was appointed to a position teaching mathematics in Graz, Austria. He was a quiet, introspective person and not greatly interested in teaching; in fact, he was probably a terrible teacher for he had the reputation of interrupting his own lectures to silently mull over some idea that had just occurred to him.

In Graz, Kepler had an idea that changed the course of his life. This idea, which became an obsession with him, was (he thought) a glimpse into the mind of God: a vision of the basic

---

[1] Much of the material in this chapter is an introduction to celestial mechanics. Many excellent celestial mechanics books are available, including A. E. Roy, *Orbital Motion*, Adam Hilger Press, Bristol, 1988, and J. M. A. Danby, *Fundamentals of Celestial Mechanics,* William-Bell Inc., Richmond, VA, 1992. A more recent book that is particularly well written is *Solar System Dynamics* by C. D. Murray and S. F. Dermott, Cambridge University Press, Cambridge, 1999.

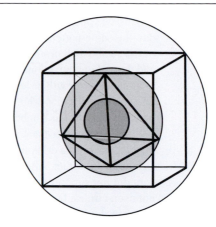

Figure 10.1 A nested tetrahedron and cube with inscribed and circumscribed spheres. In Kepler's model the nested perfect solids (from innermost outwards) were an octahedron, an icosahedron, a dodecahedron, a tetrahedron, and a cube.

structure of the universe. It was, so to speak, revealed to him why there are only five planets, and why they are in their particular orbits around the Sun. Kepler's mind-boggling inspiration was this: There are only five planets (besides Earth) because there are only five "perfect" solids, and the orbits of the planets correspond to spheres inscribed and circumscribed about the perfect solids when they are nested, one inside the other.

The "perfect" solids (or "simple polyhedrons") are the geometrical figures formed from regular polygons. A regular polygon is one with equal sides. Thus, for example, a tetrahedron (or equilateral pyramid) is made up of four equilateral triangles. A cube is made up of six squares. Similarly, an octahedron is an eight-sided solid made of equilateral triangles, an icosahedron is made of twenty equilateral triangles, and the dodecahedron is composed of twelve pentagons. Many other solids can be constructed from polygons; for example, a "soccer ball" shape can be made up of pentagons and hexagons. There are, however, only *five* "perfect" solids whose faces are a *single* type of regular polygon. There is a very neat proof that there can *only* be five such solids; this proof can be traced back to the ancient Greeks. If you are interested, it is reproduced in the book *Cosmos*, by Carl Sagan.[2]

Kepler's idea was that if the perfect solids were placed one inside the other, and each was circumscribed with a sphere, then the Sun would be at the center of the system and each of the five planets would orbit the Sun in a circular orbit whose radius would be equal to the radius of the corresponding circumscribed sphere. Kepler built models of these nested solids and their circumscribed spheres, but he could not prove his theory because he did not have enough information on the distances of the five (known) planets to the Sun. Figure 10.1 illustrates the idea behind Kepler's model.

At that time, the best astronomical data in Europe were in the observatory of an eccentric Danish astronomer named Tycho Brahe.[3] Kepler went to visit Brahe to get his data. He was certain that Brahe would be overwhelmed by his wonderful new theory. Brahe was not overwhelmed. In fact, at first Brahe would not even let Kepler see the data! However, Brahe soon realized that Kepler was an excellent mathematician and he offered to let him work on a small portion of the data to calculate the orbit of Mars. This was certainly not what Kepler had in mind, but he grudgingly agreed.

---

[2] Carl Sagan, *Cosmos*, Random House, New York, 1980 (Appendix 2).

[3] Tycho Brahe (1546–1601) set up an observatory at Uraniborg on the Danish island of Hven. A stream of young assistants from all over Europe came to Uraniborg where they carried out experiments in chemistry during the day and observed the heavens at night. Uraniborg has been described as the first research institution involved in "big science." For the full story read the book by John Robert Christianson, *On Tycho's Island: Tycho Brahe and his Assistants, 1570–1601*, Cambridge University Press, Cambridge, 2000. At the time Kepler went to visit him, Brahe had moved to Prague.

This was, perhaps, the original "graduate student–professor" relationship in science. To this day the same pattern exists. A young, aspiring scientist makes the pilgrimage to the laboratory of the established professor with the hope of being allowed to share in the professor's knowledge and data. You yourself may be doing this a few years from now. I hope your relationship with your graduate advisor will be less tempestuous than that of Kepler and Tycho. They did not get along at all. Tycho was a man who loved a party, who spent his evenings eating, drinking, and carousing, whereas Kepler was somber and rather puritanical and did not at all approve of Tycho's lifestyle.

Some years after the collaboration began, Tycho died. (Rumor has it that he died of a burst bladder while on a drinking spree.) Kepler inherited all of Tycho's data as well as Tycho's position as court astronomer. After a great deal of analysis, much to his dismay, Kepler found that Tycho's data did not support his grandiose theory. In fact, the data showed that the orbits of planets were not even circles, but rather ellipses! Kepler tried to modify his theory by slipping elliptical orbits between the inscribed perfect solids, but it did not quite work. Kepler never did know that there are more than five planets, as the discoveries of the planets Uranus and Neptune came many years after his death. By the time these were discovered, Kepler's theory on the inscribed orbits of the planets was no more than a historical oddity.

Kepler's life is full of instructive incidents for physicists. His most valuable contribution to science was his analysis and synthesis of the observations of Tycho Brahe. It is interesting to consider that he was obsessed by a beautiful theory that did not agree with experiment. No matter how beautiful a theory may be, if it does not agree with experimental measurements, it must be discarded! Physics is the study of the physical universe and it is Nature that determines the way things behave. It is to Kepler's credit that he respected and believed the data, even though the data did not agree with his theory.

10.2 Kepler's Laws

Kepler's synthesis of Tycho's planetary data can be expressed as three statements that are now called Kepler's laws of planetary motion. They are:

(1) The orbit of a planet is an ellipse with the Sun at one focus.
(2) The radius vector of a planet (the Sun–planet line) sweeps out equal areas in equal times.
(3) The period of a planet squared is proportional to the cube of the semimajor axis of its elliptical orbit.

These laws express three facts about the motion of a planet. They are "empirical laws," that is, they were obtained from the data but they had no theoretical basis. Kepler's laws describe the motion of planets but neither Kepler nor anyone else living at that time could give a reasonable explanation for this behavior. About 70 years later, Newton applied the law of universal gravitation and his laws of motion to the problem and succeeded brilliantly in showing that Kepler's laws are a consequence of some very basic physical relations. It is no wonder that other scientists of his era were in such awe of him. In this chapter, I will show you what Newton did, but of course I will use modern methods. Near the end of the chapter I will return to Kepler's laws: by that time you should have a deeper appreciation for them.

10.3 Central Forces

Any force that is directed toward or away from a fixed point (often taken as the origin of coordinates) and whose magnitude is a function *only* of the distance to the fixed point is called a *central force*. In general, a central force will have the form

$$\mathbf{F} = f(r)\hat{\mathbf{r}},$$

where $f(r)$ is the magnitude of the force (a function only of r) and $\hat{\mathbf{r}}$ is the radial unit vector. The direction of the force is along the line joining the particle and the origin. Two important central forces are the gravitational force between masses and the electrostatic force between charges. Assuming one of the bodies is fixed at the origin, the gravitational force is a central force given by

$$\mathbf{F} = -\frac{Gm_1m_2}{r^2}\hat{\mathbf{r}},$$

and the electrostatic force is

$$\mathbf{F} = \frac{Q_1Q_2}{4\pi\epsilon_or^2}\hat{\mathbf{r}}.$$

We can also *imagine* other central forces such as, for example,

$$\mathbf{F} = \frac{k}{r^5}\hat{\mathbf{r}}.$$

Such forces may or may not exist in nature, but we can analyze them mathematically anyway. Although this seems like a useless theoretical exercise, such studies may have a practical outcome. For example, to a first approximation, the force between molecules can be expressed as the sum of two central forces. This "Lennard-Jones" force can be expressed in terms of the potential energy as

$$V(r) = -\frac{a}{r^6} + \frac{b}{r^{12}}.$$

Here a and b are constants that depend on the properties of the particular molecules involved in the interaction. The force itself is obtained from $\mathbf{F} = -\nabla V$.

We are interested in determining the motion of a particle subjected to the gravitational force of another body (which we also treat as if it were a particle). For the sake of simplicity let us assume the more massive body is at rest. Strictly speaking, this cannot be true; the two objects actually orbit about their center of mass. But if one particle is much more massive than the other, then the more massive particle has a much smaller acceleration and in the limit, as the ratio of the masses goes to infinity, the more massive particle can be considered to be at rest.

To appreciate this, imagine a binary star system in which both stars have the same mass. The two stars will orbit around their center of mass, which is at a point halfway between them. But suppose one of the stars is more massive than the other. The center of mass will be closer to the more massive star. If one star is infinitely more massive than the other, the center of mass will be at the center of the larger star. Similarly, the International Space Station and the Earth are orbiting around their center of mass, but for all intents and purposes, the center of mass of this system lies at the center of the Earth and we are perfectly justified in considering the Earth to be at rest.[4]

In considering the motion of a particle of mass m in a central force field, the first thing to note is that *the particle moves in a plane*. The proof of this statement is given by the following argument. At any instant in time the particle is in the plane defined by the position vector and the velocity vector. The only way the particle can move out of this plane is if its acceleration has a component perpendicular to the plane. But for a central force, the force (and hence the acceleration) lies along $\mathbf{r}$, and therefore there is no component of the acceleration perpendicular to the plane.

---

[4] Recall the discussion of reduced mass, defined as $\mu = m_1m_2/(m_1 + m_2)$. See Equation (6.20). If the central force problem is treated in terms of the reduced mass, then no approximation is made. The relations obtained are the same as those derived here except that m is replaced by μ and $\mathbf{r}$ is the position of one body relative to the other rather than the distance from the center of mass.

Perhaps the most important aspect of the motion of a particle under the action of a central force is that the moving particle *has a constant angular momentum.* By definition, the angular momentum of a particle is $\mathbf{l} = \mathbf{r} \times \mathbf{p}$, where $\mathbf{r}$ is the position of the particle and $\mathbf{p}$ is its linear momentum. Recall from Section 7.2.2 that the time derivative of the angular momentum is

$$\frac{d\mathbf{l}}{dt} = \frac{d}{dt}(\mathbf{r} \times \mathbf{p}) = \mathbf{r} \times \mathbf{F}.$$

In the case of a central force, $\mathbf{F} = f(r)\hat{\mathbf{r}}$. Hence,

$$\frac{d\mathbf{l}}{dt} = \mathbf{r} \times f(r)\hat{\mathbf{r}} = r\hat{\mathbf{r}} \times f(r)\hat{\mathbf{r}}$$
$$= rf(r)(\hat{\mathbf{r}} \times \hat{\mathbf{r}}) = 0.$$

Consequently, for a central force, the time derivative of the angular momentum is zero. This means that the angular momentum is constant. Also recall that the time rate of change of angular momentum is equal to the torque. A central force does not exert a torque on a particle.

These simple physical arguments lead us to an important conservation law concerning the motion of *any* particle under the action of *any* central force:

The angular momentum is constant.

This means that both the *magnitude* and the *direction* of the angular momentum are constant.

For a more sophisticated proof of the constancy of angular momentum, you can write the Lagrangian for a particle in a central force field. In polar coordinates the kinetic energy is $T = \frac{1}{2}m(\dot{r}^2 + r^2\dot{\theta}^2)$ and the potential energy is $V = -\int_{r_0}^{r} f(r)dr$. The Lagrangian is

$$L = T - V = \frac{1}{2}m\dot{r}^2 + \frac{1}{2}mr^2\dot{\theta}^2 + \int_{r_0}^{r} f(r)dr.$$

Since θ is ignorable, $\partial L / \partial \dot{\theta} = \text{constant}$. That is,

$$\frac{\partial L}{\partial \dot{\theta}} = mr^2\dot{\theta} = l = \text{constant}.$$

We can use the constancy of angular momentum to generate another proof that a particle in a central force field moves in a plane. Assume the particle is at position $\mathbf{r}$ and has velocity $\mathbf{v}$. These two vectors define a plane and, of course, the particle lies in this plane. By the definition of the cross product, the vector $\mathbf{r} \times \mathbf{v}$ is perpendicular to the plane containing the vectors $\mathbf{r}$ and $\mathbf{v}$. But $\mathbf{l} = m\mathbf{r} \times \mathbf{v}$, so the vector $\mathbf{l}$ is perpendicular to the plane containing $\mathbf{r}$ and $\mathbf{v}$. Because $\mathbf{l} = \text{constant}$, the perpendicular to the plane containing $\mathbf{r}$ and $\mathbf{v}$ is constant. Therefore, the particle moves in a constant plane. See Figure 10.2.

Because the motion of the particle lies in a plane, two coordinates are sufficient to specify its position. These can be x and y or r and θ. The origin of the coordinate system is usually placed

Figure 10.2 The angular momentum vector is perpendicular to the plane containing $\mathbf{r}$ and $\mathbf{v}$. Because the angular momentum is constant, the plane is invariant.

Figure 10.3 The position of a planet referenced to "The First Point in Aries."

at the primary (assumed to be at rest).[5] To orient the coordinates, it is necessary to specify a fixed direction. Astronomers pick an imaginary line from the center of the Earth towards a position called "The First Point in Aries" which is the position of the Sun at the vernal equinox.

In Figure 10.3 the symbol Υ indicates the fixed line in space. In polar coordinates the angle θ is measured from Υ. In Cartesian coordinates one usually defines the x-axis along this fixed line. The y-axis is selected in the plane of the orbit and perpendicular to the x-axis. The z-axis is then perpendicular to the orbit and along the angular momentum vector. For many problems it is safe to assume that this coordinate system is an inertial system.

Five thousand years ago the position of the Sun at noon on the vernal equinox was in the constellation Aries ("the ram"). Because the Earth's axis of rotation precesses with a period of about 25 000 years, the position of the Sun at noon on the vernal equinox has changed and it is presently in the constellation Pisces; within a few hundred years it will enter into the constellation Aquarius. However, the name "First Point in Aries" and the symbol Υ, representing ram's horns, are still used to represent this arbitrary fixed line in space. It is amusing to note that due to the precession of the Earth's axis, the positions of the zodiacal constellations have shifted, but astrologers still use the values of 5000 years ago. Thus, people born when the Sun was in the constellation Pisces think they are Aries, those who were born when the Sun was in Aries think they are Taurus, and so on. If astrology had any validity, this horrible mix-up in the zodiacal signs would be serious indeed!

Worked Example 10.1   A particle is in a circular orbit under the action of an inverse cubed attractive central force given by $f(r) = -k/r^3$. Obtain an expression for the angular momentum and show that it is constant.

Solution By Newton's second law, $m\ddot{\mathbf{r}} = -f(r)\hat{\mathbf{r}}. = -\frac{k}{r^3}$ Recall from our discussion of plane polar coordinates in Chapter 2 that the acceleration in polar coordinates is

$$\ddot{\mathbf{r}} = (\ddot{r} - r\dot{\theta}^2)\hat{\mathbf{r}} + (r\ddot{\theta} + 2\dot{r}\dot{\theta})\hat{\boldsymbol{\theta}}.$$

For a circular orbit, the magnitude r is constant, so $\dot{r} = 0$ and $\ddot{r} = 0$ so

$$m\ddot{\mathbf{r}} = m(-r\dot{\theta}^2\hat{\mathbf{r}} + r\ddot{\theta}\hat{\boldsymbol{\theta}}) = -f(r)\hat{\mathbf{r}}.$$

Therefore, for an inverse cube force law, $m\ddot{\mathbf{r}} = -f(r)\hat{\mathbf{r}}$ yields two relations, namely

$$-mr\dot{\theta}^2 = -f(r) = -\frac{k}{r^3},$$

and

$$mr\ddot{\theta} = 0.$$

---

[5] In the language of astronomy, if one body is much more massive than the other, the massive body is called the "primary."

The second equation boils down to $\ddot{\theta} = 0$, telling us that the angular velocity, $\dot{\theta}$ is constant. The first equation yields

$$mr^4\dot{\theta}^2 = k.$$

But because $l = mr^2\dot{\theta}$ we see that the left-hand side is l^2/m and consequently

$$l^2 = mk,$$

and hence the angular momentum is constant. This result also follows from $r = $ constant and $\dot{\theta} = $ constant.

Worked Example 10.2   Consider a particle in an attractive force field whose potential energy is of the form $V(r) = kr^{n+1}$. Show that for a periodic orbit the average kinetic energy is related to the average potential energy by

$$\langle T \rangle = \frac{n+1}{2} \langle V \rangle.$$

Apply to the gravitational force. (This is a special case of the virial theorem.)

Solution   This problem asks us to determine the relationship between the average values of the kinetic and potential energy. The time average of any quantity (such as the kinetic energy) is defined as:

$$\langle T \rangle = \frac{1}{\tau} \int_0^\tau T(t)dt.$$

For this problem, it is reasonable to let τ be the orbit period.

   If a particle is in orbit, after one period it will have returned to its original location and have the same velocity as it did initially. Therefore, the quantity $\mathbf{p} \cdot \mathbf{r}$ repeats periodically. Let us define the periodic function $G(t) = \mathbf{p} \cdot \mathbf{r}$. Consider the time average of the derivative of G with respect to time:

$$\left\langle \frac{dG}{dt} \right\rangle = \frac{1}{\tau} \int_0^\tau \frac{dG}{dt} dt = \frac{1}{\tau} \int_0^\tau dG = \frac{1}{\tau} [G(\tau) - G(0)] = 0.$$

But

$$\frac{dG}{dt} = \frac{d}{dt}(\mathbf{p} \cdot \mathbf{r}) = \mathbf{p} \cdot \dot{\mathbf{r}} + \dot{\mathbf{p}} \cdot \mathbf{r}$$

$$= m\mathbf{v} \cdot \mathbf{v} + \mathbf{F} \cdot \mathbf{r} = 2T - \nabla V \cdot \mathbf{r}$$

$$= 2T - \frac{dV}{dr}r,$$

because $\nabla V \cdot \mathbf{r} = \frac{dV(r)}{dr}r$. Therefore,

$$\left\langle \frac{dG}{dt} \right\rangle = \frac{1}{\tau} \int_0^\tau \left(2T - \frac{dV}{dr}r \right) dt = 0,$$

and

$$2\langle T \rangle - \left\langle \frac{dV}{dr}r \right\rangle = 0,$$

$$\langle T \rangle = \frac{1}{2} \left\langle \frac{dV}{dr}r \right\rangle.$$

Since $V = kr^{n+1}$ we have

$$\frac{dV}{dr}r = (n+1)kr^n r = (n+1)kr^{n+1} = (n+1)V,$$

and

$$\left\langle \frac{dV}{dr}r \right\rangle = (n+1)\langle V \rangle,$$

so

$$\langle T \rangle = \frac{n+1}{2}\langle V \rangle.$$

The gravitational potential has the form $V = -\frac{k}{r}$ so $n = -2$ and

$$\langle T \rangle = -\frac{1}{2}\langle V \rangle.$$

This relationship between kinetic and potential energy is useful when solving orbital mechanics problems. It might be mentioned that the virial theorem has important applications in thermodynamics and statistical mechanics.

Exercise 10.1

Use $\mathbf{F} = -\nabla V$ to obtain an expression for the Lennard-Jones force. Determine the value of r where the force changes from attractive to repulsive. Answer: $\mathbf{F} = -\hat{\mathbf{r}}\left(\frac{6a}{r^7} - \frac{12b}{r^{13}}\right)$. ∎

Exercise 10.2

Use $V = -\int_{r_0}^{r} F(r)dr$ to obtain the potential energy for the gravitational force, the electrostatic force, and the force exerted by a spring. Select r_0 appropriately. ∎

Exercise 10.3

A planet of mass m is in a circular orbit of radius r_0 about a star of mass M. Detemine its angular momentum. Answer: $l = m\sqrt{GMr_0}$. ∎

Exercise 10.4

Assume that at some initial moment the radius vector $\mathbf{r}$ and the velocity vector $\mathbf{v}$ are perpendicular to one another – however, the angle between them is changing with time. Explain why the two vectors cannot ever be parallel to one another. Answer: Because angular momentum is constant. ∎

10.4 The Equation of Motion

Our next task is to determine the equation of motion for a planet. We will do this in two ways: first by writing down Newton's second law, and second by applying the Lagrangian technique.

The force acting on a particle (planet) of mass m is given by Newton's law of universal gravitation

$$\mathbf{F} = -\frac{GMm}{r^2}\hat{\mathbf{r}},$$

where M is the mass of the larger attracting body, assumed fixed at the origin of coordinates. The equation of motion of the satellite (mass m) is

$$m\ddot{\mathbf{r}} = -G\frac{Mm}{r^2}\hat{\mathbf{r}},$$

or

$$\ddot{\mathbf{r}} = -\frac{GM}{r^2}\hat{\mathbf{r}}. \tag{10.1}$$

We evaluated the acceleration $\ddot{\mathbf{r}}$ in polar coordinates in Chapter 2 – see Equation (2.14) – obtaining

$$\ddot{\mathbf{r}} = (\ddot{r} - r\dot{\theta}^2)\hat{\mathbf{r}} + (r\ddot{\theta} + 2\dot{r}\dot{\theta})\hat{\boldsymbol{\theta}}.$$

Substituting this expression into Equation (10.1), yields

$$(\ddot{r} - r\dot{\theta}^2)\hat{\mathbf{r}} + (r\ddot{\theta} + 2\dot{r}\dot{\theta})\hat{\boldsymbol{\theta}} = -\frac{GM}{r^2}\hat{\mathbf{r}}.$$

Separating the radial and angular components leads to the following two scalar equations

$$\ddot{r} - r\dot{\theta}^2 = -\frac{GM}{r^2}, \tag{10.2}$$

$$r\ddot{\theta} + 2\dot{r}\dot{\theta} = 0.$$

These equations are a pair of *coupled* second-order ordinary differential equations. The second equation, the "θ-equation," is easy to analyze by going back to the definition of angular momentum and recalling that

$$\mathbf{l} = \mathbf{r} \times \mathbf{p} = \mathbf{r} \times m\mathbf{v}.$$

The velocity in plane polar coordinates is $\mathbf{v} = \dot{r}\hat{\mathbf{r}} + r\dot{\theta}\hat{\boldsymbol{\theta}}$, so

$$\mathbf{l} = mr\hat{\mathbf{r}} \times (\dot{r}\hat{\mathbf{r}} + r\dot{\theta}\hat{\boldsymbol{\theta}}).$$

Now carry out the cross products, using the facts that $\hat{\mathbf{r}} \times \hat{\mathbf{r}} = 0$ and $\hat{\mathbf{r}} \times \hat{\boldsymbol{\theta}} = \hat{\mathbf{k}}$, where $\hat{\mathbf{k}}$ is perpendicular to the plane of motion. You obtain

$$\mathbf{l} = mr(r\dot{\theta})\hat{\mathbf{k}} = mr^2\dot{\theta}\hat{\mathbf{k}}. \tag{10.3}$$

This, in itself, is a useful equation. But it becomes even more useful if you take its time derivative,

$$\frac{d\mathbf{l}}{dt} = \frac{d}{dt}\left(mr^2\dot{\theta}\right)\hat{\mathbf{k}}.$$

For central forces, the angular momentum is constant, so $d\mathbf{l}/dt = 0$ and the left-hand side of this equation is zero. Therefore,

$$0 = \frac{d}{dt}\left(mr^2\dot{\theta}\right) = m\left(2r\dot{r}\dot{\theta} + r^2\ddot{\theta}\right),$$
$$= mr\left(2\dot{r}\dot{\theta} + r\ddot{\theta}\right).$$

Since neither m nor r is equal to zero, this implies that

$$r\ddot{\theta} + 2\dot{r}\dot{\theta} = 0.$$

But this is just the θ equation! Therefore, we see that the θ equation of motion, the second of Equations (10.2), is a statement that angular momentum is constant. Consequently, that equation can be replaced by the equivalent relation

$$l = mr^2\dot{\theta} = \text{constant}.$$

This equation, in turn, gives a nice expression for $\dot{\theta}$, namely,

$$\dot{\theta} = \frac{l}{mr^2}. \tag{10.4}$$

Replacing $\dot{\theta}$ in the r equation, the first of Equations (10.2), by expression (10.4) yields the following equation for the radial motion of mass m:

$$\ddot{r} - \frac{l^2}{m^2 r^3} = -\frac{GM}{r^2}. \tag{10.5}$$

This equation involves only r. Thus, conservation of angular momentum *decouples* the equations of motion. The equation has only one variable (r), so it is often called a "one-dimensional equation." But you should always keep in mind that the motion takes place in *two* dimensions. Once Equation (10.5) has been solved for $r = r(t)$, you can use it in Equation (10.4) to determine $\theta = \theta(t)$.

We shall now use the Lagrangian technique to determine the equations of motion. (We had better obtain the same result!)

As you know, the Lagrangian is $L = T - V$. In this problem V is the potential energy for a particle of mass m attracted gravitationally to a body of mass M. According to Equation (9.6) this is

$$V(r) = -\frac{GMm}{r}.$$

In polar coordinates, the kinetic energy $\left(\frac{1}{2}mv^2 = \frac{1}{2}m\mathbf{v}\cdot\mathbf{v}\right)$ is

$$T = \frac{1}{2}m\dot{r}^2 + \frac{1}{2}mr^2\dot{\theta}^2.$$

Therefore the Lagrangian is

$$L = \frac{1}{2}m\dot{r}^2 + \frac{1}{2}mr^2\dot{\theta}^2 + \frac{GMm}{r}.$$

Recall that the Lagrange equations of motion have the form

$$\frac{d}{dt}\frac{\partial L}{\partial \dot{q}_i} - \frac{\partial L}{\partial q_i} = 0,$$

where the q_is are now r and θ. So we have two equations, namely,

$$\frac{d}{dt}\frac{\partial L}{\partial \dot{r}} - \frac{\partial L}{\partial r} = 0,$$

and

$$\frac{d}{dt}\frac{\partial L}{\partial \dot{\theta}} - \frac{\partial L}{\partial \theta} = 0.$$

The partial derivatives are easily evaluated. You should prove for yourself that the two equations of motion are

$$\frac{d}{dt}(m\dot{r}) - mr\dot{\theta}^2 + \frac{GMm}{r^2} = 0, \tag{10.6}$$

and

$$\frac{d}{dt}\left(mr^2\dot{\theta}\right) = 0, \quad \text{or} \quad \frac{dl}{dt} = 0. \tag{10.7}$$

The second of these equations gives

$$\dot{\theta} = \frac{l}{mr^2}.$$

Using this expression, Equation (10.6) leads to

$$\ddot{r} - \frac{l^2}{m^2 r^3} = -\frac{GM}{r^2}. \tag{10.8}$$

These are, of course, the same as the equations of motion obtained using Newton's second law, namely Equations (10.4) and (10.5). Note, however, how much easier it is to use the Lagrangian than Newton's second law.

Exercise 10.5

Carry out the steps to show that $T = \frac{1}{2}m\dot{r}^2 + \frac{1}{2}mr^2\dot{\theta}^2$.   ∎

Exercise 10.6

Obtain Equations (10.6) and (10.7).   ∎

Exercise 10.7

Suppose the force between a particle of mass m and a fixed point is given by $\mathbf{F} = -kr\hat{\mathbf{r}}$, where k is a constant. Obtain the Lagrangian and the equations of motion. Is angular momentum conserved for this system? Answer: $m\ddot{r} - mr\dot{\theta}^2 + kr = 0$; $\frac{d}{dt}(mr^2\dot{\theta}) = 0$.   ∎

10.5 Energy and the Effective Potential Energy

We have (twice!) obtained the equation of motion for the radial coordinate – Equation (10.5) and Equation (10.8). You probably expect to proceed by integrating it to get the value of r as a function of time and initial conditions. Indeed, we shall do that in Section 10.6, but first let us consider what the conservation of energy principle tells us about our problem. Since gravity is a conservative force (and there are no other forces), the total mechanical energy of our system is constant. Thus,

$$E = \text{constant} = T + V$$
$$= \frac{1}{2}m(\dot{r}^2 + r^2\dot{\theta}^2) - G\frac{Mm}{r}.$$

But $l = mr^2\dot{\theta}$, so we can replace $\dot{\theta}$ by l/mr^2 and write

$$E = \frac{1}{2}m\dot{r}^2 + \frac{m}{2}r^2\left(\frac{l^2}{m^2r^4}\right) - \frac{GMm}{r},$$

or

$$E = \frac{1}{2}m\dot{r}^2 + \frac{l^2}{2mr^2} - \frac{GMm}{r}. \tag{10.9}$$

The first term depends on the square of the radial speed ($\dot{r}^2$) and the other terms depend only on position, so the first term *looks like* kinetic energy and the remaining terms *look like* potential energy. So it seems reasonable to associate a "radial kinetic energy" with the term $\frac{1}{2}m\dot{r}^2$ and an "effective potential energy" V_{eff} with the remaining two terms. The effective potential energy V_{eff} is given by[6]

$$V_{eff} = \frac{l^2}{2mr^2} - \frac{GMm}{r}. \tag{10.10}$$

Although V_{eff} looks and acts like a potential energy and is a function only of position, it is definitely *not* a potential energy because it actually contains a kinetic energy term, namely $l^2/(2mr^2) = \frac{1}{2}mr^2\dot{\theta}^2$.

It is instructive to draw an energy diagram in terms of the effective potential energy.[7] Note that V_{eff} is the sum of two terms, one positive and the other negative. For $r \to \infty$, the negative term in V_{eff} is the dominant term because

$$\left.\frac{1}{r}\right|_{r\to\infty} > \left.\frac{1}{r^2}\right|_{r\to\infty}.$$

As $r \to 0$, the positive term dominates because

$$\left.\frac{1}{r}\right|_{r\to 0} < \left.\frac{1}{r^2}\right|_{r\to 0}.$$

Therefore the plot of V_{eff} vs. r must have the general shape shown in Figure 10.4. Study this figure carefully and convince yourself it is qualitatively correct, specifically that V_{eff} is positive as $r \to 0$ and negative as $r \to \infty$. Notice particularly that the $l^2/2mr^2$ term bends much more sharply than $-GMm/r$. Also, don't forget that r can only take on positive values.

Figure 10.5 is also a plot of $V_{eff}(r)$ vs. r. In this plot the effective potential energy does not have quite the right shape because I drew it to make it easy for you to appreciate various aspects of the effective potential that are hard to see on a more accurate plot, such as Figure 10.4. In Figure 10.5 you see four possible values for the total energy, denoted E_0, E_1, E_2, and E_3. Consider first a particle with energy E_1. From Equations (10.9) and (10.10), we can write

$$\frac{1}{2}m\dot{r}^2 = E - V_{eff}. \tag{10.11}$$

---

[6] The terms "effective potential energy" and "effective potential" are used interchangeably, even though a "potential" is actually potential energy per unit mass. Some books use the term "fictitious potential." An older term is "centrifugal potential." In general, the effective potential energy is defined as the sum of $l^2/2mr^2$ and the potential energy. Thus for an electron orbiting about a proton, the effective potential would be

$$V_{eff} = \frac{l^2}{2mr^2} - \frac{e^2}{4\pi\epsilon_0 r}.$$

[7] At this time you may wish to review the material on energy diagrams in Section 5.6.

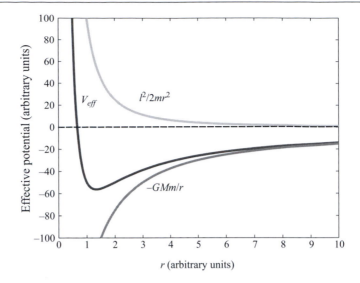

Figure 10.4 The effective potential energy $V_{eff}(r)$ is the sum of two terms, one positive and one negative.

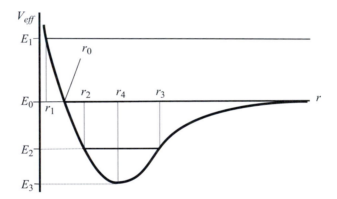

Figure 10.5 Energy diagram for the effective potential. The turning points for various values of total energy are indicated.

Because $\frac{1}{2}m\dot{r}^2$ can never be negative, the particle cannot be located at values of r for which $E < V_{eff}$. From Figure 10.5 this implies that if the energy is E_1, then $r \geq r_1$, where r_1 is the point of closest approach. You can imagine the particle starting at $r = \infty$, coming closer and closer to the primary until it reaches r_1 (the turning point), and then moving back out to infinity.

Note that according to Equation (10.11), the radial component of the velocity $\dot{r}$ is given by

$$\dot{r} = \sqrt{\frac{2}{m}\left[E - V_{eff}(r)\right]}. \tag{10.12}$$

Thus, the square of the radial speed is proportional to $E - V_{eff}$. (Note that $E - V_{eff}$ is not the kinetic energy, only the radial part of the total kinetic energy.) On the energy diagram (Figure 10.5) the distance between the horizontal line at E_1 and the heavy line representing V_{eff} is proportional to $\dot{r}^2$. At r_1 the value of V_{eff} is E_1, so $\dot{r} = 0$. That is, at the turning point the particle has zero radial velocity, which makes perfect sense. At the turning point, the angular

Figure 10.6 A comet or some other celestial object in a hyperbolic path.

component of the velocity is a maximum because $r\dot\theta = (r)(l/mr^2) = l/mr$ is greatest when r is smallest.

Let me emphasize that this not a complete description of the motion; it is only a description of the *radial* motion. Meanwhile the particle is also moving in θ with a velocity given by $\dot\theta = l/mr^2$. The angular velocity increases as the radial distance decreases, in agreement with Kepler's second law. As we shall see, for positive values of the energy, the path of the particle is a hyperbola as illustrated in Figure 10.6. This could be the path of a comet coming in from infinity, speeding up as it approaches the Sun, swooping around the Sun, and then moving out to infinity. (Here "infinity" is a place far from the Sun, such as the Oort Cloud from which many comets are believed to originate.)

Next consider a particle with zero total energy ($E = E_0 = 0$; see Figure 10.5). This means that the positive "radial kinetic energy" $\frac{1}{2}m\dot r^2$ is equal in magnitude to the negative effective potential energy V_{eff}. The motion of the particle as it comes from $r = \infty$ to the turning point at r_0 and then goes back out to $r = \infty$ is similar to the motion of the particle with energy E_1. As we shall see, the main difference is that the trajectory for energy $E_1 > 0$ is a hyperbola and the trajectory for energy $E_0 = 0$ is a parabola. For the parabolic orbit, the radial speed of the particle $\dot r$ is zero at infinity as well as at the turning point. As the particle comes in from infinity, $\dot r$ increases, reaching a maximum at r_4 where V_{eff} reaches its greatest negative value, then decreases to zero at r_0. The angular velocity is given by $\dot\theta = l/mr^2$. (You should be able to describe the angular velocity as the particle comes in from infinity to the point of closest approach and then moves back out to infinity.)

If the particle has *negative* total energy, as indicated by E_2 in Figure 10.5, the motion is quite different; there are now *two* turning points and the particle can neither reach $r = 0$ nor move out to $r = \infty$. That is, the motion is *bounded*. The particle is trapped in a potential well. As it moves back and forth radially between the two points denoted by r_2 and r_3, it is also moving azimuthally with a varying angular velocity $\dot\theta$. As we shall see shortly, this combination of radial and angular motion represents a trajectory which is an ellipse.

Finally, if the particle has energy E_3, the minimum possible total energy, the value of $\dot r$ is zero at all times and the particle is at a constant radial position $r = r_4$. The path is a circle. The angular velocity is $\dot\theta = l/mr_4^2 = $ constant. The particle is therefore moving with constant angular velocity in a circular path.

Depending on the value of the energy, the trajectory or orbit of the particle is a hyperbola, a parabola, an ellipse, or a circle. These are called conic sections because they can be generated by cutting a cone in various different ways, as shown in Figure 10.7.

We have been considering the motion of massive bodies interacting gravitationally, but the ideas and methods developed here are quite general and can be easily adapted to any central force problem, such as the motion of an electron in orbit around a proton in the Bohr model of the hydrogen atom.

Exercise 10.8

A particle is in a parabolic orbit. Where is its turning point? Answer: $r_0 = l^2/2GMm^2$. ∎

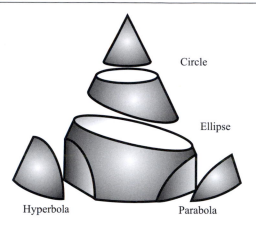

Circle

Ellipse

Hyperbola   Parabola

Figure 10.7 The conic sections. If the cone is cut as shown, the cross sections of the cuts will be a circle, an ellipse, a parabola, and one branch of a hyperbola.

Exercise 10.9

In the Bohr model of the hydrogen atom, an electron in the lowest orbit has angular momentum $l = \hbar$, where $\hbar$ is Planck's constant divided by 2π. The electrostatic potential energy of the electron and proton is $V = -e^2/4\pi\epsilon_0 r$. Show that the total energy (mechanical plus electrostatic) of the lowest orbit is $-(1/2)me^4/(4\pi\epsilon_0\hbar)^2$. Assume a circular orbit.   ∎

10.6 Solving the Radial Equation of Motion

The radial equation of motion, Equation (10.5), is a second-order ordinary differential equation of the form

$$\frac{d^2r}{dt^2} = \frac{l^2}{m^2r^3} - \frac{GM}{r^2}.$$

In previous chapters we learned two different ways of solving such an equation. In Section 3.6 we integrated directly by setting $\frac{d^2r}{dt^2} = v\frac{dv}{dr}$. In Section 5.7 we solved for the motion by using the energy integral in the form of Equation (5.15), namely,

$$\int_{x_0}^{x(t)} \frac{dx}{\sqrt{E - V(x)}} = \sqrt{\frac{2}{m}}t.$$

Applying the energy approach to the present problem, we see that Equation (10.9) leads to

$$\frac{dr}{dt} = \left[\frac{2}{m}\left(E - \frac{l^2}{2mr^2} + \frac{GMm}{r}\right)\right]^{\frac{1}{2}}. \tag{10.13}$$

This equation yields the following definite integral

$$\int_{r_0}^{r(t)} \frac{dr}{\sqrt{\frac{2}{m}\left(E - \frac{l^2}{2mr^2} + \frac{GMm}{r}\right)}} = \int_0^t dt = t.$$

Integrating gives

$$t = t(r, r_0, E, l),$$

which can, in principle, be inverted to yield

$$r = r(t, r_0, E, l).$$

In this solution, the total energy E and angular momentum l are arbitrary constants. Note the use of the term *arbitrary* constant. Such constants are arbitrary in the sense that the differential equation is satisfied by the solution no matter what values the constants happen to have. But for a particular problem, these constants are anything but arbitrary! In simple kinematics problems the "arbitrary constants" are usually the initial position and initial velocity. In more complex problems such as the one we just solved, they tend to be other constants of the motion such as the energy and angular momentum.

> **Exercise 10.10**
>
> Obtain $t = t(r, r_0, E, l)$ for a parabolic orbit (for which $E = 0$). (You will have to look up the integral.) Note that it would be very tedious to invert the expression to obtain $r = r(t)$.
> Answer:
>
> $$t = \frac{\sqrt{2m}}{3(GMm)^2} \left(GMmr + \frac{l^2}{m} \right) \sqrt{GMmr - \frac{l^2}{2m}}$$
> $$- \frac{\sqrt{2m}}{3(GMm)^2} \left(GMmr_0 + \frac{l^2}{m} \right) \sqrt{GMmr_0 - \frac{l^2}{2m}}.$$ ∎

10.7 The Equation of the Orbit

Many physicists work in the space program. In a few years you may be involved in planning a Moon shot or a planetary probe to study the clouds of Titan. If that happens, you will need to know the position of the spacecraft as a function of time. You will have to calculate $\mathbf{r} = \mathbf{r}(t)$ and you could use the technique described in the previous section.[8]

On the other hand, if you wish to describe the *orbit* of a satellite, planet or comet, you are not interested in its position at a particular time. Rather, you want a description of the *path* followed by the celestial object. Mathematically, this is given by the *equation of the orbit*, $r = r(\theta)$. Once you have this equation, you can determine r for every value of θ and thus trace out the path.

I will first present a "brute force" method and then a "sophisticated technique" for obtaining the orbit.

Brute Force Method
To determine the equation of the orbit, $r = r(\theta)$, start with the expression for $\frac{dr}{dt}$ obtained using the energy equation, Equation (10.13). Using the chain rule,

$$\frac{dr}{dt} = \frac{dr}{d\theta} \frac{d\theta}{dt} = \frac{dr}{d\theta} \dot{\theta}.$$

But recall that the angular momentum is $l = mr^2 \dot{\theta}$, so

$$\frac{dr}{dt} = \frac{l}{mr^2} \frac{dr}{d\theta}. \tag{10.14}$$

---

[8] There are other ways to determine the position of a space probe. These are described in engineering books on "astronautics."

Substituting into Equation (10.13) yields

$$\frac{dr}{d\theta} = \frac{mr^2}{l}\left[\frac{2}{m}\left(E - \frac{l^2}{2mr^2} + \frac{GMm}{r}\right)\right]^{\frac{1}{2}}$$

$$= \left[\frac{2mE}{l^2}r^4 - r^2 + \frac{2GMm^2}{l^2}r^3\right]^{\frac{1}{2}}$$

$$= \left[\alpha r^2 + \beta r^3 + \gamma r^4\right]^{\frac{1}{2}},$$

where I used the Greek letters $\alpha, \beta,$ and γ to illustrate the form of this equation. Rewriting and integrating the last equation gives

$$\int_{r_0}^{r} \frac{dr}{r(\alpha + \beta r + \gamma r^2)^{1/2}} = \int_{\theta_0}^{\theta} d\theta. \tag{10.15}$$

This integral can be found in tables of integrals.

Thus, in principle, the problem of determining the equation of the orbit is solved. The only things left to do are: (1) integrate Equation (10.15), and (2) invert the result to obtain $r = r(\theta)$.

Sophisticated Technique

There is a different way to obtain the equation of the orbit that cleverly avoids evaluating the complicated integral in Equation (10.15) and then carrying out the inversion to obtain $r = r(\theta)$. Bear with me for a little while because the math is a bit involved (but not difficult).

We will begin with Equation (10.2), the radial equation of motion:

$$\ddot{r} - r\dot{\theta}^2 = -\frac{GM}{r^2}.$$

Using $l = mr^2\dot{\theta}$ to eliminate $\dot{\theta}$ gives Equation (10.8) which is repeated here for convenience:

$$\ddot{r} - \frac{l^2}{m^2r^3} = -\frac{GM}{r^2}.$$

Now let us introduce a new variable, u, defined as the inverse of r. That is,

$$u \equiv \frac{1}{r}.$$

Then $r = u^{-1}$ and $dr = -(1/u^2)du$. Therefore,

$$\frac{dr}{dt} = -\frac{1}{u^2}\frac{du}{dt} = -\frac{1}{u^2}\frac{du}{d\theta}\frac{d\theta}{dt},$$

where the last step uses the chain rule and the fact that $u = u(\theta)$. So,

$$\frac{dr}{dt} = -\frac{1}{u^2}\frac{du}{d\theta}\dot{\theta} = -r^2\dot{\theta}\frac{du}{d\theta}.$$

But $r^2\dot{\theta} = l/m$. Replacing,

$$\frac{dr}{dt} = -\frac{l}{m}\frac{du}{d\theta}.$$

Taking the derivative with respect to time again,

$$\frac{d}{dt}\left(\frac{dr}{dt}\right) = \frac{d}{dt}\left(-\frac{l}{m}\frac{du}{d\theta}\right),$$

$$\ddot{r} = -\frac{l}{m}\frac{d}{dt}\left(\frac{du}{d\theta}\right) = -\frac{l}{m}\frac{d}{d\theta}\left(\frac{du}{d\theta}\right)\frac{d\theta}{dt},$$

$$= -\frac{l}{m}\frac{d^2u}{d\theta^2}\left(\frac{l}{mr^2}\right) = -\frac{l^2}{m^2r^2}\frac{d^2u}{d\theta^2}.$$

But $1/r^2 = u^2$ so

$$\ddot{r} = -\frac{l^2u^2}{m^2}\frac{d^2u}{d\theta^2}.$$

Substituting this into Equation (10.8) and using $1/r^3 = u^3$ we have

$$-\frac{l^2u^2}{m^2}\frac{d^2u}{d\theta^2} - \frac{l^2}{m^2}u^3 = -GMu^2,$$

or

$$\frac{d^2u}{d\theta^2} + u = \frac{GMm^2}{l^2}. \tag{10.16}$$

This is a very interesting equation. It looks like the equation for simple harmonic motion except for the additional constant term on the right-hand side. It can be made to look *exactly* like the simple harmonic motion equation by defining a new variable

$$w = u - \frac{GMm^2}{l^2}.$$

Then, since the last term is a constant,

$$\frac{dw}{d\theta} = \frac{du}{d\theta},$$

and

$$\frac{d^2u}{d^2\theta} = \frac{d^2w}{d\theta^2}.$$

Equation (10.16) can now be written in the form of the SHM equation:

$$\frac{d^2w}{d\theta^2} + w = 0.$$

As noted in Section 3.7, the solution is sinusoidal and we can write

$$w = A\cos(\theta - \theta_0),$$

where A and θ_0 are integration constants. Consequently,

$$u - \frac{GMm^2}{l^2} = A\cos(\theta - \theta_0).$$

Now u is just the inverse of r, so the derivation finally gives an equation for r in terms of θ :

$$r = \frac{1}{u} = \frac{1}{\frac{GMm^2}{l^2} + A\cos(\theta - \theta_0)}.$$

Dividing top and bottom by GMm^2/l^2 puts this in a nicer form:

$$r = \frac{l^2/GMm^2}{1 + \frac{Al^2}{GMm^2}\cos(\theta - \theta_0)}. \tag{10.17}$$

From a mathematical point of view, the problem is now solved because r has been expressed in terms of θ and other known quantities (such as the angular momentum) and two constants of integration, A and θ_0.

Equation (10.17) describes all the possible orbits for the two-body problem, i.e., circles, ellipses, parabolas, and hyperbolas. Because elliptical motion is of particular interest, let us assume that the total energy E is negative. Then the motion is bounded and the particle (planet) oscillates radially between turning points r_2 and r_3 as illustrated in Figure 10.5 for energy E_2.

Equation (10.17) contains two constants of integration, A and θ_0. We now consider their physical significance, beginning with A. At the turning points, the effective potential energy is equal to the total energy, $V_{eff}^{TP} = E$. This fact allows us to evaluate r_2 and r_3 which we write, in general, as r_{tp}. From Equation (10.10) we have

$$V_{eff}^{TP} = \frac{l^2}{2mr_{tp}^2} - \frac{GMm}{r_{tp}} = E.$$

Therefore, in terms of u,

$$\frac{l^2}{2m}u_{tp}^2 - GMmu_{tp} - E = 0.$$

The two solutions of this quadratic equation for u_{tp} are:

$$u_{\pm} = \frac{m}{l^2}\left(GMm \pm \sqrt{(GMm)^2 + \frac{4l^2E}{2m}}\right). \tag{10.18}$$

(Note that $1/u_{\pm}$ are the r_2 and r_3 turning points of Figure 10.5.)

But we have seen that

$$u = A\cos(\theta - \theta_0) + \frac{GMm^2}{l^2}.$$

The maximum and minimum values of this expression occur when $\cos(\theta - \theta_0) = \pm 1$. That is,

$$u_+ = A + \frac{GMm^2}{l^2},$$

and

$$u_- = -A + \frac{GMm^2}{l^2}.$$

Using u_+ and equating the expression above to the relation given by Equation (10.18) leads to

$$A + \frac{GMm^2}{l^2} = \frac{GMm^2}{l^2} + \frac{m}{l^2}\sqrt{(GMm)^2 + \frac{2l^2E}{m}},$$

or

$$A = \left[\frac{(GMm)^2m^2}{l^4} + \frac{2Em}{l^2}\right]^{\frac{1}{2}}. \tag{10.19}$$

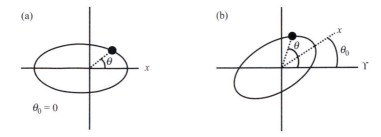

Figure 10.8 The position of a satellite relative to the major axis is given by the angle θ. (a) The major axis can be defined to be the x-axis. (b) The angle θ_0 gives the orientation of the elliptical orbit relative to a fixed line in space.

Thus A is related to the total energy and the angular momentum in a rather complicated way.

Let us now turn our attention to θ_0, the other constant of integration. Considering Equation (10.17) we note that when $\cos(\theta - \theta_0) = +1$, the denominator is maximized and r reaches its minimum value. If M is the mass of the Sun, the place where r is a minimum is called the *perihelion*. The x-axis is usually taken to be a line from M through the perihelion, as shown in Figure 10.8. If θ is measured from the x-axis, then $\theta_0 = 0$.

On the other hand, if we measure θ from a fixed direction in space, such as the "first point in Aries" (Υ) then θ_0 is the angle between Υ and the x-axis and gives the orientation of the ellipse relative to the fixed direction.

We have, then, determined A in terms of E and l and given the physical interpretation of θ_0 as a description of the orientation of the ellipse with respect to inertial axes. Now the problem is indeed fully solved. Unfortunately, as you have no doubt noticed, our expressions are rather unwieldy. In a moment I will write them in a neater form. But first we need to consider some properties of conic sections, particularly ellipses.

10.8 The Equation of an Ellipse

You probably remember the grade school method for drawing an ellipse. You place a sheet of paper on a cork board and stick two tacks in it. Then you take a piece of string, tie the ends together, and loop it around the tacks. Finally you place a pencil in the loop of string and move it around the tacks, keeping the string taut, and if you are very careful you will draw an ellipse on the paper. See Figure 10.9.

As the pencil traces out the ellipse, the distance from the pencil to either tack varies. But the sum of the distances to the two tacks remains the same because the string is of constant length and the distance between the two tacks does not change. Therefore the distance from the first tack to the pencil *plus* the distance from the pencil to the second tack is constant. This fact is used for the mathematical definition of an ellipse.

An ellipse is the locus of points whose distances from two fixed points sum to a constant.

That is a bit complicated to say, but it is expressed quite simply as an equation. First choose the two fixed points, call them focal points and denote them F and F'. The distance from a point on the ellipse to F will be called r and the distance from that same point to F' will be r'. See Figure 10.10. Then an ellipse is defined as the locus of points such that $r + r' = $ constant.

It is easy to obtain an equation for the ellipse in polar coordinates by introducing the angle θ between r and the major axis of the ellipse. The length of the semimajor axis $\left(\frac{1}{2}\overline{PP'}\right)$ will be

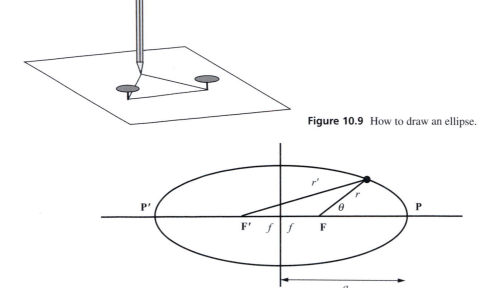

Figure 10.9 How to draw an ellipse.

Figure 10.10 An ellipse. The total distance $r + r' + \overline{F'F}$ is a constant, so $r + r'$ must be constant. Points P' and P are called the apsides.

denoted a, and the semiminor axis is denoted b. The distance from the center of the ellipse to either focal point will be $f = ea$, where e is a number less than one, called the *eccentricity*. The eccentricity e and the semimajor axis a determine the size and shape of the ellipse.

It turns out that

$$r + r' = 2a.$$

This relation may surprise you, but it is easily proved. Consider point P. There $r = \overline{FP}$ and $r' = \overline{F'P}$. By symmetry, $\overline{F'P} = \overline{FP'}$ and therefore $r + r' = \overline{FP} + \overline{FP'} = 2a$. Since $r + r'$ is a constant, if it is equal to $2a$ at one point, it is equal to $2a$ at any point.

Now consider the triangle formed by r, r', and $2f$, where $f = ea$. See Figure 10.10. According to the law of cosines, if θ is the "external" angle,

$$r'^2 = r^2 + (2ae)^2 + 2r(2ae) \cos \theta. \tag{10.20}$$

Substituting $r' = 2a - r$ into Equation (10.20) gives

$$(2a - r)^2 = r^2 + 4a^2e^2 + 4aer \cos \theta,$$

and a little bit of algebra leads to

$$r = \frac{a(1 - e^2)}{1 + e \cos \theta}. \tag{10.21}$$

This is the equation for an ellipse in plane polar coordinates. If we were to go through the same process for hyperbolas and parabolas, we would obtain similar equations for $r = r(\theta)$. In fact, the equation for any conic section can be expressed in the form

$$r = \frac{p}{1 + e \cos \theta}, \tag{10.22}$$

where $p = a(1 - e^2)$ for an ellipse, $p = a(e^2 - 1)$ for a hyperbola,[9] and $p = a$ for a parabola[10] (a is a characteristic constant for each type of curve and, in general, $p = l^2/GMm^2$). The various curves correspond to different choices of the eccentricity e. Specifically,

$$
\begin{array}{lll}
e = 0 & \text{gives} & \text{a circle,} \\
e < 1 & \text{gives} & \text{an ellipse,} \\
e = 1 & \text{gives} & \text{a parabola,} \\
e > 1 & \text{gives} & \text{a hyperbola.}
\end{array}
$$

The semimajor axis a determines the *size* of an ellipse and the eccentricity determines its *shape*. At zero eccentricity, the ellipse degenerates into a circle. As the eccentricity gets closer and closer to one, the ellipse gets flatter and flatter. That is, as e approaches unity, the ratio of the semiminor axis to the semimajor axis approaches zero. As you will prove in Exercise 10.12,

$$
\frac{b}{a} = \left(1 - e^2\right)^{1/2}. \tag{10.23}
$$

You probably know from geometry that the area of an ellipse is given by $S = \pi a b$. It is sometimes convenient to write this in terms of a only. Using Equation (10.23),

$$
S = \pi a^2 \left(1 - e^2\right)^{1/2}. \tag{10.24}
$$

Having considered the properties of conics, let's go back to the motion of planets. Comparing the equation of an ellipse (10.21) with the equation of the orbit (10.17), you can see that the orbit is an ellipse (or, more generally, a conic section). A term-by-term comparison of these two equations shows that

$$
\frac{l^2}{GMm^2} = a(1 - e^2), \tag{10.25}
$$

and that

$$
\frac{Al^2}{GMm^2} = e.
$$

But we showed in Equation (10.19) that

$$
A = \left[\frac{(GMm)^2 m^2}{l^4} + \frac{2Em}{l^2}\right]^{\frac{1}{2}},
$$

so

$$
e^2 = \frac{(GMm)^2 m^2}{l^4} \frac{l^4}{(GMm)^2 m^2} + \frac{2Em}{l^2} \frac{l^4}{(GMm^2)^2}.
$$

Therefore,

$$
e^2 = 1 + \frac{2El^2}{m(GMm)^2}. \tag{10.26}
$$

---

[9] A hyperbola has two branches and is described by the equation

$$
r = \frac{a(e^2 - 1)}{\pm 1 + e\cos\theta},
$$

where the "plus branch" corresponds to an attractive force and the "minus branch" corresponds to a repulsive force. The quantity a is one half the distance between the vertices.

[10] For a parabola $a = 2r_p$, where r_p is the distance from the focus to the nearest point on the parabolic curve.

This gives us an expression for the eccentricity in terms of constants of the motion. Note that for $E > 0$ we get $e > 1$, a hyperbola. For $E = 0$ we get $e = 1$, a parabola. For $E < 0$, we get $e < 1$, an ellipse. Finally, for a circle, $e = 0$ so $2El^2/m(GMm)^2 = -1$.

Plugging Equation (10.26) into (10.25) gives the following expression for the semimajor axis:

$$a = -\frac{GMm}{2E}. \tag{10.27}$$

The minus sign is necessary because the total energy is negative for an ellipse. Equation (10.27) shows that the semimajor axis depends only on the energy.

Worked Example 10.3 A comet of mass m starts from infinity with velocity v_0 and impact parameter b. (If undeflected, the path of the comet would be a straight line passing the Sun at a distance b.) (a) Show that the distance of closest approach to the Sun is approximately $b^2 v_0^2/2GM$. (b) Write the equation of the orbit in polar coordinates in terms of M, v_0, and b, where M is the mass of the Sun.

Solution (a) Let the Sun–comet distance at perihelion be denoted d. At that point, the speed of the comet is v_d and its velocity is perpendicular to the Sun–comet line.

By conservation of angular momentum $l_\infty = l_d$, and

$$mv_0 b = mv_d d \Rightarrow \frac{b}{d} = \frac{v_d}{v_c}.$$

By conservation of energy

$$\frac{1}{2}mv_0^2 = \frac{1}{2}mv_d^2 - \frac{GMm}{d},$$

so dividing by $mv_0^2/2$,

$$1 = \left(\frac{b}{d}\right)^2 - \frac{2GM}{dv_0^2}.$$

Multiply by d^2 and obtain the quadratic

$$d^2 + \frac{2GM}{v_0^2}d - b^2 = 0,$$

$$d = -\frac{GM}{v_0^2} \pm \sqrt{\left(\frac{GM}{v_0^2}\right)^2 + b^2}.$$

Since $d > 0$, we use the positive sign and write

$$d = \frac{GM}{v_0^2}\left(-1 + \left[1 + \frac{b^2 v_0^4}{G^2 M^2}\right]^{1/2}\right).$$

Apply the binomial expansion to get

$$d = \frac{GM}{v_0^2}\left(-1 + \left[1 + \frac{1}{2}\frac{b^2 v_0^4}{G^2 M^2} + \cdots\right]\right),$$

$$d \simeq \frac{GM}{v_0^2}\frac{1}{2}\frac{b^2 v_0^4}{G^2 M^2} = \frac{b^2 v_0^2}{2GM}.$$

(b) In polar coordinates the hyperbolic orbit under an attractive force is

$$r = \frac{a(e^2 - 1)}{1 + e \cos \theta}.$$

The problem requires us to express a and e in terms of the given parameters, namely v_0, b, and constants. Taking the derivative of r with respect to time we have

$$\dot{r} = \frac{a(e^2 - 1)}{(1 + e \cos \theta)^2} e \sin \theta \dot{\theta},$$

Now $l = mr^2 \dot{\theta}$ so

$$\dot{\theta} = \frac{l}{mr^2} = \frac{l}{m} \frac{(1 + e \cos \theta)^2}{a^2(e^2 - 1)^2},$$

and we can write

$$\dot{r}^2 = \frac{e^2 l^2 / m^2}{\left[a(e^2 - 1)\right]^2} \sin^2 \theta.$$

The square of the speed of the comet in polar coordinates is

$$v^2 = \dot{r}^2 + r^2 \dot{\theta}^2.$$

The second term in v^2 is $r^2 \dot{\theta}^2$. But,

$$r\dot{\theta} = \frac{1}{r} \frac{l}{m} = \frac{l}{m} \frac{1 + e \cos \theta}{a(e^2 - 1)},$$

$$\therefore r^2 \dot{\theta}^2 = \frac{l^2}{m^2} \frac{1 + 2e \cos \theta + e^2 \cos^2 \theta}{\left[a(e^2 - 1)\right]^2},$$

so

$$v^2 = \frac{l^2(1 + 2e \cos \theta + e^2 (\sin^2 \theta + \cos^2 \theta))}{m^2 \left[a(e^2 - 1)\right]^2}$$

$$= \frac{l^2}{m^2 a(e^2 - 1)} \left(\frac{2 + 2e \cos \theta + e^2 - 1}{a(e^2 - 1)} \right)$$

$$= \frac{l^2}{m^2 a(e^2 - 1)} \left(2\frac{1 + e \cos \theta}{a(e^2 - 1)} + \frac{e^2 - 1}{a(e^2 - 1)} \right)$$

$$= \frac{l^2}{m^2 a(e^2 - 1)} \left[\frac{2}{r} + \frac{1}{a} \right].$$

Comparing $r = \frac{a(e^2-1)}{1+e \cos \theta}$ for a hyperbolic orbit with the general form of the equation of the orbit, Equation (10.17), we appreciate that

$$\frac{l^2}{GMm^2} = a(e^2 - 1).$$

Consequently,

$$v^2 = \frac{GMm^2 a(e^2 - 1)}{m^2 a(e^2 - 1)} \left[\frac{2}{r} + \frac{1}{a} \right] = GM \left[\frac{2}{r} + \frac{1}{a} \right].$$

At $r = \infty$, the velocity is v_0 and we see that

$$v_0^2 = \frac{GM}{a},$$

so

$$a = \frac{GM}{v_0^2}.$$

Also

$$a(e^2 - 1) = \frac{l^2}{GMm^2}.$$

Solving for e we obtain

$$e = \left(1 + \left[\frac{v_0^2 b}{(GM)} \right]^2 \right)^{1/2},$$

and finally

$$r = \frac{v_0^2 b^2 / GM}{1 + \sqrt{1 + \left(\frac{b v_0^2}{GM} \right)^2} \cos\theta}.$$

Exercise 10.11

Starting with Equation (10.20), obtain Equation (10.21). ◼

Exercise 10.12

Show that for an ellipse $b/a = \left(1 - e^2\right)^{1/2}$. (Hint: Apply the Pythagorean theorem to the triangle FOP in Figure 10.11.) ◼

Exercise 10.13

A planet is in an elliptical orbit with semimajor axis a. By averaging the largest and smallest values of r, show that it leads to an "average" value of the potential energy equal to $-GMm/a$. (Compare with the virial theorem discussed in Worked Example 10.2.) ◼

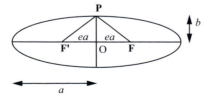

Figure 10.11 An ellipse. Note that the distance from the center to the focal point is ea and that the sum of the distances FP and $F'P$ is $2a$.

Exercise 10.14

Plot Equation (10.22) for $e = 2, e = 1.0, e = 0.5$, and $e = 0$. ■

10.9 Kepler's Laws Revisited

Having applied Newton's second law and Newton's law of universal gravitation to the problem of planetary motion, let us now return to Kepler's laws that were given in Section 10.2. You will find that Kepler's laws are simply descriptions of some aspects of the theory we have developed.

Recall that Kepler's first law states:

The orbit of a planet is an ellipse with the Sun at one of the focal points.

We applied Newton's second law to a particle of mass m in the gravitational field of M, writing $m\ddot{\mathbf{r}} = (-GMm/r^2)\hat{\mathbf{r}}$. After a significant amount of math, this led to the equation of the orbit $r = r(\theta)$, which we recognized as the equation of an ellipse. Thus Kepler's first law is a consequence of Newton's law of universal gravitation. Actually, the proof is even more general than Kepler's first law, for we have seen that objects in the solar system acting under the inverse square gravitational force will move in a trajectory which may be an ellipse, but it could also be one of the other conic sections.

Kepler studied the orbit of Mars all his life, and he deduced other important facts about its motion. For one thing, he found that the motion of Mars as a function of time was not uniform. Mars speeds up as it approaches the Sun, moves fastest at its point of nearest approach, then slows down as it moves away from the Sun. He expressed this fact in a quantitative way in his "Law of Areas" or, as we call it, Kepler's second law:

The radius vector of a planet sweeps out equal areas in equal times.

This is illustrated in Figure 10.12. If the planet moves from a to b in the same amount of time as to go from c to d then, according to Kepler's second law, the two shaded areas are equal. So another way to state Kepler's second law is, "The *areal* velocity of a planet is constant."

In point of fact, we already proved Kepler's second law, but you might not have noticed it because it was stated in quite a different way. We said, "For a central force, the angular momentum is constant." Are the two statements equivalent? Does the fact that the angular momentum is constant imply that the *areal velocity* is constant? The answer is yes. To prove that this is true, consider a planet located at vector position $\mathbf{r}$. After a time interval dt the planet will be at $\mathbf{r} + d\mathbf{r} = \mathbf{r} + \mathbf{v}dt$. See Figure 10.13. The shaded region is the area swept out. Recall that $|\mathbf{a} \times \mathbf{b}|$ is equal to the area of a parallelogram whose sides are the vectors $\mathbf{a}$ and $\mathbf{b}$. The shaded region in Figure 10.13 is one half of the parallelogram formed from $\mathbf{r}$ and $d\mathbf{r}$. Therefore, the area swept out by the planet's radius vector in time dt is

Figure 10.12 Kepler's second law states that equal areas are swept out in equal times.

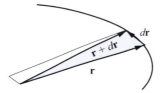

Figure 10.13 The area of the triangle is one half the area of the parallelogram. Therefore, $\Delta S = (1/2)\,|\mathbf{r} \times d\mathbf{r}| = (1/2)\,|\mathbf{r} \times \mathbf{v}\Delta t|$.

$$dS = \frac{1}{2}|\mathbf{r} \times \mathbf{v}dt| = \frac{dt}{2}|\mathbf{r} \times \mathbf{v}|.$$

But by the definition of angular momentum, $|\mathbf{r} \times \mathbf{v}| = l/m$. Therefore,

$$dS = \frac{1}{2}\frac{l}{m}dt,$$

or

$$\frac{dS}{dt} = \text{areal velocity} = \frac{l}{2m}.$$

The mass of a planet is constant, so the areal velocity is proportional to the angular momentum and $\frac{dS}{dt} = $ constant is completely equivalent to the relation $l = $ constant. Kepler's second law is simply a consequence of the fact that in central force motion, the angular momentum is conserved.

Kepler's third law states:

The period of a planet squared is proportional to its semimajor axis cubed.

In mathematical terms, if τ is the time required for a planet to orbit the Sun and if a is the semimajor axis of the orbit, the third law can be written

$$\tau^2 = Ka^3,$$

where K is some constant. Kepler discovered this law after years of studying planetary data. On the other hand, our analytical study of elliptical motion leads to it in a fairly simple way. We just note that

$$\frac{dS}{dt} = \frac{l}{2m} \Rightarrow \oint_{orbit} dS = \frac{l}{2m}\oint_{orbit} dt.$$

Let S equal the area of the ellipse. Then if τ is the period,

$$S = \frac{l}{2m}\tau.$$

But we have seen that $S = \pi ab = \pi a^2(1 - e^2)^{1/2}$ so

$$\frac{l}{2m}\tau = \pi a^2(1 - e^2)^{1/2}.$$

According to Equation (10.25), for an ellipse, $(1 - e^2) = l^2/GMm^2a$. Squaring we obtain

$$\frac{l^2}{4m^2}\tau^2 = \pi^2 a^4 \frac{l^2}{GMm^2a},$$

or

$$\tau^2 = \frac{4\pi^2}{GM}a^3.$$

This relation states that the period squared is proportional to the semimajor axis cubed. Thus, Kepler's third law is proved. Note that the constant of proportionality depends only on the mass of the Sun, so the ratio τ^2/a^3 should be the same for all the planets. Therefore, determining the period of a planet or asteroid allows us to evaluate its semimajor axis. Actually, this result is not completely accurate. There is a very small correction that is due to the fact that the Sun is not really at rest at the origin. The Sun and the planet move around their common center of mass. This point is almost, but not quite, at the center of the Sun. This introduces a small correction into Kepler's third law so that the quantity M in the denominator should be replaced by $M + m$. Once again we appreciate the technique in physics of first solving an easier idealized problem and introducing the complications later. (Problem 10.11 asks you to carry out the appropriate calculations.)

Worked Example 10.4 A particle moves in an elliptical orbit with semimajor axis a and eccentricity e. The velocity of the particle is observed to vary from a minimum value v_1 to a maximum value v_2. Determine the period.

Solution The angular momentum of the particle is constant:

$$l = mvr = \text{constant.}$$

Therefore, the maximum (and minimum) speeds occur when r is at the smallest (and greatest) distance from the force center. Call these distances r_1 and r_2. By conservation of angular momentum,

$$mv_1r_1 = mv_2r_2.$$

The points at r_1 and r_2 as well as the force center all lie on the semimajor axis, and

$$r_1 + r_2 = 2a.$$

Hence

$$l = mv_1(2a - r_2) = mv_2r_2,$$

$$\therefore r_2 = \frac{2av_1}{v_1 + v_2},$$

so the angular momentum can be written

$$l = m\frac{2av_1v_2}{v_1 + v_2}.$$

The angular momentum is related to the areal velocity by

$$\frac{l}{2m} = \frac{dS}{dt} = \text{areal velocity.}$$

The period is equal to the area divided by the areal velocity

$$\tau = \frac{\text{area}}{\text{areal velocity}} = \frac{\pi ab}{(l/2m)} = \frac{\pi ab(2m)}{m\frac{2av_1v_2}{v_1+v_2}} = \frac{\pi b(v_1 + v_2)}{v_1v_2}.$$

By Exercise 10.12 the semiminor axis, b, is equal to $a(1 - e^2)^{1/2}$, so

$$\tau = \pi a(1 - e^2)^{1/2}\frac{v_1 + v_2}{v_1v_2}.$$

Worked Example 10.5 Suppose you were familiar with Kepler's work, so you knew that the orbits of planets are ellipses and that the angular momentum is constant. This suggests to you that the force on the planet is a central force, so $F = f(r)$, but you don't know any other properties of the force. Show that the force obeys the inverse square law, that is, show that $f(r) \propto 1/r^2$.

Solution In polar coordinates the radial equation of motion is

$$\ddot{r} - r\dot{\theta}^2 = -f(r)/m.$$

The angular equation of motion leads to

$$\dot{\theta} = l/mr^2.$$

Substituting,

$$\ddot{r} - \frac{l^2}{m^2 r^3} = -\frac{f(r)}{m}.$$

In terms of $u = 1/r$, recalling that $\ddot{r} = -\frac{l^2 u^2}{m^2} \frac{d^2 u}{d\theta^2}$, we obtain

$$-\frac{l^2 u^2}{m^2} \frac{d^2 u}{d\theta^2} - \frac{l^2 u^3}{m^2} = -\frac{f(r)}{m},$$

$$\frac{d^2 u}{d\theta^2} + u = \frac{m}{l^2 u^2} f(r). \qquad (10.28)$$

Now the equation of an ellipse has the form

$$r = \frac{p}{1 + e \cos\theta},$$

$$\therefore u = \frac{1}{p} + \frac{e}{p} \cos\theta.$$

Taking the derivative with respect to θ twice yields

$$\frac{d^2 u}{d\theta^2} = -\frac{e}{p} \cos\theta,$$

and Equation (10.28) becomes

$$-\frac{e}{p} \cos\theta + \left(\frac{1}{p} + \frac{e}{p} \cos\theta \right) = \frac{m}{l^2 u^2} f(r)$$

$$\frac{1}{p} = \frac{m}{l^2 u^2} f(r) = \frac{m}{l^2} r^2 f(r).$$

That is,

$$f(r) = \frac{l^2}{mp} \frac{1}{r^2} \propto \frac{1}{r^2}.$$

Q.E.D.

Exercise 10.15

A certain asteroid has a period of four years. What is its semimajor axis in AU? (Assume you do not know the gravitational constant G or the mass of the Sun, but you do know that the Earth is at a distance of 1 AU from the Sun.) Answer: 2.52 AU. ∎

Exercise 10.16

(a)Using the form $\tau^2 = (4\pi^2/GM)a^3$ and assuming you do not know the mass of the Sun, determine the semimajor axis of Saturn in AU, given its period is 29.5 Earth years. (b)Using the correction $(M + m)$ and looking up the appropriate values, obtain a corrected value. Answers: 9.55 AU, 9.52 AU. ∎

Exercise 10.17

For a planet in a circular orbit, $F = ma$ can be written as $GMm/r^2 = mv^2/r$. Use this to derive Kepler's third law for circular orbits. ∎

Exercise 10.18

Halley's comet has an eccentricity of 0.967. Its perihelion distance is 8.81×10^{10} m. What is its period? Answer: 75.4 years. ∎

10.10 Orbital Mechanics

The first artificial satellite (Sputnik 1) was launched by the USSR on October 4th, 1957. Since then, nearly 10 000 satellites have been launched and about half of them are still in orbit. Artificial earth satellites have become an integral part of modern life. The Global Positioning System (GPS) is a constellation of satellites that can specify a location on the Earth to within an accuracy of up to 30 cm. You probably use it nearly every day. Satellites are used for navigation, weather forecasting, communications, oceanography, atmospheric science, astronomy, and military purposes.

This section is a brief consideration of some of the physics used in analyzing the motion of artificial satellites. The two principal physical concepts are energy and angular momentum.

10.10.1 Energy and Angular Momentum of a Satellite

The "specific energy" of a satellite is usually expressed as

$$\mathcal{E} = \frac{1}{2}v^2 - \frac{\mu}{r},$$

where $\mathcal{E} = E/m$ is the energy per unit mass of the satellite and $\mu = GM$ is an abbreviation commonly used in celestial mechanics.[11] M is the mass of the Earth.

The angular momentum per unit mass of an object is denoted $\mathbf{h}$. By definition

$$\mathbf{h} = \mathbf{l}/m = \mathbf{r} \times \mathbf{v}.$$

and is a constant of the motion.

Artificial satellites are placed in elliptical or circular orbits. The most important parameters of such orbits are the eccentricity (e) and the semimajor axis (a). Two other useful parameters are the distances from the center of the Earth to perigee (r_p) and to apogee (r_a). At these two points the velocity vector $\mathbf{v}$ and the position vector $\mathbf{r}$ are perpendicular to one another and we can write

$$h = r_a v_a = r_p v_p.$$

---

[11] I will use μ for GM in this section only.

We have seen that for an elliptical orbit,

$$r = \frac{a(1 - e^2)}{1 + e \cos \theta},$$

consequently, by Equation (10.25)

$$a(1 - e^2) = \frac{1}{GM} \left(\frac{l}{m}\right)^2 = \frac{h^2}{\mu}.$$

$$\therefore r = \frac{h^2/\mu}{1 + e \cos \theta}.$$

Measuring θ from perigee, we obtain

$$r_p = \frac{h^2/\mu}{1 + e} \quad \text{and} \quad r_a = \frac{h^2/\mu}{1 - e}.$$

The velocities at perigee and apogee are

$$v_p = \frac{h}{r_p} \quad \text{and} \quad v_a = \frac{h}{r_a}.$$

The energy (per unit mass) at perigee is

$$\mathcal{E} = \frac{1}{2}v_p^2 - \frac{\mu}{r_p} = \frac{h^2}{2r_p^2} - \frac{\mu}{r_p} = -\frac{\mu}{2a},$$

where the last step is obtained from Equation (10.27).

10.10.2 The Hohmann Transfer Orbit

Satellites are usually initially placed in a circular "parking orbit," and then raised to their final orbit by means of a "transfer orbit." We now consider a simple transfer orbit called the Hohmann transfer orbit. See Figure 10.14.

If a satellite in a circular orbit of radius r_c is given an impulse by turning on a rocket motor for a short burst, its velocity will increase by Δv and the orbit is transformed into an ellipse. The perigee of the ellipse will be $r_p = r_c$. The energy will have increased to $\mathcal{E}_t$. The elliptical orbit will have a semimajor axis given by $a = -\frac{\mu}{2\mathcal{E}_t}$, and the apogee will be at $r_a = 2a - r_p$. We can

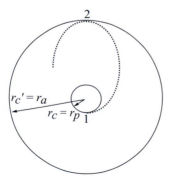

Figure 10.14 A Hohmann transfer orbit. The satellite is given an impulse at point 1 to place it in the elliptical orbit, and another impulse at 2 to place it in the circular orbit.

assume the position of the satellite has not changed during the "burn," but now its velocity will be given by

$$v_1 = v_c + \Delta v_1 = \sqrt{2\left(\frac{\mu}{r_p} + \mathcal{E}_t\right)},$$

where v_c is the velocity of the satellite in the initial circular orbit.

In a Hohmann transfer orbit, when the satellite reaches apogee it is given a second impulse to change the speed at apogee from $v_2 = v_a$ to that of a circular orbit of radius $r'_c = r_a$. Note that at the apogee of the elliptical transfer orbit, the speed is

$$v_2 = \sqrt{2\left(\frac{\mu}{r_a} + \mathcal{E}_t\right)},$$

whereas the speed of a satellite in a circular orbit of radius r_a is

$$v'_c = \sqrt{2\left(\frac{\mu}{r_a} - \frac{\mu}{2a}\right)} = \sqrt{\frac{\mu}{r_a}},$$

because $r_a = r'_c = a$. Recalling that the energy is a negative quantity, we appreciate that the satellite has to be given an additional impulse to increase its speed to that of the desired circular orbit. That is, $v'_c = v_2 + \Delta v_2$.

There are other orbit transfer schemes; the Hohmann transfer is the most economical (least amount of fuel burned) but the transfer takes a longer time than the other schemes.

Exercise 10.19

Show that $\mathcal{E} = \frac{\mu}{2a}$ implies that $r_p = a(1 - e)$.   ■

Exercise 10.20

Show that the velocity of a satellite at perigee can be expressed as $v_p = \sqrt{\frac{1+e}{r_p}}$.   ■

10.11 A Perturbed Circular Orbit

In this section we will study the *stability* of a circular orbit. When a physical system in an equilibrium state is perturbed by a small force, it will often oscillate about equilibrium. This is considered a *stable* equilibrium. On the other hand, if the perturbation causes the system to undergo a large change, the equilibrium is unstable. (The usual example of stable equilibrium is a marble at the bottom of a dip, and unstable equilibrium is a marble balanced on the tip of a bulge.) This section is both an application of the concepts studied in the previous sections and an introduction to the methods for dealing with small perturbations.

Consider a particle moving in a circular orbit under the action of a central force. For reasons that will eventually become clear, I will not require the force to obey the inverse square law. That is, although $\mathbf{F} = f(r)\hat{\mathbf{r}}$, the functional form of $f(r)$ will be left undetermined for now.

The equations of motion are given by generalizations of Equations (10.2):

$$m(\ddot{r} - r\dot{\theta}^2) = f(r), \tag{10.29}$$

$$m(r\ddot{\theta} + 2\dot{r}\dot{\theta}) = 0.$$

As before,

$$\dot{\theta} = \frac{l}{mr^2}, \tag{10.30}$$

with $l =$ constant. Inserting this expression for $\dot{\theta}$ into the radial equation of motion, Equation (10.29), leads to

$$m\ddot{r} = f(r) + \frac{l^2}{mr^3}. \tag{10.31}$$

If the particle is moving in a circular orbit of radius a, then

$$r = a = \text{constant}.$$

Consequently,

$$\ddot{r} = 0.$$

The radial equation of motion, Equation (10.31), then reduces to

$$f(a) = -\frac{l^2}{ma^3}. \tag{10.32}$$

I will come back to this equation shortly.

Now consider the question of the *stability* of a perturbed circular orbit. For example, we might be considering the motion of a planet in a perfectly circular orbit which is hit by a comet (as happened some years ago when comet Shoemaker–Levy collided with Jupiter). We want to know if the planet will continue moving in a stable orbit after a slight perturbation, or if it will behave in an erratic manner. In other words, we would like to know if a collision with a relatively small body could cause Jupiter to go flying out of the solar system.

After a collision or some other sort of perturbation, the radial position is no longer *exactly* equal to a, but it is still *nearly* equal to a, so we can write

$$r = a + \eta, \tag{10.33}$$

where $\eta \ll a$, that is, η is a very small quantity compared to a.

Inserting expression (10.33) for r into the radial equation of motion in the form of Equation (10.31) we obtain

$$m\frac{d^2}{dt^2}(a + \eta) = f(a + \eta) + \frac{l^2}{m(a + \eta)^3},$$

or

$$m\ddot{\eta} = f(a + \eta) + \frac{l^2}{ma^3(1 + \frac{\eta}{a})^3}. \tag{10.34}$$

The force at the position $r = a + \eta$ is very nearly equal to the force at $r = a$ so we are justified in expanding $f(a + \eta)$ in a Taylor's series expansion and keeping only the leading terms. Thus:

$$f(a + \eta) = f(a) + \eta\frac{df}{dr}\bigg|_{r=a} + \frac{1}{2}\eta^2\frac{d^2f}{dr^2}\bigg|_{r=a} + \cdots.$$

The term $(1 + \eta/a)^3$ in the denominator of the last term of Equation (10.34) can also be expanded. Using the binomial expansion:[12]

$$\frac{1}{(1 + \eta/a)^3} = \left(1 + \frac{\eta}{a}\right)^{-3} = 1 - 3\frac{\eta}{a} + 6\left(\frac{\eta}{a}\right)^2 + \cdots.$$

Consequently, Equation (10.34) can be written as

$$m\ddot{\eta} = f(a) + \eta\left.\frac{df}{dr}\right|_a + \cdots + \frac{l^2}{ma^3}\left(1 - 3\frac{\eta}{a} + \cdots\right),$$

or (to first order in η),

$$m\ddot{\eta} \doteq f(a) + \eta\left.\frac{df}{dr}\right|_a + \frac{l^2}{ma^3} - 3\frac{\eta}{a}\frac{l^2}{ma^3}.$$

Now recall that Equation (10.32) states that $f(a) = -l^2/ma^3$, so the first and third terms on the right-hand side cancel. This leaves

$$m\ddot{\eta} \doteq \eta\left.\frac{df}{dr}\right|_a - 3\frac{\eta}{a}\frac{l^2}{ma^3}$$

$$= \eta\left(\left.\frac{df}{dr}\right|_a - \frac{3l^2}{ma^4}\right),$$

or

$$\ddot{\eta} + \eta\left(\frac{3l^2}{m^2a^4} - \frac{1}{m}\left.\frac{df}{dr}\right|_a\right) = 0. \tag{10.35}$$

This equation has the form

$$\ddot{\eta} + K\eta = 0.$$

We will be considering equations of this form in detail later, but for now it is sufficient to note that the general form of the solution depends on whether K is positive or negative. For positive K the solution has the form

$$\eta = A\sin\sqrt{K}t + B\cos\sqrt{K}t, \tag{10.36}$$

and for negative K the solution is

$$\eta = Ce^{+\sqrt{|K|}t} + De^{-\sqrt{|K|}t}. \tag{10.37}$$

Thus, for positive K the solution is a sinusoidal oscillation, as in simple harmonic motion (see Section 3.7). On the other hand, if K is negative, the solution is the sum of an exponential decrease plus an exponential increase. The exponentially decreasing term quickly dies out so $\eta(t)$ increases with time exponentially. We can interpret this result physically to mean that for $K < 0$ the orbit

---

[12] If you have not already done so, you should immediately memorize the following extremely important series expansions.

Taylor's series:

$$F(a + \delta x) = F(a) + \delta x\left.\frac{dF}{dx}\right|_a + \frac{1}{2!}\delta x^2\left.\frac{d^2F}{dx^2}\right|_a + \frac{1}{3!}\delta x^3\left.\frac{d^3F}{dx^3}\right|_a + \cdots.$$

Binomial expansion:

$$(1 + x)^n = 1 + nx + \frac{n(n-1)}{2!}x^2 + \frac{n(n-1)(n-2)}{3!}x^3 + \cdots.$$

is unstable because the distance η from the radius a of a circular orbit grows without bounds. On the other hand, if $K > 0$, then η oscillates back and forth around zero, meaning that the particle oscillates about $r = a$ in simple harmonic motion. Therefore, the condition for a stable orbit is that $K > 0$. That is,

$$\left. \frac{3l^2}{m^2a^4} - \frac{1}{m}\frac{df}{dr} \right|_{r=a} > 0.$$

Once again we can use the fact that $l^2/ma^3 = -f(a)$, and rewrite this equation as

$$-\frac{3}{a}f(a) - \left.\frac{df}{dr}\right|_{r=a} > 0. \tag{10.38}$$

The gravitational force is $f = -GMm/r^2$ and $\frac{df}{dr} = +2GMm/r^3$, so Equation (10.38) becomes

$$\frac{3}{a}\left(\frac{GMm}{a^2}\right) - \frac{2GMm}{a^3} = +\frac{GMm}{a^3} > 0.$$

Consequently, the gravitational force leads to stable orbits.

But what if the force is not inverse square? A more general central force might have the form

$$f = -\frac{c}{r^n}.$$

Here n is a nonzero integer. (For the gravitational force, $n = 2$; notice that I explicitly incorporated the negative sign on the right-hand side because only attractive force laws lead to orbital motion.)

If $f = -\frac{c}{r^n}$, then

$$\left.\frac{df}{dr}\right|_a = \frac{cn}{a^{n+1}}.$$

Consequently, the stability condition (10.38) is

$$-\frac{3}{a}\left.\left(-\frac{c}{r^n}\right)\right|_a - \frac{cn}{a^{n+1}} > 0$$

$$-\frac{3}{a}\left(\frac{-c}{a^n}\right) - \frac{cn}{a^{n+1}} > 0$$

$$\frac{c}{a^{n+1}}(3 - n) > 0$$

$$n - 3 < 0$$

or

$$n < 3.$$

The inverse square force law ($n = 2$) leads to orbits that are stable under small perturbations. If the gravitational force had the form $F \propto 1/r^4$, a planetary system would be impossible because any tiny perturbation would cause the planets to spiral away from their orbits.

Notice that in determining the stability of circular orbits, we also solved the problem of determining the frequency of small oscillations in a perturbed orbit because Equation (10.35), the equation of motion for η, has the form

$$\ddot{\eta} + K\eta = 0,$$

and, as mentioned previously, this is the equation for simple harmonic motion. The frequency of harmonic motion is $\omega_0 = \sqrt{K}$, so

$$\omega_0^2 = \frac{3l^2}{m^2 a^4} - \frac{1}{m}\frac{df}{dr}\Big|_{r=a}$$

$$= -\frac{3}{ma}f(a) - \frac{1}{m}\frac{df}{dr}\Big|_{r=a}.$$

For a particle subjected to an inverse square force law, such as the gravitational force,

$$f = -cr^{-2},$$

$$\frac{df}{dr}\Big|_{r=a} = +2cr^{-3}\Big|_{r=a} = \frac{2c}{a^3}.$$

Also

$$f(a) = -\frac{c}{a^2}.$$

Consequently,

$$\omega_0^2 = -\frac{3}{ma}\left(-\frac{c}{a^2}\right) - \frac{1}{m}\left(\frac{2c}{a^3}\right) = \frac{c}{ma^3}. \tag{10.39}$$

This equation gives the frequency of the *radial* oscillations, Note that $2\pi/\omega_0$ is the time for the planet to go from maximum η to minimum η and back to maximum η. While carrying out this radial motion, the planet has an angular velocity given by[13]

$$\dot\theta = \frac{l}{mr^2}\Big|_{r=a} = \frac{l}{ma^2}.$$

Therefore,

$$\dot\theta^2 = \frac{l^2}{m^2 a^4} = \frac{-f(a)ma^3}{m^2 a^4} = \frac{-(c/a^2)ma^3}{m^2 a^4} = \frac{c}{ma^3}. \tag{10.40}$$

Comparing Equation (10.39) with Equation (10.40) we see that $\omega_0 = \dot\theta$, so for a $1/r^2$ force law, the radial oscillations have exactly the same frequency as the orbital motion. Therefore, in the time required for the particle to go completely around in its orbit, it will go through one complete radial oscillation. The combination of these two motions converts a circular orbit into an elliptical orbit.[14]

It can happen that an orbiting particle is perturbed in such a way that the orbital period and the period of radial oscillation are not equal. For example, the Earth is not a perfect sphere; it is an oblate spheroid and bulges slightly at the equator. This bulge exerts a force on an orbiting artificial satellite, and gives rise to a radial oscillation of frequency ω_r that is not equal to $\dot\theta$. The satellite moves in an elliptical orbit, but because the radial motion is not exactly synchronized with the angular motion, the ellipse slowly precesses. In other words, the major axis of the ellipse slowly rotates at a rate which is the difference between the two oscillations. The angular frequency of the precession of the ellipse, ω_p is given by

---

[13] Actually we should write the average value of the angular velocity, that is, $\langle\dot\theta\rangle$ rather than $\dot\theta$, but because $\langle\dot\theta\rangle \doteq \dot\theta$, I used $\dot\theta$ for simplicity in notation.

[14] You might conclude that the eccentric orbits of planets are a result of collisions with asteroids and comets. However, astronomers do not believe that this accounts for the eccentricities of planetary orbits.

$$\omega_p = \omega_r - \dot{\theta},$$

where ω_r is the frequency of the radial oscillations.

Exercise 10.21

Draw the orbit of a planet assuming the radial oscillation has a period half that of the angular motion. ∎

Exercise 10.22

By simple substitution, show that Equations (10.36) and (10.37) are solutions to the simple harmonic motion differential equation. ∎

10.12 Resonances

An orbit is usually thought of as the path of a particle that retraces its motion over and over again, as in the case of a single planet around a star. If the particle always passes through the same points, the orbit is said to be *closed*. In a simple closed orbit, the radial period and the angular period are equal. That is, the planet goes from one turning point to the other and back again in the same time as the angular displacement goes from 0 to 2π. If the radial period is slightly greater than the angular period, the planet will reach perihelion a short time after its angular position has advanced 2π. The orbit is not closed and the perihelion precesses. The orbit can actually be closed even if the two periods are not equal as long as they are in the ratio of integers, as in the middle picture of Figure 10.15. When two periods or (equivalently) two frequencies are in the ratio of small integers, they are said to be *commensurable* or *in resonance*. If the ratio of the radial frequency to the angular frequency is an irrational number such as π or $\sqrt{2}$, the orbit is not closed; it will never repeat itself. Figure 10.15 illustrates two commensurable orbits and one incommensurable orbit.

Resonances are very important in celestial mechanics. One often hears of a system being "locked" into a particular resonance. For example, the rotation of the Moon is locked into a 1:1 resonance with its orbital motion. Therefore, the same side of the Moon always faces the Earth. Similarly, the orbit of a 24 hour satellite is in a 1:1 resonance with the rotation of the Earth. An interesting case is the 2:5 resonance between the periods of Jupiter and Saturn. In the old-fashioned terminology used in celestial mechanics, this resonance is called "The Great Inequality."

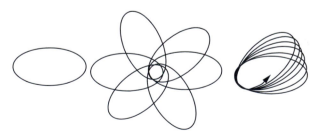

Figure 10.15 Commensurable and incommensurable orbits. The two orbits on the left are commensurable and sooner or later the planet traces out the same path. The orbit on the right is incommensurable and the planet never repeats exactly the same path.

10.13 Summary

A *central force* is directed towards or away from the origin and has a magnitude that depends only on the distance to the origin, $\mathbf{F} = f(r)\hat{\mathbf{r}}$. An example is the gravitational force acting on a planet as it orbits the Sun.

Kepler studied the motion of the planets and determined that this motion obeys three relations which we call *Kepler's laws.*

A particle moving in a central force field has constant angular momentum. The equations of motion for a particle of mass m in the field of a body of mass M are

$$\ddot{r} - \frac{l^2}{m^2 r^3} = -\frac{GM}{r^2}$$

$$\frac{d}{dt}(mr^2\dot{\theta}) = \frac{dl}{dt} = 0.$$

The effective potential energy for a particle in a gravitational field is

$$V_{eff} = \frac{l^2}{2mr^2} - \frac{GMm}{r}.$$

The radial motion of such a particle can be visualized on a plot of V_{eff} vs. r, such as Figure 10.5. Such plots indicate that the orbit of the particle (planet) is a *conic section.*

The equations of motion can be used to obtain the position of the planet as a function of time or to determine the equation of the orbit, $r = r(\theta)$. If the total energy is negative, the equation of the orbit is

$$r = \frac{l^2/GMm^2}{1 + \frac{Al^2}{GMm^2}\cos(\theta - \theta_0)},$$

which has the form of the equation for an ellipse,

$$r = \frac{a(1 - e^2)}{1 + e\cos\theta}.$$

Using these relations one can relate Kepler's laws to physical concepts as follows.

First law: Planets move in elliptical orbits. This is a consequence of the inverse square nature of the gravitational force.

Second law: Planets sweep out equal areas in equal times. This is a consequence of the conservation of angular momentum in a central force field.

Third law: The period squared is proportional to the semimajor axis cubed. This is a consequence of the conservation of energy and the fact that the magnitude of the total energy is inversely proportional to the semimajor axis.

When a planet in a circular orbit is perturbed, it will oscillate radially around the circular path, converting the orbit into an ellipse. The orbit is stable for any force law having the form $f = -cr^n$ as long as n is greater than -3.

10.14 Problems

Problem 10.1 (Bohr model of the atom.) Consider an electron in a circular orbit around a proton. Assume the angular momentum can only take on values equal to $n\hbar$ where n is an integer ($n = 1, 2, 3, \ldots$) and $\hbar$ is a constant. Determine the possible values of the radius for the orbit and the possible values for the total energy. Make up a table of the energy for the first four energy states (that is, for the electron in the four smallest orbits).

Problem 10.2 A particle of mass m is in a Lennard-Jones force field. Assume the force center is stationary. The particle has velocity $\mathbf{v}$.

(a) Write the Lagrangian.

(b) Obtain the equations of motion.

(c) Determine the angular momentum and show that it is a constant.

Problem 10.3 In a galaxy far away an advanced alien civilization has designed a spaceship that consists of two large spheres of masses m_1 and m_2 whose centers are connected by a long rod of length b and negligible mass. (The spaceship resembles a dumbbell.) In a region of space far from any material bodies the spaceship moves at constant linear velocity, while it rotates about its center of mass with angular speed ω. (a) Determine the total angular momentum. (b) Determine the linear velocities of the two masses. (c) Determine the tension in the rod connecting the two masses.

Problem 10.4 An asteroid is in a circular orbit at 5 AU from the Sun.

(a) Evaluate its speed (v_0) in m/s.

(b) For some reason, space scientists of the future decide that the asteroid should be ejected from the solar system. Show that this requires increasing its speed by a factor of $\sqrt{2}$. (Ignore the effect of the planets.)

Problem 10.5 An artificial satellite is in an elliptical orbit about Earth. Its velocity at perigee is 7000 m/s and its velocity at apogee is 6000 m/s. Determine the semimajor axis and the eccentricity of the orbit.

Problem 10.6 The perihelion of Halley's comet is 88×10^6 km and it has an eccentricity $e = 0.967$. (a) Determine the aphelion. (b) What is the speed of the comet when it is at perihelion?

Problem 10.7 Integrate Equation (10.15) directly and show that it leads to the equation of an ellipse. Assume that $\alpha = -1$, β is positive, and γ is negative.

Problem 10.8 A satellite of mass m is in an elliptical orbit around the Earth. Somehow you determine that the maximum and minimum speeds of the satellite are v_1 and v_2. Find the values of a, e, the energy E, the angular momentum l, and the period τ in terms of v_1, v_2, G, m, and M_E, the mass of the Earth.

Problem 10.9 An interplanetary spacecraft is parked in an elliptical orbit about the Sun. The period is τ. The rocket motors are fired in a short burst, and the speed of the spacecraft increases from V to $V + \Delta V$. What is the change in the period?

Problem 10.10 (a) By some magical process, the Sun suddenly loses half of its mass. Show that the Earth's orbit will be a parabola (and so the Earth will escape to infinity). (b) By some other magical process, the Sun's mass suddenly doubles. What will the Earth's orbital period be in this case?

Problem 10.11 A certain binary star system is composed of two stars of comparable masses m_1 and m_2. (For this system we cannot assume one mass is infinitely greater than the other.) Let $\mathbf{r}_1$ and $\mathbf{r}_2$ be the positions of m_1 and m_2 relative to an inertial origin and let $\mathbf{r}_1'$ and $\mathbf{r}_2'$ be their positions relative to the center of mass. (a) Show that the primed coordinates are related to the relative coordinate $\mathbf{r}$ by

$$\mathbf{r}_1' = -\frac{m_2}{m_1 + m_2}\mathbf{r},$$

$$\mathbf{r}_2' = +\frac{m_1}{m_1 + m_2}\mathbf{r},$$

where the relative coordinate $\mathbf{r}$ points from m_1 to m_2. (b) Obtain the Lagrangian in terms of the relative coordinate $\mathbf{r}$ and the position of the center of mass $\mathbf{R}$. (c) Obtain the equations of motion in terms of the relative coordinate. (You may assume the center of mass remains at rest.) (d) Express the radial equation (for the magnitude of $\mathbf{r}$) in terms of the angular momentum and the reduced mass. (e) By comparing the radial equation with Equation (10.8), obtain an expression for Kepler's third law for this situation.

Problem 10.12 Show that for a planet in an elliptical orbit about the Sun, the radial velocity at any time is given by

$$r^2 \dot{r}^2 = \left[\frac{GM}{a} (a[1+e] - r)(r - a[1-e]) \right],$$

where M is the mass of the Sun. (Hint: Use Equations (10.27) and (10.25).)

Problem 10.13 A particle of mass m is acted upon by an attractive central force given by K/r^4. The particle is placed a distance a from the force center and given an initial velocity $\sqrt{2K/3ma^3}$ at right angles to the radius vector. (a) Show that the particle spirals into the force center by deriving the equation for the orbit. (b) Determine the time for the particle to collide with the force center. (Hint: For this force, the potential energy is $V = -K/3r^3$.)

Problem 10.14 Consider a central force given by $F(r) = -K/r^3$ with $K > 0$. Plot the effective potential and discuss possible types of motion.

Problem 10.15 A particle is subjected to a central force

$$F(r) = -\frac{K}{r^2} + \frac{K'}{r^3}.$$

Assume $K > 0$ and consider both signs for K'. (a) Draw the effective potential and discuss possible types of motion. (b) Solve the orbital equation and show that the bounded orbits have the form

$$r = \frac{B}{1 + C \cos \alpha\theta}$$

as long as $K' + \frac{l}{mr^2} > 0$. Note that B, C, and α are constants depending on l, m, K, K'.

Problem 10.16 The first artificial satellite of the Earth was the Russian Sputnik I. Its perigee was 227 km above the Earth's surface. At this point its speed was 28 710 km/h. Determine its period of revolution. What is the maximum distance from the satellite to the surface of the Earth? Given: The Earth has a radius of 6.37×10^3 km.

Problem 10.17 A certain satellite has a perigee of 360 km and an apogee of 2549 km above the Earth's surface. Find its distance above the surface when it is 90° from perigee, as measured from the center of the Earth.

Problem 10.18 For the two body-problem there is a conserved quantity called the Laplace vector. (It is also called the Runge–Lenz vector.) The Laplace vector is a vector pointing towards periapsis. Its magnitude is proportional to the eccentricity. It can be expressed as:

$$\mathbf{A} = \mathbf{p} \times \mathbf{l} - GMm^2 \hat{\mathbf{r}},$$

where M is the mass of the primary, $\mathbf{p}$ is the linear momentum and $\mathbf{l}$ is the angular momentum.
(a) Show that $\mathbf{A}$ lies in the orbit plane.
(b) Show that $\mathbf{A}$ is a constant of the motion.
(c) Show that the magnitude of $\mathbf{A}$ is $GMm^2 e$. (Hint: Evaluate $\mathbf{r} \cdot \mathbf{A}$.)

Problem 10.19 Imagine a circular orbit passing through the origin. Then $r = A \cos \theta$. Show that a central force that gives rise to such an orbit has a $1/r^5$ dependence.

Problem 10.20 A particle moves under the action of a central force in a spiral described by $r = Ae^{b\theta}$. Show that the force acting on it is inversely proportional to the cube of r.

Problem 10.21 A satellite is in a circular orbit, a distance r from the center of the Earth. It has a known velocity v. Show that this satellite has an escape velocity (from r to ∞) of $\sqrt{2}v$.

Problem 10.22 Two stars of equal mass orbit in a circle around their common center of mass with period τ. Show that if they were suddenly stopped, dead still, and allowed to fall toward each other, they would collide in a time $\tau/\sqrt{32}$.

Problem 10.23 Consider a binary star system. (a) Show that Kepler's third law can be written as

$$a^3 = \tau^2(M_1 + M_2),$$

where a is in AU (astronomical units), τ is in Earth years, and $M_1 + M_2$ is the sum of the masses of the stars in solar masses. (Note: All the equations we developed in this chapter assumed $M \gg m$. Hint: Rederive Equation (10.1) and show that if the two masses are comparable, that GM should be replaced by $G(M_1 + M_2)$ everywhere.) (b) Star A and star B are members of a binary star system. An astronomer determines that the period of this system is 32 years and that the stars are separated by 16 AU. Furthermore, star A is found to be 12 AU from the center of mass of the system. Determine the masses of these stars. (Note: This is one of the ways in which astronomers evaluate the masses of stars.)

Problem 10.24 The Rutherford problem considers the hyperbolic orbit of an alpha particle interacting with a gold nucleus. The force is a repulsive force of magnitude $F = K/r^2$. (a) Show that the eccentricity can be expressed as

$$e^2 = 1 + \frac{2El^2}{mK^2}.$$

(b) Show that the distance of closest approach is

$$r_1 = a(e + 1).$$

(Hint: Recall that for a repulsive force the orbit is the "minus branch" of the hyperbola.)

Problem 10.25 A satellite is in a circular "parking" orbit of radius $2R_E$. It is desired to place it in a circular orbit of radius $6R_E$ by giving it an impulse such that it enters an elliptical orbit tangent to the two circular orbits. Determine the time required for this transfer.

Problem 10.26 Assume a space shuttle is in a circular orbit about the Earth with its nose pointing forward. The rocket motors are fired for a few moments. You would expect the shuttle to speed up. Show that, instead, it actually rises to a higher orbit and slows down. (This behavior is called the "satellite paradox.")

Problem 10.27 (Hohmann transfer.) A satellite is parked in a circular equatorial orbit of radius $r_1 = 2R_\oplus$, where $R_\oplus$ is the radius of the Earth. It is desired to "raise" the satellite to a geostationary circular orbit (having radius r_2). The transfer is accomplished by two "burns." The first burn changes the energy to that of a satellite in an elliptical orbit with perigee r_1 and apogee r_2. (The burns can be characterized by the change in velocity that they impart to the satellite.) Upon reaching apogee a second burn increases the velocity to that of a circular orbit of radius r_2.
(a) Determine the radius of the geostationary orbit.
(b) Determine the velocity increments Δv_1 and Δv_2 generated by the two burns.
(c) Determine the time required to carry out this maneuver.

Problem 10.28 Assume the force between two particles is a central attractive force proportional to the separation between the particles. That is, $\mathbf{F} = -kr\hat{\mathbf{r}}$. You can assume one of the particles is much more massive than the other one. Show that the motion of the lighter mass is an ellipse with the more massive particle at the *center* of the ellipse (rather than at a focal point).

Problem 10.29 The Moon raises tides on the Earth, causing the Earth's rotation rate to decrease and causing the radius of the Moon's orbit to increase. (See the discussion accompanying Figure 7.3.) According to Kepler's third law, the angular velocity of the Moon will decrease. Assume the Earth is a uniform sphere and that the Moon is in a circular orbit around the Earth. Ignore the Sun. The initial rotational angular velocity of the Earth is ω_0 and the initial orbital angular velocity of the Moon is Ω_0. Initially the distance between Earth and Moon is $r_0 = 3.84 \times 10^8$ m. Determine the distance between Earth and Moon when the day has become equal to the present month. (Neglect the rotational angular momentum of the Moon.)

Problem 10.30 One of the strongest arguments made for the general theory of relativity is that it predicts the correct value for the precession of the orbit of the planet Mercury. Before Einstein formulated that theory, it was suggested that the precession could be explained if the solar system were filled with dust having a low density ρ. The spherical distribution of dust would exert an additional central force on a planet given by $F' = -mKr$, where m is the mass of the planet and K is a constant equal to $(4\pi/3)\rho G$. Show that for $F' \ll F$ where F is the gravitational force of the Sun, a planet will move in an elliptical orbit whose major axis precesses at an angular velocity

$$\omega_p = 2\pi\rho r_0^{3/2}\sqrt{G/M},$$

where M is the mass of the Sun, and r_0 is the average radius of the orbit.

Problem 10.31 Consider the areal velocity of the planets in the solar system. You may assume they are in circular orbits. (a) Show that the areal velocity increases as the square root of the radius of the planet's orbit. (b) Show that the linear velocity of a planet decreases as the inverse of the square root of the radius of the orbit. (c) Determine a force law such that all the planets would have the same areal velocity.

Problem 10.32 Consider a particle that is acted upon by the central force:

$$F = -\frac{A}{r^2} - \frac{B}{r^4},$$

where A and B are constants. The orbit is a circle of radius a. Determine the condition for the orbit to be stable.

Problem 10.33 A particle is moving in a circular path under the action of an attractive central force given by

$$F = \frac{1}{r^2}e^{-r/a}.$$

Show that if the radius of the circle is greater than a the motion is stable, and if the radius of the circle is less than a, the motion is unstable.

Problem 10.34 An ideal massless string passes through a small hole in a perfectly smooth table. Two equal masses are attached to the ends of the string so that one mass is on the table a distance a from the hole, and the other mass is hanging freely. See the sketch (Figure 10.16). The mass on the table is set in motion with a velocity $\sqrt{ga}$ perpendicular to the direction of the string. Show that the mass on the table moves in a circular path. Now the hanging mass is perturbed slightly so that it oscillates up and down. Show that the period of this oscillation is $2\pi\sqrt{2a/3g}$.

Problem 10.35 Consider a particle moving in an attractive inverse cube force law. That is, $\mathbf{F} = -(c/r^3)\hat{\mathbf{r}}$, where c is a positive constant. (a) Show that if the particle is slightly perturbed, the zeroth-, first-, and second-order terms in the equation for $\ddot{\eta}$ are zero. (b) Assuming that you can generalize the results of part

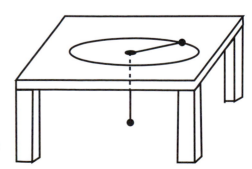

Figure 10.16 Mass on frictionless table (see Problem 10.34).

(a) to $\ddot{\eta} = 0$, describe what you expect for the subsequent motion of the perturbed particle. (c) You can get a good idea of the motion by plotting the effective potential for this problem. Show that the particle will either spiral in to the origin or spiral out to infinity. (The curve $r = r(\theta)$ is called a Cotes spiral.)

Computational Projects

Computational Project 10.1 A comet is jostled loose from the Oort Cloud and heads toward the Sun. Suppose that when it passes the orbit of Pluto it has velocity components $v_x = -0.01, v_y = 0.05$ AU/year. Plot the trajectory of the comet using three algorithms: Euler–Cromer, second-order Runge–Kutta, and fourth-order Runge–Kutta. Show that the angular momentum is constant. (The only bodies involved are the comet and the Sun. Hint: Use $GM = 4\pi^2$.)

Computational Project 10.2 This problem is the same as the previous problem, but now include the effect of Jupiter (which you can assume has a circular orbit). The trajectory of the comet lies in the plane of the orbit of Jupiter. Use RK4.

Computational Project 10.3 Write a program to trace out the positions of the planets as a function of time, using the information in the following table. You may assume the planets have zero inclination (so they all move in the same plane). Start the planets so that they are all lined up in a straight line from the Sun. (This is called a "conjunction".) (a) Determine the number of years between the conjunctions of Jupiter and Saturn. (b) What fraction of the time is Pluto nearer the Sun than Neptune? (I know Pluto is no longer a planet.)

Planet	a (AU)	e	Mass (10^{24}kg)
Mercury	0.387	0.206	0.33
Venus	0.723	0.007	4.87
Earth	1.000	0.017	5.97
Mars	1.524	0.093	0.64
Jupiter	5.203	0.048	1898.6
Saturn	9.537	0.054	568.46
Uranus	19.191	0.047	86.83
Neptune	30.069	0.009	102.43
Pluto	39.482	0.249	0.013

Computational Project 10.4 Compute and plot the motion of a particle under the action of the central force

$$F = -\frac{K}{r^3}\left(1 - \frac{\alpha}{r}\right)r,$$

where K and α are constants. Show that this orbit precesses. Show how your choice of K and α affect the motion.

Computational Project 10.5 We consider the effect of atmospheric drag on an artificial Earth satellite. The drag force acts tangentially to the orbit. This force can be expressed as

$$F = -\frac{1}{2m}C_D A\rho v^2,$$

where m = mass of the satellite, C_D = the drag coefficient, A = the cross-sectional area of the satellite, ρ = the air density, and v = the velocity of the satellite. The density of air at altitude η is given approximately by

$$\rho = \rho_0 \exp\left[-(\eta - \eta_0)/H\right],$$

where $\rho_0 = 6.5 \times 10^{-12}$ kg/m^3, $\eta_0 = 384$ km, and H is a "scale height" for which we can use 40 km at altitudes above about 100 km.

It is not difficult to show that the effect of drag on the semimajor axis and the eccentricity of the orbit are given by the expressions

$$\Delta a = -\frac{A}{m} C_D a^2 \int_0^{2\pi} \rho(\eta) \frac{(1 + e \cos E)^{3/2}}{(1 - e \cos E)^{1/2}} dE$$

$$\Delta e = -\frac{A}{m} C_D a (1 - e^2) \int_0^{2\pi} \rho(\eta) \frac{(1 + e \cos E)^{1/2}}{(1 - e \cos E)^{1/2}} \cos E \, dE,$$

where E is a parameter called the eccentric anomaly (see Computational Project 10.6). Evaluate Δa and Δe by numerically integrating the given expressions. You may assume $C_D = 1.05$, $A = 6 \text{ m}^2$, and $m = 1000 \text{ kg}$. Note: The distance from the center of the Earth to the satellite can be expressed in terms of the eccentric anomaly by $r = a(1 - e \cos E)$.

Computational Project 10.6 In Exercise 10.13 you evaluated the average potential energy of a planet in orbit to be $-GMm/a$ by simply averaging the maximum and minimum values of r. You are now asked to show that the time average of potential energy is given by that expression. You should determine the potential energy at equal time steps as the planet goes around in orbit. For simplicity let the planet be at 1 AU, have a period of 1 year and an eccentricity $e = 0.2$. The problem is complicated by the fact that the planet does not orbit at a constant angular speed. However, you can get the position of the planet at equal intervals of time by determining the *eccentric anomaly* E which is related to the average angular velocity n by Kepler's equation[15]

$$E - e \sin E = nt.$$

This is a transcendental equation so you will have to solve for E numerically. Write a program that obtains E iteratively, using the fact that e is a small quantity. That is, write Kepler's equation as

$$E = nt + e \sin E.$$

Because e is small, the first approximation to E is $E = nt$. Use this for E on the right-hand side and get the second approximation. Repeat over and over. Thus:

$$E_0 = nt$$
$$E_1 = nt + e \sin E_0$$
$$E_2 = nt + e \sin E_1,$$
$$\text{etc.}$$

To solve for the average potential energy, let the planet orbit the Sun 10 times, determine the time average potential energy and compare it with $-GMm/a$.

---

[15] The eccentric anomaly is a parameter that is well known by celestial mechanicians and astronomers. Kepler's equation is a famous equation whose analytical solution has not been found. You can read about it in Chapter 2 of C. D. Murray and S. F. Dermott, *Solar System Dynamics,* Cambridge University Press, Cambridge, 1999.

Part IV

Oscillations and Waves

11 Harmonic Motion

This chapter is an in-depth study of harmonic motion. You are familiar with two examples of simple harmonic oscillators, namely, a pendulum and a mass on a spring. Now you will study several different kinds of harmonic motion, including underdamped, overdamped, and critically damped oscillatory motion, as well as forced harmonic motion and the motion of coupled harmonic oscillators. The techniques you will learn here are used in nearly every branch of physics.

11.1 Springs and Pendulums

A very simple oscillatory system consists of a mass m connected to a spring of constant k. To avoid complications due to forces such as gravity or friction, assume the mass is on a perfectly frictionless horizontal surface and can slide back and forth, as shown in Figure 11.1.

The kinetic energy of the mass is $T = \frac{1}{2}m\dot{x}^2$ and the potential energy of the spring is $V = \frac{1}{2}kx^2$. Therefore, as described in Section 4.3 the Lagrangian for the system is

$$L = T - V = \frac{1}{2}m\dot{x}^2 - \frac{1}{2}kx^2.$$

The equation of motion is obtained from Lagrange's equation:

$$\frac{d}{dt}\frac{\partial L}{\partial \dot{x}} - \frac{\partial L}{\partial x} = 0,$$

which yields

$$\ddot{x} + \frac{k}{m}x = 0. \tag{11.1}$$

Another simple oscillatory system is a pendulum consisting of a particle of mass m (the bob) on an inextensible, massless string of length l. See Figure 11.2. The kinetic energy is

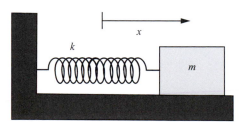

Figure 11.1 A mass on a frictionless surface connected to a spring. The distance x is measured from the equilibrium position to the center of mass of the block.

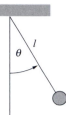

Figure 11.2 A simple pendulum.

$T = \frac{1}{2}m(\dot{x}^2 + \dot{y}^2)$, where $x = l\sin\theta$ and $y = -l\cos\theta$ (assuming the origin is located at the point where the string is attached to the support).

$$T = \frac{1}{2}m(l^2\dot{\theta}^2\cos^2\theta + l^2\dot{\theta}^2\sin^2\theta) = \frac{1}{2}ml^2\dot{\theta}^2.$$

The potential energy is

$$V = -mgl\cos\theta.$$

Consequently, see Equation (4.4),

$$L = T - V = \frac{1}{2}ml^2\dot{\theta}^2 + mgl\cos\theta.$$

The Lagrange equation of motion is:

$$\frac{d}{dt}\frac{\partial L}{\partial \dot{\theta}} - \frac{\partial L}{\partial \theta} = 0.$$

That is,

$$\frac{d}{dt}(ml^2\dot{\theta}) + mgl\sin\theta = 0,$$

or

$$ml^2\ddot{\theta} + mgl\sin\theta = 0. \tag{11.2}$$

By restricting the motion to small angles ($\theta \lesssim 20°$) you can replace $\sin\theta$ by θ (in radians). Then, dividing by ml^2,

$$\ddot{\theta} + \frac{g}{l}\theta = 0. \tag{11.3}$$

Note that Equations (11.1) and (11.3) have exactly the same *form;* they both tell us that the second derivative of the position variable is negatively proportional to the displacement from equilibrium, meaning the system is subjected to a restoring force that is always directed toward the equilibrium point. This form for the equation of motion is characteristic of oscillatory systems. Whenever you see it, you can be sure that you are faced with a case of simple harmonic motion. You can easily show by substitution that possible solutions to the two simple harmonic motion equations above are

$$x(t) = A\cos\left(\sqrt{\frac{k}{m}}t + \beta\right) \tag{11.4}$$

and

$$\theta(t) = A\cos\left(\sqrt{\frac{g}{l}}t + \beta\right). \tag{11.5}$$

These solutions describe a sinusoidal oscillation of amplitude A and angular frequency $\omega = \sqrt{k/m}$ for the spring and angular frequency $\omega = \sqrt{g/l}$ for the pendulum. At $t = 0$ the position is given by $x(0) = A \cos \beta$. The quantity β is called the phase constant and is related to the initial displacement.

Consider now the somewhat more complicated case of an oscillating system subjected to a *retarding force* such as air resistance. For the sake of a specific example, let the system be a mass on a spring acted upon by a resistive force that is proportional to the velocity.[1] Then the total force acting on the mass is not just $-kx$, but rather

$$F = -kx - b\dot{x},$$

where b is a constant of proportionality. The minus sign guarantees that the retarding force opposes the motion. The $-b\dot{x}$ term is called a *damping* force. The equation of motion will now be

$$m\ddot{x} + b\dot{x} + kx = 0. \tag{11.6}$$

This is the equation of motion for a *damped harmonic oscillator* (assuming the damping force is proportional to the first power of the velocity).

Finally, a harmonic oscillator may be subjected to some external force, F_e. In that situation we simply add the external force to the other forces (but put it on the other side of the equation) and write

$$m\ddot{x} + b\dot{x} + kx = F_e. \tag{11.7}$$

This is the equation for a *forced, damped, harmonic oscillator.*

Note that these equations of motion are all linear, ordinary, second-order differential equations with constant coefficients. You learned how to solve such equations in your course on differential equations. I will give a short review of the technique, but if you are uncertain as to any of the details please review your differential equations textbook.

Exercise 11.1

(a) Prove by substitution that Equation (11.4) and Equation (11.5) are solutions to their respective equations of motion. (b) Prove by substitution that

$$x = A \cos \sqrt{k/m}\, t + B \sin \sqrt{k/m}\, t$$

is a solution to Equation (11.1). This is one way to express the most general solution to Equation (11.1). ∎

Exercise 11.2

A certain spring stretches 6 cm when a force of 6 N is applied to it. A 5 kg mass is attached to the spring which is stretched 3 cm and released with a push so that its initial speed is -6 cm/s. The system lies on a frictionless horizontal surface. Find an equation for the position of the mass as a function of time. (Hint: Use the general form given in Exercise 11.1.) Answer: $x = 0.030 \cos(4.47t) - 0.013 \sin(4.47t)$ m. ∎

[1] Some 300 years ago, Charles Coulomb's experiments led him to conclude that kinetic friction is independent of velocity. Later more careful experiments showed that this is only approximately true. The velocity dependence of kinetic friction is still not well understood, but the frictional force for wetted surfaces is believed to be proportional to the velocity for low velocities.

> **Exercise 11.3**
>
> At time $t = 0$ a simple harmonic oscillator with angular frequency $\omega = 2$ rad/s is at $x = 3$ cm and has a speed of 1 cm/s. Obtain an expression for $x = x(t)$ and plot x vs. t. Answer: $x = 3\cos 2t + (1/2)\sin 2t$ cm.   ■

11.2 Solving the Differential Equation

An nth-order linear differential equation with constant coefficients has the form

$$a_n\frac{d^n x}{dt^n} + a_{n-1}\frac{d^{n-1}x}{dt^{n-1}} + \cdots + a_2\frac{d^2 x}{dt^2} + a_1\frac{dx}{dt} + a_o x = f(t).$$

If the quantity $f(t)$ on the right-hand side is zero, the equation is called *homogeneous*. If $f(t) \neq 0$ the equation is called *inhomogeneous*. By "linear" we mean that the dependent variable (x) and its derivatives appear only to the first power. However, the independent variable (t) can be raised to any power. Thus $\frac{dx}{dt} = 3t^2$ is linear, but $\frac{dx}{dt} = 3x^2$ and $\left(\frac{dx}{dt}\right)^2 = 3x$ are not linear.[2]

11.2.1 Homogeneous Linear Differential Equations

A homogeneous differential equation of the form given above has a solution

$$x = e^{pt}.$$

The technique for obtaining the general solution is to plug $x = e^{pt}$ into the equation and carry out the indicated mathematical operations. This yields an equation for p in terms of the constants. I will illustrate this procedure in a moment. But first I want to mention that a first-order differential equation yields a single value for p, a second-order differential equation yields two values for p, and so on. Thus, depending on the order of the differential equation, you get several different solutions. Which one of these various solutions should you use? The answer is: all of them!

Rule 1. If $x_1(t)$ and $x_2(t)$ are both solutions of a linear homogeneous differential equation then $x_1(t) + x_2(t)$ is also a solution.

For simple harmonic motion we have a *second*-order differential equation so we will obtain *two* values for p. Let us call these two values p_1 and p_2. The two solutions are:

$$x_1(t) = e^{p_1 t} \quad \text{and} \quad x_2(t) = e^{p_2 t}.$$

By Rule 1, the sum of these two solutions is also a solution. That is,

$$x(t) = e^{p_1 t} + e^{p_2 t}$$

is a solution.

We can generalize even further using another simple fact.

Rule 2. If $x(t)$ is a solution of a linear homogeneous differential equation, then $Cx(t)$, where C is an arbitrary constant, is also a solution.

Consequently,

$$x = C_1 e^{p_1 t} + C_2 e^{p_2 t}$$

is a solution to a second-order differential equation. In fact, it is the most general form for the solution of a homogeneous linear second-order differential equation with constant coefficients. Any other solution can be expressed as a linear combination of $e^{p_1 t}$ and $e^{p_2 t}$.

---

[2] The "degree" of a differential equation is the power of its highest-order derivative. Thus a linear differential equation is of the first degree. The "order" of a differential equation is the order of the highest derivative appearing in the equation.

Similarly, the general solution for a third-order differential equation is

$$x = C_1 e^{p_1 t} + C_2 e^{p_2 t} + C_3 e^{p_3 t},$$

and so on.

Note that the values of p ($p_1, p_2, \ldots$) come from the parameters of the problem (such as the masses and spring constants) whereas the values of the constants ($C_1, C_2, \ldots$) come from the initial conditions.

Let me apply these concepts to my favorite differential equation,

$$m\ddot{x} + kx = 0.$$

If $x = e^{pt}$ then $\ddot{x} = p^2 e^{pt}$. Plug in to obtain

$$mp^2 e^{pt} + ke^{pt} = 0.$$

Dividing by e^{pt} we obtain an equation for p,

$$mp^2 + k = 0.$$

This is called the *auxiliary equation*. Solving for p gives

$$p = \pm\sqrt{-\frac{k}{m}} = \pm i\sqrt{\frac{k}{m}} = \pm i\omega,$$

where $i \equiv \sqrt{-1}$ and $\omega = \sqrt{k/m}$.

Thus we have two (linearly independent) solutions, namely,

$$x = e^{+i\omega t} \quad \text{and} \quad x = e^{-i\omega t}.$$

Because the sum of solutions is also a solution and because we can multiply each solution by an arbitrary constant, the general solution to the second-order differential equation is

$$x = C_1 e^{+i\omega t} + C_2 e^{-i\omega t}. \tag{11.8}$$

This equation is in a perfectly acceptable form. However, it can also be expressed in terms of sines and cosines, which may be preferable in some cases.

Using "Euler's formula"

$$e^{\pm i\theta} = \cos\theta \pm i\sin\theta,$$

Equation (11.8) can be written

$$x = C_1(\cos\omega t + i\sin\omega t) + C_2(\cos\omega t - i\sin\omega t)$$

or

$$x = (C_1 + C_2)\cos\omega t + i(C_1 - C_2)\sin\omega t.$$

You might think that since x is a real quantity, you should discard the last term because it is imaginary; however, this would be unjustified because C_1 and C_2 are not yet determined and they could very well be complex numbers.

You now know basically everything you need for solving simple harmonic motion problems, except for one small detail: If the auxiliary equation for a second-order differential equation has two equal roots, then $p_1 = p_2 = p$ and the general solution is

$$x = C_1 e^{pt} + C_2 t e^{pt}. \tag{11.9}$$

This is easily verified by substitution. (Note the factor of t in the second term.)

Worked Example 11.1 Solve

$$\frac{d^2 x}{dt^2} - 4x = 0,$$

given the initial conditions $x(t = 0) = 0$ m and $\dot{x}(t = 0) = 3$ m/s.

Solution Note that this is not the SHM equation because the coefficient of x is negative. To solve, substitute e^{pt} into the equation and obtain

$$p^2 - 4 = 0.$$

Therefore, $p = \pm 2$ and the solution is

$$x = C_1 e^{2t} + C_2 e^{-2t}.$$

The initial condition $x(t = 0) = 0$ leads immediately to $C_2 = -C_1$ so

$$x = C_1(e^{2t} - e^{-2t}) = 2C_1 \sinh(2t).$$

The initial condition $\dot{x} = 3$ leads to

$$\dot{x} = 2C_1(2)\cosh(2t) = 4C_1\cosh(0) = 3.$$
$$\therefore C_1 = 3/4,$$

and

$$x = \frac{3}{2}\sinh(2t) \text{ m.}$$

Exercise 11.4

(a) Prove Rule 1 for a linear homogeneous second-order differential equation. (b) Prove Rule 2 for a linear homogeneous second-order differential equation. ∎

Exercise 11.5

Verify that the relation $y = Ae^{ax}\cos bx + Be^{ax}\sin bx$ satisfies the equation $[(D - a)^2 + b^2]y = 0$, where $D \equiv \frac{d}{dx}$. ∎

Exercise 11.6

Solve the equation

$$\frac{d^3 x}{dt^3} - 4\frac{d^2 x}{dt^2} + \frac{dx}{dt} + 6x = 0.$$

Answer: $x = C_1 e^{-t} + C_2 e^{2t} + C_3 e^{3t}$. ∎

Exercise 11.7

Solve $(D^2 + 2D - 3)y = 0$, where $D \equiv \frac{d}{dx}$.
Answer: $y = C_1 e^x + C_2 e^{-3x}$. ∎

Exercise 11.8

Solve $(D^2 - 6D + 9)y = 0$, where $D \equiv \frac{d}{dt}$. ∎

Exercise 11.9

Obtain formulas for $\sin 2\theta$ and $\cos 2\theta$ in terms of $\sin\theta$ and $\cos\theta$. (Hint: Note that $e^{2i\theta} = \left(e^{i\theta}\right)^2$.) ∎

11.2.2 An Undamped Harmonic Oscillator

Consider once again a block of mass m attached to a spring of constant k. Suppose you pull the mass until the spring is stretched to x_0. Now release it from rest. You are asked to determine the motion, $x = x(t)$ and to evaluate the amplitude and period of the oscillations.

As given by Equation (11.8), the solution has the form

$$x(t) = C_1 e^{i\omega t} + C_2 e^{-i\omega t},$$

with $\omega = \sqrt{k/m}$. The only thing you need to do now is determine the constants C_1 and C_2. The initial conditions are

$$x(t = 0) = x_0,$$

and

$$\dot{x}(t = 0) = 0.$$

Plugging these into the solution you obtain

$$x(t = 0) = x_0 = C_1 + C_2,$$

and

$$\dot{x}(t = 0) = 0 = i\omega C_1 - i\omega C_2.$$

The second equation requires that $C_2 = C_1$. Then the first equation implies that $C_1 = C_2 = \frac{1}{2}x_0$. Therefore, the solution is

$$x = \frac{1}{2}x_0(e^{+i\omega t} + e^{-i\omega t}) = \frac{1}{2}x_0(2\cos\omega t) = x_0 \cos\omega t.$$

This equation tells us that the mass oscillates back and forth between the values $+x_0$ and $-x_0$. The quantity x_0 is called the amplitude and is usually denoted by A. The motion takes place with a frequency

$$f = \frac{\omega}{2\pi} = \frac{1}{2\pi}\sqrt{\frac{k}{m}}.$$

A plot of position as a function of time is presented in Figure 11.3. Note that the mass returns to its original position after a time t given by

$$t = \frac{2\pi}{\omega} = \frac{1}{f} = P = \text{period}.$$

The period (P) is the time for one complete oscillation. The frequency (f) is the number of oscillations per unit time and is equal to the inverse of the period. The quantity $\omega = 2\pi f$ is called the "angular frequency" but you should be aware that often it is carelessly referred to as the frequency.

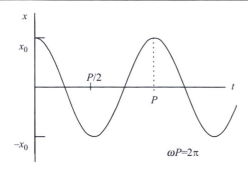

Figure 11.3 A plot of position as a function of time for a mass undergoing simple harmonic motion.

If the initial conditions are different, the form of the solution is slightly different. For example, suppose that you had been told that at $t = 0$ the block was at $x = 0$ and had a velocity v_0. That is, $x(t = 0) = 0$ and $\dot{x}(t = 0) = v_0$. Then, using Equation (11.8) again you obtain

$$0 = C_1 + C_2,$$

and

$$v_0 = i\omega(C_1 - C_2).$$

Therefore,

$$C_1 = -C_2 \quad \text{and} \quad C_1 = v_0/2i\omega.$$

Consequently,

$$x = \frac{v_0}{2i\omega}\left(e^{i\omega t} - e^{-i\omega t}\right) = \frac{v_0}{2i\omega}(2i\sin\omega t) = \frac{v_0}{\omega}\sin\omega t. \tag{11.10}$$

The solution, $x = x(t)$, also gives information on the energy of the system. When $x = 0$, $\dot{x} = v_0$, and the potential energy is zero whereas the kinetic energy is a maximum. Also, when $x = x_0$ the potential energy is a maximum but the kinetic energy is zero. Since the total energy is constant,[3]

$$\frac{1}{2}mv_0^2 = \frac{1}{2}kx_0^2.$$

You can use this equation to relate x_0 and v_0, thus:

$$x_0 = \pm\sqrt{\frac{m}{k}}\,v_0 = \pm\frac{v_0}{\omega}.$$

Therefore the result, Equation (11.10), can be written in the more familiar form

$$x = x_0 \sin\omega t.$$

The two examples you have just considered had very simple sets of initial conditions. In both cases C_1 and C_2 were replaced by a single constant. In general, however, the solution will contain two different constants. Then the general solution can be expressed in the form

$$x = x_0 \sin(\omega t + \beta).$$

---

[3] Note that this result is in agreement with results from the virial theorem discussed in Worked Example 10.2. In the present problem, $n = 1$ so $\langle T \rangle = \frac{n+1}{2}\langle V \rangle$ yields $\langle T \rangle = \langle V \rangle$.

Energy of an Undamped Oscillator

The total energy $(E = T + V)$ for an undamped oscillator is given by

$$E = \frac{1}{2}m\dot{x}^2 + \frac{1}{2}kx^2,$$

where $x = A\sin(\omega t + \beta)$ and $\dot{x} = A\omega\cos(\omega t + \beta)$. Therefore,

$$E = \frac{1}{2}m\dot{x}^2 + \frac{1}{2}kx^2,$$
$$= \frac{1}{2}mA^2\omega^2\cos^2(\omega t + \beta) + \frac{1}{2}kA^2\sin^2(\omega t + \beta),$$

but $\omega^2 = k/m$, so

$$E = \frac{1}{2}kA^2(\cos^2(\omega t + \beta) + \sin^2(\omega t + \beta)),$$
$$= \frac{1}{2}kA^2 = \frac{1}{2}m\omega^2 A^2.$$

Thus the total energy of a harmonic oscillator is proportional to the amplitude squared.

Worked Example 11.2 A mass of 0.25 kg is attached to a spring of force constant 1.0 N/m. The mass is displaced 0.15 m from its equilibrium point and released with zero initial velocity. Evaluate the total energy of the oscillator (kinetic plus potential). What is the maximum velocity? What is the period of the oscillation?

Solution The total energy is constant. The initial energy is $E = T + V = \frac{1}{2}mv_0^2 + \frac{1}{2}kx_0^2$. But $v_0 = 0$. $\therefore E = \frac{1}{2}kx_0^2 = \frac{1}{2}(1)(0.15)^2 = 1.13 \times 10^{-2} J$.

The maximum velocity occurs when $x = 0$, so $E = \frac{1}{2}v_{max}^2$ Consequently,

$$v_{max}^2 = \frac{2E}{m} = \frac{(2)(1.13 \times 10^{-2})}{0.25},$$

and $v_{max} = 0.3$ m/s.

Since $\omega = \sqrt{k/m} = \sqrt{1/0.25} = 2$, the period is $P = 2\pi/\omega = \pi$.

Exercise 11.10

Show that $x = x_0\sin(\omega t + \beta)$ is equivalent to $x = C_1\cos\omega t + C_2\sin\omega t$. ∎

Exercise 11.11

A spring of constant 10 N/m is connected to a mass of 2 kg. It is initially at rest in an unstretched position. It is given an initial speed of 3 m/s. What is the amplitude of the oscillation? Answer: 1.34 m. ∎

11.3 The Damped Harmonic Oscillator

Consider now an oscillator with a retarding force proportional to the velocity. This is a *damped harmonic oscillator*. See Figure 11.4. Common sense tells us that such an oscillator will lose energy and eventually stop.

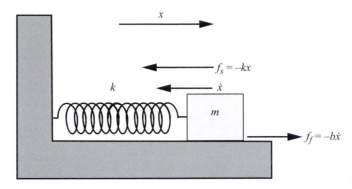

Figure 11.4 A damped harmonic oscillator. At the moment illustrated, the block is moving to the left, toward its equilibrium position. The spring force (f_s) is directed toward the left and the retarding force (f_f) is directed toward the right, opposite to the velocity.

The equation of motion of a damped harmonic oscillator is given by Equation (11.6) which I repeat here:

$$m\ddot{x} + b\dot{x} + kx = 0.$$

This is a second-order linear ordinary homogeneous differential equation with constant coefficients. As previously described, to solve it you make the substitution $x = e^{pt}$, giving

$$mp^2 e^{pt} + bpe^{pt} + ke^{pt} = 0.$$

Dividing by e^{pt} yields the auxiliary equation:

$$p^2 + \frac{b}{m}p + \frac{k}{m} = 0.$$

This is a quadratic equation with two solutions:

$$p = -\frac{b}{2m} \pm \sqrt{\left(\frac{b}{2m}\right)^2 - \frac{k}{m}}. \tag{11.11}$$

Let us call the two solutions p_1 and p_2. Since $e^{p_1 t}$ and $e^{p_2 t}$ are both solutions to the differential equation, the general solution is the sum of these:

$$x(t) = C_1 e^{p_1 t} + C_2 e^{p_2 t}, \tag{11.12}$$

where C_1 and C_2 are constants to be determined from the initial conditions.

If $\left(\frac{b}{2m}\right)^2 > \frac{k}{m}$, the quantity under the square root in Equation (11.11) is positive and the root is real. Note that the root will be smaller than $\frac{b}{2m}$ so p_1 and p_2 will be negative. Then, according to Equation (11.12), the quantity $x(t)$ will be exponentially decreasing.

On the other hand, if $\left(\frac{b}{2m}\right)^2 < \frac{k}{m}$ then the quantity under the square root is negative and p_1 and p_2 are complex. That is, they have an imaginary component (call it $i\omega$) and the solution contains terms of the form

$$e^{i\omega t} = \cos \omega t + i \sin \omega t.$$

We have reached an important conclusion. Let me repeat myself for emphasis. From the equation for p, Equation (11.11), you note that p is complex if and only if the quantity under the square root is negative, that is, iff

$$\left(\frac{b}{2m}\right)^2 < \frac{k}{m}.$$

In this case the system oscillates. On the other hand, if

$$\left(\frac{b}{2m}\right)^2 \geq \frac{k}{m},$$

then p is real and the motion dies out exponentially.

The relation of k/m to $(b/2m)^2$ characterizes the three types of motion for the damped oscillator. These are called "underdamped," "overdamped," and "critically damped." They are described in the following table.

Name	Condition	Description
underdamped	$(b/2m)^2 < k/m$	oscillatory motion
critically damped	$(b/2m)^2 = k/m$	exponential decrease
overdamped	$(b/2m)^2 > k/m$	exponential decrease

The motion of the damped oscillator depends on the relative values of the parameters b, k, m. If $b/2m$ is large, the system is subjected to a considerable retarding force and the oscillations will die out quickly. On the other hand, if $b/2m$ is small, there is little damping and the oscillator will remain in motion for a long time. Furthermore, if k/m is large, the frequency of oscillation will be large, but if k/m is small, the system will oscillate slowly.

11.3.1 The Underdamped Oscillator

Let us begin by considering the underdamped oscillator. Then, $(b/2m)^2 < k/m$ and Equation (11.11) shows that p will be a complex quantity. It is convenient to change notation at this point. Because $\sqrt{k/m}$ is the "natural" frequency of the undamped oscillator, we shall denote k/m by ω_0^2. Furthermore, $b/2m$ can be denoted by γ the "damping coefficient." The new notation is :

$$\frac{k}{m} = \omega_0^2 = \text{square of frequency of undamped oscillator,}$$

$$\frac{b}{2m} = \gamma = \text{damping coefficient.}$$

Using this notation, Equation (11.11) becomes

$$p = -\gamma \pm \sqrt{\gamma^2 - \omega_0^2}. \tag{11.13}$$

An underdamped oscillator has $\gamma^2 < \omega_0^2$ so the last term is the square root of a negative number, and p is complex. In that case it is convenient to write

$$\sqrt{(-1)(\omega_0^2 - \gamma^2)} = i\omega_1,$$

where $\omega_1 = \sqrt{\omega_0^2 - \gamma^2}$ is a real number. Then

$$p = -\gamma \pm i\omega_1,$$

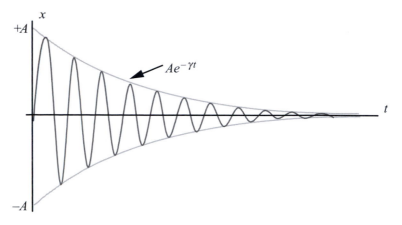

Figure 11.5 The motion of an underdamped oscillator consists of rapid oscillations (described by $\cos(\omega_1 t + \theta)$) modulated by an exponentially decreasing amplitude (described by $Ae^{-\gamma t}$).

a form that clearly shows that p is complex. Inserting the two values of p into the solution (11.12) gives

$$x(t) = C_1 e^{(-\gamma + i\omega_1)t} + C_2 e^{(-\gamma - i\omega_1)t}$$
$$= e^{-\gamma t}\left(C_1 e^{i\omega_1 t} + C_2 e^{-i\omega_1 t}\right).$$

C_1 and C_2 are complex constants related to the initial conditions. These constants can be written in a variety of different ways. You obtain a very nice-looking expression if you write them in the form

$$C_1 = \frac{1}{2}Ae^{i\theta} \text{ and } C_2 = \frac{1}{2}Ae^{-i\theta}. \tag{11.14}$$

This basically replaces the two complex constants C_1 and C_2 by the real constants A and θ. Then

$$x(t) = e^{-\gamma t}\frac{1}{2}A\left(e^{i(\omega_1 t + \theta)} + e^{-i(\omega_1 t + \theta)}\right),$$

or

$$x(t) = Ae^{-\gamma t}\cos(\omega_1 t + \theta). \tag{11.15}$$

Plotting x vs. t generates an oscillatory solution (due to the cosine term), but the amplitude of the oscillations is $Ae^{-\gamma t}$, so the amplitude decreases exponentially with time. This behavior is illustrated in Figure 11.5.

Note that the oscillations have an angular frequency ω_1, which is smaller than the natural frequency ω_0. That is, the damped system oscillates more slowly than the undamped system. You probably expected this type of behavior.

Worked Example 11.3 An underdamped harmonic oscillator has $k = 2$ N/m, $m = 1$ kg, and $b = 0.1$ kg/s. How many oscillations does the system make before the amplitude decreases to $1/e$ of its initial value?

Solution The time when the amplitude is $1/e$ of its initial value is obtained from

$$Ae^{-\gamma t} = \frac{1}{e}A \quad \therefore t = 1/\gamma = \frac{2m}{b} = 20 \text{ s}.$$

Then,

$$\omega_1 t = \left(\sqrt{\omega_0^2 - \gamma^2}\right)t = \left(\sqrt{k/m - \gamma^2}\right)t$$

$$= \left(\sqrt{2 - 0.05^2}\right)(20) = 28.26 \quad \text{rad},$$

and the number of oscillations is $28.26/2\pi = 4.5$.

Exercise 11.12

The complex constants C_1 and C_2 can be written in the form $C_1 = a_1 + ib_1$ and $C_2 = a_2 + ib_2$. But there can only be two independent constants, so a_1, b_1 and a_2, b_2 are not all independent. Using Equation (11.14) show that $a_2 = a_1$ and $b_2 = -b_1$. ∎

Exercise 11.13

Obtain the general solution of the following differential equations: (a) $m\ddot{x} + b\dot{x} - kx = 0$, (b) $m\ddot{x} - b\dot{x} + kx = 0$. Assume m, b, and k are positive real quantities. ∎

11.3.2 The Overdamped Oscillator

Now consider the *overdamped* oscillator, defined to be an oscillator for which

$$\left(\frac{b}{2m}\right)^2 > \frac{k}{m}.$$

According to Equation (11.11) or Equation (11.13), p will now have two *real* values denoted γ_1 and γ_2. Using the notation of Equation (11.13), the solutions of the auxiliary equation are

$$p_1 = -\gamma_1 = -\gamma - \sqrt{\gamma^2 - \omega_0^2},$$

and

$$p_2 = -\gamma_2 = -\gamma + \sqrt{\gamma^2 - \omega_0^2}.$$

Note that both γ_1 and γ_2 are positive and that γ_1 is greater than γ_2.

Plugging these expressions into the general solution of the differential equation (11.12) leads to

$$x(t) = C_1 e^{-\gamma_1 t} + C_2 e^{-\gamma_2 t}, \tag{11.16}$$

where γ_1 and γ_2 are real. Both terms are exponentially decreasing with time, as indicated in Figure 11.6. Because $\gamma_1 > \gamma_2$ we note that $e^{-\gamma_1 t}$ decays more rapidly than $e^{-\gamma_2 t}$.

The motion of the oscillator (that is, the value of x as a function of time) is, of course, the sum of two curves, as illustrated in Figure 11.6. The constants C_1 and C_2 depend on the initial conditions. For example, if you pull the block to some position x_0 and release it from rest ($v_0 = 0$), at $t = 0$, Equation (11.16) reduces to

$$x_0 = C_1 + C_2 \quad \Rightarrow \quad C_1 = x_0 - C_2.$$

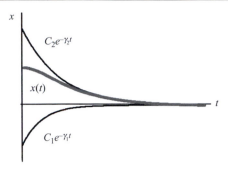

Figure 11.6 An overdamped harmonic oscillator. The motion is the sum of the two curves. The constants C_1 and C_2 depend on the initial conditions.

Since $\dot{x} = -\gamma_1 C_1 e^{-\gamma_1 t} - \gamma_2 C_2 e^{-\gamma_2 t}$,

$$v_0 = -\gamma_1 C_1 - \gamma_2 C_2.$$

But $v_0 = 0$, so $C_2 = -(\gamma_1 C_1/\gamma_2)$, and hence,

$$C_1 = x_0 - C_2 = x_0 + (\gamma_1/\gamma_2)C_1.$$

Therefore,

$$C_1 = -\frac{\gamma_2}{\gamma_1 - \gamma_2} x_0,$$

and

$$C_2 = \frac{\gamma_1}{\gamma_1 - \gamma_2} x_0.$$

Consequently,

$$x(t) = -\frac{\gamma_2}{\gamma_1 - \gamma_2} x_0 e^{-\gamma_1 t} + \frac{\gamma_1}{\gamma_1 - \gamma_2} x_0 e^{-\gamma_2 t}.$$

Depending on the initial conditions the oscillator will behave in one of three ways. The displacement $x(t)$ may rapidly decay to zero. Or $x(t)$ may increase for a time and then decay to zero. Finally, $x(t)$ could change sign, reach a local maximum or minimum, then approach zero.

Exercise 11.14

Consider the motion of an overdamped oscillator. Assume that initially $x(t = 0) = 1$ and $\dot{x}(t = 0) = 0$. Let $\gamma_1 = 3.414$ and $\gamma_2 = 0.586$. Plot $x(t)$. ■

Exercise 11.15

Obtain C_1 and C_2 for an overdamped oscillator if the initial conditions are $x(t = 0) = 0$ and $\dot{x}(t = 0) = v_0$. Answer: $C_1 = v_0/(\gamma_2 - \gamma_1)$; $C_2 = -C_1$. ■

11.3.3 The Critically Damped Oscillator

Finally, consider the case of a critically damped harmonic oscillator for which

$$\left(\frac{b}{2m}\right)^2 = \frac{k}{m}.$$

Now the term under the root in Equation (11.11) is zero.

For convenience, define the quantity γ_c as follows:

$$\gamma_c = \frac{b}{2m} = \sqrt{\frac{k}{m}} = \omega_0.$$

Then Equation (11.13) yields a single value for p namely $p = -\gamma_c$. Recall that if p has only one value, the solution to a second-order differential equation has the form of Equation (11.9), that is

$$x(t) = C_1 e^{-\gamma_c t} + C_2 t e^{-\gamma_c t},$$

or

$$x(t) = (C_1 + C_2 t) e^{-\gamma_c t}.$$

For critical damping x decays quickly to zero.

Exercise 11.16

A critically damped harmonic oscillator is initially at its equilibrium position with velocity v_0. Obtain expressions for C_1 and C_2. Answer: $C_1 = 0$, $C_2 = v_0$. ∎

Exercise 11.17

A critically damped harmonic oscillator has $b = 0.3$ kg/s and $k = 0.4$ N/m. (You are not given the mass.) It is released from rest at $x = 0.04$ m. Evaluate the constants C_1 and C_2. Answer: $C_1 = 0.04$, $C_2 = 0.107$ (in meters). ∎

11.4 The Forced Harmonic Oscillator

11.4.1 Statement of the Problem

A *forced* harmonic oscillator is an oscillator which is subjected to some sort of driving force. The external driving force is often periodic with the same frequency as the oscillator. When you push your little brother on a swing, you give the seat a shove each time it comes back to you at the top of its swing. The periodicity of your shoves is the same as the natural period of the swing. Similarly, the pendulum in a "grandfather" clock receives a small push on each swing. An old-fashioned pocket watch has an oscillating spring wheel that receives small periodic impulses from a spring and ratchet system.

The oscillator will, in general, be damped, so the equation of motion for such a system is

$$m\ddot{x} + b\dot{x} + kx = F(t), \tag{11.17}$$

where $F(t)$ is the driving force. The purpose of the driving force is to overcome the damping. If $F(t)$ is too small, the oscillations will gradually die out. On the other hand, if $F(t)$ is too large, the amplitude may grow unacceptably large. (If you push the swing too hard it will "go over the top," leading to unpleasant consequences.) As you shall see shortly, the amplitude of the oscillations also depends on the *frequency* of the driving force. This is related to the phenomenon of resonance that we will consider later.

The equation of motion of the driven harmonic oscillator Equation (11.17) is an *inhomogeneous* second-order linear differential equation. I will now make a short aside to describe how to solve such an equation.

11.4.2 Inhomogeneous Linear Ordinary Differential Equations

The title of this section describes the kind of equation we want to solve. You should understand the implication of each of the many longish words in the description. The equation is, of course, a *differential* equation because there are derivatives in it. It is *linear* because the dependent variable (x) and its derivatives are not raised to a power larger than the first. We are dealing with an *ordinary* differential equation because Equation (11.17) has no partial derivatives in it.

You probably recall that differential equations are often expressed in terms of the *differential operator* $D \equiv \frac{d}{dt}$. Then $D^2 = \frac{d^2}{dt^2}$ and so on. A linear *homo*geneous differential equation is an equation which can be put in the form

$$a_n D^n x + a_{n-1} D^{n-1} x + \cdots + a_1 Dx + a_0 x = 0.$$

Although the a_is are usually constants, they can be functions of t. Note the zero on the right-hand side.

Of course, the equation need not have a zero on the right. There could be a function of x and t on the right-hand side. Thus,

$$a_n D^n x + a_{n-1} D^{n-1} x + \cdots + a_1 Dx + a_0 x = f(x,t).$$

Such an equation is said to be *inhomo*geneous. The solution of an inhomogeneous differential equation can be determined using the following rule.

Rule 3. The solution of an *inhomogeneous* differential equation is the sum of the general solution of the corresponding *homogeneous* equation and a *particular solution* of the inhomogeneous equation.

Therefore, the first step in solving an inhomogeneous differential equation is to set the right-hand side equal to zero. This generates the corresponding homogeneous differential equation which we can solve by the method of Section 11.2. The general solution will, of course, contain as many arbitrary constants as the order of the equation. A second-order equation has a general solution with two arbitrary constants. We denote this general solution of the homogeneous equation by x_g. It is usually called the "complementary function." Next we consider the *in*homogeneous equation. We can sometimes figure out a solution by inspection. We denote this *particular* solution by x_p. The *general* solution of the inhomogeneous equation is $x_p + x_g$. It is easy to prove this assertion, and it is left as an exercise for you. We now return to the problem of the driven harmonic oscillator.

Exercise 11.18

Show that the general solution of an inhomogeneous differential equation is the sum of a particular solution and the general solution of the homogeneous differential equation.   ∎

11.4.3 Obtaining the Particular Solution

The equation of motion for the driven, damped harmonic oscillator is

$$m\ddot{x} + b\dot{x} + kx = F(t).$$

The forcing term, $F(t)$, is often periodic because that will keep the oscillator in motion, counteracting the effect of damping. Another common situation is an oscillator that has been subjected to an impulsive force. In general, the forcing term can be any function of time. It could even be a constant. In fact, for the sake of making the analysis simple, I will begin with a constant force. For example, a mass hanging from a spring in an external gravitational field (see Figure 11.7) has the equation of motion

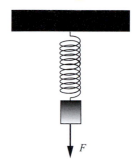

Figure 11.7 A mass hanging from a spring in a constant gravitational field.

$$m\ddot{x} + b\dot{x} + kx = mg, \qquad (11.18)$$

where the damping could be due to air resistance, for example. The mass is only allowed to oscillate vertically; it does not simultaneously swing like a pendulum.

The corresponding homogeneous equation was solved in the previous section. If the oscillation is underdamped, for example, the general solution of the homogeneous equation is given by Equation (11.15) as:

$$x_g = x(t) = Ae^{-\gamma t}\cos(\omega_1 t + \theta).$$

Now you need to find a particular solution to Equation (11.18). That is, you need to find a function of x which satisfies (11.18). After a little creative staring at that equation you will realize that if x is constant, the first two terms are zero and you are left with $kx = mg$. Therefore, $x = mg/k$ is a possible solution to the equation. It is the *particular* solution, x_p. The general solution of the inhomogeneous equation is, therefore,

$$x(t) = x_g + x_p = Ae^{-\gamma t}\cos(\omega_1 t + \theta) + \frac{mg}{k}.$$

This tells you that the effect of the gravitational force is just to stretch the spring, giving a different (lower) value for the equilibrium point.

I will now go on to a more interesting problem, namely a driving force that varies sinusoidally.[4] Let the driving frequency be denoted ω_d and assume $F(t) = F_0 \sin \omega_d t$. For simplicity, we shall begin with an *undamped* oscillator. Then the equation of motion is

$$m\ddot{x} + kx = F_0 \sin \omega_d t,$$

or

$$\ddot{x} + \omega_0^2 x = \frac{F_0}{m} \sin \omega_d t. \qquad (11.19)$$

The general solution of the homogeneous equation is

$$x_g(t) = A \cos(\omega_0 t + \theta_0),$$

where A and θ_0 depend on the initial conditions. For example, $\theta_0 = 0$ corresponds to the oscillator at position $x = A$ at time $t = 0$. (For a pendulum this means that at time zero the bob was at the top of the swing, and for a mass on a spring, it means that at time zero the spring was fully extended.)

---

[4] Recall that any periodic function can be expressed as a sum of sines and cosines (a "Fourier series") so it is reasonable to consider a sinusoidal driving force.

Having a general solution to the homogeneous equation, you still need a particular solution to the inhomogeneous equation. Previously, we obtained a particular solution by inspection. This is usually not possible, so fortunately there are techniques available for determining a particular solution.[5] One technique that always works for a second-order differential equation is to integrate the equation twice. To apply the method, first write the equation in the form

$$(a_2 D^2 + a_1 D + a_0)x = f(t). \tag{11.20}$$

This equation can be expressed as the product

$$(D - b_1)(D - b_2)x = f(t). \tag{11.21}$$

Let $u = (D - b_2)x$. Then the differential equation is

$$(D - b_1)u = f(t).$$

You now have a first-order linear differential equation that you can solve for $u(t)$. Next, going back to the definition of u, you solve

$$(D - b_2)x(t) = u(t)$$

for $x(t)$. This technique produces the general solution for the inhomogeneous equation, that is, it generates the complementary function as well as the particular solution. However, it may involve a considerable amount of mathematics, so I want to show you two other techniques which might be classified as "tricks" for finding the particular solution. (When a mathematical trick becomes very familiar we quit calling it a trick and start calling it a technique!)

Trick #1

The first trick (or technique) is used when the function on the right-hand side is of the form $f(t) \propto e^{at}$. Then, writing the second-order differential equation in the form

$$(D - b_1)(D - b_2)x = K e^{at}, \tag{11.22}$$

the particular solution is

$$
\begin{aligned}
x_p &= C e^{at} \quad \text{if} \quad a \neq b_1 \quad \text{and} \quad a \neq b_2 \quad \text{and} \ b_1 \neq b_2, \\
x_p &= C t e^{at} \quad \text{if} \quad a = b_1 \quad \text{or} \quad a = b_2 \quad \text{but} \ b_1 \neq b_2, \\
x_p &= C t^2 e^{at} \ \text{if} \quad a = b_1 = b_2,
\end{aligned}
\tag{11.23}
$$

where C is determined by plugging x_p back into the differential equation.

Trick #2

The second trick is based on the first trick and we use it when $f(t)$ (the right-hand side) is either a sine or a cosine. In that case, write the equation in one of the two following forms:

$$(D - b_1)(D - b_2)\xi_p = \left\{ \begin{array}{c} K \sin at \\ K \cos at \end{array} \right\}, \tag{11.24}$$

where ξ_p is a complex function of t. Note that $K \sin at$ is the imaginary part of $K e^{iat}$ and $K \cos at$ is the real part of $K e^{iat}$. We then solve

$$(D - b_1)(D - b_2)\xi_p = K e^{iat}$$

using formulas (11.23) and take either the real or the imaginary part of the answer. That is, the particular solution x_p is either $\mathrm{Re}(\xi_P)$ or $\mathrm{Im}(\xi_p)$.

---

[5] See, for example, Mary L. Boas, *Mathematical Methods in the Physical Sciences,* 3rd ed., John Wiley & Sons Inc., Hoboken, NJ, 2006, p. 417 ff.

Finally, we can generalize the first trick if the right-hand side is an exponential, e^{at}, times a polynomial of degree n, $P_n(t)$. Then the particular solution of

$$(D - b_1)(D - b_2)x = e^{at}P_n(t)$$

is

$$x_p = \left\{ \begin{array}{ll} e^{at}Q_n(t) & \text{if } a \neq b_1 \text{ and } a \neq b_2 \\ te^{at}Q_n(t) & \text{if } a = b_1 \text{ or } a = b_2 \text{ and } b_1 \neq b_2 \\ t^2 e^{at}Q_n(t) & \text{if } a = b_1 = b_2 \end{array} \right\},$$

where $Q_n(t)$ is a polynomial of degree n :

$$Q_n(t) = A + Bt + Ct^2 + \cdots.$$

To determine $A, B, C, \ldots$ you plug x_p into the differential equation and equate the coefficients of the different powers of t.

Worked Example 11.4 Find the general solution for

$$(D^2 - 4)x = 2 - 8t$$

when $t = 0$, $x = 0$, and $\dot{x} = 5$.

Solution We use the method described by Equation (11.21) and the discussion following it. We write the differential equation as

$$(D - 2)(D + 2)x = 2 - 8t.$$

Let $u = (D + 2)x$ and write

$$(D - 2)u = 2 - 8t.$$

For a first-order differential equation such as

$$(D + f(t))x = g(t),$$

the integrating factor is $e^{\int f(t)dt} = e^{-2t}$. Multiplying by e^{-2t} we obtain

$$e^{-2t}\frac{du}{dt} - 2ue^{-2t} = 2e^{-2t} - 8te^{-2t},$$

$$\frac{d}{dt}(ue^{-2t}) = 2e^{-2t} - 8te^{-2t},$$

$$\int d(ue^{-2t}) = \int 2e^{-2t}dt - \int 8te^{-2t}dt.$$

Therefore,

$$u = C_1 e^{2t} + 4t + 1.$$

But since $u = (D + 2)x$ we have

$$\frac{dx}{dt} + 2x = C_1 e^{2t} + 4t + 1.$$

Now the integrating factor is e^{+2t} so we multiply through by this factor to get

$$\frac{d}{dt}(xe^{2t}) = C_1 e^{4t} + 4te^{2t} + e^{2t},$$

and consequently,

$$x(t) = C_2 e^{-2t} + \frac{1}{4} C_1 e^{2t} + 2t - \frac{1}{2}.$$

Inserting the initial conditions leads to $C_1 = 4$ and $C_2 = -1/2$, and finally

$$x(t) = e^{2t} - \frac{1}{2} e^{-2t} + 2t - \frac{1}{2}.$$

Note that this method yields the complete general solution, that is, it gives us the sum of the particular solution and the complementary function all at the same time.

Exercise 11.19

Obtain the general solution to $(D^2 - D - 6)x = 8$, where $D \equiv \frac{d}{dt}$. Answer: $x(t) = C_1 e^{3t} + C_2 e^{-2t} - \frac{4}{3}$. ∎

Exercise 11.20

Obtain the general solution to $(D^2 - 9)x = 5e^{-2t}$ assuming that $x(t = 0) = 0$ and $\dot{x}(t = 0) = 1$ (in the appropriate set of units). Answer: $x(t) = \frac{1}{3}e^{3t} + \frac{2}{3}e^{-3t} - e^{-2t}$. ∎

Exercise 11.21

Obtain the general solution of $(D^2 - 4)x = \sin t$. Answer: $x(t) = C_1 e^{2t} + C_2 e^{-2t} - \frac{1}{5}\sin t$. ∎

11.4.4 The Forced Undamped Oscillator

Now that you have seen some methods for solving inhomogeneous second-order differential equations, let us return to the problem of the forced undamped harmonic oscillator.

The differential equation we set out to solve was Equation (11.19):

$$\ddot{x} + \omega_0^2 x = \frac{F_0}{m}\sin \omega_d t.$$

The right-hand side of this equation is proportional to $\sin \omega_d t$ so we use trick #2 and write

$$(D - i\omega_0)(D + i\omega_0)\xi_p(t) = \frac{F_0}{m} e^{i\omega_d t}. \tag{11.25}$$

According to trick #2 and Equation (11.23) the particular solution of Equation (11.25) is

$$\xi_p(t) = C e^{i\omega_d t}.$$

To obtain C, plug ξ_p back into the differential equation, thus:

$$\left(\frac{d^2}{dt^2} + \omega_0^2\right) C e^{i\omega_d t} = \frac{F_0}{m} e^{i\omega_d t}.$$

Hence,

$$-\omega_d^2 C + \omega_0^2 C = \frac{F_0}{m},$$

and

$$C = \frac{F_0/m}{\omega_0^2 - \omega_d^2}.$$

Recall that x_p is the imaginary part of the complex function ξ_p, so the particular solution is

$$x_p(t) = \mathcal{I}m\left(\xi_p(t)\right) = \mathcal{I}m\left(\frac{F_0/m}{\omega_0^2 - \omega_d^2}e^{i\omega_d t}\right) = \frac{F_0/m}{\omega_0^2 - \omega_d^2}\sin\omega_d t,$$

and the general solution of Equation (11.19) is

$$x(t) = A\cos(\omega_0 t + \theta_0) + \frac{F_0/m}{\omega_0^2 - \omega_d^2}\sin\omega_d t.$$

The problem is now solved, except for determining A and θ_0 from the initial conditions.

Before leaving this problem, I want you to notice that there is something quite interesting about the particular solution,

$$x_p = \frac{F_0/m}{\omega_0^2 - \omega_d^2}\sin\omega_d t.$$

This expression indicates that as the driving frequency ω_d gets closer and closer to ω_0, the natural frequency of oscillation, the amplitude of the oscillations get larger and larger. If $\omega_d = \omega_0$, the amplitude is infinite. This phenomenon is known as *resonance.* We will consider resonance further in the next section, refining our analysis and showing that the amplitude does not actually become infinite for a real physical system.

Exercise 11.22

A simple harmonic oscillator consists of a mass of 3 kg on a spring of force constant 0.15 N/m. Determine the frequency of the driving force which will cause the amplitude of oscillations to grow extremely large. Answer 0.22 rad/s. ■

11.4.5 The Forced Damped Oscillator

We now consider a *damped,* driven, harmonic oscillator and assume the driving force is, once again, $F_0 \sin \omega_d t$. The equation of motion is

$$\ddot{x} + \frac{b}{m}\dot{x} + \frac{k}{m}x = \frac{F_0}{m}\sin\omega_d t.$$

Note that $\sqrt{k/m} = \omega_0$ is the "natural" frequency of the undamped oscillator. For the sake of a specific example, we assume the oscillator is underdamped.

The technique for finding the particular solution is the same as before, so set

$$\xi_p = Ce^{i\omega_d t},$$

and plug it into the differential equation. This gives

$$C(-\omega_d^2)e^{i\omega_d t} + \frac{b}{m}C(i\omega_d)e^{i\omega_d t} + \omega_0^2 Ce^{i\omega_d t} = \frac{F_0}{m}e^{i\omega_d t},$$

where $\sin \omega_d t$ was replaced by $e^{i\omega_d t}$. (This means that at the end of the procedure we must only keep the imaginary part of the solution.)

Solving for C yields

$$C = \frac{F_0/m}{(\omega_0^2 - \omega_d^2) + i\omega_d b/m},$$

which is complex. To make this quantity easier to manipulate, the imaginary term is usually placed in the numerator. To achieve this, multiply top and bottom by the complex conjugate of the denominator and obtain

$$C = \frac{(F_0/m)\left[(\omega_0^2 - \omega_d^2) - i\omega_d b/m\right]}{\left[(\omega_0^2 - \omega_d^2)^2 + (\omega_d b/m)^2\right]}.$$

You have already seen that the general solution of the homogeneous equation for the underdamped harmonic oscillator is given in Equation (11.15) as $x = Ae^{-\gamma t}\cos(\omega_1 t + \theta)$. This is the complementary function for the inhomogeneous equation. The general solution of the forced, underdamped harmonic oscillator is, then,

$$x(t) = Ae^{-\gamma t}\cos(\omega_1 t + \theta) + \mathcal{Im}\left(\frac{F_0/m\left[(\omega_0^2 - \omega_d^2) - i\omega_d b/m\right]}{\left[(\omega_0^2 - \omega_d^2)^2 + (\omega_d b/m)^2\right]}e^{i\omega_d t}\right).$$

Eventually, the first term will die out. (This is, of course, also true for overdamped and critically damped oscillators.) Dropping the first term, you are left with

$$x(t) = \mathcal{Im}\left(\frac{F_0/m\left[(\omega_0^2 - \omega_d^2) - i\omega_d b/m\right]}{\left[(\omega_0^2 - \omega_d^2)^2 + (\omega_d b/m)^2\right]}(\cos\omega_d t + i\sin\omega_d t)\right).$$

After a bit of algebra, this expression can be written

$$x(t) = \frac{F_0}{m}\frac{(\omega_0^2 - \omega_d^2)\sin\omega_d t - (\omega_d b/m)\cos\omega_d t}{(\omega_0^2 - \omega_d^2)^2 + (\omega_d b/m)^2}. \tag{11.26}$$

At resonance, when $\omega_0 = \omega_d$, the denominator is minimized, but the damping term prevents it from going to zero. Therefore, at resonance you get a large amplitude, but it is not infinite. In real physical systems, there is always some damping that prevents the system from having infinite amplitude at resonance.

Equation (11.26) can be expressed in a more convenient way by using the trigonometric identity

$$A\sin\omega t + B\cos\omega t = \sqrt{A^2 + B^2}\cos(\omega t - \phi),$$

where $\phi = \tan^{-1}(A/B)$. So,

$$x(t) = \left[\frac{(F_0/m)^2\left[(\omega_0^2 - \omega_d^2)^2 + \omega_d^2 b^2/m^2\right]}{\left[(\omega_0^2 - \omega_d^2)^2 + \omega_d^2 b^2/m^2\right]^2}\right]^{1/2}\cos(\omega_d t - \phi)$$

$$= \frac{F_0}{\left[m^2(\omega_0^2 - \omega_d^2)^2 + \omega_d^2 b^2\right]^{1/2}}\cos(\omega_d t - \phi), \tag{11.27}$$

where, from Equation (11.26),

$$\phi = \tan^{-1}\left[-\frac{\omega_0^2 - \omega_d^2}{\omega_d b/m}\right].$$

It is interesting to note that the amplitude of $x(t)$ is not maximized at $\omega_d = \omega_0$ but rather at a nearby frequency called the resonant frequency which you can determine by evaluating the derivative of the denominator and setting it equal to zero, thus:

$$\frac{d}{d\omega_d} \left[m^2(\omega_0^2 - \omega_d^2)^2 + \omega_d^2 b^2 \right]^{1/2} = 0.$$

This leads to the following condition for a minimum

$$\omega_d^2 = \omega_0^2 - \frac{b^2}{2m^2}. \tag{11.28}$$

The greatest amplitude occurs when the driving frequency is equal to the resonant frequency which we can denote by ω'.

$$\omega' = \sqrt{\omega_0^2 - \frac{b^2}{2m^2}}.$$

This is neither the frequency of the undamped oscillator (ω_0), nor the frequency of the damped oscillator (ω_1).

When $\omega_d = \omega'$ the amplitude of the oscillation is given by

$$A = \frac{F_0}{\left[m^2(\omega_0^2 - \omega'^2)^2 + \omega'^2 b^2 \right]^{1/2}}.$$

The resonance phenomenon is often associated with the transfer of energy. The average energy of a harmonic oscillator is proportional to the amplitude squared, so an important and useful parameter in studying resonance is A^2, given by

$$A^2 = \frac{F_0^2}{m^2(\omega_0^2 - \omega'^2)^2 + \omega'^2 b^2}.$$

Worked Example 11.5 Show that the phase angle ϕ in Equation (11.27) is given by $\phi = \tan^{-1} \left[-\frac{\omega_0^2 - \omega_d^2}{\omega_d b/m} \right]$.

Solution We have shown in Equation (11.26) that

$$x = x(t) = \frac{F_0}{m} \frac{\left(\omega_0^2 - \omega_d^2\right) \sin \omega_d t - (\omega_d b/m) \cos \omega_d t}{\left(\omega_0^2 - \omega_d^2\right)^2 + (\omega_d b/m)^2}.$$

Writing this in the form $A \sin \omega t + B \cos \omega t$ we have

$$A = \frac{F_0}{m} \frac{\left(\omega_0^2 - \omega_d^2\right)}{\left(\omega_0^2 - \omega_d^2\right)^2 + (\omega_d b/m)^2},$$

and

$$B = \frac{F_0}{m} \frac{-(\omega_d b/m)}{\left(\omega_0^2 - \omega_d^2\right)^2 + (\omega_d b/m)^2}.$$

So $\phi = \tan^{-1}(\mathcal{A}/\mathcal{B})$ yields

$$\phi = \tan^{-1}\frac{\frac{(\omega_0^2-\omega_d^2)}{(\omega_0^2-\omega_d^2)^2+(\omega_d b/m)^2}}{\frac{-(\omega_d b/m)}{(\omega_0^2-\omega_d^2)^2+(\omega_d b/m)^2}} = \tan^{-1}\frac{(\omega_0^2-\omega_d^2)}{-(\omega_d b/m)}$$

$$= \tan^{-1}\frac{-(\omega_d^2-\omega_0^2)}{\omega_d b/m}.$$

The phase shift ϕ describes the lag between the applied force $F(t)$ and the response $x(t)$.

Exercise 11.23

Fill in the missing steps in the derivation of Equation (11.28).   ∎

Exercise 11.24

Show that $\mathcal{A}\sin\omega t + \mathcal{B}\cos\omega t = \sqrt{\mathcal{A}^2+\mathcal{B}^2}\cos(\omega t - \phi)$, where $\tan\phi = \mathcal{A}/\mathcal{B}$.   ∎

Exercise 11.25

A damped harmonic oscillator is driven by a force $F_0 e^{-at}$. Find the general solution. (Note: The particular solution can be expected to have the same time dependence as the driving force.)

∎

The Q Factor

The Q factor is an important property of a damped driven harmonic oscillator. It is related to both the shape of the resonance curve and the energy loss rate of the oscillator.[6]

There are several related definitions of the Q factor. These yield similar but not identical results. You should be aware that in specific applications the Q factor may be expressed differently than the way we will describe here, but the basic ideas are always the same.

Let us begin with the definition of the Q factor in terms of the *shape* of the resonance curve. An appropriate definition of Q is the ratio of the resonant frequency (ω_0) to the width of the resonance curve at half maximum of the curve. ($\Delta\omega$ is denoted "full width at half maximum.") That is, $Q = \frac{\omega_0}{\Delta\omega}$ which, as we shall show, can be written as

$$Q \equiv \frac{m}{b}\omega_0 = \sqrt{\frac{km^2}{b^2}}.$$

Note that the definition only involves the physical properties of the oscillator. As mentioned previously, a driven oscillator will have the largest amplitude when the frequency of the driving force is equal to the resonant frequency, which we have denoted ω'. For simplicity, we shall consider systems with small damping values, so $\omega' \doteq \omega_0 = \sqrt{k/m}$. A plot of the square

---

[6] There are many applications of driven harmonic oscillators. For example, in musical instruments such as trumpets or French horns the vibration of a reed or of the musician's lips produces sounds with a wide range of frequencies. The instrument has a resonance at the desired frequency, thus picking out the desired note. In optical devices, such as the resonant cavity of lasers, the Q factor is an important parameter. Electromechanical resonators such as quartz crystals in digital watches allow for highly accurate timepieces.

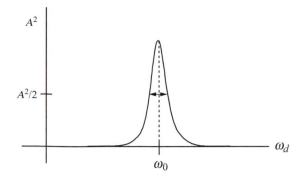

Figure 11.8 The resonance curve with full width at half maximum indicated.

of the amplitude as a function of driving frequency (ω_d) will look somewhat as illustrated in Figure 11.8.

To obtain a large response at a specific frequency the curve of Figure 11.8 should be very tall and very thin. In technical terms, it is desirable for the curve to have a small value for the width at half maximum (as shown by the double-headed arrow in the figure).

We have presented the resonance curve in terms of the amplitude squared rather than the amplitude. This is convenient because the energy of the oscillator is proportional to the square of the amplitude. Furthermore, a slightly different definition of Q is the ratio of the maximum energy stored in the oscillator to the energy lost per radian of motion. (We explore this alternative definition in Problem 11.16.)

The square of the amplitude is a maximum at $\omega_d = \omega_0$ and is given by

$$A_{\omega_0}^2 = \frac{F_0^2}{\omega_0 b^2}.$$

For some nearby driving frequency ω the amplitude is

$$A_\omega^2 = \frac{F_0^2/m^2}{[(\omega_0^2 - \omega^2)^2 + \omega^2 b^2/m^2]}.$$

If $A_\omega^2 = \frac{1}{2} A_{\omega_0}^2$ we have

$$2\omega_0^2 b^2/m^2 = (\omega_0^2 - \omega^2)^2 + \omega^2 b^2/m^2,$$

$$\frac{b^2}{m^2}(2\omega_0^2 - \omega^2) = (\omega_0^2 - \omega^2)^2,$$

and for ω close to ω_0, we can write the approximate relation

$$\frac{b}{m}\omega_0 = \omega_0^2 - \omega^2.$$

Now $(\omega_0^2 - \omega^2) = (\omega_0 - \omega)(\omega_0 + \omega) \approx \Delta\omega_{1/2}(2\omega_0)$, where $\Delta\omega_{1/2} = \omega - \omega_0$ is the half width at half maximum. The full width at half maximum is $\Delta\omega = 2\Delta\omega_{1/2}$, so

$$\frac{b\omega_0}{m} = 2\Delta\omega_{1/2} \cdot \omega_0 = \Delta\omega\omega_0,$$

and

$$Q = \frac{\omega_0}{\Delta\omega} = \frac{\omega_0}{b/m} = \frac{m}{b}\omega_0.$$

You can appreciate that large Q means a small value of $\Delta\omega$, that is, a thin peak. It might be mentioned that a very large Q is not always desirable, as for example in a musical instrument it would make it harder to "hit the note." On the other hand, for a radio to tune into one of two stations broadcasting at nearby frequencies, a high Q factor is highly desirable.

Another use of the Q factor is in determining the loss of energy of a damped oscillator. For simplicity let us consider an unforced oscillator. As you know, due to damping, the amplitude of the oscillations will decay exponentially as $e^{-\gamma t}$, where $\gamma = b/2m$; see Equation (11.15). Energy is proportional to the amplitude squared, so the decrease in energy in time t is given by $(e^{-\gamma t})^2 = e^{-2\gamma t}$. Therefore, the energy loss in the time the oscillator goes through one complete cycle is $e^{-2\gamma T}$, where T is the period ($T = \frac{2\pi}{\omega_0}$). Consequently, the ratio of final energy to initial energy is

$$\frac{Ee^{-2\gamma T}}{E} = e^{-2\gamma T} = e^{-2\frac{b}{2m}\frac{2\pi}{\omega_0}} = e^{-2\pi \frac{b}{m\omega_0}} = e^{-2\pi/Q}.$$

This tells us that an oscillator with large Q does not lose energy as quickly as an oscillator with small Q.

Resonance in Electrical Circuits (Optional)

Resonance is an interesting physical phenomenon of importance in such diverse fields as the design of electrical circuits and the motion of celestial bodies. For example, the RLC circuit you studied in your introductory electricity course is illustrated in Figure 11.9.

The charge on the capacitor plates is q. If the plates are initially charged, there will be a flow of charge (a current, I) through the inductor, setting up a back emf or voltage difference given by $L\frac{dI}{dt}$. This can also be written as $L\frac{d^2q}{dt^2}$. The voltage drop across the resistor is IR and this can be written as $R\frac{dq}{dt}$. Finally, the voltage drop across the capacitor is given by $\frac{q}{C}$. The sum of voltage drops around the circuit is zero, so

$$L\frac{d^2q}{dt^2} + R\frac{dq}{dt} + \frac{1}{C}q = 0.$$

But this is exactly the equation for a damped harmonic oscillator! If you add a "driving force" by placing some source of emf in the circuit (such as an oscillator or power supply) you will have a driven oscillator. The natural frequency of the circuit is $\omega_0 = \sqrt{1/LC}$, as you can appreciate from the form of the equation if the damping term is removed. If the driving frequency is near or equal to the natural frequency, the amplitude of the current increases greatly. This is the principle behind tuning a radio or television receiver. The "driving" voltage is supplied by the antenna picking up electromagnetic radiation. Suppose a station you want to tune to is broadcasting at a frequency ω_d. The receiver contains a variable capacitor whose capacitance you change by turning a knob or pushing a button on your receiver. Changing C will, of course, change the value of ω_0 and hence the resonant frequency ω'. When ω' is equal to ω_d, the current through the resistor reaches a much higher value than for other frequencies. This is illustrated in Figure 11.10.

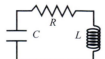

Figure 11.9 An RLC circuit.

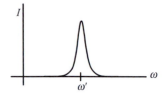

Figure 11.10 A plot of the current (I) as a function of the angular frequency ω, illustrating the resonance phenomenon in an LRC circuit. When the frequency ω is nearly equal to the resonant frequency ω', the current in the circuit increases greatly. If there were no resistance in the circuit, the current would become infinite at $\omega = \omega_0$.

Resonance in Astronomy

As mentioned in Chapter 10, celestial bodies are affected by resonances. For example, the orbital period and the rotational period of the Moon are "locked" into a 1:1 resonance. In celestial mechanics, a resonance is defined as a system in which the frequency of the "driving" force (usually the effect of a perturbing body) and the "natural" frequency of the system are in the ratio of small numbers. Resonances explain why there are gaps in the rings of Saturn and in the asteroid belts between Mars and Jupiter, and why certain satellites (including the Moon) have rotation rates equal to their orbital rates. (The resonance phenomenon in celestial mechanics is often called a "commensurability.")

Exercise 11.26

An RLC series circuit consists of a 10 ohm resistor, a 6 microfarad capacitor, and an 0.2 henry inductor. Determine the resonant frequency for the circuit. Answer: 913 rad/s. ■

11.5 Coupled Oscillators

In the real physical world, oscillators are seldom isolated from their environment and an oscillating body will set up oscillations in nearby bodies. A very good example of *coupled oscillators* is the motion of atoms in a crystal. A vibrating atom will cause neighboring atoms to start oscillating.

The analysis of coupled oscillators can be quite complicated. This section is a study of the simplest of such systems, namely two undamped oscillators connected by springs.

Figure 11.11 illustrates two masses attached to rigid walls by springs of constants k_1, k_2, and k_3. Let x_1 and x_2 be the displacements of masses m_1 and m_2 *measured from their equilibrium positions*. The equations of motion for the masses are easily determined from the Lagrangian or from Newton's second law. We obtain the following *coupled* differential equations:[7]

$$m_1\ddot{x}_1 = -k_1x_1 - k_3(x_1 - x_2), \tag{11.29}$$
$$m_2\ddot{x}_2 = -k_2x_2 - k_3(x_2 - x_1).$$

For simplicity, assume that $m_1 = m_2 = m$ and that $k_1 = k_2 = k$, but allow k_3 to be different. Then the equations of motion reduce to

$$m\ddot{x}_1 + (k + k_3)x_1 - k_3x_2 = 0,$$
$$m\ddot{x}_2 + (k + k_3)x_2 - k_3x_1 = 0.$$

---

[7] When applied to the atoms in a solid, we often assume all the masses are equal and all the "spring constants" are equal, yielding n equations of the form

$$m\ddot{x}_i = k(x_i - x_{i-1}) + k(x_{i+1} - x_i) \qquad i = 1, \ldots, n.$$

These relations reduce to Equations (11.29) for the situation pictured in Figure 11.11.

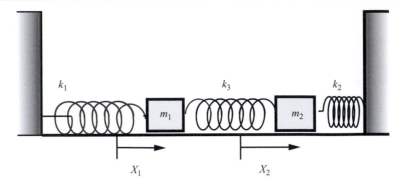

Figure 11.11 Coupled oscillators. Note that the positions of the two masses are measured from their equilibrium points.

Denote $k + k_3$ by k'. Dividing by m yields:

$$\ddot{x}_1 + (k'/m)x_1 - (k_3/m)x_2 = 0,$$
$$\ddot{x}_2 + (k'/m)x_2 - (k_3/m)x_1 = 0.$$

Now replace k'/m by ω_0^2. The reason for this notation is that ω_0 is the frequency of oscillation of either one of the masses if the other mass is held at rest. Furthermore, let us denote k_3/m by $\Delta\omega^2$, for reasons that will soon become clear.

Using this new notation the equations become

$$\ddot{x}_1 + \omega_0^2 x_1 - \Delta\omega^2 x_2 = 0, \tag{11.30}$$
$$\ddot{x}_2 + \omega_0^2 x_2 - \Delta\omega^2 x_1 = 0.$$

This system of *coupled* linear differential equations can be solved by assuming the solutions

$$x_1 = A e^{i\omega t},$$
$$x_2 = B e^{i\omega t}.$$

We assume oscillatory solutions (that is why the exponent is imaginary) with both of the variables x_1 and x_2 having the same angular frequency ω. (Except for a slight change in the notation, this is the same technique previously used in assuming solutions of the form e^{pt}.) Following our usual procedure we insert the expressions for x_1 and x_2 into the coupled equations of motion to determine the "auxiliary equations" that will give us ω. We obtain

$$A(i\omega)^2 e^{i\omega t} + \omega_0^2 A e^{i\omega t} - \Delta\omega^2 B e^{i\omega t} = 0,$$
$$B(i\omega)^2 e^{i\omega t} + \omega_0^2 B e^{i\omega t} - \Delta\omega^2 A e^{i\omega t} = 0,$$

or

$$(-\omega^2 + \omega_0^2)A - \Delta\omega^2 B = 0, \tag{11.31}$$
$$-\Delta\omega^2 A + (-\omega^2 + \omega_0^2)B = 0.$$

These two equations are a system of homogeneous coupled linear *algebraic* equations in the unknowns A and B. (Solving them will indirectly lead to a solution for ω.)

The technique for solving systems of coupled algebraic equations is to apply Cramer's rule. I believe you are familiar with this rule, but I will review it briefly anyway.

Consider the pair of linear algebraic equations

$$ax + by = e,$$
$$cx + dy = f.$$

To solve, first evaluate the determinant of the coefficients, thus

$$D = \begin{vmatrix} a & b \\ c & d \end{vmatrix}.$$

The first column of this determinant is formed from the coefficients of x and the second column contains the coefficients of y. Then, according to Cramer's rule, the solutions are:

$$x = \frac{\begin{vmatrix} e & b \\ f & d \end{vmatrix}}{D} \quad \text{and} \quad y = \frac{\begin{vmatrix} a & e \\ c & f \end{vmatrix}}{D}.$$

Note that the determinant in the numerator of the equation for x is obtained from D by replacing the column of the coefficients of x with the constants on the right-hand side of the equations. Similarly y is given by the same equation but with the column of coefficients of y replaced by the right-hand constants. (You may recall using Cramer's rule to determine the currents in an electric circuit when applying Kirchhoff's laws.)

Unfortunately, when Cramer's rule is applied to *homogenous* linear equations such as Equations (11.31), the zeros on the right-hand side mean that the determinants in the numerators have a column of zeros. This leads to the *trivial solutions* $x = 0$ and $y = 0$. However, if the *denominator* is also zero (if $D = 0$), then $x = \frac{0}{0}$ and $y = \frac{0}{0}$, both of which are undefined. You may think this is not much of an advantage, but it actually allows you to find the solution! In fact it is a general rule that coupled homogeneous algebraic equations have nontrivial solutions *if and only if* the determinant of the coefficients is zero.

Going back to Equations (11.31), the condition for a nontrivial solution is that the determinant of the coefficients is equal to zero:

$$D = \begin{vmatrix} -\omega^2 + \omega_0^2 & -\Delta\omega^2 \\ -\Delta\omega^2 & -\omega^2 + \omega_0^2 \end{vmatrix} = 0,$$

or

$$(-\omega^2 + \omega_0^2)^2 - (\Delta\omega^2)^2 = 0,$$

$$\omega^2 = +\omega_0^2 \mp \Delta\omega^2.$$

That is,

$$\omega = \pm(\omega_0^2 \mp \Delta\omega^2)^{\frac{1}{2}}.$$

This yields four possible values for ω. Two of them are

$$\omega_1 = \pm(\omega_0^2 - \Delta\omega^2)^{\frac{1}{2}},$$

and the other two are

$$\omega_2 = \pm(\omega_0^2 + \Delta\omega^2)^{\frac{1}{2}}. \tag{11.32}$$

The general solutions for x_1 and x_2 are the sum of all possible solutions:

$$x_1 = A_1 e^{i\omega_1 t} + A_{-1} e^{-i\omega_1 t} + A_2 e^{i\omega_2 t} + A_{-2} e^{-i\omega_2 t}, \tag{11.33}$$

$$x_2 = B_1 e^{i\omega_1 t} + B_{-1} e^{-i\omega_1 t} + B_2 e^{i\omega_2 t} + B_{-2} e^{-i\omega_2 t}.$$

These two expressions involve eight constants. But solving two second-order differential equations should lead to only four constants. Therefore, not all of the eight constants are independent. We now determine the four independent constant amplitudes.

Note that each term in our expression for x_1 satisfies the differential equation. Thus, for example, a possible solution involves oscillations of frequency ω_1. Then,

$$x_1 = A_1 e^{i\omega_1 t}$$

$$x_2 = B_1 e^{i\omega_1 t},$$

where $\omega_1^2 = \omega_0^2 - \Delta\omega^2$. Plugging into the equations of motion, Equations (11.30), yields

$$(\omega_0^2 - \omega_1^2)A_1 - \Delta\omega^2 B_1 = 0$$

and

$$-\Delta\omega^2 A_1 + (\omega_0^2 - \omega_1^2)B_1 = 0.$$

Either one of these yields the same result, namely,

$$\left(-\omega_0^2 - (\omega_0^2 - \Delta\omega^2)\right)A_1 - \Delta\omega^2 B_1 = 0,$$

or

$$+\Delta\omega^2 A_1 - \Delta\omega^2 B_1 = 0.$$

Therefore,

$$B_1 = A_1.$$

By considering the solution in $-\omega_1$ we obtain $B_{-1} = A_{-1}$, and $\pm\omega_2$ lead to $B_2 = -A_2$ and $B_{-2} = -A_{-2}$.

So our equations reduce to

$$x_1 = A_1 e^{i\omega_1 t} + A_{-1} e^{-i\omega_1 t} + A_2 e^{i\omega_2 t} + A_{-2} e^{-i\omega_2 t},$$

$$x_2 = A_1 e^{i\omega_1 t} + A_{-1} e^{-i\omega_1 t} - A_2 e^{i\omega_2 t} - A_{-2} e^{-i\omega_2 t}.$$

These solutions are not in a convenient format because x_1 and x_2 must be real, but $e^{\pm i\omega_1 t}$ and $e^{\pm i\omega_2 t}$ are complex, so the coefficients A_i must also be complex. These complex coefficients can be expressed in terms of real quantities, thus:

$$A_1 = \frac{1}{2}C_1 e^{i\theta_1},$$

$$A_{-1} = \frac{1}{2}C_1 e^{-i\theta_1},$$

$$A_2 = \frac{1}{2}C_2 e^{i\theta_2},$$

$$A_{-2} = \frac{1}{2}C_2 e^{-i\theta_2}.$$

Here C_1, C_2, θ_1, and θ_2 are all real. Then,

$$x_1 = \frac{1}{2}C_1 \left(e^{i\theta_1}e^{+i\omega_1 t} + e^{-i\theta_1}e^{-i\omega_1 t}\right) + \frac{1}{2}C_2 \left(e^{i\theta_2}e^{+i\omega_2 t} + e^{-i\theta_2}e^{-i\omega_2 t}\right),$$

or

$$x_1 = C_1 \cos(\omega_1 t + \theta_1) + C_2 \cos(\omega_2 t + \theta_2), \tag{11.34}$$

and similarly,

$$x_2 = C_1 \cos(\omega_1 t + \theta_1) - C_2 \cos(\omega_2 t + \theta_2). \tag{11.35}$$

For the more general case in which $k_1 \neq k_2$ and $m_1 \neq m_2$ the expressions are more complicated and lead to

$$x_1 = C_1 \cos(\omega_1 t + \theta_1) + C_2 \left(\frac{\Delta\omega^2}{\kappa^2} \sqrt{\frac{m_2}{m_1}} \right) \cos(\omega_2 t + \theta_2),$$

$$x_2 = -C_1 \left(\frac{\Delta\omega^2}{\kappa^2} \sqrt{\frac{m_1}{m_2}} \right) \cos(\omega_1 t + \theta_1) + C_2 \cos(\omega_2 t + \theta_2),$$

where $\kappa^2 = k_3 / \sqrt{m_1 m_2}$.

Exercise 11.27

(a) Derive the equations of motion (11.29) using the Lagrangian technique. (b) Derive these equations using Newton's second law. ∎

Exercise 11.28

Use Cramer's rule to evaluate x, y, z if

$$2x + 3y + z = 1,$$
$$-5x - 2y + z = 5,$$
$$x + 2y + 2z = 13.$$

Answer: $x = 3, y = -5$, and $z = 10$. ∎

11.5.1 Normal Modes

Suppose that all of the masses in a system of coupled oscillators are oscillating at the same frequency. This type of motion is called a "normal mode." You might suppose that a normal mode is a very special situation that would not occur very often. You would be correct. Nevertheless, normal modes are quite important because it turns out that *the motion of a system of oscillators can always be expressed as a linear combination of normal modes.*

As a simple example, assume the system described by Equations (11.34) and (11.35) was set in motion in such a way that $C_2 = 0$. The positions of the two masses would then be described by the equations

$$x_1 = C_1 \cos(\omega_1 t + \theta_1), \tag{11.36}$$
$$x_2 = C_1 \cos(\omega_1 t + \theta_1).$$

In this situation, both masses oscillate with the same frequency ω_1, so this is a normal mode. Furthermore, $x_1 = x_2$ at all times. What does the motion look like? The two masses are moving *in phase* with each other. The frequency of the motion is

$$\omega_1 = \left(\omega_0^2 - \Delta\omega^2 \right)^{\frac{1}{2}} = \sqrt{k/m}.$$

Notice that this is the natural frequency of the individual masses.

How could this normal mode be set up? As your physical intuition probably suggests, you could generate the motion by pulling the two masses equal distances from equilibrium and releasing them from rest simultaneously. For example, if both masses are pulled a distance b from equilibrium and released from rest, the initial values of x_1 and x_2 are both b and initial values of $\dot{x}_1$ and $\dot{x}_2$ are both 0.

Recall that in general,

$$x_1(t) = C_1 \cos(\omega_1 t + \theta_1) + C_2 \cos(\omega_2 t + \theta_2),$$
$$x_2(t) = C_1 \cos(\omega_1 t + \theta_1) - C_2 \cos(\omega_2 t + \theta_2).$$

Setting $t = 0$ we have

$$b = C_1 \cos \theta_1 + C_2 \cos \theta_2, \tag{11.37}$$
$$b = C_1 \cos \theta_1 - C_2 \cos \theta_2.$$

The velocities of the blocks are:

$$\dot{x}_1(t) = -C_1 \omega_1 \sin(\omega_1 t + \theta_1) - C_2 \omega_2 \sin(\omega_2 t + \theta_2),$$
$$\dot{x}_2(t) = -C_1 \omega_1 \sin(\omega_1 t + \theta_1) + C_2 \omega_2 \sin(\omega_2 t + \theta_2).$$

Since the blocks are released from rest at $t = 0$,

$$0 = -C_1 \omega_1 \sin \theta_1 - C_2 \omega_2 \sin \theta_2,$$
$$0 = -C_1 \omega_1 \sin \theta_1 + C_2 \omega_2 \sin \theta_2.$$

Adding these two equations shows that $\theta_1 = 0$ and subtracting them yields $\theta_2 = 0$. Equations (11.37) then become

$$b = C_1 + C_2,$$
$$b = C_1 - C_2.$$

Hence,

$$C_1 = b,$$
$$C_2 = 0.$$

Therefore the motion is described by Equations (11.36) as desired. Furthermore, $C_1 = b$ and $\theta_1 = 0$, so

$$x_1(t) = b \cos \omega_1 t, \tag{11.38}$$
$$x_2(t) = b \cos \omega_1 t,$$

where $\omega_1 = \sqrt{\omega_0^2 - \Delta \omega^2}$. This is the low-frequency normal mode.

A different normal mode can be generated by either pulling the two masses apart or pushing them together by equal amounts, and releasing them from rest. Assuming the initial displacements are equal to b' (but in opposite directions), we find $C_1 = 0$ and $C_2 = b'$.

The motion is then described by

$$x_1(t) = b' \cos \omega_2 t, \tag{11.39}$$
$$x_2(t) = -b' \cos \omega_2 t,$$

where

$$\omega_2 = (\omega_0^2 + \Delta \omega^2)^{\frac{1}{2}} = \sqrt{\frac{k + 2k_3}{m}}.$$

This means the system oscillates at a higher frequency than the "natural" frequency ω_0.

Note that the general solutions, Equations (11.34) and (11.35), are simply linear combinations of the normal mode solutions, Equations (11.38) and (11.39). As mentioned before, the motion of any system of coupled oscillators is always a linear combination of normal modes. The problem gets more complicated when the number of oscillators increases. This leads to the subject of a continuous system of oscillators that is treated in Chapter 14.

Exercise 11.29

Using the given initial conditions show that $C_2 = b$ and $\theta_2 = 0$ in Equations (11.39).   ∎

By substituting the definitions of ω_0 and $\Delta\omega$ show that $\omega_1 = \pm\sqrt{k/m}$ and $\omega_2 = \pm\sqrt{(k + 2k_3)/m}$. ∎

11.6 Summary

This chapter was a study of harmonic motion using a mass connected to a spring as an example. There were several digressions to discuss the mathematical techniques involved in solving the equations of motion. These techniques are used frequently in many areas of physics, so you should make a special effort to understand and remember them.

The analysis started with an undamped simple harmonic oscillator with equation of motion

$$m\ddot{x} + kx = 0.$$

Next we generalized to a damped harmonic oscillator in which there was a frictional force acting to slow the system down, with the equation of motion

$$m\ddot{x} + b\dot{x} + kx = 0.$$

There were three possible situations: the underdamped oscillator with $(b/2m)^2 < k/m$, the critically damped oscillator with $(b/2m)^2 = k/m$, and the overdamped oscillator with $(b/2m)^2 > k/m$. The underdamped oscillator is the only one exhibiting truly oscillatory motion with the mass moving back and forth in a regular, repetitive way. Note, however, that in all three cases the amplitude of the motion decreases as time goes on. Mathematically this is described by the amplitude being given by $Ae^{-\gamma t}$ so that the motion of an underdamped oscillator is

$$x(t) = Ae^{-\gamma t}\cos(\omega_1 t + \theta),$$

the motion of a critically damped oscillator is

$$x(t) = (C_1 + C_2 t)\, e^{-\gamma_c t},$$

and the motion of an overdamped oscillator is

$$x(t) = C_1 e^{-\gamma_1 t} + C_2 e^{-\gamma_2 t}.$$

Note also that the frequency of the motion is affected; an underdamped oscillator has a frequency smaller than the frequency of the undamped oscillator. The overdamped oscillator has a greater "frequency," but it does not really oscillate, the motion simply dies out before a single full oscillation can occur.

Next you studied the forced harmonic oscillator. In this case the motion is represented by the following inhomogeneous differential equation:

$$m\ddot{x} + b\dot{x} + kx = F(t).$$

Obtaining the motion required summing the general solution of the homogeneous differential equation and a particular solution of the inhomogeneous equation. You learned a number of "tricks" for determining the particular solutions of inhomogeneous equations.

Finally, you learned about coupled undamped harmonic oscillators whose equations of motion are

$$m_1\ddot{x}_1 = -k_1 x_1 - k_3(x_1 - x_2),$$
$$m_2\ddot{x}_2 = -k_2 x_2 - k_3(x_2 - x_1).$$

This is an important problem in physics because there are many physical systems that can be represented as coupled harmonic oscillators. Normal modes were defined and it was noted that the general motion of coupled oscillators can be expressed as a linear combination of normal mode solutions.

11.7 Problems

Problem 11.1 Determine the condition such that $x = Cte^{pt}$ is a solution of $m\ddot{x} + b\dot{x} + kx = 0$.

Problem 11.2 Determine the time average of kinetic energy and potential energy for a simple undamped harmonic oscillator and show that they are equal.

Problem 11.3 The motion of an underdamped harmonic oscillator is $x(t) = Ae^{-\gamma t} \cos(\omega_1 t + \theta)$. Assuming the initial conditions $x_0 = 1$ and $v_0 = 0$, determine A and θ.

Problem 11.4 A critically damped oscillator consists of a 5 kg block and a spring of constant 200 N/m. It receives an impulse that gives it an initial velocity of 10 m/s. The initial position is $x(t = 0) = 0$. Determine the position of the block at times 0.5 s and 2.0 s.

Problem 11.5 A bathroom scale should not oscillate. Ideally it would be critically damped. Show that if a scale is critically damped for a person of weight W it will be overdamped for a person whose weight is less than W. Suppose that when the scale is critically damped, the platform deflects 2 cm if a 70 kg person stands on it. Determine the spring constant k and the damping constant b.

Problem 11.6 In a railroad switching yard, at the end of each track there is a large wooden bar connected to heavy springs. The purpose of this bumper is to bring a rolling freight car to a stop. Clearly, it is not desirable for the freight car to bounce off the bumper and go rolling back on the track. (Note that the freight car is not attached to the bumper.) (a) Using appropriate sketches of position as a function of time explain why the car will come to a stop in contact with the bumper if it is critically damped or overdamped. (b) Assume critical damping and obtain an expression for the maximum compression of the springs in terms of γ_c and the speed of the car when it hits the bumper.

Problem 11.7 Show that the critically damped oscillator will usually approach zero faster than the overdamped oscillator. Determine the special condition such that the overdamped oscillator approaches zero faster than the critically damped oscillator.

Problem 11.8 For the critically damped harmonic oscillator obtain expressions for C_1 and C_2 for arbitrary initial conditions x_0 and v_0.

Problem 11.9 In a certain industrial process, packages fall off of the end of a conveyer belt onto a scale that is a distance h below it. It is found that if the package has mass M the scale deflects a distance δ below its equilibrium position. It is desired that the scale reach this position as rapidly as possible without overshooting. Obtain the optimum value of the damping coefficient of the scale.

Problem 11.10 An automobile has a mass of 800 kg. (This is the approximate mass of a Volkswagen Beetle.) Assume the suspension system is critically damped when a 100 kg driver is in the car because it is not desirable for the car to oscillate for a long time after going over a bump. Assume $k = 3600$ N/m. In this problem we calculate the time required for the amplitude of the oscillation to die down to $1/e$ of its initial value under the following three circumstances:
(a) The driver is the only person in the car.
(b) The driver is accompanied by three other 100 kg people.
(c) The only person in the car has a mass of 50 kg.

Problem 11.11 Consider the underdamped harmonic oscillator solution. Show that if $\omega_1 = \sqrt{\omega_0^2 - \gamma^2}$ is very small, the underdamped solution is approximately equal to the critically damped solution for a short time interval. Estimate the time (in terms of the relevant constants) when the two solutions can no longer be considered equivalent.

Problem 11.12 A 0.25 kg mass is attached to a spring of force constant 0.02 N/m. It is placed in a resistive medium and released from rest at a distance of 0.1 m from equilibrium. After 5 s, it is observed that the amplitude of the oscillations is 0.05 m. What is the damping constant b? What is the frequency of the oscillations?

Problem 11.13 A pendulum consists of a bob of mass 0.25 kg on a string of length 0.5 m. The force of air resistance varies with the first power of the speed of the bob and is given by $f_r = -bv$ with $b = 0.05$. Assume the pendulum is released at an initial angular displacement of $20°$, so the small angle approximation is appropriate.

(a) Use $N = I\alpha$ to obtain the equation of motion. Compare with Equation (11.6).
(b) Show that the motion is underdamped.
(c) Write an expression for $\theta(t)$ by comparing with the solution for a mass on a spring in Equation (11.15).
(d) Determine the period of the oscillations
(e) Determine the time required for the amplitude of the oscillations to decrease to one tenth of the initial value.

Problem 11.14 An undamped harmonic oscillator consisting of a mass and spring on a frictionless horizontal surface has a period of 1 s. Now assume the oscillator is placed on a surface with a small coefficient of sliding friction, such that the oscillator is critically damped. It is released from rest with initial displacement 0.3 m. Determine its position at time $t = 0.2$ s.

Problem 11.15 Determine the energy of a lightly damped harmonic oscillator. (You may assume $\omega_1 \doteq \omega_0$, where $\omega_0 = \sqrt{k/m}$.)

Problem 11.16 We defined the Q factor for a damped harmonic oscillator by $Q = \frac{\omega_0 m}{b}$. Obtain the expression for Q if we define it as the ratio of the maximum energy during a cycle to the energy dissipated *per radian*. Assume a lightly damped oscillator so that you can replace ω_1 with $\omega = \sqrt{k/m}$.

Problem 11.17 A long pendulum with a heavy bob and small-amplitude oscillations will swing back and forth for a long time before coming to rest. This means that the damping is small, that is, $\gamma \ll \omega_0$. Nevertheless, the pendulum is losing energy. Show that the rate of change of the energy can be expressed as

$$\frac{d}{dt}(\ln E) = -2\gamma.$$

(You can use the equations we obtained for a mass on a spring, because the mathematics is the same.)

Problem 11.18 An underdamped harmonic oscillator is subjected to a driving force given by $F_0 \cos(\omega_d t)$. Obtain an expression for the position as a function of time.

Problem 11.19 Consider a driven underdamped harmonic oscillator. Let the driving frequency ω_d be equal to ω_1, which is the frequency of the undriven oscillator and which is close to the resonance frequency ω'.
(a) Show that the ratio of the square of the maximum amplitude (the value of A^2 when $\omega_d = \omega'$) to the square of the amplitude at the nearby frequency is

$$\frac{A_{max}^2}{A^2} = \frac{4m^2(\omega_d - \omega')^2 + b^2}{b^2}.$$

(b) Show that the resonance half width is $\Delta = b/2m$. The resonance half width is the half width of the resonant peak $(\omega' - \omega_d)$ at a point where the amplitude squared has fallen to half of its maximum value.

Problem 11.20 Consider an underdamped harmonic oscillator subjected to a driving force $F_0 \sin(\omega t + \theta_0)$. Determine the rate at which work is done on the oscillator by the applied force, and evaluate the average power delivered by the force per cycle.

Problem 11.21 Determine the motion $x(t)$ for a critically damped harmonic oscillator subjected to an applied force given by $F_0 \cos \omega_d t$. You may assume that initially the mass m was displaced from the equilibrium point by a distance x_0 and that it had zero velocity.

Problem 11.22 Determine the motion for a driven damped harmonic oscillator which is acted upon by an exponentially increasing force given by $F_0(1 - e^{-at})$. The oscillator consists of a block of mass m on a horizontal surface and attached to a spring of constant $k = 4ma^2$. The damping coefficient is $b = ma$. Assume the oscillator is at rest at $t = 0$. Plot the displacement of the mass as a function of time.

Problem 11.23 Two blocks of mass m_1 and m_2 are connected by a spring with force constant k. The masses are on a frictionless horizontal surface. Therefore, the masses slide freely on the surface and oscillate back and forth relative to one another. Prove that the velocity of the center of mass of the system is constant. Determine the frequency of oscillation of the masses in terms of m_1, m_2, and k.

Problem 11.24 Figure 11.11 illustrates a system of two masses and three springs. The analysis in the text assumed that $m_1 = m_2$ and $k_1 = k_2$. Now you are asked to consider the situation when the masses are still the same, but the spring constants are different. Assume that $k_1 = 0.8k$ and $k_2 = 1.2k$, but k_3 is weaker and is given by $k_3 = 0.1k$. The initial conditions are that one of the masses is held at rest in its equilibrium position and the other is displaced from equilibrium by a distance d. Obtain expressions for $x_1(t)$ and $x_2(t)$. Plot these expressions and show they are in qualitative agreement with the sketch below (Figure 11.12).

Problem 11.25 Consider two masses coupled through a spring as in Figure 11.11. Prove the assertion that the low-frequency normal mode is generated if both masses are displaced an equal amount either to the right or the left.

Problem 11.26 For the system pictured in Figure 11.11, assume the two masses are equal but all three springs have different force constants. Find the characteristic frequencies.

Problem 11.27 A system consists of three masses connected by two springs, as shown in Figure 11.13 (This is called "coupling through a mass.") Set up the equations of motion for this system. The relaxed lengths of the two springs are l_1 and l_2. Separate the problem into two problems, one involving the motion of the center of mass and the other involving the motion of the two end masses relative to the central mass, as given by the two coordinates x_1 and x_2. Determine the normal modes for the system.

Problem 11.28 The equation of motion for the relative coordinate in the two-body problem is

$$\ddot{\mathbf{r}} = -\mu \frac{\mathbf{r}}{r^3}.$$

Show that by a change of variables from t to X, where $dt = r\,dX$, this equation can be reduced to the scalar equation of a shifted harmonic oscillator, namely,

$$r'' + \beta r = \mu,$$

where primes indicate differentiation with respect to X, and β is a constant equal to twice the negative of the total energy per unit mass. That is,

$$\beta = \frac{2\mu}{r} - v^2.$$

Note that $\dot{\mathbf{r}} = \frac{d\mathbf{r}}{dt}$ and $\mathbf{r}' = \frac{d\mathbf{r}}{dX}$, so $\mathbf{r}' = r\frac{d\mathbf{r}}{dt} = r\mathbf{v}$. Be careful not to confuse $\mathbf{r}$ with r.

Two differential equation problems now follow.

Problem 11.29 Obtain the general solution to $(D^2 + 1)x = 12\cos^2 t$. (Hint: Use the method of undetermined coefficients. You will have to look it up in your differential equations text.)

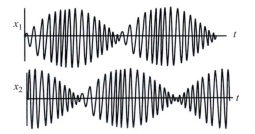

Figure 11.12 Illustration for Problem 11.24.

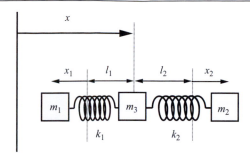

Figure 11.13 Coupling through a mass.

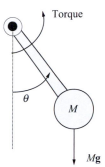

Figure 11.14 A forced physical pendulum.

Problem 11.30 Obtain the general solution to $(D^2+4)x = 4\sin^2 t$. (Hint: Use the method of undetermined coefficients.)

Computational Project

Computational Project 11.1 A physical pendulum is illustrated in Figure 11.14. Assuming that it is damped and driven by a force of frequency f, its equation of motion can be written in the (simplified) form (where T is related to torque)

$$\frac{d^2\theta}{dt^2} + b\frac{d\theta}{dt} + \sin\theta = T\sin(2\pi ft).$$

It will be easiest to solve this equation numerically if you introduce three new variables defined as follows:

$$x_1 = \frac{d\theta}{dt},$$
$$x_2 = \theta,$$
$$x_3 = 2\pi ft.$$

Then the equations of motion are

$$\frac{dx_1}{dt} = T\sin(x_3) - \sin(x_2) - bx_1,$$
$$\frac{dx_2}{dt} = x_1,$$
$$\frac{dx_3}{dt} = 2\pi f.$$

Write a computer program and plot θ (or x_2) as a function of time for $t \leq 250$. First assume $b = 0$ and $T = 0$ to be sure that your solution generates the oscillatory motion of a pendulum. Then set $b = 0.1$ and obtain the motion of a damped pendulum. Finally, let $T = 1$, and $b = 0.1$ and allow f to range from 0.001 to 0.1. Note that for $f = 0.1$ the motion is unpredictable. This is an example of chaos in a dynamical system.

12 The Pendulum

As you will soon appreciate, the physics of a pendulum is much more complex than you might imagine.[1]

The pendulum has been one of our prime examples of harmonic motion, but so far we have only considered pendulums whose motion is restricted to a vertical plane. Furthermore, we assumed that the amplitude of the swing was small so the sine of the angle could be replaced by the angle itself ($\sin \theta \approx \theta$). Finally, we assumed the pendulum was "ideal" and consisted of a point bob attached to a massless, inextensible string. In this chapter we generalize the pendulum problem in several different ways. The first generalization is to allow the pendulum swing to have an arbitrary amplitude. Next, we assume the pendulum is an extended body, not a suspended point mass. (In that case it is called a "physical" or "compound" pendulum.) Finally, we consider a pendulum whose bob is not restricted to swing in a plane but is free to trace out any motion on the surface of a sphere whose radius is equal to the length of the pendulum. (That is called a "spherical" pendulum.) If the bob of a spherical pendulum is restricted to move in a horizontal circle, it is called a "conical" pendulum (because the string sweeps out a cone). While studying these various types of pendulums you will be exposed to valuable mathematical and physical techniques; for example, you will find out how to use and evaluate elliptic integrals, as well as applying the method of successive approximations to the problem of the spherical pendulum. The Foucault pendulum, however, will be left until Chapter 15 when we consider motion in accelerated reference frames.

12.1 A Simple Pendulum with Arbitrary Amplitude

Let us begin by considering a simple (ideal) pendulum, just to get the necessary equations in their most basic form. Then I will remove the restriction of small-angle oscillations. (The subsequent analysis will lead to a consideration of two mathematical techniques, namely, the use of elliptic integrals and expansions in series.)

Figure 12.1 shows a simple pendulum made up of a point mass m and a massless inextensible string of length l. You have obtained the equation of motion for this system using Newton's second law – see Equation (4.1) – and also by the Lagrange technique (see Worked Example 4.3). In this chapter we shall be using the equations of rotational dynamics, so let us derive the equation of motion for the pendulum one more time, using the rotational form of Newton's second law:

$$\mathbf{N} = I\boldsymbol{\alpha}.$$

---

[1] Entire books have been written on pendulums. For example, G. L. Baken and J. A. Blackburn, *The Pendulum, A Case Study in Physics*, Oxford University Press, 2005, is a 300-page book on various aspects of pendulums.

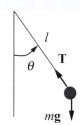

Figure 12.1 A simple pendulum. The forces on the bob are gravity (mg) and the tension (**T**) in the string.

Here **N** is the torque, I is the moment of inertia, see Equation (7.10), and $|\alpha| = \alpha = \ddot{\theta}$ is the angular acceleration. The torque is

$$N = r \times F = r \times (T + mg).$$

If **r** is measured from the point of support, $\mathbf{r} \times \mathbf{T} = 0$, leaving us with

$$N = r \times mg.$$

Since $|\mathbf{r}| = l$, we can write

$$N = |\mathbf{N}| = -lmg \sin\theta,$$

where the minus sign is due to the convention that torques producing clockwise motion are negative. It is also a reminder that this is a "restoring torque" that tends to make the angle θ smaller.

The moment of inertia is defined as $I = \sum m_i r_i^2$ so, for a simple pendulum,

$$I = ml^2.$$

Then $N = I\ddot{\theta}$ yields

$$-mgl \sin\theta = ml^2\ddot{\theta},$$

or

$$\ddot{\theta} + \frac{g}{l}\sin\theta = 0. \tag{12.1}$$

This seemingly simple equation turns out to be surprisingly difficult to solve unless we make the small angle approximation. If $\theta \lesssim 0.1\pi = 18°$, replacing $\sin\theta$ by θ introduces an error of less than 2 percent. So for small-amplitude oscillations, Equation (12.1) can be approximated by

$$\ddot{\theta} + \frac{g}{l}\theta = 0.$$

You recognize this as the equation for simple harmonic motion. As discussed in Chapter 11, the solution is

$$\theta = A\cos(\omega t + \beta), \tag{12.2}$$

where $\omega = \sqrt{g/l}$ = the angular frequency. A and β are constants that depend on the initial conditions. The period of a simple pendulum with small amplitude oscillations is $P = 2\pi/\omega = 2\pi\sqrt{l/g}$.

Now I will remove the small-angle restriction and determine the motion of a pendulum with arbitrary amplitude. Instead of integrating the equation of motion twice, I will use the energy integral described in Section 5.7. This is convenient because it introduces a "first integral" or constant of the motion, namely, the total energy E.

The total energy of a pendulum is constant (as long as there is no damping). The kinetic energy of an oscillating pendulum is $T = \frac{1}{2}I\omega^2$ or

$$T = \frac{1}{2}ml^2\dot{\theta}^2.$$

The potential energy is

$$V = -mgl\cos\theta.$$

Consequently, the total energy is

$$E = T + V = \frac{1}{2}ml^2\dot{\theta}^2 - mgl\cos\theta.$$

Solving for $\dot{\theta}$ yields

$$\dot{\theta} = \frac{d\theta}{dt} = \sqrt{\frac{2g}{l}\left(\frac{E}{mgl} + \cos\theta\right)}.$$

This is the differential equation we need to solve. The first step is to separate variables. Using $\sqrt{g/l} = \omega$, and rearranging,

$$\frac{d\theta}{\sqrt{\left(\frac{E}{mgl} + \cos\theta\right)}} = \sqrt{2}\omega dt.$$

Integrating,

$$\int_{\theta_0=0}^{\theta(t)} \frac{d\theta}{\sqrt{\left(\frac{E}{mgl} + \cos\theta\right)}} = \sqrt{2}\,\omega \int_0^t dt,$$

or

$$t = \frac{1}{\sqrt{2}\omega} \int_0^\theta \frac{d\theta}{\sqrt{\left(\frac{E}{mgl} + \cos\theta\right)}}, \tag{12.3}$$

where the lower limit, $\theta_0 = 0$, means the bob was at the bottom of the swing when the timer was started.

12.1.1 Solution in Terms of Elliptic Integrals

It may surprise you to learn that the integral in Equation (12.3) is not in your table of integrals. However, it can be put into the form of a function called an *elliptic integral*. Elliptic integrals are tabulated in all good books of mathematical tables and are included in the libraries of many computer languages. A complete elliptic integral of the first kind has the form

$$\int_0^{\pi/2} \frac{d\phi}{\sqrt{1 - k^2 \sin^2\phi}} \tag{12.4}$$

and is a function of k. Writing the integral in Equation (12.3) in the form of Equation (12.4) yields an expression for the period of a pendulum of arbitrary amplitude in terms of an elliptic integral. This may sound easy, but the algebra gets a bit complicated.

For convenience, assume the pendulum is not "going over the top" so it has a maximum angular displacement, θ_m. That is, at the top of the swing, $\theta = \theta_m$. But at the top of the swing, the kinetic energy is zero, so the total energy is

$$E = -mgl \cos \theta_m.$$

Since the total energy is constant, we can replace E/mgl in Equation (12.3) by $-\cos \theta_m$. Thus,

$$t = \frac{1}{\sqrt{2\omega}} \int_0^{\theta(t)} \frac{d\theta}{\sqrt{\cos \theta - \cos \theta_m}}.$$

The period (P) of this pendulum is the time required for the bob to go through four quarter cycles where a quarter cycle is a swing from $\theta = 0$ to $\theta = \theta_m$. Therefore,

$$P = 4 \left[\frac{1}{\sqrt{2\omega}} \int_0^{\theta_m} \frac{d\theta}{\sqrt{\cos \theta - \cos \theta_m}} \right]. \tag{12.5}$$

This is *still* not in the form of the elliptic integral of Equation (12.4). To put Equation (12.5) into the required form recall the trigonometric identity

$$\cos \theta = 1 - 2 \sin^2 \tfrac{\theta}{2}.$$

Consequently,

$$\sqrt{\cos \theta - \cos \theta_m} = \left[\left(1 - 2 \sin^2 \tfrac{\theta}{2} \right) - \left(1 - 2 \sin^2 \tfrac{\theta_m}{2} \right) \right]^{1/2},$$

$$= \left[2 \sin^2 \tfrac{\theta_m}{2} - 2 \sin^2 \tfrac{\theta}{2} \right]^{1/2},$$

$$= \left[2 \sin^2 \tfrac{\theta_m}{2} \right]^{1/2} \left[1 - \frac{\sin^2 \tfrac{\theta}{2}}{\sin^2 \tfrac{\theta_m}{2}} \right]^{1/2}.$$

For convenience, let $k = \sin \tfrac{\theta_m}{2}$ and write

$$\sqrt{\cos \theta - \cos \theta_m} = k\sqrt{2} \sqrt{1 - \frac{1}{k^2} \sin^2 \tfrac{\theta}{2}}.$$

Therefore, Equation (12.5) becomes

$$P = \frac{4}{\sqrt{2\omega}} \int_0^{\theta_m} \frac{d\theta}{\sqrt{2}k\sqrt{1 - \frac{1}{k^2} \sin^2 \tfrac{\theta}{2}}}. \tag{12.6}$$

Even this is not in the right form, but it is getting close. Define a new angle ϕ by the relation

$$\sin \phi = \frac{1}{k} \sin \frac{\theta}{2}.$$

When $\theta = \theta_m$, the value of ϕ is $\pi/2$. Now,

$$d(\sin \phi) = d \left(\frac{1}{k} \sin \frac{\theta}{2} \right) = \frac{1}{k} \cos \frac{\theta}{2} d \left(\frac{\theta}{2} \right) = \frac{1}{2k} \cos \frac{\theta}{2} d\theta.$$

Since $d(\sin \phi) = \cos \phi d\phi$ we can write

$$d\theta = \frac{2k \cos \phi d\phi}{\cos \tfrac{\theta}{2}} = \frac{2k \cos \phi d\phi}{\sqrt{1 - \sin^2 \tfrac{\theta}{2}}} = \frac{2k \cos \phi d\phi}{\sqrt{1 - k^2 \sin^2 \phi}}.$$

Plugging into Equation (12.6) we obtain

$$P = \frac{4}{\sqrt{2}\,\omega} \int_{\phi=0}^{\phi=\pi/2} \frac{2k \cos\phi\, d\phi}{\sqrt{1 - k^2 \sin^2\phi}} \frac{1}{\sqrt{2}k\sqrt{1 - \sin^2\phi}},$$

$$= \frac{8}{2\omega} \int_0^{\pi/2} \frac{\cos\phi\, d\phi}{\sqrt{1 - k^2 \sin^2\phi}} \frac{1}{\cos\phi}.$$

Hence,

$$P = \frac{4}{\omega} \int_0^{\pi/2} \frac{d\phi}{\sqrt{1 - k^2 \sin^2\phi}}, \tag{12.7}$$

which has the desired form.

This somewhat tiresome derivation yielded the expression for the period of a pendulum of arbitrary amplitude. The period is, however, expressed in terms of an elliptic integral. Let me remind you that an elliptic integral is a function like the sine or cosine in the sense that its value can be looked up in any standard set of math tables. (I do not know if any handheld calculators have elliptical integrals stored in them.)

Elliptic integrals come in various different "flavors," the primary ones being the following.

(1) Elliptic integrals of the first kind:

$$F(k, \theta) = \int_0^\theta \frac{d\phi}{\sqrt{1 - k^2 \sin^2\phi}}. \tag{12.8}$$

(2) Elliptic integrals of the second kind:

$$E(k, \theta) = \int_0^\theta \sqrt{1 - k^2 \sin^2\phi}\, d\phi. \tag{12.9}$$

(3) Elliptic integrals of the third kind:

$$\Pi(n, k, \theta) = \int_0^\theta \frac{d\phi}{(1 - n \sin^2\phi)\sqrt{1 - k^2 \sin^2\phi}}. \tag{12.10}$$

(4) Complete elliptic integrals: These are the values of F and E when $\theta = \pi/2$. Thus, for example, the *complete* elliptical integral of the first kind is:

$$F\left(k, \frac{\pi}{2}\right) = F(k) = \int_0^{\pi/2} \frac{d\phi}{\sqrt{1 - k^2 \sin^2\phi}}.$$

Equation (12.7) tells us that the period of a pendulum is given by a complete elliptic integral of the first kind.

Worked Example 12.1 A certain pendulum has a small amplitude period of two seconds. Determine its period when it swings to a maximum angle of $80°$.

Solution The angular frequency for small amplitude oscillations is $\omega = 2\pi/P = 2\pi/2 = \pi = 3.14$ rad/s. The quantity k is

$$k = \sin\frac{\theta_m}{2} = \sin 40°.$$

Looking in a table of complete elliptic integrals of the first kind we find

$\sin^{-1} k$	$F(k, \pi/2)$
$0°$	1.5708
$\vdots$	$\vdots$
$39°$	1.7748
$40°$	1.7868
$41°$	1.7992.

Therefore, the period when the amplitude is $80°$ is

$$P = \frac{4}{\pi}(1.7868) = 2.28\text{s}.$$

Exercise 12.1

Use a table to determine the values of (a) the elliptic integral of the first kind for $k = 0.5$ $\phi = 30°$, (b) the elliptic integral of the second kind for $k = 0.5$ $\phi = 30°$, and (c) the complete elliptic integral of the second kind for $k = 0.5$. (The purpose of this exercise is to get you to look at a table of elliptic integrals.) ∎

12.1.2 Solution Expressed as a Series Expansion

If you do not have a book with tabulated values of the elliptic integrals and your computer is not available, you can still solve Equation (12.7) the "old-fashioned way" by expanding the integrand. This is a particularly useful mathematical procedure. It works in this case because both k and $\sin \phi$ are less than unity.

Let

$$k^2 \sin^2 \phi = x.$$

Then

$$P = \frac{4}{\omega} \int_0^{\pi/2} \frac{d\phi}{\sqrt{1-x}} = \frac{4}{\omega} \int_0^{\pi/2} d\phi [1 - x]^{-1/2}.$$

You can expand the term in brackets using the binomial expansion because $x < 1$. Keeping the first few terms,

$$P = \frac{4}{\omega} \int_0^{\pi/2} d\phi \left[1 + \frac{1}{2}x + \frac{3}{8}x^2 + \cdots \right] \tag{12.11}$$

or

$$P \cong \frac{4}{\omega} \int_0^{\pi/2} d\phi \left[1 + \frac{1}{2}k^2 \sin^2 \phi + \frac{3}{8}k^4 \sin^4 \phi \right]$$

$$= \frac{4}{\omega} \left\{ \int_0^{\pi/2} d\phi + \frac{1}{2}k^2 \int_0^{\pi/2} \sin^2 \phi d\phi + \frac{3}{8}k^4 \int_0^{\pi/2} \sin^4 \phi d\phi \right\}.$$

The integrals over ϕ are easily evaluated, yielding

$$P = \frac{4}{\omega} \left[\frac{\pi}{2} + \frac{\pi}{4} \frac{k^2}{2} + \cdots \right] = 4\pi \sqrt{\frac{l}{g}} \left(\frac{1}{2} + \frac{k^2}{8} + \cdots \right).$$

So,

$$P \cong 2\pi \sqrt{l/g}(1 + k^2/4) = 2\pi \sqrt{l/g} \left(1 + \frac{1}{4}\sin^2 \frac{\theta_m}{2}\right).$$

Note that the first term in this expression is the period of a small amplitude ideal pendulum.

Exercise 12.2

Use the binomial expansion to carry out Equation (12.11) to one more term.   ■

Exercise 12.3

A simple pendulum has a length of 1.5 m. It swings through an arc with maximum angular displacement of 40°. Determine its period. Answer: 2.53 s.   ■

Exercise 12.4

A simple pendulum swings through an angle of 45 degrees ($\theta_m = 45°$). Show that the period is about 4 percent greater than that of small-angle oscillations.   ■

12.2 The Physical Pendulum

We have considered the motion of an ideal pendulum with a point-mass bob suspended by an inextensible, massless string. This is a good approximation for some pendulums, but everyone is familiar with oscillating objects that are not at all similar to an ideal pendulum. For example, a large chandelier or a hanging sign may undergo regular oscillations. Such objects are called "physical" or "compound" pendulums.

A compound pendulum is an extended body constrained to rotate about a fixed axis that passes through some point in the body, as illustrated in Figure 12.2. Let the point of support (through which the axis passes) be denoted by O and let G indicate the position of the center of mass. The angle between the line $\overline{OG}$ and the vertical is θ.

A rigid body is a collection of particles. The total torque on the body is, therefore,

$$\mathbf{N} = \sum (\mathbf{r}_i \times \mathbf{F}_i),$$

where $\mathbf{N}$ is the torque about O, $\mathbf{r}_i$ is the vector from O to the ith particle (which has mass m_i), and $\mathbf{F}_i = m_i \mathbf{g}$ is the gravitational force on the ith particle. Consequently,

$$\mathbf{N} = \sum \mathbf{r}_i \times (m_i \mathbf{g}) = \left(\sum m_i \mathbf{r}_i\right) \times \mathbf{g}.$$

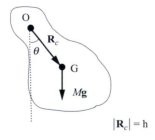

$|\mathbf{R}_c| = h$

Figure 12.2 A physical pendulum. The axis of rotation is perpendicular to the page and passes through point O. The center of mass is denoted G. The distance from the axis of rotation to the center of mass is denoted h. That is, $h = |\mathbf{R}_c|$.

By definition, the center of mass (G) is at $\mathbf{R}_c = \sum m_i \mathbf{r}_i / M$, where M is the total mass. Therefore, the torque is

$$\mathbf{N} = \mathbf{R}_c \times M\mathbf{g}.$$

This expression shows that the torque on the body is the same as if a force $M\mathbf{g}$ were acting at the center of mass of the body. That is, as far as the torque is concerned, you can assume all the mass is concentrated at the center of mass.

In Figure 12.2 the torque is a vector pointing into the page and it produces a *clockwise* rotation of the body. By convention, this is a negative torque, so the magnitude of $\mathbf{N}$ is

$$N = -Mg \, |\mathbf{R}_c| \sin \theta = -Mgh \sin \theta,$$

where $h = |\mathbf{R}_c|$ is the distance from the axis of rotation to the center of mass.

When considering the motion of rigid bodies it is convenient to define a quantity called the "radius of gyration," denoted by k. The radius of gyration is defined in terms of the moment of inertia. Recall that the dimensions of the moment of inertia are mass multiplied by the square of a distance. If I is the moment of inertia of a body relative to some given axis, and M is the mass of the body, then the radius of gyration is the distance k such that

$$I = Mk^2.$$

Although the radius of gyration does not have a geometrical meaning, it is sometimes helpful to think of it as the distance from the axis of rotation to the point where all the mass of the body would have to be concentrated to have the same moment of inertia as the actual body. The radius of gyration (k) and the distance to the center of mass (h) are generally not equal to one another.

The moment of inertia of the physical pendulum of Figure 12.2 with respect to the axis through O can be written

$$I_O = Mk_O^2.$$

Using $I\ddot{\theta} = N$, the equation of motion of the physical pendulum is

$$Mk_O^2 \ddot{\theta} = -Mgh \sin \theta.$$

Consequently,

$$\ddot{\theta} = -\frac{gh}{k_O^2} \sin \theta.$$

Recall that the equation of motion of a simple pendulum is

$$\ddot{\theta} = -\frac{g}{l} \sin \theta.$$

Comparing the two equations you see that the compound pendulum oscillates with the same period as a simple pendulum of length

$$l = \frac{k_O^2}{h}. \tag{12.12}$$

That is, a simple pendulum of length k_O^2/h will oscillate "in time" with the compound pendulum. Figure 12.3 illustrates such a pendulum. The point of suspension is O and the bob is at O$'$ (a distance $l = k_O^2/h$ from O). The point O$'$ is called the "center of oscillation" of the compound pendulum. For small oscillations, the period of a simple pendulum is $P = 2\pi \sqrt{l/g}$, and the period of a compound pendulum is

$$P = 2\pi \sqrt{k_O^2/hg}.$$

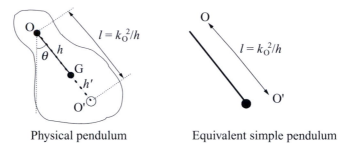

Physical pendulum   Equivalent simple pendulum

Figure 12.3 The location of the bob of an equivalent simple pendulum would be at O'. This point is called the "center of oscillation." The distance from the center of mass (G) to the center of oscillation is denoted h'.

Worked Example 12.2 A uniform hoop of radius a and mass m is free to oscillate about an axis perpendicular to the plane of the hoop and passing through the rim. (Given: The moment of inertia of the hoop about an axis through its center and perpendicular to the plane of the hoop is ma^2.) (a) Determine its period, assuming small-amplitude oscillations. (b) Locate the center of oscillation for this compound pendulum.

Solution For the hoop, the distance from the rim to the center of mass is the radius, so $h = a$:

$$P = 2\pi\sqrt{k^2/hg}.$$

Now $k = \sqrt{I/m}$ and by the parallel axis theorem the moment of inertia about a point on the rim will be

$$I = I_{CM} + MR^2 = ma^2 + ma^2 = 2ma^2$$

$$k^2 = 2a^2.$$

So

$$P = 2\pi\sqrt{2a^2/ag} = 2\pi\sqrt{2a/g}.$$

(b) The center of oscillation is a distance k^2/h from the point of suspension, that is, at a distance $2a^2/a = 2a$.

If we were to turn our physical pendulum "upside down" and hang it from an axis through O', it would oscillate with the same period as the "right-side up" pendulum. See Figure 12.3. In other words, a pendulum suspended at O' has its center of oscillation at O. This is proved in Worked Example 12.3.

Worked Example 12.3 Show that a pendulum suspended at O' has the same period as a pendulum suspended at O.

Solution From Figure 12.3 we note that $l = h + h'$, where h' is the distance from the center of oscillation (O') to the center of mass (G). From Equation (12.12)

$$k_0^2 = h^2 + hh'. \tag{12.13}$$

Now the parallel axis theorem states that $I_0 = I_G + Mh^2$ or $Mk_0^2 = Mk_G^2 + Mh^2$, so

$$k_0^2 = k_G^2 + h^2. \tag{12.14}$$

Inserting k_0^2 from Equation (12.13) into Equation (12.14) we obtain

$$k_G^2 = hh'. \tag{12.15}$$

Now hang the pendulum from O'. The center of oscillation of this upside-down pendulum will be at a distance l' from O' where

$$l' = k_{O'}^2/h' = \frac{1}{h}(h'^2 + hh') = h' + h = l.$$

Therefore, a pendulum suspended at O' has its center of oscillation at O. Consequently it will have the same period as a pendulum suspended at O.

Exercise 12.5

Determine the radius of gyration of a disk of mass M and radius R (a) relative to a perpendicular axis through its center, and (b) relative to a parallel axis tangent to the edge of the disk. Answers: (a) $R/\sqrt{2}$, (b) $R\sqrt{3/2}$. ∎

Exercise 12.6

A uniform rod of mass M and length L is suspended from one end. Determine the radius of gyration (k). Answer: $k = L/\sqrt{3}$. ∎

Exercise 12.7

A thin uniform rod of length L and mass M oscillates about an axis through one end. Determine the period of small oscillations. Answer: $(8\pi^2 L/3g)^{1/2}$. ∎

12.3 The Center of Percussion

If a rigid body is struck at an arbitrary point it will move. In general, this motion will consist of both a rotation and a translation. But it turns out that there is a point in the body where the translational and rotational motions cancel out so this point is *instantaneously* at rest. The location of this point depends on the location of the point where the force is applied. The point where the impulse is received is called the "center of percussion." In everyday terms the center of percussion is referred to as the "sweet spot" of a tennis racket or baseball bat and the place where you are holding the bat is the point where there is no acceleration of the bat, and no reaction force. If you hit the ball at any other point on the racket or bat, you will feel a stinging in your hands due to the reaction force. In other words, an impulse delivered at the sweet spot does not cause an impulsive reaction at the instantaneous axis of rotation (which is where you hold the racket or bat).[2]

---

[2] If you want to be very picky about it, the stinging in your hands is due to the vibration of the bat, not the reaction force. A purist defines the sweet spot as the location where the impulse does not cause the bat to vibrate. It turns out that the sweet spot and the center of percussion are very close to one another.

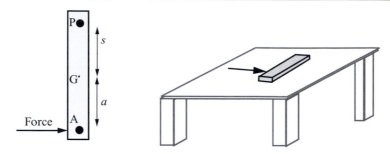

Figure 12.4 A rod on a smooth table is subjected to an impulsive force at point A. The distance between the center of mass (G) and A is a. The point P is a distance s from G. A is the center of percussion and P is the instantaneous axis of rotation.

To begin the analysis, consider a rod of mass M that receives a blow at point A. See Figure 12.4.

First assume the rod is not constrained by an axis of rotation. For example, it might be lying on a perfectly smooth table. What is the motion of the rod after receiving a blow at point A? This can be determined in a straightforward way by an application of Newton's laws.

If the force on the rod is an impulsive blow, the impulse $\mathbf{J}$ is given by

$$\mathbf{J} = \int \mathbf{F} dt.$$

But $\mathbf{F} = \frac{d\mathbf{p}}{dt}$ so

$$\mathbf{J} = \int \frac{d\mathbf{p}}{dt} dt = \int d\mathbf{p} = \mathbf{p}_f - \mathbf{p}_i.$$

Assuming the initial momentum is zero and the final momentum is Mv, the velocity of the rod immediately after the blow is

$$\mathbf{v} = \frac{\mathbf{J}}{M}.$$

To be a bit more explicit, this is the linear velocity of the *center of mass* of the rod immediately after the blow.

We now calculate the angular velocity about the center of mass.

It is clear from the physical situation that the rod will begin to rotate. The rate of change of its angular momentum is

$$\frac{d\mathbf{L}}{dt} = \mathbf{N}.$$

The moment of inertia about the center of mass is $I = Mk_G^2$. The magnitude of the torque on the rod (taken about the center of mass) due to the impulsive force $\mathbf{F}$ is

$$N = |\mathbf{r} \times \mathbf{F}| = aF,$$

where a is the distance from the center of mass to the line of action of the force. Recalling that $L = I\omega$, we have

$$aF = \frac{d}{dt}(I\omega).$$

Therefore,

$$\int a F\, dt = \int d(I\omega),$$

$$\therefore a J = I\omega.$$

Thus the angular velocity about the center of mass, immediately after the impulse, will be

$$\omega = \frac{aJ}{I}.$$

The motion of the rod is rather complicated. The velocity of a point on the rod is the linear velocity of the center of mass (given by J/M), plus the angular velocity *about* the center of mass (given by $\boldsymbol{\omega} \times \mathbf{r}$ where $\mathbf{r}$ is the vector from the center of mass). These two velocities must be added vectorially. For example, consider the motion of point P, a distance s from the center of mass, as shown in Figure 12.4. Immediately after the impulse, the velocity of this point is

$$\mathbf{v_P} = \frac{\mathbf{J}}{M} + \boldsymbol{\omega} \times \mathbf{s}. \tag{12.16}$$

The right-hand rule and Figure 12.4 tell us that $\mathbf{J}$ is directed toward the right and $\boldsymbol{\omega} \times \mathbf{s}$ is directed toward the left. The magnitude of the velocity of point P is, consequently,

$$v_P = \frac{J}{M} - \frac{aJ}{Mk_G^2} s$$

$$= \frac{J}{M}\left(1 - \frac{as}{k_G^2}\right).$$

Note that v_P will be zero if

$$\frac{as}{k_G^2} = 1.$$

We conclude that immediately after the impulse there is a point on the rod having zero velocity. Specifically, immediately after the impulse, a point P located at

$$s = \frac{k_G^2}{a} \tag{12.17}$$

will have zero velocity. Point A, where the impulse is delivered, is the *center of percussion* and point P is the instantaneous axis of rotation.

In the notation of the previous section, the distances a and s are h' and h, respectively. Inserting these quantities into Equation (12.17) gives

$$hh' = k_G^2,$$

which is just Equation (12.15).

As time goes on, the impulsive force is no longer acting and the linear and angular velocities will be constant. It is instructive to plot the positions of a few points in the rod as a function of time to convince yourself that the rod will rotate about its center of mass at a constant angular velocity, while the center of mass itself moves at a constant linear velocity.

Now consider the possibility that the rod in Figure 12.4 is *pivoted* at point P. In this case, if the rod is given a blow at point A there will be no impulsive *reaction* at the pivot. To appreciate this, assume the rod is pivoted at a distance s from the center of mass.

In general, the struck rod will tend to translate, but a pivot at P will prevent this. Point P remains at rest at all times. But if point P is to remain at rest, the net force on that point must be zero.

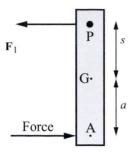

Figure 12.5 A rod pivoted at point P. The applied impulsive force is **F**. The reaction force of the pivot on the rod is $\mathbf{F}_1$.

The forces acting on P are the force due to the blow at A and the reaction force due to the pivot. Let the force exerted by the pivot be $\mathbf{F}_1$, as indicated in Figure 12.5.

The two forces (**F** and $\mathbf{F_1}$) acting on the rod, yield two simultaneous impulses, **J** and $\mathbf{J}_1$. The velocity of the center of mass after the blow will be

$$v = \frac{J - J_1}{M}.$$

Because the rod is constrained to rotate about the pivot located a distance s from the center of mass, the angular velocity of the rod is $\omega = v/s$ or

$$\omega = \frac{1}{s}\frac{J - J_1}{M}.$$

After the blow, the angular momentum is

$$L = I_p\omega,$$

where I_p is the moment of inertia about point P. But L is also given by

$$L = \int N dt = \int (s + a)F dt = (s + a)J.$$

Equating the two expressions for L, the angular velocity can be written

$$\omega = \frac{J(a + s)}{I_p}.$$

Set the two expressions for ω equal to one another to obtain

$$\frac{1}{s}\frac{J - J_1}{M} = \frac{J(a + s)}{I_p}.$$

Therefore, the reactive impulse is

$$J_1 = J\left[1 - \frac{Ms(a + s)}{I_p}\right].$$

Finally, according to the parallel axis theorem,

$$I_p = Mk_G^2 + Ms^2,$$

so

$$J_1 = J\frac{k_G^2 - as}{k_G^2 + s^2}.$$

J_1 will be zero if $s = k_G^2/a$. If an extended body is struck by an impulsive force at A, there will be zero reaction force at point P. If you are holding an extended body that happens to be a baseball bat and a ball hits it at point A, you should hope that your hands are gripping the bat at point P because otherwise you will have to exert a reaction force on the bat, and you will end up with red, stinging palms.

Worked Example 12.4 A billiard ball strikes the edge of a pool table and rebounds. It is desired that the angle of incidence be equal to the angle of reflection. To obtain this result, it is important that the ball should not start to slide on the table. The question is: What is the correct height (d) for the "bumper" on the edge of the table in Figure 12.6?

Solution Because the ball does not slide, the force at P must be zero. This means that point A should be the center of percussion relative to P. Therefore, by Equation (12.17)

$$ar = k_G^2.$$

For a sphere, $I = (2/5)Mr^2$, so $k_G^2 = (2/5)r^2$. Consequently,

$$ar = (2/5)r^2,$$

and

$$d = a + r = (7/5)r.$$

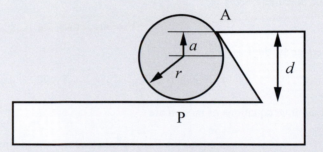

Figure 12.6 A billiard ball of radius r strikes the edge of the table at A.

12.4 The Spherical Pendulum

A spherical pendulum consists of a point mass m suspended by a string (or perhaps a massless rigid rod) of length l. The bob is constrained to remain a constant distance from the point of support, but otherwise it is free to move anywhere on the surface of a sphere of radius l.

To analyze the spherical pendulum, we shall determine the Lagrangian for the system, obtain the Lagrange equations and integrate them to determine the motion of the bob.

The Cartesian and spherical coordinates for a spherical pendulum are shown in Figure 12.7. Observe that θ is measured from the positive z-axis, so when the pendulum is at rest and the string is hanging straight down, the value of θ is π. The relations between the Cartesian and the spherical coordinates are

$$x = l \sin\theta \cos\phi,$$
$$y = l \sin\theta \sin\phi,$$
$$z = l \cos\theta.$$

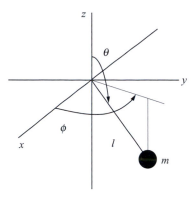

Figure 12.7 A spherical pendulum. The length of the string is l. The position of the bob is given by (l, θ, ϕ).

The kinetic energy of the bob is

$$
\begin{aligned}
T &= \frac{1}{2}m(\dot{x}^2 + \dot{y}^2 + \dot{z}^2), \\
&= \frac{1}{2}ml^2(\dot{\theta}^2 + \sin^2\theta\,\dot{\phi}^2),
\end{aligned}
\tag{12.18}
$$

and the potential energy is

$$
V = mgl\cos\theta.
$$

Consequently, the Lagrangian is

$$
L = T - V = \frac{1}{2}ml^2(\dot{\theta}^2 + \sin^2\theta\,\dot{\phi}^2) - mgl\cos\theta.
\tag{12.19}
$$

The Lagrange equations of motion are

$$
\frac{d}{dt}\left(ml^2\dot{\theta}\right) - ml^2\dot{\phi}^2\sin\theta\cos\theta - mgl\sin\theta = 0,
\tag{12.20}
$$

$$
\frac{d}{dt}\left(ml^2\sin^2\theta\,\dot{\phi}\right) = 0.
\tag{12.21}
$$

The second of these equations can be integrated immediately to give

$$
ml^2\sin^2\theta\,\dot{\phi} = \text{constant} = p_\phi.
\tag{12.22}
$$

The constant is denoted p_ϕ because it is the generalized (angular) momentum associated with the (ignorable) coordinate ϕ.

It is convenient to introduce the energy of the pendulum. Adding the kinetic and potential energies we obtain

$$
E = \frac{1}{2}ml^2(\dot{\theta}^2 + \sin^2\theta\,\dot{\phi}^2) + mgl\cos\theta.
$$

Using the expression for p_ϕ to eliminate $\dot{\phi}$ leads to

$$
E = \frac{1}{2}ml^2\dot{\theta}^2 + \frac{p_\phi^2}{2ml^2\sin^2\theta} + mgl\cos\theta.
\tag{12.23}
$$

The form of this expression suggests introducing an "effective potential" V_{eff}. (Recall the use of the effective potential for the central force problem – see Section 10.5.) For the spherical pendulum, the term $(1/2)ml^2\dot{\theta}^2$ is a kinetic energy term depending on the velocity, but the

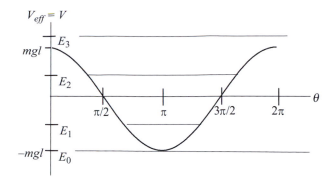

Figure 12.8 Effective potential (equal to the potential energy) for a simple planar pendulum. Note that $p_\phi = 0$.

following two terms depend only on coordinates and not velocities so they "look like" potential energies. Therefore, the effective potential is defined as

$$V_{eff} = \frac{p_\phi^2}{2ml^2 \sin^2 \theta} + mgl \cos \theta, \tag{12.24}$$

and Equation (12.23) can be written as:

$$\frac{1}{2}ml^2\dot{\theta}^2 = E - V_{eff}. \tag{12.25}$$

The effective potential depends on the value of p_ϕ. This parameter characterizes the motion of the pendulum bob with respect to the azimuthal angle ϕ.

If $p_\phi = 0$, the spherical pendulum reduces to a simple planar pendulum and the effective potential reduces to the potential energy (that is, $V_{eff} = mgl \cos \theta$). The effective potential (or potential energy) for this special case is plotted in Figure 12.8.

We now consider the motion[3] of a simple planar pendulum for different possible values of the total energy $E = T + V$. Note that T is always positive, and V varies between $\pm mgl$. The smallest energy occurs when the kinetic energy is zero and $V = -mgl$. As illustrated in Figure 12.8 this corresponds to $E = E_0 = -mgl$ and represents a pendulum which is not moving at all and is hanging straight down ($\theta = \pi$). For larger values of the total energy, such as $E = E_1 < 0$ in the figure, the pendulum is trapped in a potential well and oscillates back and forth around $\theta = \pi$. The angular amplitude of these oscillations is less than $\pi/2$; the pendulum does not rise above the $z = 0$ plane. For larger values of total energy such as E_2 the angular amplitude of the oscillations exceeds 90°. The bob rises above the $z = 0$ level on each swing. Finally, for energies such as $E_3 > +mgl$, the pendulum actually "goes over the top" and the angle increases (or decreases) continuously.

A more interesting situation arises when $p_\phi \neq 0$. The pendulum now moves azimuthally. It may or may not have motion in the θ direction.[4]

The effective potential changes drastically for $p_\phi \neq 0$ because the first term in Equation (12.24) is no longer zero. V_{eff} is plotted in Figure 12.9 for various constant nonzero values of p_ϕ. The effective potential rises to infinity at $\theta = 0$ and at $\theta = \pi$, indicating that any motion

---

[3] We are ignoring the loss of energy during the oscillations of the pendulum.

[4] If θ is constant, the pendulum bob traces out a horizontal circle of constant radius. This special case is the conical pendulum.

of the pendulum involves angles between these limiting values. However, as we shall see, the motion is actually limited to values of θ between π and $\pi/2$.

The effective potential reaches a minimum at a point whose location depends on p_ϕ. This minimum gets closer and closer to $\pi/2$ as p_ϕ increases. For a given value of p_ϕ the pendulum swings in a circle about the vertical axis, while undergoing oscillations in θ. The simplest motion of this type occurs when θ is constant and equal to its value at the minimum in V_{eff}. This is, of course, a conical pendulum. To treat the general case of the spherical pendulum, it is convenient to start the analysis with the conical pendulum.

Exercise 12.8

Fill in the missing steps to obtain Equations (12.18), (12.20), and (12.21).   ∎

Exercise 12.9

For the spherical pendulum, show that $ml^2 \sin^2 \theta \dot{\phi}$ is the generalized momentum conjugate to ϕ.   ∎

12.4.1 The Conical Pendulum

The conical pendulum is a special case of the spherical pendulum in which $p_\phi \neq 0$ and $\theta =$ constant. The bob moves in a horizontal circular path.

For a given value of p_ϕ and a constant θ, the energy of the conical pendulum will correspond to the lowest value of V_{eff}. See Equation (12.24) and Figure 12.9. (For higher energies, θ will oscillate between two values, but this is no longer a conical pendulum.) As the pendulum bob moves faster and faster, the minimum in V_{eff} is closer and closer to $\theta = \pi/2$ as can be appreciated by observing that in Figure 12.9 the minimum in V_{eff} moves towards $\pi/2$ as p_ϕ increases. That is, as the value of p_ϕ increases, the pendulum swings nearer and nearer to the horizontal plane.

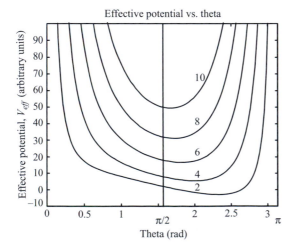

Figure 12.9 Effective potential vs. θ for various values of p_ϕ. For small values of p_ϕ (such as the $p_\phi = 2$ curve) the θ value oscillates around an angle of about 2.5 rad (nearly vertical). For large values of p_ϕ (as for $p_\phi = 10$) the value of θ oscillates around an angle of about 1.6 rad. (The string is nearly horizontal.)

Denoting the polar angle of the conical pendulum by θ_0 we can determine the relationship between p_ϕ and θ_0 by noting that the effective potential has a minimum at θ_0. Therefore,

$$\left[\frac{dV_{eff}}{d\theta}\right]_{\theta_0} = \left[\frac{d}{d\theta}\left(\frac{p_\phi^2}{2ml^2\sin^2\theta} + mgl\cos\theta\right)\right]_{\theta_0} = 0, \tag{12.26}$$

$$0 = \left[\frac{p_\phi^2}{2ml^2}\frac{d}{d\theta}\sin^{-2}\theta - mgl\sin\theta\right]_{\theta_0},$$

$$0 = -\frac{p_\phi^2\cos\theta_0}{ml^2\sin^3\theta_0} - mgl\sin\theta_0. \tag{12.27}$$

Consequently, at the minimum of V_{eff},

$$p_\phi^2 = -m^2l^3g\sin^3\theta_0\tan\theta_0. \tag{12.28}$$

As the bob moves faster and faster around in a horizontal circle, p_ϕ increases from zero to a large value and the angle θ varies from π (straight down) to nearly $\pi/2$ (horizontal) because $\tan\theta_0$ varies from 0 to $-\infty$. Therefore, the pendulum bob is always below the horizontal ($z = 0$) plane. Furthermore, squaring the definition of p_ϕ in Equation (12.22) and inserting it into Equation (12.28), leads to

$$\dot\phi = \sqrt{-g/(l\cos\theta_0)}.$$

The negative sign does not give an imaginary quantity because θ_0 lies between π and $\pi/2$. Let me repeat that as the azimuthal velocity $\dot\phi$ increases, the value of θ_0 gets closer and closer to $\pi/2$, in agreement with your childhood experiments with conical pendulums.

The energy of a conical pendulum is given by Equation (12.23) with $\dot\theta = 0$, that is,

$$E_{cp} = \frac{p_\phi^2}{2ml^2\sin^2\theta} + mgl\cos\theta, \tag{12.29}$$

$$= \frac{(ml^2\sin^2\theta\dot\phi)^2}{2ml^2\sin^2\theta} + mgl\cos\theta.$$

Setting $\theta = \theta_0$ and using the expression for $\dot\phi$ yields (after a bit of somewhat complicated algebra)

$$E_{cp} = \frac{1}{2}\frac{mgl}{\cos\theta_0}(2 - 3\sin^2\theta_0).$$

Exercise 12.10

The string of an 85 cm conical pendulum makes an angle of 20° with the negative z-axis. (That is, the vertex angle of the cone is 20°.) Determine the period of the circular motion of the bob. Answer: 1.83 s. ∎

12.4.2 The Spherical Pendulum

We now consider the spherical pendulum. We shall only consider a pendulum for which the oscillations in θ are of small amplitude. (A spherical pendulum is essentially a conical pendulum

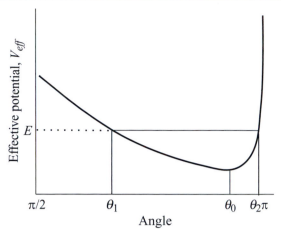

Figure 12.10 Effective potential for a spherical pendulum with a given value of p_ϕ. The total energy is E and the turning points are at θ_1 and θ_2.

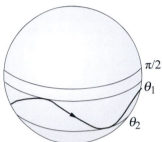

Figure 12.11 Trajectory of a spherical pendulum showing the motion between two parallel circles at different values of θ. The point of support of the pendulum is the center of the sphere.

which rises above and descends below the horizontal plane at $z = l \cos \theta_0$ as it goes around azimuthally.[5])

According to Equation (12.23) the energy of the spherical pendulum is

$$E = \frac{1}{2}ml^2\dot{\theta}^2 + V_{eff}(\theta) = \frac{1}{2}ml^2\dot{\theta}^2 + \frac{p_\phi^2}{2ml^2\sin^2\theta} + mgl\cos\theta.$$

Figure 12.10 is a plot of the effective potential for an arbitrarily selected value of p_ϕ. (Compare with the $p_\phi = 2$ curve of Figure 12.9.) Note the turning points at θ_1 and θ_2. The θ motion of the pendulum bob is an oscillation between these values as it rotates azimuthally with constant p_ϕ. This motion is illustrated in Figure 12.11, where the pendulum bob moves between the two horizontal circles denoted by θ_1 and θ_2.

The θ equation of motion is given by Equation (12.20), which can be written as

$$\ddot{\theta} - \dot{\phi}^2 \sin\theta \cos\theta - (g/l)\sin\theta = 0. \tag{12.30}$$

This differential equation can be solved in terms of elliptic integrals (see Problem 12.20) but if the θ variations are small the problem can also be solved by the technique of successive approximations, which we will now consider.

In general, the technique of successive approximations is based on finding the exact solution of a simpler system that approximates the actual system. The difference between the simple system and the actual system can be expressed in terms of a small parameter ϵ. The simple system that

---

[5] Coherent discussions of the spherical pendulum are found in H. C. Corben and Philip Stehle, *Classical Mechanics*, Wiley and Sons Inc., Hoboken, NJ, 1950 (republished by Dover Press in 1994), and in Grant Fowles and George L. Cassiday, *Analytical Mechanics*, 7th ed., Cengage Learning, Boston, MA, 2004.

can be solved exactly is called the zeroth-order approximation. For the spherical pendulum with small variations in θ, the motion is nearly that of a conical pendulum. The conical pendulum is the zeroth-order approximation and is proportional to $\epsilon^0 = 1$. The solution that is linear in the small quantity (ϵ^1) is called the first-order approximation; the solution in which we keep terms involving the small quantity squared (ϵ^2) is called the second-order approximation, and so on. Thus,

$$[\text{spherical pendulum}] = \underbrace{\underbrace{\underbrace{\epsilon^0[\text{conical pend}]}_{\text{zeroth order}} + \epsilon^1[f_1(\theta,\phi)]}_{\text{first order}} + \epsilon^2[f_2(\theta,\phi)]}_{\text{second order}} + \cdots$$

We will limit ourselves to the first-order approximation of the spherical pendulum.

Let a conical pendulum with polar angle θ_0 be slightly perturbed. Then the angle θ of the resulting spherical pendulum will be close to θ_0 and, to first order,

$$\theta = \theta_0 + \epsilon\theta_1. \tag{12.31}$$

Here ϵ is the "smallness parameter."

Since

$$p_\phi = ml^2 \sin^2\theta\,\dot\phi,$$

$\dot\phi$ can be eliminated from the equation of motion, Equation (12.30). Then defining $\beta^2 = p_\phi^2/m^2l^4$, Equation (12.30) can be written as:

$$\ddot\theta - \beta^2 \frac{\cos\theta}{\sin^3\theta} - (g/l)\sin\theta = 0. \tag{12.32}$$

Inserting Equation (12.31) into Equation (12.32) leads to

$$\epsilon\ddot\theta_1 - \beta^2 \frac{\cos(\theta_0 + \epsilon\theta_1)}{\sin^3(\theta_0 + \epsilon\theta_1)} - (g/l)\sin(\theta_0 + \epsilon\theta_1) = 0.$$

Using the sum of angles formulas[6] followed by the binomial expansion and discarding terms containing ϵ raised to powers of 2 or more, we obtain

$$\epsilon\ddot\theta_1 - \beta^2 \left[\frac{\cos\theta_0}{\sin^3\theta_0} - \left(\frac{1}{\sin^2\theta_0} + \frac{3\cos^2\theta_0}{\sin^4\theta_0}\right)\epsilon\theta_1\right] + (\sin\theta_0 + \epsilon\theta_1\cos\theta_0)\,(-g/l) = 0.$$

Next, collecting terms in various powers of ϵ we get

$$\epsilon^0\left[-\beta^2 \frac{\cos\theta_0}{\sin^3\theta_0} - \frac{g}{l}\sin\theta_0\right] + \epsilon^1\left[\ddot\theta_1 + \theta_1\beta^2\left(\frac{1}{\sin^2\theta_0} + \frac{3\cos^2\theta_0}{\sin^4\theta_0}\right) - \frac{g}{l}\theta_1\cos\theta_0\right] = 0.$$

---

[6] Note that

$$\cos(\theta_0 + \epsilon\theta_1) = \cos\theta_0\cos\epsilon\theta_1 - \sin\theta_0\sin\epsilon\theta_1$$
$$\doteq \cos\theta_0 - \epsilon\theta_1\sin\theta_0,$$

and

$$\sin(\theta_0 + \epsilon\theta_1) = \sin\theta_0\cos\epsilon\theta_1 + \cos\theta_0\sin\epsilon\theta_1$$
$$\doteq \sin\theta_0 + \epsilon\theta_1\cos\theta_0,$$

where we used $\sin\epsilon\theta_1 \doteq \epsilon\theta_1$ and $\cos\epsilon\theta_1 \doteq 1$.

Since ϵ is arbitrary, terms involving the various powers of ϵ must vanish separately. Setting the coefficients of ϵ^0 and ϵ^1 equal to zero, generates the following two relations:

$$\beta^2 \frac{\cos \theta_0}{\sin^3 \theta_0} + \frac{g}{l} \sin \theta_0 = 0, \tag{12.33}$$

and

$$\ddot{\theta}_1 + \left[\left(\frac{1}{\sin^2 \theta_0} + \frac{3\cos^2 \theta_0}{\sin^4 \theta_0} \right) \beta^2 - \frac{g}{l} \cos \theta_0 \right] \theta_1 = 0. \tag{12.34}$$

Equation (12.33) gives

$$\beta^2 = -\frac{g}{l} \frac{\sin^4 \theta_0}{\cos \theta_0}.$$

(Recall that $\cos \theta_0$ is negative because $\theta_0 > 90°$.) Inserting the expression for β^2 into Equation (12.34) yields

$$\ddot{\theta}_1 - \left(\frac{1 + 3\cos^2 \theta_0}{\cos \theta_0} \right) \frac{g}{l} \theta_1 = 0. \tag{12.35}$$

But this equation has the form $\ddot{\theta}_1 + \omega^2 \theta_1 = 0$ (simple harmonic motion), where

$$\omega = \sqrt{-\frac{g}{l \cos \theta_0} (1 + 3\cos^2 \theta_0)} = \frac{\beta}{\sin^2 \theta_0} \sqrt{1 + 3\cos^2 \theta_0}. \tag{12.36}$$

With this definition of ω, a solution of Equation (12.35) is[7]

$$\theta_1 = \cos \omega t, \tag{12.37}$$

and so the θ motion is

$$\theta = \theta_0 + \epsilon \cos \omega t. \tag{12.38}$$

We can now determine the corresponding ϕ motion. Start with the definition of β,

$$\beta^2 = \frac{p_\phi^2}{m^2 l^4} = \frac{\left(ml^2 \sin^2 \theta \dot\phi \right)^2}{m^2 l^4} = \dot\phi^2 \sin^4 \theta.$$

Therefore,

$$\dot\phi = \frac{\beta}{\sin^2 \theta} = \frac{\beta}{\sin^2 (\theta_0 + \epsilon \theta_1)}.$$

Using the sum of angles formula and the fact that ϵ is very small, we can write this as

$$\dot\phi = \frac{\beta}{(\sin \theta_0 + \epsilon \theta_1 \cos \theta_0)^2}$$

$$= \frac{\beta}{\sin^2 \theta_0} \left(1 + \epsilon \theta_1 \frac{\cos \theta_0}{\sin \theta_0} \right)^{-2}$$

$$= \frac{\beta}{\sin^2 \theta_0} (1 - 2\epsilon \theta_1 \cot \theta_0)$$

$$= \frac{\beta}{\sin^2 \theta_0} (1 - 2\epsilon \cos \omega t \cot \theta_0).$$

[7] In general, $\theta_1 = A \cos(\omega t + \gamma)$ so $\theta_1 = \cos \omega t$ is not the most general solution. Nevertheless it is *a* solution and it does satisfy the differential equation. You will soon discover that the amplitude of the oscillations is ϵ, but writing $\theta_1 = \cos \omega t$ at this stage makes the derivation less confusing.

Integrating, we obtain

$$\phi = \frac{\beta}{\sin^2 \theta_0} \left(t - \frac{2\epsilon}{\omega} \sin \omega t \cot \theta_0 \right). \tag{12.39}$$

This shows that on average ϕ is equal to its unperturbed value $(\beta t / \sin^2 \theta_0)$ but it oscillates around this value with a frequency ω.

I have not given a physical explanation for the parameter ϵ. It was treated merely as a mathematical device for carrying out the expansion. Now note that according to Equation (12.38) $\theta(t) = \theta_0 + \epsilon \theta_1 = \theta_0 + \epsilon \cos \omega t$. So ϵ is the amplitude of the oscillations of θ about θ_0. Due to our choice of constants in Equation (12.38), ϵ is the initial displacement of the bob in the θ direction:

$$\theta(t = 0) = \theta_0 + \epsilon.$$

That is, ϵ is the initial displacement of the bob from the conical angle θ_0 of the unperturbed conical pendulum. Thus, in Figure (12.11) the values of θ_1 and θ_2 are such that $\theta_2 - \theta_1 = 2\epsilon$. (If we had written Equation (12.37) more generally as $\theta_1 = A \cos(\omega_1 t + \alpha)$ then these statements would have to be modified somewhat.)

The period of the motion in the θ "direction" is $P = 2\pi/\omega$ and in time P the pendulum will have gone through an entire oscillation in θ. We can easily determine how far the azimuthal coordinate ϕ has advanced in this time. Assume that $\phi = 0$ at time $t = 0$. We determine ϕ at time $t = 2\pi/\omega$ by plugging $t = 2\pi/\omega$ into Equation (12.39). An approximate value for $\phi(t = 2\pi/\omega)$ is obtained by setting $\epsilon = 0$ in Equation (12.39), yielding

$$\phi(t = 2\pi/\omega) \doteq \frac{\beta}{\sin^2 \theta_0} \frac{2\pi}{\omega}.$$

But by Equation (12.36),

$$\omega = (\beta/\sin^2 \theta_0)\sqrt{1 + 3\cos^2 \theta_0},$$

so

$$\phi(t = 2\pi/\omega) \doteq \frac{2\pi}{\sqrt{1 + 3\cos^2 \theta_0}}.$$

Now $\cos^2 \theta_0$ must lie between zero and one. If $\cos^2 \theta_0 = 0$, then $\phi = 2\pi$, and if $\cos^2 \theta_0 = 1$, then $\phi = \pi$. Therefore,

$$\pi \leq \phi \leq 2\pi.$$

That is, while the pendulum goes through a complete oscillation in θ, it advances azimuthally through an angle between π and 2π.

This concludes the analysis of the spherical pendulum, but I would like to take a moment to consider the problem from a slightly different point of view, based on the fact that the effective potential is a minimum at θ_0. Expanding the effective potential about θ_0 in a Taylor series, we obtain

$$V_{eff}(\theta) = V_{eff}(\theta_0) + \eta \left. \frac{dV_{eff}}{d\theta} \right|_{\theta_0} + \frac{1}{2} \eta^2 \left[\frac{d^2 V_{eff}}{d\theta^2} \right]_{\theta_0} + \cdots,$$

where $\eta = \theta - \theta_0$. Let us consider the three terms on the right-hand side. Because θ_0 is constant, Equation (12.25) tells us that $V_{eff}(\theta_0) = E_0$. And, according to Equation (12.26), $\left. \frac{dV_{eff}}{d\theta} \right|_{\theta_0} = 0$. Furthermore,

$$\left. \frac{d^2 V_{eff}}{d\theta^2} \right|_{\theta_0} = \frac{-mgl}{\cos\theta_0}\left(1 + 3\cos^2\theta_0\right) = \kappa, \tag{12.40}$$

where κ is a constant. Therefore, for values of θ near θ_0,

$$V_{eff}(\theta) = E_0 + \frac{1}{2}\kappa\eta^2,$$

and the energy equation (12.25) becomes

$$\tfrac{1}{2}ml^2\dot\theta^2 = E - E_0 - \tfrac{1}{2}\kappa\eta^2.$$

Because $\dot\theta = \dot\eta$, and $E - E_0$ is just a constant, this equation is

$$\tfrac{1}{2}ml^2\dot\eta^2 + \tfrac{1}{2}\kappa\eta^2 = E',$$

where $E' = E - E_0$. Note that this has the same *form* as the energy equation for a simple harmonic oscillator. The frequency of oscillation in η is $\omega = \sqrt{\kappa/ml^2}$. Plugging in the expression (12.40) for κ this reproduces Equation (12.36). Therefore, the motion of a spherical pendulum with energy slightly greater than E_0 is circular motion around the vertical axis with small oscillations in θ of frequency ω.

Exercise 12.11

For a spherical pendulum, show that the ratio of the angular frequencies in the azimuthal (ϕ) and polar (θ) directions is $\sqrt{1 + 3\cos^2\theta_0}$. ∎

12.5 Summary

After studying this chapter you will probably agree that the "simple pendulum" is not so simple after all! One of the benefits of studying the motion of a pendulum is that it leads to a consideration of a variety of mathematical techniques.

The period of a simple pendulum with arbitrary amplitude is

$$P = \frac{4}{\omega}\int_0^{\pi/2} \frac{d\phi}{\sqrt{1 - k^2\sin^2\phi}},$$

a complete elliptic integral of the first kind. These have been tabulated but they can also be evaluated using a series expansion.

An analysis of the physical pendulum introduces a number of properties of extended bodies, including the radius of gyration k defined in terms of the moment of inertia as

$$I = Mk^2.$$

A simple pendulum oscillating with the same frequency as a physical pendulum has a length given by

$$l = k_0^2/h,$$

where h is the distance from the axis of rotation to the center of mass, and k_0 is the radius of gyration relative to the axis. The distance l is also the distance from the axis of rotation to the center of oscillation. If a physical pendulum receives an impulse at the center of percussion, there will be no reaction at the axis. The center of percussion lies at the same point as the center of oscillation (but conceptually these two quantities are quite different).

The motion of a spherical pendulum is most easily understood by introducing an effective potential

$$V_{eff} = \frac{p_\phi^2}{2ml^2 \sin^2 \theta} + mgl \cos \theta.$$

For the special case of a conical pendulum the angle $\theta = \theta_0$ is constant and the value of p_ϕ is also constant, leading to an azimuthal angular velocity of

$$\dot\phi = \sqrt{-g/(l \cos \theta_0)},$$

and a total energy of

$$E_{cp} = \frac{1}{2} \frac{mgl}{\cos \theta_0} (2 - 3 \sin^2 \theta_0).$$

The spherical pendulum presents a number of difficulties, but it warrants study because it can be treated as a conical pendulum subjected to a small perturbation about a known solution. (This technique was used in Chapter 10 to study the motion of a slightly perturbed planet. It will be used again in Chapter 15.) You will become very familiar with perturbation techniques when you study quantum mechanics. A conical pendulum which has been perturbed so that the amplitude of oscillations in θ is ϵ, has an azimuthal angular position given by

$$\phi = \frac{\beta}{\sin^2 \theta_0} \left(t - \frac{2\epsilon}{\omega} \sin \omega t \cot \theta_0 \right).$$

The oscillations in θ have an angular frequency

$$\omega = (\beta/\sin^2 \theta_0)\sqrt{1 + 3 \cos^2 \theta_0},$$

where the parameter β is defined by

$$\beta^2 = -\frac{g}{l} \frac{\sin^4 \theta_0}{\cos \theta_0}.$$

12.6 Problems

Problem 12.1 The rotational analogue to $F = \frac{dp}{dt}$ is $N = \frac{dL}{dt}$. Starting with this equation, obtain a work–energy theorem for rotational motion.

Problem 12.2 Prove that if the torque is a function only of the angle, $N = N(\theta)$, the total mechanical energy is conserved.

Problem 12.3 Consider a large amplitude simple pendulum. Estimate the fractional error in the period (relative to the elliptic integral) due to ignoring all terms beyond the first-order term in the expansion (12.11) for a pendulum whose amplitude is $60°$.

Problem 12.4 Elliptic integrals can be used to calculate the length of the arc of a curve. In this problem you are asked to obtain a formula for the circumference of an ellipse whose semimajor axis is a and semiminor axis is b. Define

$$e = \sqrt{\frac{a^2 - b^2}{a^2}}$$

and show that the circumference is given by

$$C = 4a E(e).$$

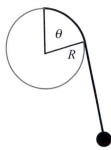

Figure 12.12 A pendulum of variable length. (Problem 12.6.)

(Hint: The parametric equations for an ellipse are $x = a \cos \theta$, and $y = b \sin \theta$. An element of length along the curve is $ds = [dx^2 + dy^2]^{1/2}$.)

Problem 12.5 A simple pendulum is attached to a device that causes the point of support to oscillate up and down according to $y' = A \cos ft$, so the y-coordinate of the bob is given by $y = y' - l \cos \theta$. Here θ is measured counterclockwise from the lowest point of the pendulum swing. (a) Obtain the Lagrangian. (b) Obtain the equation of motion. (c) Determine the angular frequency.

Problem 12.6 A pendulum of length l_0 is attached to the top of a stationary vertical disk of radius R as shown in Figure 12.12. The pendulum is long enough and the amplitude is small enough that the bob never touches the disk. (a) Show that the Lagrangian can be written as

$$L = \frac{1}{2} m (A + B\theta)^2 \dot{\theta}^2 + mg \left[(A + B\theta) \sin \theta + B \cos \theta \right]$$

and obtain explicit expressions for A and B. (b) Obtain the equation of motion. (c) The equilibrium position is $\theta_0 = \pi/2$ (the string hangs straight down). Determine the frequency of small-amplitude oscillations about θ_0.

Problem 12.7 A child stands on the seat of a playground swing. The child's body can be approximated by a cylinder of radius 10 cm, length 1 m and mass 30 kg. The swing is suspended by ropes of length 2.5 m. The mass of each rope is 2 kg. (a) Determine the period of oscillation. (b) Where is the center of oscillation of this system?

Problem 12.8 A physical pendulum oscillates about an axis 10 cm from its center of mass. The radius of gyration about the axis is 15 cm. (a) What is the period? (b) Where is the center of oscillation?

Problem 12.9 A solid sphere of radius r is cut in half and a homogeneous hemisphere of mass M is set upon a table (with its flat side up). The surface of the table is perfectly rough. The hemisphere rocks back and forth with small-amplitude excursions from equilibrium. What is the length of an equivalent simple pendulum? Justify approximations. Note that the center of mass of a hemisphere is at a distance $3r/8$ below the center of the sphere.

Problem 12.10 A crane is lowering a heavy mass at a constant rate r. The mass swings back and forth, so this is (essentially) a pendulum of increasing length. Show that the equation of motion can be written as

$$l \frac{d^2\theta}{dl^2} + 2 \frac{d\theta}{dl} + \frac{g}{r^2} \theta = 0,$$

where l is the length of the cable. Assume small oscillations.

Problem 12.11 A wheel is mounted on an axle. One end of a massless rigid rod is attached to the rim of the wheel and the other end is attached to a spring, so that the rod exerts a torque on the rim of the wheel. Assume the wheel is a disk of mass M and radius R and let the spring constant be k. Determine the period of small amplitude oscillations. See Figure 12.13.

Figure 12.13 Illustration for Problem 12.11.

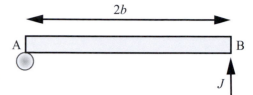

Figure 12.14 A rod of mass m and length $2b$ lies on a smooth horizontal table. (Problem 12.14.)

Figure 12.15 A bead on a rotating hoop. (Problem 12.18.)

Problem 12.12 Kater's pendulum is a physical pendulum that has the same period when supported from either of two points which lie on opposite sides of the center of mass. Show that if ω is the angular frequency of Kater's pendulum, then $\omega^2 = g/d$, where d is the distance between the two points of support. (This type of pendulum can be used to make very accurate measurements of the acceleration of gravity because one does not need an accurate determination of the moment of inertia of the body.)

Problem 12.13 Obtain an expression for the period of oscillation of a physical pendulum if the oscillations are not restricted to small amplitudes.

Problem 12.14 A rod of mass m and length $2b$ lies on a smooth horizontal table, as shown in Figure 12.14. At one end a smooth peg A is fixed into the table. The side of the rod rests against the peg. An impulse J is given to the rod at end B. Find the magnitude and direction of the impulse given to the peg.

Problem 12.15 A conical pendulum has a bob of mass 200 g and a string of length 75 cm. The string will break if the tension exceeds 2.5 N. Determine the maximum angular velocity ($\dot{\phi}$) allowed.

Problem 12.16 A conical pendulum is perturbed such that it has an initial amplitude in the θ direction of 0.01 rad. Determine how far the azimuth advances during one oscillation in θ. Assume $\theta_0 = 160°$.

Problem 12.17 Consider a spherical pendulum. Obtain an expression for the tension in the string as a function of E and θ. For given values of E and p_ϕ, determine the angle at which the string will collapse.

Problem 12.18 A bead of mass m can slide freely on a circular hoop of radius a. The hoop is upright (the plane of the hoop is vertical) and it is rotating at a constant angular speed ω about the vertical diameter. The position of the bead on the hoop is specified by the angle θ which is measured from the bottom of the vertical diameter as illustrated in Figure 12.15. (a) Obtain the Lagrangian for this system. (b) Obtain the equation of motion and show that it reduces to the equation of a simple pendulum when $\omega = 0$. (c) Determine the value of θ for which the bead is in equilibrium.

Problem 12.19 Write the Hamiltonian and Hamilton's equations of motion for a spherical pendulum.

Problem 12.20 Two constants of the motion for the spherical pendulum are the total energy E and the generalized momentum p_ϕ. Show that these first integrals lead to the following expressions:

$$\phi = \frac{p_\phi}{\sqrt{2ml^2}} \int \frac{d\theta}{\sin^2\theta \left[E - V_{eff}(\theta)\right]^{1/2}},$$

and

$$t = \sqrt{\frac{ml^2}{2}} \int \frac{d\theta}{\left[E - \frac{p_\phi^2}{2ml^2\sin^2\theta} - mgl\cos\theta\right]^{1/2}}.$$

(The first of these integrals can be put into the form of an elliptic integral of the third kind and the second can be expressed as an elliptic integral of the first kind.)

Problem 12.21 A simple pendulum has a period that depends on the amplitude of the swing. In this problem we show that if the path of the pendulum bob is a portion of a cycloid, the pendulum is *isochronous*, that is, the period of the oscillations is independent of the amplitude of the oscillations. This can be achieved by varying the length of the pendulum string as the pendulum swings back and forth. However, the analysis is easier if we replace the pendulum with the equivalent problem of a bead of mass m sliding without friction on a wire having the shape of a cycloid. Suggestion: Recall that simple harmonic motion (such as a mass on a spring) is isochronous. Write the Lagrangian for the bead on the wire in terms of the arc length s and show that the potential energy must be proportional to s^2 for isochronous motion. Then obtain expressions for x and y and show they can be expressed in the form of the parametric equations of a cycloid, $x = \rho(\phi + \sin\phi)$ and $y = \rho(1 - \cos\phi)$. See Figure 12.16.

Problem 12.22 Our analyses of pendulums have mostly ignored the effects of the environment. For example, the period of a pendulum depends on the local gravitational acceleration, and on the temperature (because the expansion and contraction of the bob changes the length). Furthermore, unless it is situated in a vacuum, a pendulum is affected by the air in various ways. Physicists have obtained formulas to account for the buoyancy of air, for the effect of air being dragged along by the pendulum and the resistance of air on the string and on the bob. Assume the pendulum consists of a massless string and a bob of mass M and volume V. In this problem, we consider the effect of the buoyancy of air which can be treated as an additional force acting opposite to gravity. (a) Show that for a pendulum in air of density D, the equation of motion is

$$I_0\ddot{\theta} = -Mgh\sin\theta + VDgh\sin\theta,$$

where $I_0 = Mk^2$ is the moment of inertia of the pendulum with respect to the point of support, h is the distance from the point of support to the center of mass of the bob, and V is the volume of the bob. (b) Show that if l is the length of an equivalent simple pendulum,

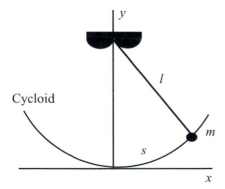

Cycloid

Figure 12.16 An isochronous pendulum can be constructed such that the length of the string varies with angle. However, the analysis requested in Problem 12.21 is easier if one assumes the bob is a bead on a cycloidal wire.

$$l = \frac{k^2}{(1 - DV/M)h}.$$

(c) Show that if the density of air is given by $D + \delta D$, where D is a standard density, the length of the equivalent pendulum changes by δl, where

$$\frac{\delta l}{l^2} = \frac{h}{k^2} \frac{V}{M} \delta D.$$

Computational Projects

Computational Project 12.1 Write a program to evaluate elliptic integrals of the first kind.

Computational Project 12.2 Use the results of Computational Project 12.1 to write a program that evaluates the period of a simple pendulum with arbitrary amplitude. Vary the amplitude and obtain a plot of period as a function of amplitude.

Computational Project 12.3 Plot position vs. time for several points on a rod undergoing an impulse, as illustrated in Figure 12.4. Assume the impulse is $J = 10$ Ns, the rod has mass $M = 2$ kg and length $L = 0.5$ m.

13 Waves

A wave is an oscillation of a medium, such as a water wave in the ocean or a wave propagating down a stretched slinky or a taut string. In general, the medium as a whole does not translate. For example, a wave in the ocean can be thought of as water molecules moving vertically up and down while the wave itself moves horizontally.[1] A strange and interesting thing about waves is that although there is no transport of mass, the wave does transport energy and momentum.

A *transverse* wave, such as a wave in a string, is one in which the material particles move perpendicular to the direction of the wave. On the other hand, a sound wave in air or the wave set up in a metal rod when you hit one end with a hammer, are examples of *longitudinal* waves. In these waves the material particles move back and forth parallel to the direction of the wave.

This chapter is an introduction to wave motion and is limited to an analysis of waves in strings.[2] However, the wave equation and the wave properties derived here are valid for all types of waves. The mathematical techniques considered in this chapter include expressing a periodic function as a Fourier series and solving the wave equation by separation of variables. You will find the material of this chapter to be particularly useful when you study electromagnetic waves.

13.1 A Wave in a Stretched String

Consider a long string. If you shake one end of the string up and down, you can set up a *traveling wave*. A single shake will generate a pulse that moves along the string; continuous shaking will produce a wave train which is often sinusoidal in form. If you have a shorter string, such as one of the strings in your guitar, and you pluck the string, you can set up a *standing wave*.

A wave is characterized by a number of physical parameters. The main ones are the following.

Amplitude (A): The maximum displacement from equilibrium.

Wavelength (λ): The distance from peak to peak or between any two corresponding points on the wave.

Period (τ): The time it takes for a point on the wave to go through one complete oscillation.

Frequency (f): The number of oscillations per unit time. Note that $f = 1/\tau$.

Speed (v): How fast the waveform is displaced in the direction of its motion, $v = \lambda/\tau = \lambda f$.

Angular frequency (ω): Another measure of frequency, defined by $\omega = 2\pi f = 2\pi/\tau$.

---

[1] This is not quite true. A molecule of water also has a small horizontal oscillation and traces out an elliptical path – but you get the general idea.

[2] An ideal string is perfectly flexible and linearly elastic. This means the tension is everywhere the same and always directed tangentially. Linear elasticity means the tension depends linearly on the amount the string is stretched (Hooke's law).

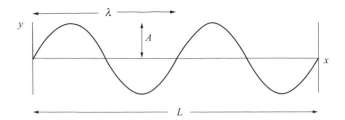

Figure 13.1 A wave in a string of length L. The wavelength and amplitude are indicated on the figure. The displacement from equilibrium is $y = y(x)$.

Wave number (k): A parameter related to wavelength, defined by $k = 2\pi/\lambda$. Because $\lambda = v\tau$ the wave number can also be expressed as $k = \omega/v$. Note that the wave number is analogous to the angular frequency: one of them is an inverse time and the other is an inverse length.

Consider a string of length L with fixed end points. Suppose you generated a wave in the string and took a photograph of the wave. In your photograph the string might look somewhat like the sketch in Figure 13.1. The vertical displacement of a particle in the string is y. This displacement depends on the horizontal position, x, so $y = y(x)$. (There is no time dependence yet because this is just a description of the instantaneous photograph of the wave.) The curve repeats after a distance $x = \lambda$. Suppose you measured the vertical displacement (y) along the string and found it satisfied the following sinusoidal curve[3]

$$y(x) = A \sin\left(\frac{2\pi x}{\lambda}\right) = A \sin(kx).$$

When x is any multiple of $\lambda/2$ the quantity in parentheses is a multiple of π and there is zero displacement at that point. Such points are called the *nodes* of the waveform.

The wave in Figure 13.1 has a wavelength equal to one half the length of the string, L. The only waves that will "fit" onto the string are those with wavelengths $\lambda = 2L/n$ where $n = 1, 2, 3, \ldots$.

Now imagine that instead of an instantaneous photograph you have a video of the oscillating string. If you look at a particular point on the string you will see it moving up and down. Thus the displacement y is also a function of the time t. As you might suspect, the wave can be expressed mathematically by an expression involving sines and cosines of the quantities kx and ωt.

To derive the expression for $y = y(x, t)$ we need to determine the equation of motion and then integrate it. Consider a string of length L under tension F. The two ends of the string are at $x = 0$ and $x = L$. Ignore gravity. If you pluck the string you can set up a "standing wave" as in a guitar string. If you look at a plucked guitar string you will notice that the amplitude of the wave is much smaller than the length of the string. That is, y is relatively a small quantity. Furthermore, the slope dy/dx is also small. The tension in the string is the same at every point.

Figure 13.2 shows an infinitesimal portion (dx) of the string as a pulse propagates along it.

The portion of the string extends from x to $x + dx$. The mass of this element is $dm = \rho dx$, where ρ is the mass per unit length. The two ends are displaced vertically by $y(x, t)$ and $y(x + dx, t)$. The element of the string is subjected to a tension $F(x)$ acting to the left and down at x (at angle θ) and a tension $F(x + dx)$ acting to the right and up at $x + dx$ (at angle ϕ). The horizontal components of these two tensions must cancel because the element of string is not accelerating to

---

[3] As you will see, a wave can generally be expressed as a sum of sines and cosines. For simplicity, at the moment, we assume a single sinusoid.

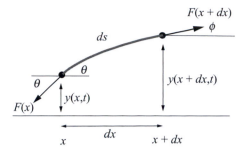

Figure 13.2 A small portion of a vibrating string.

the right or the left. (The tension in the string is the same at all points so there actually is a small oscillating horizontal force that we shall consider in Section 13.7.) There is, however, a vertical component of force bringing the string back down to its equilibrium position.

The magnitude of the tension is the same at both ends, that is,

$$F(x + dx) = F(x) = F.$$

The vertical acceleration is $a_y = \partial^2 y / \partial t^2$, so applying Newton's second law gives[4]

$$\rho dx \frac{\partial^2 y(x,t)}{\partial t^2} = F(x + dx) \sin \phi - F(x) \sin \theta,$$

$$= F(\sin \phi - \sin \theta).$$

Assuming the angles θ and ϕ are small (because both y and dy/dx are small)

$$\sin \theta \doteq \tan \theta = \frac{\partial y(x,t)}{\partial x},$$

$$\sin \phi \doteq \tan \phi = \frac{\partial y(x + dx,t)}{\partial x}.$$

So,

$$\rho dx \frac{\partial^2 y(x,t)}{\partial t^2} = F \left[\frac{\partial y(x + dx,t)}{\partial x} - \frac{\partial y(x,t)}{\partial x} \right].$$

Dividing both sides by ∂x and taking the limit as $\partial x \to 0$ leads to

$$\rho \frac{\partial^2 y}{\partial t^2} = F \frac{\partial}{\partial x} \left[\frac{\partial y}{\partial x} \right],$$

or

$$\frac{\partial^2 y}{\partial x^2} = \frac{\rho}{F} \frac{\partial^2 y}{\partial t^2}.$$

If this last equation looks familiar to you it is because it has the form of the well-known relation called the "wave equation" which is usually expressed as

$$\frac{\partial^2 y}{\partial x^2} = \frac{1}{v^2} \frac{\partial^2 y}{\partial t^2}, \tag{13.1}$$

where v is the speed of the wave.

---

[4] We use partial derivatives because the acceleration depends on x as well as t.

Thus we have not only obtained the equation of motion for the wave in a taut string, we have also shown that the wave propagates in the x direction with speed

$$v = \sqrt{F/\rho}.$$

The wave equation (13.1) is a very important, fundamental relation. It is a second-order partial differential equation giving the displacement of the string as a function of time and position. I strongly suggest you memorize it.

Exercise 13.1

A string of length 6 m is fixed at both ends. Its mass is 0.1 kg and the tension in the string is 50 N. (a) What is the wavelength of the longest possible standing wave in this string? (b) What is the frequency of that wave? Answer: (a) 12 m, (b) 4.56 Hz. ∎

Exercise 13.2

If n is the number of half wavelengths, show that the frequency of a standing wave in a string is $f = nv/2L$. ∎

13.2 Direct Solution of the Wave Equation

Our next task is to solve Equation (13.1), the wave equation, to obtain an expression for $y = y(x,t)$. Many second-order partial differential equations (including this one) can be solved by a technique called *separation of variables*.

The basic idea behind the separation of variables method is to express the unknown function (in our case $y = y(x,t)$) as the product of two functions, each depending on only one variable. That is, you *assume* that you can write $y(x,t)$ in the following form:

$$y(x,t) = X(x)T(t),$$

where X is a function only of x and T is a function only of t. Substitute this expression into the differential equation (13.1) to get

$$\frac{\partial^2 X(x)T(t)}{\partial t^2} = v^2 \frac{\partial^2 X(x)T(t)}{\partial x^2},$$

$$X\frac{\partial^2 T}{\partial t^2} = v^2 T \frac{\partial^2 X}{\partial x^2},$$

$$\frac{1}{T}\frac{\partial^2 T}{\partial t^2} = \frac{v^2}{X}\frac{\partial^2 X}{\partial x^2}.$$

Now the left-hand side depends only on t and the right-hand side depends only on x. Since x and t can be varied independently, the only way the equation can be satisfied is if the right-hand side and the left-hand side are equal to the same constant. Call it $-\omega^2$. The partial differential equation has then been *separated* into the following two *ordinary* differential equations:

$$\frac{v^2}{X}\frac{d^2 X}{dx^2} = -\omega^2,$$

$$\frac{1}{T}\frac{d^2 T}{dt^2} = -\omega^2,$$

or

$$\frac{d^2 X}{dx^2} + \frac{\omega^2}{v^2} X = 0,$$

$$\frac{d^2 T}{dt^2} + \omega^2 T = 0.$$

Both of these equations have the form of the simple harmonic oscillator equation. You have seen the SHM equation many times, so I will simply write down the general solutions

$$X(x) = A \cos \frac{\omega}{v} x + B \sin \frac{\omega}{v} x = A \cos kx + B \sin kx,$$

$$T(t) = C \cos \omega t + D \sin \omega t.$$

The solution for $y = y(x,t)$ is the product of these two functions, thus:

$$y(x,t) = X(x)T(t) = (A \cos kx + B \sin kx)(C \cos \omega t + D \sin \omega t). \qquad (13.2)$$

Note that the partial differential equation (13.1) is satisfied by the four expressions

$$y(x,t) \propto \begin{cases} \cos kx \cos \omega t \\ \cos kx \sin \omega t \\ \sin kx \cos \omega t \\ \sin kx \sin \omega t \end{cases}.$$

The general solution, Equation (13.2), is the sum of all of these. The coefficients, A, B, C, D are determined from the boundary conditions.

Exercise 13.3

Show that an expression of the form

$$y = A \sin kx + B \cos kx$$

can be expressed as

$$y = C \cos(kx + \alpha).$$

Express C and α in terms of A and B. (The quantity α is called the "phase constant.") Answer: $C = \sqrt{A^2 + B^2}$, $\alpha = \tan^{-1}(-A/B)$. ∎

13.2.1 Fourier Series (Optional)

You are probably familiar with Fourier series from your mathematics courses, so this discussion is intended to refresh your memory. If you have not been exposed to Fourier series previously, I strongly suggest you read about them in a "Math Methods" text.[5]

Any periodic function (such as a wave) or a portion of a periodic function (such as a standing wave) can be represented as a sum of sines and cosines. Consider, for example, the portion of a "square wave" shown in Figure 13.3.

The function we are approximating with a Fourier series is $f(x) = -\pi/2$ for $-\pi < x < 0$, $f(x) = +\pi/2$ for $0 < x < +\pi$. This is a portion of the square wave which alternates between plus and minus $\pi/2$ with a periodicity of π. In the middle panel of Figure 13.3, I plotted $2 \sin x$

---

[5] An excellent source is Chapter 7 in *Mathematical Methods in the Physical Sciences* by Mary L. Boas (Wiley and Sons, 2006).

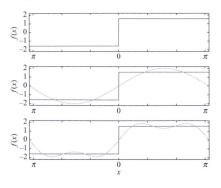

Figure 13.3 The top panel shows the square wave $f(x) = -\pi/2$ for $-\pi < x < 0$, $f(x) = +\pi/2$ for $0 < x < +\pi$. The middle panel also shows $f(x) = 2\sin x$, a rather poor approximation to the square wave. The bottom panel shows $f(x) = 2\sin x + \frac{2}{3}\sin 3x$, a somewhat better approximation.

Figure 13.4 A string plucked at its center. The displacement at $L/2$ is b.

which has the appropriate periodicity but does not approximate the original function very well. In the bottom panel the additional term $\frac{2}{3}\sin 3x$ has been added to our highly truncated Fourier series, and you can appreciate that the approximation is getting better. The next approximation (not plotted) is $f(x) = 2\sin x + \frac{2}{3}\sin 3x + \frac{2}{5}\sin 5x$. The Fourier series more and more closely approximates the square wave as more and more terms are included.

Before getting into the details of how to obtain a Fourier series, let us consider why they are useful. That is, if we already know the function $f(x)$, why would we want to express it as an infinite sum of sines and cosines? To answer this question let us note that the solution to the wave equation (13.2) is expressed in terms of sines and cosines which are multiplied by constant coefficients which we do not know. These coefficients will be determined from the initial and/or boundary conditions. In the problem of an oscillating mass on a spring or a pendulum this was easily done because the initial conditions were the position and the velocity of what was essentially a point mass. But the initial conditions for a vibrating string involve the positions and velocities of all the points of the string. Suppose that we know the initial configuration of a plucked string, and it looks like Figure 13.4, which shows us two straight lines. How can we match this initial condition to the solution that looks like Equation (13.2), i.e., a sum of sines and cosines? The answer is to express a function such as that of Figure 13.4 as a Fourier series and thus determine the coefficients in the solution.

In general a function $f(x)$ defined in the interval $-\pi < x < +\pi$ can be expressed as a Fourier series having the form

$$f(x) = \frac{1}{2}a_0 + a_1 \cos x + a_2 \cos 2x + \cdots + b_1 \sin x + b_2 \sin(2x) + \cdots$$

or

$$f(x) = \frac{a_0}{2} + \sum_{n=1}^{\infty} a_n \cos nx + \sum_{n=1}^{\infty} b_n \sin nx. \tag{13.3}$$

The problem now reduces to determining the coefficients a_n and b_n. This is accomplished by using the orthogonality properties of sines and cosines which can be expressed by the following three integrals:

$$\int_{-\pi}^{+\pi} \sin nx \cos mx\, dx = 0 \text{ for all } m \text{ and } n$$

$$\int_{-\pi}^{+\pi} \sin nx \sin mx\, dx = \pi \text{ for } n = m, \text{ and } 0 \text{ for } n \neq m$$

$$\int_{-\pi}^{+\pi} \cos nx \cos mx\, dx = \pi \text{ for } n = m, \text{ and } 0 \text{ for } n \neq m$$

To determine the coefficients a_n and b_n we multiply Equation (13.3) by $\cos mx$ and integrate from $-\pi$ to $+\pi$:

$$\int_{-\pi}^{+\pi} f(x) \cos mx\, dx = \frac{a_0}{2} \int_{-\pi}^{+\pi} \cos mx\, dx$$

$$+ \int_{-\pi}^{+\pi} \sum_{n=1}^{\infty} a_n \cos nx \cos mx\, dx$$

$$+ \int_{-\pi}^{+\pi} \sum_{n=1}^{\infty} b_n \sin nx \cos mx\, dx.$$

For $m = 0$ this reduces to $\int_{-\pi}^{+\pi} f(x)dx = \int_{-\pi}^{+\pi} 1\, dx = \frac{a_0}{2}(\pi + \pi)$, or

$$a_0 = \frac{1}{\pi} \int_{-\pi}^{+\pi} f(x)dx$$

and for $m \neq 0$

$$\int_{-\pi}^{+\pi} f(x) \cos mx\, dx = \frac{a_0}{2} \int_{-\pi}^{+\pi} \cos mx\, dx + \sum_{n=1}^{\infty} a_n \pi \delta_{mn} + \sum_{n=1}^{\infty} b_n \int_{-\pi}^{+\pi} \sin nx \cos mx\, dx$$

$$= 0 + a_m \pi + 0.$$

Therefore,

$$a_m = \frac{1}{\pi} \int_{-\pi}^{+\pi} f(x) \cos mx\, dx.$$

Similarly, if we multiply Equation (13.3) by $\sin mx$ and integrate we obtain

$$b_m = \frac{1}{\pi} \int_{-\pi}^{+\pi} f(x) \sin mx\, dx.$$

We assumed the function $f(x)$ was defined in the interval $-\pi$ to $+\pi$. But this assumption was only made so that we could keep the notation simple. For a function defined on the arbitrary interval $-l$ to $+l$ we simply change the variable from x to $\frac{\pi x}{l}$ and obtain the coefficients from

$$a_m = \frac{1}{l} \int_{-l}^{+l} f(x) \cos \frac{m\pi}{l} x\, dx \quad m = 0, 1, 2, 3, \ldots$$

$$b_m = \frac{1}{l} \int_{-l}^{+l} f(x) \sin \frac{m\pi}{l} x\, dx \quad m = 1, 2, 3, \ldots.$$

A labor saving fact is that if $f(x)$ is an odd function of x ($f(x) = -f(-x)$), then only sine terms appear in the Fourier series, and if $f(x)$ is an even function of x ($f(x) = f(-x)$), then only cosine terms appear in the series.

Exercise 13.4

Show that if $f(x)$ is defined in the interval $0 < x < l$, the coefficients in the Fourier expansion are

$$a_m = \frac{2}{l} \int_0^{+l} f(x) \cos \frac{m\pi}{l} x \, dx,$$

$$b_m = \frac{2}{l} \int_0^{+l} f(x) \sin \frac{m\pi}{l} x \, dx.$$

∎

13.3 Standing Waves

As an example of how to determine the coefficients in the general solution, Equation (13.2), consider a standing wave, as in the string illustrated in Figure 13.1. The string is tied down at the end points so

$$y(x = 0) = y(x = L) = 0.$$

Then

$$y(0, t) = 0 = (A \cos 0 + B \sin 0)(C \cos \omega t + D \sin \omega t),$$

or

$$0 = A(C \cos \omega t + D \sin \omega t).$$

This requires that $A = 0$. The general equation has now been reduced to

$$y(x, t) = B \sin(kx)(C \cos \omega t + D \sin \omega t).$$

The boundary condition at $x = L$ leads to

$$y(L, t) = 0 = B \sin(kL)(C \cos \omega t + D \sin \omega t).$$

This condition is satisfied if $B = 0$, but in that case $y = 0$ everywhere and at all times! Although it satisfies the differential equation, this is obviously not a satisfactory solution. (It is the "trivial" solution.) However, there is another way to ensure that this expression is zero; if

$$\sin(kL) = 0,$$

the condition is satisfied. This means that kL must be an integer multiple of π because if $\sin \theta = 0$, then $\theta = 0, \pi, 2\pi, 3\pi \ldots$. That is,

$$kL = n\pi, \quad n = 1, 2, 3, \ldots.$$

Note that $n = 0$ is not included because that would require $k = 0$ and again $y(x, t)$ would be zero for all values of x. If k can take on the values $n\pi/L$, we must also require that ω take on a set of allowed values because $\omega = kv = k\sqrt{\frac{F}{\rho}}$. Thus, ω has the values

$$\omega_n = \frac{n\pi}{L} \sqrt{\frac{F}{\rho}} = \frac{n\pi v}{L},$$

and Equation (13.2) reduces to

$$y(x, t) = B \sin \frac{n\pi x}{L} (C \cos \omega_n t + D \sin \omega_n t).$$

Finally, combining the constants BC and BD and calling them A_n and B_n we write

$$y(x,t) = (A_n \cos \omega_n t + B_n \sin \omega_n t) \sin \frac{n\pi x}{L}. \tag{13.4}$$

The differential equation and the boundary conditions are satisfied for any integer value of n. But note that the coefficients A_n and B_n are generally different for different values of n. The general solution to the differential equation is, of course, the sum of all possible solutions, that is,

$$y(x,t) = \sum_{n=1}^{\infty} (A_n \cos \omega_n t + B_n \sin \omega_n t) \sin \frac{n\pi x}{L}, \tag{13.5}$$

$$= \sum_{n=1}^{\infty} \left(A_n \cos \frac{n\pi v}{L} t + B_n \sin \frac{n\pi v}{L} t \right) \sin \frac{n\pi x}{L}.$$

To determine the coefficients A_n and B_n we use the initial conditions, as illustrated in the following example.

Worked Example 13.1 A string of length L is pulled up at its midpoint a distance b and then released. The displacement of the string at any position $(0 \le x \le L)$ and time $(t \ge 0)$ is given by Equation (13.5). Determine the coefficients A_n and B_n. See Figure 13.4.

Solution The waveform at $t = 0$ is given by the two straight lines, $y = (2b/L)x$ for $0 \le x \le L/2$ and $y = (-2b/L)x + 2b$ for $L/2 \le x \le L$.

At time $t = 0$ the string is not moving, so $\partial y/\partial t = 0$. From Equation (13.5)

$$0 = \frac{\partial}{\partial t} \left[\sum_{n=1}^{\infty} A_n \sin \frac{n\pi x}{L} \cos \omega_n t + \sum_{n=1}^{\infty} B_n \sin \frac{n\pi x}{L} \sin \omega_n t \right]_{t=0},$$

$$0 = \left[\sum_{n=1}^{\infty} A_n \sin \frac{n\pi x}{L} (-\omega_n \sin \omega_n t) + \sum_{n=1}^{\infty} B_n \sin \frac{n\pi x}{L} (\omega_n \cos \omega_n t) \right]_{t=0},$$

$$0 = \sum_{n=1}^{\infty} B_n \sin \frac{n\pi x}{L}.$$

This is zero for all values of x iff $B_n = 0$ for all n. So we are left with

$$y(x,t) = \sum_{n=1}^{\infty} A_n \sin \frac{n\pi x}{L} \cos \omega_n t,$$

$$y(x,0) = \sum_{n=1}^{\infty} A_n \sin \frac{n\pi x}{L}.$$

But $y(x,0)$ has the shape illustrated in Figure 13.4. As you will show in Problem 13.1, this can be expressed as the following Fourier series:

$$y(x,0) = \frac{8b}{\pi^2} \left(\sin \frac{\pi x}{L} - \frac{1}{3^2} \sin \frac{3\pi x}{L} + \frac{1}{5^2} \sin \frac{5\pi x}{L} \cdots \right). \tag{13.6}$$

That is,

$$y(x,0) = \frac{8b}{\pi^2} \left(\sin \frac{\pi x}{L} - \frac{1}{3^2} \sin \frac{3\pi x}{L} + \frac{1}{5^2} \sin \frac{5\pi x}{L} \cdots \right)$$

$$= \sum_{n=1}^{\infty} A_n \sin \frac{n\pi x}{L},$$

where

$$A_n = \frac{8b}{n^2\pi^2} \sin \frac{n\pi}{2} \qquad n = 1,3,5,\ldots,$$

and

$$A_n = 0 \qquad n = 2,4,6,\ldots.$$

The factor $\sin(n\pi/2)$ is included to give the sign of the odd n terms and to delete all the even n terms. Finally, the waveform of the plucked string is

$$y(x,t) = \sum_{n=odd}^{\infty} \left(\frac{8b}{n^2\pi^2} \sin \frac{n\pi}{2} \right) \cos \omega_n t \sin \frac{n\pi x}{L}.$$

Exercise 13.5

Equation (13.6) is a Fourier series. Plotting the terms in this series, we find the integer n is equal to the number of *loops* in the standing wave when only the y_n term is excited. The various modes, $n = 1,2,3,\ldots$ are called the first, second, third ... harmonics. Plot the first four harmonics for a string of length L. ∎

13.4 Traveling Waves

In the previous section we considered a standing wave. Now consider a traveling wave, such as a pulse traveling down an infinite string (or an electromagnetic wave propagating in empty space). Imagine a very long (perhaps infinite) string that is plucked at some point, or perhaps shaken at one end. Your experience with ropes and slinkies tells you that a pulse will move down the string, as pictured in Figure 13.5. If there is no loss of energy, the shape will be repeated at a later time at a further position, as shown in the lower panel of the figure. Suppose the top panel gives the waveform at time $t = 0$, so the plot is $y = y(x,0)$. The bottom panel shows the wave at a later time t. But $y(x,t) = y(x - vt,0)$, that is, the displacement at point x at time t is the same as the displacement at $x - vt$ at the earlier time. Note that during the time interval t, the wave has traveled a distance vt. Therefore, v is the speed at which the waveform propagates. This is called the *phase speed*. (Recall that we identified v with the quantity $\sqrt{F/\rho}$.)

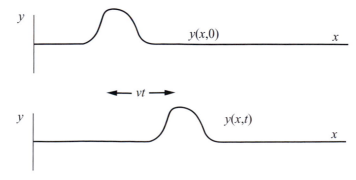

Figure 13.5 A pulse in a string moving to the right. It does not change shape, so the displacement obeys the condition $y(x,t) = y(x + dx, t + dt)$.

The fact that the displacement at time t is the same as the displacement at $x - vt$ at an earlier time suggests that the waveform for a wave traveling at speed v will have the functional form

$$y(x,t) = f(x - vt).$$

In Figure 13.5, the lower panel represents a later time, so the wave is moving toward the right, that is, toward positive x. A wave traveling in the opposite direction would have the functional form

$$y(x,t) = g(x + vt).$$

For example, a sinusoidal traveling wave moving toward the right can be represented as

$$y(x,t) = A \sin k(x - vt).$$

Here I introduced k simply as a constant to make the argument of the sine dimensionless; however, if you think about it, you will realize that k is actually the wave number, so the expression becomes

$$y(x,t) = A \sin(kx - \omega t).$$

You might plot this expression for a few nearby values of time and verify for yourself that this indeed is a representation of a traveling sine wave moving in the positive x-direction.

Although the expression

$$y = f(x - vt)$$

seems a reasonable way to describe a traveling wave, it remains to be shown that it satisfies the wave equation for any function f that depends on x and t in the combination $x - vt$. For example, a number of different possible forms of f would be

$$y(x,t) = A \cos k(x - vt)$$
$$y(x,t) = A e^{ik(x-vt)}$$
$$y(x,t) = A \sin\left(k^2(x - vt)^2\right).$$

The assumption that any function of $(x - vt)$ satisfies the wave equation can be proved by plugging $y = f(x - vt)$ into the equation and noting that if $u = x - vt$, then $\frac{\partial f}{\partial t} = \frac{\partial f}{\partial u}\frac{\partial u}{\partial t} = -v\frac{\partial f}{\partial u}$, so

$$\frac{\partial^2 y}{\partial t^2} = \frac{\partial}{\partial t}\left[\frac{\partial f(u)}{\partial t}\right] = -v\frac{\partial}{\partial t}\left[\frac{\partial f}{\partial u}\right]$$
$$= -v\left[\left(\frac{\partial}{\partial u}\frac{\partial f}{\partial u}\right)\frac{\partial u}{\partial t}\right] = +v^2\frac{\partial^2 f}{\partial u^2},$$

and since $\frac{\partial f}{\partial x} = \frac{\partial f}{\partial u}\frac{\partial u}{\partial x} = \frac{\partial f}{\partial u}(1)$, we can write

$$v^2\frac{\partial^2 y}{\partial x^2} = v^2\frac{\partial}{\partial x}\left[\frac{\partial f}{\partial x}\right] = v^2\frac{\partial}{\partial x}\left[\frac{\partial f}{\partial u}\right]$$
$$= v^2\frac{\partial}{\partial u}\left[\frac{\partial f}{\partial u}\right] = v^2\frac{\partial^2 f}{\partial u^2}.$$

The right-hand sides of the two expressions are equal, proving the wave equation is satisfied for *any* function of $x - vt$.

A particularly useful function of $x - vt$ that is often used to represent a traveling wave is

$$y(x,t) = Ae^{ik(x-vt)} = Ae^{i(kx-\omega t)}.$$

Note that this can be expressed in terms of sines and cosines using the Euler relation:

$$y(x,t) = Ae^{i(kx-\omega t)} = A\left[\cos(kx - \omega t) + i\sin(kx - \omega t)\right].$$

Although this expression satisfies the wave equation, it is not obvious how to interpret the imaginary part. Therefore, we avoid the problem by writing

$$y(x,t) = \mathrm{Re}\left[Ae^{i(kx-\omega t)}\right],$$

that is, only taking the real part of the expression. This formulation is particularly useful when studying electromagnetic waves.

Exercise 13.6

Argue that the constant k introduced above as a proportionality constant must actually be the wave number. ∎

Exercise 13.7

Plot $y = A\sin(kx - \omega t)$ for four nearby values of time to show that this does indeed represent a wave traveling to the right. (Select any reasonable values for the parameters, but make sure that the four values of t are less than a quarter of the period.) ∎

13.5 Standing Waves as a Special Case of Traveling Waves

You have just seen that the solution to the wave equation is any function of $(kx \pm \omega t)$. But the standing wave of Figure 13.1 that was considered earlier was also a solution of the wave equation and it had the form

$$y(x,t) = A\sin kx \cos \omega t + B\sin kx \sin \omega t. \tag{13.7}$$

This expression is a very slightly modified version of Equation (13.4). It does not look much like a function of $(kx \pm \omega t)$! Are the two expressions equivalent? The answer is yes, and the underlying reason is that a standing wave can be considered to be the sum of two traveling waves moving in opposite directions. We now show that this is indeed true.

To be a bit less general but to make the analysis easier to follow, assume the traveling waves are sinusoidal. The general form of two sinusoidal waves traveling in opposite directions is a combination of sine and cosine terms, thus,

$$y(x,t) = A\sin(kx - \omega t) + B\sin(kx + \omega t) + C\cos(kx - \omega t) + D\cos(kx + \omega t).$$

The boundary conditions require that $y = 0$ at $x = 0$ and at $x = L$. If $x = 0$

$$0 = -A\sin \omega t + B\sin \omega t + C\cos \omega t + D\cos \omega t,$$
$$= (B - A)\sin \omega t + (C + D)\cos \omega t.$$

This condition holds at all times if

$$B = A,$$
$$C = -D.$$

and the solution reduces to

$$y = A \left[\sin(kx - \omega t) + \sin(kx + \omega t)\right] + C \left[\cos(kx - \omega t) - \cos(kx + \omega t)\right].$$

Imposing the second boundary condition ($y = 0$ at $x = L$) gives

$$0 = A \left[\sin(kL - \omega t) + \sin(kL + \omega t)\right] + C \left[\cos(kL - \omega t) - \cos(kL + \omega t)\right].$$

This expression is valid at all times, including $t = 0$, in which case it is easy to appreciate that k is equal to any of the values

$$k = \frac{n\pi}{L}, \qquad n = 1, 2, 3 \ldots.$$

(Recall that the standing wave satisfied this condition.)

The sum of the two traveling waves (with $y = 0$ at $x = 0, L$) is, therefore,

$$y(x,t) = A \left[\sin(kx - \omega t) + \sin(kx + \omega t)\right] + C \left[\cos(kx - \omega t) - \cos(kx + \omega t)\right],$$

with $k = n\pi/L$ and $\omega = kv = n\pi v/L$.

Using the trigonometric identities for the sums and differences of angles,[6]

$$y(x,t) = [2A \sin(kx)] \cos \omega t + [2C \sin(kx)] \sin \omega t.$$

Recalling that the standing wave was given by

$$y(x,t) = A \sin(kx) \cos \omega t + B \sin(kx) \sin \omega t,$$

we see that the two expressions are identical except for the notation of the coefficients. Thus, we have shown that the standing wave can be considered the superposition of two traveling waves moving in opposite directions.

Exercise 13.8

Show that $y = 0$ when $x = L$.   ■

13.6 Energy

Let us determine the energy in a wave. As an example, consider a standing wave in a string of length L.

The kinetic energy of the mass element ρdx as it oscillates up and down is

$$dT = \frac{1}{2}\rho dx \left(\frac{dy}{dt}\right)^2.$$

To determine the potential energy note that the element of unstretched length dx has stretched length ds, so the displacement produces a net stretch in the string of $ds - dx$. (See Figure 13.2.) The tension in the string is F, so the work done to stretch the string (which is equal to the increase in its potential energy) is

$$dV = F(ds - dx).$$

---

[6] These are:

$$\sin \alpha + \sin \beta = 2 \sin \frac{1}{2}(\alpha + \beta) \cos \frac{1}{2}(\alpha - \beta)$$

$$\cos \alpha - \cos \beta = -2 \sin \frac{1}{2}(\alpha + \beta) \sin \frac{1}{2}(\alpha - \beta).$$

But

$$ds = \sqrt{dx^2 + dy^2} = \sqrt{1 + \left(\frac{dy}{dx}\right)^2}\, dx,$$

so

$$dV = F\left[\sqrt{1 + \left(\frac{dy}{dx}\right)^2} - 1\right] dx.$$

The binomial expansion of the square root leads to

$$dV = F\,dx\left[1 + \frac{1}{2}\left(\frac{dy}{dx}\right)^2 + \cdots - 1\right] \doteq F\frac{1}{2}\left(\frac{dy}{dx}\right)^2 dx.$$

Consequently, the total energy in a vibrating string of length L is

$$E = T + V = \int_0^L (dT + dV) = \int_0^L \left[\frac{1}{2}\rho\left(\frac{\partial y}{\partial t}\right)^2 + \frac{1}{2}F\left(\frac{\partial y}{\partial x}\right)^2\right] dx.$$

For example, if the wave is described by

$$y(x,t) = A\sin\frac{\pi x}{L}\cos\omega t,$$

then

$$\frac{\partial y}{\partial t} = -\omega A\sin\frac{\pi x}{L}\sin\omega t,$$

and

$$\frac{\partial y}{\partial x} = \frac{\pi}{L}A\cos\frac{\pi x}{L}\cos\omega t.$$

Consequently,

$$E = \int_0^L \left[\frac{1}{2}\rho\omega^2 A^2\left(\sin^2\frac{\pi x}{L}\sin^2\omega t\right) + \frac{1}{2}F\frac{\pi^2}{L^2}A^2\left(\cos^2\frac{\pi x}{L}\cos^2\omega t\right)\right] dx,$$

$$= \frac{1}{2}A^2\left[\rho\omega^2\sin^2\omega t\int_0^L \sin^2\frac{\pi x}{L}dx\right] + \frac{1}{2}A^2\left[F\frac{\pi^2}{L^2}\cos^2\omega t\int_0^L \cos^2\frac{\pi x}{L}dx\right],$$

$$= \frac{1}{2}A^2\frac{L}{2}\left[\rho\omega^2\sin^2\omega t + F\frac{\pi^2}{L^2}\cos^2\omega t\right].$$

Now recall that $F/\rho = v^2$ and $\omega = \pi v/L$, so $F\pi^2/L^2 = \rho v^2\pi^2/(\pi^2 v^2/\omega^2) = \rho\omega^2$ and

$$E = \frac{1}{2}A^2\frac{L}{2}\rho\omega^2(\sin^2\omega t + \cos^2\omega t)$$

$$= A^2\rho\omega^2\frac{L}{4}.$$

Note that the energy is proportional to the amplitude squared.

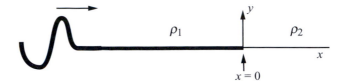

Figure 13.6 Two strings of different linear densities are connected at $x = 0$.

13.6.1 Energy Flow

Consider a wave passing from one medium into another. An example is the sunlight incident on the window of your room. As you know, some of the incident light will be transmitted and some will be reflected. Reflection is a characteristic property of any kind of wave when it is incident upon the interface of two materials in which the wave travels at different speeds. For the light wave, glass has a larger index of refraction than air, and the wave travels more slowly in glass. For a wave in a string, the speed depends on the mass per unit length of the string (ρ).

An analogy to a light wave incident on glass would be a traveling wave in a string of linear density ρ_1 incident on a second string of different linear density ρ_2. This is illustrated in Figure 13.6.

Imagine a wave coming in from the left (from $x = -\infty$). Call this the *incident* wave. At the junction of the two strings (at $x = 0$), the incident wave is split into two waves, the *reflected* wave that moves back toward $x = -\infty$, and the *transmitted* wave that continues on toward $x = +\infty$. Let us denote these three waves by y_i, y_r, and y_t. These are all traveling waves, so they can be represented as follows:

$$
\begin{aligned}
y_i &= A_i \cos(k_1 x - \omega t), \\
y_r &= A_r \cos(k_1 x + \omega t), \\
y_t &= A_t \cos(k_2 x - \omega t).
\end{aligned}
\tag{13.8}
$$

Note the sign on the ωt term in the expression for y_r and the subscript on k for y_t.

The speed of the wave is different in the two mediums. Because $v = \omega/k$, a change in the speed could be due to a change in either ω or k (or both). But it turns out that the waves have the same value of ω. You may wonder why k changes but ω stays the same. The reason is that the frequency of the wave is determined by the physical mechanism that is generating the wave. If you are standing at one end of the string and shaking it up and down, you are controlling the frequency of the wave by how quickly (or slowly) you move the end of the rope. This you can control. However, you have no control over the wavelength. The wavelength depends on the properties of the medium. When the properties of the medium change the wavelength changes, and so the wave number $k = 2\pi/\lambda$ also changes.

The energy of a wave is proportional to the square of its amplitude. An interesting and important question is how much energy is reflected at an interface and how much is transmitted into the second medium. The question can be answered by determining the amplitudes of the reflected and transmitted waves relative to the incident amplitude. In other words, it is desirable to determine the ratios A_r/A_i and A_t/A_i. This can be done by realizing that both y and dy/dx must be continuous at the junction between the two strings. The value of y is continuous because the strings are attached to one another. The reason why dy/dx is the same on both sides of the junction is that the tension is the same on both sides. These two boundary conditions can be expressed as

$$[y_i + y_r]_{x=0} = [y_t]_{x=0},$$

$$\left[\frac{\partial y_i}{\partial x} + \frac{\partial y_r}{\partial x}\right]_{x=0} = \left[\frac{\partial y_t}{\partial x}\right]_{x=0}.$$

Substituting the expressions from Equations (13.8) we obtain

$$A_i \cos(-\omega t) + A_r \cos(+\omega t) = A_i \cos(-\omega t),$$

$$A_i + A_r = A_t,$$

and

$$k_1(A_i - A_r) = k_2 A_t,$$

or

$$\frac{1}{v_1}(A_i - A_r) = \frac{1}{v_2}A_t.$$

Therefore,

$$\frac{A_r}{A_i} = \frac{v_2 - v_1}{v_1 + v_2},$$

$$\frac{A_t}{A_i} = \frac{2v_2}{v_1 + v_2}.$$

Note that, since $v = \sqrt{F/\rho}$, if the second string is heavier than the first, $\rho_2 > \rho_1$ and $v_2 < v_1$ and A_r is negative. This means the reflected wave is inverted ("upside down"). Going to the extreme case, if the second string is infinitely massive, $v_2 = 0$ and $A_t = 0$. This is equivalent to string 1 being "nailed down" at $x = 0$.

To determine the energy flow through the junction consider the rate at which work is done by a particle on one side of $x = 0$ on a particle on the other side. Assume the particle on the left (at $x = 0^-$) exerts a downward force on the particle at $x = 0^+$, given by

$$F_y = -F \sin\theta \doteq -F \tan\theta = -F\frac{\partial y}{\partial x}.$$

The rate at which work is done is the power and is given by $P = \mathbf{F} \cdot \mathbf{v}$. Therefore,

$$\left[\frac{dW}{dt}\right]_{x=0^+} = \left[F_y\frac{\partial y}{\partial t}\right]_{x=0}$$

$$= \left[\left(-F\frac{\partial y_t}{\partial x}\right)\frac{\partial y_t}{\partial t}\right]_{x=0}$$

$$= -F\left[(k_2 A_t \sin(k_2 x - \omega t))(A_t(-\omega)\sin(k_2 x - \omega t))\right]_{x=0}$$

$$= F A_t^2 k_2 \omega \sin^2(\omega t) = F A_t^2(\omega^2/v_2)\sin^2(\omega t),$$

where we used $k = \omega/v$. The average of $\sin^2(\omega t)$ over one period is $1/2$. The average energy flow into string number 2 is equal to the average work done, so

$$\left\langle\frac{dE}{dt}\right\rangle_{x=0^+} = \frac{F A_t^2 \omega^2}{2v_2}.$$

An equivalent analysis for the energy leaving string number 1 gives

$$\left\langle\frac{dE}{dt}\right\rangle_{x=0^-} = \frac{F(A_i^2 - A_r^2)\omega^2}{2v_1}.$$

These expressions must be equal if there is no energy loss at the junction.

Worked Example 13.2 Obtain an expression for the reflection coefficient R defined as the ratio of the intensity of the reflected wave to the intensity of the incident wave ($R = I_r/I_i$).

Solution The intensity of a wave is defined as the energy transported across a normal unit area per unit time, that is, power per unit area. Because energy is proportional to the amplitude squared

$$R = \frac{I_r}{I_i} = \frac{A_r^2/(\text{area} \times \text{time})}{A_i^2/(\text{area} \times \text{time})} = \frac{A_r^2}{A_i^2} = \left(\frac{v_2 - v_1}{v_2 + v_1}\right)^2$$

$$= \left(\frac{k_1 - k_2}{k_1 + k_2}\right)^2.$$

Exercise 13.9

Show that if the tension is continuous at the junction, the slope dy/dx is continuous. ■

Exercise 13.10

Obtain the appropriate boundary condition for two strings joined by a knot of mass m. Answer:

$$F\left(\left.\frac{\partial y}{\partial x}\right|_{0^+} - \left.\frac{\partial y}{\partial x}\right|_{0^-}\right) = m\left.\frac{\partial^2 y}{\partial t^2}\right|_0. \qquad ■$$

13.7 Momentum (Optional)

The particles in a transverse wave move perpendicular to the motion of the wave. Thus, it would seem that a transverse wave would not transport momentum longitudinally. However, it is a known fact that waves in strings *do* transport momentum. This means that our assumption of purely transverse motion for a particle in a string must be wrong. There must be some longitudinal motion as well as transverse motion.[7] This can be easily appreciated conceptually by (mentally) replacing the string with a series of particles of mass m connected by massless springs of constant k, as shown in Figure 13.7. As indicated in the figure, if one particle is plucked so that it is displaced vertically, it causes the particles on either side to be displaced both vertically *and* horizontally.

When the plucked particle is released, it sets up a (large amplitude) transverse wave as well as a (small amplitude) longitudinal wave. Rowland and Pask[8] carried out numerical studies using

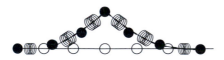

Figure 13.7 A string modeled as a series of particles connected by springs. The open circles represent the equilibrium positions of the particles. The filled circles represent their positions when a particle is displaced vertically.

[7] Electromagnetic waves in a vacuum are transverse. The momentum density associated with these waves is $\epsilon_0 \mathbf{E} \times \mathbf{B}$, which is perpendicular to the transverse oscillations of $\mathbf{E}$ and $\mathbf{B}$. For a string, however, perfectly transverse waves would not have a longitudinal component of momentum.

[8] David R. Rowland and Colin Pask, "The Missing Wave Momentum Mystery," *American Journal of Physics*, **67**, 378–388 (1999) and David R. Rowland, "Comment on 'What Happens to Energy and Momentum When Two Oppositely-Moving Wave Pulses Overlap?' by N. Gauthier," *American Journal of Physics*, **72**, 1425–1429 (2004). It is interesting to note that the theory of a vibrating string was first developed by Lagrange in 1759 but there are still

this model of a string and showed the existence of the longitudinal waves. (A particle in the string actually traces out a tall, thin figure eight, moving up and down in a large amplitude oscillation while oscillating horizontally with a small amplitude oscillation of the same frequency.)

Consider a mass element dm of a string. In the absence of a wave or pulse, this element of mass will be located at (x_0, y_0). However, if there is a wave in the string, the mass element will move vertically and will also undergo a small horizontal displacement. For convenience, we shall denote the horizontal and vertical displacements from equilibrium by (ξ, η). Because ξ is very small ($\xi \ll \eta$) we are justified in setting $\eta = y$.

Assume the ends of the string are located at fixed points separated by a distance L_0. When a wave (or pulse) is generated in the string, the string will be *stretched* by a small amount, and the length of the string will now be

$$L = \int_0^{L_0} ds = \int_0^{L_0} \sqrt{dx^2 + dy^2} = \int_0^{L_0} \sqrt{1 + \left(\frac{dy}{dx}\right)^2} \, dx$$

$$\simeq L_0 + \int_0^{L_0} \frac{1}{2}\left(\frac{dy}{dx}\right)^2 dx,$$

where we used the binomial expansion.

The "stretch" of the string is

$$\Delta L = \int_0^{L_0} \frac{1}{2}\left(\frac{dy}{dx}\right)^2 dx = \int_0^{L_0} \frac{1}{2}\eta'^2 dx$$

where we replaced y with η. Note that the prime indicates differentiation with respect to x.

Now let us express the wave as $y = f(x - vt)$ which we can write as $\eta = f(u)$ where $u = x - vt$. Then

$$\eta' = \frac{d\eta}{dx} = \frac{d\eta}{du}\frac{du}{dx} = \frac{d\eta}{du}(1). \tag{13.9}$$

Similarly, the time derivative of η, denoted $\dot{\eta}$, is

$$\dot{\eta} = \frac{d\eta}{dt} = \frac{d\eta}{du}\frac{du}{dt} = \frac{d\eta}{du}(-v). \tag{13.10}$$

Equating the expressions for $\frac{d\eta}{du}$ from Equations (13.9) and (13.10) we obtain

$$\eta' = -\frac{1}{v}\dot{\eta}.$$

Therefore,

$$\Delta L = \int_{wave} \frac{1}{2}\eta'^2 dx = \int_{wave} \frac{1}{2}\eta'\eta' dx = \int_{wave} \frac{1}{2}\eta'\left(\frac{-1}{v}\dot{\eta}\right) dx,$$

but $v = \frac{dx}{dt}$ so

$$\Delta L = \int_{wave} -\frac{1}{2}\eta'\dot{\eta}\frac{dt}{dx} dx = \int_{wave} -\frac{1}{2}\eta'\dot{\eta} dt. \tag{13.11}$$

Note that we changed the limits on the integral to simply *"wave"* because changing the variable of integration requires changing the limits.

interesting aspects of the problem to be explored. For example, Eugene Butikov discusses the potential energy associated with the small longitudinal stretching of a transverse wave in the paper "Peculiarities in the Energy Transfer by Waves on Strained Strings," *Physica Scripta*, **88**, 6 (2013).

But ΔL can also be expressed as

$$\Delta L = \int_{wave} \dot{\xi} \, dt. \tag{13.12}$$

Equating the expressions for ΔL from Equations (13.11) and (13.12) yields

$$\dot{\xi} = -\frac{1}{2} \eta' \dot{\eta}.$$

Thus, we have obtained an expression for the horizontal velocity of an element of the string in terms of the transverse quantities η' and $\dot{\eta}$. Consequently, the horizontal component of momentum density (momentum per unit length) is just

$$g = \rho \dot{\xi} = -\frac{1}{2} \rho \eta' \dot{\eta}.$$

Although our analysis is correct, it might be criticized as not being complete because, depending on initial conditions, there may also be a small amplitude purely longitudinal wave generated in the string. We have ignored this effect. Also, it might be noted that many older textbooks and papers in physical journals erroneously left off the factor of $1/2$ for subtle reasons that are described by Rowland and Pask.[9]

13.8 Summary

In this chapter you were exposed to an analysis of waves in strings. The relations obtained, however, are applicable (with slight modifications) to other kinds of waves.

Considering the motion in an element of a string and applying Newton's second law, one can derive the relation

$$\frac{\partial^2 y}{\partial t^2} = \frac{F}{\rho} \frac{\partial^2 y}{\partial x^2}.$$

For a traveling wave, the speed of propagation (phase speed) is $v = \sqrt{F/\rho}$ and the relation above becomes

$$\frac{\partial^2 y}{\partial t^2} = v^2 \frac{\partial^2 y}{\partial x^2},$$

which is called the *wave equation.*

The wave equation is solved by the technique of *separation of variables*, leading to the general solution

$$y(x,t) = (A \cos kx + B \sin kx)(C \cos \omega t + D \sin \omega t).$$

The "arbitrary" constants A, B, C, D are determined from the boundary conditions. For a standing wave in a string tied down at both ends, the solution can be expressed as

$$y(x,t) = \sum_{n=1}^{\infty} A_n \sin \frac{n\pi x}{L} \cos \omega t + \sum_{n=1}^{\infty} B_n \sin \frac{n\pi x}{L} \sin \omega t.$$

The coefficients A_n and B_n can be determined if the waveform is known at some initial time by expanding in a Fourier series.

---

[9] Rowland and Pask, *American Journal of Physics*, **67**, 378–388 (1999).

A traveling wave can always be expressed as a function of $(x - vt)$ and $(x + vt)$, or $(kx \pm \omega t)$. For example, a wave moving toward positive x can be expressed as

$$y(x,t) = \text{Re}\left[Ae^{i(kx-\omega t)}\right].$$

A standing wave is the sum of two traveling waves moving in opposite directions.

The energy in a wave is proportional to the square of the amplitude of the wave. For example, a standing wave in a stretched string of length L has energy

$$E = A^2 \rho \omega^2 \frac{L}{4}.$$

When a wave is incident on the interface between two mediums it will, in general, give rise to both a reflected wave and a transmitted wave. The relative amplitudes of these waves are:

$$\frac{A_r}{A_i} = \frac{v_1 - v_2}{v_1 + v_2},$$
$$\frac{A_t}{A_i} = \frac{2v_2}{v_1 + v_2},$$

where v_1 and v_2 are the speeds of the waves in medium 1 and medium 2. The average energy transmitted per unit time is:

$$\left\langle\frac{dW}{dt}\right\rangle_t = \frac{F\omega^2}{2v_2}A_t^2.$$

13.9 Problems

Problem 13.1 Obtain expression (13.6) for the shape of the string in Figure 13.4.

Problem 13.2 A massless elastic string of length l and constant k is loaded with n evenly spaced particles of mass m. The string is stretched to $l + \Delta l$. Determine the speed of the transverse oscillations.

Problem 13.3 A string of linear mass density ρ and length L is fixed at both ends and subjected to a tension F. Somehow it is stretched into a parabolic shape described by $y = a(L^2/4 - x^2)$, where $x = 0$ corresponds to the center of the string and a is a constant having units of inverse length. (a) Plot $y = y(x,0)$ for $-L/2 \le x \le L/2$. (b) Express $y(x,0)$ as a Fourier series. (c) The string is released at time $t = 0$. Write an expression for $y = y(x,t)$.

Problem 13.4 A string of length L under tension F_0 is released from rest. Assume that initially the shape of the string was given by $y(x,0) = A \sin \frac{\pi x}{L}$. Determine $y(x,t)$.

Problem 13.5 A string of length 1 m and mass 50 g is fixed at both ends. It is subjected to a tension of 50 N. The midpoint of the string is pulled upwards 2 cm and released. Determine the motion.

Problem 13.6 A uniform string of length L and mass density ρ is subjected to a tension F_0. The end points are fixed. Somehow it is given an initial shape that is the arc of a circle. The string is then released from rest. Because the string is not displaced very far from its horizontal equilibrium position, the radius of the circular arc is considerably longer than the length of the chord, so terms smaller than $(L/R)^2$ can be ignored. Determine the motion.

Problem 13.7 The three-dimensional wave equation is $\nabla^2 \Phi - \frac{1}{v^2}\frac{\partial^2 \Phi}{\partial t^2} = 0$, where $\Phi = \Phi(x, y, z, t)$. Solve by separation of variables.

Problem 13.8 Consider the traveling wave represented by $y(x,t) = Ae^{i(kx-\omega t)}$. Show that if k is complex and ω is real, this represents a damped wave.

Problem 13.9 A vibrating string is oscillating in a viscous medium. Assume that all portions of the string are subjected to a retarding force proportional to the speed given by $-b\frac{\partial y}{\partial t}dx$. (a) Write the equation of motion. (b) Assume $y = \sum_{n=1}^{\infty} \phi_n \sin n\pi x/L$ (the "normal" solution). Plug into the equation of motion, multiply by $\sin m\pi x/L$, and integrate to obtain a differential equation for $\phi_m = \phi_m(t)$. (c) Solve for ϕ_m and write the solution $y = y(x,t)$.

Problem 13.10 In Worked Example 13.1 we expressed the displacement of a plucked string as the Fourier series

$$y(x,t) = \sum_{n \text{ odd}}^{\infty} \frac{8b}{n^2\pi^2}\left(\sin\frac{n\pi}{2}\right)\cos\frac{n\pi ct}{L}\sin\frac{n\pi x}{L}.$$

Write this as the sum of traveling waves. Write out the first three terms explicitly.

Problem 13.11 Show that the Lagrangian for a transversely vibrating string is $L = \frac{1}{4}\sum_{n=1}^{\infty}\left(\rho l \dot{Q}_n^2 - \frac{F_0\pi^2}{l}n^2 Q_n^2\right)$, where the "normal" coordinates Q_n are given by $Q_n = A_n\cos\omega_n t + B_n\sin\omega_n t$.

Problem 13.12 The Lagrangian for the vibrating string is given as the solution of Problem 13.11. (a) Obtain the equation of motion. (b) Solve for the "normal" coordinates Q_n.

Problem 13.13 A stretched string has one fixed end, but the other end (at $x = L$) is tied to a massless ring that slides on a frictionless vertical rod. Determine the boundary conditions and obtain an expression for a wave in this string.

Problem 13.14 One end of a string of length L is connected to a mechanism that makes it oscillate according to $y(0,t) = A\sin\omega t$. The other end ($x = L$) is fixed. Determine the motion.

Problem 13.15 A string of linear density ρ_1 is tied to a second string of smaller linear density ρ_2. Assume the massless knot joining the strings is at $x = 0$, and that the strings are infinitely long. A pulse y_i propagates in the first string in the $+x$ direction. When it reaches the knot, it produces a reflected pulse y_r and a transmitted pulse, y_t. (a) Write the boundary conditions at the knot using the continuity of the strings and the slopes of the strings at that point. (b) Show that the amplitudes of the reflected and transmitted waves are related to the amplitude of the incident wave by

$$A_r = \frac{v_2 - v_1}{v_2 + v_1}A_i, \quad \text{and} \quad A_t = \frac{2v_2}{v_2 + v_1}A_i.$$

Problem 13.16 Two long strings with linear mass densities ρ_1 and ρ_2 with $\rho_1 > \rho_2$ are joined by a massless knot. See Figure 13.6.
(a) Show that the transmitted energy per unit time (averaged over a period) is $(F\omega^2/2v_2)A_t^2$.
(b) Show that the rate energy is supplied to the junction is

$$(F\omega^2/2v_2)(A_i^2 - A_r^2).$$

(c) Prove that these two terms are equal.

Problem 13.17 In Worked Example 13.1 we solved the problem of a plucked string and obtained a standing wave. Now solve the problem by assuming the resultant standing wave is the sum of two traveling waves $f(x - vt)$ and $g(x + vt)$. Make sure initial and boundary conditions are satisfied.

Problem 13.18 A string of length L is fixed at both ends. At time $t = 0$ the displacement at any point is given by $u(x)$ and the vertical speed is given by $v(x)$. Recall that the general solution can be written as

$$y(x,t) = \sum_{n=1}^{\infty}(A_n\sin\omega_n t + B_n\cos\omega_n t)\sin\frac{n\pi x}{L}.$$

Show that

$$A_n = \frac{2}{\omega_n L} \int_0^L v(x) \sin \frac{n\pi x}{L} dx,$$

and

$$B_n = \frac{2}{L} \int_0^L u(x) \sin \frac{n\pi x}{L} dx.$$

Problem 13.19 A simple model of a one-dimensional crystal consists of a straight line of n particles of equal mass, initially equally spaced along a light elastic string that is clamped at both ends (Figure 13.8). The particles can move longitudinally, that is, along the string. Let q_i be the displacement of the ith particle from its equilibrium position, noting that i ranges from 1 to n and that $q_0 = q_{n+1} = 0$. (a) Show that the Lagrangian for this system is $L = \frac{1}{2} \sum_{i=1}^n \left[m\dot{q}_i^2 - k(q_{i+1} - q_i)^2 \right]$, where k is the elastic constant. (b) Obtain the equations of motion. (c) Show that if $q_i = A_i \cos \omega t$, then the amplitudes of the oscillations are related by the recursion relation $A_i = \frac{k}{m\omega^2 + 2k}(A_{i+1} + A_{i-1})$.

Problem 13.20 A normal mode consists of an oscillation at a single frequency. The general motion of a vibrating string is

$$y(x,t) = \sum_{n=1}^{\infty} (A_n \sin \omega_n t + B_n \cos \omega_n t) \sin \frac{n\pi x}{L}.$$

This can be expressed as a sum of normal modes by defining the normal coordinate ϕ_n as

$$\phi_n = A_n \sin \omega_n t + B_n \cos \omega_n t,$$

leading to

$$y(x,t) = \sum_{n=1}^{\infty} \phi_n \sin \frac{n\pi x}{L}.$$

(a) Express the total energy in terms of the normal coordinates ϕ_n. (b) Express the Lagrangian in terms of the normal coordinates and obtain the equations of motion in terms of ϕ_n.

Problem 13.21 Consider a transversely vibrating string of length l and linear mass density ρ. The string is fixed at both ends and is subjected to a tension F. Derive the Lagrangian and the Hamiltonian for the string by carrying out the indicated steps. First note that the wave can be expressed as

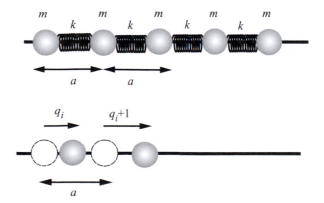

Figure 13.8 A string modeled as a series of particles connected by springs. In the lower diagram, the open circles represent the equilibrium positions of the particles. The filled circles represent their positions when a particle is displaced longitudinally. (Problem 13.19.)

$$y(x,t) = \sum_n Z_n \sin k_n x,$$

where $Z_n = Z_n(t) = A_n \sin \omega_n t + B_n \cos \omega_n t$. We can treat Z_n as a generalized coordinate. (a) Show that the kinetic and potential energies can be written as

$$T = \frac{1}{4}\rho l \sum_n \dot{Z}_n^2$$

$$V = \frac{\pi^2 F}{4l} \sum_n Z_n^2.$$

(b) Write the Lagrangian using Z_n as the generalized coordinate.
(c) Obtain the Lagrange equations of motion.
(d) Write the Hamiltonian, noting that the conjugate generalized momentum is $p_n = \frac{\partial L}{\partial \dot{Z}_n}$.

Computational Projects

Computational Project 13.1 Using the expression for $y(x,0)$ obtained in Worked Example 13.1, sum the first five terms and plot. Then sum the first ten terms and plot.

Computational Project 13.2 Given: $y(x,t) = A \sin(kx - \omega t)$. Plot $y(x,t)$ as a function of x for five nearby values of t and verify that this represents a wave moving in the positive x-direction. Choose reasonable values for the constants A, k, ω.

Computational Project 13.3 Consider the two traveling waves $y_1(x,t) = A \sin(kx - \omega t)$ and $y_2(x,t) = A \sin(kx + \omega t)$. Show that the sum of these waves is a standing wave. To do so, generate a three-panel ("movie") plot showing the two traveling waves and the standing wave as functions of time.

Computational Project 13.4 Using the model of Figure 13.7 as a basis and displacing a particle vertically, show the development of a traveling wave. Compare your results with Figure 5 of Rowland and Pask, *American Journal of Physics*, **67**, 383 (1999).

Small Oscillations (Optional)

14.1 Introduction

I denoted this chapter as "optional" because it contains essentially no new physics. However, it does introduce some useful mathematical concepts and techniques. The methods introduced here are applied in several areas of physics including quantum mechanics and solid-state physics.

We shall limit our considerations to the study of small oscillations of coupled systems of particles about their equilibrium positions (such as the oscillations of the molecules in a solid). Idealized examples of such systems are the particles connected by massless springs in Figure 14.1 and the coupled pendulums shown in Figure 14.2. These systems were treated in a more elementary manner in Chapter 11; you may wish to refer to that chapter as we go along.[1]

14.2 Statement of the Problem

Figure 14.1 illustrates a one-dimensional system consisting of n particles whose positions might be given by the Cartesian coordinates $x_1, x_2, \ldots$ referred to an arbitrary origin. It will not surprise you that the problem is best formulated in a terms of a set of generalized coordinates, such as η_1 and η_2 as shown in Figure 14.2. But before we specify a particular set of generalized coordinates, let us consider a one-dimensional system consisting of n particles whose positions are given by $q_1, q_2, \ldots$. Let us assume the system is conservative and has a potential energy whose value depends only on the coordinates, thus:

$$V = V(q_1, \ldots, q_n). \tag{14.1}$$

Next we need to determine the kinetic energy. We shall further assume the Lagrangian for the system is a function only of the qs and the $\dot{q}$s, and does not depend on time:

$$L = L(q_1, \ldots, q_n; \dot{q}_1, \ldots, \dot{q}_n). \tag{14.2}$$

14.2.1 Static Equilibrium

As pointed out in Section 1.6, in equilibrium all accelerations are zero and in static equilibrium all velocities are also zero. Thus, if the system is in static equilibrium, $\ddot{q}_i = 0$ and $\dot{q}_i = 0$ for all i. We shall denote the value of q_i in a state of static equilibrium by q_i^0.

In terms of Cartesian coordinates, measured from an arbitrary origin, the kinetic energy is $T = \sum_i \frac{1}{2} m_i \dot{x}_i^2$. Because the transformation equations can be written as $x_i = x_i(q_1, \ldots, q_n)$, the total time derivative of x_i is

---

[1] A very good discussion of small oscillations is found in the text by A. L. Fetter and J. D. Walecka, *Theoretical Mechanics of Particles and Continua*, McGraw Hill, New York, 1980, Chapter 4.

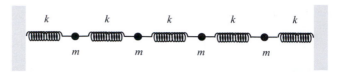

Figure 14.1 Coupled masses. The masses are equal and are connected by springs having equal spring constants.

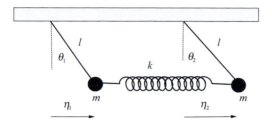

Figure 14.2 Coupled pendulums. The bobs have equal masses and the massless strings are of equal length. The bobs are connected by a massless spring.

$$\dot{x}_i = \frac{d}{dt}x_i = \sum_{j=1}^{n} \frac{\partial x_i}{\partial q_j}\frac{dq_j}{dt} = \sum_{j=1}^{n} \frac{\partial x_i}{\partial q_j}\dot{q}_j.$$

Consequently, the kinetic energy can be written as follows:

$$T = \sum_{i=1}^{n} \frac{1}{2}m_i \dot{x}_i^2 = \sum_{i=1}^{n} \frac{1}{2}m_i \left(\sum_{j=1}^{n} \frac{\partial x_i}{\partial q_j}\dot{q}_j\right)\left(\sum_{k=1}^{n} \frac{\partial x_i}{\partial q_k}\dot{q}_k\right)$$

$$= \frac{1}{2}\sum_{j=1}^{n}\sum_{k=1}^{n}\left\{\sum_{i=1}^{n} m_i \frac{\partial x_i}{\partial q_j}\frac{\partial x_i}{\partial q_k}\right\}\dot{q}_j\dot{q}_k$$

$$= \frac{1}{2}\sum_{j=1}^{n}\sum_{k=1}^{n} M_{kj}\dot{q}_j\dot{q}_k, \tag{14.3}$$

where M_{kj} is *not* the mass of a particle. Equation (14.3) gives the expression for kinetic energy in terms of generalized coordinates for time-independent transformation equations.[2] Note that the expression for the kinetic energy is a quadratic form.

14.2.2 The Mass Matrix

The quantity M_{kj} in Equation (14.3) is defined by

$$M_{kj} = \sum_{i} m_i \left(\frac{\partial x_i}{\partial q_j}\right)_{q^0}\left(\frac{\partial x_i}{\partial q_k}\right)_{q^0}, \tag{14.4}$$

where the subscripts q^0 indicate that the expressions are evaluated at points of static equilibrium. Note that $M_{kj} = M_{jk} = M_{kj}^*$.

The quantities M_{kj} are the elements of a real, constant, symmetric matrix called the *mass matrix* which is denoted by $\mathcal{M}$. We shall be using this later when we formulate the problem in terms of matrices.

---

[2] A more general (and more complicated) expression is given in K. Symon, *Mechanics*, 3rd ed., Addison Wesley, Reading, MA, 1971, p. 357, Equation (9.9). However, Equation (14.3) is applicable to most problems and is sufficient for our present purposes.

When the system is in static equilibrium, $\dot{q}_i = 0$ and, according to Equation (14.3), $T = 0$. Note also that $\frac{\partial T}{\partial \dot{q}_i} = 0$ because T depends only on the $\dot{q}_i$s, and furthermore $\frac{\partial T}{\partial \dot{q}_i} = 0$ in static equilibrium because $\dot{q}_i = 0$. That is,

$$\frac{\partial T}{\partial \dot{q}_i} = \frac{1}{2} \sum_{j=1}^{n} M_{ij} \dot{q}_j = 0.$$

Thus, in static equilibrium, the coordinates are all constant and the kinetic energy is zero. The demonstration of this result may have been a bit abstract, but the result itself merely states a reasonable fact. Furthermore, as shown in Exercise 14.1, the generalized force vanishes.

Exercise 14.1

Write the Lagrange equation in the Nielsen form as in Equation (4.13) and show that the generalized force is zero. ∎

14.2.3 The Potential Matrix

Figure 14.3 shows three possible shapes for the potential energy near an equilibrium point.

Stable equilibrium implies that if the system is displaced from the equilibrium point by a small amount it will tend to return to the equilibrium point. From the figure we note that only for case (a) in which the potential is a minimum at the equilibrium point is the system in *stable* equilibrium. For cases (b) and (c) a similar displacement (even by an infinitesimal amount) does not result in the system returning to the original position.

A small displacement from equilibrium can be represented by

$$q_i = q_i^0 + \eta_i, \qquad i = 1, \ldots, n, \tag{14.5}$$

where η_i is a small quantity because we are interested in the behavior of a system in a near-equilibrium situation. Equation (14.5) implies that $\dot{q}_i = \dot{\eta}_i$, a result we will be using shortly.

Using Equation (14.5) the potential energy near an equilibrium point can be written as

$$V(q_1, \ldots, q_n) = V(q_1^0 + \eta_1, q_2^0 + \eta_2, \ldots, q_n^0 + \eta_n). \tag{14.6}$$

Carrying out a Taylor's series expansion about q_i^0 leads to

$$V(q_1, \ldots, q_n) = V(q_1^0, \ldots, q_n^0) + \sum_{j=1}^{n} \eta_j \left(\frac{\partial V}{\partial q_j} \right)_{q^0}$$

$$+ \frac{1}{2} \sum_{j=1}^{n} \sum_{k=1}^{n} \eta_j \eta_k \left(\frac{\partial^2 V}{\partial q_j \partial q_k} \right)_{q^0} + \cdots . \tag{14.7}$$

Now the first term on the right-hand side is $V(q_1^0, \ldots, q_n^0)$, which is just a constant and can be set equal to zero by redefining the zero point of potential energy. The next term on the right involves

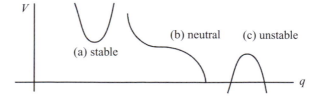

Figure 14.3 Three types of equilibrium: (a) stable, (b) neutral, and (c) unstable.

$\left(\frac{\partial V}{\partial q_j}\right)_{q_j^0}$. But at the equilibrium point the slope of the potential energy curve is zero, so this term is equal to zero. Ignoring terms of order higher than η^2, Equation (14.7) reduces to

$$V = \frac{1}{2} \sum_{j=1}^{n} \sum_{k=1}^{n} \left(\frac{\partial^2 V}{\partial q_j \partial q_k}\right)_{q^0} \eta_j \eta_k,$$

$$= \frac{1}{2} \sum_{j=1}^{n} \sum_{k=1}^{n} V_{kj} \eta_j \eta_k, \tag{14.8}$$

where

$$V_{kj} = \left(\frac{\partial^2 V}{\partial q_j \partial q_k}\right)_{q^0}. \tag{14.9}$$

This expression defines the elements of a constant, real, symmetric matrix called the *potential matrix*, which we will denote $\mathcal{V}$.

14.2.4 The Lagrange Equations

For a system displaced slightly from stable equilibrium $\dot{q}_i = \dot{\eta}_i$. The kinetic energy T is given by Equation (14.3) and the potential energy is given by Equation (14.8) so the Lagrangian is

$$L = T - V = \frac{1}{2} \sum_{kj} \left(M_{kj} \dot{\eta}_j \dot{\eta}_k - V_{kj} \eta_j \eta_k\right). \tag{14.10}$$

The Lagrange equations of motion are

$$\frac{d}{dt} \frac{\partial L}{\partial \dot{\eta}_k} - \frac{\partial L}{\partial \eta_k} = 0.$$

From Equation (14.10)

$$\frac{d}{dt} \frac{\partial L}{\partial \dot{\eta}_k} = \frac{d}{dt} \frac{1}{2} \sum_j M_{kj} \dot{\eta}_j = \frac{1}{2} \sum_j M_{kj} \ddot{\eta}_j,$$

and

$$\frac{\partial L}{\partial \eta_k} = -\frac{1}{2} \sum_j V_{kj} \eta_j,$$

so the equations of motion are

$$\sum_{j=1}^{n} \left(M_{kj} \ddot{\eta}_j + V_{kj} \eta_j\right) = 0 \qquad k = 1, \ldots, n. \tag{14.11}$$

Note that M_{kj} and V_{kj} are symmetric so the order of the subscripts makes no difference. Equations (14.11) are a set of n linear, homogeneous, coupled, second-order differential equations with real, constant coefficients. Each equation has $2n$ terms involving the elements of the $n \times n$ matrices we have named the *mass matrix* $\mathcal{M}$ and the *potential matrix* $\mathcal{V}$. If you can find functions $\eta_i(t)$ that satisfy Equations (14.11), you have solved the dynamical problem.

Equation (14.11) can be written in matrix form as

$$\mathcal{M} \ddot{\eta} + \mathcal{V} \eta = 0,$$

where $\mathcal{M}$ and $\mathcal{V}$ are the $n \times n$ mass and potential matrices and η is an n-component column vector giving the displacements of all the particles from their equilibrium positions.

Worked Example 14.1   Determine the potential matrix and the mass matrix for the system of two masses connected by springs as shown in Figure 11.11.

Solution Note that the spring between the two masses has spring constant k_3 and the springs attached to the walls and to masses m_1 and m_2 have spring constants k_1 and k_2 respectively. The positions of the masses (relative to equilibrium) are x_1 and x_2 (measured positively to the right). Therefore, the potential energy is

$$V = \frac{1}{2}k_1 x_1^2 + \frac{1}{2}k_2 x_2^2 + \frac{1}{2}k_3(x_2 - x_1)^2.$$

We now evaluate

$$V_{kj} = \left(\frac{\partial^2 V}{\partial q_j \partial q_k} \right)_{q^0},$$

where $q_1 = x_1$ and $q_2 = x_2$. The various elements are:

$$V_{11} = \frac{\partial^2 V}{\partial x_1 \partial x_1} = \frac{\partial}{\partial x_1}[k_1 x_1 - k_3(x_2 - x_1)] = k_1 + k_3,$$

$$V_{12} = \frac{\partial^2 V}{\partial x_1 \partial x_2} = \frac{\partial}{\partial x_1}[k_2 x_2 + k_3(x_2 - x_1)] = -k_3,$$

$$V_{21} = \frac{\partial^2 V}{\partial x_2 \partial x_1} = \frac{\partial}{\partial x_2}[k_1 x_1 - k_3(x_2 - x_1)] = -k_3,$$

$$V_{22} = \frac{\partial^2 V}{\partial x_2 \partial x_2} = \frac{\partial}{\partial x_2}[k_2 x_2 + k_3(x_2 - x_1)] = k_2 + k_3,$$

and so the potential matrix is

$$\mathcal{V} = \begin{pmatrix} k_1 + k_3 & -k_3 \\ -k_3 & k_2 + k_3 \end{pmatrix}.$$

The mass matrix is given by

$$M_{kj} = \sum_i m_i \left(\frac{\partial x_i}{\partial q_j} \right)_{q^0} \left(\frac{\partial x_i}{\partial q_k} \right)_{q^0},$$

so the elements are

$$M_{11} = m_1 \frac{\partial x_1}{\partial x_1} \frac{\partial x_1}{\partial x_1} + m_2 \frac{\partial x_2}{\partial x_1} \frac{\partial x_2}{\partial x_1} = m_1,$$

$$M_{12} = m_1 \frac{\partial x_1}{\partial x_1} \frac{\partial x_1}{\partial x_2} + m_2 \frac{\partial x_2}{\partial x_1} \frac{\partial x_2}{\partial x_2} = 0,$$

$$M_{21} = m_1 \frac{\partial x_1}{\partial x_2} \frac{\partial x_1}{\partial x_1} + m_2 \frac{\partial x_2}{\partial x_2} \frac{\partial x_2}{\partial x_1} = 0,$$

$$M_{22} = m_1 \frac{\partial x_1}{\partial x_2} \frac{\partial x_1}{\partial x_2} + m_2 \frac{\partial x_2}{\partial x_2} \frac{\partial x_2}{\partial x_2} = m_2.$$

The mass matrix is

$$\mathcal{M} = \begin{pmatrix} m_1 & 0 \\ 0 & m_2 \end{pmatrix}.$$

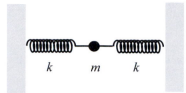

Figure 14.4 One mass and two springs.

Consider a block of mass m connected to a spring of constant k. The block is allowed to slide on a frictionless horizontal surface, as illustrated in Figure 4.3. Determine the mass matrix and the potential matrix for this one-dimensional system. Answer: $M_{11} = m$; $V_{11} = k$. ■

Determine the mass matrix and the potential matrix for a system of one mass and two springs arranged as shown in Figure 14.4. (Assume that both springs have the same spring constant.) Answer: $M_{11} = m$; $V_{11} = 2k$. ■

14.3 Normal Modes

We now obtain solutions to Equations (14.11). This is a rather formal procedure, but in the course of arriving at the answer, you will encounter a number of important concepts such as the *normal coordinates*, the *modal matrix*, and the determination of eigenvalues.

14.3.1 One-Dimensional Motion: A Simple Pendulum

Let us begin with a one-dimensional problem, that is, a system characterized by a single generalized coordinate q. For the sake of an explicit example we will determine the equation of motion for a simple pendulum. The generalized coordinate is the angle the string makes with the vertical, i.e., $q = \theta$. From Figure 14.5 you can appreciate that the transformation equations are:

$$x = x_1 = l \sin \theta,$$
$$y = x_2 = l \cos \theta.$$

The kinetic energy is given by Equation (14.3) with $q_i = q_i^0 + \eta_i$, where η is the displacement from equilibrium. For the pendulum of Figure 14.5, equilibrium corresponds to the pendulum hanging straight down, so $\theta_0 = 0$. Because there is only one generalized coordinate ($\eta = \theta$), the equation for the kinetic energy reduces to

$$T = \frac{1}{2} M_{11} \dot{\eta}_1 \dot{\eta}_1 = \frac{1}{2} M_{11} \dot{\theta}^2.$$

We can determine M_{11} from the definition in Equation (14.4) noting that there are two Cartesian coordinates, x and y, so

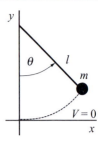

Figure 14.5 A simple pendulum. The potential energy V is arbitrarily set equal to zero when the pendulum is hanging vertically.

$$M_{11} = \sum_{i=1}^{2} m_i \left(\frac{\partial x_i}{\partial q_j}\right)_{q^0} \left(\frac{\partial x_i}{\partial q_k}\right)_{q^0},$$

$$= m \left[\frac{\partial x}{\partial \theta}\frac{\partial x}{\partial \theta} + \frac{\partial y}{\partial \theta}\frac{\partial y}{\partial \theta}\right]_{\theta=0},$$

$$= m \left[(l \cos \theta)^2 + (l \sin \theta)^2\right],$$

$$= ml^2.$$

Thus, $T = \frac{1}{2}ml^2\dot{\theta}^2$.

Similarly, we can determine the potential energy. From Figure 14.5, $V = mgl \cos \theta$. But we want to express it in the form of Equation (14.8),

$$V = \frac{1}{2}V_{11}\eta_1\eta_1 = \frac{1}{2}V_{11}\theta^2.$$

We can evaluate V_{11} from Equation (14.8), obtaining

$$V_{11} = \left(\frac{\partial^2 V}{\partial q_j \partial q_k}\right)_{q^0} = -\frac{\partial}{\partial \theta}\left(\frac{\partial}{\partial \theta}(mgl \cos \theta)\right)_{\theta=0},$$

$$V_{11} = mgl \cos \theta \,|_{\theta=0} = mgl,$$

and consequently, $V = \frac{1}{2}mgl\theta^2$.

The Lagrangian is

$$L = T - V = \frac{1}{2}ml^2\dot{\theta}^2 - \frac{1}{2}mgl\theta^2,$$

and the Lagrange equation of motion is

$$0 = \frac{d}{dt}\frac{\partial l}{\partial \dot{\theta}} - \frac{\partial L}{\partial \theta} = \frac{d}{dt}ml^2\dot{\theta} + mgl\theta,$$

or

$$\ddot{\theta} = -\frac{g}{l}\theta,$$

as expected.

For this problem, the mass matrix and the potential matrix are 1×1 matrices having a single element. In general, if the system is characterized by a single coordinate, you will find that $M_{kj} = m$ and $V_{kj} = \kappa$, where m and κ are real numbers (but not necessarily equal to the mass and the force constant) and the equation of motion (14.11) reduces to

$$m\ddot{\eta} + \kappa\eta = 0, \tag{14.12}$$

or

$$\ddot{\eta} = -\frac{\kappa}{m}\eta.$$

(I am using κ for the force constant to avoid confusing it with k, the subscript index.)

14.3.2 One-Dimensional Motion: A Particle on a Spring

The equation of motion of a particle of mass m attached to a spring of constant κ is $\ddot{\eta} = -\frac{\kappa}{m}\eta$. I realize you know this is the equation for simple harmonic motion whose solution you have seen many times, but bear with me while I solve this simple differential equation in a rather complicated way which will make our work easier when we generalize to a system of many particles.

We can solve $\ddot{\eta} = -\frac{\kappa}{m}\eta$ in our usual way by assuming $\eta = Ae^{pt}$. This leads to $p = \pm i\omega$, where $\omega = \sqrt{\kappa/m}$. In Chapter 11 we simply expressed the solution as

$$\eta = A\cos(\omega t + \beta),\tag{14.13}$$

and noted that substitution into the differential equation showed this expression satisfied the differential equation.

A Complex Solution

Although it will initially complicate the analysis, it turns out to be convenient to express the solution in terms of a complex function denoted $z(t)$. This will be the general solution of the differential equation

$$\ddot{z} = -\frac{\kappa}{m}z.$$

The real part of z will be the actual displacement from equilibrium. That is, $\eta = \mathrm{Re}(z)$.

Since there are two values of $p = \pm i\sqrt{\kappa/m} = \pm i\omega_0$ there are two possible solutions. The general solution is the sum of all possible solutions. Let us express it as

$$z(t) = z_+^{(0)}e^{i\omega_0 t} + z_-^{(0)}e^{-i\omega_0 t},\tag{14.14}$$

where the superscripts (0) on the amplitudes z_+ and z_- remind us that these amplitudes correspond to the frequencies $\pm\omega_0$.

The ultimate quantity of interest is the real physical displacement $\eta = \mathrm{Re}(z)$. Recalling that the real part of a complex number z is given by $\frac{1}{2}(z + z^*)$, we can express η as

$$\eta = \mathrm{Re}(z) = \frac{1}{2}(z + z^*)$$
$$= \frac{1}{2}\left\{\left[z_+^{(0)} + z_-^{(0)*}\right]e^{i\omega_0 t} + \left[z_+^{(0)*} + z_-^{(0)}\right]e^{-i\omega_0 t}\right\}.\tag{14.15}$$

This is certainly cumbersome, but we can express our solution much more simply by introducing another complex number in polar form, defined as

$$\rho_0 e^{i\phi_0} = z_+ + z_-^*.$$

Note that ρ_0 and ϕ_0 are real numbers. In terms of ρ_0 and ϕ_0 the solution (14.15) can be written as

$$\eta = \frac{1}{2}\rho_0\left[e^{i(\phi_0 + \omega_0 t)} + e^{-i(\phi_0 + \omega_0 t)}\right] = \rho_0\cos(\omega_0 t + \phi_0),\tag{14.16}$$

regaining the form of Equation (14.13).

Note that we have expressed our solution in three different ways, namely: (1) in terms of η, the real physical distance from equilibrium, (2) as the sum of complex quantities $z_\pm$, and (3) as a complex number in polar form, $\rho e^{i\phi}$. These various forms will be useful to us when we consider the oscillatory motion of a system of many particles.

Exercise 14.4

Show that if $z = z^0 e^{i\omega_0 t}$, then $\omega_0 = \pm\sqrt{\kappa/m}$.   ∎

Exercise 14.5

Determine the relationships between ρ, ϕ of Equation (14.16) and the real and imaginary parts of z_+, z_- of Equation (14.14). Answers: $\rho^2 = [\text{Re}(z_+) + \text{Re}(z_-^*)]^2 + [\text{Im}(z_+) + \text{Im}(z_-^*)]^2$ and $\tan\phi = \frac{\text{Im}(z_+) + \text{Im}(z_-^*)}{\text{Re}(z_+) + \text{Re}(z_-^*)}$. ∎

14.3.3 Solving the Coupled Equations

Having solved the problem for a single particle, we now assume n particles such as the coupled oscillators in Figure 14.1. We will describe the system by a set of n coupled equations having the form of Equation (14.11) but with η replaced by the complex quantity z:

$$\sum_{j=1}^{n}(M_{kj}\ddot{z}_j + V_{kj}z_j) = 0, \quad k = 1, \ldots, n. \tag{14.17}$$

For greater generality we formulated the problem in terms of the complex variable z, but don't forget that the physical solution is $\eta_k = \text{Re}(z_k)$.

I will now assume that all the particles are oscillating with exactly the *same* frequency. As you may remember from Chapter 11, this type of motion is called a *normal mode*. And, as shown in that chapter, the general motion of a system of particles can be expressed as a linear combination of normal modes.

If the assumed frequency is ω, the normal mode solution has the form

$$z_j = z_j^0 e^{i\omega t}. \tag{14.18}$$

The problem is now reduced to determining the coefficients z_j^0.

Plugging Equation (14.18) into the equations of motion (14.17) gives

$$\sum_{j=1}^{n}(M_{kj}(-\omega^2)z_j^0 e^{i\omega t} + V_{kj}z_j^0 e^{i\omega t}) = 0, \quad k = 1, 2, \ldots, n$$

or

$$\sum_{j=1}^{n}(V_{kj} - \omega^2 M_{kj})z_j^0 = 0 \quad k = 1, 2, \ldots, n. \tag{14.19}$$

In matrix form we can write this equation as

$$(\mathcal{V} - \omega^2 \mathcal{M})z^0 = 0,$$

This has the form of an eigenvalue equation, a fact that we will exploit a bit later.

Equations (14.19) are a set of n linear homogeneous coupled *algebraic* equations. As you have seen previously (in Section 11.5), such a set of equations has a nontrivial solution if and only if the determinant of the coefficients is zero. That is, a nontrivial solution exists iff

$$\det\left|V_{kj} - \omega^2 M_{kj}\right| = 0. \tag{14.20}$$

Evaluating the determinant of Equation (14.20) leads to an nth-order polynomial in ω^2 so there will be n values of ω^2 that satisfy the condition. Let us denote these n allowed values of ω^2 by ω_s^2, where $s = 1, 2, \ldots, n$.

The frequencies ω_s^2 must be real because Equation (14.18) represents a stable solution only if the ωs are real. If ω_s is zero or complex, then according to Equation (14.18) the solution z_j will be constant or exponentially increasing or decreasing. These conditions correspond to the neutral or unstable situations depicted in Figure 14.3.

Worked Example 14.2 Prove that the frequencies, the ωs in Equation (14.19), are real and determine the condition for stable oscillations.

Solution Consider a particular frequency, say ω_s. Let $z_j^{(s)}$ be the corresponding solutions to the set of equations (14.19). Multiply each equation in the set by $\left(z_k^{(s)}\right)^*$ and sum over k to obtain

$$\sum_{k,j}^{n} (z_k^{(s)*} V_{kj} - \omega_s^2 z_k^{(s)*} M_{kj}) z_j^{(s)} = 0,$$

or

$$\sum_{k,j}^{n} z_k^{(s)*} V_{kj} z_j^{(s)} = \omega_s^2 \sum_{k,j}^{n} z_k^{(s)*} M_{kj} z_j^{(s)}.$$

Consequently

$$\omega_s^2 = \frac{\sum_{k,j}^{n} z_k^{(s)*} V_{kj} z_j^{(s)}}{\sum_{k,j}^{n} z_k^{(s)*} M_{kj} z_j^{(s)}}. \tag{14.21}$$

Take the complex conjugate of this expression and use the fact that the V_{kj} and M_{kj} are real to obtain

$$\left(\omega_s^2\right)^* = \frac{\sum_{k,j}^{n} z_k^{(s)} V_{kj} z_j^{(s)*}}{\sum_{k,j}^{n} z_k^{(s)} M_{kj} z_j^{(s)*}}.$$

But V_{kj} and M_{kj} are elements of symmetrical matrices so

$$\left(\omega_s^2\right)^* = \frac{\sum_{k,j}^{n} z_k^{(s)} V_{kj} z_j^{(s)*}}{\sum_{k,j}^{n} z_k^{(s)} M_{kj} z_j^{(s)*}} = \frac{\sum_{k,j}^{n} z_j^{(s)*} V_{kj} z_k^{(s)}}{\sum_{k,j}^{n} z_j^{(s)*} M_{kj} z_k^{(s)}} = \omega_s^2, \tag{14.22}$$

which proves that ω_s^2 is real. If ω_s^2 is real, then ω_s is also real as long as $\omega_s^2 \geq 0$. Therefore, the condition for a stable solution is

$$\frac{\sum_{k,j}^{n} z_k^{(s)*} V_{kj} z_j^{(s)}}{\sum_{k,j}^{n} z_k^{(s)*} M_{kj} z_j^{(s)}} \geq 0. \tag{14.23}$$

The elements of the mass matrix (M_{kj}) are positive, as is evident from the definition, Equation (14.4), so the denominator in (14.23) is positive. The numerator will be positive or negative depending on the value of V_{kj}, which is defined in Equation (14.9):

$$V_{kj} = V_{kj} = \left(\frac{\partial^2 V}{\partial q_j \partial q_k}\right)_{q^0}.$$

If the potential energy V is a minimum at equilibrium point q^0 then its second derivative is positive, so $V_{kj} > 0$ and condition (14.23) is met. Therefore, stable oscillations can only occur around points where the potential energy is a minimum – a result that we knew all along.

Returning to the problem of determining the motion of the system, let us pick a particular frequency (ω_s^2) and consider the form of the solutions of Equations (14.19) which we can write as

$$\sum_{j=1}^{n}(V_{kj} - \omega_s^2 M_{kj})z_j^{(s)} = 0 \quad k = 1, 2, \ldots, n. \tag{14.24}$$

As noted previously, these equations have a nontrivial solution iff the determinant of the coefficients vanishes; see Equation (14.20). Recall that if the determinant is zero, the equations are not all linearly independent. The nonindependent equation(s) can be discarded. This can be done as follows: Renumber the equations so that the nth equation is the one you plan to discard.[3] Your system of equations now looks like this:

$$a_{11}z_1 + a_{12}z_2 + \cdots + a_{1n}z_n = 0, \tag{14.25}$$
$$a_{21}z_1 + a_{22}z_2 + \cdots + a_{2n}z_n = 0,$$

$$\vdots$$

$$a_{n1}z_1 + a_{n2}z_2 + \cdots + a_{nn}z_n = 0.$$

Divide all the equations by z_n and discard the last equation. You will now have a set of $n - 1$ equations having the form

$$a_{11}z_1/z_n + a_{12}z_2/z_n + \cdots = -a_{1n}, \tag{14.26}$$
$$a_{21}z_1/z_n + a_{22}z_2/z_n + \cdots = -a_{2n},$$

$$\vdots$$

$$a_{n-1,1}z_1/z_n + a_{n-1,2}z_2/z_n + \cdots = -a_{n-1,n}.$$

This is a set of $n - 1$ *inhomogeneous* equations that can be solved using Cramer's rule for the ratios $z_1/z_n, z_2/z_n, \ldots$.

The coefficients a_{kj} in Equations (14.26) are

$$a_{kj} = V_{kj} - \omega_s^2 M_{kj},$$

and they are *real*. Thus, Equations (14.26) are a set of linear *algebraic* equations with real coefficients. Consequently, the Cramer's rule solutions z_j/z_n are real. But the z_j were assumed to be complex – see Equation (14.18)! How can the ratio of two complex numbers be real? This can only happen if the complex part of the two numbers is the same. To appreciate this, write the complex z_js in polar form

$$z_j^{(s)} = \rho_j^{(s)}e^{i\phi_s}. \tag{14.27}$$

The value of ϕ_s must be the same for all the z_k^ss. But if $e^{i\phi_s}$ is a common factor, you can divide all of the equations by it and write Equations (14.24) in the form

$$\sum_{j=1}^{n}V_{kj}\rho_j^{(s)} = \omega_s^2\sum_{j=1}^{n}M_{kj}\rho_j^{(s)}, \quad k = 1, 2, \ldots, n. \tag{14.28}$$

Equation (14.28) is one way to write the *eigenvector equation*. Note that the $\rho_j^{(s)}$ are a set of n real quantities. They can be ordered into a column matrix, thus,

---

[3] If there is more than one nonindependent equation, you can go through the process over and over.

$$\boldsymbol{\rho}^{(s)} = \begin{pmatrix} \rho_1^{(s)} \\ \rho_2^{(s)} \\ \vdots \\ \rho_n^{(s)} \end{pmatrix}. \tag{14.29}$$

Here $\boldsymbol{\rho}^{(s)}$ is an *eigenvector*, and the frequency ω_s^2 is an *eigenvalue*.

In like manner, the eigenvector equation corresponding to a different frequency ω_t^2 is

$$\sum_{k=1}^{n} V_{kj} \rho_k^{(t)} = \omega_t^2 \sum_{k=1}^{n} M_{kj} \rho_k^{(t)}. \tag{14.30}$$

Let us operate on Equation (14.28) with $\sum_k \rho_k^{(t)}$ and on Equation (14.30) with $\sum_j \rho_j^{(s)}$ to obtain

$$\sum_{k,j} \rho_k^{(t)} V_{kj} \rho_j^{(s)} = \omega_s^2 \sum_{k,j} \rho_k^{(t)} M_{kj} \rho_j^{(s)}, \tag{14.31}$$

and

$$\sum_{k,j} \rho_j^{(s)} V_{kj} \rho_k^{(t)} = \omega_t^2 \sum_{k,j} \rho_j^{(s)} M_{kj} \rho_k^{(t)}. \tag{14.32}$$

The left-hand sides of Equations (14.31) and (14.32) are identical, so subtracting the two equations yields

$$0 = (\omega_s^2 - \omega_t^2) \sum_{k,j} \rho_k^{(t)} M_{kj} \rho_j^{(s)}. \tag{14.33}$$

If the eigenvalues are different, that is, if $\omega_s^2 \neq \omega_t^2$, then

$$\sum_{k,j} \rho_k^{(t)} M_{kj} \rho_j^{(s)} = 0, \qquad s \neq t. \tag{14.34}$$

This is an orthogonality condition on the eigenvectors.[4] You can multiply $\rho^{(s)}$ and $\rho^{(t)}$ by appropriate constants such that the following *orthonormality* condition is met[5]

$$\sum_{k,j} \rho_k^{(t)} M_{kj} \rho_j^{(s)} = \delta_{ts}. \tag{14.35}$$

The ρ vectors are then said to be normalized.

Considering the form of the differential equation (14.17), you can appreciate that nothing is changed if you multiply the solution, Equation (14.27), by an arbitrary constant. Therefore, the general solution to (14.17) is the sum of terms of the form

$$z_j^{(s)} = C^{(s)} \rho_j^{(s)} e^{i\phi_s}, \qquad j = 1, \ldots, n. \tag{14.36}$$

There are n solutions of this form for *each* of the eigenvalues ω_s. Each such solution has (in general) a different value of $\rho_j^{(s)}$ but the factors $C^{(s)}$ and ϕ_s are common to all the solutions corresponding to the same eigenvalue. For example, for the system illustrated in Figure 14.1, the displacements of particles $1, 2, 3, \ldots, j, \ldots, n$ correspond to the real parts of

---

[4] For Cartesian coordinates the orthogonality condition is $\sum_{jk} \rho_k^{(t)} \rho_j^{(s)} = 0$ so Equation (14.34) is the orthogonality condition in a non-Cartesian space described by the metric M_{kj}.

[5] A set of vectors is orthonormal if all the vectors have unit length and are perpendicular to one another.

$z_1^{(s)}, z_2^{(s)}, \ldots, z_j^{(s)}, \ldots z_n^{(s)}$ where all the particles are assumed to oscillate at the same (normal) frequency ω_s.

The solution of the coupled problem is obtained as a linear superposition of all the particular solutions, thus the generalized form of Equation (14.14) is

$$z_j(t) = \sum_{s=1}^{n} \left[z_{+j}^{(s)} e^{i\omega_s t} + z_{-j}^{(s)} e^{-i\omega_s t} \right], \qquad j = 1, \ldots, n, \tag{14.37}$$

where

$$z_{+j}^{(s)} = C_+^{(s)} \rho_j^{(s)} e^{i\phi_s^+}, \qquad z_{-j}^{(s)} = C_-^{(s)} \rho_j^{(s)} e^{i\phi_s^-}. \tag{14.38}$$

The general solution is the sum over all the normal modes. If the normal mode solutions satisfy the differential equation, then a sum of normal mode solutions will also satisfy the differential equation. This is the principle of superposition.[6]

As you can prove (see Problem 14.16), the actual physical motion, η_j, is obtained by taking the real part of Equation (14.37), yielding

$$\eta_j(t) = \sum_{s=1}^{n} C^{(s)} \rho_j^{(s)} \cos(\omega_s t + \phi_s). \tag{14.39}$$

Exercise 14.6

Show that the condition for stable equilibrium is not met at a maximum of the potential energy. ∎

Exercise 14.7

Show that Equation (14.39) is the real part of Equation (14.37). ∎

14.4 Matrix Formulation

In the previous section we solved the problem of coupled oscillators algebraically. We did mention that some of the equations can be formulated in terms of matrices and we did employ the terminology of matrix analysis, defining eigenvalues and eigenvectors. Nevertheless, we carried out the analysis without the benefit of matrix notation. As you will now see, the problem we are considering can be formulated in an elegant and very concise way by using matrices. The drawback is that the analysis tends to be rather abstract.

Suppose $\mathcal{A}$ is an $n \times n$ matrix and $\mathbf{x}$ is a column matrix having n elements. Then $\mathcal{A}\mathbf{x}$ is a column matrix, $\mathbf{x}^T$ is a row matrix, and $\mathbf{x}^T \mathcal{A}\mathbf{x}$ is a scalar.

As defined in Sections 14.2.2 and 14.2.3, $\mathcal{M}$ is the mass matrix (with elements M_{kj}) and $\mathcal{V}$ is the potential matrix (with elements V_{kj}). In terms of $\mathcal{V}$ and $\mathcal{M}$, the eigenvector equation (14.28) is

$$\left(\mathcal{V} - \omega_s^2 \mathcal{M} \right) \rho^{(s)} = 0. \tag{14.40}$$

---

[6] For a differential equation, the general solution is the sum over all the independent solutions. Because the normal modes are linearly independent, the general solution can be expressed as a sum over all normal modes, as in Equation (14.37).

The orthonormality condition (14.35) is

$$\left(\boldsymbol{\rho}^{(s)}\right)^T \mathcal{M} \boldsymbol{\rho}^{(t)} = \delta_{st},$$

and the physical displacements (14.39) can be expressed as a linear combination of the normal mode solutions,

$$\boldsymbol{\eta}(t) = \sum_s \boldsymbol{\rho}^{(s)} C^{(s)} \cos(\omega_s t + \phi_s). \tag{14.41}$$

The constants $C^{(s)}$ and ϕ_s can be associated with the initial conditions $\boldsymbol{\eta}(0)$ and $\dot{\boldsymbol{\eta}}(0)$ as obtained from Equation (14.41) and its derivative:

$$\boldsymbol{\eta}(0) = \sum_s \boldsymbol{\rho}^{(s)} C^{(s)} \cos \phi_s,$$

$$\dot{\boldsymbol{\eta}}(0) = -\sum_s \boldsymbol{\rho}^{(s)} C^{(s)} \omega_s \sin \phi_s.$$

You are usually given $\boldsymbol{\eta}(0)$ and $\dot{\boldsymbol{\eta}}(0)$ and asked to find $C^{(s)}$ and ϕ_s. These are determined by multiplying by $\left(\boldsymbol{\rho}^{(t)}\right)^T \mathcal{M}$ from the left and using the orthogonality conditions:

$$\left(\boldsymbol{\rho}^{(t)}\right)^T \mathcal{M} \boldsymbol{\eta}(0) = \sum_s \left(\boldsymbol{\rho}^{(t)}\right)^T \mathcal{M} \boldsymbol{\rho}^{(s)} C^{(s)} \cos \phi_s = C^{(t)} \cos \phi_t,$$

and

$$\left(\boldsymbol{\rho}^{(t)}\right)^T \mathcal{M} \dot{\boldsymbol{\eta}}(0) = -\sum_s \left(\boldsymbol{\rho}^{(t)}\right)^T \mathcal{M} \boldsymbol{\rho}^{(s)} \omega_s C^{(s)} \sin \phi_s = -\omega_t C^{(t)} \sin \phi_t.$$

14.4.1 The Modal Matrix

Solving the problem of n coupled oscillators introduced n eigenvectors $\boldsymbol{\rho}^{(s)}$, each corresponding to a different frequency ω_s. Further, each eigenvector had n elements, $\rho_j^{(s)}$, $j = 1, \dots, n$. Therefore, you can construct an $n \times n$ matrix using the eigenvectors as columns, thus:

$$\mathcal{A} = \begin{pmatrix} \rho_1^{(1)} & \rho_1^{(2)} & \rho_1^{(3)} & \cdots & \rho_1^{(n)} \\ \rho_2^{(1)} & \rho_2^{(2)} & \rho_2^{(3)} & \cdots & \rho_2^{(n)} \\ & & \vdots & & \\ \rho_n^{(1)} & \rho_n^{(2)} & \rho_n^{(3)} & \cdots & \rho_n^{(n)} \end{pmatrix} = \begin{pmatrix} \boldsymbol{\rho}^{(1)} & \boldsymbol{\rho}^{(2)} & \boldsymbol{\rho}^{(3)} & \cdots & \boldsymbol{\rho}^{(n)} \\ \downarrow & \downarrow & \downarrow & & \downarrow \\ \downarrow & \downarrow & \downarrow & & \downarrow \\ \downarrow & \downarrow & \downarrow & & \downarrow \end{pmatrix}. \tag{14.42}$$

$\mathcal{A}$ is called the *modal matrix*. If the elements of $\mathcal{A}$ are denoted a_{ij}, then $a_{ij} = \rho_i^{(j)}$. The modal matrix has some interesting and useful properties.

Property 1 The matrix product $\mathcal{A}^T \mathcal{M} \mathcal{A}$ is the unit matrix:

$$\left(\mathcal{A}^T \mathcal{M} \mathcal{A}\right)_{ij} = \sum_{kl} a_{ik}^T M_{kl} a_{lj} = \sum_{kl} a_{ki} M_{kl} a_{lj} = \sum_{kl} \rho_k^{(i)} M_{kl} \rho_l^{(j)} = \delta_{ij}, \tag{14.43}$$

or

$$\mathcal{A}^T \mathcal{M} \mathcal{A} = \mathbf{1}. \tag{14.44}$$

Consequently, $\mathcal{A}^T \mathcal{M} \mathcal{A}$ is an $n \times n$ diagonal matrix with ones along the diagonal. This means that the modal matrix diagonalizes the mass matrix.

Property 2 The matrix product $\mathcal{A}^T \mathcal{V} \mathcal{A}$ is a diagonal matrix whose elements are the normal mode frequencies squared:

$$\left(\mathcal{A}^T \mathcal{V} \mathcal{A}\right)_{ij} = \sum_{kl} a_{ik}^T V_{kl} a_{lj} = \sum_{kl} \rho_k^{(i)} V_{kl} \rho_l^{(j)} = \sum_k \left\{ \rho_k^{(i)} \sum_l V_{kl} \rho_l^{(j)} \right\}.$$

But the eigenvector equation (14.28) allows us to write the summation over l as $\omega_j^2 \sum_l M_{kl} \rho_l^{(j)}$, so

$$\left(\mathcal{A}^T \mathcal{V} \mathcal{A}\right)_{ij} = \sum_k \left\{ \rho_k^{(i)} \omega_j^2 \sum_l M_{kl} \rho_l^{(j)} \right\} = \omega_j^2 \sum_{kl} \rho_k^{(i)} M_{kl} \rho_l^{(j)} = \omega_j^2 \delta_{ij},$$

or

$$\mathcal{A}^T \mathcal{V} \mathcal{A} = \begin{pmatrix} \omega_1^2 & 0 & 0 & \cdots \\ 0 & \omega_2^2 & 0 & \cdots \\ 0 & 0 & \omega_3^2 & \cdots \\ \vdots & \vdots & \vdots & \vdots \end{pmatrix} \equiv \omega_D^2. \tag{14.45}$$

Thus, the modal matrix also diagonalizes the potential matrix, generating a matrix whose only nonzero elements are the squares of the normal mode frequencies.

14.5 Normal Coordinates

I expressed the positions of the oscillating mass points in various ways. The Cartesian coordinate, x_i, is the distance from the *origin* to m_i. The distance of a mass point from its *equilibrium position* is denoted η_i and it is the first of several generalized coordinates. (I referred to the η_i as the "physical" coordinates.) The $z_i^{(s)}$ are a set of complex generalized coordinates. They are the position coordinates obtained from the equation of motion assuming a normal mode solution, i.e., one in which all the particles oscillate at the same frequency ω_s. The physical interpretation of the z_i is a bit difficult to appreciate. Since the z_i are complex, they can be expressed in polar form as

$$z_i^{(s)} = \rho_i^{(s)} e^{i\phi_s}.$$

The $\rho_i^{(s)}$ are elements of the eigenvector $\rho^{(s)}$ which describes the position of all the particles in a particular normal mode. The relationship between the η_i and the eigenvectors for any general motion of the system is the real part of a superposition of all the normal modes, as given by Equation (14.39).

Given this multiplicity of ways to express the positions of the particles, it may seem a bit ludicrous that I should introduce yet another set of generalized coordinates. However, as you will appreciate shortly, the *normal coordinates* (denoted by ζ_i) are a very convenient way to formulate the problem of coupled oscillators. The normal coordinates are related to the generalized physical coordinates η_i by the following defining equation involving the modal matrix:

$$\eta(t) = \mathcal{A} \zeta(t). \tag{14.46}$$

Multiplying both sides by $\mathcal{A}^T \mathcal{M}$ isolates ζ; see Equation (14.44). The result is

$$\zeta(t) = \mathcal{A}^T \mathcal{M} \eta(t). \tag{14.47}$$

Since $\mathcal{A}^T$ and $\mathcal{M}$ are matrices with constant, real elements, the normal coordinates ζ_i are linear combinations of the original generalized coordinates η_i.

Using the normal coordinates, the Lagrange equations of motion take on a particularly simple form. Equation (14.10) for the Lagrangian was

$$L = T - V = \frac{1}{2} \sum_{ij} \left(M_{ij} \dot{\eta}_i \dot{\eta}_j - V_{ij} \eta_i \eta_j \right).$$

This can be expressed as

$$2L = \sum_{ij} \dot{\eta}_i M_{ij} \dot{\eta}_j - \sum_{ij} \eta_i V_{ij} \eta_j,$$

or

$$2L = \dot{\eta}^T M \dot{\eta} - \eta^T V \eta. \tag{14.48}$$

But $\eta(t) = A \zeta(t)$ and $\dot{\eta}(t) = A \dot{\zeta}(t)$, so

$$2L = \left(A \dot{\zeta} \right)^T M \left(A \dot{\zeta} \right) - \left(A \zeta \right)^T V \left(A \zeta \right),$$
$$= \dot{\zeta}^T \left(A^T M A \right) \dot{\zeta} - \zeta^T \left(A^T V A \right) \zeta,$$
$$= \dot{\zeta}^T \dot{\zeta} - \zeta^T \omega_D^2 \zeta, \tag{14.49}$$

or, in component form,

$$L = \frac{1}{2} \sum_{i=1}^{n} \left(\dot{\zeta}_i^2 - \omega_i^2 \zeta_i^2 \right). \tag{14.50}$$

Note that using the modal matrix to define the new generalized coordinates has, in a sense, diagonalized the Lagrangian. Note also that the ω_is in Equation (14.50) are the set of normal frequencies for the system.

In terms of the normal coordinates, ζ_i, the Lagrange equations of motion are

$$\frac{d}{dt} \frac{\partial}{\partial \dot{\zeta}_i} \left(\frac{1}{2} \sum_{i=1}^{n} \left(\dot{\zeta}_i^2 - \omega_i^2 \zeta_i^2 \right) \right) - \frac{\partial}{\partial \zeta_i} \left(\frac{1}{2} \sum_{i=1}^{n} \left(\dot{\zeta}_i^2 - \omega_i^2 \zeta_i^2 \right) \right) = 0, \tag{14.51}$$

or

$$\ddot{\zeta}_i = -\omega_i^2 \zeta_i, \tag{14.52}$$

which is the equation for simple harmonic motion. Therefore diagonalizing the Lagrangian decoupled the equations of motion and reduced the problem to a set of n independent, decoupled, simple harmonic oscillator equations. Each normal coordinate ζ_i oscillates independently with angular frequency ω_i.

The derivation has been perfectly general and implies that any mechanical system undergoing small amplitude oscillations around points of stable equilibrium can be described in terms of normal coordinates.

The solution to Equation (14.52) can be expressed in component form as

$$\zeta_i = C^{(i)} \cos(\omega_i t + \phi_i), \qquad i = 1, \ldots, n \tag{14.53}$$

or in the form of a column vector as

$$\zeta = \begin{pmatrix} C^{(1)} \cos(\omega_1 t + \phi_1) \\ C^{(2)} \cos(\omega_2 t + \phi_2) \\ \vdots \\ C^{(n)} \cos(\omega_n t + \phi_n) \end{pmatrix}. \tag{14.54}$$

The physical generalized coordinates are obtained from the definition $\eta(t) = \mathcal{A}\zeta(t)$, which can be written as

$$
\eta = \begin{pmatrix} \rho^{(1)} & \rho^{(2)} & \rho^{(3)} & \cdots & \rho^{(n)} \\ \downarrow & \downarrow & \downarrow & & \downarrow \\ \downarrow & \downarrow & \downarrow & & \downarrow \\ \downarrow & \downarrow & \downarrow & & \downarrow \end{pmatrix} \begin{pmatrix} \zeta \\ \downarrow \\ \downarrow \\ \downarrow \end{pmatrix}.
$$

Therefore,

$$
\eta_i = \sum_j \rho_i^{(j)} C^{(j)} \cos(\omega_j t + \phi_j), \tag{14.55}
$$

which is just Equation (14.39).

Comparing Equations (14.55) and (14.53) you can see that the normal coordinates ζ_i are the coefficients of the eigenvectors ρ_i in the expansion of η_i.

14.6 Coupled Pendulums: An Example

Consider the two pendulums connected by a spring of constant k illustrated in Figure 14.6. Note that $\sin\theta = \eta/l$. For small-amplitude oscillations, $\sin\theta \doteq \theta$. The kinetic energy of the bobs is

$$
T = \frac{1}{2}m\left[(l\dot{\theta}_1)^2 + (l\dot{\theta}_2)^2\right] = \frac{1}{2}m\left[\dot{\eta}_1^2 + \dot{\eta}_2^2\right]. \tag{14.56}
$$

The potential energy is the gravitational potential energy V_G plus the potential energy in the spring V_S. In terms of η_1 and η_2, we can express V_G as

$$
V_G = mgl\left[(1 - \cos\theta_1) + (1 - \cos\theta_2)\right],
$$

$$
= mgl\left[\left(1 - \left\{1 - \sin^2\theta_1\right\}^{\frac{1}{2}}\right) + \left(1 - \left\{1 - \sin^2\theta_2\right\}^{\frac{1}{2}}\right)\right],
$$

$$
\doteq mgl\frac{1}{2}\left(\sin^2\theta_1 + \sin^2\theta_2\right) \simeq \frac{1}{2}\frac{mg}{l}(l\theta_1^2 + l\theta_2^2).
$$

Therefore, to second order in small quantities,

$$
V_G = \frac{mg}{2l}(\eta_1^2 + \eta_2^2). \tag{14.57}
$$

The potential energy of the spring is $\frac{1}{2}k(d - d_0)^2$, where d is the distance between the masses and d_0 is the unstretched length of the spring. These quantities are illustrated in Figure 14.6 from which you can appreciate that

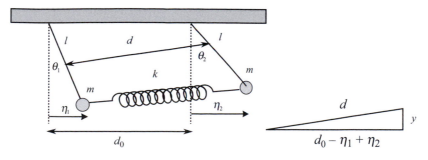

Figure 14.6 Coupled pendulums.

$$d = \left[y^2 + (d_0 - \eta_1 + \eta_2)^2\right]^{\frac{1}{2}},$$

$$= d_0 \left(\frac{y^2}{d_0^2} + 1 + \frac{2(\eta_2 - \eta_1)}{d_0} + \frac{(\eta_2 - \eta_1)^2}{d_0^2}\right)^{\frac{1}{2}}.$$

Discarding the higher-order terms $\frac{y^2}{d_0^2}$ and $\frac{(\eta_2-\eta_1)^2}{d_0^2}$ and using the binomial expansion we obtain

$$d - d_0 \doteq \eta_2 - \eta_1,$$

Therefore,

$$V_S = \frac{1}{2}k(d - d_0)^2 = \frac{1}{2}k(\eta_2 - \eta_1)^2. \tag{14.58}$$

Having obtained the kinetic and potential energies, the Lagrangian is

$$L = T - V = \frac{1}{2}m\left[\dot{\eta}_1^2 + \dot{\eta}_2^2\right] - \frac{mg}{2l}(\eta_1^2 + \eta_2^2) - \frac{1}{2}k(\eta_2 - \eta_1)^2. \tag{14.59}$$

The equations of motion are

$$m\ddot{\eta}_1 + \left(k + \frac{mg}{l}\right)\eta_1 - k\eta_2 = 0, \tag{14.60}$$

$$m\ddot{\eta}_2 + \left(k + \frac{mg}{l}\right)\eta_2 - k\eta_1 = 0.$$

To obtain a solution, we look for normal modes, i.e., a solution in which both bobs oscillate at the same frequency ω. That is, we assume that

$$\eta_i = C\rho_i \cos(\omega t + \phi), \quad i = 1, 2. \tag{14.61}$$

Substituting into the equation of motion yields the following (eigenvalue) equations:

$$\left(k + \frac{mg}{l} - m\omega^2\right)\rho_1 - k\rho_2 = 0, \tag{14.62}$$

$$-k\rho_1 + \left(k + \frac{mg}{l} - m\omega^2\right)\rho_2 = 0.$$

Remember that the eigenvector equation has the form

$$\left(\mathcal{V} - \mathcal{M}\omega^2\right)\rho = 0. \tag{14.63}$$

For this problem ρ is the two-component column vector

$$\rho = \begin{pmatrix} \rho_1 \\ \rho_2 \end{pmatrix}. \tag{14.64}$$

By comparing Equations (14.62) and (14.63) we see that

$$\mathcal{M} = m \begin{pmatrix} 1 & 0 \\ 0 & 1 \end{pmatrix}, \tag{14.65}$$

and

$$\mathcal{V} = \begin{pmatrix} k + \frac{mg}{l} & -k \\ -k & k + \frac{mg}{l} \end{pmatrix}. \tag{14.66}$$

The eigenvalue equations (14.62) are a set of coupled homogeneous algebraic equations having a nontrivial solution iff the determinant of the coefficients vanishes. That is,

$$\begin{vmatrix} k + \frac{mg}{l} - m\omega^2 & -k \\ -k & k + \frac{mg}{l} - m\omega^2 \end{vmatrix} = 0.$$

This yields a quadratic equation for ω^2 with solutions

$$\omega_1 = \sqrt{\frac{g}{l}}, \tag{14.67}$$

$$\omega_2 = \sqrt{\frac{g}{l} + 2\frac{k}{m}}. \tag{14.68}$$

To obtain the eigenvectors corresponding to these two eigenvalues, plug them into Equation (14.62) and obtain

$$\rho_1^{(1)} = \rho_2^{(1)} \qquad \text{for } \omega = \omega_1, \tag{14.69}$$

and

$$\rho_1^{(2)} = -\rho_2^{(2)} \qquad \text{for } \omega = \omega_2. \tag{14.70}$$

Equation (14.69) represents motion in which the bobs oscillate in phase with equal amplitudes at the frequency of a free pendulum. Equation (14.70) represents the bobs oscillating out of phase with a frequency larger than that of a free pendulum. These *normal modes* are easily generated with the simple apparatus illustrated in Figure 14.6. They are not, however, the most general type of motion. To describe the general motion go back to the equations of motion, Equations (14.60). Note that the vector

$$\rho = \begin{pmatrix} \rho_1 \\ \rho_2 \end{pmatrix}$$

can be multiplied by a constant without affecting anything except the amplitude. This allows you to normalize the eigenvectors so that the orthonormality condition

$$\left(\rho^{(s)} \right)^T \mathcal{M} \rho^{(t)} = \delta_{st}$$

is satisfied. Using Equation (14.69) one can show that the normalization constant is $1/\sqrt{2m}$. Thus,

$$\rho^{(1)} = \frac{1}{\sqrt{2m}} \begin{pmatrix} 1 \\ 1 \end{pmatrix}, \tag{14.71}$$

$$\rho^{(2)} = \frac{1}{\sqrt{2m}} \begin{pmatrix} 1 \\ -1 \end{pmatrix}.$$

Now construct the modal matrix $\mathcal{A}$ using these normalized eigenvectors as the columns of the matrix:

$$\mathcal{A} = \frac{1}{\sqrt{2m}} \begin{pmatrix} 1 & 1 \\ 1 & -1 \end{pmatrix}.$$

The normal coordinates are obtained from $\boldsymbol{\zeta} = A^T M \boldsymbol{\eta}$ and are given by the following linear combinations of the ηs:

$$\zeta_1 = \sqrt{\frac{m}{2}}(\eta_1 + \eta_2), \tag{14.72}$$

$$\zeta_2 = \sqrt{\frac{m}{2}}(\eta_1 - \eta_2).$$

According to Equation (14.50), the Lagrangian is

$$L = \frac{1}{2}\sum_{i=1}^{2}(\dot{\zeta}_i^2 - \omega_i^2 \zeta_i^2),$$

and the equations of motion are

$$\ddot{\zeta}_i = -\omega_i^2 \zeta_i, \qquad i = 1, 2. \tag{14.73}$$

The general solution is

$$\zeta_1 = C^{(1)} \cos(\omega_1 t + \phi_1), \tag{14.74}$$

$$\zeta_2 = C^{(2)} \cos(\omega_2 t + \phi_2).$$

The constants $C^{(1)}, C^{(2)}, \phi_1$, and ϕ_2 are determined from the initial conditions. For example, if bob 1 is displaced a distance α and released from rest, $\phi_1 = \phi_2 = 0$ and $C^{(1)} = C^{(2)} = \alpha\sqrt{m/2}$. Therefore,

$$\boldsymbol{\zeta} = \alpha\sqrt{\frac{m}{2}}\begin{pmatrix} \cos\omega_1 t \\ \cos\omega_2 t \end{pmatrix}.$$

The original ("physical") generalized coordinates (obtained from $\boldsymbol{\eta} = A\boldsymbol{\zeta}$) are

$$\eta_1(t) = \frac{\alpha}{2}(\cos\omega_1 t + \cos\omega_2 t) \tag{14.75}$$

$$\eta_2(t) = \frac{\alpha}{2}(\cos\omega_1 t - \cos\omega_2 t)$$

or, using a trigonometric relation,

$$\eta_1(t) = \left[\alpha\cos\left\{\frac{1}{2}(\omega_2 - \omega_1)t\right\}\right]\cos\left\{\frac{1}{2}(\omega_2 + \omega_1)t\right\}, \tag{14.76}$$

$$\eta_2(t) = \left[\alpha\sin\left\{\frac{1}{2}(\omega_2 - \omega_1)t\right\}\right]\cos\left\{\frac{1}{2}(\omega_2 + \omega_1)t\right\}.$$

The first term (in square brackets) is a slowly varying amplitude and the second term (involving $\omega_1 + \omega_2$) is a rapidly varying term. The functions η_1 and η_2 are plotted as functions of time in Figure 14.7.

Exercise 14.8

The system illustrated in Figure 14.6 has bobs of mass 0.1 kg and a spring whose force constant is 5 N/m. The pendulum string and the unstretched spring length are both 0.3 m. Determine the normal mode frequencies. Answers: 5.72/s and 11.52/s. ∎

Exercise 14.9

Verify Equations (14.67) and (14.68). ∎

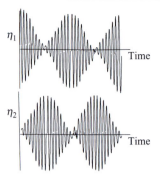

η_1 — Time

η_2 — Time

Figure 14.7 Oscillations of coupled pendulums.

Figure 14.8 A massless string with point masses separated by equal distances.

Exercise 14.10

Show that $\rho^{(1)}$ and $\rho^{(2)}$, as given by Equations (14.71), satisfy the orthonormality condition.

■

14.7 Many Degrees of Freedom

As a final example, consider a system with many degrees of freedom: specifically, the transverse oscillations of mass points on a stretched massless string, as illustrated in Figure 14.8.

14.7.1 Statement of the Problem

Assume the N particles have mass m and are uniformly distributed on the string. Let a be the horizontal distance between the particles and let y_j be the vertical displacement of particle j from its equilibrium position. Let τ be the constant, uniform tension in the string.[7] See Figure 14.9.

The equation of motion of the jth particle can be obtained by elementary means, that is, by applying $F = ma$. (See Section 13.1.) This yields

$$m\ddot{y}_j = \tau \sin\phi - \tau \sin\theta. \tag{14.77}$$

But

$$\sin\phi \simeq \tan\phi = \frac{1}{a}\left(y_{j+1} - y_j\right),$$

and

$$\sin\theta \simeq \tan\theta = \frac{1}{a}\left(y_j - y_{j-1}\right).$$

Hence,

$$m\ddot{y}_j + \frac{2\tau}{a}y_j - \frac{\tau}{a}\left(y_{j+1} + y_{j-1}\right) = 0, \qquad j = 1, \ldots, N. \tag{14.78}$$

[7] Recall the discussion in Section 13.1.

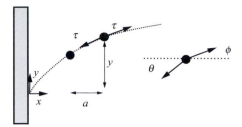

Figure 14.9 Forces acting on a point mass on a string. The angles between the string and the horizontal at the mass point are θ and ϕ.

At the two end points the string is attached to the walls and not allowed to move, so

$$y_0 = y_{N+1} = 0. \tag{14.79}$$

Normal Modes: The jth particle is a horizontal distance ja from the origin. It is convenient to define $x_j = ja$ and to denote y_j by $y(x_j)$. This is simply a useful change in notation. The system will oscillate and traveling plane waves can be set up in it. (Recall that a stationary or *standing* wave can be considered as two traveling waves moving in opposite directions.) A traveling plane wave is described mathematically by

$$y(x_j,t) = Ae^{i(kx_j - \omega t)}. \tag{14.80}$$

Plugging Equation (14.80) into the equations of motion (14.78) leads to the following expression for ω^2 :

$$\omega^2 = \frac{4\tau}{ma} \sin^2 \frac{ka}{2}. \tag{14.81}$$

An expression such as this which expresses ω as a function of k, $\omega = \omega(k)$ is called a *dispersion relation*.

In Equation (14.80) and Equation (14.81) both ω and k could be continuous, and in many applications they are. For the problem at hand, however, the system is constrained by boundary conditions that render ω and k discrete.

Equation (14.80) represents a plane wave traveling in the positive x-direction. A plane wave traveling in the opposite direction can be represented by $Be^{-i(kx_j + \omega t)}$ and this expression also satisfies the equations of motion. Therefore, the general solution is the sum of these two expressions:

$$y(x_j,t) = Ae^{i(kx_j - \omega t)} + Be^{-i(kx_j + \omega t)}. \tag{14.82}$$

The boundary condition at $x_j = 0$ is $y(0,t) = 0$. This can be met by requiring that $B = -A$. The other end point is also stationary so

$$y([N+1]a, t) = 0.$$

Therefore,

$$0 = A \left\{ e^{ik(N+1)a} e^{-i\omega t} - e^{-ik(N+1)a} e^{-i\omega t} \right\},$$

$$= Ae^{-i\omega t} \left[e^{ik(N+1)a} - e^{-ik(N+1)a} \right] = \left(Ae^{-i\omega t} \right) 2i \sin(k(N+1)a).$$

So,

$$\sin(k(N+1)a) = 0,$$

and

$$k(N+1)a = n\pi, \qquad n = 1, 2, \ldots, N. \tag{14.83}$$

Consequently, from Equation (14.81),

$$\omega^2 = \frac{4\tau}{ma} \sin^2 \left(\frac{1}{2} \frac{n\pi}{N+1} \right), \qquad n = 1, 2, \ldots, N. \tag{14.84}$$

The solution (14.82) for a particular ω_n is

$$
\begin{aligned}
y(x_j, t) &= A_n e^{-i\omega_n t} \left[e^{ikx_j} - e^{-ikx_j} \right], \\
&= A_n e^{-i\omega_n t} 2i \sin(kx_j), \\
&= 2i A_n \sin \left(\frac{n\pi x_j}{a(N+1)} \right) [\cos \omega_n t - i \sin \omega_n t].
\end{aligned}
$$

Because $y(x_j, t)$ is real, we only keep the real part of this solution. Assuming A_n is real,

$$y(x_j, t) = 2A_n \sin \left(\frac{n\pi x_j}{a(N+1)} \right) \sin \omega_n t, \tag{14.85}$$

where ω_n^2 is given by Equation (14.84).

Exercise 14.11

Show that $\omega^2 = (4\tau/ma) \sin^2(ka/2)$ follows from Equations (14.78) and (14.80). ∎

Worked Example 14.3 Consider a massless string with three particles on it. Determine the normal mode frequencies (eigenfrequencies), the allowed wavelengths, and the maximum frequency.

Solution For this system, $N = 3$. The three values of ω_n are obtained from Equation (14.84) and are:

$$\omega_1 = 0.765 \sqrt{\frac{\tau}{ma}},$$

$$\omega_2 = 1.414 \sqrt{\frac{\tau}{ma}},$$

$$\omega_3 = 1.848 \sqrt{\frac{\tau}{ma}}.$$

These are the normal mode frequencies. According to Equation (14.85), the amplitudes of the motion for the various normal modes are functions of position given by $2A_n \sin(n\pi x_j/4a)$. Plotting these amplitudes gives a good qualitative feeling for the motion of the system. See Figure 14.10.

It is useful to introduce a "characteristic velocity" by

$$v = \sqrt{\frac{\tau}{m/a}}. \tag{14.86}$$

In terms of v, the eigenfrequencies are

$$\omega_n = \frac{2v}{a} \sin \frac{n\pi}{2(N+1)}. \tag{14.87}$$

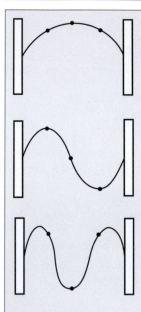

Figure 14.10 Positions of the three point masses.

By definition, the wavenumber is

$$k = \frac{2\pi}{\lambda},$$

(14.88)

where λ is the wavelength. But $k = n\pi / [2(N+1)a]$, so

$$\lambda = \frac{2(N+1)a}{n}.$$

If the length of the string is l then $l = (N+1)a$ and

$$\lambda = \frac{2l}{n}.$$

The largest value of n is N, hence the shortest wavelength is

$$\lambda_{\min} = \frac{2l}{N} = \frac{2(N+1)a}{N} \gtrsim 2a.$$

For large N the minimum wavelength is twice the spacing between the particles, which is an expected conclusion. However, this also means there is a maximum frequency, which may not be quite as obvious. It is

$$\omega_{\max} = \frac{2v}{a} \sin \frac{N\pi a}{(N+1)2a} \approx \frac{2v}{a} \sin \frac{\pi}{2} = \frac{2v}{a}.$$

14.8 Transition to Continuous Systems

The previous section treated a system composed of discrete particles of mass m separated by distances a. If the interparticle spacing is very small compared with the characteristic scale of the phenomenon, the system can be considered continuous. In this section we let $a \to 0$ so that the system of the previous section more and more closely resembles a continuous string. We first describe how the analysis presented above is carried to the continuous limit. Then we consider the motion of a string, first from an elementary point of view, then from an advanced point of view using the Lagrangian approach.

The Continuum Limit. To go from the discrete particle case to the continuous case, let $N \to \infty$ and $a \to 0$, but still require that the length of the string remain unchanged. That is,

$$l = (N + 1)a = \text{constant}.$$

Furthermore, we require that the linear mass density, σ, is constant.

$$\sigma = \frac{m}{a} = \text{constant}. \tag{14.89}$$

(Note that I am using σ for the linear mass density of the string because in this chapter λ is used to represent wavelength and ρ is used to represent the eigenvectors.)

By Equation (14.84) the normal mode frequencies can be written

$$\omega_n^2 = \lim_{N \to \infty} \left\{ \frac{4v^2}{a^2} \sin^2 \left(\frac{1}{2} \frac{n\pi}{N+1} \right) \right\} = \frac{4v^2}{a^2} \left(\frac{n\pi}{2N} \right)^2 = v^2 \left(\frac{n\pi}{l} \right)^2. \tag{14.90}$$

($N \to \infty$ implies that any normal mode under consideration will have $n \ll N$.)

The solutions are, of course, of the form

$$y(x,t) = A e^{i(kx - \omega t)},$$

where $k = 2\pi n / l$ and ω is any one of the normal frequencies ω_n.

A Continuous String. In Chapter 13 the wave equation was derived from first principles. For a string of mass density σ under tension τ we found by Equation (13.1):

$$\frac{\partial^2 y}{\partial t^2} = v^2 \frac{\partial^2 y}{\partial x^2}, \tag{14.91}$$

where $v^2 = \tau / \sigma$.

Worked Example 14.4 Solve the wave equation (14.91) using the techniques developed in this chapter. (Recall that in Chapter 13 the wave equation was solved by separation of variables.)

Solution Assume normal mode solutions of the form

$$y(x,t) = C\rho(x) \cos(\omega t + \phi). \tag{14.92}$$

Plugging this expression into the differential equation (14.91) yields

$$\frac{d^2 \rho(x)}{dx^2} + k^2 \rho(x) = 0, \tag{14.93}$$

where $k = \omega / v$. Equation (14.93) is an ordinary differential equation that is also an eigenvalue equation. Solutions exist only for specified allowed values of k^2. (The solutions are the eigenfunctions.) Because Equation (14.93) has the form of the SHM equation, we can write a solution immediately:

$$\rho(x) = \sqrt{\frac{2}{l\sigma}} \sin kx, \tag{14.94}$$

where the coefficient $\sqrt{2/l\sigma}$ was selected for normalization.

For a string tied at both ends, the boundary conditions are $\rho(0) = 0$ and $\rho(l) = 0$. The solution (14.94) meets the first boundary condition, but the second condition is met iff $\sin kl = 0$, i.e., iff k is one of the set

$$k_n = \frac{n\pi}{l}, \quad n = 1, 2, \ldots, \infty. \tag{14.95}$$

Since $k = \omega/v$, the normal mode frequencies are

$$\omega_n = k_n v = \frac{n\pi v}{l}.$$

The allowed wavelengths are

$$\lambda = \frac{2\pi}{k} = \frac{2l}{n}, \quad n = 1, 2, \ldots, \infty. \tag{14.96}$$

Thus, the string must contain an integer number of half wavelengths. The general solution is obtained as a superposition of normal modes and is given by

$$y(x,t) = \sum_{n=1}^{\infty} C_n \rho^{(n)}(x) \cos(\omega_n t + \phi), \tag{14.97}$$

or, equivalently,

$$y(x,t) = \sum_{n=1}^{\infty} \sqrt{\frac{2}{l\sigma}} \sin k_n x \, (a_n \cos \omega_n t + b_n \sin \omega_n t). \tag{14.98}$$

Here $a_n = C_n \cos \phi_n$ and $b_n = -C_n \sin \phi_n$. Each term in this sum satisfies the wave equation and the boundary conditions. The values of the constants can be determined from the *initial conditions:*

$$y(x,0) \equiv \sum_{n=1}^{\infty} a_n \sqrt{\frac{2}{l\sigma}} \sin k_n x,$$

and

$$\dot{y}(x,0) \equiv \sum_{n=1}^{\infty} \omega_n b_n \sqrt{\frac{2}{l\sigma}} \sin k_n x.$$

These expressions are Fourier series so you can use the orthonormality property of these series to determine the coefficients a_n and $\omega_n b_n$. They are:

$$a_n = \sqrt{\frac{2\sigma}{l}} \int_0^l y(x,0) \sin(k_n x) \, dx, \tag{14.99}$$

and

$$\omega_n b_n = \sqrt{\frac{2\sigma}{l}} \int_0^l \dot{y}(x,0) \sin(k_n x) \, dx. \tag{14.100}$$

Exercise 14.12

Using condition (14.96), plot the first four waves in a string with fixed end points.   ∎

14.8.1 The Lagrangian Density: A Continuous String

In advanced work one often encounters the *Lagrangian density.* In this section we introduce this concept by relating it to transverse oscillations in a continuous string. Consider an element of a string, such as that illustrated in Figure 14.9. To determine the Lagrangian evaluate the kinetic and potential energies of the element and then integrate over the entire string.

The kinetic energy of an element of length dx is

$$dT = \frac{1}{2}(\sigma dx)\left(\frac{\partial y}{\partial t}\right)^2,$$

and for the entire string,

$$T = \frac{1}{2}\int_0^l \left(\frac{\partial y(x,t)}{\partial t}\right)^2 \sigma dx. \tag{14.101}$$

The potential energy is equal to the work required to stretch the element from its original length dx to its final length ds. If τ is the tension in the string, this work is given by

$$dW = \tau(ds - dx) = \tau\left[(dx^2 + dy^2)^{\frac{1}{2}} - dx\right] = \frac{1}{2}\tau\left(\frac{\partial y}{\partial x}\right)^2 dx,$$

and for the entire string,

$$V = \frac{1}{2}\int_0^l \tau\left(\frac{\partial y(x,t)}{\partial x}\right)^2 dx. \tag{14.102}$$

The Lagrangian follows immediately from $L = T - V$. If σ and τ are constant,

$$L = \frac{\sigma}{2}\int_0^l \left(\frac{\partial y(x,t)}{\partial t}\right)^2 dx - \frac{\tau}{2}\int_0^l \left(\frac{\partial y(x,t)}{\partial x}\right)^2 dx. \tag{14.103}$$

For a continuous one-dimensional system of length l, the *Lagrangian density* $\mathcal{L}$ is defined by

$$L = \int_0^l \mathcal{L}dx. \tag{14.104}$$

Hamilton's Principle for a Continuous System. Recall from Section 4.8.1 that Hamilton's principle states that the variation of the action integral is zero:

$$\delta \int_{t_1}^{t_2} Ldt = 0. \tag{14.105}$$

In terms of $\mathcal{L}$, Hamilton's principle is

$$\delta \int_{t_1}^{t_2} dt \left\{\int_0^l dx \mathcal{L}\left(y, \frac{\partial y}{\partial x}, \frac{\partial y}{\partial t}; x, t\right)\right\} = 0. \tag{14.106}$$

Consider a virtual displacement,

$$y(x,t) \rightarrow y(x,t) + \delta y(x,t),$$

subject to the condition of fixed end points in time, i.e.,

$$\delta y(x,t_1) = \delta y(x,t_2) = 0, \quad \text{for all } x.$$

This describes a system that has the same configuration at times t_1 and t_2, such as a string oscillating with period $t_2 - t_1$.

The string is also subject to the spatial boundary condition of fixed end points, so

$$\delta y(x = 0,t) = \delta y(x = l,t) = 0 \quad \text{for all } t.$$

Equation (14.106) can be written as

$$0 = \int_{t_1}^{t_2} dt \left\{ \int_0^l dx \left[\frac{\partial \mathcal{L}}{\partial y} \delta y + \frac{\partial \mathcal{L}}{\partial \left(\frac{\partial y}{\partial x} \right)} \delta \left(\frac{\partial y}{\partial x} \right) + \frac{\partial \mathcal{L}}{\partial \left(\frac{\partial y}{\partial t} \right)} \delta \left(\frac{\partial y}{\partial t} \right) \right] \right\}. \tag{14.107}$$

We can write

$$\delta \left(\frac{\partial y}{\partial x} \right) = \frac{\partial}{\partial x} \delta y,$$

and

$$\delta \left(\frac{\partial y}{\partial t} \right) = \frac{\partial}{\partial t} \delta y.$$

Carrying out integrations by parts as usual, leads to

$$0 = \int_{t_1}^{t_2} dt \left\{ \int_0^l dx \left[\frac{\partial \mathcal{L}}{\partial y} - \frac{\partial}{\partial x} \frac{\partial \mathcal{L}}{\partial \left(\frac{\partial y}{\partial x} \right)} - \frac{\partial}{\partial t} \frac{\partial \mathcal{L}}{\partial \left(\frac{\partial y}{\partial t} \right)} \right] \delta y \right\}.$$

Consequently, Lagrange's equation for a continuous system is

$$\frac{\partial}{\partial t} \frac{\partial \mathcal{L}}{\partial \left(\frac{\partial y}{\partial t} \right)} + \frac{\partial}{\partial x} \frac{\partial \mathcal{L}}{\partial \left(\frac{\partial y}{\partial x} \right)} - \frac{\partial \mathcal{L}}{\partial y} = 0. \tag{14.108}$$

Worked Example 14.5   Derive the wave equation for a wave on a string using the Lagrange density and Equation (14.108)

Solution   For a string, from Equation (14.103).

$$\mathcal{L} = \frac{\sigma}{2} \left(\frac{\partial y}{\partial t} \right)^2 - \frac{\tau}{2} \left(\frac{\partial y}{\partial x} \right)^2, \tag{14.109}$$

so

$$\frac{\partial \mathcal{L}}{\partial \left(\frac{\partial y}{\partial t} \right)} = \sigma \frac{\partial y}{\partial t},$$

and

$$\frac{\partial \mathcal{L}}{\partial \left(\frac{\partial y}{\partial x} \right)} = -\tau \frac{\partial y}{\partial x}.$$

Lagrange's equation (14.108) reduces to

$$\frac{\partial}{\partial t} \left(\sigma \frac{\partial y}{\partial t} \right) + \frac{\partial}{\partial x} \left(-\tau \frac{\partial y}{\partial x} \right) - 0 = 0,$$

or, using $v = \sqrt{\tau/\sigma}$,

$$\frac{\partial^2 y}{\partial t^2} = v^2 \frac{\partial^2 y}{\partial x^2}.$$

Thus, we obtained the one-dimensional wave equation.

14.9 Summary

Interacting particles often exhibit small oscillations about equilibrium. Two classical examples of such systems are illustrated in Figures 14.1 and 14.2. The problem could be set up in terms of Cartesian coordinates (x_i), but it is much more convenient to introduce a set of generalized coordinates q_i defined as $q_i = q_i^0 + \eta_i$, where q_i^0 is the equilibrium position of the ith particle and η_i is its instantaneous displacement from equilibrium.

The kinetic energy can be expressed in terms of the physical coordinates (η_i) as

$$T = \frac{1}{2} \sum_{j=1}^{n} \sum_{k=1}^{n} M_{kj} \dot{q}_j \dot{q}_k = \frac{1}{2} \sum_{j=1}^{n} \sum_{k=1}^{n} M_{kj} \dot{\eta}_j \dot{\eta}_k.$$

The potential energy is given by

$$V = \frac{1}{2} \sum_{j=1}^{n} \sum_{k=1}^{n} \left(\frac{\partial^2 V}{\partial q_j \partial q_k} \right)_{q^0} \eta_j \eta_k = \frac{1}{2} \sum_{j=1}^{n} \sum_{k=1}^{n} V_{kj} \eta_j \eta_k.$$

The elements M_{kj} and V_{kj} define the *mass matrix* and the *potential matrix*. They are given by

$$M_{kj} = \sum_i m_i \left(\frac{\partial x_i}{\partial q_j} \right)_{q^0} \left(\frac{\partial x_i}{\partial q_k} \right)_{q^0},$$

and

$$V_{kj} = \left(\frac{\partial^2 V}{\partial q_j \partial q_k} \right)_{q^0}.$$

Knowing the kinetic energy and the potential energy allows one to determine the Lagrangian and hence the equations of motion. The Lagrangian is

$$L = T - V = \frac{1}{2} \sum_{kj} \left(M_{kj} \dot{\eta}_j \dot{\eta}_k - V_{kj} \eta_j \eta_k \right),$$

and the equations of motion are

$$\sum_j \left(M_{kj} \ddot{\eta}_j + V_{kj} \eta_j \right) = 0, \qquad k = 1, \ldots, n.$$

To solve the equations of motion it is convenient to introduce a set of complex quantities z_j related to the η_j by

$$\eta_j = \mathrm{Re}\left(z_j \right).$$

The equations of motion are, then

$$\sum_j \left(M_{kj} \ddot{z}_j + V_{kj} z_j \right) = 0, \qquad k = 1, \ldots, n.$$

The solution is obtained by making the usual assumption that all the particles oscillate at the same frequency ω (normal-mode assumption) so that

$$z_j = z_k^0 e^{i\omega t}.$$

Plugging this into the equations of motion leads to a set of n coupled algebraic equations that have a nontrivial solution iff

$$\det \left| V_{kj} - \omega^2 M_{kj} \right| = 0.$$

The normal mode frequencies ω_s can then be determined. Once the frequencies are determined, the complex coordinates z_k can be evaluated using Cramer's rule. The coefficients in the Cramer's rule procedure are

$$a_{kj} = V_{kj} - \omega_s^2 M_{kj},$$

and the solution gives n values of $z_k^{(s)}$ for each frequency ω_s. Here the superscript (s) indicates the particular normal mode. (Actually, we obtain $n - 1$ ratios of the form $z_k^{(s)}/z_n^{(s)}$.)

The complex coordinates $z_k^{(s)}$ can be expressed in polar form, thus,

$$z_k^{(s)} = \rho_k^{(s)} e^{i\phi_s}.$$

There is a different $\rho_k^{(s)}$ for each value of k, but the ϕ_s are all the same (for a given s). The set of ρ values for a given normal mode (s) is conveniently expressed as a column vector, thus:

$$\boldsymbol{\rho}^{(s)} = \begin{pmatrix} \rho_1^{(s)} \\ \rho_2^{(s)} \\ \vdots \\ \rho_n^{(s)} \end{pmatrix}.$$

This is called the eigenvector for the normal mode. The various eigenvectors are orthogonal. Once the eigenvectors have been determined, the physical coordinates can be obtained from

$$\eta_j(t) = \sum_{s=1}^{n} C^{(s)} \rho_j^{(s)} \cos(\omega_s t + \phi_s),$$

where $C^{(s)}$ and ϕ_s are obtained from the initial conditions.

The expressions obtained are probably better expressed as matrix relations. The mass matrix $\mathcal{M}$ has elements M_{kj} and the potential matrix $\mathcal{V}$ has elements V_{kj}. Assuming the normal mode frequencies have been determined, the eigenvectors $\boldsymbol{\rho}^{(s)}$ can be evaluated from

$$\left(\mathcal{V} - \omega_s^2 \mathcal{M} \right) \boldsymbol{\rho}^{(s)} = 0.$$

The physical coordinates can also be expressed as a column vector,

$$\boldsymbol{\eta}(t) = \sum_s \boldsymbol{\rho}^{(s)} C^{(s)} \cos(\omega_s t + \phi_s).$$

The modal matrix $\mathcal{A}$ is the matrix generated by using the eigenvectors $\boldsymbol{\rho}^{(s)}$ as columns.

It is convenient to introduce a set of coordinates called "normal coordinates" and denoted by ζ. These are defined by

$$\boldsymbol{\eta}(t) = \mathcal{A} \boldsymbol{\zeta}(t).$$

Consequently,

$$\boldsymbol{\zeta}(t) = \mathcal{A}^T \mathcal{M} \boldsymbol{\eta}(t).$$

The normal coordinates are useful because they decouple the equations of motion and lead to the set of equations

$$\ddot{\zeta}_s = -\omega_s^2 \zeta_s, \qquad s = 1, \ldots, n.$$

The solutions can be written immediately as

$$\zeta_s = C^{(s)} \cos(\omega_s t + \phi_s), \qquad s = 1, \ldots, n.$$

Examples of the application of this theory are coupled pendulums and transverse oscillations of mass points on a massless string. The second example can be generalized to a continuous string.

The final topic was the Lagrangian of a continuous system, such as a string. We obtained

$$L = \int_0^l \mathcal{L} dx = \frac{\sigma}{2} \int_0^l \left(\frac{\partial y(x,t)}{\partial t} \right)^2 dx - \frac{\tau}{2} \int_0^l \left(\frac{\partial y(x,t)}{\partial x} \right)^2 dx,$$

where σ is the linear mass density, and τ is the tension in the string.

14.10 Problems

Problem 14.1 Determine the mass matrix and the potential matrix for a system of two equal masses and three equal springs arranged as in Figure 14.1.

Problem 14.2 A particle of mass m finds itself in a region of space where the potential is given by

$$V(x, y, z) = \frac{V_0}{a^2} \left(4x^2 + 5y^2 + 6z^2 - 2ax - 5ay \right).$$

V_0 and a are positive constants. (a) Determine the location and value of V at its minimum. (b) Evaluate the mass matrix and the potential matrix. (c) Determine the normal mode oscillation frequencies about the equilibrium point.

Problem 14.3 A massless spring of constant k is fixed to the ceiling and is hanging vertically. A particle of mass m is attached to the bottom of the spring. Then another identical spring is attached to the particle and finally a second particle (also having mass m) is attached to the second spring. The system only moves in the vertical direction. Determine the normal frequencies and the eigenvectors $\rho^{(1)}$ and $\rho^{(2)}$.

Problem 14.4 For the problem of two coupled pendulums described in Section 14.6, show that $\mathcal{A}^T \mathcal{M} \mathcal{A} = 1$ and $\mathcal{A}^T \mathcal{V} \mathcal{A} = \omega_D^2$. Also show that the Lagrangian $L = \frac{1}{2} \sum_{j=1}^2 (\dot{\zeta}_j^2 - \omega_j^2 \zeta_j^2)$ reduces to the form of Equation (14.10).

Problem 14.5 Three equal masses m are connected by four equal springs of force constant k and length a, as in the system of Figure 14.1. Somehow the masses are constrained to oscillate transversely, that is, perpendicular to the line of the springs. (Perhaps the masses are mounted on thin vertical rods.) The oscillations all take place in a single plane. Find the normal mode frequencies.

Problem 14.6 Three equal springs are placed in a frictionless circular trough of radius a. At the junctions of the springs are masses $m, m,$ and $2m$. Determine the normal coordinates and frequencies of oscillation.

Problem 14.7 Assume the system of Figure 14.1 consists of two identical masses and three identical springs. (a) Find the Lagrangian and Lagrange's equations. (b) Determine the normal mode frequencies and the eigenvectors. (c) Construct the modal matrix.

Problem 14.8 Consider the system of Figure 14.1 as described in the previous problem. Let one mass be displaced from equilibrium by a small distance a in the direction of the other mass. Obtain the motion.

Problem 14.9 A system like the one in Figure 14.1 consists of two equal masses and three equal springs. The system is initially at rest. Then the mass on the right is subjected to an additional force $F = A \sin \omega t$. Determine the motion.

Problem 14.10 Consider a linearly symmetrical triatomic molecule. In equilibrium two atoms of mass m are located on either side of an atom of mass M. All three atoms lie along a straight line and in equilibrium the distance between them is b. The interatomic forces can be assumed to obey Hooke's law. Assume vibrations take place along the line of the molecule. Determine the normal mode frequencies. Describe the normal modes. (Note, the zero-frequency normal mode corresponds to a translation of the entire molecule.)

Problem 14.11 Consider again the linearly symmetrical triatomic molecule described in the previous problem. Show that it can be reduced to a problem with only two degrees of freedom by using the distances between the atoms as coordinates. Write the 2×2 mass and potential matrices and obtain the normal mode frequencies.

Problem 14.12 A system such as that illustrated in Figure 14.1 consists of N identical particles with masses m and $N + 1$ identical massless springs with force constant k. In equilibrium each spring is stretched so that its length (a) is greater than the unstretched length (a_0). Note that oscillations in the longitudinal mode and the transverse mode are allowed, but assume all oscillations take place in a single plane. (a) Determine the Lagrangian for the system. (b) By evaluating the equations of motion, show that the longitudinal and transverse modes are decoupled.

Problem 14.13 A large number of pendulums of length l and mass m are coupled by springs of constant k. When the system is in static equilibrium, all the pendulums hang vertically and the distance a between the bobs is equal to the equilibrium length of the springs. (a) Write the Lagrangian for the system in terms of η_i, the displacement from equilibrium of the ith pendulum. (b) Obtain the equations of motion. (c) Assume a traveling wave solution and obtain an expression for ω.

Problem 14.14 A rod of length l and mass m is suspended from its ends by two springs of equilibrium length b and constant k. The springs are hanging from the ceiling and are suspended from two points a distance L apart. Since $L > l$, the springs are not vertical, but form an angle θ with the vertical. Assume the rod oscillates only in the vertical plane formed by rod and springs. Determine the normal modes, assuming small oscillations.

Problem 14.15 Use the orthogonality of the Fourier series to determine the expressions given in Equations (14.99) and (14.100).

Problem 14.16 Show that the real part of $z_j(t)$, as given by Equation (14.37) is $\eta_j(t)$ in the form given by Equation (14.39). Note that $z_+^{(s)}$ and $z_-^{(s)}$ are complex numbers and $\eta = \mathrm{Re}(z) = \frac{1}{2}(z + z*)$.

Computational Projects

Computational Project 14.1 Plot Equation (14.76) using reasonable values for the parameters. Study the effect of varying the parameters.

Computational Project 14.2 Assume the system of Figure 14.1 consists of 10 equal masses (and 11 springs). The masses are all 0.1 kg and the springs all have $k = 5$ N/m. Write a computer program to determine the normal mode frequencies.

Part V

Rotation

15 Accelerated Reference Frames

By now you have heard me say many times that Newton's laws of motion are applicable only in inertial (nonaccelerated) frames of reference. The question arises: How do we deal with motion in an *accelerated* reference frame? After all, most real reference frames are accelerating and, for many of them, the acceleration cannot be neglected. Since we live on the surface of a large rotating sphere, it is important for us to be able to solve physics problems in noninertial systems.

In this chapter you will learn how to express Newton's second law in a noninertial (accelerating) reference frame. We are, of course, particularly interested in rotating coordinate systems, which are a prime example of accelerating reference frames. Our study of these systems will introduce you to "fictitious" forces such as the centrifugal force and the Coriolis force. The Coriolis force, as you may know, is responsible for a number of important geophysical processes such as hurricanes and ocean currents. Finally, we will consider the motion of a very long simple pendulum and show that the plane of this "Foucault" pendulum precesses due to the rotation of the Earth.

15.1 A Linearly Accelerating Reference Frame

Although the most interesting noninertial reference frame for our purposes is a rotating reference frame such as the Earth, I would like to begin by looking at a reference frame that is accelerating *linearly* relative to an inertial reference frame. Figure 15.1 shows a reference frame O that is at rest "with respect to the fixed stars" and a second reference frame, O', that is accelerating at a constant acceleration $\ddot{\mathbf{r}}$ relative to O.[1] In the figure, the position of a particle of mass m with respect to the inertial frame is $\mathbf{r}_O$ and the position of the particle with respect to the accelerating frame is $\mathbf{r}_{O'}$. The position of the origin O' with respect to O is $\mathbf{r}$. By tip to tail addition of vectors,

$$\mathbf{r}_O = \mathbf{r} + \mathbf{r}_{O'}. \tag{15.1}$$

Because O is an inertial coordinate system, Newton's second law holds in that frame and we can write

$$m\ddot{\mathbf{r}}_O = \mathbf{F}, \tag{15.2}$$

where $\mathbf{F}$ is the force acting on the particle. The acceleration of the particle relative to the accelerating frame is $\ddot{\mathbf{r}}_{O'}$. The force on the particle and the mass of the particle are assumed independent of the reference frame. (I am specifically excluding relativistic effects.)

---

[1] Please ignore the annoying fact that eventually O' will be moving at the speed of light!

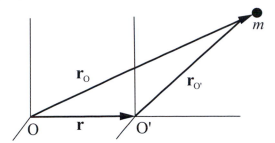

Figure 15.1 Coordinate system O' is accelerating at a constant rate $\ddot{\mathbf{r}}$ with respect to coordinate system O.

Taking the second time derivative of Equation (15.1) yields

$$\ddot{\mathbf{r}}_O = \ddot{\mathbf{r}}_{O'} + \ddot{\mathbf{r}},$$

where $\ddot{\mathbf{r}}$ is the acceleration of reference frame O' relative to the inertial frame O. Newton's law (15.2) then reads

$$\mathbf{F} = m\ddot{\mathbf{r}}_O = m(\ddot{\mathbf{r}}_{O'} + \ddot{\mathbf{r}}) = m\ddot{\mathbf{r}}_{O'} + m\ddot{\mathbf{r}}.$$

Therefore, from the point of view of the observer in the *accelerated reference frame,* Newton's second law has the form

$$m\ddot{\mathbf{r}}_{O'} = \mathbf{F} - m\ddot{\mathbf{r}}.$$

That is, there appears to be an *additional* force $\mathbf{f} = -m\ddot{\mathbf{r}}$ acting on the particle. If asked to determine the force on the particle, the observer in the accelerated reference frame would answer $\mathbf{F} + \mathbf{f}$.

Thus the observer in the accelerated system will believe that an additional force is acting on the particle. This is called a *fictitious force* because it is really just a mass times acceleration term which has been moved to the other side of the equation. Well-known fictitious forces include the centrifugal force and the Coriolis force.

15.2 A Rotating Coordinate Frame

Consider an ant standing still on the surface of a basketball. If the ball is rotating, the ant is moving in a circle around the axis of rotation. (The ant may *think* it is at rest. You may *think* you are sitting still in this room, but you are actually moving around in a large circle at over 600 miles per hour.) An object moving in a circle has a continuously changing velocity. Therefore, it is accelerating. The acceleration is directed towards the center of the circle. A coordinate system attached to a rotating sphere is an accelerated coordinate system. A particle at rest relative to this coordinate system is accelerating relative to inertial space.

As a very important special case, consider a coordinate system (x', y', z') rigidly connected to the rotating Earth. Suppose the origin of this coordinate system is at the center of the Earth, as shown in Figure 15.2. The Earth and the coordinate system rotate together about the z'-axis at an angular velocity $\boldsymbol{\Omega}$ (assumed constant). Figure 15.2 also shows an inertial coordinate system (x, y, z) that is at rest with respect to the fixed stars. (I am ignoring the motion of the Earth around the Sun and the motion of the Sun around the center of the Galaxy.)

The common origin of those two coordinate systems is the center of the Earth. The z- and z'-axes coincide and lie along the axis of rotation of the Earth. (You can imagine these axes pointing from the center of the Earth out through the North Pole.) The x, y and x', y' axes all lie in the

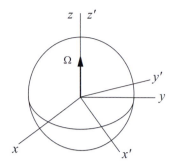

Figure 15.2 A rotating coordinate system (x', y', z') fixed in the Earth and an inertial system (x, y, z) at rest with respect to the fixed stars.

equatorial plane. Again, the coordinate axes (x, y, z) are fixed in space and the coordinate axes (x', y', z') are rigidly attached to the Earth and rotating with it.

A person standing still on the surface of the Earth is at rest with respect to the rotating coordinate system. The position of this person is given by the vector $\mathbf{r}$ from the origin to the point P where the person is standing. The components of $\mathbf{r}$ can be specified in either the inertial or the rotating coordinate system. Because P is at rest in the rotating (Earth or primed) system,

$$\left[\frac{d\mathbf{r}}{dt}\right]_{\text{rot}} = 0.$$

On the other hand, according to an observer in the inertial system, the point P is moving with angular velocity $\boldsymbol{\Omega}$. The linear velocity of the point in the inertial system is

$$\left[\frac{d\mathbf{r}}{dt}\right]_{\text{inertial}} = \boldsymbol{\Omega} \times \mathbf{r}.$$

Next assume the person has a velocity $\mathbf{v}_r$ with respect to the Earth, so that

$$\left[\frac{d\mathbf{r}}{dt}\right]_{\text{rot}} = \mathbf{v}_r.$$

Then, from the point of view of the inertial frame, the person's velocity is

$$\left[\frac{d\mathbf{r}}{dt}\right]_{\text{inertial}} = \mathbf{v}_r + \boldsymbol{\Omega} \times \mathbf{r}.$$

This last equation can be written

$$\left[\frac{d\mathbf{r}}{dt}\right]_{\text{inertial}} = \left[\frac{d\mathbf{r}}{dt}\right]_{\text{rot}} + \boldsymbol{\Omega} \times \mathbf{r}.$$

It is easy to generalize this relationship to the time derivative of any vector $\mathbf{U}$. The relation between the rate of change of $\mathbf{U}$ in the inertial system and the rate of change of $\mathbf{U}$ in the rotating system is

$$\left[\frac{d\mathbf{U}}{dt}\right]_{\text{inertial}} = \left[\frac{d\mathbf{U}}{dt}\right]_{\text{rot}} + \boldsymbol{\Omega} \times \mathbf{U}. \tag{15.3}$$

This can be expressed as an *operator equation*

$$\left[\frac{d}{dt}\right]_{\text{inertial}} = \left[\frac{d}{dt}\right]_{\text{rot}} + \boldsymbol{\Omega} \times, \tag{15.4}$$

where it is understood that the various terms are all operating on the same vector.

Exercise 15.1

Somewhere out in space there is an inertial reference frame and (strange to say) in it there is a merry-go-round that is rotating at angular velocity $\boldsymbol{\Omega}$. A child is sitting on a wooden horse a distance R from the axis of the merry-go-round. (a) What is the velocity of the child in the rotating frame? (b) What is the velocity of the child in the inertial frame? (c) What is the apparent acceleration of the child in the rotating frame? (d) Determine the acceleration of the child in the inertial frame by elementary methods, that is, by using the definition of centripetal acceleration. (e) Determine the acceleration of the child in the inertial frame using Equation (15.3). Answers (a) zero, (b) $\Omega R \hat{\boldsymbol{\phi}}$, (c) zero, (d) $-\Omega^2 R \hat{\boldsymbol{\rho}}$. ∎

Exercise 15.2

Prove that the rate of change of the angular velocity of the rotating system is the same in both the inertial and the rotating reference frame. ∎

15.3 Fictitious Forces

Let us continue ignoring the motion of the Earth around the Sun and imagine the Earth to be a rotating sphere isolated in space. It is rotating at a nearly constant angular velocity of 2π rad/day. The relationship between the time rate of change of the position vector ($\mathbf{r}$) of a particle in the inertial reference frame and its time rate of change in the rotating reference frame is, by Equation (15.3),

$$\left[\frac{d\mathbf{r}}{dt}\right]_{\text{inertial}} = \left[\frac{d\mathbf{r}}{dt}\right]_{\text{rot}} + \boldsymbol{\Omega} \times \mathbf{r}.$$

The velocity of the particle with respect to inertial space is $\left[\frac{d\mathbf{r}}{dt}\right]_{\text{inertial}}$. Call it $\mathbf{v}_i$. Also, $\left[\frac{d\mathbf{r}}{dt}\right]_{\text{rot}}$ is the velocity of the particle with respect to the rotating frame. Call it $\mathbf{v}_r$. Consequently,

$$\mathbf{v}_i = \mathbf{v}_r + \boldsymbol{\Omega} \times \mathbf{r}.$$

This equation indicates that if a person on the Earth perceives a particle to be at rest, an observer in inertial space will perceive the particle to have a velocity $\mathbf{v}_i = \boldsymbol{\Omega} \times \mathbf{r}$.

Acceleration is defined as the time rate of change of velocity. Hence,

$$\mathbf{a}_{\text{inertial}} = \left[\frac{d\mathbf{v}_i}{dt}\right]_i.$$

This is the acceleration in the inertial reference frame and is, of course, the acceleration that is expressed in Newton's second law. However, we live in a rotating reference frame, so we want to express the second law in terms of quantities we can measure, that is, $\mathbf{a}_r$ and $\mathbf{v}_r$.

Applying operator equation (15.4) yields

$$\mathbf{a}_i = \left[\frac{d}{dt}(\mathbf{v}_r + \boldsymbol{\Omega} \times \mathbf{r})\right]_{\text{inertial}}$$

$$= \left[\frac{d}{dt}(\mathbf{v}_r + \boldsymbol{\Omega} \times \mathbf{r})\right]_{\text{rot}} + \boldsymbol{\Omega} \times (\mathbf{v}_r + \boldsymbol{\Omega} \times \mathbf{r}),$$

$$= \left[\frac{d}{dt}\mathbf{v}_r\right]_{\text{rot}} + \left[\frac{d}{dt}(\boldsymbol{\Omega} \times \mathbf{r})\right]_{\text{rot}} + \boldsymbol{\Omega} \times \mathbf{v}_r + \boldsymbol{\Omega} \times (\boldsymbol{\Omega} \times \mathbf{r}),$$

$$= \mathbf{a}_r + \frac{d\boldsymbol{\Omega}}{dt} \times \mathbf{r} + \left[\left(\boldsymbol{\Omega} \times \frac{d\mathbf{r}}{dt}\right)\right]_{\text{rot}} + \boldsymbol{\Omega} \times \mathbf{v}_r + \boldsymbol{\Omega} \times (\boldsymbol{\Omega} \times \mathbf{r}).$$

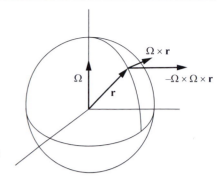

Figure 15.3 The centrifugal acceleration, $-\boldsymbol{\Omega} \times (\boldsymbol{\Omega} \times \mathbf{r})$, points away from the axis of rotation.

If the Earth's rotation rate is assumed constant, $\frac{d\boldsymbol{\Omega}}{dt} = 0$. Using the notation $\left[\frac{d\mathbf{r}}{dt}\right]_{\text{rot}} = \boldsymbol{v}_{\text{r}}$, the inertial acceleration is given by

$$\mathbf{a}_{\text{i}} = \mathbf{a}_{\text{r}} + 2\boldsymbol{\Omega} \times \boldsymbol{v}_{\text{r}} + \boldsymbol{\Omega} \times (\boldsymbol{\Omega} \times \mathbf{r}).$$

Newton's second law is valid in the inertial frame, so $\mathbf{a}_{\text{i}} = \mathbf{F}/m$, where $\mathbf{F}$ is the net external ("physical") force acting on the particle. Replacing $\mathbf{a}_{\text{i}}$ and rearranging leads to

$$\mathbf{a}_{\text{r}} = \mathbf{F}/m - 2\boldsymbol{\Omega} \times \boldsymbol{v}_{\text{r}} - \boldsymbol{\Omega} \times (\boldsymbol{\Omega} \times \mathbf{r}). \tag{15.5}$$

This equation shows that the acceleration *as measured in the rotating reference frame* will depend not only on the force acting on the particle, but also on two other terms. These additional terms are a consequence of considering the motion in a noninertial reference frame.

One often writes Equation (15.5) in the suggestive form

$$\mathbf{F} - 2m\boldsymbol{\Omega} \times \boldsymbol{v}_{\text{r}} - m\left[\boldsymbol{\Omega} \times (\boldsymbol{\Omega} \times \mathbf{r})\right] = m\mathbf{a}_{\text{r}}. \tag{15.6}$$

This equation *looks* like $F = ma$. The additional terms on the left-hand side of the equation have the units of force and they are referred to as "fictitious forces." Keep in mind, however, that they really are just mass times acceleration terms that have been brought over from the other side of the equation. A "real" force is an interaction between bodies. A "fictitious" force is a consequence of the acceleration of the coordinate system.

The term

$$-2m\boldsymbol{\Omega} \times \boldsymbol{v}_{\text{r}},$$

is called the "Coriolis force" and the term

$$-m\boldsymbol{\Omega} \times (\boldsymbol{\Omega} \times \mathbf{r}),$$

is called the "centrifugal force." The reason for the name is that this "force" is directed away from axis of rotation, as illustrated in Figure 15.3

Worked Example 15.1 An old-fashioned record player is rotating with an angular velocity $\boldsymbol{\Omega}_0 = \Omega_0\hat{\mathbf{k}}$. A bug is crawling with constant speed v_b relative to the record along a groove in the record, a distance r from the axis of rotation. Assume the bug is moving in a circular path, so there is no radial component to its velocity. (a) Determine the velocity of the bug relative to the room (that is the "inertial space"). (b) Determine the acceleration of the bug by elementary means, using the definition of centripetal acceleration. (c) Determine the acceleration of the bug using Equation (15.4) and compare your answer to part (b). (Comment: This example will help you learn the mechanics of using Equation (15.4).) In particular, you will appreciate that the same vector must be used on both sides of that equation. That is, if you are calculating

$\left[\frac{dv_i}{dt}\right]_{\text{inertial}}$ you get the wrong answer if you put $\left[\frac{dv_{\text{rot}}}{dt}\right]_{\text{rot}}$ on the other side. This is fairly obvious, but it is an easy error to make.)

Solution Let $\hat{\rho}, \hat{\phi}, \hat{k}$ be cylindrical unit vectors in the rotating frame. (a) The velocity of the bug in the rotating coordinate frame is

$$\left[\frac{d\mathbf{r}}{dt}\right]_{\text{rot}} = v_b\hat{\phi}.$$

The velocity of the bug in inertial space by Equation (15.4) is

$$\left[\frac{d\mathbf{r}}{dt}\right]_{\text{inertial}} = \left[\frac{d\mathbf{r}}{dt}\right]_{\text{rot}} + \mathbf{\Omega}_0 \times \mathbf{r}$$

$$\mathbf{v}_i = \mathbf{v}_{\text{rot}} + \mathbf{\Omega}_0 \times \mathbf{r}$$

$$= v_b\hat{\phi} + \Omega_0 r(\hat{k} \times \hat{\rho})$$

$$= v_b\hat{\phi} + \Omega_0 r\hat{\phi}.$$

For convenience, let us write $v_b\hat{\phi}$ as $\Omega_b r\hat{\phi}$. Then

$$\mathbf{v}_i = (\Omega_b + \Omega_0)r\hat{\phi}.$$

(b) By the definition of centripetal acceleration

$$\mathbf{a}_i = \frac{v_i^2}{r}(-\hat{\rho}) = \frac{-\hat{\rho}}{r}[(\Omega_b + \Omega_0)r]^2$$

$$= -\hat{\rho}r(\Omega_b + \Omega_0)^2.$$

(c) Using Equation (15.4) to determine $\mathbf{a}_i$ we have

$$\mathbf{a}_i = \left[\frac{dv_i}{dt}\right]_{\text{inertial}} = \left[\frac{d}{dt}(\Omega_b + \Omega_0)r\hat{\phi}\right]_{\text{inertial}}$$

$$= \left[\frac{d}{dt}(\Omega_b + \Omega_0)r\hat{\phi}\right]_{\text{rot}} + \mathbf{\Omega}_0 \times (\Omega_b + \Omega_0)r\hat{\phi}.$$

Since $v_b = \Omega_b r$ and because the rate of change of $\hat{\phi}$ in the rotating frame is only due to the motion of the bug, we have $\left[\frac{d\hat{\phi}}{dt}\right]_{\text{rot}} = -\Omega_b\hat{\rho}$. Furthermore, keeping in mind that $|\mathbf{r}| = r = $ constant, we obtain

$$\mathbf{a}_i = (\Omega_b + \Omega_0)r(-\Omega_b)\hat{\rho} + \Omega_0(\Omega_0 + \Omega_b)r(\hat{k} \times \hat{\phi})$$

$$= -\hat{\rho}r[(\Omega_b + \Omega_0)\Omega_b + \Omega_0(\Omega_b + \Omega_0)]$$

$$= -\hat{\rho}r[\Omega_b^2 + 2\Omega_b\Omega_0 + \Omega_0^2]$$

$$= -\hat{\rho}r(\Omega_b + \Omega_0)^2,$$

in agreement with the result of part (b).

15.4 Centrifugal Force and the Plumb Bob

Assume the Earth is a sphere. Most people probably assume that a hanging plumb bob points directly towards the center of the Earth. However, that would only be true if the Earth were not

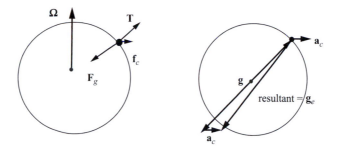

Figure 15.4 A plumb bob in the Northern Hemisphere points below the center of a spherical Earth. The sketch on the left shows the "forces" acting on the bob. The vector $\mathbf{f}_c$ represents the centrifugal force. The sketch on the right illustrates the vector sum of $\mathbf{g}$ and $\mathbf{a}_c = -\mathbf{\Omega} \times (\mathbf{\Omega} \times \mathbf{r})$.

rotating. On a rotating Earth, the centrifugal force causes the plumb bob to be deflected slightly away from the line to the center. The proof is fairly simple. You may want to try it on your own before reading my derivation.

As shown in Figure 15.4, the "real" forces acting on the bob are the tension in the string and the gravitational attraction of the Earth. That is,

$$\mathbf{F} = \mathbf{T} + m\mathbf{g}.$$

Inserting this expression for the force into the "$F = ma$" equation in the form of Equation (15.6) we get

$$\mathbf{T} + m\mathbf{g} - 2m\mathbf{\Omega} \times \mathbf{v}_r - m\left[\mathbf{\Omega} \times (\mathbf{\Omega} \times \mathbf{r})\right] = m\mathbf{a}_r,$$

or

$$\mathbf{T} + m\mathbf{g} - m\left[\mathbf{\Omega} \times (\mathbf{\Omega} \times \mathbf{r})\right] = 0,$$

where I used the fact that the velocity and acceleration of the plumb bob in the rotating frame are zero.

Now $\mathbf{g}$ points towards the center of the Earth, but the equation for the plumb bob written in the form

$$\mathbf{T} + m\mathbf{g} = m\left[\mathbf{\Omega} \times (\mathbf{\Omega} \times \mathbf{r})\right],$$

indicates that the tension is not parallel to $m\mathbf{g}$ and therefore it is not directed along the line to the center of the Earth.

It is usually convenient to include the centrifugal acceleration into the gravitational acceleration. The "effective" gravitational acceleration is defined by

$$\mathbf{g}_e = \mathbf{g} - \mathbf{\Omega} \times (\mathbf{\Omega} \times \mathbf{r}),$$

and the equation reduces to

$$\mathbf{T} + m\mathbf{g}_e = 0.$$

That is, the tension is directed opposite to $\mathbf{g}_e$ rather than opposite to $\mathbf{g}$. See Figure 15.4. In the Northern Hemisphere $\mathbf{g}_e = \mathbf{g} - \mathbf{\Omega} \times (\mathbf{\Omega} \times \mathbf{r})$ points slightly below the center of the Earth while in the Southern Hemisphere it points slightly above the center.

The centrifugal force is responsible for the fact that the Earth is not a sphere; it has a "bulge" at the equator. To see how this comes about, consider a particle on the surface of a perfectly spherical Earth. This particle is subjected to the gravitational force, a normal force and the centrifugal force. If you draw a force diagram for this situation, you will find that the centrifugal force is an unbalanced force. Resolving the centrifugal force into components parallel and perpendicular to the Earth's surface, you will find that the perpendicular component simply adds to the normal force, but the unbalanced parallel component points toward the equator. Thus, the centrifugal force is responsible for the fact that the Earth is an oblate spheroid rather than a sphere. You can convince yourself that if the Earth's equatorial bulge is taken into consideration there will be no unbalanced forces on a particle on the surface. This argument also tells you that although a plumb line does not point towards the center of the Earth, it is perpendicular to the Earth's *surface*, so it does define a local perpendicular.

Exercise 15.3

Draw the force diagram for a particle on the surface of a perfectly smooth, spherical, rotating planet and show that the particle accelerates towards the equator. Do this for both the Northern and Southern Hemispheres. Draw the same sort of sketch for a planet with an equatorial bulge and demonstrate that in this case there is no net force on the particle.   ■

15.5 The Coriolis Force

It is well known (at least to physicists) that a projectile fired near the surface of the Earth will veer towards the right in the Northern Hemisphere and towards the left in the Southern Hemisphere. This is due to the so-called Coriolis force, which is also responsible for oceanic circulation, cyclones, hurricanes, and other geophysical phenomena.

The Coriolis force is the term $-2m\mathbf{\Omega} \times \mathbf{v}_\mathrm{r}$ in Equation (15.6). Note that the direction of this fictitious force depends on the direction of the velocity vector.

When studying the effect of the Coriolis force on objects moving near the surface of the Earth it is convenient to introduce a Cartesian coordinate system in which the z-axis points up from the surface (i.e., opposite to the direction of the effective gravitational acceleration $\mathbf{g}_e$), the x-axis points East, and the y-axis points North.[2] See Figure 15.5. Note that the angular velocity of the Earth ($\mathbf{\Omega}$) has components along the y-axis (North) and along the $+z$-axis (up). Thus:

$$\mathbf{\Omega} = \Omega \cos \lambda \hat{\mathbf{k}} + \Omega \sin \lambda \hat{\mathbf{j}},$$

where λ is the "colatitude."

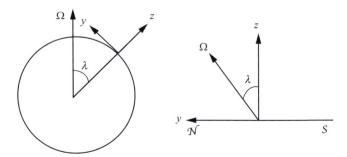

Figure 15.5 A reference frame fixed to the rotating Earth. The z-axis is along the local vertical, the y-axis points North and the x-axis points East (into the page). The angle λ is the colatitude.

---

[2] Some books use a different convention with the x-axis pointing South. The equations for the motion will be slightly different, so be careful when comparing equations from different books.

Draw force diagrams to determine the direction of the Coriolis force on the following objects. (a) A falling stone. (b) A projectile fired due North in the Northern Hemisphere. (c) A projectile fired due East in the Northern Hemisphere. (d) A projectile fired due East in the Southern Hemisphere. ∎

15.5.1 A Falling Body

An object of mass m is dropped from a height h. Where does it land? It turns out that the object hits the ground a tiny bit to the East of the spot directly below the point where it was released. The reason is that the falling object has a downward velocity so there is a Coriolis force acting on it, given by $-2m\mathbf{\Omega} \times \mathbf{v_r}$. (In the rotating reference frame of the Earth, the forces acting on the particle are the gravitational force, the Coriolis force, and the centrifugal force.)

For simplicity, we incorporate the centrifugal acceleration into $\mathbf{g}$ and write $\mathbf{g}_e$ for the sum of these two terms. The equation of motion of the falling body is, then

$$m\mathbf{a_r} = m\mathbf{g}_e - 2m\mathbf{\Omega} \times \mathbf{v_r}.$$

If you think about it, you will soon realize we cannot solve this equation for $\mathbf{a_r}$ because we do not know $\mathbf{v_r}$, and we cannot determine $\mathbf{v_r}$ until we solve the equation! It would seem that we are faced with an unsolvable problem.

The way out of this dilemma is to use the fact that the Coriolis acceleration is much smaller than the gravitational acceleration, so $\mathbf{v_r}$ is almost (but not quite) equal to the velocity of the falling body in the absence of the Coriolis force. That is, the Coriolis effect can be treated as a *perturbation.* The technique is as follows: First solve the unperturbed problem and determine the unperturbed velocity and time for the object to fall. This is called the "zeroth-order approximation." Then use the unperturbed velocity in the expression for the Coriolis acceleration to obtain the "first-order approximation." This process can be repeated as many times as desired to get better and better approximations to the correct answer, but usually the first- or second-order approximation is quite adequate.[3]

Let us begin by solving the problem for the *unperturbed* velocity and position of the particle as a function of time (the zeroth-order approximation). The only force acting on the body is the (effective) gravitational force. Consequently,

$$\mathbf{a_r} = -g_e\hat{\mathbf{k}}.$$

Integrating once,

$$\mathbf{v_r} = -g_e t\hat{\mathbf{k}},$$

and integrating again,

$$\mathbf{z} = h\hat{\mathbf{k}} - \tfrac{1}{2}g_e t^2\hat{\mathbf{k}}.$$

The time required for the particle to reach the ground, t_f, is the value of t when $z = 0$, that is,

$$t_f = \sqrt{2h/g_e}.$$

Note that none of the above relations include the Coriolis effect.

Now include the Coriolis acceleration, given by:

$$\mathbf{a}_c = -2\mathbf{\Omega} \times \mathbf{v_r}.$$

---

[3] This technique is known as the method of successive approximations.

You cannot evaluate $\mathbf{a}_c$ exactly because you do not know the value of $\mathbf{v}_r$ which itself depends on the Coriolis acceleration. However, you know the zeroth-order approximation to the velocity. It is $\mathbf{v}_r = -g_e t \hat{\mathbf{k}}$. The *first-order* correction is obtained by including the Coriolis term in the acceleration but using the zeroth-order velocity. That is, $\mathbf{a} = g_e \hat{\mathbf{k}} + a_c$ where

$$\mathbf{a}_c = -2\mathbf{\Omega} \times \mathbf{v}_r = -2 \left[\Omega \cos \lambda \hat{\mathbf{k}} + \Omega \sin \lambda \hat{\mathbf{j}} \right] \times \left[-g_e t \hat{\mathbf{k}} \right]$$

$$= 2 g_e t \Omega \sin \lambda \hat{\mathbf{i}}.$$

Note that for the falling body, the Coriolis acceleration is in the positive x-direction, that is, towards the East. The eastward component of the velocity (v_x) is initially zero and at any later time it is obtained by integrating the expression above. Therefore,

$$v_x = g_e t^2 \Omega \sin \lambda.$$

The eastward deflection (x) is obtained by integrating again.

$$x = \frac{1}{3} g_e t^3 \Omega \sin \lambda.$$

To determine the total eastward deflection when the object is dropped from height h, substitute the time of flight ($t = \sqrt{2h/g_e}$) into this last expression.

The results obtained are the first-order solution. As mentioned above, these results can be plugged back into the equation of motion to obtain the second-order solution, and so on.

Students are sometimes mystified by this result. The eastward deflection is not due to the Earth rotating under the falling body – anyway, that would give a westward deflection! A physical explanation for the horizontal (eastward) deflection of a dropped body comes from the conservation of angular momentum. Let $\mathbf{l}$ be the angular momentum of the particle in the inertial frame. There is no torque acting on it, so $\mathbf{l}$ is constant. Initially the object is at position $\mathbf{r}$ (where $\mathbf{r}$ is a vector from the center of the Earth). In the inertial frame the body has an eastward linear velocity $\mathbf{v}$. Because the body is at rest with respect to the surface of the Earth, this initial velocity is $\mathbf{v} = \mathbf{\Omega} \times \mathbf{r} \doteq \Omega R_e \sin \lambda \hat{\mathbf{i}}$. The angular momentum of the particle is

$$\mathbf{l} = m\mathbf{r} \times \mathbf{v}.$$

For $\mathbf{l}$ to remain constant while $\mathbf{r}$ decreases (as the particle falls), the velocity $\mathbf{v}$ must increase. Therefore, in the Earth-based reference frame, the object acquires a velocity towards the East.[4]

Another strange and interesting effect of Coriolis forces is the fact that if you throw an object straight up, it does not land at the point from which it started. (You will prove this in Problem 15.6.)

Worked Example 15.2   To first order, a falling particle is not deflected towards the South. However, there is a southward deflection in the second-order approximation. Estimate this deviation and compare it with the first-order deviation towards the East.

Solution   To first order, the velocity is

$$\mathbf{v}_r = v_z \hat{\mathbf{k}} + v_x \hat{\mathbf{i}} = -g_e t \hat{\mathbf{k}} + g_e t^2 \Omega \sin \lambda \hat{\mathbf{i}}.$$

---

[4] A slightly different explanation for the eastward deflection of a dropped object is based on the fact that if you hold an object above a point on the Earth's surface, both the object and the surface point are moving towards the East with the same angular velocity (Ω). But because $v = \Omega r$ the object has a slightly greater eastward velocity than the surface point. This greater eastward velocity is not affected by the fall and therefore the object lands to the East of the surface point.

The Coriolis acceleration is

$$\mathbf{a}_c = -2\boldsymbol{\Omega} \times \mathbf{v_r}$$

$$= -2\left[\Omega \cos\lambda \hat{\mathbf{k}} + \Omega \sin\lambda \, \hat{\mathbf{j}}\right] \times \left[-g_e t \hat{\mathbf{k}} + g_e t^2 \Omega \sin\lambda \hat{\mathbf{i}}\right].$$

That is,

$$a_x = 2\left(\Omega \sin\lambda\right) g_e t,$$

$$a_y = -2\left(\Omega^2 \cos\lambda \sin\lambda\right) g_e t^2,$$

$$a_z = 2\left(\Omega^2 \sin^2\lambda\right) g_e t^2 - g_e.$$

Consequently,

$$v_y = \int a_y dt = -\frac{2}{3}\left(\Omega^2 \cos\lambda \sin\lambda\right) g_e t^3,$$

and

$$y = -\frac{1}{6}\left(\Omega^2 \cos\lambda \sin\lambda\right) g_e t^4.$$

Using the first-order value for the time of flight we obtain

$$y = -\frac{2}{3}\frac{h^2}{g_e}\Omega^2 \cos\lambda \sin\lambda.$$

Note that the time of flight will actually be slightly longer because the Coriolis acceleration opposes the gravitational acceleration.

Comparing the southward and eastward deflections we have

$$\frac{y}{x} = \frac{-(1/6)\left(\Omega^2 \cos\lambda \sin\lambda\right) g_e t^4}{(1/3)\left(\Omega \sin\lambda\right) g_e t^3} = -\frac{1}{2}\left(\Omega \cos\lambda\right) t = -\Omega \cos\lambda\sqrt{h/2g_e}.$$

Since Ω is about 10^{-5} and $\cos\lambda$ is near unity and the time of flight will be in the range of 1–10 s, we appreciate that the southward deflection is significantly smaller than the eastward deflection.

Exercise 15.5

A stone is dropped from a height of 50 m. How far is it deflected towards the East? Assume the latitude is 60° N. Answer: 0.39 cm. ■

15.5.2 A Projectile

To determine the motion of a projectile fired from the surface of the Earth with arbitrary initial speed and direction, let us return to the equation for the acceleration in a rotating coordinate frame. A slight modification of Equation (15.5) gives

$$\mathbf{a} = \mathbf{g} - 2(\boldsymbol{\Omega} \times \mathbf{v}).$$

I dropped the subscript "e" on $\mathbf{g}$, but keep in mind that the centrifugal force is incorporated into $\mathbf{g}$. I also dropped the subscript r on $\mathbf{v_r}$ and on $\mathbf{a_r}$. The velocity $\mathbf{v}$ at any time can be expressed as

$$\mathbf{v} = v_x\hat{\mathbf{i}} + v_y\hat{\mathbf{j}} + v_z\hat{\mathbf{k}},$$

and

$$\boldsymbol{\Omega} = \Omega \sin \lambda \hat{\mathbf{j}} + \Omega \cos \lambda \hat{\mathbf{k}}.$$

Hence,

$$\mathbf{a} = -g\hat{\mathbf{k}} + 2\left(v_y \Omega \cos \lambda - v_z \Omega \sin \lambda\right)\hat{\mathbf{i}} - 2v_x \Omega \cos \lambda \hat{\mathbf{j}} + 2v_x \Omega \sin \lambda \hat{\mathbf{k}}, \tag{15.7}$$

or

$$a_x = 2v_y \Omega \cos \lambda - 2v_z \Omega \sin \lambda,$$
$$a_y = -2v_x \Omega \cos \lambda,$$
$$a_z = 2v_x \Omega \sin \lambda - g.$$

Recall that you cannot determine the motion by integrating the acceleration terms twice because the accelerations depend on the velocity which is unknown. So once again it is necessary to use the method of successive approximations. That is, you obtain the first-order approximation by using the velocity components of the zeroth-order approximation, and similarly calculating the higher-order approximations. However, even without doing any calculations, the functional dependence of the acceleration components give a good qualitative idea of the motion. For example, in the Northern Hemisphere, if the projectile is fired due North, then v_x (the eastward component of the velocity) is initially zero and there is no Coriolis acceleration in the y- or z-directions. There is, however, an *eastward* component of the acceleration and the projectile will be deflected towards the East (to the right of its direction of motion). To actually evaluate this deflection you must incorporate the zeroth-order approximations of v_y and v_z into the expression for a_x and then integrate.

Similarly, the expressions for the acceleration components indicate that a projectile fired due East (x-direction) is deflected towards the South. The velocity component v_x slightly changes a_z, making it less negative, therefore slightly increasing the time of flight of the projectile resulting in an increase in range. Furthermore, a_x is nonzero, which also affects the range of the projectile. Although the vertical velocity v_z is positive during the first part of the trajectory and negative during the second part, you cannot assume that the second term in a_x averages out to zero.

In solving for the motion of a projectile, you must use perturbation techniques, but fortunately a reasonably accurate solution is usually given by the lowest order approximation that includes the Coriolis force.

You should convince yourself that in the Northern Hemisphere a projectile will be deflected toward its right and in the Southern Hemisphere it will be deflected toward its left.

Worked Example 15.3   A large cannon is located at colatitude λ in the Northern Hemisphere. It fires a shell with velocity $\mathbf{v}_0$. The cannon is aimed at an angle ϕ East of North and has an elevation angle θ. (a) Write expressions for the initial (unperturbed) velocity components. (Call them v_{0x}, v_{0y}, v_{0z}.) (b) Write expressions for a_x, a_y, a_z in terms of the unperturbed velocity components. (c) Obtain first-order expressions for the velocity components as functions of time and the unperturbed values of the velocity. Ignore air resistance. (The results of this example will be found useful when solving the problems at the end of the chapter.)

Solution   (a) By inspection of Figure 15.6 the initial velocity components are

$$v_{0x} = v_0 \cos \theta \sin \phi,$$
$$v_{0y} = v_0 \cos \theta \cos \phi,$$
$$v_{0z} = v_0 \sin \theta.$$

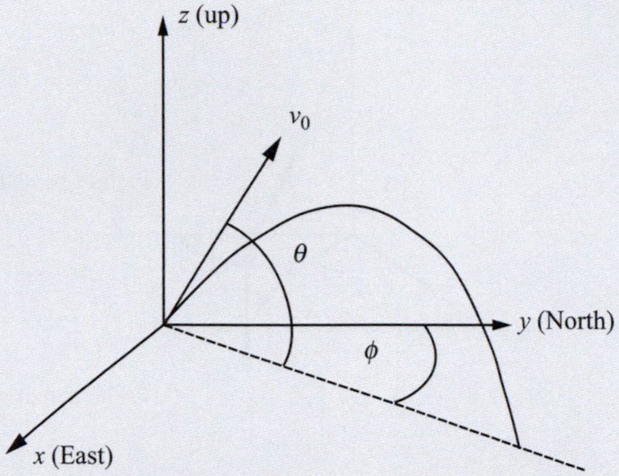

Figure 15.6 A projectile fired East of North. Note that ϕ is measured from North. The elevation angle θ is measured from the horizontal plane.

(b) The Coriolis acceleration is

$$\mathbf{a}_c = -2\mathbf{\Omega} \times \mathbf{v} = -2\left[\Omega \sin \lambda \hat{\mathbf{j}} + \Omega \cos \lambda \hat{\mathbf{k}}\right] \times \left[v_x \hat{\mathbf{i}} + v_y \hat{\mathbf{j}} + v_z \hat{\mathbf{k}}\right],$$

$$= -2\left[-v_x \Omega \sin \lambda \hat{\mathbf{k}} + v_z \Omega \sin \lambda \hat{\mathbf{i}} + v_x \Omega \cos \lambda \hat{\mathbf{j}} - v_y \Omega \cos \lambda \hat{\mathbf{i}}\right].$$

Since $v_x = v_{0x}, v_y = v_{0y}$ and $v_z = v_{0z} - gt$ we obtain

$$a_x = 2v_0 \Omega \cos \lambda \cos \theta \cos \phi - 2\Omega \sin \lambda (v_0 \sin \theta - gt),$$
$$a_y = -2v_0 \Omega \cos \lambda \cos \theta \sin \phi,$$
$$a_z = 2v_0 \Omega \sin \lambda \cos \theta \sin \phi - g.$$

(c) The velocity is determined by integrating the acceleration components to obtain:

$$v_x = v_{0x} + \int a_x dt$$
$$= v_0 \cos \theta \sin \phi + (2v_0 \Omega \cos \lambda \cos \theta \cos \phi)\, t$$
$$- (2v_0 \Omega \sin \lambda \sin \theta)\, t + gt^2 \Omega \sin \lambda,$$

$$v_y = v_{0y} + \int a_y dt = v_0 \cos \theta \cos \phi - (2v_0 \Omega \cos \lambda \cos \theta \sin \phi)\, t,$$

$$v_z = v_{0z} + \int a_z dt = v_0 \sin \theta - gt + (2v_0 \Omega \sin \lambda \cos \theta \sin \phi)\, t.$$

Exercise 15.6

Carry out the operation $\mathbf{\Omega} \times \mathbf{v}$ to verify Equation (15.7). ∎

Exercise 15.7

A projectile located at 60°N is fired due East with an initial velocity of 300 m/s at an angle 25° above the horizontal. Evaluate the initial values of the components of the Coriolis acceleration. (Note: $\lambda = 30°$.) Answer: $a_x(t = 0) = -9.22 \times 10^{-3}$ m/s^2. ∎

15.6 The Foucault Pendulum

The Foucault pendulum is an important example of the effect of a rotating coordinate system on the dynamics of a physical system. This pendulum consists of a massive bob suspended by a very

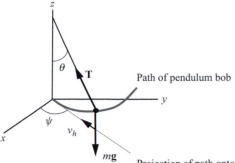

Path of pendulum bob

Projection of path onto xy-plane   **Figure 15.7** The Foucault pendulum.

long string or thin wire. Foucault pendulums range from ten to sixty meters in length. When set in motion the bob swings back and forth and the plane of motion of the pendulum slowly precesses. The precession rate is $\Omega \cos \lambda$, where Ω is the angular velocity of the Earth (2π rad/day) and λ is the colatitude. A long pendulum goes through many oscillations before the motion is damped out, and the precession is quite noticeable. When Foucault suspended a pendulum inside the Pantheon in Paris large crowds showed up to "observe the rotation of the Earth."

The behavior of the Foucault pendulum is easy to understand if you imagine a pendulum suspended above the North Pole. The Earth rotates under the pendulum and an observer standing on the surface sees the plane of motion precessing at a rate of one revolution per day. On the other hand, at the equator, a pendulum does not precess at all. The motion at an arbitrary latitude is not intuitively obvious and requires an analysis to understand it.

Consider a long pendulum at an arbitrary latitude. Let the z-axis be oriented along the local vertical. See Figure 15.7, which shows the projection of the path of the pendulum onto the xy-plane. The projected path makes an angle ψ with the x-axis. The precession of the pendulum causes ψ to change as a function of time. The basic problem consists in demonstrating that the rate of change of ψ is

$$\frac{d\psi}{dt} = \Omega \cos \lambda.$$

The pendulum bob is acted upon by two real forces: the tension in the string ($\mathbf{T}$) and gravity ($m\mathbf{g}$). It is also acted upon by two fictitious forces: the Coriolis force ($-2m\mathbf{\Omega} \times \mathbf{v}$) and the centrifugal force. For convenience assume the centrifugal force is absorbed into $m\mathbf{g}$.

Therefore,

$$m\mathbf{a} = \mathbf{T} + m\mathbf{g} - 2m(\mathbf{\Omega} \times \mathbf{v}).$$

In component form this relationship becomes

$$ma_x\hat{\mathbf{i}} + ma_y\hat{\mathbf{j}} + ma_z\hat{\mathbf{k}} = -T \sin\theta \cos\psi\hat{\mathbf{i}} - T \sin\theta \sin\psi\hat{\mathbf{j}} + T \cos\theta\hat{\mathbf{k}} - mg\hat{\mathbf{k}}$$
$$- 2m\left[\Omega\cos\lambda\hat{\mathbf{k}} + \Omega\sin\lambda\hat{\mathbf{j}}\right] \times \left[v_x\hat{\mathbf{i}} + v_y\hat{\mathbf{j}} + v_z\hat{\mathbf{k}}\right].$$

That is,

$$-T \sin\theta \cos\psi + 2mv_y\Omega\cos\lambda - 2mv_z\Omega\sin\lambda = ma_x, \tag{15.8}$$
$$-T \sin\theta \sin\psi - 2mv_x\Omega\cos\lambda = ma_y, \tag{15.9}$$
$$T \cos\theta - mg + 2mv_x\Omega\sin\lambda = ma_z. \tag{15.10}$$

Consider Equations (15.8) and (15.9). Because the vertical displacement is small, the vertical velocity v_z is much smaller than the horizontal velocity. Therefore, we are justified in ignoring the term containing v_z in Equation (15.8), leading to the set of coupled differential equations

$$-T\sin\theta\cos\psi + 2mv_y\Omega\cos\lambda = ma_x,$$
$$-T\sin\theta\sin\psi - 2mv_x\Omega\cos\lambda = ma_y.$$

The pendulum is very long so the angle θ is quite small and the string is essentially vertical at all times (Figure 15.7 is significantly out of scale). To a high degree of accuracy the tension in the string is equal to mg. Furthermore, $x = l\sin\theta\cos\psi$ and $y = l\sin\theta\sin\psi$. Therefore, the coupled equations above can be formulated as:

$$-\left(\frac{g}{l}\right)x + 2v_y\Omega\cos\lambda = a_x,$$
$$-\left(\frac{g}{l}\right)y - 2v_x\Omega\cos\lambda = a_y.$$

Noting that $g/l = \omega^2$, where ω is the oscillatory frequency of the pendulum, and letting $K = \Omega\cos\lambda$, these equations become

$$\ddot{x} - 2K\dot{y} + \omega^2 x = 0, \tag{15.11}$$
$$\ddot{y} + 2K\dot{x} + \omega^2 y = 0.$$

The problem now is to solve these coupled linear differential equations. I will do this in three ways. First I will use a "trick" then I will solve the equations in a straightforward, but more tedious manner. Then, in Worked Example 15.5, I will show you a third way to find the solution. The reason for doing this is because I want you to see several techniques for solving coupled differential equations.

A Neat Trick

The set of coupled equations (15.11) can be solved very nicely by introducing the complex quantity

$$\zeta = x + iy.$$

Multiplying the second equation by i ($= \sqrt{-1}$) and adding the two equations gives

$$\ddot{\zeta} + 2iK\dot{\zeta} + \omega^2\zeta = 0,$$

which resembles the equation of motion for a damped harmonic oscillator. This linear differential equation is solved in the usual way, assuming $\zeta = e^{pt}$ and obtaining the "secular" or "characteristic" equation

$$p^2 + 2iKp + \omega^2 = 0,$$

with solution

$$p = -iK \pm i\sqrt{K^2 + \omega^2}.$$

Now $K = \Omega\cos\lambda$, where Ω is the rotation rate of the Earth, whereas $\omega = \sqrt{g/l}$ is the angular frequency of the pendulum. So $K \ll \omega$ and the K^2 under the root can be neglected. There are two possible values for p, so the solution is

$$\zeta = Ae^{-i(K+\omega)t} + Be^{-i(K-w)t}, \tag{15.12}$$

where A and B are complex constants. Writing the complex constants in the form $(a + ib)$ and $(c + id)$ the solution is

$$\zeta(t) = (a + ib)e^{-i(K+\omega)t} + (c + id)e^{-i(K-w)t}. \tag{15.13}$$

This complex solution is expressed in terms of real quantities, but it is still not obvious what the motion looks like. The following example will help you to visualize the motion described by Equation (15.13).

Worked Example 15.4 At time $t = 0$ the Foucault pendulum is released from rest at position $x = \mathcal{A}$, $y = 0$. Obtain expressions for $x(t)$ and $y(t)$.

Solution Let us write the constants in Equation (15.12) as $\frac{1}{2}Ce^{\pm i\theta}$, thus

$$\begin{aligned}
\zeta &= \frac{1}{2}Ce^{-i\theta}e^{-i(\omega+K)t} + \frac{1}{2}Ce^{i\theta}e^{i(\omega-K)t} \\
&= \frac{1}{2}Ce^{-iKt}\left[e^{-i(\omega t+\theta)} + e^{+i(\omega t+\theta)}\right] \\
&= \frac{1}{2}C\left[\cos Kt - i\sin Kt\right]\left[2\cos(\omega t + \theta)\right] \\
&= C\left[\cos Kt\cos(\omega t + \theta) - i\sin Kt\cos(\omega t + \theta)\right].
\end{aligned}$$

Because $x = \text{Re}(\zeta)$ and $y = \text{Im}(\zeta)$ we conclude that

$$\begin{aligned}
x &= C\cos(t\Omega\cos\lambda)\cos(\omega t + \theta), \\
y &= -C\sin(t\Omega\cos\lambda)\cos(\omega t + \theta).
\end{aligned}$$

The constants C and θ are now easily obtained from the initial conditions, yielding $C = \mathcal{A}$ and $\theta = 0$, so

$$\begin{aligned}
x(t) &= [\mathcal{A}\cos(t\Omega\cos\lambda)]\cos\omega t, \\
y(t) &= -[\mathcal{A}\sin(t\Omega\cos\lambda)]\cos\omega t.
\end{aligned}$$

The motion is a rapid oscillation of frequency ω, modulated by a low-frequency variation in amplitude ($\Omega\cos\lambda$). The modulated x and y oscillations have time varying amplitudes that are $180°$ out of phase. The resultant motion is a rotation of the plane of motion with a frequency $\Omega\cos\lambda$. This is illustrated in Figure 15.8.

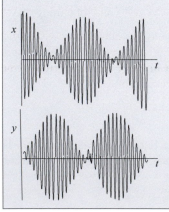

Figure 15.8 Amplitudes of oscillations in the x and y directions for the Foucault pendulum.

Exercise 15.8

A Foucault pendulum is observed to have a velocity v_0 in the $+x$-direction at time $t = 0$ when it passes through the origin. Obtain expressions for x and y as functions of t. Answer: $x = \text{Re}(\zeta) = -\frac{v_0}{K} \sin \omega t \cos Kt$, $y = \text{Im}(\zeta) = \frac{v_0}{K} \sin \omega t \sin Kt$. ∎

A More General Way to Solve the Problem

Return again to Equations (15.11). They are a pair of coupled second-order linear differential equations whose solution is desired. By this time you are very familiar with the technique of assuming a solution of the form Ae^{pt}. In this case there are two unknowns and the procedure is to assume two solutions, but to let both of them have the same time dependence.[5] In other words, assume the solutions

$$x = Ae^{pt},$$
$$y = Be^{pt}.$$

Plugging these solutions into Equations (15.11) leads to the following set of secular equations

$$p^2 A - 2KpB + \omega^2 A = 0,$$
$$p^2 B + 2KpA + \omega^2 B = 0.$$

or

$$A(p^2 + \omega^2) - B(2Kp) = 0,$$
$$A(2Kp) + B(p^2 + \omega^2) = 0.$$

This is a system of homogeneous linear algebraic equations in the unknowns A and B. Recall that the condition for a nontrivial solution is

$$\begin{vmatrix} p^2 + \omega^2 & -2Kp \\ 2Kp & p^2 + \omega^2 \end{vmatrix} = 0.$$

This leads to

$$p = -iK \pm i\sqrt{K^2 + \omega^2}. \tag{15.14}$$

The rest of the solution is straightforward.[6]

Worked Example 15.5 Solve for the precession rate of a Foucault pendulum by considering the problem in a rotating reference frame in which the pendulum does not precess.

Solution Using a rotating reference frame makes the problem easier to solve. Imagine looking down on the pendulum from a coordinate system rotating at an angular velocity $K = \Omega \cos \lambda$, equal to the precession rate. In this coordinate frame the bob moves back and forth in a straight line. Let the axes in this coordinate frame be x^* and y^* and define the x^*-axis to lie along the path of the bob. Then y^* is always zero. Denoting the distance from the origin to the bob along the x^*-axis by s, Figure 15.9 shows that the Earth-based coordinates x and y are

$$x = s \cos Kt$$

---

[5] You probably remember that this is the procedure we used to solve the coupled oscillators problem. (See Section 11.5.)

[6] It is interesting to note that the Foucault pendulum problem has many parallels to a seemingly completely unrelated problem, namely the violation of CP symmetry in the decay of an elementary particle called the neutral kaon. To explore further, see the paper "Classical Illustration of CP Violation in Kaon Decays" by Jonathan L. Rosner and Scott A. Slezak, *American Journal of Physics*, **69**, 44–49 (2001).

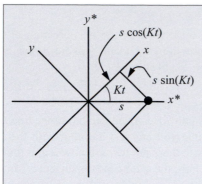

Figure 15.9 The starred coordinate system rotates at an angular velocity $K = \Omega \cos \lambda$ relative to the unstarred system. The motion of the pendulum bob is along the x^*-axis.

and

$$y = -s \sin Kt.$$

The sum of the equations of motion, Equations (15.11), is

$$\ddot{x} + \ddot{y} + 2K(\dot{x} - \dot{y}) + \omega^2(x + y) = 0 \qquad (15.15)$$

Plugging in the expressions for x and y in terms of s gives

$$\ddot{s} + (K^2 + \omega^2)s = 0. \qquad (15.16)$$

Again, $K \ll \omega$ so, to an excellent approximation,

$$\ddot{s} + \omega^2 s = 0.$$

This is just the SHM equation. Thus, in the rotating coordinate system the Foucault pendulum simply oscillates back and forth, or, from the point of view of the observer on the surface of the Earth, the plane of the pendulum rotates at a constant rate $K = \Omega \cos \lambda$.

Exercise 15.9

Fill in the missing steps between Equations (15.15) and (15.16).   ■

Exercise 15.10

Show that Equation (15.14) leads to the same result as obtained in Worked Example 15.5.   ■

Exercise 15.11

Suppose that a 20 m Foucault pendulum is mounted at San Francisco, California (latitude 37.77°). Determine the precession rate of its plane of motion. Answer: 11.9°/h.   ■

15.7 Application: The Tidal Force (Optional)

The tidal force is an interesting and important astronomical phenomenon. Tides are responsible for the formation of rings around large planets, tidal locking (or resonances) between bodies, the stretching of bodies falling into black holes ("spaghetization"), the frictional heating of Jupiter's

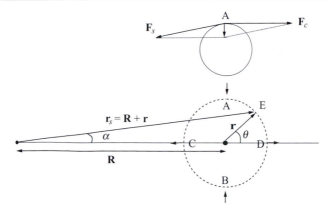

Figure 15.10 The lower figure gives the geometrical parameters. The upper figure shows that forces at point A do not cancel.

moon Io, and moonquakes, to mention a few tidal effects. In Chapter 7 we noted that the ocean tides on the Earth, due primarily to the Moon, are slowing down the rotation rate of the Earth.

A particularly interesting aspect of tidal forces is that they are due to the inhomogeneity of the gravitational field acting on a body and consequently are given by an inverse cube force law.

To analyze the tidal force, we shall consider the tides raised by the Sun on the Earth's oceans. Although the lunar tides are nearly twice as big as the solar tides, the solar tides are easier to analyze (because the center of mass of the system is essentially at the center of the Sun) and the physical principles are the same. The equations we will derive are valid for both situations, although the numerical values will be different.

We begin by assuming a solid rigid Earth that is not rotating about its axis, but it is revolving about the Sun in a circle at angular velocity Ω. Thus a coordinate system fixed in the Earth has its axes pointing towards the same fixed stars as it orbits the Sun. This has been described as the same motion a cook uses in moving a frying pan in a circle while holding the handle. Of course, the Earth really is rotating and we will include that fact a little later, but the picture of the nonrotating Earth can be helpful in understanding why there are two tidal bulges. The geometry of the system is shown in the bottom part of Figure 15.10.

In this coordinate system the forces acting on a particle *at the center of the Earth* are the attractive force of the Sun ($\mathbf{F}_s = -G\frac{Mm}{R^2}\hat{\mathbf{R}}$) and the (fictitious) centrifugal force ($\mathbf{F}_c = -m\,\mathbf{\Omega} \times (\mathbf{\Omega} \times \mathbf{R})$), where $\mathbf{R}$ is the vector from Sun to Earth, M is the mass of the Sun and m is the mass of the particle. These forces are equal and opposite.

The Earth is assumed to be perfectly rigid so all points in the Earth are accelerating at the same rate. That is, every part of the Earth feels the same centrifugal force. But not all points feel the same gravitational attraction toward the Sun. As indicated in the top part of Figure 15.10, at point A the two forces are not perfectly antiparallel and there is a small unbalanced component directed toward the center of the globe. Such forces, due to the difference between $\mathbf{F}_s$ and $\mathbf{F}_c$ are called "tidal forces." Similarly, at point B there is a net force toward the center. Point C is nearer to the Sun than the center of the planet, so the solar attraction is greater than the centrifugal force, resulting in a small unbalanced (tidal) component directed *away* from the center. At point D the centrifugal force is greater than the solar force so once again the tidal force is away from the center. In other words, at points A and B there is a compression and at points C and D the tidal force exerts a stretching force on the Earth. (It might be helpful to consider the behavior of a spherical gas cloud falling towards the Sun. The same forces are acting on the gas cloud, but it is not rigid so the cloud is squeezed and stretched into a long thin shape. This would be an example of spaghetization.)

Let us now determine the tidal force at an arbitrary point on Earth's surface such as point E in the bottom part of Figure 15.10. Because the centrifugal force at the center of the Earth is equal and opposite to the Sun's gravitational attraction we can write

$$\mathbf{F}_c = \frac{GMm}{R^3}\mathbf{R}.$$

At point E the gravitational force depends on the distance to the Sun, $\mathbf{r}_s$.

$$\mathbf{F}_s = -\frac{GMm}{r_s^3}\mathbf{r}_s.$$

Therefore, the tidal force, being the difference between these, is

$$\mathbf{F}_{Tide} = \mathbf{F}_c - \mathbf{F}_s = -GMm\left(\frac{\mathbf{r}_s}{r_s^3} - \frac{\mathbf{R}}{R^3}\right).$$

Let us now determine r_s^{-3}.

Note that $r_s^2 = (\mathbf{R}+\mathbf{r})\cdot(\mathbf{R}+\mathbf{r}) = R^2 + r^2 + 2(\mathbf{R}\cdot\mathbf{r})$. Since $R^2 \gg r^2$ we can write

$$r_s^2 \approx R^2\left(1 + 2\frac{\mathbf{R}\cdot\mathbf{r}}{R^2}\right)$$

$$\therefore 1/r_s^3 = (r_s^2)^{-3/2} = \left[R^2\left(1 + 2\frac{\mathbf{R}\cdot\mathbf{r}}{R^2}\right)\right]^{-3/2}$$

$$\doteq \frac{1}{R^3}\left(1 - 3\frac{\mathbf{R}\cdot\mathbf{r}}{R^2}\right),$$

so

$$\mathbf{F}_{Tide} = -GMm\left[\frac{\mathbf{r}_s}{R^3}\left(1 - 3\frac{\mathbf{R}\cdot\mathbf{r}}{R^2}\right) - \frac{\mathbf{R}}{R^3}\right]$$

$$= -\frac{GMm}{R^3}\left[(\mathbf{R}+\mathbf{r})\left(1 - 3\frac{\mathbf{R}\cdot\mathbf{r}}{R^2}\right) - \mathbf{R}\right]$$

$$= -\frac{GMm}{R^3}\left(\mathbf{R}+\mathbf{r} - 3\frac{\mathbf{R}\cdot\mathbf{r}}{R^2}(\mathbf{R}+\mathbf{r}) - \mathbf{R}\right).$$

But $\mathbf{R}+\mathbf{r} \approx \mathbf{R}$, so

$$\mathbf{F}_{Tide} = -\frac{GMm}{R^3}\left(\mathbf{r} - 3\mathbf{R}\frac{\mathbf{R}\cdot\mathbf{r}}{R^2}\right).$$

Observe that the main contributions of $\mathbf{F}_s$ and $\mathbf{F}_c$ have cancelled, leaving only their difference. At points A and B we have $\mathbf{R} \perp \mathbf{r}$ so $\mathbf{F}_{Tide} = -\frac{GMm}{R^3}\mathbf{r}$ which is pointing towards the center of the Earth, i.e., a compression. At point C the angle θ is π so and we obtain $\mathbf{F}_{Tide} = +2\frac{GMm}{R^3}\mathbf{r}$ which points away from the surface. A similar expression is obtained at D.

The tidal forces at various points on the surface of the globe are shown in Figure 15.11.

At a general point, such as E in Figure 15.11, the tidal force can be resolved into components perpendicular to the surface (along $\hat{\mathbf{r}}$) and parallel to the surface (along $\hat{\boldsymbol{\theta}}$). The components perpendicular to the surface (such as at ABCD) simply add to the Earth's gravitational force and are not of interest. But the tangential components can affect particles on the Earth's surface that are not rigidly attached to it and these are the forces that produce the tides in the ocean and atmosphere.

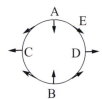

Figure 15.11 Equilibrium solar tidal forces on the Earth.

We shall denote the tangential component of $\mathbf{F}_{Tide}$ at an arbitrary point, such as E, by $\mathbf{F}_\theta$. Then since

$$\mathbf{F}_{Tide} = -\frac{GMm}{R^3}\left(r\hat{\mathbf{r}} - 3\frac{\mathbf{R}\cdot\mathbf{r}}{R^2}R\hat{\mathbf{R}}\right),$$

and $\hat{\mathbf{R}} = \cos\theta\hat{\mathbf{r}} - \sin\theta\hat{\boldsymbol{\theta}}$, we have

$$\mathbf{F}_\theta = -\left(\frac{GMm}{R^3}\right)3\frac{Rr\cos\theta}{R^2}R\sin\theta\hat{\boldsymbol{\theta}}$$

$$= -3\frac{GMm}{R^3}r\sin\theta\cos\theta\hat{\boldsymbol{\theta}}$$

$$= -3\frac{GMm}{R^3}r\sin 2\theta\hat{\boldsymbol{\theta}}.$$

Note that $\mathbf{F}_\theta = 0$ at ABCD.

We now include the fact that the Earth is rotating at $\omega = 2\pi/\text{day}$. This means that the angle θ can be replaced by ωt and the tangential component of the tidal force can be expressed as

$$\mathbf{F}_\theta = -3\frac{GMm}{R^3}r\sin(2\omega t)\hat{\boldsymbol{\theta}}.$$

This result suggests that the tidal forces cause the waters of the oceans to form bulges at C and D. That is, that high tides would occur on the Sun–Earth line, occurring locally at noon and midnight. We know this is not true, therefore the static (or "equilibrium") theory described above fails to explain a crucial fact about tides. Nevertheless, it does explain why there are two tides each day and it has led to the interesting fact that the tidal force is inversely proportional to the *cube* of the distance to the driving body. (Thus the lunar tides are twice as large as the solar tides although the gravitational force of the Sun on the Earth is much greater than that of the Moon on the Earth.) Another failure of the static theory described above is that it predicts tidal bulges of about 54 cm; generally the tides are much greater than that.

Nevertheless, the static theory does provide a basis for a more general theory, usually referred to as a dynamical theory of the tides.[7] This theory considers the interaction between the driving force and the response of the oceans. We will not go into this theory in any detail because it involves some fairly advanced analyses and because it does not actually predict the tides at any particular place. But it is a reasonable attempt at developing a general theory. A usual starting point to this theory is to assume the entire Earth is covered by an ocean of constant depth. An even simpler model assumes a wide frictionless water filled canal along the Earth's equator. The Sun is assumed to lie in the Earth's equatorial plane. According to the static theory, the water in the canal should form bulges at the subsolar points C and D. If the tidal force would suddenly disappear, water would flow away from the bulge at C and the surface of the water would flatten out. But because the canal is frictionless, the water would build up again on the opposite side of

---

[7] The dynamical theory was suggested by Laplace and later developed by the mathematician G. B. Airy and the astronomer G. H. Darwin.

the Earth. Meanwhile, the bulge at D would also flatten out at first and then build up at C. That is, we would have two waves moving in opposite directions. From hydrodynamics, the speed of these waves is $\sqrt{gh}$, where h is the depth of the canal. For a canal of depth 3.5 km, the waves would oscillate with a period of about 30 hours and the bulges would lag the Sun by about 6 hours. But even this analysis is far too simple. One needs to include the inclination of the Earth's axis to its orbit plane as well as the eccentricity and inclination of the Moon's orbit. Finally, a more realistic model would include frictional forces as well as the effect of continents.

Although tidal tables are available for most ports and people have developed empirical formulas that are applicable to a particular location, these relations typically consist of sums of some 40 sinusoidal terms. A general theory of ocean tides on our planet has not been developed.

15.8 Summary

A rotating reference frame is an accelerated reference frame. The time derivative of a vector in a rotating frame is related to the time derivative in an inertial frame by the operator equation

$$\left[\frac{d}{dt} \right]_{\text{inertial}} = \left[\frac{d}{dt} \right]_{\text{rot}} + \boldsymbol{\Omega} \times .$$

The acceleration $\mathbf{a}_r$ relative to a rotating frame is obtained by applying the operator equation twice to the position vector, yielding:

$$\mathbf{a}_r = \mathbf{F}/m - 2\boldsymbol{\Omega} \times \mathbf{v}_r - \boldsymbol{\Omega} \times (\boldsymbol{\Omega} \times \mathbf{r}),$$

that is often written as

$$\mathbf{F} - 2m\boldsymbol{\Omega} \times \mathbf{v_r} - m\left[\boldsymbol{\Omega} \times (\boldsymbol{\Omega} \times \mathbf{r}) \right] = m\mathbf{a}_r.$$

The term $-2m\boldsymbol{\Omega} \times \mathbf{v}_r$ is called the Coriolis force and $-m\left[\boldsymbol{\Omega} \times (\boldsymbol{\Omega} \times \mathbf{r}) \right]$ is called the centrifugal force.

The centrifugal force causes a plumb bob to point slightly away from the center of the Earth.

The Coriolis force causes a falling body to be deflected towards the East and a projectile to be deflected to the right of its initial velocity. Solving for the motion under the action of the Coriolis force usually involves applying the method of successive approximations.

The Foucault pendulum is a long simple pendulum that oscillates for enough time to allow one to appreciate the precession of the plane of oscillation at $\Omega \cos \lambda$.

15.9 Problems

Problem 15.1 A child runs on a merry-go-round with velocity $\mathbf{V}$ with respect to the merry-go-round. Show that

$$\left(\frac{d\mathbf{r}}{dt} \right)_I = \mathbf{V} + (\omega_e + \omega_{mgr}) \times \mathbf{r},$$

where ω_e is the angular velocity of the Earth and ω_{mgr} is the angular velocity of the merry-go-round.

Problem 15.2 A wedge of mass M and angle α is resting on a frictionless horizontal plane. A rectangular box of mass m is on the wedge. Since all surfaces are frictionless, the box will slide down the wedge and the wedge will slide in the opposite direction on the plane. So far, this is the same as Problem 4.11, which you may have solved. There you used the Lagrangian technique to solve the problem and found that the horizontal acceleration of the wedge is

$$\ddot{X} = -\frac{m}{m+M}\left(\frac{g\sin\alpha}{1 - \frac{m}{m+M}\cos^2\alpha}\right)\cos\alpha.$$

Here you are asked to find the horizontal component of the acceleration of the box relative to the wedge. Solve the problem by applying Newton's second law directly, but using the fact that the wedge is accelerating so the motion of the box relative to the wedge is motion in an accelerated reference frame.

Problem 15.3 Obtain an equation analogous to Equation (15.5) but including the orbital motion of the Earth around the Sun. You may assume the two rotation vectors are constant and parallel and that the Earth's orbit is circular. Compare the magnitude of any additional term to the Coriolis term in Equation (15.5).

Problem 15.4 Assume the Earth is an inertial coordinate system. Consider a child on a rotating platform. (a) The child runs radially towards the edge. (b) The child runs in the azimuthal direction. Determine the fictitious forces acting on the child in both situations.

Problem 15.5 A wheel of radius a rolls at constant speed V in a circular path of radius R. Assume the wheel is perpendicular to the plane of the path and that the surface of the Earth is an inertial reference frame. Consider a point at the top of the wheel. Determine the Coriolis acceleration, and the centrifugal acceleration of this point. See Figure 15.12.

Problem 15.6 A particle is projected vertically upward with an initial velocity v_0. Determine, to first order, the point where it lands relative to the point where it was thrown.

Problem 15.7 A NASA research center in Cleveland, Ohio has an underground vacuum drop tank in which an object can fall 132 meters. The latitude of Cleveland is 41.5°N. Determine the Coriolis deflection to second order.

Problem 15.8 A 5 kg stone is dropped from a height of 100 meters above the surface of the Earth at a location where the latitude is 40°. Assume the radius of the Earth is 6378 km. (a) What is the linear velocity of the stone just before it is dropped? (b) What is its angular momentum at that moment? (c) Determine the difference in its eastward velocity by the time it hits the ground by noting that the angular momentum of the stone does not change.

Problem 15.9 A projectile is fired due East with an initial velocity of 2000 m/s and at an elevation of 25°. Due to Coriolis effects it misses its target. Determine how far away from the target it lands to first order, assuming that the target is at the zeroth-order location. Obtain the correction to the time of flight. Assume $\lambda = 30°$.

Problem 15.10 A low-pressure area develops over the North Atlantic. Draw a vector diagram to show that the air mass will tend to move in a circle centered on the low-pressure region. Is the air moving clockwise or counterclockwise? Do the same thing for an air mass in the South Atlantic.

Problem 15.11 A low-pressure region in the North Atlantic is located at 30°N. A weather ship is located a distance of 80 km from the low. The atmospheric pressure at the ship is 1 atm, but it decreases towards the low with a pressure gradient of 10^{-2} Pa/m. Evaluate the wind velocity at the position of the ship. (The density of air is 1.3 kg/m^3.)

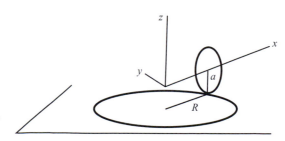

Figure 15.12 Wheel rolling on a circular path. (Problem 15.5.)

Problem 15.12 An object is dropped from 2 km above the Earth's surface at colatitude $45°$. Determine how far it has deviated horizontally from a line directed towards the center of the Earth by the time it hits the surface due solely to the effect of the centrifugal force. Although they are much greater influences, neglect both the Coriolis force and air resistance.

Problem 15.13 John and Mary (who live at $45°$N) go sky diving. They jump out of the airplane together. His free-fall velocity is 8 m/s and hers is 10 m/s. If the jump started at an altitude of 5 km, how far apart will they be when they land? (Assume the Coriolis force is the only factor determining this difference in the final position.)

Problem 15.14 An airplane is flying due East along the $30°$N latitude circle at 800 km/h. (a) Determine the Coriolis acceleration acting on it. (b) If the Coriolis force is not compensated for, how far off course will the airplane be after one hour?

Problem 15.15 A stationary pendulum is hanging from the ceiling of an airplane flying at a constant velocity of 800 km/h, following a meridian of longitude. Determine the deviation of the string of the pendulum relative to a line to the center of the Earth at the moment the airplane flies over the North Pole.

Problem 15.16 Suppose the Earth is perfectly spherical. An eccentric millionaire in Seattle builds an ice rink that is perfectly flat and tangent to the surface of the Earth. Furthermore, the ice is perfectly frictionless. The millionaire is surprised to find that a puck set in motion at the point of tangency slides back and forth like the bob of a Foucault pendulum. Find the frequency of oscillation and the frequency of precession of the trajectory of the puck.

Problem 15.17 It has been suggested that a "space habitat" in the shape of a giant rotating doughnut (a torus) would allow people to live inside in an environment with artificial gravity due to the centrifugal force. (This would be analogous to being in the inner tube of a large rotating bicycle wheel.) If the torus were big enough (say 3 km radius) and rotated fast enough (at about 30 rev/h), the centrifugal force on the "outside" wall would be approximately equal to g. Suppose the inhabitants of such a space habitat decided to play basketball. In the toroidal cavity the playing court would be perpendicular to the radial direction. (That is, "down" is radially outward and "up" is radially inward.) Define the z-axis along the direction of $\mathbf{\Omega}$, the rotational velocity of the torus, and ϕ perpendicular to r and z such that $\hat{\mathbf{r}} \times \hat{\boldsymbol{\phi}} = \hat{\mathbf{k}}$. Determine the Coriolis force on a ball that is thrown at speed v in the following directions: (a) "up," (b) the azimuthal direction, and (c) parallel to $\mathbf{\Omega}$. (d) Derive an expression for the point where a ball will land if it is thrown upward at speed v_0. (Note: This means the initial velocity is $-v_0\hat{\mathbf{r}}$.)

Problem 15.18 A dirigible crossing the North Atlantic at $50°$ latitude was flying at 200 mph due East. The navigator went to sleep and the Coriolis force caused the dirigible to move in a circular path. Determine the radius of the circle.

Problem 15.19 During the Falkland Islands war, the Argentine Air force fired a number of Exocet missiles at British ships. Assume an airplane flying horizontally due East at 1000 km/h at an altitude of 6 km fires a missile. The rocket motor of the missile accelerates it at 10 m/s^2 horizontally. Determine how far the Coriolis force will have caused the missile to deviate from "straight ahead" by the time it hits the water. You may assume the incident took place at a latitude of $55°$ S. Ignore air resistance.

Problem 15.20 A particle of mass m is on a frictionless rotating turntable. It is acted upon by a force directed towards the axis of rotation and given by $F = -kr$. We wish to obtain an expression for the position of the particle as a function of time in the rotating reference frame. To do so, first find the rotation rate Ω for the turntable such that the particle remains at rest. Next perturb the body radially ($r = a + \eta$, where $a =$ unperturbed value of r). Obtain expressions for $r(t)$ and $\theta(t)$ and then for $x(t)$ and $y(t)$ in the rotating reference frame. Describe the motion. (Hint: The θ equation can be easily integrated.)

Problem 15.21 A mass m is acted upon by a force directed towards the origin and given by $F = -kr$. Analyze the motion in a rotating coordinate system. (a) Determine the rotation rate of a coordinate system such that the unperturbed body is at rest. (b) Allow the body to be perturbed so that now it undergoes radial oscillations. Obtain expressions for $r = r(t)$ and $\theta = \theta(t)$, where r and θ are measured in the rotating coordinate system.

Computational Projects

Computational Project 15.1 Write a computer program to determine the impact point for a shell from a long-range cannon. Include the Coriolis effect to third order. (Neglect air resistance.)

Computational Project 15.2 Generate Figure 15.8 assuming a pendulum of length 40 m at 40°N.

Computational Project 15.3 A stone is dropped from a height of 100 m at latitude 40°N. Determine the distance it is deflected horizontally by the Coriolis force in the absence of air resistance. Next include air resistance and compare the results.

16 Rotational Kinematics

This chapter and the next are a study of the general motion of a rigid body. This is a fairly complicated topic which involves mathematical concepts that you may not have encountered before.

We have already considered the rotation of a symmetrical body about a fixed *axis*. We now generalize to rotations about a fixed *point*. For example, consider a spinning top. A top is a symmetrical body that rotates about its axis of symmetry. When it is spinning very fast, the top remains upright, and as long as the top is not slipping, its center of mass is at rest. But as the top slows down, it leans over. The center of mass traces a circle around the vertical. The axis of rotation precesses about this vertical. Eventually, the top will begin to nutate or "nod" up and down as it precesses. If the top is not sliding, it has one fixed point, namely the point that is in contact with the ground. Thus, this is rotation about a fixed point, and not about a fixed axis.

Another example of rotation about a fixed point is illustrated in Figure 16.1, which shows a disk with an axle passing through it. In this case, the axle is not perpendicular to the disk. If the system is not supported in any way (imagine it is a poorly designed satellite) this nonsymmetrical body will wobble while it rotates. The axis of rotation does not have a fixed direction. If the center of mass is at rest, this is a rotation about a fixed point.

Analyzing the motion of a body rotating about a fixed point involves developing a system of *rotational kinematics*. Recall that in Chapter 2 you studied the kinematics of a point particle. This entailed learning how to describe the position (as well as the velocity and acceleration) of a particle in terms of Cartesian, cylindrical, and spherical coordinates. In a similar manner, you will now learn how to describe the *orientation* of a rigid body and how this changes as the body rotates.

From a mathematical point of view, the rotation of a rigid body can be considered an *orthogonal transformation*. The analysis of orthogonal transformations, as you will see, leads to a consideration of Euler angles and Euler's theorem. Euler's angles represent three rotations that carry a body from any given initial orientation to any final orientation. Euler's theorem goes one better and tells us that we can carry a body from any initial orientation to any final orientation with a *single* rotation. (I feel obliged to warn you that some of this material is pretty tough going!)

16.1 Orientation of a Rigid Body

All the points in a rigid body maintain fixed relative distances from one another. This means that the distance between any two points i and j is given by the constraint equation

$$\left| \mathbf{r}_j - \mathbf{r}_i \right| = r_{ij} = \text{constant.}$$

Figure 16.1 A body that is not symmetric about an axis will tend to wobble when it rotates.

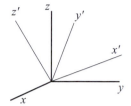

Figure 16.2 An inertial reference frame (x, y, z) and a reference frame (x', y', z') fixed in the body.

Suppose the rigid body is rotating about a fixed point. That is, there is one point in the body that is fixed in space and all the other points in the body can move about it in any way that is consistent with the set of constraints above.[1] Our first task is to describe the orientation of the body relative to a set of fixed axes in terms of the smallest set of independent (generalized) coordinates.

Six coordinates are sufficient to completely describe the position and orientation of a rigid body. (I will now show you that this statement is true; it might help to draw a figure while going through my explanation.) To specify the orientation of the body you need to know the relative positions of three noncollinear points. Three coordinates are required to locate a particular point P in the body – this point will often be the center of mass. You need to know the positions of two more points relative to P. The second point is at some given distance d from P so no matter how the body is rotated about P, the second point must lie somewhere on the surface of a sphere of radius d with center at P. Two coordinates are sufficient to specify the position of a point on the surface of the sphere so you need two coordinates to determine the position of this second point. You still need to specify the position of a third point relative to the first two. Draw a line between the first two points. Now the body can be rotated about an axis through these points without changing their positions. Therefore, the third point must lie on a circle centered on this line; to locate it requires a single coordinate. Consequently, a total of six independent coordinates are sufficient to specify both the position and the orientation of the body.

If some point, such as the center of mass, is *fixed* at some known position, you only need *three* coordinates to describe the orientation of the body. Let the center of mass be the origin of an inertial Cartesian coordinate system fixed in space with axes denoted by x, y, z. Let x', y', z' be another Cartesian coordinate system that is *fixed in the body* and rotates with the body. The two systems are illustrated in Figure 16.2. The orientation of the body is given by the orientation of the rotating (primed) system relative to the inertial (unprimed) system. The orientation of x', y', z' relative to x, y, z can be described using just three independent quantities. Another way to say this is to point out that x, y, z can be transformed into x', y', z' with three rotations. If the rigid body is rotating, the orientation of the body axes relative to the inertial axes will be changing with time. To keep our analysis simple, we will consider an instant frozen in time so that x', y', z' are not moving.

---

[1] For a body rotating about a *fixed* axis, there is the additional constraint that all the points on the axis are fixed in space and all off-axis points move in circles about the axis. However, the present problem considers a body with only one fixed point (possibly the center of mass).

Denote the unit vectors for the two systems by $\hat{\mathbf{i}}$, $\hat{\mathbf{j}}$, $\hat{\mathbf{k}}$ and $\hat{\mathbf{i}}'$, $\hat{\mathbf{j}}'$, $\hat{\mathbf{k}}'$. The direction of the x'-axis with respect to the unprimed axes is given by its direction cosines $\alpha_1, \alpha_2, \alpha_3$, defined by

$$\alpha_1 = \cos(x', x) = \hat{\mathbf{i}}' \cdot \hat{\mathbf{i}}, \tag{16.1}$$
$$\alpha_2 = \cos(x', y) = \hat{\mathbf{i}}' \cdot \hat{\mathbf{j}},$$
$$\alpha_3 = \cos(x', z) = \hat{\mathbf{i}}' \cdot \hat{\mathbf{k}}.$$

Similarly, the directions of the y'- and z'-axes are determined by their direction cosines. The nine direction cosines (three for each of the primed axes) give the orientation of the primed (or "body") axes with respect to the fixed (or inertial) axes. However, *nine* quantities are not needed to describe the relative orientation of the two sets of axes; as you know, *three* are sufficient. This means that not all of the direction cosines are independent.

Just as α_i ($i = 1, 2, 3$) denotes the direction cosines for the x'-axis, the directions cosines for the y'- and z'-axes can be denoted by β_i and γ_i with definitions analogous to Equations (16.1). Using the orthogonality of the unit vectors, you can show that

$$\alpha_l \alpha_m + \beta_l \beta_m + \gamma_l \gamma_m = \delta_{lm} \quad (l, m = 1, 2, 3), \tag{16.2}$$

where δ_{lm} is the Kronecker delta and is equal to unity if $l = m$ and equal to zero otherwise. Equations (16.2) are known as the "orthogonality conditions."[2]

Exercise 16.1

Write expressions analogous to Equation (16.1) for β_i and γ_i. ∎

Exercise 16.2

Prove relations (16.2). ∎

16.2 Orthogonal Transformations

Consider the axes illustrated in Figure 16.2. The position of a point in space is given by (x, y, z) in the inertial system and by (x', y', z') in the rotated system. The relation between these coordinates is

$$x' = \alpha_1 x + \alpha_2 y + \alpha_3 z, \tag{16.3}$$
$$y' = \beta_1 x + \beta_2 y + \beta_3 z,$$
$$z' = \gamma_1 x + \gamma_2 y + \gamma_3 z.$$

To make the notation easier to handle, it is traditional to write x, y, z, as x_1, x_2, x_3, and instead of x', y', z', one writes x_1', x_2', x_3'. Furthermore, instead of α_1 we write a_{11}, instead of α_2 we write a_{12}, instead of β_3 we write a_{23}, and so on, so that the transformation equations (16.3) are expressed as

$$x_1' = a_{11} x_1 + a_{12} x_2 + a_{13} x_3, \tag{16.4}$$
$$x_2' = a_{21} x_1 + a_{22} x_2 + a_{23} x_3,$$
$$x_3' = a_{31} x_1 + a_{32} x_2 + a_{33} x_3.$$

---

[2] The word "orthogonality" implies that one thing is perpendicular to another. In this case, we are just noting that the unit vectors are mutually perpendicular. In the case of orthogonal matrices the geometric concept of perpendicular quantities is harder to visualize.

It is convenient to use the Einstein summation convention to express Equations (16.4) in the very compact form

$$x_i' = a_{ij}x_j.$$ (16.5)

The Einstein summation convention states that if a term contains the same subscript twice, a summation over that subscript is implied. Thus

$$a_{ij}x_j = \sum_j a_{ij}x_j = a_{i1}x_1 + a_{i2}x_2 + a_{i3}x_3.$$

The orthogonality relationships, Equations (16.2), are then simply

$$a_{ij}a_{ik} = \delta_{jk}.$$ (16.6)

Note that the transformation from the unprimed to the primed coordinates is carried out using the coefficients a_{ij}. The a_{ij} are related to each other by the orthogonality conditions given by Equation (16.6). Therefore, the transformation (16.4) is called an "orthogonal transformation."

You will show in Problem 16.1 that the orthogonality relation (16.6) leads to

$$x_i'x_i' = x_ix_i.$$ (16.7)

But

$$x_ix_i = x_1x_1 + x_2x_2 + x_3x_3 = x^2 + y^2 + z^2$$

is the square of the length of the vector $x\hat{\mathbf{i}} + y\hat{\mathbf{j}} + z\hat{\mathbf{k}}$. So Equation (16.7) tells us that an orthogonal transformation conserves the length of a vector.

The coefficients a_{ij} lend themselves quite naturally to being expressed as a matrix, thus

$$\mathcal{A} = \begin{pmatrix} a_{11} & a_{12} & a_{13} \\ a_{21} & a_{22} & a_{23} \\ a_{31} & a_{32} & a_{33} \end{pmatrix}.$$ (16.8)

It is convenient at this time to express vectors as matrices. For example, the position vector in the inertial system is denoted $\mathbf{x}$ and is expressed as a column matrix as follows:

$$\mathbf{x} = \begin{pmatrix} x \\ y \\ z \end{pmatrix} = \begin{pmatrix} x_1 \\ x_2 \\ x_3 \end{pmatrix}.$$

In the rotated system the position vector is denoted by $\mathbf{x}'$ where

$$\mathbf{x}' = \begin{pmatrix} x' \\ y' \\ z' \end{pmatrix} = \begin{pmatrix} x_1' \\ x_2' \\ x_3' \end{pmatrix}.$$

Note that $\mathbf{x}$ and $\mathbf{x}'$ refer to the same vector, but expressed in terms of two different coordinate systems. That is, the components of $\mathbf{x}$ are (x, y, z) whereas the components of $\mathbf{x}'$ are (x', y', z').

The transformation equation (16.4) relating $\mathbf{x}$ and $\mathbf{x}'$ can now be written in the compact matrix form,

$$\mathbf{x}' = \mathcal{A}\mathbf{x}.$$ (16.9)

Worked Example 16.1   Determine the transformation matrix $\mathcal{A}$ for a two-dimensional rotation.

Solution Consider point P in Figure 16.3(a). The coordinates of P in the unprimed system are (x, y) and the coordinates of P in the primed system are (x', y'). Panel (b) of the figure shows that x' can be constructed as the sum of the segments $x \cos\theta + y \sin\theta$ and y' can be constructed as the difference $y \cos\theta - x \sin\theta$. That is, the transformation equations are

$$x' = x \cos\theta + y \sin\theta, \tag{16.10}$$
$$y' = -x \sin\theta + y \cos\theta.$$

If we write $\mathbf{x}' = \mathcal{A}\mathbf{x}$, then

$$\mathcal{A} = \begin{pmatrix} \cos\theta & \sin\theta \\ -\sin\theta & \cos\theta \end{pmatrix}. \tag{16.11}$$

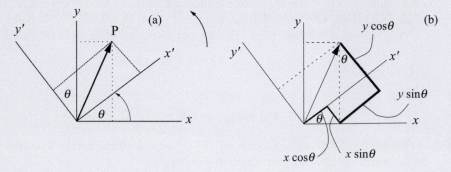

Figure 16.3 (a) The coordinates of point P in the two reference frames. (b) A construction showing that x' is the sum of two segments and y' is the difference between two segments.

Exercise 16.3

Carefully draw axes such as those shown in Figure 16.3 and determine the inverse relations to Equation (16.10), that is, obtain $x = x(x', y')$ and $y = y(x', y')$. ∎

Exercise 16.4

Show that if we represent vectors by row and column matrices, the dot product of **A** and **B** can be written

$$(A_x \quad A_y \quad A_z) \begin{pmatrix} B_x \\ B_y \\ B_z \end{pmatrix}.$$

∎

Interpretation

The matrix $\mathcal{A}$ describes the transformation from the unprimed system to the primed system. If **r** is a vector from the origin to some point P, then (x, y) and (x', y') represent the components of the vector **r** in the two coordinate systems. The matrix $\mathcal{A}$ tells us how the two sets of components

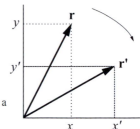

Figure 16.4 The "active" interpretation assumes the *vector* rotates in a clockwise sense rather than the *coordinate system* rotating in the counterclockwise sense.

are related. That is, $\mathcal{A}$ describes the result of a counterclockwise rotation of the *coordinate system* through the angle θ. This is called the "passive" interpretation of $\mathcal{A}$, and is illustrated in Figure 16.3.

There is, however, a different interpretation in which $\mathcal{A}$ is considered to be an *operator* that rotates the vector $\mathbf{r}$ in a *clockwise* manner to generate a new vector $\mathbf{r}'$. In this interpretation, the coordinate system is fixed and (x', y') are the components of the new vector, as shown in Figure 16.4. This is, of course, called the "active" interpretation.

16.2.1 Orthogonal Matrices

Let us consider in more detail the transformation matrix $\mathcal{A}$. Start with

$$\mathbf{r}' = \mathcal{A}\mathbf{r}. \tag{16.12}$$

This can be written as

$$x_i' = \sum_j a_{ij} x_j, \tag{16.13}$$

or, using the summation convention,

$$x_i' = a_{ij} x_j. \tag{16.14}$$

(I will be using the summation convention frequently in what follows: watch for repeated indices.) The inverse transformation from $\mathbf{r}'$ to $\mathbf{r}$ can be obtained by multiplying both sides of Equation (16.12) by the inverse of $\mathcal{A}$, denoted $\mathcal{A}^{-1}$. By definition,

$$\mathcal{A}\mathcal{A}^{-1} = \mathcal{A}^{-1}\mathcal{A} = \mathbf{1},$$

where

$$\mathbf{1} = \begin{pmatrix} 1 & 0 & 0 \\ 0 & 1 & 0 \\ 0 & 0 & 1 \end{pmatrix}$$

is the unit matrix. Thus

$$\mathcal{A}^{-1}\mathbf{r}' = \mathcal{A}^{-1}(\mathcal{A}\mathbf{r}) = \mathcal{A}^{-1}\mathcal{A}\mathbf{r} = \mathbf{r},$$

or

$$\mathbf{r} = \mathcal{A}^{-1}\mathbf{r}'.$$

Now by the rules of matrix multiplication, if $\mathcal{C} = \mathcal{A} \cdot \mathcal{B}$, then

$$c_{jk} = a_{ji} b_{ik}. \tag{16.15}$$

Consequently, if

$$A^{-1}A = 1,$$

then

$$a_{ji}^{-1}a_{ik} = \delta_{jk}. \tag{16.16}$$

But recall that the orthogonality condition, Equation (16.6), can be written as

$$a_{ij}a_{ik} = \delta_{jk}. \tag{16.17}$$

Comparing Equation (16.17) with Equation (16.16) you will appreciate that

$$a_{ji}^{-1} = a_{ij}.$$

That is, the elements of the inverse matrix are obtained by transposing the rows and columns of the original matrix. The transposed matrix is denoted by A^T. Thus, *for an orthogonal matrix,*

$$A^T = A^{-1}, \tag{16.18}$$

and hence,

$$A^T A = 1. \tag{16.19}$$

In physics we are often interested in the matrix obtained by taking the transpose of the original matrix and then replacing each element by its complex conjugate. The resulting matrix is called the *adjoint* matrix and is usually denoted by $A^\dagger$.

$$A^\dagger = \left(A^T\right)^*. \tag{16.20}$$

If $A^\dagger$ is the inverse of A, that is, if

$$A^\dagger A = 1, \tag{16.21}$$

then A is said to be *unitary.*

If $A^\dagger$ is equal to A, that is, if

$$A^\dagger = A, \tag{16.22}$$

then A is said to be *hermitian* or "self-adjoint."

The table below summarizes the names and properties of these various kinds of orthogonal matrices.

Orthogonal Matrices

Name	Symbol	Element	Property
Orthogonal	A	a_{ij}	$|A|^2 = 1$
Inverse	A^{-1}	$a_{ij}^{-1} = a_{ji}$	$A^{-1}A = 1$
Transpose	A^T	$a_{ij}^T = a_{ji}$	$A^T A = 1$
Adjoint	$A^\dagger$	$a_{ij} = a_{ji}^*$	
Unitary			$A^\dagger = A^{-1}$
Hermitian			$A^\dagger = A$

One of the properties of the orthogonal matrices we are using to represent rotations is the fact that they have a determinant of $+1$. This important property is easily proved. Taking the determinant of Equation (16.19) yields[3]

$$\left| \mathcal{A}^T \right| |\mathcal{A}| = |\mathbf{1}|.$$

But the determinant of the transpose is the same as the determinant of the original matrix because the determinant of a matrix is not affected by the interchange of rows and columns. Therefore,

$$|\mathcal{A}|^2 = 1,$$

and hence $|\mathcal{A}| = \pm 1$.

The determinant of $\mathcal{A}$ is either plus or minus unity. The value $+1$ corresponds to a rotation and the value -1 corresponds to an inversion of the coordinate axes. (An inversion is a reflection of all the axes.) This can be appreciated by considering a simple matrix $\mathcal{S}$ that has a determinant -1 :

$$\mathcal{S} = \begin{pmatrix} -1 & 0 & 0 \\ 0 & -1 & 0 \\ 0 & 0 & -1 \end{pmatrix}.$$

Then, if $\mathbf{r}' = \mathcal{S}\mathbf{r}$,

$$\begin{pmatrix} x' \\ y' \\ z' \end{pmatrix} = \begin{pmatrix} -1 & 0 & 0 \\ 0 & -1 & 0 \\ 0 & 0 & -1 \end{pmatrix} \begin{pmatrix} x \\ y \\ z \end{pmatrix} = \begin{pmatrix} -x \\ -y \\ -z \end{pmatrix}.$$

That is,

$$x' = -x; \quad y' = -y; \quad z' = -z.$$

Thus, the operator $\mathcal{S}$ inverts the axes, changing a right-handed coordinate system into a left-handed coordinate system. Any matrix with determinant -1 generates an inversion because it can be written as the product of $\mathcal{S}$ and a matrix with determinant $+1$. Therefore, an orthogonal matrix representing a rotation must have a determinant equal to $+1$.

If the matrix $\mathcal{B}$ is considered an operator that generates a rotation of the coordinate system, the representation of a vector $\mathbf{V}$ in the rotated coordinate system is

$$\mathbf{V}' = \mathcal{B}\mathbf{V}.$$

In the passive interpretation, this is the transformation law giving the relationship between the components of the vector in the rotated coordinate system and the components of the same vector in the original system.

A matrix (call it $\mathcal{A}$) is described by a set of elements a_{ij} and these will, in general, be different in the rotated coordinate system than in the original system. Let us determine the transformation law for a *matrix* under the rotation of the coordinate system.

Assume that in some coordinate system the matrix $\mathcal{A}$ acting on the vector $\mathbf{V}$ produces the vector $\mathbf{W}$:

$$\mathbf{W} = \mathcal{A}\mathbf{V}. \tag{16.23}$$

---

[3] The determinant of a product of matrices is the product of the determinants of the matrices. (Say that three times quickly!) In other words,

$$|\mathcal{A}\mathcal{B}| = |\mathcal{A}||\mathcal{B}|.$$

Let $\mathcal{B}$ represent a rotation of the coordinate system. In terms of the rotated set of axes, the vector **W** is given by

$$\mathbf{W}' = \mathcal{B}\mathbf{W} = \mathcal{B}\mathcal{A}\mathbf{V}. \tag{16.24}$$

But in the rotated coordinate system **V**, is given by $\mathbf{V}' = \mathcal{B}\mathbf{V}$, so it is desirable to multiply **V** by $\mathcal{B}$. This can be accomplished by introducing the unit matrix $\mathcal{B}^{-1}\mathcal{B}$ into Equation (16.24) thus:

$$\mathbf{W}' = \mathcal{B}\mathcal{A}\mathbf{V} = \mathcal{B}\mathcal{A}(\mathcal{B}^{-1}\mathcal{B})\mathbf{V} = (\mathcal{B}\mathcal{A}\mathcal{B}^{-1})\mathcal{B}\mathbf{V} = (\mathcal{B}\mathcal{A}\mathcal{B}^{-1})\mathbf{V}'. \tag{16.25}$$

This equation shows that the matrix $\mathcal{B}\mathcal{A}\mathcal{B}^{-1}$ transforms the vector $\mathbf{V}'$ into $\mathbf{W}'$. That is, Equation (16.25) describes the same operation as Equation (16.23), but referred to the rotated coordinate system. Therefore, the operator $\mathcal{A}$ in the rotated system is represented by

$$\mathcal{A}' = \mathcal{B}\mathcal{A}\mathcal{B}^{-1}. \tag{16.26}$$

This is the transformation law for matrices. It is called a *similarity transformation*. In general, you can consider a similarity transformation as a way to generate a new matrix that has certain properties; there is no need to interpret it in geometrical terms as the result of a rotation. For example, you can use a similarity transformation to transform a matrix into diagonal form. It is left as a problem to show that a similarity transformation preserves both the determinant and the trace of a matrix.

Worked Example 16.2   Consider a matrix $\mathcal{B}$ that represents a rotation about the z-axis through an angle ϕ. Using $\mathcal{B}$ carry out a similarity transformation of the matrix $\mathcal{A}$ of Equation (16.11) and interpret the result.

Solution Write

$$\mathcal{B} = \begin{pmatrix} \cos\phi & \sin\phi \\ -\sin\phi & \cos\phi \end{pmatrix},$$

and

$$\mathcal{A} = \begin{pmatrix} \cos\theta & \sin\theta \\ -\sin\theta & \cos\theta \end{pmatrix}.$$

Then,

$$\mathcal{A}' = \mathcal{B}\mathcal{A}\mathcal{B}^{-1}$$

$$= \begin{pmatrix} \cos\phi & \sin\phi \\ -\sin\phi & \cos\phi \end{pmatrix} \begin{pmatrix} \cos\theta & \sin\theta \\ -\sin\theta & \cos\theta \end{pmatrix} \begin{pmatrix} \cos\phi & -\sin\phi \\ \sin\phi & \cos\phi \end{pmatrix},$$

where we used the fact that $\mathcal{B}$ is orthogonal so $\mathcal{B}^{-1} = \mathcal{B}^T$. Carrying out the matrix multiplications and simplifying we end up with

$$\mathcal{A}' = \begin{pmatrix} \cos\theta & \sin\theta \\ -\sin\theta & \cos\theta \end{pmatrix},$$

which tells us that the rotation matrix is unchanged by this rotation of the axes.

Exercise 16.5

Consider the transformation matrix given by Equation (16.11). (a) Is this matrix orthogonal? (b) Is it unitary? (c) Is it hermitian?   ∎

16.3 The Euler Angles

We now return to the question of describing the orientation of a rigid body. As usual we let x', y', z' represent a coordinate system embedded in the body and rotating with it, and we let x, y, z represent an inertial coordinate system. For convenience, we assume the two systems have a common fixed origin.

The orientation of x', y', z' with respect to x, y, z can be described in terms of the nine direction cosines denoted a_{ij}. These are related by the orthogonality conditions (16.6), so they are not all independent. As discussed above, if the origins of the two systems coincide, there are only three independent coordinates. The Euler angles are a set of three independent coordinates that describe the orientation of x', y', z' with respect to x, y, z. Specifically, these angles describe three rotations that carry the axes x, y, z into x', y', z'.

The first Euler angle is a rotation about the z-axis through ϕ. This causes x and y to be transformed to a new pair of axes, call them ξ and η. This rotation is illustrated in Figure 16.5. This procedure defines a new coordinate system ξ, η, ζ. Note that ζ lies along the original z-axis and ξ and η lie in the same plane as x, y.

Next, carry out a rotation through θ about the ξ-axis. This causes the ζ-axis and the η-axis to change direction as indicated in Figure 16.6. The axes resulting from this rotation are denoted ξ', η', ζ'. The old xy-plane is now the $\xi'\eta'$-plane and is inclined to the original plane by an angle θ. (The line from the origin along the ξ (or ξ') axis is called the line of nodes.)

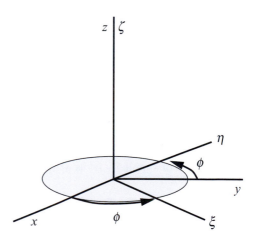

Figure 16.5 The first Euler angle: a rotation through an angle ϕ about the z-axis.

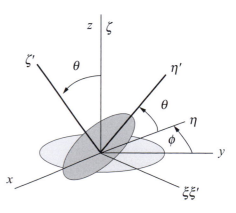

Figure 16.6 The second Euler angle: a rotation through θ about the ξ-axis.

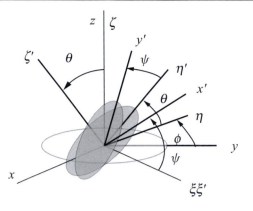

Figure 16.7 The third Euler angle: A rotation through ψ about the ζ'-axis.

Finally carry out a rotation about the ζ'-axis through ψ. This operation leaves ζ' unchanged but carries ξ' to x' and η' to y', as indicated in Figure 16.7. The new coordinate system is x', y', z'.

It is easy to appreciate that the three Euler angles ϕ, θ, ψ can be used to carry the original axes x, y, z to any desired final orientation x', y', z'.

I described the three Euler rotations in words. Figures 16.5, 16.6, and 16.7 illustrate the rotations graphically. Now I will show you how to represent these rotations mathematically. A rotation about a fixed axis is described by a matrix such as given by Equation (16.8). Therefore, the first rotation (through ϕ about z) can be expressed mathematically as

$$\mathcal{D} = \begin{pmatrix} \cos\phi & \sin\phi & 0 \\ -\sin\phi & \cos\phi & 0 \\ 0 & 0 & 1 \end{pmatrix}. \tag{16.27}$$

Note the similarity to the rotation matrix of Equation (16.11). You can easily verify that

$$\begin{pmatrix} \xi \\ \eta \\ \zeta \end{pmatrix} = \mathcal{D} \begin{pmatrix} x \\ y \\ z \end{pmatrix}. \tag{16.28}$$

For simplicity let us write this as

$$\xi = \mathcal{D}\mathbf{x}.$$

The next rotation is through θ about ξ. This rotation leaves the ξ-axis unchanged, and it is expressed by the matrix

$$\mathcal{C} = \begin{pmatrix} 1 & 0 & 0 \\ 0 & \cos\theta & \sin\theta \\ 0 & -\sin\theta & \cos\theta \end{pmatrix}. \tag{16.29}$$

Letting

$$\xi' = \begin{pmatrix} \xi' \\ \eta' \\ \zeta' \end{pmatrix},$$

we can write

$$\xi' = \mathcal{C}\xi.$$

Finally, the last rotation is through ψ about ζ'. This has the same structure as $\mathcal{D}$, so we write

$$\mathcal{B} = \begin{pmatrix} \cos\psi & \sin\psi & 0 \\ -\sin\psi & \cos\psi & 0 \\ 0 & 0 & 1 \end{pmatrix}. \tag{16.30}$$

Again, letting

$$\mathbf{x}' = \begin{pmatrix} x' \\ y' \\ z' \end{pmatrix},$$

we have

$$\mathbf{x}' = \mathcal{B}\xi'.$$

The transformation from x, y, z to x', y', z' is a sequence of three rotations described by the three matrices $\mathcal{D}, \mathcal{C}, \mathcal{B}$. Defining the matrix $\mathcal{A}$ by

$$\mathcal{A} = \mathcal{B}\mathcal{C}\mathcal{D}$$

the total transformation is given by

$$\mathbf{x}' = \mathcal{A}\mathbf{x}. \tag{16.31}$$

The transformation matrix $\mathcal{A}$ is the matrix product of three matrices and is rather complicated. As you can verify by some rather tedious matrix multiplication, it is given by[4]

$$\mathcal{A} = \begin{pmatrix} \cos\phi\cos\psi - \sin\phi\cos\theta\sin\psi & \sin\phi\cos\psi + \cos\phi\cos\theta\sin\psi & \sin\theta\sin\psi \\ -\cos\phi\sin\psi - \sin\phi\cos\theta\cos\psi & -\sin\phi\sin\psi + \cos\phi\cos\theta\cos\psi & \sin\theta\cos\psi \\ \sin\phi\sin\theta & -\cos\phi\sin\theta & \cos\theta \end{pmatrix} \tag{16.32}$$

Note that $\mathcal{A}$ involves only the independent coordinates ϕ, θ, ψ.

Worked Example 16.3 A student is asked to determine the orientation of the coordinate axes if the Euler angles are $30°, 45°$, and $60°$. The student does the rotations in reverse order. What is the difference in the rotation matrices and in the directions of the rotated axes?

Solution The appropriate angles are $\phi = 30°$, $\theta = 45°$, and $\psi = 60°$. So Equation (16.32) reduces to

$$\mathcal{A} = \begin{pmatrix} 0.13 & 0.78 & 0.61 \\ -0.93 & -0.13 & 0.35 \\ 0.35 & 0.61 & 0.71 \end{pmatrix}.$$

However, if the rotations are taken in reverse order, then $\phi = 60°$, $\theta = 45°$, and $\psi = 30°$, and

$$\mathcal{A}' = \begin{pmatrix} 0.13 & 0.93 & 0.35 \\ -0.78 & -0.13 & 0.61 \\ 0.61 & -0.35 & 0.71 \end{pmatrix}.$$

---

[4] An alternate set of angles describing rotations about three different axes is often used in aeronautical engineering. Those three angles are called roll, pitch, and yaw.

Thus, the correct directions for the new coordinates axes are

$$x' = 0.13x + 0.78y + 0.61z,$$
$$y' = -0.61x - 0.13y + 0.35z,$$
$$z' = 0.35x + 0.61y + 0.71z,$$

whereas the student obtains

$$x' = 0.13x + 0.93y + 0.35z,$$
$$y' = -0.78x - 0.13y + 0.61z,$$
$$z' = 0.61x - 0.35y + 0.71z.$$

Exercise 16.6

Verify Equation (16.28).   ■

Exercise 16.7

Verify Equation (16.32).   ■

Exercise 16.8

Describe in words the three rotations denoted roll, pitch, and yaw.   ■

Exercise 16.9

Write the 3×3 matrix for a rotation about the z-axis through $30°$, followed by a rotation about the new x-axis through $45°$.

$$\text{Answer:} \begin{pmatrix} 0.866 & 0.500 & 0 \\ -0.354 & 0.612 & 0.707 \\ 0.354 & -0.612 & 0.707 \end{pmatrix}.$$ ■

16.4 Euler's Theorem

So far we have been studying the rotation of a rigid body with one fixed point and you have seen that such a rotation is described by the matrix $\mathcal{A}$ where

$$\mathbf{x}' = \mathcal{A}\mathbf{x}.$$

In other words, if the position of a point P in the body is given by $\mathbf{x}$ before the rotation, it is given by $\mathbf{x}'$ after the rotation.[5]

As we have seen, any desired orientation of a rigid body can be achieved through three rotations (the Euler angles). But experience suggests the possibility of reaching any given orientation by *a rotation through a single angle* about some appropriately chosen axis. This concept is formulated more precisely by Euler's theorem of rigid body motion:

---

[5] Note that $\mathcal{A}$ can be a function of time so $\mathcal{A} = \mathcal{A}(t)$ and $\mathcal{A}(t = 0) = \mathbf{1}$. Thus, $\mathcal{A}(t)$ leads to a description of the orientation of the body at any instant of time.

Theorem 16.1 (**Euler's Theorem**) *The general displacement of a rigid body with one fixed point is a rotation about some axis.*

   This theorem states that as long as one point in the body is fixed, it is possible to find some axis such that a single rotation about that axis will carry the body to the desired orientation. (Note that you still need three parameters to describe the rotation. Two of these give the *direction* of the axis of rotation and the third gives the *angle* through which the body is rotated.)

   A rotation has two important properties: (1) The length of a vector is unchanged by a rotation, and (2) a vector lying on the axis of rotation is unchanged in direction and magnitude. If the matrix $\mathcal{A}$ describes a rotation, then property (1) will be satisfied by requiring $\mathcal{A}$ to be an orthogonal matrix as seen from Equation (16.7). To find the conditions for property (2) to be satisfied, consider a vector $\mathbf{R}$ lying along the axis of rotation. Then $\mathbf{R}' = \mathbf{R}$, where $\mathbf{R}'$ is the vector in the rotated system. But $\mathbf{R}' = \mathcal{A}\mathbf{R}$. Consequently, for this vector,

$$\mathcal{A}\mathbf{R} = \mathbf{R}. \tag{16.33}$$

The matrix $\mathcal{A}$ represents a rotation from an initial orientation to some final orientation. $\mathcal{A}$ may have been generated by the three rotations through Euler angles, for example. What Euler's theorem states is that the same final orientation can be achieved by a single rotation about some axis. If this is so, then a vector that is unchanged by the rotation must lie on the axis of rotation. That is, there exists a vector $\mathbf{R}$ satisfying Equation (16.33).

   But Equation (16.33) is just a special case of the more general equation

$$\mathcal{A}\mathbf{R} = \lambda\mathbf{R}, \tag{16.34}$$

where λ is a real or complex scalar constant. Equation (16.34) is called an *eigenvalue equation;* λ is called the *eigenvalue* and $\mathbf{R}$ is called the *eigenvector*. The eigenvalue problem consists in finding the eigenvalues and eigenvectors for a given matrix (or operator) $\mathcal{A}$.

   Euler's theorem requires that $\mathcal{A}\mathbf{R} = \mathbf{R}$. Therefore, *the theorem is proved if we can show that any orthogonal matrix representing a rotation will have at least one eigenvalue equal to* +1.

   Let us write Equation (16.34) in the equivalent form

$$(\mathcal{A} - \lambda\mathbf{1})\,\mathbf{R} = 0, \tag{16.35}$$

and express $\mathbf{R}$ as a column vector,

$$\mathbf{R} = \begin{pmatrix} X \\ Y \\ Z \end{pmatrix}.$$

Expanding Equation (16.35),

$$(a_{11} - \lambda)X + a_{12}Y + a_{13}Z = 0, \tag{16.36}$$
$$a_{21}X + (a_{22} - \lambda)Y + a_{23}Z = 0,$$
$$a_{31}X + a_{32}Y + (a_{33} - \lambda)Z = 0,$$

where X, Y, Z are the components of the vector $\mathbf{R}$. The problem of finding a vector that is unchanged by the rotation is solved by determining a vector $\mathbf{R}$ corresponding to the eigenvalue +1. Actually, it is not necessary to find the vector itself, but only its direction. This means that it is sufficient to find the relative values of the components of $\mathbf{R}$. Note from Equation (16.34) that any multiple of an eigenvector is also an eigenvector. Therefore, the *magnitude* of the eigenvector is not a particularly important property. However, the *direction* of the eigenvector is important.

Homogeneous Algebraic Equations

Equations (16.36) for the components of **R** are a set of three coupled linear homogeneous algebraic equations. You ran into the problem of solving for a set of two such equations while studying coupled oscillators. I will review some of the main points in obtaining a solution, but if you are unclear on the concepts you might reread the material in Section 11.5.

Using Cramer's rule to solve coupled homogeneous equations leads to the so-called "trivial solution," $X = 0, Y = 0, Z = 0$, unless the determinant of the coefficients in Equation (16.36) is zero. That is, a nontrivial solution is possible if and only if

$$D = \begin{vmatrix} a_{11} - \lambda & a_{12} & a_{13} \\ a_{21} & a_{22} - \lambda & a_{23} \\ a_{31} & a_{32} & a_{33} - \lambda \end{vmatrix} = 0.$$

If the determinant of a matrix is zero, it means that one (or more) of the rows or columns is not linearly independent of the others. So $D = 0$ indicates that one of the three homogeneous equations can be obtained from the other equations. For example, one equation might simply be a multiple of another equation or it might be the sum of the other two equations. Then one of the equations does not give any information that is not contained in the other two equations. That equation can be discarded. Suppose, for the sake of argument that the last equation in the set is the one which is not linearly independent, and that it has been discarded. You started out with three homogeneous equations in three unknowns. You now have two equations in three unknowns (X, Y, Z). This is not a good situation to be in! If, however, you divide each equation by Z, you end up with the following two equations

$$(a_{11} - \lambda)\frac{X}{Z} + a_{12}\frac{Y}{Z} = -a_{13}, \tag{16.37}$$

$$a_{21}\frac{X}{Z} + (a_{22} - \lambda)\frac{Y}{Z} = -a_{23}.$$

This is a pair of two *in*homogeneous equations in the two unknown ratios that you can solve using Cramer's rule! (This technique also works for much larger systems of linear homogeneous coupled algebraic equations. If the new set of equations are also homogenous just repeat the process.)

What is the point of all this? The point is that the equations only have the trivial solution $X = Y = Z = 0$ *unless* the determinant of the coefficients is zero. That is, a nonzero solution requires that

$$\begin{vmatrix} a_{11} - \lambda & a_{12} & a_{13} \\ a_{21} & a_{22} - \lambda & a_{23} \\ a_{31} & a_{32} & a_{33} - \lambda \end{vmatrix} = 0. \tag{16.38}$$

This condition guarantees that a nontrivial solution for the ratios X/Z and Y/Z can be obtained. Furthermore, Equation (16.38) can be solved for λ. The cubic equation for λ is called the *secular equation*. There will, of course, be three roots; let us denote them λ_1, λ_2, and λ_3. These can each be substituted into Equations (16.37) to solve for the ratios X/Z and Y/Z. Note that in general you will obtain three solutions, each corresponding to one of the three values of λ. (Remember that you want to show that one of these eigenvalues is $+1$.)

Knowing the ratios X/Z and Y/Z (or equivalently $X:Y:Z$) does not actually specify the vector with components (X, Y, Z), but it does give the *direction* of this vector and this is all you really need to know. Specifically, you have determined a vector with components $(X/Z, Y/Z, 1)$. This vector may not have the same magnitude as the vector you set out to find and it might even be pointing in the opposite *sense*, but at least it lies in the correct direction, and all you want to determine is the axis about which to rotate the body.

For each of the three roots λ_1, λ_2, and λ_3 (the three *eigenvalues*) you will get a different set of components. Therefore, in general, you get three different vectors corresponding to the three different eigenvalues. The vector obtained using λ_1, for example, is called the *eigenvector* corresponding to λ_1.

Because the ratios of the components give the direction of the eigenvector, you can use this information to generate a unit vector $\hat{\mathbf{e}}_i$ in that direction. Similarly, you can obtain unit vectors in the other two directions. These unit vectors can be denoted by

$$\hat{\mathbf{e}}_1, \hat{\mathbf{e}}_2, \hat{\mathbf{e}}_3,$$

and they specify the directions of the eigenvectors. The three lines that lie in these directions are called the *principal axes* of the matrix $\mathcal{A}$.

The eigenvectors of $\mathcal{A}$ can be used to generate a useful matrix by defining $\mathcal{S}$ to be the matrix whose columns are the eigenvectors of $\mathcal{A}$. For example, suppose that the eigenvector corresponding to λ_1, the first eigenvalue of $\mathcal{A}$, is

$$\mathbf{R}^{(1)} = \begin{pmatrix} X^{(1)} \\ Y^{(1)} \\ Z^{(1)} \end{pmatrix} = \begin{pmatrix} R_1^{(1)} \\ R_2^{(1)} \\ R_3^{(1)} \end{pmatrix}.$$

Similarly, let $\mathbf{R}^{(2)}$ and $\mathbf{R}^{(3)}$ be the eigenvectors corresponding to eigenvalues λ_2 and λ_3. Then

$$S = \begin{pmatrix} R_1^{(1)} & R_1^{(2)} & R_1^{(3)} \\ R_2^{(1)} & R_2^{(2)} & R_2^{(3)} \\ R_3^{(1)} & R_3^{(2)} & R_3^{(3)} \end{pmatrix} = \begin{pmatrix} \uparrow & \uparrow & \uparrow \\ \mathbf{R}^{(1)} & \mathbf{R}^{(2)} & \mathbf{R}^{(3)} \\ \downarrow & \downarrow & \downarrow \end{pmatrix}. \tag{16.39}$$

The matrix $\mathcal{S}$ (called the "modal matrix") itself describes a rotation, or more generally, an orthogonal transformation. Using $\mathcal{S}$ to transform $\mathcal{A}$ via a similarity transformation generates a transformed matrix $\mathcal{A}'$. This transformed matrix is diagonal and the elements along the diagonal are the eigenvalues. That is,

$$\mathcal{A}' = \mathcal{S}^{-1} \mathcal{A} \mathcal{S} = \begin{pmatrix} \lambda_1 & 0 & 0 \\ 0 & \lambda_2 & 0 \\ 0 & 0 & \lambda_3 \end{pmatrix}. \tag{16.40}$$

Thus, the modal matrix $\mathcal{S}$ "diagonalizes" the matrix $\mathcal{A}$. (This will be proved later.)

Application

The preceding discussion has been rather abstract, so let us consider a specific example using the matrix that represents a rotation of $45°$ about the z-axis. This matrix is

$$\mathcal{A} = \begin{pmatrix} \cos 45° & \sin 45° & 0 \\ -\sin 45° & \cos 45° & 0 \\ 0 & 0 & 1 \end{pmatrix} = \frac{\sqrt{2}}{2} \begin{pmatrix} 1 & 1 & 0 \\ -1 & 1 & 0 \\ 0 & 0 & 2/\sqrt{2} \end{pmatrix}.$$

Our first task is to find the eigenvalues and the eigenvectors of this matrix. The eigenvalue equation is

$$\mathcal{A}\mathbf{C} = \lambda \mathbf{C},$$

where $\mathbf{C}$ is an eigenvector that can be written in the form of a column matrix, thus

$$\mathbf{C} = \begin{pmatrix} c_1 \\ c_2 \\ c_3 \end{pmatrix}.$$

That is,

$$\begin{pmatrix} 1 & 1 & 0 \\ -1 & 1 & 0 \\ 0 & 0 & 2/\sqrt{2} \end{pmatrix} \begin{pmatrix} c_1 \\ c_2 \\ c_3 \end{pmatrix} = \frac{2}{\sqrt{2}}\lambda \begin{pmatrix} c_1 \\ c_2 \\ c_3 \end{pmatrix}.$$

Carrying out the matrix multiplications we obtain the following three equations:

$$(1 - \sqrt{2}\lambda)c_1 + c_2 = 0, \tag{16.41}$$
$$-c_1 + (1 - \sqrt{2}\lambda)c_2 = 0,$$
$$(1 - \lambda)c_3 = 0.$$

The third equation yields $\lambda = 1$. The first two are coupled homogeneous algebraic equations and have a nontrivial solution if and only if the determinant of their coefficients is zero:

$$\begin{vmatrix} (1 - \sqrt{2}\lambda) & 1 \\ -1 & (1 - \sqrt{2}\lambda) \end{vmatrix} = 0.$$

This leads to the secular equation for λ, namely

$$2\lambda^2 - 2\sqrt{2}\lambda + 2 = 0.$$

Solving for λ yields

$$\lambda = \frac{1}{\sqrt{2}}(1 \pm i).$$

The three eigenvalues are

$$\lambda^{(1)} = 1,$$
$$\lambda^{(2)} = \frac{1}{\sqrt{2}}(1 + i),$$
$$\lambda^{(3)} = \frac{1}{\sqrt{2}}(1 - i).$$

There is one eigen*vector* associated with each eigen*value*. Inserting eigenvalue $\lambda^{(1)} = 1$ into the first of Equations (16.41) gives (for the components of $\mathbf{C}^{(1)}$)

$$(1 - \sqrt{2})c_1^{(1)} + c_2^{(1)} = 0,$$

from which

$$c_2^{(1)} = 0.41 c_1^{(1)},$$

where the superscripts are reminders of which eigenvalue is being used. The second equation of (16.41) gives

$$-c_1^{(1)} + (1 - \sqrt{2})c_2^{(1)} = 0,$$

which yields

$$c_2^{(1)} = \frac{1}{0.41}c_1^{(1)}.$$

The two relations between $c_1^{(1)}$ and $c_2^{(1)}$ are contradictory. They can only be satisfied if $c_1^{(1)} = 0$ and $c_2^{(1)} = 0$. Now insert eigenvalue $\lambda = 1$ into the third equation of (16.41) to obtain

$$c_3^{(1)}(1 - \lambda) = 0,$$

which is satisfied for any value of $c_3^{(1)}$. We usually set $c_3^{(1)} = 1$ for convenience. This completes the determination of the first eigenvector, giving

$$\mathbf{C}^{(1)} = \begin{pmatrix} 0 \\ 0 \\ 1 \end{pmatrix}.$$

Next use $\lambda^{(2)} = \frac{1}{\sqrt{2}}(1 + i)$ in the three equations. The first equation of (16.41) then reads

$$\left(1 - \sqrt{2}\left[\frac{1}{\sqrt{2}}(1 + i)\right]\right)c_1 + c_2 = 0$$

from which

$$c_2^{(2)} = ic_1^{(2)}.$$

Plugging $\lambda^{(2)}$ into the second equation of (16.41) gives the same relation, so the equations only give the *relative* values of the components. At this point, one usually simply lets $c_1^{(1)} = 1$. Lastly, plugging $\lambda^{(2)}$ into the third equation gives $c_3^{(2)} = 0$. So the second eigenvector is

$$\mathbf{C}^{(2)} = \begin{pmatrix} 1 \\ i \\ 0 \end{pmatrix}.$$

Similarly, using $\lambda^{(3)}$ leads to the following expression for the third eigenvector

$$\mathbf{C}^{(3)} = \begin{pmatrix} 1 \\ -i \\ 0 \end{pmatrix}.$$

Once you have obtained the eigenvalues and the related eigenvectors you can write down the modal matrix S by remembering that the columns of S are the eigenvectors. That is, the columns of S are $\mathbf{C}^{(1)}, \mathbf{C}^{(2)}$, and $\mathbf{C}^{(3)}$ thus:

$$S = \begin{pmatrix} 0 & 1 & 1 \\ 0 & i & -i \\ 1 & 0 & 0 \end{pmatrix}.$$

Now I want to show you that the similarity transformation $S^{-1}AS$ will generate a diagonal matrix with the eigenvalues along the diagonal. But to do so, I first need to determine S^{-1}, the inverse of S. You may not recall the technique for obtaining the inverse of a matrix. One method is to first generate a matrix whose elements are the cofactors of the elements of S. (The cofactor of an element of a matrix is the determinant formed from what is left after scratching out the row and column of the element. Thus, the cofactor of S_{11} is 0.) The cofactor of S_{12} is $-i$ (because of the sign change associated with alternating rows and columns of determinants). The matrix of the cofactors of S is

$$\mathcal{D} = \begin{pmatrix} 0 & -i & -i \\ 0 & -1 & 1 \\ -2i & 0 & 0 \end{pmatrix}.$$

Next we take the transpose of $\mathcal{D}$. Exchanging rows and columns gives

$$\mathcal{D}^T = \begin{pmatrix} 0 & 0 & -2i \\ -i & -1 & 0 \\ -i & 1 & 0 \end{pmatrix}.$$

Now the inverse of $\mathcal{S}$ is given by

$$\mathcal{S}^{-1} = \frac{1}{\det \mathcal{S}} \mathcal{D}^T,$$

where

$$\det \mathcal{S} = |\mathcal{S}| = \begin{vmatrix} 0 & 1 & 1 \\ 0 & i & -i \\ 1 & 0 & 0 \end{vmatrix} = -2i.$$

Consequently,

$$\mathcal{S}^{-1} = \frac{1}{-2i} \begin{pmatrix} 0 & 0 & -2i \\ -i & -1 & 0 \\ -i & 1 & 0 \end{pmatrix}.$$

Finally, we can evaluate $\mathcal{S}^{-1}\mathcal{A}\mathcal{S}$ to find

$$\mathcal{S}^{-1}\mathcal{A}\mathcal{S} = \begin{pmatrix} 1 & 0 & 0 \\ 0 & \frac{1}{\sqrt{2}}(1+i) & 0 \\ 0 & 0 & \frac{1}{\sqrt{2}}(1-i) \end{pmatrix},$$

and we see that the diagonal elements are indeed the eigenvalues of $\mathcal{A}$.

Exercise 16.10
Obtain $\mathbf{C}^{(3)}$.

Exercise 16.11
Verify that $\mathcal{S}^{-1}\mathcal{S} = \mathbf{1}$, where $\mathbf{1}$ is the unit matrix.

Exercise 16.12
Verify the expression obtained for $\mathcal{S}^{-1}\mathcal{A}\mathcal{S}$.

Proof of Euler's Theorem

After that rather long discussion of homogeneous algebraic equations, we return to Euler's theorem. I mentioned previously that the theorem is proved if you can show that the real orthogonal matrix representing the rotation of a rigid body with one point fixed always has at least one eigenvalue equal to $+1$.

Although I showed you in the example above that the matrix $\mathcal{S}$ diagonalizes $\mathcal{A}$ and that the diagonal elements are the eigenvalues, I will now prove it in general. Recall that

$$\mathcal{A}\mathbf{R} = \lambda\mathbf{R}$$

represents the three equations (one for each eigenvalue/eigenvector pair),

$$\mathcal{A}\mathbf{R}^{(1)} = \lambda^{(1)}\mathbf{R}^{(1)}, \quad \mathcal{A}\mathbf{R}^{(2)} = \lambda^{(2)}\mathbf{R}^{(2)}, \quad \mathcal{A}\mathbf{R}^{(3)} = \lambda^{(3)}\mathbf{R}^{(3)}. \tag{16.42}$$

These equations are expressed more neatly by defining the matrix λ as

$$\lambda = \begin{pmatrix} \lambda^{(1)} & 0 & 0 \\ 0 & \lambda^{(2)} & 0 \\ 0 & 0 & \lambda^{(3)} \end{pmatrix}. \tag{16.43}$$

Then, since the modal matrix $\mathcal{S}$ is

$$\mathcal{S} = \begin{pmatrix} \uparrow & \uparrow & \uparrow \\ \mathbf{R}^{(1)} & \mathbf{R}^{(2)} & \mathbf{R}^{(3)} \\ \downarrow & \downarrow & \downarrow \end{pmatrix},$$

the three equations (16.42) can be expressed as the single matrix equation,

$$\mathcal{A}\mathcal{S} = \mathcal{S}\lambda. \tag{16.44}$$

Multiplying Equation (16.44) on the left by $\mathcal{S}^{-1}$ you obtain

$$\mathcal{S}^{-1}\mathcal{A}\mathcal{S} = \mathcal{S}^{-1}\mathcal{S}\lambda = \lambda.$$

Therefore, the similarity transformation $\mathcal{S}^{-1}\mathcal{A}\mathcal{S}$ does indeed generate a diagonal matrix whose elements are the eigenvalues. Furthermore, the determinant of λ is +1 because λ is obtained from the orthogonal matrix $\mathcal{A}$ by a similarity transformation (which conserves the determinant). As proved in Section 16.2.1 the determinant of $\mathcal{A}$ is +1. Consequently,

$$|\lambda| = +1. \tag{16.45}$$

From Equation (16.43) you can see that this means

$$\lambda^{(1)}\lambda^{(2)}\lambda^{(3)} = +1. \tag{16.46}$$

The proof of Euler's theorem is almost complete, but it is convenient at this point to consider three simple facts about the nature of the eigenvalues of orthogonal rotation matrices. These are:

Fact 1. All the eigenvalues of an orthogonal rotation matrix have unit magnitude. This is a consequence of the fact that a rotation preserves the length of a vector. That is, $\left|\mathbf{R}'\right| = |\mathbf{R}|$, or

$$x'^2 + y'^2 + z'^2 = x^2 + y^2 + z^2.$$

However, things are not quite as simple as they might appear because the secular equation could have imaginary or complex roots. In that case the eigenvectors are complex. The magnitude of a complex vector is obtained from

$$|\mathbf{R}|^2 = \mathbf{R}^* \cdot \mathbf{R}.$$

Therefore, the "rotated" vector $\mathbf{R}'$ has the property

$$\mathbf{R}'^* \cdot \mathbf{R}' = \mathbf{R}^* \cdot \mathbf{R}.$$

But $\mathbf{R}' = \lambda\mathbf{R}$, so

$$\mathbf{R}'^* \cdot \mathbf{R}' = \lambda^*\lambda\mathbf{R}^* \cdot \mathbf{R}.$$

Consequently, complex eigenvalues, must satisfy the condition

$$\lambda^*\lambda = +1.$$

In other words, if λ is real, then $\lambda = +1$ but if λ is complex, then $\lambda^*\lambda = +1$.

Fact 2. A real orthogonal 3×3 matrix has at least one real eigenvalue. This is proved from the fact that the secular equation is *cubic* so it has the form

$$f(\lambda) = \lambda^3 + a\lambda^2 + b\lambda + c.$$

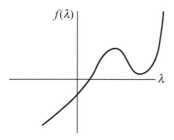

Figure 16.8 A plot of the cubic secular equation.

This function is illustrated in Figure 16.8.

Note that as $\lambda \to -\infty$, $f(\lambda)$ is negative, and as $\lambda \to +\infty$, $f(\lambda)$ is positive. Therefore, the function crosses the axis at least once and there must be at least one real root. You have seen that a rotation requires that $|\mathcal{A}| = +1$ so the real root must be $\lambda = +1$.

Fact 3. The complex conjugate of an eigenvalue is also an eigenvalue. This is a consequence of Fact 1. Also, it is a consequence of the fact that the elements of $\mathcal{A}$ are real because in that case the coefficients of the secular equation are real. So if λ is a solution of $\lambda^3 + a\lambda^2 + b\lambda + c = 0$, we can take the complex conjugate of the equation and see that λ^* is also a solution. Therefore, if λ_1 is complex, then $\lambda_2 = \lambda_1^*$.

The proof of Euler's theorem is now in hand. You know that $|\boldsymbol{\lambda}| = \lambda_1\lambda_2\lambda_3 = +1$. There are two possibilities, namely that all the roots are real or that one root is real and the two complex roots are conjugates.

Consider the first possibility. All the roots are real. Recall that all *real* roots must equal $+1$. Then $\mathcal{A} = \mathbf{1}$. This represents a *null* rotation.[6]

That leaves the second possibility. One root is real and equal to $+1$. The other two roots are complex conjugates. We can write

$$\lambda^{(1)} = e^{i\phi}, \qquad \lambda^{(2)} = e^{-i\phi}, \qquad \lambda^{(3)} = +1.$$

This corresponds to a rotation through ϕ about the z-axis.

Conclusion. For any nontrivial situation, one and only one of the eigenvalues is $+1$. Euler's theorem is proved.

You can use Euler's theorem to determine the direction of the axis of rotation and the angle of the single rotation that takes a rigid body from any given initial orientation to any desired final orientation. Begin with the matrix $\mathcal{A}$ using, for example, the Euler angles, as in Equation (16.32). Assume $\lambda = +1$ and solve the eigenvalue equations (16.36) for X, Y, Z. This gives you the direction of the desired axis of rotation. The magnitude of the single rotation about this axis is obtained by evaluating the trace of the matrix $\mathcal{A}$. To appreciate this, note that you can always use a similarity transformation to transform $\mathcal{A}$ to a set of axes in which the single rotation is about the z-axis. Call the angle of rotation ϕ. The transformed matrix will have the form

$$\mathcal{A}' = \begin{pmatrix} \cos\phi & \sin\phi & 0 \\ -\sin\phi & \cos\phi & 0 \\ 0 & 0 & 1 \end{pmatrix}.$$

The trace of $\mathcal{A}'$ is

$$\mathrm{Tr}\,\mathcal{A}' = 1 + 2\cos\phi.$$

---

[6] Another possibility is $\lambda_1\lambda_2\lambda_3 = (-1)(-1)(+1)$. This represents a rotation through π about one axis or an inversion of the other two axes. The conclusion that only one eigenvalue is $+1$ still holds.

(Recall that the trace is the sum of the diagonal elements.) The trace of an orthogonal matrix is invariant under a similarity transformation, so the trace of the original matrix $\mathcal{A}$ is also $1+2\cos\phi$. Therefore, you can determine the rotation angle by simply evaluating the trace of $\mathcal{A}$.

Exercise 16.13

Show that the two eigenvalues of $\mathcal{A}$ obey the relation $\lambda^*\lambda = +1$.   ∎

Exercise 16.14

Assume a, b, c are real. Show that if λ is a solution of $\lambda^3 + a\lambda^2 + b\lambda + c = 0$, then λ^* is also a solution.   ∎

16.5 Infinitesimal Rotations

According to Euler's theorem, any rotation or series of rotations can be described as a single rotation about a suitably selected axis. Furthermore, as you know, a rotation can be described in terms of three independent quantities. A vector has three independent components. Therefore, it would seem that a rotation could be described by a vector. Indeed, we already did this in previous chapters when we described rotations about a fixed axis in terms of the vector $\boldsymbol{\Omega}$.

There is, however, a problem in using a vector such as $\boldsymbol{\Omega}$ to describe a finite rotation about a fixed point (rather than a fixed axis). The problem is that finite rotations about different axes do not commute. The reason for this is that rotations are described by matrices and in general $\mathcal{A}\mathcal{B} \neq \mathcal{B}\mathcal{A}$. For example, if an airplane undergoes a 90° roll followed by a 90° yaw, it ends up in a different orientation than if it undergoes a 90° yaw followed by a 90° roll. If these two rotations are represented by two vectors, say $\boldsymbol{\Omega}_1$ and $\boldsymbol{\Omega}_2$, then, because vector addition is commutative, you would expect that $\boldsymbol{\Omega}_1 + \boldsymbol{\Omega}_2 = \boldsymbol{\Omega}_2 + \boldsymbol{\Omega}_1$, which is clearly not true for the situation just described. Finite rotations cannot, in general, be represented by vectors.

Nevertheless, *infinitesimal* rotations do commute. An infinitesimal rotation only changes the components of a vector slightly so that, for example,

$$x_1' = x_1 + \epsilon_{11}x_1 + \epsilon_{12}x_2 + \epsilon_{13}x_3.$$

In this equation the ϵ_{ij} are infinitesimal so we are justified in keeping only first-order terms in the ϵs. Using the summation convention, the infinitesimal rotation is expressed as

$$x_i' = (\delta_{ij} + \epsilon_{ij})x_j$$

or in matrix notation

$$\mathbf{x}' = (\mathbf{1} + \boldsymbol{\epsilon})\mathbf{x}. \tag{16.47}$$

The matrix $(\mathbf{1} + \boldsymbol{\epsilon})$ is the matrix describing an infinitesimal rotation. It is almost equal to the identity transformation.

It is now easy to prove that infinitesimal rotations commute. If $\mathbf{1} + \boldsymbol{\epsilon}_1$ and $\mathbf{1} + \boldsymbol{\epsilon}_2$ represent two different infinitesimal rotations, then

$$(\mathbf{1} + \boldsymbol{\epsilon}_1)(\mathbf{1} + \boldsymbol{\epsilon}_2) = \mathbf{1} + \boldsymbol{\epsilon}_1 + \boldsymbol{\epsilon}_2 + \boldsymbol{\epsilon}_1\boldsymbol{\epsilon}_2 \doteq \mathbf{1} + \boldsymbol{\epsilon}_1 + \boldsymbol{\epsilon}_2.$$

Here $\boldsymbol{\epsilon}_1\boldsymbol{\epsilon}_2$ has been neglected because it is a second-order infinitesimal. The product of the matrices taken in reverse order gives the same result.

Other Properties of Infinitesimal Rotations

If $\mathcal{A} = 1 + \epsilon$, then $\mathcal{A}^{-1} = 1 - \epsilon$ because

$$\mathcal{A}\mathcal{A}^{-1} = (1 + \epsilon)(1 - \epsilon) = 1 + \epsilon - \epsilon = 1.$$

Furthermore, since the matrix representing a rotation is orthogonal, $\mathcal{A}^T = \mathcal{A}^{-1}$ and therefore

$$\mathcal{A}^T = 1 + \epsilon^T = 1 - \epsilon,$$

so

$$\epsilon^T = -\epsilon.$$

This means that ϵ is an antisymmetric matrix. That is, the elements on the diagonal are zero and the off-diagonal elements are such that $\epsilon_{ij} = -\epsilon_{ji}$. Therefore the matrix ϵ has the form

$$\epsilon = \begin{pmatrix} 0 & d\Omega_3 & -d\Omega_2 \\ -d\Omega_3 & 0 & d\Omega_1 \\ d\Omega_2 & -d\Omega_1 & 0 \end{pmatrix}. \tag{16.48}$$

Going back to Equation (16.47), note that the differential change in the vector $\mathbf{x}$ due to the rotation is

$$d\mathbf{x} = \mathbf{x}' - \mathbf{x} = \epsilon \mathbf{x}.$$

Therefore, using Equation (16.48), the components of $d\mathbf{x}$ are

$$dx_1 = x_2 d\Omega_3 - x_3 d\Omega_2,$$
$$dx_2 = x_3 d\Omega_1 - x_1 d\Omega_3,$$
$$dx_3 = x_1 d\Omega_2 - x_2 d\Omega_1.$$

These equations are the familiar expressions for the components of the cross product of the vectors $\mathbf{x}$ and $d\mathbf{\Omega}$, where

$$d\mathbf{\Omega} = \begin{pmatrix} d\Omega_1 \\ d\Omega_2 \\ d\Omega_3 \end{pmatrix}.$$

That is,

$$d\mathbf{x} = \mathbf{x} \times d\mathbf{\Omega}.$$

The vector $d\mathbf{\Omega}$ is a vector lying along the axis of rotation.

Exercise 16.15

Draw a picture of an airplane. Draw the same airplane after it has undergone a pitch of $90°$ followed by a yaw of $90°$. Now draw the same airplane after it has undergone a yaw of $90°$ followed by a pitch of $90°$. ∎

16.6 Summary

The orientation of a rigid body is described in terms of the orientation of a set of axes (x', y', z') fixed in the body relative to a set of inertial axes (x, y, z). Mathematically, the orientation is represented by the set of nine direction cosines a_{ij} giving the angles between the primed and the unprimed axes. Because three parameters are sufficient to describe the orientation of a rigid body with one fixed point, the direction cosines are not all independent; indeed they are related by the six orthogonality conditions,

$$a_{ij}a_{ik} = \delta_{jk}.$$

The rotation (or transformation) matrix $\mathcal{A}$ is a 3×3 matrix with elements a_{ij}; see Equation (16.8). The position vector in the rotated frame is $\mathbf{x}'$ and is related to the position vector in the inertial frame by Equation (16.9).

$$\mathbf{x}' = \mathcal{A}\mathbf{x}.$$

The transformation matrix $\mathcal{A}$ can be interpreted in an *active* sense as a rotation of the vector $\mathbf{x}$, or in the *passive* sense as an opposite rotation of the coordinate system. The transformation matrix is *orthogonal*, that is,

$$\mathcal{A}^T = \mathcal{A}^{-1}.$$

If $\mathcal{A}$ represents a rotation, its determinant is $+1$.

If $\mathcal{B}$ represents a rotation of a coordinate system, then the elements of $\mathcal{A}$ will change. The change in $\mathcal{A}$ under such a rotation is given by the *similarity transformation*,

$$\mathcal{A}' = \mathcal{B}\mathcal{A}\mathcal{B}^{-1}.$$

A body can be transformed from any initial orientation to any final orientation by applying three successive rotations through the Euler angles as follows: (1) about z through ϕ, (2) about the new x-axis through θ, and (3) about the new z-axis through ψ. These rotations are represented by the 3×3 matrices $\mathcal{D}, \mathcal{C}, \mathcal{B}$; these are given in Equations (16.27), (16.29), and (16.30). The total transformation matrix $\mathcal{A} = \mathcal{B}\mathcal{C}\mathcal{D}$ is given in Equation (16.32).

Euler's theorem states that the transformation from any given initial orientation to any given final orientation can be achieved by a *single* rotation about an appropriately chosen axis. Points on the axis will not be affected by the rotation, so if $\mathbf{R}$ lies along the axis

$$\mathcal{A}\mathbf{R} = \mathbf{R},$$

or, in terms of an eigenvalue equation,

$$\mathcal{A}\mathbf{R} = \lambda\mathbf{R}.$$

To prove Euler's theorem requires showing that one eigenvalue of $\mathcal{A}$ is $+1$. This involves defining the eigenvectors $\mathbf{R}^{(1)}, \mathbf{R}^{(2)}$, and $\mathbf{R}^{(3)}$ associated with the three eigenvalues λ_1, λ_2, and λ_3. The modal matrix $\mathcal{S}$ whose columns are the eigenvectors, diagonalizes the transformation matrix $\mathcal{A}$ by the similarity transformation of Equation (16.40):

$$\mathcal{A}' = \mathcal{S}^{-1}\mathcal{A}\mathcal{S} = \begin{pmatrix} \lambda_1 & 0 & 0 \\ 0 & \lambda_2 & 0 \\ 0 & 0 & \lambda_3 \end{pmatrix} = \lambda.$$

This implies that $|\lambda| = \lambda_1\lambda_2\lambda_3 = +1$. By showing that two of the eigenvalues are complex (and complex conjugates) it follows that one eigenvalue is $+1$, thus proving Euler's theorem.

Rotations about a fixed point are represented by matrices. They cannot, in general, be represented by vectors because vectors commute whereas finite rotations do not commute. However, infinitesimal rotations do commute and these can be represented by an infinitesimal vector $\mathbf{d}\boldsymbol{\Omega}$.

16.7 Problems

Problem 16.1 Show that the length of a vector is unchanged by an orthogonal transformation; that is, prove that

$$x_i' x_i' = x_i x_i.$$

Problem 16.2 Show that the dot product of two vectors can be represented as the matrix product of a row vector with a column vector.

Problem 16.3 Prove that the matrix $\mathcal{A} = \begin{pmatrix} -1 & 1 & 3 \\ 1 & 2 & 0 \\ 3 & 0 & 2 \end{pmatrix}$ is not orthogonal by showing that its transpose is not its inverse.

Problem 16.4 Show that the determinant of the Euler transformation matrix, Equation (16.32), is +1.

Problem 16.5 Consider a rotation through the two Euler angles, ϕ and θ. (a) Write the rotation matrix $\mathcal{A}$ from Equation (16.32). (b) Obtain the inverse matrix $\mathcal{A}^{-1}$. (c) Show that $\mathcal{A}^{-1}\mathcal{A} = \mathbf{1}$.

Problem 16.6 Prove that the determinant and the trace of a matrix are unchanged by a similarity transformation.

Problem 16.7 Show that each term in a rotation matrix is equal to its cofactor.

Problem 16.8 A "natural" coordinate system for the Earth has the origin at the center of the planet, the z-axis passing through the North Pole, and the x-axis in the equatorial plane and going through the Greenwich meridian. (Where is the y-axis?) Suppose that you wanted to change to a system with the z-axis going through London ($0°$ E, $51°31'$ N), and the x-axis also passing through the Greenwich meridian. Determine the Euler angles to generate this transformation.

Problem 16.9 People in Fresno felt neglected and asked us to determine an axis of rotation such that a single rotation would put Fresno on the top of the world, that is, at the North Pole. Determine the direction of the axis of rotation as well as the angle required to put Fresno where it wants to be. At present, Fresno is at $119°47'$ W and $36°44'$ N.

Problem 16.10 The Earth is magically rotated through the following Euler angles: $\phi = 30°$, $\theta = 45°$, and $\psi = 30°$. Determine the latitude and longitude of the new directions of the x-, y-, z-axes. At present, the z-axis points through the North Pole, the x-axis goes through ($0°$ E, $0°$ N), and the y-axis goes through ($90°$ E, $0°$ N).

Problem 16.11 A Cartesian coordinate system undergoes a rotation of $\pi/2$ about the z-axis, followed by a rotation of $\pi/4$ about the new x-axis. (That is, $\phi = \pi/2$ and $\theta = \pi/4$.) Determine a single rotation that will yield the same final orientation. That is, give the direction of the axis of rotation and the magnitude of the single angle of rotation.

Problem 16.12 Determine the eigenvalues and eigenvectors for the matrix

$$\begin{pmatrix} 1 & 3 & 0 \\ 3 & -2 & -1 \\ 0 & -1 & 1 \end{pmatrix}.$$

Problem 16.13 Prove that the product of two orthogonal matrices is an orthogonal matrix.

Problem 16.14 Assume the following matrices act on a two dimensional vector $\mathbf{R} = x\hat{\mathbf{i}} + y\hat{\mathbf{j}}$.

(a) Describe the effect on $\mathbf{R}$ of $\mathcal{A}\mathbf{R}$, where $\mathcal{A} = \begin{pmatrix} 1 & 0 \\ 0 & -1 \end{pmatrix}$.

(b) Describe the effect on $\mathbf{R}$ of $\mathcal{B}\mathbf{R}$, where $\mathcal{B} = \begin{pmatrix} -1 & 0 & 0 \\ 0 & -1 & 0 \\ 0 & 0 & 1 \end{pmatrix}$.

Problem 16.15 A vector $\mathbf{r}$ is reflected in the xy-plane. (a) Write a vector equation for $\mathbf{r}'$, the vector after reflection. (b) Generalize to a reflection in a plane perpendicular to the unit vector $\hat{\mathbf{n}}$. (c) Write the transformation as a matrix equation. (d) Show that the transformation matrix is an improper orthogonal matrix.

Problem 16.16 Consider the matrix $\mathcal{T} = \begin{pmatrix} 1 & 4 \\ 1 & 1 \end{pmatrix}$. Obtain its eigenvalues and eigenvectors. Obtain the modal matrix $\mathcal{S}$. Show that $\mathcal{S}^{-1}\mathcal{T}\mathcal{S}$ diagonalizes $\mathcal{T}$ and that the diagonal elements are the eigenvalues.

Problem 16.17 Determine the eigenvectors for the matrix

$$\begin{pmatrix} \cos 60° & \sin 60° & 0 \\ -\sin 60° & \cos 60° & 0 \\ 0 & 0 & 1 \end{pmatrix}.$$

Problem 16.18 Assume $\mathcal{A}$ is an orthogonal matrix. Let each column of $\mathcal{A}$ represent a vector. Show that these vectors are orthogonal.

Problem 16.19 Consider the angular velocity vector $\boldsymbol{\omega}$. Its components along the x-, y-, and z-axes are ω_x, ω_y, and ω_z. Show that these components can be expressed in terms of the Euler angles as follows:

$$\omega_x = \dot{\theta}\cos\phi + \dot{\psi}\sin\theta\sin\phi,$$
$$\omega_y = \dot{\theta}\sin\phi - \dot{\psi}\sin\theta\cos\phi,$$
$$\omega_z = \dot{\psi}\cos\theta + \dot{\phi}.$$

Problem 16.20 (a) Using Equation (16.32) determine the rotation matrix corresponding to three Euler rotations if the angles are all $90°$. (b) Determine the eigenvalues using

$$\mathcal{A}\mathbf{C} = \lambda\mathbf{C}.$$

(c) Determine the modal matrix and verify that the diagonal elements of $\mathcal{S}^{-1}\mathcal{A}\mathcal{S}$ are the eigenvalues. Note that this problem has degenerate eigenvalues so there can be two linearly independent eigenvectors corresponding to the same eigenvalue.

Problem 16.21 A Cartesian coordinate system undergoes an infinitesimal rotation $d\boldsymbol{\Omega} = d\Omega\hat{\mathbf{k}}$.
(a) Determine the change in the components of a vector $\mathbf{x} = \alpha\hat{\mathbf{i}} + \beta\hat{\mathbf{j}} + \gamma\hat{\mathbf{k}}$.
(b) Write an expression for the vector in component form after the rotation.

Computational Projects

Computational Project 16.1 Develop a program to determine the eigenvalues of a 2×2 matrix.

Computational Project 16.2 Write a computer program to find the inverse of a 2×2 matrix.

17 Rotational Dynamics

In the preceding chapter we described the *kinematics* of a rigid body. In this chapter we study the *dynamics* of a rigid body that is rotating in an arbitrary manner about a fixed point. To make the analysis easier I will introduce tensors and dyads, physical quantities that may be new to you. (You may be using tensors for the rest of your life, so this is a good chance to learn them!) Once the necessary mathematical tools have been developed we will use them to obtain expressions for the energy and angular momentum of rotating bodies. Next we develop the equations of motion for rotating bodies; these are called Euler's equations. Finally, we use these concepts to study two problems, a freely rotating body and a rotating body acted upon by a torque. As examples we consider the Earth and a spinning top. The Earth is approximately a torque-free spinning body and a top (or gyroscope) is an example of a spinning body subjected to an external torque.[1]

It is convenient to split the dynamical problem into two parts, one corresponding to the translation of the body and the other corresponding to the rotation of the body about some point. I will not prove it, but it can be shown that one can always do this.[2] Therefore, in the remainder of this chapter I will assume that the body is rotating about a *fixed* point. As you saw in Chapter 16, the generalization from rotation about a fixed axis to rotation about a fixed point leads to a much more complicated formulation of the problem. (That is why we will need to use tensors.) The study of rotational dynamics is of particular value to physics students because many of the mathematical techniques carry over into quantum theory.

17.1 Angular Momentum

In our previous consideration of the angular momentum of rotating bodies (in Chapter 7), we assumed not only that the rotation was about a fixed axis, but also that the body was *symmetrical* about the axis of rotation. We are now going to generalize and obtain an expression for the angular momentum of a body of *arbitrary* shape rotating about a fixed *point*.

The angular moment $\mathbf{l}_i$ of a particle of mass m_i and linear momentum $\mathbf{p}_i$ is

$$\mathbf{l}_i = \mathbf{r}_i \times \mathbf{p}_i = m_i \left(\mathbf{r}_i \times \mathbf{v}_i \right),$$

where $\mathbf{r}_i$ is the position of the particle with respect to the origin. (The origin is usually the center of mass, or a point in the body that is fixed in space.) The total angular momentum of a rigid body with respect to the origin will be

---

[1] Rotational motion is a particularly rich field of study. Klein and Sommerfeld wrote a four-volume study on the motion of a spinning top (F. Klein and A. Sommerfeld, *Uber die Theorie des Kreisels*, B. G. Teubner, Leipzig, 1897–1910, vols. 1–4). As you might expect, the book is wide ranging, discussing rotating bodies in general with applications to geophysics, astronomy, and technology.

[2] Chasles' theorem states that the most general displacement of a rigid body is a translation plus a rotation. This is actually a corollary of Euler's theorem of Chapter 16.

$$L = \sum_i l_i = \sum_i m_i (\mathbf{r}_i \times \mathbf{v}_i).$$

The summation is over all the particles in the body and $\mathbf{v}_i$ is the linear velocity of particle i due to the rotation of the body. The (instantaneous) linear velocity of particle i in a body rotating with angular velocity $\boldsymbol{\omega}$ is

$$\mathbf{v}_i = \boldsymbol{\omega} \times \mathbf{r}_i,$$

so

$$L = \sum_i m_i \left[\mathbf{r}_i \times (\boldsymbol{\omega} \times \mathbf{r}_i) \right].$$

Then, using the BAC – CAB rule,

$$L = \sum_i m_i \left[\boldsymbol{\omega} (\mathbf{r}_i \cdot \mathbf{r}_i) - \mathbf{r}_i (\mathbf{r}_i \cdot \boldsymbol{\omega}) \right]. \tag{17.1}$$

By expanding this expression, you can show that the x-component of the angular momentum of the body is

$$L_x = \sum_i m_i \left[r_i^2 \omega_x - x_i (x_i \omega_x + y_i \omega_y + z_i \omega_z) \right],$$
$$= \sum_i m_i \left[\omega_x \left(r_i^2 - x_i^2 \right) - \omega_y x_i y_i + \omega_z x_i z_i \right],$$
$$= \left(\sum_i m_i (r_i^2 - x_i^2) \right) \omega_x - \left(\sum m_i x_i y_i \right) \omega_y - \left(\sum m_i x_i z_i \right) \omega_z,$$

and similarly for L_y and L_z. Therefore, the components of $\mathbf{L}$ have the form

$$L_x = I_{xx} \omega_x + I_{xy} \omega_y + I_{xz} \omega_z, \tag{17.2}$$
$$L_y = I_{yx} \omega_x + I_{yy} \omega_y + I_{yz} \omega_z,$$
$$L_z = I_{zx} \omega_x + I_{zy} \omega_y + I_{zz} \omega_z.$$

The coefficients of the angular velocity components are called *moments* of inertia if the subscripts are the same (I_{xx}, I_{yy}, I_{zz}) and *products* of inertia if the subscripts are different ($I_{xy}, I_{xz}, \ldots$). The moments of inertia are defined by expressions of the form

$$I_{xx} = \sum_i m_i (r_i^2 - x_i^2) \quad \text{or} \quad I_{xx} = \int_V \rho(\mathbf{r})(r^2 - x^2) d\tau, \tag{17.3}$$

and the products of inertia are defined by expressions of the form

$$I_{xy} = -\sum_i m_i x_i y_i \quad \text{or} \quad I_{xy} = -\int_V \rho(\mathbf{r}) xy d\tau. \tag{17.4}$$

These expressions can be written very economically as follows:

$$I_{jk} = \int_V \rho(\mathbf{r})(r^2 \delta_{jk} - r_j r_k) d\tau,$$

where the subscripts j, k take on the values $1, 2, 3$, and δ_{jk} is the Kronecker delta. Note that $I_{jk} = I_{kj}$ so there are only six independent elements in our expressions for the moments and products of inertia.[3]

The nine quantities I_{jk} defined above can be represented as a 3×3 array. This array is called the *inertia tensor*. It is denoted $\mathcal{I}$.

---

[3] In this context it is somewhat more natural to write x, y, z rather than x_1, x_2, x_3 as we did in the previous chapter. The expression $r_i^2 - x_i^2$ is often written in the equivalent form $y_i^2 + z_i^2$, and similarly for the other two moments of inertia.

Finally, using matrix multiplication, the angular momentum can be expressed as the product of the inertia tensor and the angular velocity

$$\mathbf{L} = \mathcal{I}\boldsymbol{\omega}. \tag{17.5}$$

Exercise 17.1

Expand Equation (17.1) to obtain an expression for L_y. ■

Exercise 17.2

Consider a ring of mass M and radius R. Determine the moments and products of inertia. Assume the z-axis goes through the center of the ring and the x- and y-axes are in the plane of the ring. Answer: $I_{xx} = \frac{1}{2}MR^2$. ■

Tensors

A *scalar* (mass, charge, etc.) is a physical quantity having one component. If you rotate the coordinate system, the value of a scalar does not change. A scalar is a tensor of rank zero. Examples are mass, time, energy, etc.

A *vector* (position, velocity, etc.) is a physical quantity having three components. If you rotate the coordinate system, the components of a vector change according to the rule $V_i' = a_{ij}V_j$, where the as obey condition $a_{ij}a_{ik} = \delta_{jk}$; see Equation (16.6). A vector is a tensor of rank one. Examples are momentum, force, velocity, etc.

A *tensor of rank two* is a physical quantity having nine elements[4] that transform according to the rule,

$$T_{ij}' = a_{ik}a_{jl}T_{kl}.$$

The coefficients a_{ik} can be considered elements of a matrix $\mathcal{A}$ that transforms T into T' by the similarity transformation

$$\mathcal{T}' = \mathcal{A}\mathcal{T}\mathcal{A}^T.$$

A rank-two tensor can always be represented as a square matrix. Note, however, that a tensor and a matrix are not the same thing. A matrix is simply an array of numbers. A matrix usually has no physical significance, whereas a tensor is a physical quantity that can be represented as an orthogonal matrix. A second-rank tensor is often represented by a 3×3 matrix. For *computational* purposes there is no difference between a tensor and a matrix; if you represent a tensor by a matrix then the tensor obeys all the rules of matrix algebra. (There are, however, ways to represent tensors that do not involve matrices.)

A tensor of rank N is a quantity having 3^N elements which transform in a particular way under orthogonal transformations. The general transformation law for a tensor undergoing an orthogonal transformation is

$$T_{ijk\ldots}'(\mathbf{r}') = a_{il}a_{jm}a_{kn}\cdots T_{lmn\ldots}(\mathbf{r}),$$

where the summation convention is implied. Here $T'(\mathbf{r}')$ is the tensor represented in terms of the primed coordinate system and $T(\mathbf{r})$ is the tensor represented in terms of the unprimed coordinate system. In mathematical terms, a tensor is often defined as a quantity that transforms according to this relation.

---

[4] I am assuming three-dimensional space. In a four-dimensional space a rank-two tensor has sixteen elements.

You may be thinking that this is getting rather complicated, but you will find that in practice the only tensors we shall use are those of rank zero, rank one, and rank two (and you are very familiar with tensors of rank zero and rank one).

To summarize, consider the following three definitions.

A tensor of rank zero ($N = 0$) has one component and it is invariant under a coordinate transformation. We call such an object a *scalar.*

A tensor of rank one ($N = 1$) has three components and transforms according to $T_i' = a_{il} T_l$. Such an object is called a *vector.*

A tensor of rank two ($N = 2$) has nine components and transforms according to $T_{ij}' = a_{il} a_{jm} T_{lm}$.

When we call an object a tensor, we are usually referring to a rank-two tensor.

Dyads and Dyadics

Another useful quantity is a *dyad,* defined in terms of two vectors. It is the object you would obtain if you multiplied two vectors together in the most naive way possible. For example, the dyad formed from the vectors **A** and **B** and denoted **AB** is defined as[5]

$$\mathbf{AB} = \left(A_x \hat{\mathbf{i}} + A_y \hat{\mathbf{j}} + A_z \hat{\mathbf{k}} \right) \left(B_x \hat{\mathbf{i}} + B_y \hat{\mathbf{j}} + B_z \hat{\mathbf{k}} \right)$$

$$= A_x B_x \hat{\mathbf{i}}\hat{\mathbf{i}} + A_x B_y \hat{\mathbf{i}}\hat{\mathbf{j}} + A_x B_z \hat{\mathbf{i}}\hat{\mathbf{k}}$$

$$+ A_y B_x \hat{\mathbf{j}}\hat{\mathbf{i}} + A_y B_y \hat{\mathbf{j}}\hat{\mathbf{j}} + A_y B_z \hat{\mathbf{j}}\hat{\mathbf{k}}$$

$$+ A_z B_x \hat{\mathbf{k}}\hat{\mathbf{i}} + A_z B_y \hat{\mathbf{k}}\hat{\mathbf{j}} + A_z B_z \hat{\mathbf{k}}\hat{\mathbf{k}}.$$

A *dyadic* is a linear polynomial of dyads, as for example

$$\mathbf{AB} + \mathbf{CD} + \cdots .$$

The *unit dyadic* is

$$\mathbf{1} = \hat{\mathbf{i}}\hat{\mathbf{i}} + \hat{\mathbf{j}}\hat{\mathbf{j}} + \hat{\mathbf{k}}\hat{\mathbf{k}}.$$

A dyad operates on a vector by the dot product, thus

$$\mathbf{AB} \cdot \mathbf{C} = \mathbf{A}(\mathbf{B} \cdot \mathbf{C}),$$

and

$$\mathbf{C} \cdot \mathbf{AB} = (\mathbf{C} \cdot \mathbf{A})\mathbf{B}.$$

In dyadic notation the inertia tensor can be written as

$$\mathcal{I} = m_i (r_i^2 \mathbf{1} - \mathbf{r}_i \mathbf{r}_i),$$

where the summation is implied, or as

$$\mathcal{I} = \int \rho(\mathbf{r}) \left[r^2 \mathbf{1} - \mathbf{rr} \right] d\tau.$$

Referring to Equation (17.3), we appreciate that in tensor notation the angular momentum is given by

$$\mathbf{L} = \mathcal{I} \cdot \boldsymbol{\omega}. \tag{17.6}$$

---

[5] This kind of product of two vectors is also called the "outer product," the "direct product," and the "tensor product."

This differs from Equation (17.5) only in the dot between $\mathcal{I}$ and $\boldsymbol{\omega}$, indicating that we are considering $\mathcal{I}$ as a tensor rather than as a matrix. (This makes no practical difference.)

If $\mathcal{I}$ were a scalar, then the angular momentum $\mathbf{L}$ would have the same direction as the angular velocity $\boldsymbol{\omega}$, but since $\mathcal{I}$ is a tensor, the angular momentum and angular velocity will, in general, not lie along the same line.

Worked Example 17.1 Obtain an expression for $\mathbf{L}$ from $\mathbf{L} = \mathcal{I} \cdot \boldsymbol{\omega}$ assuming $\mathcal{I}$ is (a) a matrix and (b) a dyadic.

Solution As a matrix

$$\mathbf{L} = \begin{pmatrix} L_x \\ L_y \\ L_z \end{pmatrix} = \begin{pmatrix} I_{xx} & I_{xy} & I_{xz} \\ I_{yx} & I_{yy} & I_{yz} \\ I_{zx} & I_{zy} & I_{zz} \end{pmatrix} \begin{pmatrix} \omega_x \\ \omega_y \\ \omega_z \end{pmatrix}$$

$$= \begin{pmatrix} I_{xx}\omega_x + I_{xy}\omega_y + I_{xz}\omega_z \\ I_{yx}\omega_x + I_{yy}\omega_y + I_{yz}\omega_z \\ I_{zx}\omega_x + I_{zy}\omega_y + I_{zz}\omega_z \end{pmatrix}.$$

As a dyadic

$$\mathbf{L} = \mathcal{I} \cdot \boldsymbol{\omega} = (I_{xx}\hat{\mathbf{i}}\hat{\mathbf{i}} + I_{xy}\hat{\mathbf{i}}\hat{\mathbf{j}} + I_{xz}\hat{\mathbf{i}}\hat{\mathbf{k}} + I_{yx}\hat{\mathbf{j}}\hat{\mathbf{i}} + I_{yy}\hat{\mathbf{j}}\hat{\mathbf{j}} + I_{yz}\hat{\mathbf{j}}\hat{\mathbf{k}}$$

$$+ I_{zx}\hat{\mathbf{k}}\hat{\mathbf{i}} + I_{zy}\hat{\mathbf{k}}\hat{\mathbf{j}} + I_{zz}\hat{\mathbf{k}}\hat{\mathbf{k}}) \cdot (\omega_x\hat{\mathbf{i}} + \omega_y\hat{\mathbf{j}} + \omega_z\hat{\mathbf{k}})$$

$$= I_{xx}\omega_x\hat{\mathbf{i}} + I_{xy}\omega_y\hat{\mathbf{i}} + I_{xz}\omega_z\hat{\mathbf{i}} + I_{yx}\omega_x\hat{\mathbf{j}} + I_{yy}\omega_y\hat{\mathbf{j}} + I_{yz}\omega_z\hat{\mathbf{j}}$$

$$+ I_{zx}\omega_x\hat{\mathbf{k}} + I_{zy}\omega_y\hat{\mathbf{k}} + I_{zz}\omega_z\hat{\mathbf{k}}$$

$$= \left(I_{xx}\omega_x + I_{xy}\omega_y + I_{xz}\omega_z \right)\hat{\mathbf{i}} + \left(I_{yx}\omega_x + I_{yy}\omega_y + I_{yz}\omega_z \right)\hat{\mathbf{j}}$$

$$+ \left(I_{zx}\omega_x + I_{zy}\omega_y + I_{zz}\omega_z \right)\hat{\mathbf{k}}.$$

Exercise 17.3

(a) Evaluate $\mathbf{1} \cdot \hat{\mathbf{i}}$. (b) Evaluate $\mathbf{1} \cdot (5\hat{\mathbf{i}} + 6\hat{\mathbf{j}} + 7\hat{\mathbf{k}})$. Answer: (a) $\hat{\mathbf{i}}$. ■

Exercise 17.4

Evaluate $\mathbf{AB} \cdot \mathbf{C}$ if $\mathbf{A} = 3\hat{\mathbf{i}}$, $\mathbf{B} = 2\hat{\mathbf{i}} + 5\hat{\mathbf{j}}$, and $\mathbf{C} = \hat{\mathbf{i}} + 2\hat{\mathbf{j}} + 3\hat{\mathbf{k}}$. Answer: $36\hat{\mathbf{i}}$. ■

17.2 Kinetic Energy

Having determined that the angular momentum of a rotating body is $\mathbf{L} = \mathcal{I} \cdot \boldsymbol{\omega}$, let us now obtain an expression for the kinetic energy of a rigid body rotating about a fixed point. The kinetic energy of a particle is

$$T_i = \frac{1}{2}m_i v_i^2.$$

Therefore, the kinetic energy of a rigid body is

$$T = \sum_i \frac{1}{2}m_i v_i^2.$$

But $v_i^2 = \mathbf{v}_i \cdot \mathbf{v}_i$ and $\mathbf{v}_i = \boldsymbol{\omega} \times \mathbf{r}_i$, so

$$T = \sum_i \frac{1}{2} m_i (\boldsymbol{\omega} \times \mathbf{r}_i) \cdot \mathbf{v}_i$$

$$= \sum_i \frac{1}{2} m_i \boldsymbol{\omega} \cdot (\mathbf{r}_i \times \mathbf{v}_i)$$

$$= \frac{1}{2} \boldsymbol{\omega} \cdot \sum_i \mathbf{r}_i \times m_i \mathbf{v}_i$$

$$= \frac{1}{2} \boldsymbol{\omega} \cdot \mathbf{L}.$$

Using Equation (17.6) for the angular momentum, we obtain

$$T = \frac{1}{2} \boldsymbol{\omega} \cdot \mathcal{I} \cdot \boldsymbol{\omega}. \tag{17.7}$$

Worked Example 17.2 Prove that the work–energy theorem for a rotating body is given by $\frac{dT}{dt} = \boldsymbol{\omega} \cdot \mathbf{N}$.

Solution The work–energy theorem states that the net work done on a body is equal to the body's increase in kinetic energy. For linear motion,

$$dT = dW = \mathbf{F} \cdot d\mathbf{x},$$

so

$$\frac{dT}{dt} = \mathbf{F} \frac{d\mathbf{x}}{dt} = \mathbf{F} \cdot \mathbf{v}$$

$$= \mathbf{F} \cdot (\boldsymbol{\omega} \times \mathbf{r}) = (\boldsymbol{\omega} \times \mathbf{r}) \cdot \mathbf{F} = \boldsymbol{\omega} \cdot (\mathbf{r} \times \mathbf{F})$$

$$\frac{dT}{dt} = \boldsymbol{\omega} \cdot \mathbf{N}. \qquad \text{Q.E.D.} \tag{17.8}$$

17.3 Properties of the Inertia Tensor

17.3.1 Eigenvalues and Principal Axes

The elements of the inertia tensor depend on the orientation of the body relative to the coordinate axes. For any set of coordinate axes, the tensor can be represented by a *symmetric* matrix because by the definition of the products of inertia, $I_{ij} = I_{ji}$. This also tells us that only six elements are independent. Furthermore, all the elements are real. But for a real, symmetric matrix,

$$I^\dagger = (I^T)^* = I,$$

and therefore the inertia tensor is hermitian. (See Section 16.2.1.) As I will prove in a bit, by making an appropriate choice of the axes you can obtain an expression for $\mathcal{I}$ in which all of the off-diagonal terms are zero. I will denote this diagonalized inertia tensor by $\mathcal{I}'$. In matrix form it looks like this:

$$\mathcal{I}' = \begin{pmatrix} I_1 & 0 & 0 \\ 0 & I_2 & 0 \\ 0 & 0 & I_3 \end{pmatrix}. \tag{17.9}$$

Note that I_1, I_2, and I_3 are the elements of the *diagonalized* tensor. Generally speaking they will not be equal to I_{xx}, I_{yy}, and I_{zz}.

The diagonalized inertia tensor can also be written in dyadic form, thus:

$$\mathcal{I}' = I_1 \hat{\mathbf{i}}\hat{\mathbf{i}} + I_2 \hat{\mathbf{j}}\hat{\mathbf{j}} + I_3 \hat{\mathbf{k}}\hat{\mathbf{k}}. \tag{17.10}$$

It is always helpful to express the inertia tensor in diagonalized form. It is a clean and simple way to express the tensor and it leads to very nice expressions for the angular momentum and the kinetic energy, namely:

$$L_1 = I_1 \omega_1 \qquad L_2 = I_2 \omega_2 \qquad L_3 = I_3 \omega_3,$$

$$T = \frac{1}{2} I_1 \omega_1^2 + \frac{1}{2} I_2 \omega_2^2 + \frac{1}{2} I_3 \omega_3^2.$$

Exercise 17.5

Show that Equations (17.9) and (17.10) are equivalent. ∎

We now consider how to diagonalize the inertia tensor. When a matrix is diagonalized, the diagonal elements are its eigenvalues; see Equation (16.40). Thus, the quantities I_1, I_2, and I_3 are the eigenvalues of the inertia tensor. Once you have found the eigenvalues, you can find the eigenvectors. The directions of the eigenvectors give the three axes of the coordinate system in which $\mathcal{I}$ is diagonal. (These are called the principal axes.)

Except for the trivial case, the orthogonal rotation matrices considered in Chapter 16 had only one real eigenvalue. On the other hand, for the inertia tensor, all three eigenvalues are real and the three corresponding eigenvectors are mutually orthogonal. While proving that this is true we will formulate the procedure for finding the principal axes.

To prove that the eigenvalues of $\mathcal{I}$ are real, we allow them to be complex. We also allow the components of the eigenvectors to be complex.

Let the three eigenvectors of $\mathcal{I}$ be denoted $\mathbf{R}_j$ ($j = 1, 2, 3$) and let I_j be the corresponding eigenvalues. The eigenvalue equations are:

$$\mathcal{I} \cdot \mathbf{R}_j = I_j \mathbf{R}_j \qquad j = 1, 2, 3. \tag{17.11}$$

Multiply from the left by $\mathbf{R}_l^*$. (In matrix notation $\mathbf{R}_l^*$ will have to be a row vector; otherwise you could not carry out the matrix multiplication.)

$$\mathbf{R}_l^* \cdot \mathcal{I} \cdot \mathbf{R}_j = I_j \mathbf{R}_l^* \cdot \mathbf{R}_j. \tag{17.12}$$

The complex conjugate of Equation (17.11) is

$$\mathcal{I}^* \cdot \mathbf{R}_l^* = I_l^* \mathbf{R}_l^*. \tag{17.13}$$

(Note the change in the dummy index.) By the rules of matrix multiplication, $\mathcal{I} \cdot \mathbf{R} = \mathbf{R} \cdot \mathcal{I}^T$, so Equation (17.13) can be written as

$$\mathbf{R}_l^* \cdot \mathcal{I}^\dagger = I_l^* \mathbf{R}_l^*.$$

Multiplying from the right with $\mathbf{R}_j$ gives

$$\mathbf{R}_l^* \cdot \mathcal{I}^\dagger \cdot \mathbf{R}_j = I_l^* \mathbf{R}_l^* \cdot \mathbf{R}_j. \tag{17.14}$$

But $\mathcal{I} = \mathcal{I}^\dagger$ because $\mathcal{I}$ is hermitian. Therefore, the left-hand sides of Equations (17.12) and (17.14) are identical and subtracting these equations gives us

$$(I_j - I_l^*)\mathbf{R}_l^* \cdot \mathbf{R}_j = 0. \tag{17.15}$$

Either $l = j$ or $l \neq j$. If $l = j$,

$$(I_j - I_j^*)\mathbf{R}_j^* \cdot \mathbf{R}_j = 0.$$

Now $\mathbf{R}_j^* \cdot \mathbf{R}_j = |R_j|^2 > 0$, is not equal to zero, so $I_j = I_j^*$, thus proving the eigenvalues are real. This result followed from the hermitian nature of $\mathcal{I}$, and it implies that in general the eigenvalues of a hermitian matrix are real. (This fact is very important in quantum mechanics where operators are represented by matrices and observables are represented by the eigenvalues of these operators. Because observables must be real quantities, the operators in quantum mechanics must be represented by hermitian matrices.)

Next consider the possibility $l \neq j$. In that case, if the eigenvalues are not equal, $\mathbf{R}_l^* \cdot \mathbf{R}_j = 0$ proving that the eigenvectors corresponding to different eigenvalues are orthogonal.

However, if $I_l = I_j$ this proof does not hold. In that case, the eigenvalues are identical and are said to be *degenerate*. Nevertheless, we can still show that there are three mutually perpendicular eigenvectors. One way to prove this is to assume, for example, that eigenvalues 2 and 3 are degenerate, i.e., $I_2 = I_3$, but that I_1 is different. Then there is an eigenvector $\mathbf{R}_1$ corresponding to I_1 and at least one eigenvector (call it $\mathbf{R}_2$) corresponding to the double eigenvalue. $\mathbf{R}_2$ is orthogonal to $\mathbf{R}_1$ by Equation (17.15). If the coordinate axes are selected with the x-axis along $\mathbf{R}_1$ and the y-axis along $\mathbf{R}_2$, then the eigenvalue equations (17.11) for $j = 1, 2$ reduce to

$$\mathcal{I} \cdot \hat{\mathbf{i}} = I_1 \hat{\mathbf{i}},$$

and

$$\mathcal{I} \cdot \hat{\mathbf{j}} = I_2 \hat{\mathbf{j}}.$$

If you write these equations out in full you will find that if $\mathbf{R}_1 = I_1 \hat{\mathbf{i}}$ then I_{21} and I_{31} are zero, and if $\mathbf{R}_2 = I_2 \hat{\mathbf{j}}$ then I_{12} and I_{32} are zero. And since $\mathcal{I}$ is a symmetrical tensor it follows that I_{31} and I_{32} are also zero. Thus, $\mathcal{I}$ is diagonalized. Furthermore, $\hat{\mathbf{k}}$ is also an eigenvector with eigenvalue I_2. (The principal axes are along $\hat{\mathbf{i}}, \hat{\mathbf{j}}, \hat{\mathbf{k}}$.)

A second way of dealing with degenerate eigenvalues is to note that the equality of two eigenvalues implies a symmetry of the rigid body about the axis with the distinct eigenvalue (say I_1). A plane that is perpendicular to $\mathbf{R}_1$ must contain the other two eigenvectors. Because Equations (17.11) are linear, any linear combination of eigenvectors in that plane is also a solution. Therefore, any two orthogonal directions in the plane perpendicular to $\mathbf{R}_1$ can serve as the other two principal axes.[6]

A third approach to the problem of degenerate eigenvalues is familiar to you if have already studied quantum mechanics. There you used the Gram–Schmidt procedure to generate a set of orthogonal eigenvectors from a set of nonorthogonal eigenvectors.[7]

In conclusion, if the three eigenvalues are different, the three eigenvectors are all mutually perpendicular. If two eigenvalues are the same, we can define a principal axis in the direction of the eigenvector with the different eigenvalue and the other two eigenvectors will lie in a perpendicular plane. Finally, if all three of the eigenvalues of the inertia tensor are equal, then

---

[6] See, for example, Florian Scheck, *Mechanics: From Newton's Laws to Deterministic Chaos,* Springer Verlag, Berlin, 1990, pp. 172–173.

[7] See, for example, D. J. Griffiths and D. F. Schroeter, *Introduction to Quantum Mechanics*, 3rd ed., Cambridge University Press, Cambridge, 2018, p. 98.

all directions in space are eigenvectors and any set of three orthogonal axes can be considered principal axes. In that case the inertia tensor will be diagonal to begin with.

So far we have shown that the eigenvalues of the inertia tensor are real and that there are three orthogonal eigenvectors. But we have not shown how to evaluate them. We will develop the method for obtaining the eigenvectors in Section 17.3.2. Let us now consider the eigenvalues (or "principal moments of inertia"). These are determined by recalling that the eigenvalue equation (17.11) can be written as

$$(\mathcal{I} - I\mathbf{1}) \cdot \mathbf{R} = 0.$$

A nontrivial solution can be found if and only if the determinant of the coefficients vanishes, that is, iff

$$\begin{vmatrix} I_{11} - I & I_{12} & I_{13} \\ I_{21} & I_{22} - I & I_{23} \\ I_{31} & I_{32} & I_{33} - I \end{vmatrix} = 0.$$

This "secular equation" is a cubic in I and yields the three roots which are the principal moments of inertia. Each one of these roots can then be substituted into Equation (17.11) to obtain the directions of the principal axes.

Exercise 17.6

Show by writing the equation in matrix form that if $\mathcal{I} \cdot \hat{\mathbf{i}} = I_1 \hat{\mathbf{i}}$ then I_{21} and I_{31} are zero.   ∎

17.3.2 Evaluating the Inertia Tensor

To illustrate the methods for evaluating the eigenvectors (or principal axes) and eigenvalues of the inertia tensor I will go through a few examples in detail. However, to learn this material you should solve a number of more complicated problems on your own. Some are given at the end of the chapter.

Evaluating the inertia tensor basically involves nothing more than doing some (often unpleasant) integrations of Equations (17.3) and (17.4) to determine the moments and products of inertia. There are, however, some rather nice labor saving devices that you should use whenever possible. The four labor saving devices you may find most useful are as follows.

Labor Saving Device 1 (LSD 1): The Parallel Axis Theorem

If you know the inertia tensor about the center of mass, you can determine the inertia tensor with respect to parallel axes through a different origin O by using the relation

$$\mathcal{I}_O = \mathcal{I}_{CM} + M\left(R^2\mathbf{1} - \mathbf{RR}\right), \tag{17.16}$$

where $\mathbf{R}$ is the position of the center of mass relative to O.

Labor Saving Device 2 (LSD 2): Rotating the Tensor

If it is easy to evaluate the tensor about one set of axes (say xyz) then you can obtain the expression for the inertia tensor with respect to another set of axes ($x'y'z'$) having the same origin by a similarity transformation, thus:

$$\mathcal{I}' = \mathcal{A}\mathcal{I}\mathcal{A}^T. \tag{17.17}$$

The elements of $\mathcal{A}$ are the direction cosines ($a_{ij} = \hat{\mathbf{e}}'_i \cdot \hat{\mathbf{e}}_j$).

Labor Saving Device 3 (LSD 3): If the body is composed of several parts, the inertia tensor for the whole body is the sum of the tensors for the various parts (as long as these are calculated for the same origin and the same set of axes).

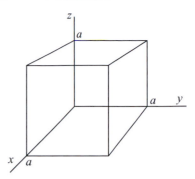

Figure 17.1 A cube of mass M and side of length a.

Labor Saving Device 4 (LSD 4): Any plane of symmetry of a body is perpendicular to a principal axis.

As a simple example of evaluating the inertia tensor, let us calculate the inertia tensor for a cube of side a and mass M with respect to coordinates parallel to the sides of the cube and with origin at one corner, as illustrated in Figure 17.1.

To determine the elements of the inertia tensor $\mathcal{I}$ we need to evaluate the terms in the following formula

$$\mathcal{I} = \begin{pmatrix} I_{xx} & I_{xy} & I_{xz} \\ I_{yx} & I_{yy} & I_{yz} \\ I_{zx} & I_{zy} & I_{zz} \end{pmatrix} = \int \int \int_{\text{Vol}} \rho \begin{pmatrix} y^2 + z^2 & -xy & -xz \\ -yx & x^2 + z^2 & -yz \\ -zx & -zy & x^2 + y^2 \end{pmatrix} d\tau. \quad (17.18)$$

For a cube, these integrals are quite simple. I will evaluate I_{xx} just to illustrate the process. Note that the density is $\rho = M/a^3$.

$$I_{xx} = \int \int \int_{\text{Vol}} \rho \, d\tau (y^2 + z^2)$$

$$= \frac{M}{a^3} \left[\int_{x=0}^{a} \int_{y=0}^{a} \int_{z=0}^{a} y^2 dx dy dz + \int_{x=0}^{a} \int_{y=0}^{a} \int_{z=0}^{a} z^2 dx dy dz \right]$$

$$= \frac{2}{3} M a^2.$$

From the symmetry of the body you can appreciate that I_{yy} and I_{zz} are also equal to $(2/3)Ma^2$. It is left as an exercise to show that the products of inertia are equal to $-(1/4)Ma^2$.

Therefore,

$$\mathcal{I} = \begin{pmatrix} \frac{2}{3} & -\frac{1}{4} & -\frac{1}{4} \\ -\frac{1}{4} & \frac{2}{3} & -\frac{1}{4} \\ -\frac{1}{4} & -\frac{1}{4} & \frac{2}{3} \end{pmatrix} Ma^2$$

where it is understood that Ma^2 multiplies every term in the matrix.

Exercise 17.7

Determine the products of inertia for the cube of Figure 17.1. ■

The evaluation of this particular inertia tensor was fairly straightforward. If you were asked to determine the inertia tensor of the cube about a point at the *center* of the cube you could use the parallel axis theorem (LSD 1), writing it as

$$\mathcal{I}_{CM} = \mathcal{I}_O - M(R^2\mathbf{1} - \mathbf{RR}).$$

The vector $\mathbf{R}$ from O to CM is

$$\mathbf{R} = \frac{a}{2}\hat{\mathbf{i}} + \frac{a}{2}\hat{\mathbf{j}} + \frac{a}{2}\hat{\mathbf{k}},$$

so

$$|\mathbf{R}| = \frac{\sqrt{3}a}{2},$$

and consequently

$$R^2\mathbf{1} = \begin{pmatrix} \frac{3a^2}{4} & 0 & 0 \\ 0 & \frac{3a^2}{4} & 0 \\ 0 & 0 & \frac{3a^2}{4} \end{pmatrix}.$$

Also,

$$\mathbf{RR} = \left(\frac{a}{2}\hat{\mathbf{i}} + \frac{a}{2}\hat{\mathbf{j}} + \frac{a}{2}\hat{\mathbf{k}}\right)\left(\frac{a}{2}\hat{\mathbf{i}} + \frac{a}{2}\hat{\mathbf{j}} + \frac{a}{2}\hat{\mathbf{k}}\right)$$

$$= \frac{a^2}{4}\left(\hat{\mathbf{i}}\hat{\mathbf{i}} + \hat{\mathbf{i}}\hat{\mathbf{j}} + \hat{\mathbf{i}}\hat{\mathbf{k}} + \hat{\mathbf{j}}\hat{\mathbf{i}} + \hat{\mathbf{j}}\hat{\mathbf{j}} + \hat{\mathbf{j}}\hat{\mathbf{k}} + \hat{\mathbf{k}}\hat{\mathbf{i}} + \hat{\mathbf{k}}\hat{\mathbf{j}} + \hat{\mathbf{k}}\hat{\mathbf{k}}\right)$$

$$= \frac{a^2}{4}\begin{pmatrix} 1 & 1 & 1 \\ 1 & 1 & 1 \\ 1 & 1 & 1 \end{pmatrix}.$$

Therefore

$$R^2\mathbf{1} - \mathbf{RR} = a^2\begin{pmatrix} \frac{1}{2} & -\frac{1}{4} & -\frac{1}{4} \\ -\frac{1}{4} & \frac{1}{2} & -\frac{1}{4} \\ -\frac{1}{4} & -\frac{1}{4} & \frac{1}{2} \end{pmatrix}.$$

Finally, the inertia tensor about parallel axes with origin at the center is

$$\mathcal{I}_{CM} = Ma^2\begin{pmatrix} \frac{2}{3} & -\frac{1}{4} & -\frac{1}{4} \\ -\frac{1}{4} & \frac{2}{3} & -\frac{1}{4} \\ -\frac{1}{4} & -\frac{1}{4} & \frac{2}{3} \end{pmatrix} - Ma^2\begin{pmatrix} \frac{1}{2} & -\frac{1}{4} & -\frac{1}{4} \\ -\frac{1}{4} & \frac{1}{2} & -\frac{1}{4} \\ -\frac{1}{4} & -\frac{1}{4} & \frac{1}{2} \end{pmatrix}$$

$$= Ma^2\begin{pmatrix} \frac{1}{6} & 0 & 0 \\ 0 & \frac{1}{6} & 0 \\ 0 & 0 & \frac{1}{6} \end{pmatrix} = \frac{Ma^2}{6}\begin{pmatrix} 1 & 0 & 0 \\ 0 & 1 & 0 \\ 0 & 0 & 1 \end{pmatrix}.$$

You are probably not surprised that the matrix is diagonal. From the simplicity of Figure 17.1 (and by LSD 4) you can infer that axes parallel to the cube sides and with origin at the CM will be the principal axes for the inertia tensor. Note that the cube has triply degenerate eigenvalues of $Ma^2/6$.

The principal axes for the cube were found almost accidentally, but it is not always that simple. To illustrate, let us determine the principal axes for the object formed when you slice the cube along a diagonal. As shown in Figure 17.2 this generates a tetrahedron with three faces that are right triangles and one face that is an equilateral triangle.[8] We will first determine the moment of inertia with respect to the (xyz) axes indicated in the figure, and then determine the principal axes.

[8] A *regular* tetrahedron or "triangular pyramid" has four identical triangular faces. The tetrahedron we have constructed is actually called a "trirectangular tetrahedron."

The inertia tensor is given by Equation (17.18). To illustrate the procedure, I will evaluate a moment of inertia. Let the mass of the tetrahedron be M. The volume of this kind of tetrahedron is $(1/6)a^3$. The density is $\rho = 6M/a^3$ and consequently,

$$I_{xx} = \int \int_{\text{Vol}} \int \rho \left(y^2 + z^2 \right) d\tau = \frac{6M}{a^3} \int_{x=0}^{a} dx \int_{y=0}^{a-x} dy \int_{z=0}^{a-x-y} (y^2 + z^2) dz$$

$$= \frac{6M}{a^3} \int_{x=0}^{a} dx \int_{y=0}^{a-x} dy \left[y^2(a - x - y) + \frac{1}{3}(a - x - y)^3 \right]$$

$$= \frac{6M}{a^3} \int_{x=0}^{a} \frac{1}{6}(a - x)^4 dx = \frac{M}{a^3} \frac{a^5}{5} = \frac{Ma^2}{5}.$$

(The last two integrals are rather messy and it is easiest to evaluate them with Maple or some other computer algebra system.)

By symmetry, $I_{yy} = I_{zz} = I_{xx} = \frac{1}{5}Ma^2$. It is left as a problem to show that the products of inertia are all equal to $-Ma^2/20$. Therefore, the inertia tensor for the tetrahedron about the (xyz) axes is

$$I = Ma^2 \begin{pmatrix} \frac{1}{5} & -\frac{1}{20} & -\frac{1}{20} \\ -\frac{1}{20} & \frac{1}{5} & -\frac{1}{20} \\ -\frac{1}{20} & -\frac{1}{20} & \frac{1}{5} \end{pmatrix} = \frac{Ma^2}{20} \begin{pmatrix} 4 & -1 & -1 \\ -1 & 4 & -1 \\ -1 & -1 & 4 \end{pmatrix}.$$

Exercise 17.8

Show that the volume of the tetrahedron of Figure 17.2 is $a^3/6$.   ∎

Having obtained the inertia tensor about the xyz axes of Figure 17.2 let us now determine the principal axes. Using LSD 4, note that the plane of symmetry indicated in the right-hand panel of Figure 17.2 will be perpendicular to a principal axis. Rotating the coordinate system by $45°$ in a clockwise direction around the z-axis generates a new set of axes, $x'y'z$ in which the x'-axis is a principal axis. To carry out the rotation use LSD 2, that is,

$$\mathcal{I}' = \mathcal{A}\mathcal{I}\mathcal{A}^T.$$

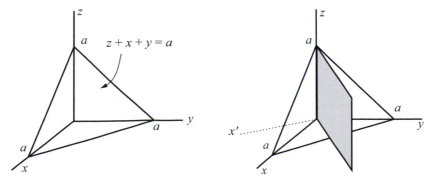

Figure 17.2 A tetrahedron formed by slicing a cube diagonally. In the sketch on the right a plane of symmetry is indicated. The x'-axis is perpendicular to the plane of symmetry.

The elements of the rotation matrix $\mathcal{A}$ are given by the dot products of unit vectors along the old and the new coordinates, and you will find that[9]

$$\mathcal{A} = \begin{pmatrix} 1/\sqrt{2} & -1/\sqrt{2} & 0 \\ 1/\sqrt{2} & 1/\sqrt{2} & 0 \\ 0 & 0 & 1 \end{pmatrix},$$

and therefore,

$$\mathcal{I}' = \begin{pmatrix} 1/\sqrt{2} & -1/\sqrt{2} & 0 \\ 1/\sqrt{2} & 1/\sqrt{2} & 0 \\ 0 & 0 & 1 \end{pmatrix} \begin{pmatrix} 4\gamma & -1\gamma & -1\gamma \\ -1\gamma & 4\gamma & -1\gamma \\ -1\gamma & -1\gamma & 4\gamma \end{pmatrix} \begin{pmatrix} 1/\sqrt{2} & 1/\sqrt{2} & 0 \\ -1/\sqrt{2} & 1/\sqrt{2} & 0 \\ 0 & 0 & 1 \end{pmatrix},$$

where $\gamma = Ma^2/20$. Carrying out the matrix multiplications generates

$$\mathcal{I}' = \frac{Ma^2}{20} \begin{pmatrix} 5 & 0 & 0 \\ 0 & 3 & -\sqrt{2} \\ 0 & -\sqrt{2} & 4 \end{pmatrix}.$$

This tensor is not completely diagonalized but you can appreciate that $\frac{Ma^2}{20}(5)$ is an eigenvalue and its eigenvector is along the x'-axis. Because the principal axes lie along the eigenvectors, this tells us that x' is indeed a principal axis. (Actually, the reason I carried out the rotation was not to show that x' was a principal axis, but just to give a practical example of rotating the inertia tensor.)

Although I obtained one principal axis by a simple rotation, it is not obvious where the other two principal axes are, so let's go back to basics and determine the eigenvalues and eigenvectors for $\mathcal{I}'$. The eigenvalue equation is

$$\mathcal{I}'\mathbf{R} = \lambda\mathbf{R},$$

or

$$(\mathcal{I}' - \lambda\mathbf{1})\mathbf{R} = 0,$$

which, written in matrix form, is

$$\begin{pmatrix} 5\gamma - \lambda & 0 & 0 \\ 0 & 3\gamma - \lambda & -\gamma\sqrt{2} \\ 0 & -\gamma\sqrt{2} & 4\gamma - \lambda \end{pmatrix} \begin{pmatrix} \uparrow \\ \mathbf{R} \\ \downarrow \end{pmatrix} = \begin{pmatrix} 0 \\ 0 \\ 0 \end{pmatrix}. \tag{17.19}$$

There are three eigenvectors, so there are three eigenvalues and three eigenvector equations. Because these will be expressed in terms of the primed coordinates, we will express them as

$$\mathbf{R_1} = \begin{pmatrix} x'_1 \\ y'_1 \\ z'_1 \end{pmatrix}, \quad \mathbf{R_2} = \begin{pmatrix} x'_2 \\ y'_2 \\ z'_2 \end{pmatrix}, \quad \mathbf{R_3} = \begin{pmatrix} x'_3 \\ y'_3 \\ z'_3 \end{pmatrix}.$$

Equation (17.19) is a system of three coupled homogeneous algebraic equations and they have a nontrivial solution if and only if the determinant of the coefficients is zero. Consequently, we require

$$\begin{vmatrix} 5\gamma - \lambda & 0 & 0 \\ 0 & 3\gamma - \lambda & -\gamma\sqrt{2} \\ 0 & -\gamma\sqrt{2} & 4\gamma - \lambda \end{vmatrix} = 0,$$

[9] This is the first Euler rotation. Compare to Equation (16.27) with $\phi = -45°$.

or

$$(5\gamma - \lambda)(3\gamma - \lambda)(4\gamma - \lambda) - (5\gamma - \lambda)(-\gamma\sqrt{2})(-\gamma\sqrt{2}) = 0,$$

$$(5\gamma - \lambda)\left[(3\gamma - \lambda)(4\gamma - \lambda) - 4\gamma^2/2\right] = 0.$$

So

$$5\gamma - \lambda = 0,$$

and

$$\lambda^2 - 7\gamma\lambda + 10\gamma^2 = 0.$$

Therefore, one eigenvalue is

$$\lambda_1 = 5\left(\frac{Ma^2}{20}\right) = 5\gamma,$$

and the other two (obtained from the quadratic for λ) are

$$\lambda_2 = 5\frac{Ma^2}{20} = 5\gamma,$$

$$\lambda_3 = 2\frac{Ma^2}{20} = 2\gamma.$$

Note that the eigenvalue 5γ appears twice, indicating a degeneracy. The first eigenvector is obtained by inserting $\lambda^{(1)}$ into Equations (17.19), yielding

$$0x_1' + 0y_1' + 0z_1' = 0$$

$$0x_1' - 2\gamma y_1' - \sqrt{2}\gamma z_1' = 0$$

$$0x_1' - \sqrt{2}y_1' - \gamma z_1' = 0.$$

The first of these equations is satisfied by any values of x_1', y_1', z_1', the components of the eigenvector corresponding to λ_1. The second and third of these equations both yield $z_1' = -\sqrt{2}y_1'$. Because we know that one eigenvector points along x', we select $y_1' = z_1' = 0$ and $x_1' = 1$. That is,[10]

$$\mathbf{R_1} = \begin{pmatrix} 1 \\ 0 \\ 0 \end{pmatrix}.$$

The next eigenvector, corresponding to the degenerate eigenvalue $\lambda_2 = 5\gamma$ must lie in the $y'z'$-plane, so now we select $x_2' = 0$.[11] Since we have $z_2' = -\sqrt{2}y_2'$, an eigenvector of unit magnitude is

$$\mathbf{R_2} = \begin{pmatrix} 0 \\ 1/\sqrt{3} \\ -\sqrt{2/3} \end{pmatrix}.$$

[10] It is convenient for the eigenvectors to have unit magnitude.
[11] For a degenerate eigenvalue we determine one eigenvector by the usual technique, and note that the other eigenvector will be orthogonal to the first one.

We now consider the third eigenvalue $\lambda_3 = 2\gamma$. Inserting this value into Equations (17.19) gives

$$3\gamma x_3' = 0,$$
$$\gamma y_3' - \sqrt{2}\gamma z_3' = 0,$$
$$-\sqrt{2}\gamma y_3' + 2\gamma z_3' = 0.$$

The first equation requires $x_3' = 0$, so the x' component of this eigenvector is zero. The second equation states that $z_3' = y_3'/\sqrt{2}$ and the third equation yields the same relation. Thus, the third eigenvector could be written as $\mathbf{R_3} = (0, -1, -1/\sqrt{2})$, but this is not one unit long so we set it to $\mathbf{R_3} = (0, -\sqrt{2/3}, -\sqrt{1/3})$ giving us a vector in the same direction as $(0, -1, -1/\sqrt{2})$ but with magnitude unity.

And so, at last, the principal axes of the body sketched in Figure 17.2 have been determined. It required a significant amount of labor but I hope you appreciated the procedure and did not get lost in the mathematics.

Exercise 17.9

For the principal axes of the tetrahedron show that x_2' and x_3' must be equal to zero. ∎

Exercise 17.10

For the tetrahedron of Figure 17.2 show that $\mathbf{R_3} = (0, -\sqrt{1/3}, \sqrt{2/3})$ is a vector of unit magnitude and is perpendicular to both $\mathbf{R_1}$ and $\mathbf{R_2}$. ∎

As a final example, let us determine the inertia tensor for an idealized gyroscope as shown in Figure 17.3. Assume the axis is massless, so the system is just a disk of radius a and mass M. To make the problem simpler, assume the disk has negligible thickness and its mass is given by $M = \int\int \sigma dA$, where σ is the mass per unit area. The left-hand side of Figure 17.3 shows a set of body axes, x', y', z'. It is natural to select z' along the axis of the gyroscope and let x' and y' lie in the plane of the disk. The inertia tensor in terms of these body axes will be denoted $\mathcal{I}'$.

The moments and products of inertia can be evaluated using Equation (17.18). For example, the value of $I_{x'x'}'$ is

$$I_{x'x'}' = \int\int \sigma(y'^2 + z'^2)dA.$$

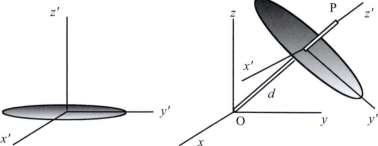

Figure 17.3 An idealized gyroscope made of a massless axis and a thin disk. The bottom end of the gyroscope is fixed at O.

Here z' is the distance in the z'-direction to the mass element, but we are assuming a zero thickness disk so we will set it equal to zero. Then, since $\sigma = M/\pi a^2$,

$$I'_{x'x'} = \frac{M}{\pi a^2} \int \int y'^2 dA.$$

Changing to polar coordinates and noting that $y' = r \sin \phi$,

$$I'_{x'x'} = \frac{M}{\pi a^2} \int_{r=0}^{a} \int_{0}^{2\pi} (r \sin \phi)^2 r dr d\phi = \frac{Ma^2}{4}.$$

Similarly, $I'_{y'y'} = I'_{x'x'} = \frac{Ma^2}{4}$, while $I'_{z'z'} = \frac{Ma^2}{2}$. (You could have obtained these results by elementary means using the perpendicular axis theorem.) All of the products of inertia are zero. (You will show this in an exercise.) The inertia tensor with respect to the body axes with origin at the center of mass of the disk is

$$\mathcal{I}' = Ma^2 \begin{pmatrix} \frac{1}{4} & 0 & 0 \\ 0 & \frac{1}{4} & 0 \\ 0 & 0 & \frac{1}{2} \end{pmatrix}.$$

Next, let us determine the inertia tensor $\mathcal{I}$ relative to the inertial axes x, y, z illustrated on the right in Figure 17.3. The method is essentially the same as for the triangular tetrahedron. In general, the gyroscope is spinning, precessing, and nutating. Point P at the free tip of the axis of rotation can move in any way consistent with the bottom end remaining fixed at O. The body axes x', y' are rotating with the disk. In general, the inertia tensor relative to the space axes x, y, z will be a complicated function of time. To simplify the problem, we shall artificially "freeze" the motion of the gyroscope at an instant of time when the body axis x' is parallel to the inertial axis x.

The first step is to translate the origin of the body axes along the z'-axis to the origin of the space axes (x, y, z). The inertia tensor with respect to these parallel translated axes can be determined using LSD 1: $\mathcal{I}'_O = \mathcal{I}_{CM} + M(R^2 \mathbf{1} - \mathbf{RR})$, where $\mathbf{R}$ is the vector from the origin to the center of mass. That is,

$$\mathbf{R} = (d)\,\hat{\mathbf{z}}',$$

where d is the distance from O to the center of the disk along the axis of rotation, and $\hat{\mathbf{z}}'$ is a unit vector in the z'-direction, pointing along the axis of the gyroscope. So $R^2 = d^2$ and $\mathbf{RR} = d^2 \hat{\mathbf{k}}\hat{\mathbf{k}}$. Therefore,

$$\mathcal{I}'_O = \begin{pmatrix} \frac{Ma^2}{4} & 0 & 0 \\ 0 & \frac{Ma^2}{4} & 0 \\ 0 & 0 & \frac{Ma^2}{2} \end{pmatrix} + \begin{pmatrix} Md^2 & 0 & 0 \\ 0 & Md^2 & 0 \\ 0 & 0 & Md^2 \end{pmatrix} - \begin{pmatrix} 0 & 0 & 0 \\ 0 & 0 & 0 \\ 0 & 0 & Md^2 \end{pmatrix}$$

$$= \begin{pmatrix} M(\frac{a^2}{4} + d^2) & 0 & 0 \\ 0 & M(\frac{a^2}{4} + d^2) & 0 \\ 0 & 0 & \frac{Ma^2}{2} \end{pmatrix}.$$

The next step is to rotate the inertia tensor to obtain an expression in terms of the space axes xyz. In general, we would have to rotate the body axes through the three Euler angles, ϕ, θ, ψ using the rotation matrix $\mathcal{A}$ given by Equation (16.32). Consequently, the inertia tensor in terms of the space axes is

$$\mathcal{I} = \mathcal{A}\mathcal{I}'_O\mathcal{A}^T,$$

where the Euler angles are functions of time. Note that in terms of the body axes, the inertia tensor is constant, but in terms of the space axes, it is a function of time. There is little point in writing out the general result of the matrix multiplication above, but to illustrate the method, let us return to our assumption of a frozen instant of time when the x'-axis and the x-axis are parallel, as illustrated in Figure 17.3. Then a single rotation about x' will line up z' and z, and also y' and y. Thus, the transformation to the space axes is a counterclockwise rotation about x' through θ, where θ is the angle between the z- and z'-axes. So,

$$\mathcal{A} = \begin{pmatrix} 1 & 0 & 0 \\ 0 & \cos\theta & -\sin\theta \\ 0 & \sin\theta & \cos\theta \end{pmatrix}.$$

Consequently, the inertia tensor in terms of the space axes is

$$\mathcal{I} = \mathcal{A}\mathcal{I}_0'\mathcal{A}^T$$

$$= \begin{pmatrix} 1 & 0 & 0 \\ 0 & \cos\theta & -\sin\theta \\ 0 & \sin\theta & \cos\theta \end{pmatrix} \begin{pmatrix} M(\frac{a^2}{4}+d^2) & 0 & 0 \\ 0 & M(\frac{a^2}{4}+d^2) & 0 \\ 0 & 0 & \frac{Ma^2}{2} \end{pmatrix} \begin{pmatrix} 1 & 0 & 0 \\ 0 & \cos\theta & \sin\theta \\ 0 & -\sin\theta & \cos\theta \end{pmatrix}$$

$$= \begin{pmatrix} M(\frac{a^2}{4}+d^2) & 0 & 0 \\ 0 & \cos^2\theta\, M(\frac{a^2}{4}+d^2)+\sin^2\theta\frac{Ma^2}{2} & \sin\theta\cos\theta\, M(d^2-\frac{a^2}{4}) \\ 0 & \sin\theta\cos\theta\, M(d^2-\frac{a^2}{4}) & \sin^2\theta\, M(\frac{a^2}{4}+d^2)+\cos^2\theta\frac{Ma^2}{2} \end{pmatrix}$$

Exercise 17.11

Show that in the body axes, the idealized gyroscope of our example has all the products of inertia equal to zero. ∎

17.4 The Euler Equations of Motion

We now determine the equation of motion of a rotating body that is subjected to an external torque. We mentioned that the motion of a rotating rigid body can be resolved into the translational motion of the center of mass and the rotational motion about the center of mass. The translational motion of the center of mass can be treated using $F = ma$. In this section we analyze the rotational motion about the center of mass. The results are applicable to the general motion of a body with one point fixed, in which case the rotation is taken about the fixed point.

The most straightforward approach to the problem is to start with Newton's second law as applied to a rotating body, i.e.,

$$\left(\frac{d\mathbf{L}}{dt}\right)_{\text{inertial}} = \mathbf{N}.$$

The time derivative referred to an inertial coordinate system can be replaced with one referred to a set of axes fixed in the rotating body by applying the usual prescription from Equation (15.3),

$$\left(\frac{d\mathbf{L}}{dt}\right)_{\text{inertial}} = \left(\frac{d\mathbf{L}}{dt}\right)_{\text{body}} + \boldsymbol{\omega} \times \mathbf{L}.$$

Replacing $\left(\frac{d\mathbf{L}}{dt}\right)_{\text{inertial}}$ by $\mathbf{N}$ and dropping the subscripts, this equation becomes

$$\mathbf{N} = \left(\frac{d\mathbf{L}}{dt}\right) + \boldsymbol{\omega} \times \mathbf{L}.$$

This equation is valid in the body (rotating) reference frame. Now $\mathbf{L} = \mathcal{I} \cdot \boldsymbol{\omega}$ and in the rotating frame the inertia tensor $\mathcal{I}$ is constant, so

$$\mathbf{N} = \mathcal{I} \cdot \frac{d\boldsymbol{\omega}}{dt} + \boldsymbol{\omega} \times (\mathcal{I} \cdot \boldsymbol{\omega}). \tag{17.20}$$

Equation (17.20) is the equation of motion for a rotating body. An interesting consequence of this equation is that for a freely rotating body (no torques), the angular velocity will be constant only if $\boldsymbol{\omega} \times (\mathcal{I} \cdot \boldsymbol{\omega}) = 0$, which means that $\mathcal{I} \cdot \boldsymbol{\omega}$ is parallel to $\boldsymbol{\omega}$. But $(\mathcal{I} \cdot \boldsymbol{\omega})$ is parallel to $\boldsymbol{\omega}$ only if $\boldsymbol{\omega}$ lies along a principal axis. This may seem like a useless theoretical fact, but just think about the last time you had your car wheels spin balanced. The purpose of the dynamical balancing is to make sure that a principal axis of the inertia tensor is aligned with the angular velocity vector. This ensures that the wheel spins without any torques being exerted on the axis. (Explain that to your mechanic!)

Assuming the body axes are principal axes, Equation (17.20) can be expressed in component form as

$$\begin{aligned} I_1\dot{\omega}_1 - \omega_2\omega_3(I_2 - I_3) &= N_1, \\ I_2\dot{\omega}_2 - \omega_3\omega_1(I_3 - I_1) &= N_2, \\ I_3\dot{\omega}_3 - \omega_1\omega_2(I_1 - I_2) &= N_3. \end{aligned} \tag{17.21}$$

These are called **Euler's equations of motion**. They are rotational analogues of Newton's second law of motion. In the next two sections we describe applications of Euler's equations.

Exercise 17.12

Show that for a symmetrical body, Euler's equations (17.21) reduce to the elementary relation $\mathbf{N} = I\boldsymbol{\alpha}$ in Equation (7.12). ∎

17.5 Torque-Free Motion

The Moon and the Sun exert small but nonzero torques on Earth's equatorial bulge, resulting in a very slow precession of the Earth's rotation vector. (We shall consider a similar mechanism in the next section.) But even if we discount this effect and assume that no external torques act on the Earth, we find that according to Euler's equations, there will be a "free" precession of Earth's rotation vector. So, for the moment, let us assume the Earth is a freely rotating oblate spheroid.

If there are no torques acting on a rotating body, Euler's equations of motion, Equations (17.21), reduce to

$$\begin{aligned} I_1\dot{\omega}_1 &= \omega_2\omega_3(I_2 - I_3), \\ I_2\dot{\omega}_2 &= \omega_3\omega_1(I_3 - I_1), \\ I_3\dot{\omega}_3 &= \omega_1\omega_2(I_1 - I_2). \end{aligned}$$

If the body is a spheroid, two of the moments of inertia will be equal. Let $I_1 = I_2$. Then Euler's equations further reduce to

$$I_1\dot{\omega}_1 = \omega_2\omega_3(I_1 - I_3),$$
$$I_1\dot{\omega}_2 = \omega_3\omega_1(I_3 - I_1),$$
$$I_3\dot{\omega}_3 = 0.$$

The third equation establishes that ω_3 is constant. The first two equations can be written as

$$\dot{\omega}_1 = \omega_3\frac{I_1 - I_3}{I_1}\omega_2,$$

$$\dot{\omega}_2 = -\omega_3\frac{I_1 - I_3}{I_1}\omega_1.$$

Denoting the constant $\omega_3 (I_1 - I_3) / I_1$ by Ω, these become

$$\dot{\omega}_1 = +\Omega\omega_2, \tag{17.22}$$
$$\dot{\omega}_2 = -\Omega\omega_1. \tag{17.23}$$

Taking the time derivative of the first of these equations and inserting it into the second equation yields

$$\ddot{\omega}_1 = -\Omega^2\omega_1.$$

But this is the equation for simple harmonic motion! A solution is

$$\omega_1 = A \cos \Omega t.$$

Similarly

$$\omega_2 = A \sin \Omega t.$$

In conclusion, ω_3 is constant but ω_1 and ω_2 vary sinusoidally out of phase with each other. This means that the total angular velocity vector, $\boldsymbol{\omega} = \omega_1\hat{\mathbf{i}} + \omega_2\hat{\mathbf{j}} + \omega_3\hat{\mathbf{k}}$, precesses about $\omega_3\hat{\mathbf{k}}$ with a precession rate Ω. (See Figure 17.4.) Note that this free precession is due to the fact that the angular momentum vector and the axis of rotation are not aligned. (An extreme situation is illustrated in Figure 16.1.) There is no external torque acting on the body so its angular momentum is constant in magnitude and direction and the rotation axis precesses around it. (Analytically, this is the same as a nutation that is discussed below in Section 17.6.2.)

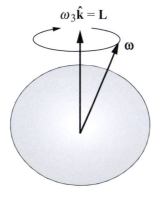

Figure 17.4 Illustrating the torque-free motion of an oblate spheroid. The angular momentum vector is constant and the rotation axis precesses about it at Ω.

This "free precession" for the Earth should have a period of $\Omega = \omega_3 (I_1 - I_3) / I_1 \approx 306$ days or approximately ten months. It may or may not have been detected. There appears to be a very small deviation of the axis of rotation of the Earth having this period, causing a wandering of the poles with an amplitude of about ten meters. However, there is a much greater (but still small) precession with a period of 420 days called the Chandler wobble which might be the free precession. The reason for the disagreement in period is believed by some geophysicists to be due to the fact that the Earth is not really a rigid body.

The *free precession* should not be confused with the *precession of the equinoxes* due to the torques on the Earth's equatorial bulge. This is responsible for the gradual precession of the first point in Aries through the zodiac with a period of 26 000 years.

Exercise 17.13

Show that $\omega_2 = A \sin \Omega t$. ∎

Exercise 17.14

Assume the total angular velocity vector $\boldsymbol{\omega}$ makes an angle α with the z-axis. Show that $\Omega = \frac{I - I_3}{I} \omega \cos \alpha$. ∎

17.6 The Spinning Top (Gyroscope)

You may recall the elementary analysis of the gyroscope that was presented in Section 7.6. (You might spend a few moments reviewing it at this time.) That analysis showed that the gravitational torque acting on the top led to a precession of the axis of rotation with an angular speed $\Omega_P = dmg/I\omega$, where d is the distance from the point of support to the center of mass, I is the moment of inertia about the axis of rotation, and ω is the spin velocity of the top. This is a correct solution, but it is not the general solution. In this section we consider the symmetrical[12] spinning top (or gyroscope) using the techniques developed in this chapter.

Figure 17.5 shows the orientation of the top in terms of the Euler angles, ϕ, θ, ψ. The spinning motion of the top is given by the rotation about the z'-axis, and the spin angular velocity is $\dot{\boldsymbol{\psi}} = \dot{\psi}\hat{\mathbf{z}}'$. If the top is precessing, the rotation vector traces out a cone around the vertical or z-axis. As the rotation axis traces out this trajectory, the angle ϕ is changing, and the angular velocity associated with precession is $\dot{\boldsymbol{\phi}} = \dot{\phi}\hat{\mathbf{z}}$. Finally, the top may be nutating, that is, the axis may be nodding up and down as it precesses. From the figure, it is clear that this type of motion represents a change in the angle θ. The angular velocity associated with nutation would be a vector along the line of nodes (ξ) and can be expressed as $\dot{\theta}\hat{\boldsymbol{\xi}}$. Recall that the Euler angles form a set of three independent coordinates so they are an appropriate set of generalized coordinates for describing the motion of the system.

Using the Euler angles of Figure 17.5, the angular velocity $\boldsymbol{\omega}$ is written in terms of its components thus

$$\boldsymbol{\omega} = \dot{\theta}\hat{\boldsymbol{\xi}} + \dot{\phi}\hat{\mathbf{z}} + \dot{\psi}\hat{\mathbf{z}}',$$

---

[12] The *asymmetrical* top is a more complicated problem and is usually reserved for graduate courses in mechanics. If you would like to look at the solution see L. D. Landau and E. M. Lifshitz, *Mechanics*, 3rd ed. Pergamon Press, Oxford, 1976, Section 37.

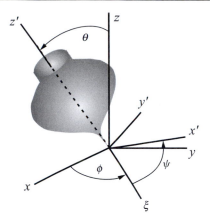

Figure 17.5 A spinning top. The orientation of the top relative to the inertial axes x, y, z is described by the Euler angles ϕ, θ, ψ.

where the unit vectors are directed along the lines ξ, z, z'. Let $\hat{\mathbf{e}}_1, \hat{\mathbf{e}}_2$, and $\hat{\mathbf{e}}_3$ be unit vectors directed along the body axes, x', y', z'. (By symmetry, these are the principal axes.) In terms of $\hat{\mathbf{e}}_1, \hat{\mathbf{e}}_2$, and $\hat{\mathbf{e}}_3$ the unit vectors along ξ, z, z' are given by

$$\hat{\boldsymbol{\xi}} = \hat{\mathbf{e}}_1 \cos \psi - \hat{\mathbf{e}}_2 \sin \psi,$$
$$\hat{\mathbf{z}} = \hat{\mathbf{e}}_1 \sin \theta \sin \psi + \hat{\mathbf{e}}_2 \sin \theta \cos \psi + \hat{\mathbf{e}}_3 \cos \theta,$$
$$\hat{\mathbf{z}}' = \hat{\mathbf{e}}_3.$$

Consequently, the components of the velocity along the body axes are

$$\omega_1 = \boldsymbol{\omega} \cdot \hat{\mathbf{e}}_1 = \dot{\theta} \cos \psi + \dot{\phi} \sin \theta \sin \psi, \tag{17.24}$$
$$\omega_2 = \boldsymbol{\omega} \cdot \hat{\mathbf{e}}_2 = -\dot{\theta} \sin \psi + \dot{\phi} \sin \theta \cos \psi,$$
$$\omega_3 = \boldsymbol{\omega} \cdot \hat{\mathbf{e}}_3 = \dot{\phi} \cos \theta + \dot{\psi}.$$

Note that ω_3 is not simply $\dot{\psi}$, the spin rate of the top about the z'-axis; the precession of the axis of rotation adds to the number of revolutions of the top. (This is similar to the fact that the Earth orbiting the Sun once a year means that the year has 365.25 days rather than 364.25.)

Let the top have mass m and assume that the center of mass lies a distance d along the axis from the point of contact on the floor. Because the top is symmetrical, the center of mass lies on the z'-axis. The torque acting on the top is $\mathbf{N} = d\hat{\mathbf{z}}' \times m\mathbf{g}$, a vector pointing along the line of nodes (ξ) and having magnitude $mgd \sin \theta$.[13]

The top is spinning about its axis, but it is also precessing and nutating. To develop a mathematical description of precession and nutation, let us begin by writing the Lagrangian for the system.

The kinetic energy is given by $T = \frac{1}{2} \boldsymbol{\omega} \cdot \mathcal{I} \cdot \boldsymbol{\omega}$, so

$$T = \frac{1}{2} \left(I_1 \omega_1^2 + I_2 \omega_2^2 + I_3 \omega_3^2 \right), \tag{17.25}$$

where the ωs are given by Equation (17.24). This leads to a fairly long and complicated expression. But for a *symmetrical* body with $I_2 = I_1$ it reduces to

$$T = \frac{1}{2} \left[I_1 \left(\dot{\theta}^2 + \dot{\phi}^2 \sin^2 \theta \right) + I_3 \left(\dot{\psi} + \dot{\phi} \cos \theta \right)^2 \right]. \tag{17.26}$$

---

[13] This torque is often said to be the source or cause of the precession. That is a bit misleading. The torque controls the precession, but does not cause it. In a similar manner, the gravitational force of the Sun on the Earth controls the orbital motion, but does not cause it. As we shall see in Section 17.6.2 the cause of the precession is the tendency of the top to tip over and fall in the gravitational field.

The potential energy for the spinning top is simply the gravitational potential energy, $V = mgd\cos\theta$. Consequently,

$$L = T - V = \frac{1}{2}\left[I_1\left(\dot{\theta}^2 + \dot{\phi}^2\sin^2\theta\right) + I_3\left(\dot{\psi} + \dot{\phi}\cos\theta\right)^2\right] - mgd\cos\theta. \tag{17.27}$$

There is a great deal of information in this Lagrangian, as well as in what is *not* in the Lagrangian. In particular, you will note that the coordinates ϕ and ψ are not in the Lagrangian. They are ignorable. Consequently the momenta conjugate to these variables are constant. To be specific, because ϕ is ignorable,

$$p_\phi = \frac{\partial L}{\partial\dot{\phi}} = I_1\dot{\phi}\sin^2\theta + I_3\cos\theta\left(\dot{\psi} + \dot{\phi}\cos\theta\right) = \text{constant}. \tag{17.28}$$

Similarly, because ψ is ignorable,

$$p_\psi = \frac{\partial L}{\partial\dot{\psi}} = I_3(\dot{\psi} + \dot{\phi}\cos\theta) = \text{constant}. \tag{17.29}$$

Furthermore, because time does not appear explicitly in the Lagrangian or in the transformation equations, the total energy is constant, that is,

$$E = \frac{1}{2}\left[I_1\left(\dot{\theta}^2 + \dot{\phi}^2\sin^2\theta\right) + I_3\left(\dot{\psi} + \dot{\phi}\cos\theta\right)^2\right] + mgd\cos\theta = \text{constant}. \tag{17.30}$$

In terms of the constant quantities p_ϕ and p_ψ, the energy is

$$E = \frac{1}{2}I_1\dot{\theta}^2 + \frac{(p_\phi - p_\psi\cos\theta)^2}{2I_1\sin^2\theta} + \frac{p_\psi^2}{2I_3} + mgd\cos\theta. \tag{17.31}$$

This expression gives the energy in terms of the *single* coordinate, θ.

The form of the total energy in Equation (17.31) suggests defining an effective potential involving all the terms with a θ. For convenience the constant term $p_\psi^2/2I_3$ is also included in the effective potential. Then,

$$E = \frac{1}{2}I_1\dot{\theta}^2 + V'(\theta), \tag{17.32}$$

where

$$V'(\theta) = \frac{(p_\phi - p_\psi\cos\theta)^2}{2I_1\sin^2\theta} + mgd\cos\theta + \frac{p_\psi^2}{2I_3}. \tag{17.33}$$

Solving Equation (17.32) for $\dot{\theta}$ yields

$$\dot{\theta} = \frac{d\theta}{dt} = \sqrt{\frac{2}{I_1}(E - V')}. \tag{17.34}$$

Given values for the constants E, p_ϕ, and p_ψ this equation can (in principle) be integrated to obtain $\theta = \theta(t)$. Inserting $\theta(t)$ in Equations (17.28) and (17.29) will then yield expressions for $\phi = \phi(t)$ and $\psi = \psi(t)$. The problem is thus solved. (It may not be easy to do this, but conceptually it presents no difficulty.)

However, even without actually solving Equation (17.34) we have learned much about the behavior of the system. For example, note that the rotation of the top around its axis of rotation is ω_3. The third of Equations (17.24) and the expression for p_ψ given in Equation (17.29) lead to

$$\omega_3 = \frac{p_\psi}{I_3}.$$

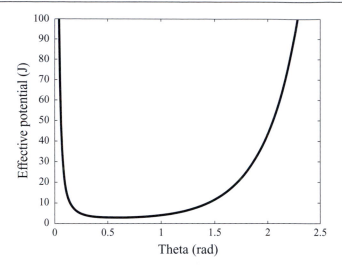

Figure 17.6 The effective potential as a function of θ for a spinning top.

Thus the rotational velocity of the top about its axis of symmetry is a constant. (This assumes, of course, that the top does not slow down due to friction with the air or at the point of contact with the ground.)

Figure 17.6 is a plot of the effective potential $V'(\theta)$ as a function of θ. The $\sin^2 \theta$ term in the denominator of Equation (17.33) means the effective potential goes to infinity at 0 and π. The minimum in $V'(\theta)$ can be found by setting the derivative of V' with respect to θ equal to zero. Denoting the angle at which V' is minimized by θ_0, this yields

$$0 = -mgd \sin \theta_0 + \frac{(p_\phi - p_\psi \cos \theta_0) p_\psi}{I_1 \sin \theta_0} - \frac{(p_\phi - p_\psi \cos \theta_0)^2 \cos \theta_0}{I_1 \sin^3 \theta_0}. \tag{17.35}$$

After some fairly complicated algebra we can write this in the interesting form

$$mgd I_1 \sin^4 \theta_0 - (p_\phi - p_\psi \cos \theta_0)(p_\psi - p_\phi \cos \theta_0) = 0. \tag{17.36}$$

Solving gives the value of θ_0 for the particular values of the constant parameters.

If we define $\beta = p_\phi - p_\psi \cos \theta_0$ we can write Equation (17.35) in the form

$$\beta^2 - \frac{p_\psi \sin^2 \theta_0}{\cos \theta_0} \beta + \frac{mgd I_1 \sin^4 \theta_0}{\cos \theta_0} = 0, \tag{17.37}$$

which is a quadratic in β whose solution is

$$\beta = \frac{p_\psi \sin^2 \theta_0}{2 \cos \theta_0} \left[1 \pm \sqrt{1 - \frac{4mgd I_1 \cos \theta_0}{p_\psi^2}} \right]. \tag{17.38}$$

We will return to this equation shortly.

Exercise 17.15

Obtain Equations (17.24).   ∎

Exercise 17.16

Obtain the general expression for the kinetic energy of a rotating body by evaluating Equation (17.25). Show that for a symmetrical body this reduces to Equation (17.26). ∎

Exercise 17.17

Obtain the expressions in Equations (17.28) and (17.29) for the generalized momenta p_ϕ and p_ψ. ∎

Exercise 17.18

Carry out the required substitutions to convert Equation (17.30) into (17.31). ∎

17.6.1 Precession without Nutation

If the total energy E is equal to $V'(\theta_0)$ the top has a constant value of θ equal to θ_0, meaning that the top is not nutating. This is the minimum value of E and the top is at the minimum of the effective potential curve. If, however, the top has somewhat more energy, the value of θ can vary between two limits (call them θ_1 and θ_2) and the top nutates, with the rotation axis oscillating between these limits.

We shall first consider the minimum energy top with $E = V'(\theta_0)$ and no nutation taking place. The precession rate of the top is $\dot\phi(\theta_0)$. To determine the value of this precession rate we insert θ_0 into Equations (17.28) and (17.29) for p_ϕ and p_ψ. From Equation (17.29),

$$\dot\psi = \frac{p_\psi}{I_3} - \dot\phi \cos\theta_0,$$

so Equation (17.28) reads

$$p_\phi = I_1 \dot\phi \sin^2\theta_0 + I_3 \cos\theta_0 \left(\frac{p_\psi}{I_3} - \dot\phi \cos\theta_0 + \dot\phi \cos\theta_0 \right).$$

Consequently,

$$\dot\phi = \frac{p_\phi - p_\psi \cos\theta_0}{I_1 \sin^2\theta_0} = \frac{\beta}{I_1 \sin^2\theta_0}. \tag{17.39}$$

Inserting the value of β from Equation (17.38) we obtain

$$\dot\phi = \frac{p_\psi}{2I_1 \cos\theta_0} \left[1 \pm \sqrt{1 - \frac{4mgdI_1 \cos\theta_0}{p_\psi^2}} \right].$$

But $p_\psi = I_3\omega_3$ so

$$\dot\phi = \frac{I_3\omega_3}{2I_1 \cos\theta_0} \left[1 \pm \sqrt{1 - \frac{4mgdI_1 \cos\theta_0}{I_3^2\omega_3^2}} \right]. \tag{17.40}$$

Since $\dot\phi$ is an observable, it must be real. Thus, the quantity under the root must be positive. That is,

$$1 - \frac{4mgdI_1 \cos\theta_0}{I_3^2\omega_3^2} \geq 0,$$

from which

$$\omega_3^2 \geq \frac{I_1}{I_3^2}\,(4mgd\cos\theta_0)\,.$$

For precession to take place, ω_3 must be at least as great as this minimum value. If the minimum is exceeded, there are two possible values for $\dot{\phi}$. These can be determined by expanding the square root in a binomial expansion. Then Equation (17.40) yields

$$\dot{\phi} \doteq \frac{I_3\omega_3}{2I_1\cos\theta_0}\left\{1 \pm \left[1 - \frac{1}{2}\frac{4mgd\,I_1\cos\theta_0}{[I_3\omega_3]^2}\right]\right\}.$$

The positive sign gives the fast precession rate,

$$\dot{\phi} \doteq \frac{I_3\omega_3}{2I_1\cos\theta_0}(2) = \frac{I_3\omega_3}{I_1\cos\theta_0},$$

and the negative sign gives the slow precession rate,

$$\dot{\phi} \doteq \frac{I_3\omega_3}{2I_1\cos\theta_0}\frac{1}{2}\frac{4mgd\,I_1\cos\theta_0}{[I_3\omega_3]^2} = \frac{mgd}{I_3\omega_3} = \Omega_P.$$

The slow precession rate Ω_P is the same as derived in Chapter 7. However, that derivation did not allow us to determine the fast precession rate. (The slow rate is the one most commonly observed.)

The derivation in Chapter 7 made the approximation that the angular momentum of the gyroscope was directed along the axis of rotation, so $\omega_3 = \dot{\psi}$. Now, however, we know that $\omega_3 = \dot{\psi} + \dot{\phi}\cos\theta$. The total angular momentum is $\mathbf{L} = \mathbf{p}_\psi + \mathbf{p}_\phi$ where $\mathbf{p}_\psi$ is directed along the principal axis $\hat{\mathbf{e}}_3 = \hat{z}'$ and $\mathbf{p}_\psi$ is along the vertical $\hat{z}$. See Figure 17.7. The vector $\mathbf{L}$ precesses about the vertical, as does the axis of rotation (along $\mathbf{z}'$).

Exercise 17.19

Determine the ratio of the slow precession to the fast precession. ∎

17.6.2 Nutation

A pure precession, such as described in the previous section, requires releasing the gyroscope with a specific set of initial conditions. A more common situation is for you to hold the free

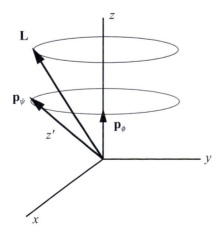

Figure 17.7 The angular momentum vector is the sum of $\mathbf{L} = \mathbf{p}_\psi + \mathbf{p}_\phi$. The gravitational force generates a torque that results in $\mathbf{L}$ precessing about the z-axis. The axis of rotation is along z' and also precesses about z.

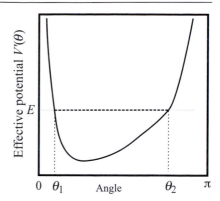

Figure 17.8 Qualitative sketch of the effective potential for a symmetrical top. If the energy E is greater than the minimum value of the effective potential V', the value of θ will vary between θ_1 and θ_2. (The top nutates.)

end of the gyroscope with your fingers, tip it to an angle θ and release it. Suppose the axis of rotation at the instant of release is over the x-axis. At this instant, p_ϕ is zero and **L** is constant. The gyroscope starts to fall. As the gyroscope falls, it acquires an angular momentum along the y-axis of Figure 17.7. The sum of p_ψ and this "new" angular momentum gives a total angular momentum with a component perpendicular to the z, z', x-plane. That is, the gyroscope axis starts to precess. As it falls further, this precession speeds up and becomes larger than Ω_P, so the axis rises and the precession rate decreases. In other words, the precession rate oscillates around the value Ω_P. As Feynman put it, "It has to go down a little in order to go around."[14] The resulting up and down oscillation in θ is called nutation.

Recall that in the preceding section we assumed that the energy of the gyroscope was $E = V'(\theta_0)$, so that the system was at the minimum of the effective potential well. Then θ remained constant and equal to θ_0. However, if the energy of the top is greater than this minimum value, θ will oscillate between values θ_1 and θ_2. See Figure 17.8.

As θ varies between θ_1 and θ_2, the axis of the gyroscope "nods" up and down. (A similar phenomenon was observed for the spherical pendulum. See Section 12.4.2.) The value of the energy at the "turning points" θ_1 and θ_2 is equal to the effective potential at those points and θ_1 and θ_2 are solutions of the equation

$$E = \frac{(p_\phi - p_\psi \cos \theta)^2}{2I_1 \sin^2 \theta} + \frac{p_\psi^2}{2I_3} + mgd \cos \theta. \tag{17.41}$$

This is the same as Equation (17.31) except that the term involving $\dot{\theta}$ is left out because $\dot{\theta}$ is zero at the turning points. But Equation (17.41) can be written as

$$2I_1 \left(E - \frac{p_\psi^2}{2I_3} \right) \sin^2 \theta = p_\phi^2 - 2p_\phi p_\psi \cos \theta + p_\psi^2 \cos^2 \theta + 2I_1 mgd \cos \theta \sin^2 \theta.$$

Replacing $\sin^2 \theta$ by $1 - \cos^2 \theta$,

$$2I_1 \left(E - \frac{p_\psi^2}{2I_3} \right) (1 - \cos^2 \theta) = p_\phi^2 - 2p_\phi p_\psi \cos \theta + p_\psi^2 \cos^2 \theta + 2I_1 mgd \cos \theta (1 - \cos^2 \theta),$$

which is a cubic in $\cos \theta$. There are three roots to this equation. Two of them are θ_1 and θ_2. The third root is greater than $+1$ and does not represent a physically possible situation.

[14] R. P. Feynman, R. B. Leighton and M. Sands, *The Feynman Lectures on Physics*, Vol. I, p. 20-7, Addison Wesley, 1963.

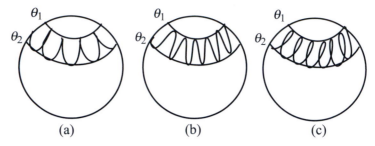

Figure 17.9 The path traced out by the rotation axis during nutation on the surface of a sphere centered on the point of contact of the spinning top with the floor.

As mentioned previously, the nutation of a spinning gyroscope depends on the initial conditions, specifically on the value of $\dot{\phi}$ when the tip of the gyroscope axis is released. If $\dot{\phi} = 0$ at $t = 0$, the tip of the axis will trace out the curve shown in Figure 17.9(a). If the axis is given a small shove in the direction of precession ($\dot{\phi}(t = 0) > 0$) the tip traces out the curve of Figure 17.9(b), and if the axis is given a shove in the opposite direction ($\dot{\phi}(t = 0) < 0$) the curve traced out is shown in Figure 17.9(c).

During nutation, the direction of the axis of rotation (θ) will vary between θ_1 and θ_2. But $\dot{\phi}$ depends on θ so the precession rate varies as well. We found from Equation (17.39) that

$$\dot{\phi} = \frac{p_\phi - p_\psi \cos\theta}{I_1 \sin^2\theta}.$$

This indicates that the sign of $\dot{\phi}$ depends on the relative values of p_ϕ and p_ψ. Specifically, if $\cos\theta < p_\phi/p_\psi$, $\dot{\phi}$ is positive, meaning that the velocity of precession is in the same sense as ω_3, the spin velocity of the top. Otherwise, $\dot{\phi}$ is negative. To express this mathematically, let us define an angle $\Theta = \cos^{-1} p_\phi/p_\psi$; then $\dot{\phi}$ is positive for $\theta > \Theta$ and negative for $\theta < \Theta$. In Problem 17.11 you will show that Θ is smaller than θ_0, so Θ is smaller than θ_2. The range of values for Θ is $0 < \Theta < \theta_0$. If $\Theta < \theta_1$ then $\dot{\phi}$ is always positive and the axis of the nutating top traces out the curve indicated in Figure 17.9(b). On the other hand, if $\Theta > \theta_1$, then as θ varies from θ_1 to θ_2, the precession rate $\dot{\phi}$ changes sign. The path traced out by the axis then has the form shown in Figure 17.9(c).

An interesting problem involves the motion of a spinning disk. If, for example, you hold a coin upright and set it spinning with a flick of your finger, you find that it spins upright for a while, then it leans over, and rotates faster and faster, then suddenly falls. The classical analysis (as we have done) would have the coin spinning forever, but actually air resistance rather quickly stops the motion.[15]

Exercise 17.20

Show explicitly that if $\theta < \Theta$, then $\dot{\phi}$ is negative. That is, $\dot{\phi}$ has the opposite sign from ω_3. ∎

Exercise 17.21

Show that if $\Theta < \theta_1$, then $\dot{\phi}$ is positive. ∎

---

[15] A spinning coin will slow down and lean over, so that it will be spinning about a point on its circumference. The axis of rotation does not pass through the coin. The point of contact between the coin and the surface is not fixed (it moves around the circumference) but at any instant it can be considered as the point about which the coin is rotating. A highly polished disk on a smoothly machined flat plate is sold as a toy called "Euler's Disk." An interesting analysis of the motion is given in the paper by H. K. Moffatt, "Euler's Disk and its Finite-time Singularity," *Nature*, **404**, 833 (2000).

17.7 Summary

The angular momentum of a body of arbitrary shape that is rotating about a fixed point is

$$\mathbf{L} = \mathcal{I} \cdot \boldsymbol{\omega},$$

where $\mathcal{I}$ is the inertia tensor. The elements of the inertia tensor are

$$I_{jk} = \int \rho(\mathbf{r})(r^2 \delta_{jk} - r_j r_k) d\tau,$$

so

$$\mathcal{I} = \int \rho(\mathbf{r})(r^2 \mathbf{1} - \mathbf{rr}) d\tau.$$

The kinetic energy of the rotating body is

$$T = \frac{1}{2} \boldsymbol{\omega} \cdot \mathcal{I} \cdot \boldsymbol{\omega}.$$

The inertia tensor can be represented by a real, hermitian matrix. This formulation is most useful when the matrix is expressed in diagonalized form. Then the axes of the coordinate system lie along the principal axes of the body.

It is often necessary to determine the inertia tensor in a rotated coordinate system. This is accomplished by the similarity transformation

$$\mathcal{I}' = \mathcal{A} \mathcal{I} \mathcal{A}^T,$$

where $\mathcal{A}$ is the appropriate rotation matrix.

To diagonalize $\mathcal{I}$ we determine the eigenvalues λ_i and the eigenvectors $\mathbf{R}_i$. (To assist in this task, four labor saving devices were described.) The eigenvalues and eigenvectors of the inertia tensor are obtained from the eigenvalue equation:

$$(\mathcal{I} - \lambda \mathbf{1}) \mathbf{R} = 0.$$

A nontrivial solution requires that the determinant of the coefficients is zero, that is,

$$|\mathcal{I} - \lambda \mathbf{1}| = 0.$$

This relation yields three equations that can be solved for the three eigenvalues, λ_i, $i = 1, 2, 3$. Plugging these eigenvalues into the eigenvalue equation leads to expressions for the components of the eigenvectors, x_i, y_i, z_i, $i = 1, 2, 3$, where i specifies the eigenvector. (Note that we actually obtain the ratio of components, such as y_2/z_2. The third component (such as x_2) is chosen such that the eigenvector has unit magnitude.)

The Euler equation of motion for a rotating body is

$$\mathbf{N} = \mathcal{I} \cdot \frac{d\boldsymbol{\omega}}{dt} + \boldsymbol{\omega} \times (\mathcal{I} \cdot \boldsymbol{\omega}),$$

where $\mathcal{I}$ is the inertia tensor in the body coordinate system. If the torque is zero (a freely rotating body) the Euler equation yields

$$I_1 \dot{\omega}_1 = \omega_2 \omega_3 (I_2 - I_3),$$
$$I_2 \dot{\omega}_2 = \omega_3 \omega_1 (I_3 - I_1),$$
$$I_3 \dot{\omega}_3 = \omega_1 \omega_2 (I_1 - I_2).$$

If the torque is not zero, the analysis is much more complicated. We illustrated the process with the example of a spinning symmetrical top. Using the Euler angles as coordinates, the Lagrangian is

$$L = \frac{1}{2}\left[I_1\left(\dot{\theta}^2 + \dot{\phi}^2\sin^2\theta\right) + I_3\left(\dot{\psi} + \dot{\phi}\cos\theta\right)^2\right] - mgd\cos\theta.$$

The two generalized momenta associated with ϕ and ψ are constants, that is,

$$p_\phi = \frac{\partial L}{\partial\dot{\phi}} = I_1\dot{\phi}\sin^2\theta + I_3\cos\theta\left(\dot{\psi} + \dot{\phi}\cos\theta\right) = \text{constant},$$

$$p_\psi = \frac{\partial L}{\partial\dot{\psi}} = I_3(\dot{\psi} + \dot{\phi}\cos\theta) = \text{constant}.$$

The total energy can then be written

$$E = \frac{1}{2}I_1\dot{\theta}^2 + \frac{(p_\phi - p_\psi\cos\theta)^2}{2I_1\sin^2\theta} + \frac{p_\psi^2}{2I_3} + mgd\cos\theta,$$

or

$$E = \frac{1}{2}I_1\dot{\theta}^2 + V'(\theta),$$

where the effective potential $V'(\theta)$ is

$$V'(\theta) = \frac{(p_\phi - p_\psi\cos\theta)^2}{2I_1\sin^2\theta} + mgd\cos\theta + \frac{p_\psi^2}{2I_3}.$$

The precession of the top consists of a variation in ϕ at a rate

$$\dot{\phi} = \frac{p_\phi - p_\psi\cos\theta_0}{I_1\sin^2\theta_0}.$$

Precession can only occur if the top is spinning at a rate

$$\dot{\psi} \geq \frac{4mgd I_1}{I_3^2}\cos^2\theta_0\sin^2\theta_0.$$

Nutation is a variation in θ, the angle between the axis of the top and the vertical. It can vary from θ_1 to θ_2, two angles whose values depend on the total energy. They can be evaluated as two of the roots of the equation

$$E = \frac{(p_\phi - p_\psi\cos\theta)^2}{2I_1\sin^2\theta} + \frac{p_\psi^2}{2I_3} + mgd\cos\theta.$$

17.8 Problems

Problem 17.1 Prove that for a symmetrical body, the products of inertia are zero.

Problem 17.2 For the tetrahedron of Figure 17.2 show that the products of inertia are $-Ma^2/20$.

Problem 17.3 A system is made up of three particles of mass M located at $(2a, 0, 0)$, $(0, 2a, 0)$, and $(0, 0, a)$. Find the principal moments of inertia about the origin. Determine a set of principal axes.

Problem 17.4 A Cartesian coordinate system has its origin at the center of mass of a sphere of radius a and mass M. (a) Determine the inertia tensor of the sphere, relative to this coordinate system. Assume the density of the sphere is constant. Express the inertia tensor as a matrix and as a dyadic. (b) Now assume the sphere is rotating with angular velocity $\boldsymbol{\omega} = (\omega/\sqrt{3})(\hat{\mathbf{i}} + \hat{\mathbf{j}} + \hat{\mathbf{k}})$. Evaluate the kinetic energy using matrix multiplication and dyadic multiplication of $T = \frac{1}{2}\boldsymbol{\omega} \cdot \mathcal{I} \cdot \boldsymbol{\omega}$. (c) Show that your result agrees with the elementary expression for the kinetic energy of a rotating sphere ($T = \frac{1}{2}I\omega^2$).

Problem 17.5 Determine the inertia tensor for the ammonia molecule about its center of mass. Determine the principal axes. The ammonia molecule (NH_3) is made up of one nitrogen atom ($MW = 14$) and three hydrogen atoms ($MW = 1$) arranged in a pyramidical shape with the nitrogen at the apex and the three hydrogens forming an equilateral triangular base. The distance between the hydrogen atoms and the nitrogen atom is 1.03 Å and the angles between the bonds are $36.4°$ and $107.2°$.

Problem 17.6 Let $\mathcal{A}_n$ be an antisymmetric matrix whose elements are given by $A_{ij} = \epsilon_{ijk}x_k$, where the x_k are the coordinates of the kth mass point of a body. Show that the matrix of the inertia tensor of the body can be written as

$$\mathcal{I} = -m_n \left(\mathcal{A}_n\right)^2.$$

By definition the Levi-Civita density tensor ϵ_{ijk} is zero if any two of the indices ijk are equal, $+1$ if $ijk = 1, 2, 3$ or any even permutation of $1, 2, 3$ and -1 for odd permutations of $1, 2, 3$.

Problem 17.7 Show that using the Einstein summation convention the kinetic energy can be expressed as

$$T = \frac{1}{2}\sum_a \omega_i I_{ij} \omega_j,$$

where

$$I_{ij} = \sum_a m_a (x_k x_k \delta_{ij} - x_i x_j).$$

(The subscript a identifies the mass particle. Hint: Keep in mind that repeated indices imply summation. Suggestion: First write $L_i = I_{ij}\omega_j$.)

Problem 17.8 Show that eigenvectors corresponding to different eigenvalues are orthogonal.

Problem 17.9 Prove Labor Saving Device 1, Equation (17.16).

Problem 17.10 A dumbbell is made up of two point masses m at the ends of a massless rod of length $2d$. The center of mass is at the origin of a Cartesian coordinate system (x, y, z) and the rod and masses lie in the xy-plane. The rod makes an angle θ with the x-axis. (a) Determine the inertia tensor of the dumbbell in this coordinate system. First obtain a general expression for $\mathcal{I}$ in terms of θ, then for the rest of the problem assume $\theta = 30°$. This will make the matrix multiplications much simpler. (b) Using a labor saving device, determine the inertia tensor in a coordinate system (x', y', z') where the primed axes are parallel to (x, y, z) but the origin is displaced by $\mathbf{R} = d(\hat{\mathbf{i}} + \hat{\mathbf{j}} + \hat{\mathbf{k}})$. (c) Using a labor saving device determine the inertia tensor in a reference frame that is rotated through an angle $-\theta$ relative to the original reference frame, so that the point masses now lie along the x-axis.

Problem 17.11 Show that $\Theta = \cos^{-1}(p_\phi/p_\psi)$ is smaller than θ_0.

Problem 17.12 A square flat metal plate of side a lies in the xy-plane. The origin of a Cartesian coordinate system is at one corner and the positive x- and y-axes extend along the sides of the plate. (a) Determine the inertia tensor relative to the origin. (b) Assume the plate is rotating about the z-axis ($\boldsymbol{\omega} = \omega\hat{\mathbf{k}}$). Determine the angular momentum and the kinetic energy.

Problem 17.13 Consider a plane lamina of arbitrary shape. Prove that relative to an arbitrary origin there always exists a set of principal axes. (This is also true of three-dimensional bodies, but the proof is simpler for a two-dimensional object.) (Hint: Place the object in the first quadrant of a Cartesian coordinate system and by rotating the axes prove that there is a an orientation for which the products of inertia vanish.)

Problem 17.14 A uniform right circular cone of height h and base radius R has a mass M. It is set on its side and it rolls without slipping in such a way that the tip of the cone remains fixed. Note that instantaneously the axis of rotation is along the line of contact between the cone and the horizontal surface. The angle between this line and a fixed line on the horizontal plane is θ so the angular velocity of the center of mass of the cone is $\dot{\theta}$. (a) Obtain an expression for the kinetic energy. (b) Obtain an expression for the angular momentum. (Hint: First determine the inertia tensor relative to a set of principal axes.)

Problem 17.15 A gyroscope is both rotating about its axis and precessing about the z-axis and has an angular velocity $\boldsymbol{\omega} = \alpha\hat{z} + \beta\hat{z}'$. Determine the angular momentum about the body axes and the kinetic energy. Assume $I_1 = I_2$ and that I_3 and θ are known.

Problem 17.16 A gyroscope is hanging from a fixed point (so $\theta_0 > \pi/2$). Show that there is one positive and one negative value for $\dot{\phi}(\theta_0)$.

Problem 17.17 A spinning top is held with a fixed polar angle $\theta = \theta_1$. Therefore, initially, $\dot{\theta} = 0$, $\dot{\phi} = 0$, and $\dot{\psi} = \omega_3$. It is then released and allowed to precess and nutate. (a) Write expressions for p_ϕ, p_ψ, V', and E. (b) The turning points for the nutation are θ_1 and θ_2 ($\theta_1 \leq \theta \leq \theta_2$). Obtain an expression for $\cos\theta_2$. (c) Assume the top is spinning rapidly so that the quantity $\alpha = (2I_1mgd)/(I_3^2\omega_3^2) \ll 1$. Show that, for this case, $\cos\theta_2 \simeq \cos\theta_1 - \alpha\sin^2\theta_1$.

Problem 17.18 The spin axis of a rapidly spinning gyroscope lies in the horizontal plane. That is, $\theta = \pi/2$. It is precessing about the vertical axis. (a) Show that elementary considerations lead to a precession rate of $\dot{\phi} = mgd/I_3\omega_3$. (b) Show that a more sophisticated analysis based on Equation (17.36) yields the same expression for the average precession rate $\langle\dot{\phi}\rangle$.

Problem 17.19 A symmetrical top is approximated by a cone of mass 200 g, height 4 cm, and base diameter 3 cm. It spins on its point at 400 rad/s. Determine the (slow) precession rate.

Problem 17.20 The disk illustrated in Figure 16.1 is spun up to an angular velocity ω while the axis on which it is mounted is constrained to remain fixed in direction. Suddenly the axis is released and system (disk plus axis) is free to wobble. (a) Determine the half-angle of the cone in space described by the axis. You may ignore the mass of the axis. Assume the angle between the axis and a line perpendicular to the plane of the disk is α. (b) Determine the (precession) period of the wobble.

Problem 17.21 A set of axes (x, y, z), fixed in a body, are principal axes. A line $\overline{OQ}$ passes through the origin (O) and has direction cosines α, β, γ relative to the axes x, y, z. Show that the moment of inertia of the body about $\overline{OQ}$ is given by

$$I_{OQ} = \alpha^2 I_x + \beta^2 I_y + \gamma^2 I_z.$$

Hint: It may be helpful to note that the angle between two vectors with direction cosines $\alpha_1, \beta_1, \gamma_1$ and $\alpha_2, \beta_2, \gamma_2$ is given by

$$\cos\theta = \alpha_1\alpha_2 + \beta_1\beta_2 + \gamma_1\gamma_2.$$

Problem 17.22 A set of axes (x, y, z) are fixed in a body. Relative to these body axes, the moments of inertia are I_x, I_y, I_z and the products of inertia are P_{xy}, P_{xz}, P_{yz}. (a) Show that the moment of inertia about a line $\overline{OQ}$ passing through the origin is

$$I = \alpha^2 I_x + \beta^2 I_y + \gamma^2 I_z - 2\alpha\beta P_{xy} - 2\beta\gamma P_{yz} - 2\alpha\gamma P_{xz},$$

where α, β, γ are the direction cosines of $\overline{OQ}$. (b) Draw a vector $\mathbf{R}$ in the direction of $\overline{OQ}$ and let $\mathbf{R} \cdot \mathbf{R}$ be inversely proportional to the moment of inertia about $\overline{OQ}$, that is,

$$R^2 = \frac{1}{I}.$$

Show that the tip of $\mathbf{R}$ generates the surface of an ellipsoid. (This is called "Poinsot's ellipsoid of inertia.")

Problem 17.23 A top rotates about a vertical axis. Show that the motion is stable if $\omega_3^2 > 4mgdI_1/I_3^2$. (Hint: $\partial^2 E/\partial\theta^2 > 0$ for stability.)

Problem 17.24 A frisbee will usually "wobble" when thrown. This wobble is the precession of the angular velocity vector $\boldsymbol{\omega}$ about the angular momentum vector $\mathbf{L}$ (assuming these two vectors are not perfectly aligned). Consider this an example of torque-free motion and carry out the following calculations to show that the precession rate $\dot{\phi}$ is given by $\dot{\phi} \cong 2\omega$. (The approximation is due to assuming the angle between $\boldsymbol{\omega}$ and the symmetry axis is small.)

(a) Consider the ξ', η', ζ' coordinate system illustrated in Figure 16.6. Show that $\omega_\xi = \dot{\theta}, \omega_{\eta'} = \dot{\phi} \sin\theta, \omega_{\zeta'} = \dot{\phi}\cos\theta + \dot{\psi}$.

(b) Assume the angular momentum vector $\mathbf{L}$ is oriented along $\hat{\mathbf{z}}$ and the angular velocity vector $\boldsymbol{\omega}$ lies in the $\hat{\mathbf{z}} - \hat{\boldsymbol{\zeta}}'$-plane at an angle α to $\hat{\boldsymbol{\zeta}}'$. Because the frisbee is essentially a flat plate, $I_{\xi'\xi'} = I_{\eta'\eta'} = I$ and $I_{\zeta'\zeta'} = I_s$. Show that $\hat{\boldsymbol{\zeta}}'$

$$\frac{L_{\eta'}}{L_{\zeta'}} = \tan\theta = \frac{I}{I_s}\tan\alpha.$$

(c) Show that $\dot{\phi} = \omega\frac{\sin\alpha}{\sin\theta} = \omega\left[1 + \left(\frac{I_s^2}{I^2} - 1\right)\cos^2\alpha\right]^{1/2}$.

(d) Show that $I_s = 2I$ and hence $\dot{\phi} \cong 2\omega$.

Problem 17.25 A "sleeping top" is one that is rotating rapidly in an upright mode and is not precessing. As the top slows down, it begins to precess, indicating that the stability has been lost as the rotation rate decreases. Using Equation (17.40), show that the criterion for the stability of a sleeping top is

$$I_3^2\omega_3^2 \geq 4mgI_1d.$$

Computational Projects

Computational Project 17.1 Write a program to solve for θ_0 using Equation (17.36). You may assume $m = 1, d = 0.1, p_\phi = 3, p_\psi = 2, I_1 = 3$, and $I_2 = 6$.

Computational Project 17.2 Use a computer algebra system to verify that Equation (17.35) can be expressed as Equation (17.36).

Part VI

Special Topics

18 Statics

This chapter treats several advanced concepts in statics, well beyond the brief summary of statics in Section 1.6. We will begin with a few definitions and two simple theorems concerning systems of forces acting on rigid bodies, then go on to analyze the statics of freely deformable bodies such as a string or cable hanging from stationary supports. This is followed by definitions of stress and strain and a generalization of Hooke's law. The last topic is d'Alembert's principle and the concept of virtual work. You will see how this principle can be used to derive Lagrange's equations. An important application is an investigation of the properties of a fluid in equilibrium (hydrostatics), but we will leave that for Chapter 19 where we consider fluids in general.

18.1 Basic Concepts

Statics is the study of bodies that are not accelerating, that is, bodies that are at rest or moving at constant velocity. It is always possible to find an inertial coordinate system in which such a body is at rest, so the linear velocity is usually assumed to be zero.

If a body is not accelerating, the net force and the net torque acting on it must be zero. Consequently, the basic conditions for equilibrium are:

$$\sum \mathbf{F}_i = 0,$$

$$\sum \mathbf{N}_i = 0.$$

There are two useful theorems related to the equilibrium of a rigid body.

Theorem 18.1   *If a body is in equilibrium under the action of three forces then the lines of action of the three forces all lie in the same plane and intersect at one point.*

Proof   Two force vectors define a plane. If the third force does not lie in that plane, the tip-to-tail sum of the vectors do not form a closed triangle. Therefore, for equilibrium, the third force must also lie in the plane. Furthermore if two of the forces intersect at some point, they contribute zero torque about that point. For the body to be in equilibrium, the third force must also contribute zero torque about that point and consequently its line of action must also pass through the point.

Theorem 18.2   *If a force $\mathbf{F}_c$ can be considered the sum of two other forces, $\mathbf{F}_c = \mathbf{F}_a + \mathbf{F}_b$, then the torque about a point due to $\mathbf{F}_c$ is the sum of the torques due to $\mathbf{F}_a$ and $\mathbf{F}_b$ if they act at the same point.*

Proof   The torque due to $\mathbf{F}_c$ is $\mathbf{N}_c = \mathbf{r} \times \mathbf{F}_c$. But

$$\mathbf{N}_c = \mathbf{r} \times \mathbf{F}_c = \mathbf{r} \times (\mathbf{F}_a + \mathbf{F}_b) = \mathbf{N}_a + \mathbf{N}_b,$$

and the theorem is proved. This is sometimes referred to as the Theorem of Varignon.

The following examples and simple exercises will refresh your knowledge on how to solve statics problems.

Worked Example 18.1   A rod of length L and mass M is supported from the ceiling by two massless vertical strings of lengths $L/2$ and L, as shown in Figure 18.1. A weight of mass $2M$ hangs from the lower end of the rod. Determine the tensions in the strings.

Solution   The net force and the net torque must be zero. All the forces act vertically so

$$\sum \mathbf{F} = 0 \implies T_1 + T_2 = 3Mg.$$

Taking torques about the lower end of the rod,

$$\sum \mathbf{N} = 0 \implies T_1 L \cos \theta = Mg \frac{L}{2} \cos \theta,$$

so

$$T_1 = \frac{1}{2} Mg,$$

and consequently

$$T_2 = 3Mg - T_1 = 2.5Mg.$$

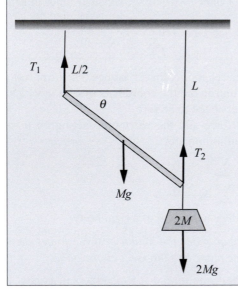

Figure 18.1   A system in equilibrium.

Worked Example 18.2   A spool of thread rests on its side on an inclined plane. See Figure 18.2. It does not roll down the plane because the string is wrapped over the top of the spool and the thread cannot simply unwind. (In fact, when the spool slips, it will have to rotate clockwise to unwind the string.) Assume the higher end of the inclined plane is slowly raised until the spool slips. Let the coefficient of static friction between spool and plane be $\mu = 0.2$. Determine the angle at which the spool begins to slip.

Solution   Forces up the plane equal forces down the plane so if F_N is the normal force,

$$mg \sin \theta = T + \mu F_N.$$

The forces perpendicular to the plane also add to zero,

$$F_N = mg \cos\theta.$$

Clockwise torques equal counterclockwise torques. Taking torques about the point of contact between spool and plane:

$$T(2R) = mgR \sin\theta,$$

$$T = \frac{1}{2}mg \sin\theta,$$

where R is the radius of the spool. Since $T = mg \sin\theta - \mu F_N = mg \sin\theta - \mu mg \cos\theta$ we have

$$\frac{1}{2}mg \sin\theta = mg \sin\theta - \mu mg \cos\theta,$$

$$\tan\theta = 2\mu = 0.4,$$

$$\theta = 21.8°.$$

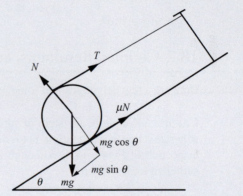

Figure 18.2 A spool of thread lies on its side on an inclined plane. As seen in the figure, the thread is wrapped around the top of the spool so it cannot roll without slipping.

Exercise 18.1

Prove that a building such as the leaning Tower of Pisa will not topple over if a vertical line from its center of mass passes through its base. ∎

Exercise 18.2

A yo-yo of mass m and radius r hangs from a nail on a wall such that its curved side is in contact with the wall. The nail is a distance L above the point of contact. Determine the force exerted by the wall on the yo-yo. You may assume the string is attached to the center of mass of the yo-yo, as shown in Figure 18.3. Answer: $N = Mgr/L$. ∎

Exercise 18.3

A ladder of length 2 m and mass 10 kg rests against a smooth wall. The coefficient of static friction between ladder and floor is 0.2. How far from the wall is the foot of the ladder when it begins to slip? Answer: 0.74 m. ∎

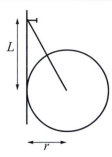

Figure 18.3 A yo-yo hangs from a nail in the wall. Assume the inner axle has diameter zero so the string is attached to the center of mass. (Exercise 18.2.)

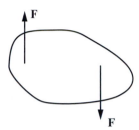

Figure 18.4 A couple.

18.2 Couples, Resultants, and Equilibrants

Here are a few definitions and some concepts that are useful when treating statics problems.

Couple A couple is a system of forces whose sum is zero, i.e., $\sum \mathbf{F}_i = 0$. A couple cannot generate a linear acceleration, but it can exert a torque on a body, as you can appreciate from an inspection of Figure 18.4. Although a couple may be a system of many forces, two forces are sufficient to describe it.

Equivalent Systems of Forces If two systems of forces exert the *same total force* and the *same total torque* about every point, then the two systems of forces are called *equivalent*.

Resultant If a force $\mathbf{R}$ acting at point P is equivalent to a system of forces $\mathbf{F}_i$ acting at points P_i, then $\mathbf{R}$ is called the *resultant* of the system of forces.

Equilibrant If a force $\mathbf{R}$ acting at point P is the resultant of the system of forces $\mathbf{F}_i$ acting at several points P_i, then applying the force $\mathbf{R}$ at P will cause the same linear and angular acceleration as all of the forces $\mathbf{F}_i$ acting at points P_i. Consequently, if one applies a force $-\mathbf{R}$ at P, the system of forces is balanced. For this reason, $-\mathbf{R}$ acting at P is called the *equilibrant*. A system that is not in static equilibrium is brought into static equilibrium by applying an equilibrant. In designing a complicated structure, an engineer will be interested in determining the equilibrant.

A useful fact about couples is given by the following theorem.

Theorem 18.3   *A couple exerts the same torque about every point.*

Proof   The torque exerted about point Q by a couple is

$$\mathbf{N}_Q = \sum \left(\mathbf{r}_{Qi} \times \mathbf{F}_i \right),$$

where $\mathbf{r}_{Qi}$ is a vector from Q to the point of application of force $\mathbf{F}_i$. The torque exerted by this couple about a different point Q' is given by

$$N_{Q'} = \sum \left(r_{Q'i} \times F_i \right).$$

But $r_{Q'i} = r_{Qi} + d$, where d is the vector from Q to Q'. So

$$N_{Q'} = \sum \left((r_{Qi} + d) \times F_i \right) = N_Q + d \times \sum_i F_i.$$

By the definition of a couple, $\sum_i F_i = 0$. Therefore $N_{Q'} = N_Q$, and the theorem is proved.

Exercise 18.4

A bridge is collapsing when Superman flies under it and holds it up by applying a single force at one point. This is possible because Superman is very good at determining the equilibrant of a system. Suppose the bridge has a mass of 100 000 kg, and a length of 20 m. Other forces acting on the bridge (just before Superman appeared) were 300 000 N at the left end and 150 000 N at the right end, both acting vertically upward. The weight of the collapsing bridge acts at its geometric center. Determine the equilibrant. Answer: $R = 5.3 \times 10^5$ N at 12.8 m from the left end. ∎

18.3 Reduction to the Simplest Set of Forces

It is often convenient to reduce a complicated set of forces to the smallest number of forces that are equivalent to the original system, that is, to the smallest number of forces that yield the same total force and same total torque.

Theorem 18.4 *Any system of forces is equivalent to a single force applied at an arbitrary point plus a couple.*

Proof Consider a system of forces, F_i, $i = 1, 2, \ldots, n$, applied at points $P_1, P_2, \ldots, P_n$. Let $F = \sum F_i$, so F is the net force exerted by the system of forces. Let N_Q be the torque about some given point Q due to the system of forces. Assume the net force F acts at an arbitrary point P. This gives rise to a torque about Q, but in general it is not equal to N_Q. Adding a couple to the system does not change the net force, but it allows one to obtain the desired net torque N_Q. Let the couple be composed of the two equal and opposite forces F_c and $-F_c$. Let $+F_c$ act at P (the same place where F is acting). Then the original system of forces is equivalent to a force $F + F_c$ acting at P plus a force $-F_c$ acting at some other point such that the torque is equal to that of the original system. Thus the system of forces is reduced to one force and a couple and the theorem is proved.

Exercise 18.5

A 1 m rod lying along the x-axis is acted upon by the following three forces: a force $F_1 = 10\hat{j}$ N at the left end, a force $F_2 = -20\hat{j}$ N at the midpoint, and a force $F_3 = 5\hat{i}$ N at the right end. Determine an equivalent system of two forces. ∎

18.4 The Hanging Cable

A string (and to a lesser extent, a cable or a chain) is a "freely deformable" body. This section is an analysis of the physics of an ideal rope or cable that is suspended from two fixed points. We will study two problems: (1) a "bridge" suspended from cables, and (2) a rope hanging under its own weight.

18.4.1 A Suspension Bridge

We begin by considering a suspension bridge, such as shown in Figure 18.5. The suspension cables are assumed massless and the only load is the weight of the roadway which is distributed uniformly in the horizontal direction. The points of support (A, B) are at the same level. It is desired to find the shape of the curve formed by the cables.

In a problem of this type it is convenient to consider a portion of a cable starting with the lowest point (C), as illustrated in Figure 18.6. Because C is the lowest point of the sagging cable, the tension at C, (T_c), is horizontal. The tension at a nearby point D, (T_d), is at an angle θ with respect to the horizontal. The section of roadway supported by this section of the cable has length x and weight wx where w is the weight per unit length of the roadway.

The lines of action of the three forces $(T_c, T_d,$ and $wx)$ pass through the same point (by Theorem 18.1). By symmetry this point is at $x/2$. The sum of the forces is zero so

$$T_d \sin \theta = wx,$$
$$T_d \cos \theta = T_c.$$

Dividing one equation by the other,

$$\tan \theta = \frac{wx}{T_c}.$$

In the limit of D being very close to C, this reduces to

$$\frac{dy}{dx} = \frac{wx}{T_c}.$$

Integrating this differential equation gives an expression for y as a function of x. But $y = y(x)$ is the equation for the curve formed by the sagging cable. Integrating:

$$y = \frac{wx^2}{2T_c} + c.$$

This is the equation of a parabola.

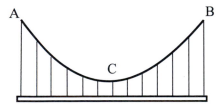

Figure 18.5 An idealized suspension bridge.

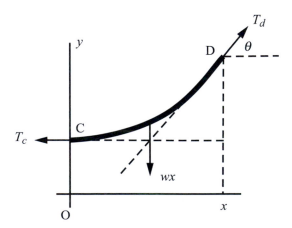

Figure 18.6 A portion of the cable of a suspension bridge.

Exercise 18.6

Show that the tension at any point in the cable is given by $T = \sqrt{T_c^2 + w^2 x^2}$. ∎

18.4.2 A Hanging Rope

Next we consider a slightly more difficult problem, namely a rope hanging under its own weight. Consider a rope (or cable or chain) suspended from two ends which may or may not be at the same height. Again, we want to determine the shape of the rope.

Consider a segment $\overline{AB}$ of length s as shown in Figure 18.7. For convenience s is measured from the lowest point in the sagging rope. The segment is subjected to three forces, namely the horizontally directed tension $\mathbf{T_0}$ at A, the tension $\mathbf{T}$ acting at angle θ at B, and the weight $\mathbf{W} = -ws$ acting downward at the center of mass of the segment. Note that if the force of gravity on the rope is expressed as a force per unit length $\mathbf{w}$ then the total "body force"[1] acting on the piece of rope of length s will be $\int_0^s \mathbf{w} ds$. In equilibrium, the vector sum of the three forces acting on the segment is zero. Equating the vertical and the horizontal components of these forces leads to

$$T \sin \theta = ws,$$
$$T \cos \theta = T_0.$$

Therefore

$$\tan \theta = \frac{w}{T_0} s,$$

or

$$s = \frac{T_0}{w} \tan \theta = c \tan \theta, \tag{18.1}$$

where $c = w/T_0$ and has the units of length. (In a little while I will give you a geometric interpretation of c as the vertical distance from the origin to the lowest point in the rope.)

If you are very good at analytical geometry, you might have recognized Equation (18.1) as the intrinsic equation of a *catenary*. In this sense, the problem is solved, but most people would prefer to have a formula for the curve in Cartesian coordinates x, y.

Note that the Cartesian coordinates shown in Figure 18.7 have the y-axis passing through the lowest point of the curve, but the origin is displaced vertically from this lowest point.

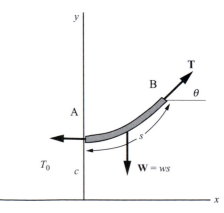

Figure 18.7 A portion of a rope sagging under its own weight.

---

[1] A "body force" is a force that acts throughout the volume of a body. It can be expressed as a density. Thus, for example, the tension in the rope would not be a body force, but the weight of the rope is a body force.

Equation (18.1) can be written as

$$\tan \theta = \frac{dy}{dx} = \frac{s}{c}.$$

Consequently,

$$\frac{d^2 y}{dx^2} = \frac{1}{c}\frac{ds}{dx} = \frac{1}{c}\frac{\sqrt{dx^2 + dy^2}}{dx},$$

$$= \frac{1}{c}\sqrt{1 + \left(\frac{dy}{dx}\right)^2}.$$

Let $p = \frac{dy}{dx}$ and write this relation as

$$\frac{dp}{dx} = \frac{1}{c}\sqrt{1 + p^2},$$

$$\frac{dp}{\sqrt{1 + p^2}} = \frac{1}{c}dx.$$

Integrating

$$\frac{x}{c} = \sinh^{-1} p + \text{constant.}$$

Because the slope (p) is zero at $x = 0$, the constant of integration is zero. Therefore,

$$p = \sinh\left(\frac{x}{c}\right).$$

But $p = dy/dx$, so

$$\frac{dy}{dx} = \sinh\left(\frac{x}{c}\right).$$

Integrating again leads to

$$y = c\cosh\left(\frac{x}{c}\right) + \text{ constant.}$$

If the origin is placed a distance c below the lowest point, then $y = c$ at $x = 0$ and the constant of integration is zero. Thus the equation of the curve formed by the hanging rope is

$$y = c\cosh\left(\frac{x}{c}\right), \tag{18.2}$$

which is the equation of a catenary in Cartesian coordinates.[2]

The tension at any point in the rope is obtained from

$$T = \frac{T_0}{\cos\theta} = \frac{T_0}{\frac{dx}{ds}} = T_0\frac{ds}{dx} = T_0\sqrt{1 + (dy/dx)^2}$$

$$= T_0\sqrt{1 + \sinh^2(x/c)}.$$

Therefore,

$$T = T_0\cosh\left(\frac{x}{c}\right).$$

---

[2] More generally, if the lowest point is at (c_1, c_2), then $y = c_2\cosh\frac{x - c_1}{c_2}$.

But, $y = c \cosh \left(\frac{x}{c} \right)$ so

$$T = \frac{T_0}{c} y,$$

and finally, since $c = w / T_0$,

$$T = wy.$$

That is, the tension at any point in the rope is directly proportional to the height of that point. The maximum tension occurs at the highest points, that is, at the end points. For a rope whose end points are at the same height and are separated by a horizontal distance a (called the "span") the end points are at $x = \pm a/2$, and $y = h + c$, where h (the "sag") is the vertical distance from the points of support to the lowest point in the rope. The maximum tension is then

$$T_m = wc \cosh \left(\frac{a}{2c} \right) = w(h + c).$$

Worked Example 18.3 Obtain an expression for the length of a hanging rope in terms of the parameters c and a.

Solution Assume the rope is hanging between points 1 and 2. If L is the length of the rope, then

$$L = \int_1^2 ds = s_2 - s_1.$$

But

$$s = c \tan \theta = c \left(\frac{dy}{dx} \right) = c \frac{d}{dx} \left(c \cosh \left(\frac{x}{c} \right) \right) = c \sinh \left(\frac{x}{c} \right).$$

Consequently,

$$L = c \left[\sinh \left(\frac{x_2}{c} \right) - \sinh \left(\frac{x_1}{c} \right) \right].$$

If points 1 and 2 are not at the same height then the difference in their heights is

$$H = y_2 - y_1 = c \left[\cosh \left(\frac{x_2}{c} \right) - \cosh \left(\frac{x_1}{c} \right) \right].$$

Consequently,

$$L^2 = H^2 + 2c^2 \left(\cosh \left(\frac{a}{c} \right) - 1 \right) = H + 4c^2 \sinh^2 \left(\frac{a}{2c} \right).$$

This relationship can be inverted to determine the parameter c of a catenary in terms of its length L, its span a, and the vertical displacement of the end points H.

Exercise 18.7

Show that the length of a sagging rope is given by $L = c \left[\sinh \left(\frac{x_2}{c} \right) - \sinh \left(\frac{x_1}{c} \right) \right]$. ■

Exercise 18.8

A rope of length 3 m and mass per unit length of 0.2 kg/m is hanging under its own weight. The end points are located at $(-1, 1.6)$ and $(1, 1.6)$. Numerically determine the parameter c. Determine the maximum value of the tension. Answer: $c = 0.62$. ■

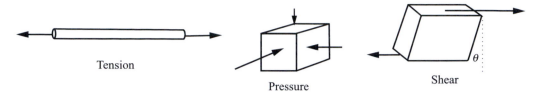

Tension   Pressure   Shear

Figure 18.8 Examples of three types of stress. A tension is represented by a force exerted across the end of a rod, tending to stretch it. A pressure is represented by forces perpendicular to the faces of a cube, tending to squeeze it. A shear is represented by forces parallel to the top and bottom faces of a cube, tending to change its shape.

18.5 Stress and Strain

Let us now turn our attention to another aspect of real bodies – they are not completely rigid. When acted upon by forces, real bodies undergo deformations. These deformations are due to forces that act across surfaces. For example, to describe what happens when you squeeze a tennis ball, you need to specify not only the magnitude and the direction of the force, but also the area and orientation of the surface across which it is transmitted. A force that acts across a surface is called a *stress*. A stress is always expressed as a force divided by an area (F/A). Stresses are characterized by their effect. If you squeeze something, you tend to compress it. The stress is then called a *pressure*. If the force tends to stretch the body upon which it acts, as when you pull on the ends of a rubber band, the stress is called a *tension*. If the stress tends to change the shape of the body as, for example, when a force acts parallel to a surface of the body, it is called a *shear*. These three types of stress are illustrated in Figure 18.8.

The deformation of the body is called the *strain*. As with the stress, the strain depends on the situation. Thus, a rod of length l subjected to a force across an end (a tension) will undergo a change in its length (Δl). The deformation will be characterized by $\Delta l / l$. A cube subjected to a compression will undergo a change in volume. This deformation is characterized by $\Delta V / V$. A parallelepiped subjected to a shear might have its top surface slide relative to the bottom surface. This is characterized by the tangent of the angle θ shown in Figure 18.8.

The relation between stress and strain is given by an empirical relation known as Hooke's law:

Strain is proportional to stress.

This law is true for most material objects as long at the strain is small. The proportionality constant is a characteristic of the material and is tabulated in various places, such as the *CRC Tables*.[3] Hooke's law for a long thin rod of length l is

$$\frac{F}{A} = Y\frac{\Delta l}{l},$$
(18.3)

where F is the tension, A is the cross-sectional area of the rod, and Y is a quantity characteristic of the composition of the rod called *Young's modulus*. Δl is the increase in length of a rod of length l when subjected to a stress F/A. If the rod is already under tension, then the left-hand side is the additional stress applied to it.

For a compression, Hooke's law is

$$\frac{F}{A} = -B\frac{\Delta V}{V},$$
(18.4)

[3] *CRC Tables of Chemistry and Physics*, 100th ed., CRC Press, Boca Raton, FL, 2019.

where B is called the *bulk modulus*. The quantity F/A is usually called the pressure, as it corresponds to the idea that an increase in the pressure will produce a compression. The negative sign in the law emphasizes that an increase in pressure causes a decrease in volume. Therefore, the bulk modulus is usually defined in term of the pressure, thus $B \equiv -V \frac{dp}{dV}$.

Finally, for a shear, Hooke's law is

$$\frac{F}{A} = n \tan \theta, \tag{18.5}$$

where n is called the *shear modulus*.

The Stress Tensor

You have certainly noticed that the left-hand side of the Hooke's law equations all have the same form although they represent different situations, specifically the direction of the force and the orientation of the surface relative to the force. It turns out that we can put all of these into a single expression by defining a quantity called the *stress tensor*.

We are considering forces that act across surfaces. In general, the material on one side of a real or imaginary surface exerts a force on the material on the other side. For example, the molecules on one side of an imaginary plane in a material body will exert forces on the molecules on the other side. Of course, we are usually interested in real surfaces with different materials on either side. The forces exerted across a surface, are called stresses. They can be a tension, a compression, or a shear.

The force acting across a surface can have an arbitrary direction and in general is given by

$$\mathbf{F} = F_x \hat{\mathbf{i}} + F_x \hat{\mathbf{j}} + F_z \hat{\mathbf{k}}.$$

The *orientation* of a surface is usually specified by a unit vector perpendicular to the surface. How can we describe a force in the x-direction acting across a surface oriented perpendicular to the y-direction? As you might guess, this could be written in a variety of ways, such as $(F_x)_y$, or some such convention. Because the stress is a force divided by an area let us use the letter P to represent the possible stresses and denote them by P_{xy}. This will be the force per unit area acting in the x-direction across a surface whose normal is in the y-direction. There are nine such terms and they can be expressed quite nicely in matrix form, thus:

$$\begin{pmatrix} P_{xx} & P_{xy} & P_{xz} \\ P_{yx} & P_{yy} & P_{yz} \\ P_{zx} & P_{zy} & P_{zz} \end{pmatrix}.$$

This is called the *stress tensor*. A matrix is an array of numbers. It need not have any physical significance. A tensor can be represented as a matrix, and as we shall see, a tensor obeys the rules of matrix algebra. But a tensor is not a matrix. A tensor is a physical quantity which is often *represented* as a matrix, but it can be represented in other ways as well. A tensor has much in common with a vector. In fact, a vector is a special kind of tensor. Note that a vector is a physical quantity with three terms. A tensor is a physical quantity with nine terms. We considered tensors in some detail in Chapter 17, but for now, I just wanted to mention the concept and make you aware that stress is a tensor.

18.6 The Centroid (Optional)

A quantity closely related to center of mass is the *centroid* of a geometrical figure. For an object of constant density, the two points coincide. However, the centroid is a purely geometrical concept and does not depend on the density distribution of the body. The centroid depends only on the

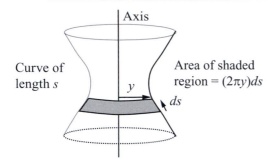

Figure 18.9 A curve of length s is revolved about an axis to generate a surface of revolution.

size and shape of the geometrical figure. If **R** represents the location of the centroid, then it is defined for a volume, a surface, and a curve as follows:

For a volume,

$$\mathbf{R} = \frac{1}{V} \int \mathbf{r} d\tau.$$

The centroid of a surface of area A is given by

$$\mathbf{R} = \frac{1}{A} \int \mathbf{r} d A,$$

and the centroid of a curve of length S is given by

$$\mathbf{R} = \frac{1}{S} \int \mathbf{r} d s.$$

In each of the preceding equations, **r** is the vector from the origin to the element of volume, area, or length.

Some theorems concerning centroids have come down from antiquity and are known as the *Theorems of Pappus*.

Theorem 18.5 *A "surface of revolution" is generated by revolving a plane curve about a line. If a plane curve is revolved about an axis in its own plane and does not intersect the axis, the area of the surface of revolution is equal to the product of the length of the curve and the distance traced out by the centroid of the curve.*

Proof Consider a small portion of the curve, ds, a distance y from the axis. See Figure 18.9. As the curve is revolved about the axis, ds sweeps out an area $d A = 2\pi y ds$. The total area swept out is $A = \int 2\pi y ds = 2\pi \int y ds$. But the centroid is at $Y = (1/s) \int y ds$, so $A = 2\pi Y s$, and the theorem is proved.

Theorem 18.6 *A "solid of revolution" is generated by revolving a planar surface around an axis in its own plane. If the surface does not intersect the axis, the generated solid has a volume equal to the product of the area of the surface and the distance traced out by the centroid of the surface.*

The proof of this theorem is left as an exercise.

As an example of the use of the theorems of Pappus consider the problem of determining the centroid of a semicircular disk. Let the disk be rotated about an axis lying on the straight side of the disk, as shown in Figure 18.10. The volume of revolution generated is a sphere of volume $V = (4/3)\pi a^3$. But according to Theorem 18.6, this volume is equal to the area of the disk times $2\pi Y$, where Y is the distance from the axis to the centroid of the shaded disk. That is

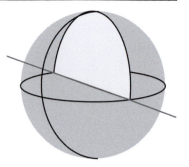

Figure 18.10 A semicircular disk is rotated about an axis to generate a sphere.

$$(4/3)\pi a^3 = (\text{area of disk})\, 2\pi Y = \left(\frac{1}{2}\pi a^2\right) 2\pi Y,$$

$$\therefore Y = \frac{4a}{3\pi}.$$

Exercise 18.9

Prove Pappus' second theorem (Theorem 18.6). ∎

18.7 The Center of Gravity (Optional)

The center of gravity is often confused with the center of mass. In a uniform gravitational field, the center of gravity and the center of mass lie at the same point. Often engineering drawings will have a point denoted "CG" but almost invariably this point is the center of mass rather than the center of gravity. The difference between these two concepts becomes clear when you consider their definitions.

You know the definition of the center of mass from Equation (5.16). It is related to how the mass is distributed in an extended body. On the other hand, the center of gravity is defined to be that point at which all of the mass of the body can be concentrated to experience the same *gravitational force* as the extended body.

As an example, think of a very long rod. The center of mass of the rod does not depend on its orientation. But the center of gravity of the rod will depend on whether the rod is held horizontally, parallel to the surface of the Earth, or vertically, perpendicular to the surface of the Earth. If the rod is horizontal, the center of mass and the center of gravity will coincide; if the rod is vertical, the center of gravity will be a little bit below the center of mass. This is because the gravitational force decreases with distance from the center of the Earth and those parts of the rod nearest the Earth are attracted a bit more strongly than the points of the rod that are further away. Of course, the rod would have to be extremely long for the distance between the two points (CG and CM) to be at all noticeable.

The center of gravity of an extended body of mass M can be defined in terms of its interaction with a particle (say m). The extended body and the particle are illustrated in Figure 18.11. Suppose the extended body shrinks to a point. Where should this point mass be located to experience the same force as the extended body? The location of this hypothetical point mass is called the center of gravity of M relative to m. In Figure 18.11 it is indicated by the symbol CG.

To determine the location of the center of gravity of M, note that the force vectors (call them $\mathbf{F}_i$) on all the mass elements in M are directed along lines that go through m. This is illustrated

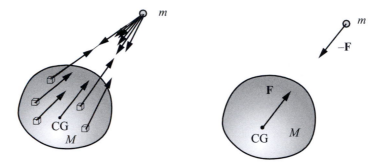

Figure 18.11 The center of gravity of the extended body is at the point CG. The left-hand panel shows that every element in M is attracting m. The net force is the sum of all the force vectors. The right-hand panel shows the action–reaction forces assuming that the extended body can be replaced by a point mass M at CG.

in the left-hand side of Figure 18.11. The net force on m is

$$-\sum \mathbf{F}_i = -Gm \sum \frac{dM_i}{r_1^2}\hat{\mathbf{r}}_i,$$

where dM_i is the mass of a mass element in M and $\hat{\mathbf{r}}_i$ is the unit vector pointing from dM_i to m.

By Newton's third law, the force exerted by m on M is

$$\mathbf{F} = Gm \int \frac{\rho d\tau}{r^2}\hat{\mathbf{r}}.$$

If all of the mass of M were concentrated at the center of gravity, the force on it would be

$$\mathbf{F} = Gm\frac{M}{r_{cg}^2}\hat{\mathbf{r}}_{cg}.$$

Equating the expressions for $\mathbf{F}$ we obtain the following formula for r_{cg}, the position of the center of gravity:

$$r_{cg} = \left[\frac{1}{M}\int \frac{\rho dV}{r^2}\right]^{-1/2}.$$

The point CG will coincide with the center of mass of M only if the body has spherical symmetry.

We have defined the center of gravity of an extended body relative to a *particle*. For two arbitrarily shaped bodies one cannot, in general, define a unique center of gravity for either body.

18.8 D'Alembert's Principle and Virtual Work

An important technique used by mechanical engineers to determine equilibrium conditions and to evaluate forces of constraint is based on the principle developed by the French mathematician and philosopher, Jean d'Alembert (1717–1783).

Recall that a constraint is an equation relating generalized coordinates (see Section 4.5). A constraint that involves only the generalized coordinates (and not the generalized velocities or any other variables) is called a "holonomic" constraint. Such a constraint can be used to express one coordinate in terms of the others, thus reducing the number of coordinates required to describe the motion (and, consequently, reducing by one the number of Lagrange equations that need to be solved).[4]

The principle of d'Alembert is, in a sense, a restatement of Newton's second law, but it is expressed in a way that makes it very useful in advanced mechanics. For a system of N particles, d'Alembert's principle is

$$\sum_{i=1}^{N} \left(\mathbf{F}_i^{ext} - \dot{\mathbf{p}}_i\right) \cdot \delta\mathbf{r}_i = 0, \tag{18.6}$$

where $\mathbf{F}_i^{ext}$ is the external force acting on particle i and $\dot{\mathbf{p}}_i$ is the rate of change of its momentum. By Newton's second law it is clear that the term in parenthesis is zero. Therefore, multiplying this term by $\delta\mathbf{r}_i$ has no effect. Furthermore, summing over all particles is just a sum of terms that are all equal to zero. Thus the fact that the expression is equal to zero is certainly true. What is not obvious (yet) is why this particular formulation is of any value.

It is convenient to express d'Alembert's principle in terms of Cartesian coordinates. Then d'Alembert principle in scalar form is

$$\sum_{i=1}^{n=3N} \left(F_i^{ext} - \dot{p}_i\right) \delta x_i = 0. \tag{18.7}$$

The quantity $\delta\mathbf{r}_i$ in Equation (18.6) or δx_i in Equation (18.7) is called a *virtual displacement*. (See Section 4.7.) A virtual displacement is one that has the following properties: It is instantaneous (that is, $dt = 0$), it is infinitesimal, and it is consistent with any constraints acting on the system. This means that time is kept constant and the forces and constraints acting on the system are "frozen" during the virtual displacement. The work done during a virtual displacement is called the "virtual work." The *principle of virtual work* states that *for a system in equilibrium,* the work done in any virtual displacement is zero, that is,

$$\delta W = \sum_i Q_i \delta q_i = 0, \tag{18.8}$$

where Q_i is the generalized force, defined by[5]

$$Q_j = \sum_i F_i \frac{\partial x_i}{\partial q_j}.$$

Because δW can also be expressed as $\Sigma F_i \delta x_i$, we appreciate that $\Sigma F_i \delta x_i = \Sigma Q_j \delta q_j$. This is a useful relationship for determining the equilibrium conditions for complicated systems.

Worked Example 18.4 Two smooth inclined planes are joined at the top. The planes form angles of θ_1 and θ_2 with the horizontal, as shown in Figure 18.12. Two masses, m_1 and m_2 are connected by a string over an ideal pulley. Using the principle of virtual work, determine the equilibrium condition.

Solution Assume a virtual displacement δx in which mass m_1 moves up one plane and mass m_2 moves down the other plane. The force of gravity acting on m_1 in this direction is $-m_1 g \sin\theta_1$ and the force on the second mass is $+m_2 g \sin\theta_2$. According to the principle of

---

[4] If the constraints acting on a system are holonomic they can be determined by using a technique called the method of Lagrange multipliers. This topic is fairly advanced and we will not consider it further.
[5] See Section 4.7.

virtual work,

$$-m_1 g \sin \theta_1 \delta x + m_2 g \sin \theta_2 \delta x = 0.$$

Therefore, equilibrium requires that

$$\frac{m_1}{m_2} = \frac{\sin \theta_2}{\sin \theta_1}.$$

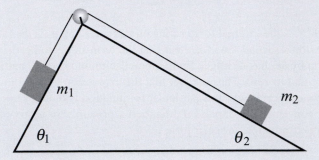

Figure 18.12 Two masses connected by a string on inclined planes. Find the equilibrium condition.

Exercise 18.10

Consider an Atwood's machine with two equal masses. Show that Equation (18.8) holds for this system.   ■

18.8.1 A Derivation of Lagrange's Equations

An interesting aspect of d'Alembert's principle is that it leads to a different way of deriving Lagrange's equations. We begin by considering the second term in d'Alembert's principle, namely $\dot{p}_i \delta x_i$ If $x_i = x_i(q_1, \ldots, q_k)$ then

$$\delta x_i = \frac{\partial x_i}{\partial q_1} \delta q_1 + \frac{\partial x_i}{\partial q_2} \delta q_2 + \cdots + \frac{\partial x_i}{\partial q_k} \delta q_k = \sum_{j=1}^{k} \frac{\partial x_i}{\partial q_j} \delta q_j,$$

so the sum of $\dot{p}_i \delta x_i$ over all particles is

$$\sum_{i=1}^{n} \dot{p}_i \delta x_i = \sum_{i=1}^{n} \dot{p}_i \sum_{j=1}^{k} \frac{\partial x_i}{\partial q_j} \delta q_j = \sum_{i=1}^{n} \sum_{j=1}^{k} m_i \ddot{x}_i \frac{\partial x_i}{\partial q_j} \delta q_j. \tag{18.9}$$

Note that

$$\frac{d}{dt} \left(m_i \dot{x}_i \frac{\partial x_i}{\partial q_j} \right) = m_i \ddot{x}_i \frac{\partial x_i}{\partial q_j} + m_i \dot{x}_i \frac{d}{dt} \frac{\partial x_i}{\partial q_j}. \tag{18.10}$$

Consequently,

$$\sum_{i,j} m_i \ddot{x}_i \frac{\partial x_i}{\partial q_j} \delta q_j = \sum_{i,j} \left[\frac{d}{dt} \left(m_i \dot{x}_i \frac{\partial x_i}{\partial q_j} \right) - m_i \dot{x}_i \frac{d}{dt} \frac{\partial x_i}{\partial q_j} \right] \delta q_j. \tag{18.11}$$

The last term can be written as

$$m_i x_i \frac{d}{dt} \frac{\partial x_i}{\partial q_j} = m_i x_i \left[\sum_j \left(\sum_k \frac{\partial^2 x_i}{\partial q_j \partial q_k} \dot{q}_k \right) + \frac{\partial^2 x_i}{\partial q_j \partial t} \right]$$

$$= m_i x_i \sum_j \frac{\partial}{\partial q_j} \left[\left(\sum_k \frac{\partial x_i}{\partial q_k} \dot{q}_k \right) + \frac{\partial x_i}{\partial t} \right]$$

$$= m_i x_i \sum_j \frac{\partial \dot{x}_i}{\partial q_j},$$

where we used the fact that

$$\dot{x}_i = \sum_k \frac{\partial x_i}{\partial q_k} \dot{q}_k + \frac{\partial x_i}{\partial t}. \tag{18.12}$$

Using these relations, Equation (18.9) becomes

$$\sum_i \dot{p}_i \delta x_i = \sum_{i,j} \left[\frac{d}{dt} \left(m_i \dot{x}_i \frac{\partial x_i}{\partial q_j} \right) - m_i \dot{x}_i \frac{\partial \dot{x}_i}{\partial q_j} \right] \delta q_j,$$

$$= \sum_{i,j} \left[\frac{d}{dt} \left(m_i \dot{x}_i \frac{\partial \dot{x}_i}{\partial \dot{q}_j} \right) - m_i \dot{x}_i \frac{\partial \dot{x}_i}{\partial q_j} \right] \delta q_j. \tag{18.13}$$

In the last equation we applied the useful relation

$$\frac{\partial x_i}{\partial q_j} = \frac{\partial \dot{x}_i}{\partial \dot{q}_j}.$$

Now the kinetic energy is $T = \sum (1/2) m \dot{x}_i^2$, so

$$\frac{\partial T}{\partial q_j} = \frac{\partial}{\partial q_j} \sum_i \left(\frac{1}{2} m_i \dot{x}_i^2 \right) = \sum_i m_i \left(\dot{x}_i \frac{\partial \dot{x}_i}{\partial q_j} \right),$$

and

$$\frac{\partial T}{\partial \dot{q}_j} = \frac{\partial}{\partial \dot{q}_j} \sum_i \left(\frac{1}{2} m_i \dot{x}_i^2 \right) = \sum_i m_i \left(\dot{x}_i \frac{\partial \dot{x}_i}{\partial \dot{q}_j} \right).$$

Comparing with the right-hand side of Equation (18.13) we appreciate that

$$\sum_i \dot{p}_i \delta x_i = \sum_j \left[\frac{d}{dt} \frac{\partial T}{\partial \dot{q}_j} - \frac{\partial T}{\partial q_j} \right] \delta q_j.$$

Finally, d'Alembert's principle reads

$$\sum_{i=1}^N \left(F_i^{ext} - \dot{p}_i \right) \delta x_i = \sum_j Q_j \delta q_j - \sum_j \left[\frac{d}{dt} \frac{\partial T}{\partial \dot{q}_j} - \frac{\partial T}{\partial q_j} \right] \delta q_j = 0,$$

$$0 = \sum_j \left(Q_j - \frac{d}{dt} \frac{\partial T}{\partial \dot{q}_j} + \frac{\partial T}{\partial q_j} \right) \delta q_j.$$

Using the fact that the δq_j are independent it is clear that each term in the summation must individually be equal to zero, so

$$\frac{d}{dt}\frac{\partial T}{\partial \dot{q}_j} - \frac{\partial T}{\partial q_j} = Q_j. \tag{18.14}$$

This derivation gives Lagrange's equations in the Nielsen form involving the generalized forces and the kinetic energy. (You have seen this alternate form of Lagrange's equations previously in Worked Example 4.6.)

Exercise 18.11

Starting with Equation (18.14) and assuming the generalized forces are derivable from a potential that is independent of the generalized velocities, derive the Lagrange equations in their usual form, as shown in Equation (4.9). ∎

18.9 Summary

This chapter on the statics of extended bodies started with a simple analysis of the conditions for equilibrium, namely that the sum of the forces is zero and the sum of the torques is zero:

$$\sum \mathbf{F}_i = 0,$$
$$\sum \mathbf{N}_i = 0.$$

You were then exposed to a number of theorems and definitions to familiarize you with the concepts and terms used in the study of statics. Thus, for example, you were given the definitions of a couple, a resultant, and an equilibrant. You found out that any system of forces can be reduced to a single force plus a couple.

The next topic was more advanced, being a study of a hanging cable. There were two applications, namely, a suspension bridge and a rope hanging under its own weight. The shape of the cable suspending a bridge is a parabola,

$$y = \frac{w}{2T_c}x^2 + c,$$

and the shape of a rope hanging under its own weight is a catenary

$$y = c \cosh\left(\frac{x}{c}\right).$$

Hooke's law states that stress and strain are proportional. Defining stress as force per unit area, leads to the three expressions in the table.

Law	Stress	Proportionality constant
$F/A = Y\,\Delta l/l$	tension	$Y =$ Young's modulus
$F/A = -B\,\Delta V/V$	pressure	$B =$ Bulk modulus
$F/A = n\tan\theta$	shear	$n =$ Shear modulus

A consideration of the orientation of the surface across which the force acts led to the concept of the stress tensor which was described but not utilized in the analysis.

The centroid is a geometrical concept similar to the physical concept of center of mass. The centroids of a volume, a surface, and a line were defined. Some properties of centroids were summarized in the theorems of Pappus.

The *center of gravity* is the point in an extended body of mass M at which a point mass M would exert the same gravitational force on a second point mass m as does the extended body.

Finally, you were exposed to d'Alembert's principle and virtual work, a rather advanced topic in the study of statics. We defined virtual work and noted that for a system in equilibrium the virtual work done in a virtual displacement is zero. We also showed how d'Alembert's principle can be used to derive Lagrange's equations.

18.10 Problems

Problem 18.1 A work of art consists of a rod of length L and mass M. It is suspended by thin wires at either end. The wires are connected to two walls and rod "floats" horizontally between the walls. The angles formed by the wires with the horizontal are θ and ϕ. Determine the angle formed by the rod with the horizontal. Determine the tensions in the wires.

Problem 18.2 A rigid rod of mass M and length L is supported from a single point by two strings attached to the ends of the rod. Each string has length L. A weight, also having mass M, is hung from one end of the rod. Find the angle the rod makes with the horizontal and determine the tensions in the strings. See Figure 18.13.

Problem 18.3 A smooth rod that is nailed to the floor and to the wall, makes an angle of $30°$ to the horizontal, as shown in Figure 18.14. A ring of mass m can slide on the frictionless rod. A string is passed through the ring; one end of the string is attached to a point on the floor and the other end of the string supports a weight of mass $3\,m$. Determine the angle ϕ between the two segments of the string. Note, the rod does not move, but the ring can slide along it.

Problem 18.4 A rigid flat plate 2 meters on a side is acted upon at its corners by the following forces: $\mathbf{F}_1 = 10\hat{\jmath}$ acting at $(-1,1)$, $\mathbf{F}_2 = 5\hat{\imath} - 5\hat{\jmath}$ acting at $(1, -1)$, $\mathbf{F}_3 = 10\hat{\jmath}$ acting at $(-1, -1)$, and $\mathbf{F}_4 = 10\hat{\imath}$ at $(1,1)$. The forces are in newtons. The center of the plate is at $(0,0)$. Determine the resultant and the equilibrant.

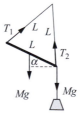

Figure 18.13 A suspended rod. (Problem 18.2.)

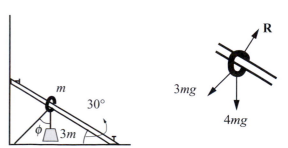

Figure 18.14 The sketch on the left shows the setup and the sketch on the right shows the forces on the ring of Problem 18.3.

Figure 18.15 Billiard balls in a cylindrical pipe. (Problem 18.5.)

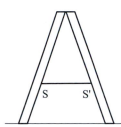

Figure 18.16 A stepladder.

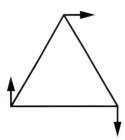

Figure 18.17 An equilateral triangle subjected to three forces of equal magnitude.

Problem 18.5 An open cylindrical pipe of mass M and radius a is set upright on a table. Two billiard balls of mass m and radius r are dropped into the pipe. Assume $r > a/2$. Show that the pipe will not tip over if $M \geq 2m(1 - \frac{r}{a})$. See Figure 18.15.

Problem 18.6 A stepladder is made from two ladders of length l and mass m by joining them at the top with a frictionless pivot. A horizontal string of length s is attached at points S, S', a third of the way up the ladders. The system rests on a smooth floor. What is the tension in the string? See Figure 18.16.

Problem 18.7 Consider the rod of Figure 18.13. If the string with tension T_2 breaks, the rod will rotate and translate until it is hanging straight down. Determine the equilibrant of the forces acting on the rod at the instant the string breaks.

Problem 18.8 A ladder of weight W and length l stands on a smooth floor and leans against a smooth wall. The angle between the ladder and the floor is α. As you might expect, because there is no friction, the ladder is slipping. What is the equilibrant? (The expression you will obtain includes the normal forces.)

Problem 18.9 A rod of length 1 meter has strings attached at either end and at the middle. The string attached to the left end makes an angle of $30°$ to the rod and exerts a force of 10 N on it. The string attached to the middle makes an angle of $270°$ to the rod and exerts a force of 15 N. The string on the right end makes an angle of $90°$ and exerts a force of 10 N. All the forces lie in the same plane. Determine the resultant and the equilibrant.

Problem 18.10 An equilateral triangle, 1 meter on a side, is subjected to three 10 N forces, as shown in Figure 18.17. (a) Describe the motion of the object. (b) Find the equilibrant.

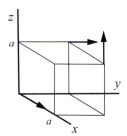

Figure 18.18 A cube subjected to forces of equal magnitude but different directions at some of its corners. (Problem 18.11.)

Problem 18.11 A cube of side a has one corner at the origin and the diagonally opposite corner at (a,a,a). The sides are oriented along the axes. The forces acting on the cube are all equal in magnitude to F. The force on the corner at $(0,0,0)$ is $F\hat{\mathbf{i}}$, the force on the corner at $(0,a,a)$ is $F\hat{\mathbf{j}}$ and the force on the corner at (a,a,a) is $F\hat{\mathbf{k}}$. Find an equivalent single force and couple. (See Figure 18.18.)

Problem 18.12 Expanding the equation for a catenary (Equation (18.2)) show that if the tension is large, $y \simeq c + x^2/2c$.

Problem 18.13 Consider a suspended cable in which the cross-sectional area of the cable varies, being smallest at the midpoint and greatest at the points of support (assumed to be at the same height). The thickness of the cable is designed such that the tension per unit area of the cross section is constant along the cable. That way it will have equal probability of breaking at all points. Note that for such a cable the tension is proportional to the area and hence to the weight per unit length, w. Hence, $T = kw$, where k is a constant of proportionality and w varies along the length of the cable. (a) Show that such a cable forms a curve given by

$$y = k \log \sec \frac{x}{k} + c,$$

where c is a constant of integration. (b) Show the maximum possible span for such a cable is πk.

Problem 18.14 Using the Taylor's series expansion for the potential, show that it leads to Hooke's law for small displacements from equilibrium. (For simplicity, you may assume a one-dimensional deformation.)

Problem 18.15 Obtain a formula for the volume of a torus. (Hint: Use a theorem of Pappus.)

Problem 18.16 Two uniform spheres of mass M are located at $x = \pm a$. Determine the center of gravity relative to a point P on the y-axis. Show that the center of gravity and the center of mass approach one another as the distance to P increases.

Problem 18.17 A satellite consists of two spheres (which we approximate as particles) each having mass m. They are connected by a massless rod of length l. The satellite is in a circular orbit of radius r about the Earth with both masses lying along the same radial line from the center of the Earth. Determine the distance h between the center of mass and the center of gravity of this satellite.

Problem 18.18 A cylindrical rod of length 20 km is held perpendicular to the surface of the Earth. Determine the distance between its center of mass and center of gravity. (Assume the Earth is a point mass located at its center.) (Hint: Use the binomial expansion).

Problem 18.19 Prove that if $x_i = x_i(q_1, q_2, \ldots, q_n, t)$ then

$$\frac{\partial x_i}{\partial q_j} = \frac{\partial \dot{x}_i}{\partial \dot{q}_j}.$$

Problem 18.20 Show that

$$\sum_i F_i \delta x_i = \sum_j Q_j \delta q_j.$$

Problem 18.21 Assume that the two tilted surfaces of Figure 18.12 are not smooth, and the coefficient of friction is μ. The system is in equilibrium. Use the principle of virtual work to obtain an expression for the coefficient of friction.

Computational Problems

Computational Techniques: Solving a System of Linear Equations

Statics problems often involve solving a system of linear equations. Computer programming languages, such as Matlab and Python, usually have built-in functions that allow one to solve such equations very simply. For example, consider the system of three equations in three unknowns

$$a_{11}x_1 + a_{12}x_2 + a_{13}x_3 = b_1,$$
$$a_{21}x_1 + a_{22}x_2 + a_{23}x_3 = b_2,$$
$$a_{31}x_1 + a_{32}x_2 + a_{33}x_3 = b_3.$$

Here the a's and b's are known quantities and one wishes to determine the x's. Using matrix notation, these equations can be written

$$\begin{pmatrix} a_{11} & a_{12} & a_{13} \\ a_{21} & a_{22} & a_{23} \\ a_{31} & a_{32} & a_{33} \end{pmatrix} \begin{pmatrix} x_1 \\ x_2 \\ x \end{pmatrix} = \begin{pmatrix} b_1 \\ b_2 \\ b_3 \end{pmatrix}$$

or

$$\mathcal{A}\mathbf{x} = \mathbf{B}.$$

Then, in the syntax of Matlab,

$$\mathbf{x} = \mathcal{A} \backslash \mathbf{B}.$$

(Note the backslash. Without going into the details, the process of solving for the vector $\mathbf{x}$ uses a technique called Gaussian elimination that basically solves the first equation for x_1 and substitutes it into the remaining equations. Then it solves the second equation for x_2 and substitutes it into the remaining equations, and so on, until the last x is obtained in terms of the coefficients. Finally, "back substitution" generates the values of all the $x's$.)

Computational Project 18.1 In Figure 18.13 assume the lengths of the rod and strings are 80 cm. Let the hanging rod have mass 12 kg. Write a program to determine T_1, T_2, and the hanging mass by solving the three coupled equations for different (input) values of α.

Computational Project 18.2 We found the equation for a rope hanging under its own weight is $y = c \cosh\left(\frac{x}{c}\right)$. Recall that

$$\cosh u = \frac{1}{2}\left(e^u + e^{-u}\right).$$

Plot e^u and e^{-u} and $\frac{1}{2}\left(e^u + e^{-u}\right)$ to generate a catenary. Show that the lowest point in the catenary is a distance c above the origin.

19 Fluid Dynamics and Sound Waves (Optional)

19.1 Introduction

In this chapter we consider the basic concepts of the statics and dynamics of fluids. As the name indicates, a fluid is any substance that flows, such as liquids and gases.[1] The general categories of our study are *fluid statics* or *hydrostatics* concerning the behavior of fluids at rest, and *fluid dynamics* which is a study of the motion of fluids and of objects moving with respect to fluids. This is further subdivided into studies of the dynamics of liquids and gases or *hydrodynamics* and *gas dynamics*.

Liquids are very nearly incompressible, whereas gases are easily compressed. What they have in common, however, is the ability to transmit a stress, specifically a pressure. In liquids this is due to intermolecular forces whereas in a gas, pressure is transmitted by collisions between the molecules. This ability to transmit a pressure allows one to treat a fluid as a unified system rather than as a collection of independent particles.

The viscosity of a fluid is an important property that measures the "stickiness" of the fluid and affects its ability to flow. As you know, a highly viscous fluid (such as honey) does not flow easily, whereas a low viscosity fluid like water or alcohol flows readily. However, the static and dynamical properties of a fluid can be most easily analyzed by assuming an ideal nonviscous fluid, so we will often ignore viscosity. (A nonviscous fluid cannot wet a surface, so Feynman referred to a nonviscous fluid as "dry water."[2])

19.2 Equilibrium of Fluids (Hydrostatics)

For fluids, the definition of *equilibrium* is more restrictive than for solids, because we require that a fluid in equilibrium be *at rest*. By definition, a fluid (such as a gas or liquid) is a substance that cannot support a shear. Even a very viscous fluid like tar will slowly deform under a shear until it reaches an equilibrium state in which the shear is zero. Therefore the only stress forces acting on a fluid in equilibrium are forces perpendicular to the surface, namely, pressures. Because hydrostatics refers to fluids at rest, the results are valid for viscous as well as nonviscous fluids.[3]

---

[1] Some substances that we think of as solids also flow slowly (amorphous solids) when subjected to shear stresses. You sometimes hear that the glass in medieval cathedrals is thicker at the bottom because it is a supercooled liquid, however, this appears to be false. See "Fact or Fiction? Glass Is a (Supercooled) Liquid?" by Ciara Curtin, *Scientific American,* February 2007, www.scientificamerican.com/article/fact-fiction-glass-liquid/.

[2] R. Feynman, *The Feynman Lectures on Physics*, Vol. 2, Chapter 40, Addison-Wesley, Reading, MA, 1964.

[3] Stress and strain are defined and discussed in Section 18.5.

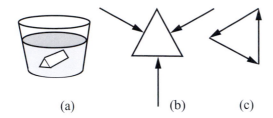

Figure 19.1 (a) An infinitesimal volume element of fluid in the shape of an equilateral prism. (b) The forces on the rectangular faces of the prism are perpendicular to the surfaces. (c) The forces add up to zero, so the magnitudes of the forces are all the same.

When I said that in a fluid in equilibrium the forces are perpendicular to the surface, I did not just mean the exposed surfaces of the fluid, but *any* surface, including surfaces entirely within the fluid. For example, you can imagine a surface lying between two layers of molecules in the fluid. The material on one side of this (imaginary) surface exerts a force on the material on the other side, and the force is perpendicular to the surface. If not, there would be a shear and by definition the fluid would not be in equilibrium.

Imagine a large tank full of some fluid, say water. You could determine the pressure at any point in the tank by lowering a pressure sensor gauge into the tank. The fluid exerts a force on the surface of the pressure sensor and the force per unit area is the pressure at that point in the fluid. It is an interesting fact that the pressure in a fluid is the same in all directions. If you rotate the pressure sensor so that it faces a different direction, it reads the same value. That is, the pressure is the same regardless of the orientation of the surface. To appreciate why this is so, consider an infinitesimal portion of the fluid. Let this portion have the shape of an elongated prism, as shown in Figure 19.1(a). The prism-shaped fluid element is at rest so the net force acting on it must be zero. That is, the forces acting normal to the rectangular surfaces of the fluid element must sum vectorially to zero, as shown in Figures 19.1(b) and (c). These three forces are equal in magnitude, and since the surfaces are all the same, the pressure on each surface is the same. Thus, the pressure is independent of the orientation of the faces.

The prism in the previous argument was assumed infinitesimal to avoid including the weight of the fluid in the prism.[4] However, if gravity (or indeed any body force) acts on the fluid in equilibrium, then in addition to the pressure there is a force on a *finite* volume element ΔV given by

$$\mathbf{F} = m\mathbf{g} = (\rho \Delta V)\mathbf{g},$$

where ρ is the density of the fluid. The body force $\mathbf{f}$, defined as the force per unit volume, is

$$\mathbf{f} = \rho \mathbf{g} = -\rho g \hat{\mathbf{k}}.$$

To obtain an expression for the pressure as a function of position, you can construct an (imaginary) upright cylinder at some arbitrary point in the fluid. Let the axis of the cylinder be vertical, that is, in the direction of the gravitational force. The top and bottom faces have area dA and the distance between the faces is dz. There are three forces acting on the fluid in the cylinder, namely: (1) the downward force due to the pressure on the top surface, (2) the upward force due to the pressure on the bottom surface, and (3) the downward force of gravity on the fluid enclosed. Let the bottom of the cylinder be at z and the top at $z + dz$. See Figure 19.2. Then the equilibrium condition

force up = force down

---

[4] The body force (weight) is proportional to the volume whereas the surface force is proportional to the area. As the fluid element shrinks to zero, the volume decreases faster than the area (volume $\propto r^3$, and area $\propto r^2$). Therefore, the body force can be neglected for an infinitesimal volume element.

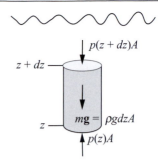

Figure 19.2 The cylinder immersed in a fluid is acted upon by stresses (pressure) on the top and bottom and by the body force (weight).

is expressed by

$$p(z)A = p(z + dz)A + \rho g\,dz\,A.$$

Using the definition of derivative, this expression can be written

$$\frac{dp}{dz} = -\rho g. \tag{19.1}$$

That is,

$$\frac{dp}{dz}\hat{\mathbf{k}} = \mathbf{f}.$$

It is easy to generalize this expression to obtain an expression for the change in pressure with respect to a displacement in an arbitrary direction, leading to

$$\nabla p = \mathbf{f}. \tag{19.2}$$

That is, the gradient of the pressure is equal to the body force. This implies that the surfaces of constant pressure are perpendicular to the body force.

Using the vector definition of differential, we have

$$\int \nabla p \cdot d\mathbf{r} = \int dp = \int \mathbf{f} \cdot d\mathbf{r},$$

so

$$p(r) = p(r_0) + \int_{r_0}^{r} \mathbf{f} \cdot d\mathbf{r}, \tag{19.3}$$

where the integral is along any path from r_0 to r in the fluid. Note that for a fluid in equilibrium, Equation (19.3) defines a pressure field (the pressure is defined at every point in the fluid). Furthermore, the equation states that pressure changes are balanced by the body force, $\mathbf{f}$.

Equation (19.3) also tells us that if the body force is constant, increasing the pressure at one point in a fluid increases the pressure by an equal amount at all other points in the fluid. This is called Pascal's law. For example, if $p(r_0)$ is the pressure at the surface of the fluid, then an increase in the surface pressure is communicated to all points in the fluid.

Exercise 19.1

Sea-level atmospheric pressure is 1.01×10^5 Pa. At the Golden Gate of San Francisco Bay, the water depth is about 100 m. Determine the pressure at that depth. The density of sea water is 1024 kg/m^3. Answer: 1.10×10^6 Pa. ∎

Exercise 19.2

A beaker is filled to the 2 cm mark with mercury and then an equal amount of water is added. What is the pressure at the bottom? (The specific gravity of mercury is 13.6.) Answer 1.04×10^5 Pa. ∎

Exercise 19.3

Show that for a fluid with constant density the pressure increases with depth as $p = p_0 + \rho g h$, where h is the depth. Determine the pressure a distance 3 m below the surface of a lake. Answer: 1.3×10^5 Pa. ∎

Exercise 19.4

A cylinder of radius 50 cm with an open nozzle on the top, is filled with water to a height of 1.5 m. A plug of radius 1 cm is at the bottom of the cylinder. (a) Determine the net force on the plug if the air pressure is 1 atmosphere (1 atm = 1.01×10^5 Pa and 1 Pa = 1 N/m^2.) (b) The nozzle is now attached to a pump and air is pumped in until the pressure reaches 3 atm. What is the force on the plug now? Answer: (a) 36.4 N. ∎

Archimedes' Principle

Consider a fluid in equilibrium. In the interior of the fluid, draw an arbitrary closed surface S enclosing a volume V of the fluid. The gravitational force acting on the fluid in this volume is

$$\mathbf{F}_W = \int_V \mathbf{f} dV = \int_V \mathbf{g}\rho dV = m\mathbf{g} = \text{weight.}$$

$\mathbf{F}_W$ is the weight of the fluid in the volume. The only other force acting on this element of the fluid is due to the pressure exerted on the surface of the element by the fluid outside of V. This can be obtained by integrating over the surface S enclosing V, thus

$$\mathbf{F}_P = - \oint_S \hat{\mathbf{n}} p dA,$$

where $\hat{\mathbf{n}}$ is the outward normal. (That is the reason for the minus sign.)

For equilibrium

$$\mathbf{F}_P = -\mathbf{F}_W.$$

This does not tell you anything unexpected. The weight of the fluid (down) is equal to the (upward) pressure force on it. Now assume the volume of fluid inside S is replaced by a solid body of the same shape and size. Then $\mathbf{F}_W$ is the *weight of the solid body,* rather than the weight of the fluid. The pressure force on the volume is the same as before, as it is due to the fluid *outside* of the body. (The force exerted by the fluid does not depend on the nature of the material in the volume.) Therefore the upward (buoyant) force on the body is equal to the weight of the displaced fluid. This is Archimedes' principle.

Exercise 19.5

Prove that the pressure is independent of direction by adding the Cartesian components of the forces acting on the rectangular faces of the prism of Figure 19.1. ∎

Exercise 19.6

Generalize the argument above to show that if the pressure is a function of position it is related to the body force by $\nabla p = f$. That is, derive Equation (19.2). ∎

Exercise 19.7

Show that Equation (19.2) implies that the body force is conservative. ∎

Exercise 19.8

Consider a cubical iceberg of side a floating in sea water. Determine the height of the exposed portion in terms of a. (Density of sea water is 1024 kg/m^3, $\rho_{ice} = 917$ kg/m^3.) ∎

An Ideal Gas

A liquid is essentially incompressible, but the volume of a gas depends on the pressure exerted on it. The basic relation, based on Hooke's law (stress $\propto$ strain), is given by Equation (18.4) which can be expressed in terms of the pressure as

$$-\frac{dV}{V} = \frac{dp}{B},$$

where B is the bulk modulus of the gas. Let us formulate this relation in terms of density rather than volume. Because $\rho = m/V$, for constant mass,

$$d\rho = -mV^{-2}dV,$$

so

$$\frac{d\rho}{\rho} = -\frac{mdV}{V^2}\frac{V}{m} = -\frac{dV}{V}.$$

Consequently

$$\frac{d\rho}{\rho} = \frac{dp}{B}. \tag{19.4}$$

Integrating yields

$$\rho = \rho_0 \exp\left[\int_{p_o}^{p} \frac{dp}{B}\right].$$

Most gases under standard conditions of temperature and pressure are in reasonably good agreement with the ideal gas law,

$$pV = nRT,$$

where T is the temperature, n is the number of moles, and R is the universal gas constant. Because ρ is defined as the mass divided by the volume and the mass is nM_W (where M_W is the molecular weight), you can replace n/V by ρ/M_W and write the ideal gas law as

$$\rho = \frac{M_W}{RT}p.$$

Inserting this expression into Equation (19.1) leads to the *barometric equation*

$$\frac{dp}{dz} = -\frac{gM_W}{RT}p. \tag{19.5}$$

As an example, consider an "isothermal atmosphere," that is, an atmosphere in which the temperature is constant. This is, of course, not at all like the real atmosphere of the Earth which has a rather complicated temperature structure. Nevertheless it is a "zeroth-order" approximation that gives some useful information. Integrating the equation for dp/dz leads to

$$p = p_0 \exp\left[-\frac{g M_W}{RT}(z - z_0)\right].$$

Here z_0 is usually taken to be sea level and set equal to zero, and p_0 is the pressure at sea level (usually assumed to be 1 atm).

Worked Example 19.1 Suppose the temperature of the atmosphere decreases linearly with altitude: $T = T_0 - \alpha z$. (This is somewhat more realistic than assuming an isothermal atmosphere.) Obtain an expression for the pressure as a function of altitude (z).

Solution Given the barometric equation:

$$\frac{dp}{dz} = -\frac{g M_W p}{RT} = \frac{g M_W p}{R}\frac{1}{T_0 - \alpha z}.$$

So

$$\frac{dp}{p} = -\frac{g M_W}{R}\frac{dz}{T_0 - \alpha z},$$

$$\int_{p_0}^{p}\frac{dp}{p} = -\frac{g M_W}{R}\int_{0}^{z}\frac{dz}{T_0 - \alpha z},$$

$$\ln(p/p_0) = \frac{M_W g}{\alpha R}\ln\left(\frac{T_0 - \alpha z}{T_0}\right) = \frac{M_W g}{\alpha R}\ln\left(1 - \frac{\alpha}{T_0}z\right),$$

and consequently

$$p = p_0\left(1 - \frac{\alpha}{T_0}z\right)^{M_W g/\alpha R}.$$

Exercise 19.9

Integrate the barometric equation to obtain the expression for the pressure as a function of altitude in an isothermal atmosphere. ∎

19.3 Fluid Kinematics

19.3.1 Viscosity, Laminar Flow, and Turbulence

Viscosity in a fluid is similar to friction between two rigid bodies. Fluids with high values of viscosity (honey, heavy oils) do not flow easily whereas fluids with low viscosity (air, water) flow readily. Consider a fluid between two parallel plates, as shown in Figure 19.3, assuming the bottom plate is at rest and the top plate is moving at speed v. The friction between the top plate and the fluid will tend to make the fluid in a thin layer next to the top plate move at speed v, whereas the fluid in contact with the stationary bottom plate will have zero speed. (This fact can be appreciated by noticing that a thin layer of dust on the blade of a fan does not get "blown off" when the fan is turned on.)

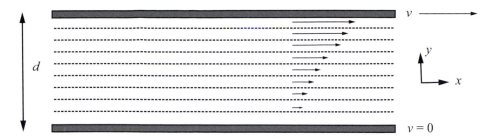

Figure 19.3 Laminar flow. The top layer of fluid moves with the top plate and the bottom layer of fluid is at rest. This is an illustration of viscous drag between two parallel plates.

Experimentally, we find that the force per unit area required to keep the upper plate moving at constant velocity is proportional to the velocity divided by the separation of the plates (d). That is,

$$\frac{F}{A} = \eta \frac{v}{d},$$

or, in the limit,

$$\frac{F}{A} = \eta \frac{dv}{dy}, \tag{19.6}$$

where η is a constant of proportionality depending on the properties of the fluid and called the *coefficient of viscosity*. Equation (19.6) is called Newton's law of viscosity.[5]

Note that F/A is a shear stress, but the strain $\propto$ stress relation has taken on a different form than in Chapter 18 where we defined the shear modulus for solids.

Fluids obeying the relation given by Equation (19.6) are called "true fluids" or "Newtonian fluids." For highly viscous liquids (that is, non-Newtonian fluids) the strain is not simply proportional to the stress, but depends on other factors. For example, ketchup flows more easily when shaken. We will not consider such fluids.

It is convenient to consider the fluid to be composed of flat *layers* with speed varying as shown in Figure 19.3. Adjacent layers exert a viscous force on each other. The fluid above tending to speed up a particular layer, while the fluid below tends to slow it down. If the fluid in a particular layer flows smoothly past the adjacent layers with little or no mixing, the flow is called "laminar."

In laminar flow, the velocity, pressure, and density at each point in the fluid remains constant in time (but varies from one location to another). Laminar flow is a somewhat idealized situation as true laminar flow usually only occurs in high-viscosity fluids at low velocities.

The other extreme type of fluid flow is *turbulent flow* that is characterized by eddies and swirls with mixing and irregular fluctuations. That is, disordered and irregular motion. The criterion for distinguishing between laminar and turbulent flow is the value of a unitless parameter called the "Reynold's number." For flow in a circular pipe the Reynolds number is given by $\mathrm{Re} \equiv \bar{v} D \rho / \eta$, where $\bar{v}$ is the average velocity, D is the diameter of the pipe, ρ is the density of the fluid, and η is the viscosity. Generally speaking, a Reynold's number less than about 2000 is characteristic of laminar flow and Re greater than about 2000 means the flow is turbulent. We will not consider turbulent flow.

---

[5] Just to make things more confusing, in the tables you will find two different values for the viscosity of a fluid. One of these will be called the *dynamic* or *absolute* viscosity. This is usually denoted by the symbol η and is the quantity defined by Equation (19.6). The other quantity is denoted the *kinematic* viscosity or *specific* viscosity and is usually denoted by v. The two quantities are related by $v = \eta/\rho$ where ρ is the density of the fluid.

> **Exercise 19.10**
>
> Determine the speed at which the flow becomes turbulent for water flowing in a 2 cm diameter pipe. The viscosity of water is 10^{-3} N s/m^2. Answer: 0.1 m/s.   ∎

19.3.2 Eulerian and Lagrangian Formulations

Consider a fluid in motion. You can imagine the fluid (a liquid or a gas) to be made up of "fluid particles." This is fairly natural in the case of a gas, but we usually think of a liquid as a continuous substance.[6]

There are two approaches to studying fluid flow called "Lagrangian" and "Eulerian." The Lagrangian approach follows the motion of a fluid particle giving its position as a function of time. If you toss a cork into a river, the equation of motion of the cork would be an example of the Lagrangian approach. On the other hand, the Eulerian approach focuses on a particular location and specifies the properties of the fluid as it flows past this point. An example would be to sit on the bank of a river and watch the behavior of the water at a particular point.

We could write the Lagrange equations for the fluid particles and follow a set of representative particles as they move in the stream. However, it is often easier to analyze the situation from the Eulerian point of view and concentrate on fluid properties such as velocity and density at each point in the fluid. This means that we are treating the fluid properties as *fields*, and at every point in the fluid we define a velocity, a density, a pressure, and so on.[7] Thus, for example, we describe the fluid in terms of the density field, $\rho(x, y, z, t)$, the velocity field, $\mathbf{v}(x, y, z, t)$, the pressure field, $p(x, y, z, t)$, etc.

We will be formulating our analysis mainly in terms of the Eulerian specification; nevertheless it is sometimes necessary to consider a fluid particle because, after all, it is the particles (and not points in space) to which the laws of mechanics apply.

In the Lagrangian description it is useful to think of a fluid particle as a small volume containing a fixed number of molecules of the fluid. As time goes on, this volume element will move with the fluid and it may change shape and size, but its mass will be constant.

In the Eulerian description we consider the fluid as a whole and define the *fields* (pressure, velocity, density, etc.) at every point in the fluid.

Before continuing our analysis we have to consider two very important concepts: the *convective derivative* and the *equation of continuity*.

19.3.3 The Convective Derivative

To determine the convective derivative we note that the properties of a fluid (pressure, density, and so on) will change in time. The rate of change will be due to two processes: (1) changes that take place at a particular fixed point, and (2) changes that are due to the translation of a fluid element in the moving fluid. Consider, for example, the change in pressure as a function of time. Because $p = p(x, y, z, t)$, its time derivative will be given by

$$\frac{dp}{dt} = \frac{\partial p}{\partial t} + \frac{\partial p}{\partial x}\frac{dx}{dt} + \frac{\partial p}{\partial y}\frac{dy}{dt} + \frac{\partial p}{\partial z}\frac{dz}{dt}.$$

---

[6] As we continue our analysis we shall sometimes find it convenient to treat the fluid as a collection of particles and sometimes as a continuous medium.

[7] Recall the definition of field as given in Section 9.2.

But $\frac{dx}{dt}, \frac{dy}{dt}, \frac{dz}{dt}$ are the components of the fluid velocity. So

$$\frac{dp}{dt} = \frac{\partial p}{\partial t} + \frac{\partial p}{\partial x}v_x + \frac{\partial p}{\partial y}v_y + \frac{\partial p}{\partial z}v_z,$$

or

$$\frac{dp}{dt} = \frac{\partial p}{\partial t} + \mathbf{v} \cdot \nabla p. \tag{19.7}$$

The same relationship will hold for any of the fields describing the flow. Consequently, it is helpful to define the convective derivative by the operator equation

$$\frac{d}{dt} = \frac{\partial}{\partial t} + \mathbf{v} \cdot \nabla.$$

You can think of the left-hand side of Equation (19.7) as the rate of change of p in the Lagrangian representation and the term $\frac{\partial p}{\partial t}$ on the right as the rate of change of p in the Eulerian representation. The term $\mathbf{v} \cdot \nabla p$ is the "bridge" between the two representations.

19.3.4 The Equation of Continuity

The equation of continuity is another important and useful relation. In its simplest form it is essentially an expression of the conservation of mass. (We are excluding relativistic effects and nuclear reactions!)

To derive the equation of continuity for matter we consider a small *fixed* volume δV which might as well be a cube of sides dx, dy, dz. Let the mass enclosed by δV be δm. Then $\frac{d}{dt}(\delta m)$ is the rate of change of the enclosed mass. We can write $\delta m = \rho \delta V$. We are interested in obtaining an expression for ρ. Note: We are assuming that δV is a constant, but ρ is not.

It may be helpful for you to visualize a fluid flowing through the fixed volume, as illustrated in Figure 19.4. The mass "entering" the volume through the face $dydz$ at $x = 0$ in time dt is

$$(\rho v_x)_x dydzdt,$$

and the mass "leaving" through the opposite face (at $x + dx$) in time dt is

$$(\rho v_x)_{x+dx} dydzdt.$$

(The velocities can be negative so the terms "entering" and "leaving" are not to be taken literally.)

We also need to consider the mass flowing in and out through the other faces, so the total change of mass enclosed is

$$d(\delta m) = \left[(\rho v_x)_x - (\rho v_x)_{x+dx}\right] dydzdt$$
$$+ \left[(\rho v_y)_y - (\rho v_y)_{y+dy}\right] dxdzdt$$
$$+ \left[(\rho v_z)_z - (\rho v_z)_{z+dz}\right] dxdydt.$$

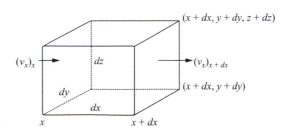

Figure 19.4 A volume fixed in space.

Dividing by dt and rearranging slightly we can write

$$-\frac{d}{dt}\delta m = \frac{(\rho v_x)_{x+dx} - (\rho v_x)_x}{dx}dxdydz$$
$$+ \frac{(\rho v_y)_{y+dy} - (\rho v_y)_y}{dy}dxdydz$$
$$+ \frac{(\rho v_z)_{z+dz} - (\rho v_z)_z}{dz}dxdydz.$$

Using the definition of derivative and noting that $\delta V = dxdydz$ we have

$$-\frac{d}{dt}\delta m = \left[\frac{\partial(\rho v_x)}{\partial x} + \frac{\partial(\rho v_y)}{\partial y} + \frac{\partial(\rho v_z)}{\partial z}\right]\delta V,$$

or, recognizing the divergence,

$$-\frac{d}{dt}\left(\frac{\delta m}{\delta V}\right) = \nabla \cdot (\rho \mathbf{v}).$$

Since $\frac{\delta m}{\delta V} = \rho$, we can rearrange to get

$$\frac{\partial \rho}{\partial t} + \nabla \cdot (\rho \mathbf{v}) = 0. \tag{19.8}$$

This is called the *equation of continuity*. It might be mentioned that there are other equations of continuity having exactly the same form but with ρ representing the density of different conserved physical quantities such as charge, momentum, and energy.

Worked Example 19.2 We obtained the equation of continuity using the Eulerian representation. Show that we obtain the same result using the Lagrangian representation.

Solution In our derivation of the equation of continuity, we assumed the volume element was constant but the mass enclosed by the volume could change. Thus, our derivation was formulated in the Eulerian representation. The same equation can be derived in the Lagrangian representation, in which we follow a group of particles enclosed by a surface that becomes distorted as the particles move with different velocities. In this derivation the mass is constant but the volume changes. In Figure 19.5 we illustrate an initially cubical volume element which is stretched in the x direction because the particles at $x+dx$ are moving faster than the particles at x. Two particles are shown, one on the "back" of the cube at x and the other one on the "front" of the cube at $x + dx$. Let the particle on the back have a velocity $v_x(x)$ and let the particle on the front have a velocity $v_x(x + dx)$.

The initial "length" of the cube (at time t) is $\delta x(t)$. Note that this is simply dx. But at a later time $t + dt$ the length is $\delta x(t + dt)$ where now the "back" particle is at

$$x + v_x(x)dt, \tag{A}$$

and the "front" particle is at

$$(x + dx) + v_x(x + dx)dt. \tag{B}$$

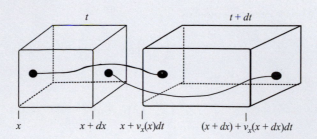

Figure 19.5 The stretching of a volume element that contains all the particles. In the figure only the displacement along x is illustrated.

Subtracting Equation (A) from Equation (B) we obtain

$$\delta x(t + dt) = \mathrm{B} - \mathrm{A}$$
$$= (x + dx) + v_x(x + dx)dt - x + v_x(x)dt$$
$$= dx + v_x(x + dx)dt - v_x(x)dt$$
$$= \delta x(t) + [v_x(x + dx) - v_x(x)]dt.$$

We can write this as

$$\delta x(t + dt) - \delta x(t) = \frac{v_x(x + dx)dt - v_x(x)}{dx}dx\,dt.$$

Using the definition of derivative on both sides,

$$\left(\frac{d}{dt}\delta x\right) = \left(\frac{\partial v_x}{\partial x}\right)\delta x\,dt,$$

or

$$d(\delta x) = \left(\frac{\partial v_x}{\partial x}\right)\delta x\,dt.$$

Consequently, the increase in volume due to this stretching along the x-axis is

$$d(\delta V) = \left(\frac{\partial v_x}{\partial x}\delta x\right)\delta y\delta z\,dt.$$

The total increase in volume is due to motion of the particles in all three directions. That is,

$$d(\delta V) = \left(\frac{\partial v_x}{\partial x} + \frac{\partial v_y}{\partial y} + \frac{\partial v_z}{\partial z}\right)\delta x\delta y\delta z\,dt = (\nabla \cdot \mathbf{v})\delta V\,dt,$$

or

$$\frac{d}{dt}\delta V = \nabla \cdot \mathbf{v}\delta V. \tag{19.9}$$

In the limit $\delta V \to 0$ we replace δV with dV and integrate to obtain

$$\frac{dV}{dt} = \int_V (\nabla \cdot \mathbf{v})dV.$$

The mass enclosed ($\delta m = \rho\delta V$) is constant, so

$$\frac{d}{dt}\delta m = \frac{d}{dt}(\rho\delta V) = 0. \tag{19.10}$$

But $\frac{d}{dt}(\rho\delta V)$ can be written as

$$\left(\frac{d\rho}{dt}\right)\delta V + \rho\frac{d}{dt}(\delta V) = 0,$$

or, using Equation (19.9)

$$\left(\frac{d\rho}{dt}\right)\delta V + \rho(\nabla \cdot \mathbf{v})\delta V = 0.$$

Applying the convective derivative $\frac{d\rho}{dt} = \frac{\partial\rho}{\partial t} + \mathbf{v}\cdot\nabla\rho$,

so

$$\frac{\partial\rho}{\partial t} + \mathbf{v}\cdot\nabla\rho + \rho\nabla\cdot\mathbf{v} = 0,$$

or

$$\frac{\partial\rho}{\partial t} + \nabla\cdot\rho\mathbf{v} = 0.$$

This is the equation of continuity.

19.3.5 Application: The Flow of Fluid through a Surface

Consider a surface dS fixed at some point in space, as shown in Figure 19.6. The fluid flows with a velocity $\mathbf{v}$, so all of the fluid in the cylinder of length vdt will pass through the surface in time dt.

As indicated in Figure 19.6, the surface need not be perpendicular to the fluid flow. The angle between $\hat{\mathbf{n}}$ and $\mathbf{v}$ is θ. Now the volume of the cylinder is $(dS\cos\theta)vdt = dSdt(\mathbf{v}\cdot\hat{\mathbf{n}})$ and the mass flowing through dS in time dt will be

$$dm = \rho(\mathbf{v}\cdot\hat{\mathbf{n}})dSdt,$$

so

$$\frac{dm}{dt} = \rho(\mathbf{v}\cdot\hat{\mathbf{n}})dS.$$

Integrating we find that the total mass flow per unit time through a finite surface S is

$$\frac{dM}{dt} = \oint_S \rho\mathbf{v}\cdot\hat{\mathbf{n}}dS = \oint_S \hat{\mathbf{n}}\cdot(\rho\mathbf{v})dS.$$

Here $M = \int dm$ and the quantity $\rho\mathbf{v}$ is the mass current. (It is also equal to the momentum per unit volume or momentum density.)

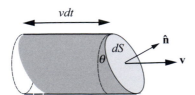

Figure 19.6 The fluid contained in the cylindrical volume element will pass through the surface dS.

19.3.6 Solenoidal and Irrotational Fluid Flow

We now consider two different types of fluid flow called "solenoidal" and "irrotational." Let us begin by defining a solenoidal flow. Consider an incompressible fluid. If the fluid cannot be compressed, it must have a constant density. From the equation of continuity, Equation (19.8), if ρ is constant, the divergence of the velocity is zero:

$$\nabla \cdot \mathbf{v} = 0. \tag{19.11}$$

A vector field with zero divergence is called a "solenoidal" field. (Other names that you might run into include "incompressible vector field," and "transverse vector field.") Although we used the symbol $\mathbf{v}$ for velocity, in this case it can stand for any vector field. Applying the divergence theorem, Equation (9.14), we obtain

$$\oint_S \mathbf{v} \cdot d\mathbf{S} = 0,$$

which tells us that the net flux of a solenoidal field through a closed surface is zero.

A geometric way of describing a solenoidal field is to state that the field has no sources or sinks. The field lines are either closed curves or terminate at infinity.

We now consider a flow whose velocity has zero curl. By definition, if the curl of the velocity at some point is zero, the flow is called "irrotational" or (sometimes) "longitudinal." That is, if at some point

$$\nabla \times \mathbf{v} = 0, \tag{19.12}$$

the field is irrotational at that point. To visualize the concept of an irrotational flow, imagine floating a small paddle wheel in the flow. If the paddle wheel does not rotate, the flow is irrotational. This has nothing to do with the overall motion of the fluid. For example, the fluid as a whole might be spiraling around in a vortex (like water going down a drain), but if the speed of the water varies inversely as the distance from the axis, then the flow is irrotational. If the fluid rotates like a rigid body, as in a bucket of water on a rotating turntable, the paddle wheel will turn about its axis as the bucket rotates. Then $\nabla \times \mathbf{v} \neq 0$, and the flow is not irrotational.

A related concept is the vorticity of a fluid given by

$$\boldsymbol{\Omega} = \frac{1}{2}\nabla \times \mathbf{v}.$$

The vorticity is the rotational velocity of a fluid *at a specific point*, not the rotational motion of the fluid as a whole. For example, a Ferris wheel rotates at some angular velocity, but the people riding in the gondolas are not rotating about their center of mass.

While we are on the topic of terminology, it might be mentioned that if the density is a function only of pressure, the fluid is said to be "homogeneous." Note that if a fluid is homogeneous and incompressible, the density must be constant.

Exercise 19.11

Show that for a solenoidal flow $\oint_S \mathbf{v} \cdot \hat{\mathbf{n}} \, dS = 0$.   ∎

Exercise 19.12

Show that if $\mathbf{v} = \nabla \times \mathbf{A}$ then $\mathbf{v}$ is solenoidal.   ∎

Exercise 19.13

Show that the vorticity field is solenoidal. ∎

Exercise 19.14

Show that if a fluid rotates like a rigid body, so that the velocity is proportional to the distance from the axis, then $\nabla \times \mathbf{v} \neq 0$. ∎

19.4 Equation of Motion: Euler's Equation

An ideal fluid is a fluid with no viscosity and there are no exchanges of heat between different parts of the system.[8] (An ideal fluid is the "dry water" mentioned by Feynman.) Although there is no such thing as an ideal fluid, it is an extremely useful fiction allowing us to easily derive the basic equations of fluid dynamics. (Honey is far from being an ideal fluid, whereas air at room temperature is essentially an ideal fluid.) An ideal fluid cannot resist a shear stress nor a tension (although it is hard to imagine a way of exerting a tension on a fluid[9]).

We shall assume the only stress that acts on an ideal fluid is pressure. Let $\mathbf{f}$ be the body force density (force per unit volume) acting on a fluid element δV. Then $\mathbf{f}\delta V$ is the force exerted by the body force. (The best example of such a body force is the gravitational force for which $\mathbf{f} = m\mathbf{g}/V = \rho\mathbf{g}$.)

Consider an element of an ideal fluid acted upon by a body force such as gravity and by pressure exerted from outside the element. Assume the mass of this element is δm and is constant. Let the fluid element have the shape of a rectangular box with sides $\delta x, \delta y$, and δz. Assume the surface perpendicular to the x-axis on the left side face at x is acted upon by a pressure p_l and the surface on the right at $x + \delta x$ is acted upon by a pressure p_r. This pressure difference produces a net force in the x direction given by

$$\delta F_x = p_l \delta y \delta z - p_r \delta y \delta z,$$

which (in the limit) gives

$$\delta F_x = -\frac{\partial p}{\partial x}\delta x \delta y \delta z = -\frac{\partial p}{\partial x}\delta V.$$

Including the other two dimensions we write

$$\delta \mathbf{F}_{\text{press}} = -\left(\hat{\mathbf{i}}\frac{\partial p}{\partial x} + \hat{\mathbf{j}}\frac{dp}{dy} + \hat{\mathbf{k}}\frac{dp}{dz}\right)\delta V = -\nabla p \delta V.$$

The total force (pressure differential plus body force) is equal to mass times acceleration. We can write $m\mathbf{a} = \mathbf{F}_{\text{tot}}$ as

$$(\delta m)\mathbf{a} = \delta\mathbf{F}_{\text{tot}} = \delta\mathbf{F}_{\text{press}} + \mathbf{f}\delta V = (-\nabla p + \mathbf{f})\delta V.$$

Divide by δV to get the following equation of motion:

$$\rho\frac{d\mathbf{v}}{dt} + \nabla p = \mathbf{f}. \tag{19.13}$$

---

[8] In an ideal fluid all processes are adiabatic.
[9] Perhaps pulling taffy might be an example of a fluid under a tension.

The equation of continuity, Equation (19.8), tells us that

$$\frac{d\mathbf{v}}{dt} = \frac{\partial\mathbf{v}}{\partial t} + (\mathbf{v} \cdot \nabla)\mathbf{v},$$

so

$$\frac{\partial\mathbf{v}}{\partial t} + (\mathbf{v} \cdot \nabla)\mathbf{v} + \frac{1}{\rho}\nabla p = \frac{\mathbf{f}}{\rho}. \tag{19.14}$$

This is called *Euler's equation.* It is the equation of motion expressed in terms appropriate to fluids.

If the fluid is homogeneous, the density depends only on pressure. The relationship between density and pressure is given by Equation (19.4) repeated here:

$$\frac{d\rho}{\rho} = \frac{dp}{B},$$

where B is the bulk modulus. One usually assumes the bulk modulus is determined under static conditions and at a constant temperature. This is the "isothermal" bulk modulus. However, if there are rapid density changes, as in a sound wave, thermal equilibrium is not reached and one must use the "adiabatic" bulk modulus. (For air the isothermal bulk modulus is about 101 kPa and the adiabatic bulk modulus is about 142 kPa, so you can see it definitely makes a difference.)

For a homogeneous fluid the properties of the fluid can be expressed as functions of velocity ($\mathbf{v}$) and pressure (p). Because velocity is a vector, we have four unknown quantities. Therefore, in general, we need four equations, or equivalently, one scalar equation and one vector equation. The scalar equation is the equation of continuity and the vector equation is Euler's equation:

$$\frac{\partial\rho}{\partial t} + \nabla \cdot (\rho\mathbf{v}) = 0,$$

$$\frac{\partial\mathbf{v}}{\partial t} + (\mathbf{v} \cdot \nabla)\mathbf{v} + \frac{1}{\rho}\nabla p = \frac{\mathbf{f}}{\rho}.$$

To determine p and $\mathbf{v}$ we need to solve these coupled partial differential equations. (It is usually assumed that the relationship between ρ and p is known. For example, if the density of the fluid is constant, then $p = -\rho gz +$ constant.) The solutions to these differential equations will depend on the boundary conditions and on the initial conditions.

If the fluid is inhomogeneous then ρ depends on the temperature as well as the pressure and we need an additional differential equation to determine $(p, \mathbf{v}, \rho)$ or $(p, \mathbf{v}, T)$ in the fluid. I will not introduce this complication because I don't want to get into the thermodynamics of the fluid system. Nevertheless, as you can imagine, when a fluid is heated, currents are generated in the fluid which tend to mix the fluid and equalize the temperature. This motion is called *convection.* It can be shown that for an ideal gas in a uniform gravitational field, convection is suppressed if $\frac{dT}{dz} > -\frac{g}{c_p}$, where c_p is the specific heat at constant pressure.[10] We will assume in our analysis that the fluid is either in thermal equilibrium or the condition that convection is absent has been met.

19.4.1 Steady Flow

By "steady flow" we mean that the quantities associated with the fluid at a given point are constant in time. Thus, at (x, y, z), the velocity, pressure, density, etc., are all constant. Hence, all partial derivatives with respect to time are zero and instead of $\frac{d}{dt} = \frac{\partial}{\partial t} + \mathbf{v} \cdot \nabla$ we simply have $\frac{d}{dt} = \mathbf{v} \cdot \nabla$.

---

[10] A bubble of hot air will cool adiabatically as it rises. If the environmental cooling rate is less than adiabatic, the bubble will soon be colder than the environment and will rise no further.

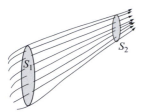

Figure 19.7 Tube of flow.

The path traced out by a fluid element is called a "streamline" and at every point it is tangent to the velocity. Let s be the distance as measured along a streamline. Then in steady flow the properties of a fluid are functions only of s. The component of ∇ along a streamline is $\frac{d}{ds}$ and so

$$\frac{d}{dt} = \mathbf{v} \cdot \hat{\mathbf{s}} \frac{d}{ds}.$$

The equation of continuity reduces to

$$\nabla \cdot (\rho \mathbf{v}) = 0.$$

The divergence theorem tells us that $\int_V \nabla \cdot (\rho\mathbf{v})d\tau = \oint_S (\rho\mathbf{v}) \cdot \hat{\mathbf{n}} dS$, so integrating over a fixed volume leads to

$$\oint_S \hat{\mathbf{n}} \cdot (\rho\mathbf{v})dS = 0,$$

which states that the total mass flowing through a closed surface is zero. (The amount that enters equals the amount that leaves.)

Consider an open surface S_1. All the streamlines passing through S_1 form a "tube of flow." In the sketch (Figure 19.7), the open surfaces S_1 and S_2 are two cross sections of the tube of flow. No fluid crosses in or out of such a tube because the velocity is always tangent to the streamlines. Consequently, we can write

$$\oint_{S_1} \rho_1 v_1 dS_1 - \oint_{S_2} \rho_2 v_2 dS_2 = 0,$$

or

$$I = \oint_S \rho v dS = \text{constant} = \text{current}.$$

This equation tells us that as the cross-sectional surface decreases, the velocity of an incompressible fluid increases. The spacing between the streamlines decreases, as illustrated in Figure 19.7. That is, the spacing between the streamlines is an indication of the speed of the fluid.

Our definition of current as a measure of fluid flow is completely analogous to the definition of electrical current if ρ represents electric charge density.

Exercise 19.15

Show that

$$(\mathbf{v} \cdot \nabla)\mathbf{v} = \hat{\mathbf{i}}\left[v_x \frac{\partial v_x}{\partial x} + v_y \frac{\partial v_x}{\partial y} + v_z \frac{\partial v_x}{\partial z} \right] + \hat{\mathbf{j}}\left[v_x \frac{\partial v_y}{\partial x} + v_y \frac{\partial v_y}{\partial y} + v_z \frac{\partial v_y}{\partial z} \right]$$

$$+ \hat{\mathbf{k}}\left[v_x \frac{\partial v_z}{\partial x} + v_y \frac{\partial v_z}{\partial y} + v_z \frac{\partial v_z}{\partial z} \right] = \sum_{i=1}^{3} \hat{\mathbf{e}}_i \left(\sum_{j=1}^{3} v_j \frac{\partial v_i}{\partial x_j} \right).$$

19.5 Conservation of Mass, Momentum, and Energy

As you know, the conservation laws of physics (conservation of momentum, energy, charge, etc.) are fundamental statements about the nature of the physical universe as well as extremely useful tools for solving physics problems. In this section we consider the conservation laws that apply to the motion of a fluid.

19.5.1 Conservation of Mass and a Generalized Equation of Continuity

The most important conservation law for fluid motion is probably the law of conservation of mass which we expressed as the equation of continuity

$$\frac{\partial \rho}{\partial t} + \nabla \cdot (\rho \mathbf{v}) = 0.$$

This relationship was derived in the Lagrangian representation in Worked Example 19.2 by expressing the conservation of mass as $\frac{d}{dt}(\rho \delta V) = 0$, where ρ was the mass density. However, the same arguments apply to other conserved quantities. That is, ρ can represent physical quantities other than mass density. For example, it can be charge or energy density or momentum density. But unlike mass, the energy and momentum of a fluid element can change. So we need to generalize our basic relation when dealing with these quantities and write

$$\frac{d}{dt}(\rho \delta V) = Q \delta V,$$

where Q is the rate at which the quantity ρ is being produced. Consequently, the generalized form of the equation of continuity is

$$\frac{\partial \rho}{\partial t} + \nabla \cdot (\rho \mathbf{v}) = Q. \tag{19.15}$$

Integrating this equation and using Gauss's divergence theorem we obtain

$$\frac{d}{dt} \int_V \rho \, dV + \oint_S \hat{\mathbf{n}} \cdot \rho \mathbf{v} \, dS = \int_V Q \, dV.$$

If the volume element is assumed to move with the fluid, the surface term disappears.

19.5.2 Conservation of Momentum

We now derive the conservation law for linear momentum. The *momentum density* is $\rho \mathbf{v}$. The equation of motion, Equation (19.13), is $\rho \frac{d\mathbf{v}}{dt} + \nabla p = \mathbf{f}$. In deriving this relation we assumed the mass of the fluid element was constant. That is, $\delta m = \rho \delta V = $ constant. So we can multiply the equation of motion by δV and bring the quantity $\rho \delta V$ in under the derivative, thus:

$$\frac{d}{dt}(\rho \mathbf{v} \delta V) = (\mathbf{f} - \nabla p) \, \delta V. \tag{19.16}$$

This is, of course, the same as $\frac{d}{dt}(\rho \delta V) = Q \delta V$ so we appreciate that the source for momentum in the fluid (per unit volume) is

$$Q = \mathbf{f} - \nabla p.$$

(This is essentially Newton's second law. It tells us the rate of change of momentum is equal to the force, but expressed in terms that are appropriate for fluid flow.)

It is interesting to express the integral of Equation (19.16), thus,

$$\frac{d}{dt}\int_V \rho \mathbf{v} dV = \int_V \mathbf{f} dV - \oint_S \hat{\mathbf{n}} p dS,$$

where we used the fact that $\int_V \nabla p dV = \oint_S \hat{\mathbf{n}} p dS$. Note that the internal pressure has been eliminated, leaving only the external pressure acting across the surface S.

The momentum conservation laws apply not only to ideal fluids, but also to viscous fluids (when suitably formulated). However, energy is not conserved for viscous fluids because viscosity is an internal friction. Nevertheless, Equation (19.16) applies in either case.

19.5.3 Conservation of Energy and Bernoulli's Equation

Conservation of *energy* is obtained by taking the dot product of $\mathbf{v}$ and Equation (19.16)

$$\mathbf{v} \cdot \frac{d}{dt}(\rho \mathbf{v} \delta V) = \mathbf{v} \cdot (\mathbf{f} - \nabla p) \delta V. \tag{19.17}$$

Because $\rho \delta V$ is constant,

$$\mathbf{v} \cdot \frac{d}{dt}(\rho \mathbf{v} \delta V) = \frac{d}{dt}\left(\frac{1}{2}\rho v^2 \delta V\right).$$

This states that the left-hand side of Equation (19.17) is the rate of change of the kinetic energy density. The right-hand side must be equal to the rate at which kinetic energy density is being produced (or depleted). Usually the body force is the gravitational force per unit volume on the fluid parcel; that is, $\mathbf{f} = \rho \mathbf{g}$. It is convenient to express this in terms of the gravitational potential per unit mass Φ which we shall relate to potential energy density:

$$\mathbf{f} = -\rho \nabla \Phi.$$

Therefore, we can write Equation (19.17) as

$$\frac{d}{dt}\left(\frac{1}{2}\rho v^2 \delta V\right) = \mathbf{v} \cdot \frac{d}{dt}(\rho \mathbf{v} \delta V) = \mathbf{v} \cdot (\mathbf{f} - \nabla p)\delta V$$

$$= \mathbf{v} \cdot (-\rho \nabla \Phi - \nabla)\delta V = -\mathbf{v} \cdot \rho \nabla \Phi \delta V - \mathbf{v} \cdot \nabla p \delta V$$

Note that

$$\frac{d}{dt}(p \delta V) = \frac{dp}{dt}\delta V + p\frac{d\delta V}{dt} = \frac{\partial p}{\partial t}\delta V + (\mathbf{v} \cdot \nabla p)\delta V + p\nabla \cdot \mathbf{v}\delta V.$$

Therefore,

$$-\mathbf{v} \cdot \nabla p \delta V = -\frac{d}{dt}(p\delta V) + \frac{\partial p}{\partial t}\delta V + p\nabla \cdot \mathbf{v}\delta V, \text{ so,}$$

$$\frac{d}{dt}\left(\frac{1}{2}\rho v^2 \delta V\right) = -\mathbf{v} \cdot \nabla \Phi \rho \delta V - \frac{d}{dt}(p\delta V) + \frac{\partial p}{\partial t}\delta V + p\nabla \cdot \mathbf{v}\delta V$$

$$= -\frac{d}{dt}(\rho \Phi \delta V) + \frac{\partial}{\partial t}(\Phi \rho \delta V) - \frac{d}{dt}(p\delta V) + \frac{\partial p}{\partial t}\delta V$$

$$+ p\nabla \cdot \mathbf{v}\delta V.$$

Collecting all total derivatives we obtain

$$\frac{d}{dt}\left[\left(\frac{1}{2}\rho v^2 + \rho \Phi + p\right)\delta V\right] = \rho \delta V \frac{\partial \Phi}{\partial t} + \frac{\partial p}{\partial t}\delta V + p\nabla \cdot \mathbf{v}\delta V. \tag{19.18}$$

We shall simplify this expression somewhat later, but for now it will be helpful to analyze the various terms in Equation (19.18).

Consider the left-hand side of Equation (19.18). Since $\frac{1}{2}\rho v^2$ is the kinetic energy density and $\Phi\rho$ is the gravitational potential energy density (usually equal to $\rho g z$, as shown in Figure 19.2), we appreciate that p can be considered another energy density term. Taking its negative gradient gives the force density due to the pressure, so p can be considered a potential energy per unit volume, and the expression on the left is the time rate of change of total energy. (Note that the energy involves three terms, namely: kinetic energy, potential energy due to gravity, and potential energy due to pressure.)

The rate of change of energy is equal to the work done on the fluid element. We expect the gravitational potential (Φ) at a given point to be constant with respect to time. Furthermore, if the pressure at a given point is constant in time so that $\frac{\partial p}{\partial t} = 0$, then the first two terms on the right-hand side of Equation (19.18) are zero.[11] If the fluid is incompressible[12] so that $\nabla \cdot \mathbf{v} = 0$, the right-hand side of Equation (19.18) is zero and the energy is constant. On the other hand, if $\nabla \cdot \mathbf{v} \neq 0$ (compressible fluid) the equation suggests that $p\nabla \cdot \mathbf{v}\delta V$ is the work related to the compression or expansion of the fluid element δV. We can validate this suggestion by calculating the work done during an expansion. Because $W_{\text{ex}} = \int \mathbf{F} \cdot d\mathbf{x} = \int (pA)dx = \int p\,dV$ we can write

$$\delta W_{\text{ex}} = p\delta V,$$

so, using Equation (19.9),

$$\frac{d(\delta W_{\text{ex}})}{dt} = p\frac{d}{dt}(\delta V) = p\nabla \cdot \mathbf{v}\delta V,$$

and the suggestion is verified. (Keep in mind that we are assuming constant pressure at a point during the infinitesimal expansion.)

It is convenient to introduce a potential energy term $u\delta m$ associated with changes in the volume. If the volume changes, the surface of the fluid element does work against the surrounding fluid. Because $F = -\frac{dU}{dx}$ we can write

$$\delta W_{\text{ex}} = -\frac{dU}{dx}\delta x = -\delta U = -u\delta m = -u\rho\delta V.$$

Therefore,

$$\frac{d}{dt}\delta W_{\text{ex}} = -\frac{d}{dt}(u\rho\delta V),$$

and consequently

$$p\nabla \cdot \mathbf{v}\delta V = -\frac{d}{dt}(u\rho\delta V).$$

Finally, replacing the last term in Equation (19.18) with $-\frac{d}{dt}(u\rho\delta V)$ and moving it to the right-hand side we obtain the somewhat simpler relationship

$$\frac{d}{dt}\left[\left(\frac{1}{2}v^2 + \Phi + \frac{p}{\rho} + u\right)\rho\delta V\right] = \left(\frac{\partial p}{\partial t} + \rho\frac{\partial \Phi}{\partial t}\right)\delta V. \tag{19.19}$$

Dividing by $\rho\delta V$ yields *Bernoulli's theorem:*

$$\frac{d}{dt}\left[\frac{1}{2}v^2 + \frac{p}{\rho} + \Phi + u\right] = \left(\frac{1}{\rho}\frac{\partial p}{\partial t} + \frac{\partial \Phi}{\partial t}\right). \tag{19.20}$$

---

[11] The pressure at a point is not constant for sound waves.
[12] This is solenoidal flow. See Equation (19.11).

A "steady flow" (as considered in Section 19.4.1) has $\frac{\partial \rho}{\partial t}$ and $\frac{\partial \Phi}{\partial t}$ equal to zero, so for that situation we can integrate to get

$$\frac{1}{2}v^2 + \Phi + \frac{p}{\rho} + u = \text{constant.} \tag{19.21}$$

This is called *Bernoulli's equation.*

The gravitational potential Φ can usually be expressed as $\Phi = gz$, where z is measured from some reference height. If the fluid is incompressible, ρ and u are constant (as is Φ) and Bernoulli's equation can be expressed in the form usually found in introductory physics texts:

$$\frac{1}{2}\rho v^2 + \rho gz + p = \text{constant.}$$

This equation yields a useful fact: for flow at constant height, the pressure decreases as the velocity increases. This is called the Venturi effect.

For steady flow, Bernoulli's theorem can be expressed in the alternate form

$$\frac{d}{ds}\left(\frac{1}{2}v^2 + \frac{p}{\rho} - \Phi + u\right) = 0,$$

where s is measured along a streamline.

19.5.4 A Velocity Potential Function

In addition to the conditions for steady flow, let us assume the flow is irrotational ($\nabla \times \mathbf{v} = 0$). Then we can define a velocity potential function $\phi(x, y, z)$ by

$$\phi(\mathbf{r}) = \int_{\mathbf{r}_s}^{\mathbf{r}} \mathbf{v} \cdot d\mathbf{r},$$

so

$$\mathbf{v} = \nabla \phi.$$

But for steady flow, $\nabla \cdot (\rho \mathbf{v}) = 0$, so $\nabla \cdot (\rho(\nabla\phi)) = 0$ or $\nabla \cdot (\rho\nabla\phi) = 0$. If the fluid is incompressible ($\rho = \text{constant}$), we obtain

$$\nabla^2 \phi = 0,$$

which you will recognize as Laplace's equation.

Exercise 19.16

Fill in the missing steps in the derivation of Equation (19.18). (Hint: Recall that $\frac{d}{dt}\delta V = \nabla \cdot \mathbf{v}\delta V$. See Equation (19.9).) ∎

Exercise 19.17

Prove that if the velocity can be obtained from a velocity potential, the flow is irrotational. ∎

Exercise 19.18

Show that

$$\mathbf{v} \cdot \frac{d}{dt}(\rho\mathbf{v}\delta V) = \frac{d}{dt}\left(\frac{1}{2}\rho v^2 \delta V\right).$$

∎

Exercise 19.19

If the velocity potential for some flow is given by $\phi = a/r$, where a is a constant, determine the velocity. Answer: $\mathbf{v} = -\frac{a}{r^2}\hat{\mathbf{r}}$. ∎

19.6 Sound Waves

We now consider the production of sound waves in a fluid. These waves consist of small, rapid changes in pressure that can be detected by the resulting oscillation of the ear drum. The transport of energy by sound waves involves changes in the density of the fluid, so we are now dealing with a *compressible* fluid.

Unlike waves in a string or electromagnetic waves, sound waves are longitudinal waves, that is, the oscillations are in the same direction as the wave velocity.

Our analysis involves two tasks. The first task is to determine the wave equation for a sound wave. The second task is to solve the equation and interpret the solution.

To determine the wave equation we use Euler's equation and an equation relating the bulk modulus to the pressure and density.

It may be helpful to consider how a guitar generates a sound wave. When you pluck the string it vibrates and this vibration is transmitted to the face of the guitar which generates waves in the fluid (air) inside the body of the instrument. The sound wave escapes to the environment through the hole in the face and propagates outward at the speed of sound. Note that there are two velocities involved in the sound wave, namely: the speed of sound in air and the speed with which the air molecules are vibrating back and forth. To keep from confusing these two speeds I will denote the velocity of the fluid elements by $\mathbf{v}$ and the speed of the sound wave by c.

We begin the analysis by considering a fluid *at rest* with pressure p_0 and density ρ_0 which can vary from point to point in the fluid, but which do not vary with time at a specific point. Recall that the equation of motion in terms of derivatives at a fixed point is Euler's equation (19.14):

$$\frac{\partial \mathbf{v}}{\partial t} + (\mathbf{v} \cdot \nabla)\mathbf{v} + \frac{1}{\rho}\nabla p = \frac{\mathbf{f}}{\rho}.$$

Because the fluid is assumed to be at rest, $\mathbf{v} = 0 = $ constant, and this equation reduces to

$$\frac{1}{\rho_0}\nabla p_0 = \frac{1}{\rho_0}\mathbf{f}_0.$$

Now let the fluid be perturbed so that $\rho = \rho_0 + \rho'$ and $p = p_0 + p'$, where ρ' and p' are small relative to ρ_0 and p_0. We can no longer assume $\mathbf{v} = 0$ so the equation of motion becomes

$$\frac{\partial \mathbf{v}}{\partial t} + (\mathbf{v} \cdot \nabla)\mathbf{v} + \frac{1}{\rho_0 + \rho'}\nabla(p_0 + p') = \frac{\mathbf{f}}{\rho_0 + \rho'}.$$

As long as the velocity of the fluid elements ($\mathbf{v}$) is much smaller than the speed of the sound wave (c) the term $\mathbf{v} \cdot \nabla \mathbf{v}$ can be neglected. Keeping only first-order terms in $\mathbf{v}$, ρ, and p we obtain

$$\frac{\partial \mathbf{v}}{\partial t} + \frac{1}{\rho_0}\nabla p_0 + \frac{1}{\rho_0}\nabla p' = \frac{1}{\rho_0}\nabla p_0,$$

or

$$\frac{\partial \mathbf{v}}{\partial t} = -\frac{1}{\rho_0}\nabla p'. \tag{19.22}$$

This is the first equation we will use to develop the wave equation. The second relation is the equation for the bulk modulus which we write as

$$\frac{d\rho}{\rho} = \frac{dp}{B}.$$

Then, for a perturbed fluid

$$\frac{d(\rho_0 + \rho')}{(\rho_0 + \rho')} = \frac{d(p_0 + p')}{B},$$

$$\frac{1}{\rho_0} d\rho' = \frac{dp'}{B}.$$

Since ρ_0 and B are constants,

$$\frac{\rho'}{\rho_0} = \frac{p'}{B},$$

or

$$\rho' = \rho_0 \frac{p'}{B}. \tag{19.23}$$

The equation of continuity is $\frac{\partial \rho}{\partial t} + \nabla \cdot \rho\mathbf{v} = 0$, so

$$\frac{\partial(\rho_0 + \rho')}{\partial t} + \nabla \cdot (\rho_0 + \rho')\mathbf{v} = 0,$$

$$\frac{\partial \rho'}{\partial t} = -\rho_0 \nabla \cdot \mathbf{v} - \mathbf{v} \cdot \nabla\rho_0.$$

If ρ_0 is uniform (or at least, approximately uniform) we can drop the last term and write

$$\frac{\partial \rho'}{\partial t} = -\rho_0 \nabla \cdot \mathbf{v}. \tag{19.24}$$

Inserting Equation (19.23) in Equation (19.24) we get

$$\frac{\partial}{\partial t}\left(\rho_0 \frac{p'}{B}\right) = -\rho_0 \nabla \cdot \mathbf{v},$$

or

$$\frac{\partial p'}{\partial t} = -B\nabla \cdot \mathbf{v}. \tag{19.25}$$

Recall that we have shown in Equation (19.22) that

$$\frac{\partial \mathbf{v}}{\partial t} = -\frac{1}{\rho_0}\nabla p'. \tag{19.26}$$

Equations (19.25) and (19.26) are the fundamental equations for sound waves. In one dimension they reduce to

$$\frac{\partial p'}{\partial t} = -B\frac{\partial v}{\partial x},$$

and

$$\frac{\partial v}{\partial t} = -\frac{1}{\rho_0}\frac{\partial p'}{\partial x}.$$

We can eliminate either p' or $\mathbf{v}$ from (19.25) and (19.26). To eliminate $\mathbf{v}$ we take the divergence of (19.26), as follows:

$$\nabla \cdot \left(\frac{\partial \mathbf{v}}{\partial t}\right) = -\frac{1}{\rho_0}\nabla \cdot (\nabla p'),$$

$$\frac{\partial}{\partial t}(\nabla \cdot \mathbf{v}) = -\frac{1}{\rho_0}\nabla^2 p'. \tag{19.27}$$

The time derivative of Equation (19.25) is

$$\frac{\partial^2 p'}{\partial t^2} = -B\frac{\partial}{\partial t}(\nabla \cdot \mathbf{v}),$$

so replacing $\frac{\partial}{\partial t}\nabla \cdot \mathbf{v}$ with $-\frac{1}{B}\frac{\partial^2 p'}{\partial t^2}$ in Equation (19.27), we obtain

$$\nabla^2 p' = \frac{\rho_0}{B}\frac{\partial^2 p'}{\partial t^2},$$

which we can write in the form of the three-dimensional wave equation

$$\nabla^2 p' - \frac{1}{c^2}\frac{\partial^2 p'}{\partial t^2} = 0, \tag{19.28}$$

where c represents the speed of the wave and is given by

$$c = \sqrt{\frac{B}{\rho_0}}. \tag{19.29}$$

Thus, we have shown that our analysis leads to the three-dimensional wave equation.[13]

Similarly, we can express the three-dimensional wave equation for sound waves in terms of the velocity:

$$\nabla^2 \mathbf{v} - \frac{1}{c^2}\frac{\partial^2 \mathbf{v}}{\partial t^2} = 0. \tag{19.30}$$

I emphasize once again that $\mathbf{v}$ is *not* the velocity of the wave. That is given by c. The quantity $\mathbf{v}$ is the speed of the fluid packets (or the net velocity of the air molecules if you prefer to think of a sound wave in air).

We have actually obtained two wave equations. Equation (19.28) is a scalar wave equation for the pressure as a function of position and time, and Equation (19.30) is a vector equation for the velocity of the fluid particles as a function of position and time. Because p and $\mathbf{v}$ both refer to the same wave, there must be a relationship between them, as we now show.

Consider the one-dimensional version of Equation (19.28):

$$\frac{\partial^2 p'}{\partial t^2} - \frac{1}{c^2}\frac{\partial^2 p'}{\partial x^2} = 0.$$

As discussed in Section 13.4, this partial differential equation is satisfied by any function of the form

$$p' = f(x - ct),$$

---

[13] Because the pressure variations in a sound wave are so rapid, there is no opportunity for heat to be dissipated. That is, heat is neither absorbed nor emitted by the fluid. Consequently, in calculating the speed of the sound wave one must use the adiabatic bulk modulus rather than the isothermal bulk modulus.

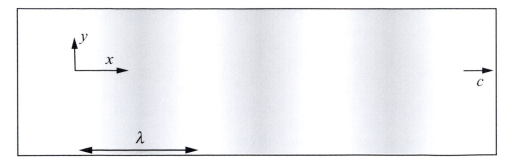

Figure 19.8 The pattern moves to the right at speed c. You can assume the shading represents either pressure or density.

representing a plane wave with values of p' constant on any plane perpendicular to the x-axis, as illustrated in Figure 19.8.

We can express the plane wave solution to the three-dimensional equation, Equation (19.28), as

$$p' = f(\hat{\mathbf{n}} \cdot r - ct),$$

where $\hat{\mathbf{n}}$ gives the direction of propagation of the wave. Note that f is a scalar function.

Similarly, the plane wave solution of Equation (19.30) for the velocity can be expressed as

$$\mathbf{v} = \mathbf{g}(\hat{\mathbf{m}} \cdot \mathbf{r} - ct),$$

where $\mathbf{g}$ is a vector function and $\hat{\mathbf{m}}$ is the direction of the velocity of the fluid particle. Comparing the two plane wave solutions we inquire into the relationships between $\hat{\mathbf{n}}$ and $\hat{\mathbf{m}}$ and the relationship between f and $\mathbf{g}$.

The unit vectors $\hat{\mathbf{n}}$ and $\hat{\mathbf{m}}$ give the direction of the propagation of the wave ($\hat{\mathbf{n}}$) and the direction of the velocity of the fluid particles ($\hat{\mathbf{m}}$). For transverse waves, such as a wave in a string or an electromagnetic wave, the oscillations are perpendicular to the direction of the motion of the wave. It is possible to introduce transverse sound waves in a solid, in which case $\hat{\mathbf{n}} \perp \hat{\mathbf{m}}$. However, a fluid cannot support a shear stress, so for a fluid $\hat{\mathbf{n}} = \hat{\mathbf{m}}$. This simply expresses the fact that sound waves in a fluid are longitudinal.

Next, let us consider the relationship between f and $\mathbf{g}$. To generate a vector that has the same direction as the motion of the wave, the function $\mathbf{g}$ must be proportional to $\hat{\mathbf{n}}$. Furthermore, we saw in Equation (19.22) that

$$\frac{\partial \mathbf{v}}{\partial t} = -\frac{1}{\rho_0} \nabla p'.$$

It is left as a problem (Problem 19.16) to show that this leads to the following relation between f and $\mathbf{g}$:

$$\mathbf{g} = \frac{\hat{\mathbf{n}}}{\sqrt{B\rho_0}} f,$$

and consequently

$$\mathbf{v} = \frac{\hat{\mathbf{n}}}{\sqrt{B\rho_0}} p'.$$

Although any function of the form $f(\hat{\mathbf{n}} \cdot r - ct)$ will satisfy the wave equation, a sinusoidal function is particularly useful. Because the argument of a sine or cosine must be dimensionless, we introduce the wave vector $\mathbf{k}$ related to the wavelength and/or the angular frequency thus:

$$\mathbf{k} = \frac{2\pi}{\lambda}\hat{\mathbf{n}} = \frac{\omega}{c}\hat{\mathbf{n}}.$$

Then, the sound wave can be expressed variously as

$$p' = A\cos(\mathbf{k}\cdot\mathbf{r} - \omega t) \quad \text{or} \quad p' = \text{Re}\left[Ae^{i(\mathbf{k}\cdot\mathbf{r}-\omega t)}\right].$$

Sound waves transport energy. Imagine a monk pulling on a bell rope. The mechanical vibrations of the bell generate a sound wave that is transported through the air until it is incident upon some surface, such as the eardrum of a listener. Let us determine the energy per unit time transmitted by a sound wave to a unit area of receiver. Assuming the surface is perpendicular to $\hat{\mathbf{n}}$, we can write

$$\frac{\text{energy}}{\text{time} \times \text{area}} = \frac{\text{power}}{\text{area}} = \frac{\text{force} \times \text{velocity}}{\text{area}} = \text{pressure} \times \text{velocity}.$$

For a sound wave in a stationary fluid the velocity v oscillates around the average value of zero and p' oscillates around the constant p_0. The average power is

$$P_{av} = \langle p'v\rangle = \left\langle p'\frac{p'}{\sqrt{B\rho_0}}\right\rangle = \frac{\langle p'^2\rangle}{\sqrt{B\rho_0}}.$$

Our analysis has been focused on plane waves in a fluid. But the three-dimensional wave equation has other solutions. An important one is a spherical wave. A small patch of the spherical wave is approximately plane and the energy per unit time passing through it will be proportional to p'^2. But the energy flow per unit area should drop off as $1/r^2$, so $p'^2 \propto 1/r^2$ and this suggests that

$$p = \frac{1}{r}f(r - ct).$$

This can be shown to satisfy the wave equation.

Worked Example 19.3 An oscillator (such as a tuning fork) is the source of a sound wave. Assume the oscillator vibrates with an amplitude A_{osc} and a frequency f. (a) Derive a formula for the excess pressure p' in a one-dimensional compression wave.[a] (b) Determine the maximum amplitude of the compression wave in your ear canal due to a tuning fork with vibrational amplitude 10^{-9} m and frequency 760 Hz. (The density of air at 20 °C is 1.2 kg/m^3 and the speed of sound is 343 m/s.) Express your answer in atmospheres.

Solution (a) Our analysis is based on the definition of bulk modulus which we can express as

$$p' = -B\frac{\delta V}{V}.$$

Consider a small volume element V of the fluid as illustrated in Figure 19.9 where we are assuming the fluid is at rest. The volume is $V = Al = A(x_2 - x_1)$, where A is the area of the plane surfaces at the equilibrium positions x_1 and x_2.

When the vibrating source is activated the planes at x_1 and x_2 will oscillate back and forth about the equilibrium positions with amplitudes equal to A_{osc}. The displacements δx of these planes will depend on the position and time according to

$$\delta x = A_{osc}\cos\left(\frac{2\pi x}{\lambda} - 2\pi ft\right).$$

At some given instant of time the first plane will be at $x_1 + \delta x_1$ and the second plane will be at $x_2 + \delta x_2$. The distance between the two planes will now be

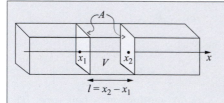

Figure 19.9 The volume of the fluid parcel is $V = Al = A(x_2 - x_1)$.

$$(x_2 + \delta x_2) - (x_1 + \delta x_1) = (x_2 - x_1) + (\delta x_2 - \delta x_1) = l + \delta l.$$

The volume of the fluid element has increased by $\delta V = A\delta l$ so

$$\frac{\delta V}{V} = \frac{\delta l}{l}.$$

Consequently, we can write

$$p' = -B\frac{\delta l}{l}.$$

Here $\delta l = \delta x_2 - \delta x_1$. Note that at any given time t, δx_1 is a function only of x_1 and δx_2 is a function only of x_2. If the planes are separated by a small distance dx (so $l = dx$) we have

$$\delta x_1 = \delta x(x),$$

$$\delta x_2 = \delta x(x + dx).$$

Therefore,

$$\delta l = \delta x_2 - \delta x_1 = \delta x(x + dx) - \delta x(x)$$

and, using the definition of derivative,

$$\delta l = \frac{\partial(\delta x)}{\partial x}dx.$$

Because $dx = l$ we can write

$$\frac{\delta l}{l} = \frac{\partial(\delta x)}{\partial x}.$$

We know that

$$\delta x = A_{\text{osc}} \cos\left(\frac{2\pi x}{\lambda} - 2\pi ft\right),$$

so

$$\frac{\partial(\delta x)}{\partial x} = -A_{\text{osc}}\frac{2\pi}{\lambda} \sin\left(\frac{2\pi x}{\lambda} - 2\pi ft\right).$$

Therefore

$$p' = -B\frac{dV}{V} = -B\frac{\delta l}{l} = -B\frac{\partial(\delta x)}{\partial x} = +BA_{\text{osc}}\frac{2\pi}{\lambda} \sin\left(\frac{2\pi x}{\lambda} - 2\pi ft\right).$$

But by Equation (19.29) $B = \rho_0 c^2$, so finally the excess pressure as a function of position and time is given by

$$p' = \rho_0 c^2 A_{\text{osc}}\frac{2\pi}{\lambda} \sin\left(\frac{2\pi x}{\lambda} - 2\pi ft\right). \tag{19.31}$$

(b) The maximum value of p' according to Equation (19.31) will occur when the sine term is 1. Recall that $\lambda = c/f$, so

$$p'_{max} = 2\pi \rho_0 c A_{osc} f = (2\pi)(1.2)(343)(1 \times 10^{-9})(760) = 0.002 \text{ N/m}^2,$$

or 1.95×10^{-8} atm. (The human ear can detect pressure differences as low as 2×10^{-10} atm.)

[a] For example, a compression wave in a water-filled pipe can be considered one dimensional.

Exercise 19.20

Derive the wave equation for velocity, Equation (19.30). ∎

Exercise 19.21

Show that $p = \frac{1}{r} f(r - ct)$ satisfies the three-dimensional wave equation. ∎

19.7 Solving the Wave Equation by Separation of Variables

Imagine a fluid in a closed container. Now the fluid vibrations are subject to *boundary conditions* that describe the behavior of the fluid at the walls of the container. For simplicity suppose the container is a rectangular box with rigid walls.

We will use separation of variables to obtain the general solution of the three-dimensional wave equations then use the boundary conditions to determine the constants of integration in our solution.[14]

The wave equation is

$$\nabla^2 p' - \frac{1}{c^2}\frac{\partial^2 p'}{\partial t^2} = 0.$$

Using separation of variables we assume

$$p'(x, y, z, t) = U(x, y, z)\Theta(t).$$

Inserting this expression into the wave equation yields

$$\Theta \nabla^2 U - \frac{1}{c^2}U\frac{\partial^2 \Theta}{\partial t^2} = 0.$$

Dividing by ΘU and moving the second term to the right-hand side we obtain

$$\frac{1}{U}\nabla^2 U = \frac{1}{c^2}\frac{1}{\Theta}\frac{\partial^2 \Theta}{\partial t^2}.$$

Because one side of this equation depends only on U, a function of x, y, z, and the other side depends only on Θ, a function of t, they can only be equal if both sides are equal to the same constant. Let us denote the constant by $-\omega^2/c^2$. That is,

$$\frac{1}{U}\nabla^2 U = \frac{1}{c^2}\frac{1}{\Theta}\frac{\partial^2 \Theta}{\partial t^2} = \text{constant} = -\frac{\omega^2}{c^2}.$$

[14] See Section 13.2 in which we solve the one-dimensional wave equation.

Therefore the Θ equation is just

$$\frac{d^2\Theta}{dt^2} + \omega^2\Theta = 0.$$

This is the simple harmonic motion equation. Consequently the solution of the Θ equation is

$$\Theta(t) = A\cos\omega t + B\sin\omega t.$$

(There is also a traveling wave solution, but that is not of interest at the moment.)

The other side of our separated equation is

$$\nabla^2 U + \frac{\omega^2}{c^2}U = 0,$$

which is the equation for three-dimensional simple harmonic motion.

The solution of the U equation can be obtained by a further separation of variables to generate three uncoupled ordinary differential equations. This is accomplished by substituting

$$U(x, y, z) = X(x)Y(y)Z(z)$$

into the differential equation for U. Plugging in leads to equations of the form

$$\frac{1}{X}\frac{d^2 X}{dx^2} = k_x,$$

with solution

$$X(x) = C_x\cos k_x x + D_x\sin k_x x,$$

and similarly for Y and Z. Note that

$$k_x^2 + k_y^2 + k_z^2 = \frac{\omega^2}{c^2}.$$

The solution to the wave equation is, consequently,

$$p' = p'(x, y, z, t) = X(x)Y(y)Z(z)\Theta(t).$$

However, our problem has just begun because we now need to consider the boundary conditions at the walls. Assume the fluid is contained in a rectangular box with sides L_x, L_y, L_z. The boundaries are at $x = 0$ and $x = L_x$ and similarly for y and z. Although we solved the wave equation for p', we could just as well have expressed it in terms of $\mathbf{v}$ because, as we have seen, $\mathbf{v} = \frac{\hat{\mathbf{n}}}{\sqrt{B\rho_0}}p'$. It turns out that the boundary conditions are more easily expressed in terms of velocity than in terms of pressure. (It is intuitively clear that at the walls the velocity of a fluid element in the direction perpendicular to the wall must be zero. It is not very obvious that at a wall the pressure will be at its maximum amplitude.) So let us reformulate the problem in terms of velocity. We begin with Equation (19.22) which states that $\frac{\partial \mathbf{v}}{\partial t} = -\frac{1}{\rho_0}\nabla p'$. Since $p = XYZ\Theta$, the x component of this equation is

$$\begin{aligned}
\frac{\partial v_x}{\partial t} &= -\frac{1}{\rho_0}\frac{\partial}{\partial x}(XYZ\Theta) \\
&= -\frac{YZ\Theta}{\rho_0}\frac{\partial X}{\partial x} \\
&= -\frac{YZ}{\rho_0}(A\cos\omega t + B\sin\omega t)\frac{\partial}{\partial x}[(C_x\cos k_x x + D_x\sin k_x x)] \\
&= -\frac{YZ}{\rho_0}(A\cos\omega t + B\sin\omega t)(-C_x k_x\sin k_x x + D_x k_x\cos k_x x).
\end{aligned}$$

Integrating to get v_x we obtain

$$v_x = -\frac{YZk_x}{\rho_0\omega}(A\sin\omega t - B\cos\omega t)(-C_x\sin k_x x + D_x\cos k_x x).$$

We require this to vanish at $x = 0$, so $D_x = 0$ and

$$v_x = \frac{YZk_x}{\rho_0\omega}(A\sin\omega t - B\cos\omega t)C_x\sin k_x x.$$

Similar expressions can be obtained for v_y and v_z.

There is also a wall at L_x and the boundary condition there is $v_x = 0$ at $x = L_x$. Therefore, $\sin k_x L_x = 0$ which implies that $k_x = l\pi/L_x$, where l is an integer, $l = 0, 1, 2, \ldots$, and

$$v_x = \frac{l\pi}{L_x\rho_0\omega}Y(y)Z(z)\sin\frac{l\pi x}{L_x}(A\sin\omega t + B\cos\omega t).$$

Similar relations hold for k_y and k_z, specifically, $k_y = m\pi/L_y$ and $k_z = n\pi/L_z$, with m and n integers. The corresponding velocity components have the form

$$v_x = \frac{l\pi}{L_x\rho_0\omega_{lmn}}(A\sin\omega_{lmn} - B\cos\omega_{lmn}t)\sin\frac{l\pi x}{L_x}\cos\frac{m\pi y}{L_y}\cos\frac{n\pi z}{L_z}.$$

Each set of integral values of l, m, n defines a *normal mode* oscillation. The frequencies of the normal modes are:

$$\omega_{lmn} = \pi c\left(\frac{l^2}{L_x^2} + \frac{m^2}{L_y^2} + \frac{n^2}{L_z^2}\right)^{\frac{1}{2}}.$$

(Note, however, that l, m, n cannot all be zero because then $\omega_{lmn} = 0$ and this does not correspond to a vibration.)

The pressure differential is $p' = U(x, y, z)\Theta(t)$. We can now write an expression for p' for a given normal mode, thus:

$$p' = (A\cos\omega_{lmn}t + B\sin\omega_{lmn}t)\cos\frac{l\pi x}{L_x}\cos\frac{m\pi y}{L_y}\cos\frac{n\pi z}{L_z}.$$

This quantity oscillates at maximum amplitude at the walls. We conclude that at the walls, the perpendicular component of the velocity has a node and the pressure has an antinode.

19.7.1 Sound Waves in Pipes (The Organ)

Recall that for a string the frequencies of the normal modes were $v_n = n(c/2l)$ so $v_2 = 2v_1, v_3 = 3v_1$, etc. They are "harmonically" related to one another. This fact allows us to obtain musical tones from a string.[15] But the normal-mode frequencies, ω_{lmn}, that we derived above are not harmonically related to one another. How then, can we get musical tones from an organ pipe? The answer is that in an organ pipe, L_x is much greater than L_y or L_z. (That is, an organ pipe is a long thin pipe.) The lowest frequencies correspond to $m = n = 0$ and l is a small integer. Consequently, the first few normal frequencies are multiples of the lowest frequency.

An organ pipe closed at one end is illustrated in Figure 19.10.

---

[15] Musical instruments generate sound waves of a given frequency (called the "fundamental") and smaller amplitude waves of frequencies that are multiples of the fundamental frequency. (These are called "overtones.") A mixture of a large number of different frequency sound waves produces seemingly random oscillations of the eardrum and is called noise.

Figure 19.10 Air at pressure somewhat greater than atmospheric pressure is blown into the tube. The air passes a flexible flap of wood or metal called a reed which oscillates, opening and closing the air passage and setting up alternate puffs of air at higher pressure.

Consider a long pipe oriented along the z-axis with a rectangular cross section of dimensions L_x and L_y. Assume, for the sake of this analysis, that the pipe is open at both ends, that is, at $z = 0$ and at $z = L_z$.[16] The zero-velocity boundary condition will apply at the surfaces at $x = 0$ and $x = L_x$ and $y = 0$ and $y = L_y$. The sound wave will have the form of a standing wave in the x- and y-directions, but the form of a traveling wave in the z-direction.

That is, in general,

$$X(x) \propto \cos k_x x \quad \text{and} \quad Y(y) \propto \cos k_y y,$$

and

$$Z(z) \propto e^{ik_z z} \quad \text{and} \quad \Theta(t) = A e^{i\omega t},$$

so

$$p' = \text{Re} \left[A e^{i(k_z z - \omega t)} \right] \cos \frac{l\pi x}{L_x} \cos \frac{m\pi y}{L_y}$$

$$= A \cos \frac{l\pi x}{L_x} \cos \frac{m\pi y}{L_y} \cos(k_z z - \omega t).$$

Each choice of l, m corresponds to a "mode of propagation."

Since $k_x^2 + k_y^2 + k_z^2 = \omega^2/c^2$, and since $k_x = l\pi/L_x$ and $k_y = m\pi/L_y$, we have

$$k_z = \left[\frac{\omega^2}{c^2} - \left(\frac{l\pi}{L_x} \right)^2 - \left(\frac{m\pi}{L_y} \right)^2 \right]^{1/2}.$$

If $l = m = 0$, $k_z = \omega/c$. But if $l \neq 0$ and $m \neq 0$ the speed of the wave depends on the values of l and m, that is,

$$c_{lm} = \frac{\omega}{|k_z|} = c \left[1 - \left(\frac{l\pi c}{\omega L_x} \right)^2 - \left(\frac{m\pi c}{\omega L_y} \right)^2 \right]^{-1/2}.$$

This means that the wave is a combination of waves having somewhat different velocities, c_{lm}, which are referred to as the "phase velocities." These sum to give the "group" which travels at c. As time goes on, the faster waves move ahead and the slower waves lag behind, so the group spreads out. This is called "dispersion." Dispersion is also observed for electromagnetic waves, in which waves of different frequencies will travel through a medium (such as glass) at different speeds.

---

[16] The opening through which slightly compressed air is blown past a reed to generate fluctuations in the air column would be placed on the side of the tube rather than on one end as in Figure 19.10.

(a) Write the equation for the speed of a shear wave in a solid. (b) The density of aluminum is 2.7×10^3 kg/m^3 and its bulk and shear moduli are $B = 7 \times 10^{10}$ Pa and $n = 2.5 \times 10^{10}$ Pa. Determine the velocities of compression waves and shear waves in aluminum. Answer: (b) 5092 m/s and 3043 m/s. ∎

What is the lowest-frequency sound wave in a box of dimensions 5 cm $\times$ 10 cm $\times$ 20 cm? (Use $c = 340$ m/s.) Answer: 5.3×10^3 rad/s = 850 Hz. ∎

Determine the dominant frequency of the sound produced by a square cross section organ pipe of dimensions 2 m $\times$ 0.05 m $\times$ 0.05 m. Use $c = 340$ m/s. ∎

19.8 Summary

A study of the equilibrium state of a fluid at rest (hydrostatics), shows that the pressure field is related to the body force $\mathbf{f}$ by

$$\mathbf{f} = \nabla p.$$

Two important applications are Pascal's law and Archimedes' principle which we applied to fluids in equilibrium and to the atmosphere of the Earth. Pascal's law states that for constant body force, increasing the pressure at one point in the fluid increases the pressure at all other points in the fluid by an equal amount. Archimedes' principle states that the buoyant force on a body is equal to the weight of the displaced fluid.

Next came a consideration of some introductory concepts in fluid dynamics. We described the Eulerian and the Lagrangian approaches to the study of fluid flow. We defined the convective derivative, described by the operator relation

$$\frac{d}{dt} = \frac{\partial}{\partial t} + \mathbf{v} \cdot \nabla,$$

and derived the equation of continuity

$$\frac{\partial \rho}{\partial t} + \nabla \cdot (\rho \mathbf{v}) = 0.$$

Solenoidal flow ($\nabla \cdot \mathbf{v} = \mathbf{0}$) and irrotational flow ($\nabla \times \mathbf{v} = \mathbf{0}$) were defined.

The equations of motion for fluid flow were derived, leading to Euler's equation

$$\frac{\partial \mathbf{v}}{\partial t} + (\mathbf{v} \cdot \nabla)\mathbf{v} + \frac{1}{\rho}\nabla p = \frac{\mathbf{f}}{\rho}.$$

The conservation laws for mass, momentum, and energy in a fluid were developed.

These concepts were applied to a study of sound waves. Sound waves are longitudinal waves for which the pressure and the velocity of the fluid elements obey the wave equation. The wave equation in three dimensions was solved using separation of variables. As an application, sound waves in an organ pipe were considered.

19.9 Problems

Problem 19.1 A hydraulic automobile lift is made up of two interconnected cylindrical reservoirs filled with hydraulic fluid. One of the cylinders, of radius 5 cm, is attached to an air pump so the surface pressure can be increased from 14.7 lb/in^2 to 44.1 lb/in^2. The other cylinder has a radius of 25 cm. How heavy an automobile can be supported by the hydraulic fluid in the second cylinder?

Problem 19.2 A solid object of weight mg is suspended by a light thread and submerged in a liquid of density ρ_L. The tension in the thread is found to be $m'g$. Show that the density of the object is given by

$$\rho_O = \frac{m\rho_L}{\Delta m},$$

where $\Delta m = m - m'$,

Problem 19.3 I think everyone knows the story about Archimedes and the crown of the king of Syracuse. (In this problem I am going to make up some numbers because the story does not give us any quantitative information.) Suppose the king gave the goldsmith 1 kg of gold. The dishonest goldsmith made a crown of mass 1 kg but substituted some of the gold with lead. The suspicious king asked Archimedes to determine if the crown was pure gold. Archimedes submerged the crown in water and found that it had an apparent mass of 0.93 kg. (a) Evaluate the density of the crown and show it was not pure gold. (b) What was the mass of gold stolen by the goldsmith? Data: density of water = 1000 kg/m^3, density of gold = 19 320 kg/m^3, density of lead = 11 343 kg/m^3.

Problem 19.4 Under some circumstances the lower atmosphere can be considered adiabatic. Assuming an adiabatic atmosphere, the relation between temperature and pressure is

$$\frac{dT}{dP} = \frac{2}{7}\frac{T}{P}.$$

(Roughly speaking, air is an ideal diatomic gas with molecular weight 29.) Obtain an expression for the *lapse rate,* that is the rate at which the temperature decreases with altitude.

Problem 19.5 The compressibility (κ) is the inverse of the bulk modulus. Determine the isothermal compressibility of an ideal gas.

Problem 19.6 A graduated cylinder of radius 10 cm is filled to the 6 cm mark with mercury and then to the 12 cm mark with water. Mercury has a density 13.6 times that of water. Determine the total (integrated) force these liquids exert on the wall of the cylinder.

Problem 19.7 The potential energy $u\delta m$ associated with volume changes is given by the negative of the work done by the pressure on the surrounding fluid as the pressure increases from p_0 to p. Show that $u\delta m = \int_{p_0}^{p} \frac{p}{\rho B} dp$.

Problem 19.8 It is easy to observe that as water flows out of a faucet the stream gets narrower. I recently measured the diameter of the stream for my kitchen faucet and found that at the nozzle the stream had a diameter of 1 cm, but after falling for 20 cm the stream diameter had decreased to 0.5 cm. Assuming water is an ideal fluid, use this information to determine the flow rate (cm^3/s).

Problem 19.9 In a drinking fountain the water shoots straight up to a height h_1 with a velocity v_1 from a spout of area A_1. If the spout area is decreased to $A_2 = \frac{1}{2}A_1$, the water reaches a greater height h_2. Determine the ratio of the heights, h_2/h_1.

Problem 19.10 A cubical container of side a has a small spout located at a bottom corner. When the container is full, the speed of the water coming out of the spout is v_1. Determine the speed through the spout when the container is half empty. Express your answer in terms of v_1.

Problem 19.11 Draw the streamlines for two different flow fields: (a) the irrotational flow of water in a bucket that is on a rotating turntable, and (b) the rotational vortex flow of water going down a drain (imagine

Figure 19.11 Determine the speed of the outflowing fluid. (Problem 19.14.)

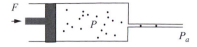

a circular tub with a drain at the center). Show that for (a) $v \propto r$ and for (b) $v \propto 1/r$. Note that for a bucket of water on a turntable the sides of the bucket exert a torque on the water.

Problem 19.12 Someday, in the far future, there will be a large "zero-gravity" park in outer space. The swimming pool in this park is a huge spherical water drop of radius R. When you jump in and swim towards the center, the pressure increases (just as on the Earth when you swim to the bottom of a swimming pool). In this problem, you are asked to derive an expression for the pressure inside the drop as a function of r, the distance from the center. For simplicity, assume the pressure at the surface of the drop (at $r = R$) is zero. Recall from Chapter 9 that $\nabla \cdot \mathbf{g} = -4\pi G\rho$, where $\mathbf{g}$ is defined in terms of the body force by $\rho\mathbf{g} = \mathbf{f}$.

Problem 19.13 D'Alembert's paradox states that an object moving at constant velocity in an ideal infinite fluid will not feel a drag force. He showed that this was a consequence of Euler's equation. Starting with Euler's equation show how d'Alembert arrived at this (erroneous) conclusion. Note that one can formulate the problem as an object at rest and a fluid in motion. Ignore gravity. (Comment: The difficulty lies in the concept of an infinite fluid at rest. Because the fluid has zero velocity far from the moving body, there is no way for energy to be dissipated. The resolution of the paradox is that for a finite fluid, waves are generated at the surface and energy is lost.)

Problem 19.14 A syringe with body inner radius 0.5 cm has a needle with orifice radius 0.16 mm. The syringe is filled with water and the plunger is pressed with a force of 210 N. Determine the speed of the water squirting out of the needle. See Figure 19.11.

Problem 19.15 Bernoulli's equation as described in Section 19.5.3, can be expressed as

$$p + \frac{1}{2}\rho v^2 - f = \text{constant}$$

along a streamline. Show that for steady, irrotational flow of an incompressible fluid, the constant is the same for all streamlines in a tube of flow.

Problem 19.16 Show that for a longitudinal sound wave the relationship between f and $\mathbf{g}$ is

$$\mathbf{g} = \frac{\hat{\mathbf{n}}}{\sqrt{B\rho_0}}f.$$

Problem 19.17 Consider an open-ended pipe and a pipe closed at one end but having the same dimensions. Show that the two pipes cannot have modes of the same frequency.

Problem 19.18 We assumed that in a sound wave the fluid density is given by $\rho = \rho_0 + \rho'$ with ρ' being much smaller than ρ_0. Show that this implies that the velocity of the fluid elements is much less than the speed of the wave. That is, show that if $\rho' \ll \rho_0$ then $v \ll c$.

Problem 19.19 An oscillator generates a sound wave in a long pipe filled with water. The oscillator generates compression waves with an amplitude of 10^{-7} m at a frequency of 1000 Hz. Determine the maximum pressure in the wave.

20 The Special Theory of Relativity

This chapter presents the basic ideas of the special theory of relativity. Relativity theory involves serious modifications to classical mechanics as well as to many of our basic notions about time, space, and causality.

Relativity often contradicts "common sense" concepts, as does quantum mechanics, the other great modification of classical physics. It may help you to think of classical mechanics as the study of objects that surround us in our familiar world. Balls and springs and falling objects all obey classical mechanics. When things are moving very, very fast, close to the speed of light, then classical physics must be modified. This is the realm of special relativity. Also, when things are very, very small (such as atoms and electrons) then there is a different set of modifications to classical physics. This is the realm of quantum mechanics. Very small things that are moving very fast fall into the realm of relativistic quantum mechanics. Figure 20.1 illustrates the realms of these theories.

20.1 Albert Einstein (Historical Note)

Albert Einstein was born in Ulm, Germany, in 1879 and died in the United States in 1955. When he was a high school student his family moved to Italy, but he attended university in Zurich, Switzerland. After graduation he found a position in the Swiss patent office, examining patent applications. During this time he developed the theory of relativity. In 1905 he published a paper describing what is now known as the "special" theory of relativity. The conclusions were not readily accepted by the scientific community. During the same year he also published important

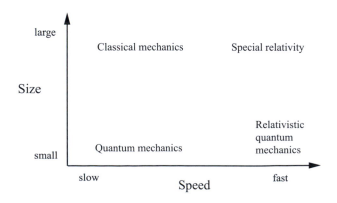

Figure 20.1 Different realms of physics. Classical mechanics applies to objects that are "large" and "slow."

papers on the specific heat of solids and a paper explaining the photoelectric effect. In 1915 he published the first paper on "general" relativity.[1] Special relativity concerns itself with reference frames moving at constant velocities and is of particular importance in electrodynamics. General relativity includes accelerated reference frames and is a theory of gravitation.

Einstein thus contributed to both electromagnetic theory and gravitation. His lifelong ambition was to develop a unified theory of gravitation and electromagnetism, but this goal eluded him.[2]

Although Einstein was one of the founding fathers of quantum mechanics there were many aspects of quantum theory that did not satisfy him. In particular, he was bothered by the probabilistic interpretation of quantum mechanics; he once declared "God does not play dice!" He also did not like to think that the state of a photon (for example) could be determined by what happens to another photon a great distance away. He called this "Spooky action at a distance." However, experimental evidence seems to indicate that Einstein was wrong in his dismissal of this interpretation of quantum mechanics.

Einstein became a world-famous personage, particularly after measurements during an eclipse bore out the prediction of the general theory of relativity that light from a star would be bent in the gravitational field of the Sun. Newspapers declared that only five people in the world could understand Einstein's theory. Of course that was not true, but it certainly added to his mystique. His opinions were sought out, even on subjects far removed from physics.

When the Nazi party came into power in Germany and began to persecute Jews, Einstein left Berlin and emigrated to the United States where he took a position at the Princeton Institute for Advanced Studies. He was convinced that German scientists would develop nuclear weapons and he signed a letter to President Roosevelt urging him to begin a project to develop the atomic bomb even though this went against his pacifist philosophy. Eventually he took out American citizenship. His last years were a model of a tranquil academic life. After he died a medical examiner removed his brain. The story of Einstein's brain and its subsequent travels is a fascinating but gruesome tale.[3]

20.2 Experimental Background

In the latter part of the nineteenth century, physicists were confronted with a paradoxical experimental fact. The work of Michelson and Morley in the years from 1881 to 1887 showed that the speed of light was the same in all reference frames. Whether the *source* of the light was approaching or receding from the observer, the speed of light was always the same, namely $c = 3 \times 10^8$ m/s (186 000 miles per second). Furthermore, regardless of whether the *observer* was approaching or receding from the source of the light, the speed of light was still 3×10^8 m/s. For example, as the Earth moves in its orbit, at one time of the year it will be approaching a distant star at a speed of about 30 km/s. Six months later, the Earth will be moving away from the star at 30 km/s. Thus, one would expect the speed of light measured at those two times to differ by 60 km/s, a difference that could be detected experimentally. However, when the measurement was performed, it was found that there was no difference in the speed of light, whether the Earth was approaching the star or moving away from it.

This experimental fact caused a great deal of concern among physicists. If you are in a car and someone throws a rock at it, the speed of the rock relative to the car depends on how fast the car

---

[1] Discussed briefly in Section 9.8.

[2] A readable biography is by Walter Isaacson, *Einstein, His Life and Universe*, Simon and Schuster, New York, 2007.

[3] *Driving Mr. Albert: A Trip Across America with Einstein's Brain*, Michael Paterniti, Dial Press, New York, 2001.

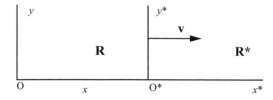

Figure 20.2 Two inertial reference frames, **R** and **R***
The relative velocity **v** is directed along the common
x-axes.

is going. But the Michelson–Morley results showed that the speed of light relative to the car does
not depend at all on the speed of the car!

Understanding the phenomenon of the constancy of the speed of light and finding an
explanation for it, requires an analysis of measurements made in two different reference frames.
The reference frames are quite simple – they are two inertial reference frames in relative motion.
Call the reference frames **R** and **R***. Both are moving at constant velocity. To make it easy to
visualize the situation, you can imagine that **R** is at rest with respect to the fixed stars and that **R***
is moving at velocity **v** relative to **R**. Assume that at some initial time ($t = 0$) the origins of the
two coordinate systems coincide and that the velocity **v** is directed along the common x-axes, as
shown in Figure 20.2.

Let us consider the relationship between the starred and the unstarred coordinates. Physicists
use the name "Galilean Transformation" to denote the classical relations. Note that in reference
frame **R** a particular *event* (such as the occurrence of a super nova, the ringing of a bell, or
any such well-localized event) is characterized by its position and the time it happened. That is,
an event is characterized by x, y, z, t. This involves the concepts of *space* and *time*. For centuries,
philosophers have debated the meaning of space and time, asking, "What is space? What is time?"
Einstein decided to cut through all the complications and give simple, operational definitions for
these two quantities. Basically, Einstein said "Space is what you measure with a ruler. Time is
what you measure with a clock."

An event that is located at x, y, z, t in **R** will be located at x^*, y^*, z^*, t^* in **R***. What is the
relation between these sets of coordinates? The Galilean transformations between x, y, z, t and
x^*, y^*, z^*, t^* are

$$x^* = x - vt, \tag{20.1}$$
$$y^* = y,$$
$$z^* = z,$$
$$t^* = t,$$

where the rulers and clocks in both reference frames are identical in all respects.

Although the transformations given by Equations (20.1) are quite satisfactory and work
perfectly well for most problems in mechanics, they cannot be correct because they do not give
the same speed for light in both reference frames. If x^* is the position of an object in **R*** at some
particular time, then the x component of its velocity in **R*** is $\dot{x}^*$, and the x component of its
velocity in **R** is $\dot{x}$ and is given by the time derivative of the first of Equations (20.1) as

$$\dot{x} = \dot{x}^* + v.$$

If the speed of light as measured in **R*** is c, then the speed of light as measured in **R** will be $c + v$.
But this is not the case. The speed of light is c in both reference frames.

Einstein realized that the problem lay in the Galilean transformations. He derived a new set of
transformations that reduced to the Galilean transformations for objects moving slowly compared
to the speed of light but which yielded the same value for the speed of light in both reference
frames. These transformation equations are called the "Lorentz" transformations because they

had been derived earlier by Lorentz. Einstein was not aware of Lorentz's work and he derived the equations himself independently. We shall derive the Lorentz transformation equations shortly, but first it is helpful to state the postulates of special relativity.

20.3 The Postulates of Special Relativity

The easiest approach to special relativity is to accept Einstein's two postulates; that is, to accept as unproved assumptions the following two statements.

Postulate 1 All inertial reference frames are equivalent.
Postulate 2 The speed of light is the same in all inertial reference frames.

The first postulate, which is often referred to as the "Principle of Relativity," implies that the laws of physics are the same in all inertial reference frames. The first postulate also implies there is no experiment that can demonstrate that a particular reference frame is at absolute rest. In fact, the concept of "absolute rest" is meaningless.

The second postulate simply states the results of experiments such as those of Michelson and Morley.

20.4 The Lorentz Transformations

Our goal is to obtain a set of transformation equations that will reduce to Equations (20.1) under the conditions of classical mechanics but will not contradict Postulate 2.

First, however, it is necessary to do away with preconceived notions of time and space. If you have two identical clocks, you expect them to run at the same rate. If you put one of the clocks in a rocket ship moving at a constant velocity, you probably do not expect the moving clock to run slow. Yet this is actually what you would observe. For example, if John is in the "stationary" reference frame $\mathbf{R}$ and his friend Mary is in the "moving" reference frame $\mathbf{R}^*$, John will claim that Mary's clock is running slow. Of course, from Mary's point of view, she is at rest and John is moving at a velocity $-\mathbf{v}$, and she will say that his clock is running slower than hers. Because John and Mary will never meet again, the question as to whose clock is actually running slow cannot be answered. (Later we will consider the "twin paradox" in which one of the observers turns around and comes back to compare clocks. However, if the two reference frames are at all times inertial, they will never meet again.)

Furthermore, if John uses telescopes and other devices to measure the length of a meter stick in Mary's reference frame, he will decide that her meter stick is shorter than his!

If all of this sounds like utter rot it may be because you believe that space and time are absolutes and you have not yet accepted that space and time are quantities that depend on the motion of the observer. If you accept Einstein's operational definition that time is what you measure with a clock and you concede the possibility that two clocks in relative motion might not run at the same rate, then you are close to accepting relativity theory.[4]

20.4.1 The Light Clock: A Gedanken Experiment

Einstein was fond of "gedanken" experiments, that is "thought" experiments, in which he applied the laws of physics to an imaginary situation and determined the logical outcome.

[4] I will often refer to a reference frame "at rest" or "moving." Perhaps each time I use these words I should add, "with respect to the other reference frame," but I will let you supply that phrase mentally.

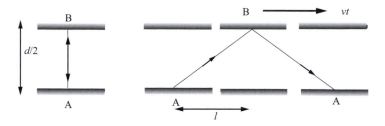

Figure 20.3 A light clock at rest in reference frame $\mathbf{R}^*$. The sketch on the right is the moving clock as seen by an observer in reference frame $\mathbf{R}$.

A well-known gedanken experiment describes the operation of a "light clock" which is made up of two mirrors. A pulse of light is emitted at one mirror, bounces off the other mirror, and returns to the first mirror. Some sort of detector lets us know when the light returns to the first mirror, and the clock ticks. (Nobody said this was a *practical* clock, but it certainly is a possible one.) The left panel of Figure 20.3 illustrates the clock. Assume Mary has the clock in her rocket ship (reference frame $\mathbf{R}^*$). The clock is at rest in this reference frame. The mirrors are separated by a distance $d/2$, so the time required for the light pulse to go from A to B and back to A is d/c, where c is the speed of light. The time between ticks (according to Mary) is $\Delta t_M = d/c$.

John, the observer in reference frame $\mathbf{R}$, sees Mary's clock as her rocket speeds past him. From his point of view, the pulse of light that goes from A to B to A and makes Mary's clock tick is now traveling a longer distance as illustrated in the sketch on the right in Figure 20.3. During the time the light pulse goes from A to B and back to A, the clock has moved to the right a distance $2l$. The distance traveled by the light ray, according to John, is $2s$ where $s^2 = l^2 + (d/2)^2$. John says the time between ticks on Mary's clock is $2s/c$. Let us call this Δt_J. Note that $s = c\,(\Delta t_J/2)$. Also $l = v\,(\Delta t_J/2)$. Consequently, $s^2 = l^2 + (d/2)^2$ can be written

$$c^2\left(\frac{\Delta t_J}{2}\right)^2 = v^2\left(\frac{\Delta t_J}{2}\right)^2 + \left(\frac{d}{2}\right)^2.$$

But we have seen that $d = c\Delta t_M$, so

$$c^2\left(\frac{\Delta t_J}{2}\right)^2 = v^2\left(\frac{\Delta t_J}{2}\right)^2 + c^2\left(\frac{\Delta t_M}{2}\right)^2.$$

Solving for Δt_J we find

$$\Delta t_J = \frac{\Delta t_M}{\sqrt{1 - v^2/c^2}}. \tag{20.2}$$

The denominator in this equation is less than unity so $\Delta t_J > \Delta t_M$. John claims that Mary's clock is ticking more slowly than it would at rest in his reference frame. He says her clock is running slow. This effect is known as "time dilation."

Remember that Δt_M is the time between ticks on a clock at rest with respect to Mary and Δt_J is the time between ticks on that same clock as determined by John. Mary's heart is a kind of clock. She says her heart is beating at 60 beats per minute. But John says that Mary's heart is beating more slowly, say, at 40 beats per minute. He decides she is aging more slowly than he is.

The time measured on the clock at rest in $\mathbf{R}^*$ is usually denoted t^*. Time intervals measured by the clocks in the two reference frames are related by

$$t_2^* - t_1^* = (t_2 - t_1)\sqrt{1 - v^2/c^2}. \tag{20.3}$$

Mary is at rest with respect to the clock in the rocket ship. If she observes the clock in John's reference frame, she will say it is his clock that is running slow. (After all, this is the theory of *relativity*!)

Having determined that moving clocks run slow (as observed from a stationary frame), let us consider the length of a moving object as observed from a stationary frame. Imagine Mary has a table in her rocket ship and the table is aligned along the common x-axis of the two systems. To determine the length of the table, she sets up a mirror at one end and a lamp at the other. She (somehow) measures the time for a light signal to go from the lamp to the mirror and back to the lamp. This time is measured on her clock and is denoted Δt_M (or Δt^*). She concludes that the length of the table is $l^* = (1/2)c\Delta t^*$. Note that the time of flight is $\Delta t^* = \Delta t_M = 2l^*/c$.

John observes this experiment. He notes that the light ray that left the lamp had to travel a distance $l + v\Delta t_1$ to get to the mirror, because the rocket moved the mirror a distance $v\Delta t_1$ while the light was traveling. The time for the light signal to get to the mirror is

$$\Delta t_1 = \frac{1}{c}(l + v\Delta t_1),$$

and solving for Δt_1 we find

$$\Delta t_1 = \frac{l}{c - v}.$$

The light signal now must return to the position of the lamp, but this time (according to John) it only has to travel a distance $l - v\Delta t_2$, where Δt_2 is the time of flight. So,

$$\Delta t_2 = \frac{1}{c}(l - v\Delta t_2),$$

and solving for Δt_2 we obtain

$$\Delta t_2 = \frac{l}{c + v}.$$

The total time for the light signal to propagate from lamp to mirror and back to lamp, according to John, is

$$\Delta t_J = \Delta t_1 + \Delta t_2 = \frac{l}{c - v} + \frac{l}{c + v} = \frac{2l/c}{1 - v^2/c^2}.$$

Finally, recalling that

$$\Delta t_J = \frac{\Delta t_M}{\sqrt{1 - v^2/c^2}} = \frac{2l^*/c}{\sqrt{1 - v^2/c^2}},$$

we equate the two expressions for Δt_J to get

$$\frac{2l/c}{1 - v^2/c^2} = \frac{2l^*/c}{\sqrt{1 - v^2/c^2}}.$$

Consequently,

$$l = l^*\sqrt{1 - v^2/c^2}. \tag{20.4}$$

That is, the measured length of a ruler is *longer* in a reference frame in which the ruler is at rest than in a frame in which it is moving. This effect is called "length contraction." It is sometimes convenient to think of l^* as the length of an object in a coordinate frame in which it is at rest and call it l_0, the "rest length." Then the length of the object, as seen by an observer in another reference frame is

$$l = l_0(1 - v^2/c^2)^{1/2}.$$

Exercise 20.1

The Space Shuttle is 60 meters in length. It flies past you at 27 000 miles per hour (1.2×10^4 m/s). How much shorter is it in your reference frame? Answer: 4.8×10^{-8} m. ∎

Exercise 20.2

Mary's rocket ship is moving at $0.92c$ relative to John. Mary's pulse rate, as measured by her, is 60 beats per minute. She puts a stethoscope over her heart and transmits her heartbeats over the radio to John. What is her pulse rate, according to John? Answer: 23.5/min. ∎

20.4.2 The Lorentz Transformations

Having determined the relativistic properties of clocks and meter sticks, we can now obtain the relations between positions and times in two coordinate systems in motion with respect to one another. That is, we will write the transformation equations relating (x^*, y^*, z^*, t^*) to (x, y, z, t). Before beginning, it is important to have a very clear idea what these coordinates refer to. Using again the John and Mary example, imagine that some event (E) occurs in Mary's rocket ship. For example, a light bulb burns out with a sudden pop. This event occurs at a well-defined place and time. Mary can measure the distance from her origin (O*) to the event and she can determine the time the event took place by looking at her clock. She characterizes E by the space-time coordinates (x^*, y^*, z^*, t^*).

John also observes the event, and he characterizes it by specifying the values of (x, y, z, t).

What is the relationship between the two sets of coordinates for the event E? (See Figure 20.4.)

Mary's rocket ship is moving along the common x, x^*-coordinate axes. Relativistic effects are due to the relative motion of the two coordinate systems. Since there is no relative motion along the y- or z-directions, it is clear that these components are unchanged. Hence the coordinates of E in these directions are related by

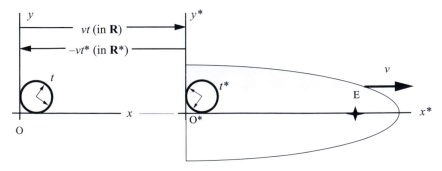

Figure 20.4 Event E is observed from two coordinate systems. The event is at (x, t) in the stationary system and at (x^*, t^*) in the moving system.

$$y^* = y,$$
$$z^* = z.$$

Suppose that John and Mary's clocks are at the origins of their coordinate systems and suppose that they synchronized their clocks, setting them to zero, at the instant the two origins coincided.

According to John, at the time of the event E the origin of Mary's coordinate system (O*) was at vt. According to Mary, at the time of the event, the origin of John's coordinate system was at $-vt^*$.

Mary uses a meter stick to measure the distance from O*, her origin, to E and she calls it l_0. John sees everything in the rocket ship as contracted and he says the distance from O* to E is

$$l = l_0\sqrt{1 - v^2/c^2}.$$

Therefore, according to Mary, the position of the event is $x^* = l_0$ and according to John it is $x = vt + l_0\sqrt{1 - v^2/c^2}$. Replacing l_0 by x^*,

$$x = vt + x^*\sqrt{1 - v^2/c^2}.$$

Solving for x^*,

$$x^* = \frac{x - vt}{\sqrt{1 - v^2/c^2}}. \tag{20.5}$$

Our job is half done, as we have determined the relation between x and x^*. Now we need to obtain the relation between t and t^*. This is quite easily done by noting that Mary will obtain the same sort of transformation equation between the starred and unstarred x-coordinates. She finds that

$$x = \frac{x^* + vt^*}{\sqrt{1 - v^2/c^2}}. \tag{20.6}$$

Plugging Equation (20.5) into (20.6) yields

$$x = \frac{1}{\sqrt{1 - v^2/c^2}}\left[\frac{x - vt}{\sqrt{1 - v^2/c^2}} + vt^*\right].$$

Solving for t^* gives

$$t^* = \frac{t - (xv/c^2)}{\sqrt{1 - v^2/c^2}}. \tag{20.7}$$

This completes the derivation of the Lorentz transformations. Gathering the equations together:

$$x^* = \frac{x - vt}{\sqrt{1 - v^2/c^2}}, \tag{20.8}$$
$$y^* = y,$$
$$z^* = z,$$
$$t^* = \frac{t - (xv/c^2)}{\sqrt{1 - v^2/c^2}}.$$

In relativity theory one often uses the shorthand expression

$$\gamma = \frac{1}{\sqrt{1 - v^2/c^2}},$$

so the transformation equations have the form

$$x^* = \gamma(x - vt), \tag{20.9}$$

$$t^* = \gamma\left(t - \frac{xv}{c^2}\right).$$

To obtain the "inverse" Lorentz relations, you can solve Equations (20.9) for x and t, obtaining

$$x = \gamma(x^* + vt^*), \tag{20.10}$$

$$t = \gamma\left(t^* + \frac{x^*v}{c^2}\right).$$

The Lorentz transformation equations yield the length contraction and time dilation relations; this is no surprise, because they were based on those concepts. However, these equations have to be applied with care, as it is easy to generate a wrong answer. The following worked examples may be helpful.

Worked Example 20.1 Mary measures the length of her rocket ship and finds it to be l_0. (That is, $x_2^* - x_1^* = l_0$.) John, who is at rest on the surface of a stationary asteroid, decides to measure the length of the rocket as it speeds past him. He calls this length l. Use the transformation equations to determine the relation between l and l_0.

Solution John must simultaneously determine the positions of the two end points of the rocket, x_1 and x_2. That is, the determinations of the locations of the end points are two events that take place at the same time. So $t_1 = t_2$. Then the first of Equations (20.8) yields

$$x_2^* - x_1^* = \frac{x_2 - x_1}{\sqrt{1 - v^2/c^2}},$$

or

$$l_0 = \frac{x_2 - x_1}{\sqrt{1 - v^2/c^2}},$$

leading to the usual length contraction relation,

$$l = l_0\sqrt{1 - v^2/c^2}.$$

(Note that the argument depended on our realizing that $t_1 = t_2$.)

Worked Example 20.2 Now suppose that Mary looks at the flashing red light on the tip of the wing of her rocket. She looks at the clock in her rocket ship and measures the time interval between two flashes (or events) and calls it $\tau_0(= t_2^* - t_1^*)$. John also sees the two events. According to his clock the time between the events is $\tau = t_2 - t_1$. What is the relation between the time measured by Mary and the time measured by John?

Solution Note that relative to Mary the two events occur at the same place, so $x_2^* = x_1^*$. The time interval in John's reference frame is calculated using the inverse relation, the last equation of (20.10). The result is

$$t_2 - t_1 = \frac{t_2^* - t_1^*}{\sqrt{1 - v^2/c^2}}.$$

Consequently,

$$\tau = \frac{\tau_0}{\sqrt{1 - v^2/c^2}}.$$

Thus $\tau > \tau_0$. The time interval in the rest frame (τ_0) is shorter than the time interval in the reference frame that is moving with respect to the events. Although John is at rest on the asteroid, in this example, his is the moving clock and moving clocks run slow.

It is interesting to consider the time read on a distant clock by a moving observer. Suppose that John has set out three synchronized clocks at positions $x = -l, 0, +l$. Mary is passing by in her rocket ship and at the instant she passes the origin of John's reference frame, she checks the times on the clocks. (Actually, it might take her a long time to make the determination, but she can always figure out later what the clocks read at the instant she passed John's origin.) In Problem 20.3 you are asked to show that according to Mary, the three clocks will read times of $+lv/c^2, 0, -lv/c^2$. Thus, the clock she is approaching reads an earlier time than the clock in her reference frame by $-lv/c^2$.

Exercise 20.3

Obtain Equation (20.6). Note the change in the sign in the numerator. ∎

Exercise 20.4

Carry out the steps to obtain Equation (20.7). ∎

Exercise 20.5

Obtain the inverse Lorentz transformation equations (20.10). ∎

Exercise 20.6

A person in a rocket ship with speed $0.92c$ relative to Earth measures the time between two events to be five hours. Both events happened at the same location on Earth. What is the interval between these events as measured by an observer on Earth? Answer: 1.96 hours. ∎

20.5 The Addition of Velocities

The theory of relativity predicts that no material object can be accelerated to a velocity greater than the speed of light. (You will prove this contention in Problem 20.21.)

But if the speed of light is the upper limit – the maximum speed for any material body – then we will have to drastically revise our idea of how velocities add. Imagine the following scenario: Professor Einstein (A) is at rest on the Earth. He is observing Captain Buck Rogers (B) who is flying away from the Earth in a rocket ship at a speed $v = 0.75c$. Buck Rogers fires a torpedo (C) in the forward direction at a speed of $0.5c$ relative to his space ship. See Figure 20.5. The question is, how fast is the torpedo going with respect to Earth? That is, what will Professor Einstein say is the speed of the torpedo?

Figure 20.5 Assume A is at rest. B is moving at a speed $v = 0.75c$ relative to A, and C is moving at a speed of $0.5c$ relative to B. What is the speed of C relative to A?

Your first impulse may be to simply add the velocity of B ($0.75c$) and the velocity of C ($0.5c$) and state that the speed of the torpedo relative to the Earth is $1.25c$. But this is wrong since the speed of the torpedo must be less than c in any reference frame. So how should you add the velocities?

It turns out to be fairly easy to derive the appropriate transformation law for velocities. Let the speed of the torpedo relative to the rest frame be u and its speed relative to the moving frame be u^*. The speed of the rocket ship with respect to the rest frame is v, because it is the speed of the moving frame with respect to the rest frame. For convenience, assume that A, B, and C all coincide at time $t = 0$. The problem consists in determining an expression for u in terms of u^*.

The position of the torpedo with respect to the moving frame is

$$x^* = u^* t^*.$$

Recall the transformation rules

$$x = \gamma(x^* + vt^*),$$
$$t = \gamma(t^* + vx^*/c^2).$$

Replace x^* by $u^* t^*$ and x by ut to obtain

$$u = \frac{u^* + v}{1 + (u^* v/c^2)}. \tag{20.11}$$

This is the relativistic "addition of velocities" law.

Worked Example 20.3 Refer to Figure 20.5. Buck Rogers fires a torpedo at $0.7c$ at an angle of $60°$ to the direction of motion. Determine the x- and y-components of the velocity of the torpedo as measured by Professor Einstein.

Solution Let us determine the addition of velocities law when the object has a velocity component in the y-direction. From Equation (20.11) the velocity component in the x-direction is

$$u_x = \frac{u_x^* + v}{1 + (u_x^* v/c^2)}.$$

The velocity in the y-direction is obtained from the Lorentz transformation equations that tell us that

$$y = y^*.$$

Although the y-components of *position* are the same in the two reference frames, the y-components of *velocity* are not the same because of time dilation.

Write $u_y^* t^*$ for y^* and $u_y t$ for y. That is,

$$u_y t = u_y^* t^*,$$

$$u_y \gamma (t^* + vx^*/c^2) = u_y^* t^*,$$

$$u_y = \frac{u_y^* t^*}{\gamma (t^* + vx^*/c^2)} = \frac{u_y^* t^*}{\gamma (t^* + vu_x^* t^*/c^2)}.$$

So,

$$u_y = \frac{u_y^*}{\gamma (1 + u_x^* v/c^2)}.$$

For the problem at hand, $v = 0.75c$, $u_x^* = 0.70c \cos 60°$, and $u_y^* = 0.70c \sin 60°$. Therefore,

$$u_x = \frac{0.70c \cos 60° + 0.75c}{1 + (0.70c \cos 60°)(0.75c)/c^2}$$

$$= \frac{1.10c}{1 + 0.263} = 0.87c,$$

and

$$u_y = \frac{u_y^*}{\gamma (1 + u_x^* v/c^2)} = \frac{0.70c \sin 60° \sqrt{1 - (0.75c)^2/c^2}}{(1 + (0.70c \cos 60°)(0.75c)/c^2)}$$

$$= \frac{0.10}{1.263} = 0.08c.$$

Exercise 20.7

Carry out the algebra steps to obtain Equation (20.11). ■

Exercise 20.8

Show that Equation (20.11) reduces to the expected relationships in the limits of velocities that are (a) much less than the speed of light, and (b) equal to the speed of light. ■

Exercise 20.9

Obtain the inverse relationship $u^* = u^*(u, v)$. ■

Exercise 20.10

A radioactive material is at rest in a physics laboratory. Two electrons are ejected in opposite directions. Each has a speed of $0.67c$ as measured in the laboratory frame. Determine the speed of one electron relative to the other (a) classically and (b) relativistically. Answer (b): $0.92c$. ■

20.6 Simultaneity and Causality

When two things happen at the same time, they are *simultaneous*. But events that are simultaneous for one observer will not, in general, be simultaneous for another observer.

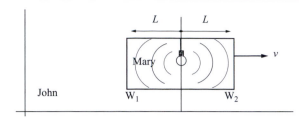

Figure 20.6 Mary turns on a light at the center of her rocket ship. She says the light reaches the two end walls simultaneously. But John sees the light reach wall W_1 before it reaches wall W_2.

For example, as shown in Figure 20.6, Mary turns on a light that is located at the exact center of her rocket ship. The light reaches the end walls (W_1 and W_2) of her ship simultaneously. But John, in his stationary reference frame, sees the light reach wall W_1 before it reaches W_2, because the light only travels a distance $L - vt$ to reach W_1, but it travels $L + vt$ to reach W_2.

If the light reaching the walls defines events E_1 and E_2, then Mary says that E_1 and E_2 are simultaneous, but John says E_1 occurred before E_2. A third stationary observer, toward whom Mary is moving, would say event E_2 occurred before E_1.

If one event *causes* another event, then it must precede the second event. In other words, causes must come before effects in *any* coordinate system. In our example, event E_1 can come before or after event E_2, depending on the observer. Therefore, event E_1 cannot in any way cause E_2.

For E_1 to cause E_2 there must be some sort of a signal from E_1 to E_2. For example, E_1 might be a bell that rings at W_1 when the light reaches it, and E_2 might be a bell that is rung at W_2 when a person at W_2 hears the bell from W_1. In this sense, you can imagine that E_1 causes E_2. As you will see, when two events are thus causally related, then E_1 *always* precedes E_2 in *all* coordinate systems.

To better appreciate this concept, let us note that according to Mary, events E_1 and E_2 were simultaneous. They occurred at positions $x_1^* = -L$, $x_2^* = +L$, and at times $t_1^* = L/c$ and $t_2^* = L/c$.

According to John, assuming the light was switched on just as the light bulb flew past him, the positions and times for the two events are (x_1, t_1) and (x_2, t_2). These are related to (x_1^*, t_1^*) and (x_2^*, t_2^*) as given in the following table.

Event = E_1	Event = E_2
$x_1 = \gamma(x_1^* + vt_1^*) = \gamma(-L + vt_1^*)$	$x_2 = \gamma(x_2^* + vt_2^*) = \gamma(L + vt_2^*)$
$t_1 = \gamma(t_1^* + \frac{v}{c^2}x_1^*) = \gamma\left(\frac{L}{c} + \frac{v}{c^2}(-L)\right)$	$t_2 = \gamma(t_2^* + \frac{v}{c^2}x_2^*) = \gamma\left(\frac{L}{c} + \frac{v}{c^2}L\right)$

Note that, although $t_2^* - t_1^* = 0$,

$$t_2 - t_1 = \gamma\left(\frac{L}{c} + \frac{vL}{c^2} - \frac{L}{c} + \frac{vL}{c^2}\right) = 2\gamma\frac{vL}{c^2}.$$

Thus, according to John, E_2 happened *after* E_1, as expected.

E_1 and E_2 cannot be causally related. In Mary's coordinate system, E_1 and E_2 occurred at the same time, so for E_1 to cause E_2, a signal from E_1 to E_2 would have to be propagated instantaneously, that is, at infinite speed. From John's point of view, the time between the events was not zero, but rather $2\gamma vL/c^2$. The spatial separation of the two events was

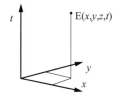

Figure 20.7 The event E represented as a point in a three-dimensional projection of four-dimensional space-time.

$$x_2 - x_1 = \gamma(L + vt_2^*) - \gamma(-L + vt_1^*)$$
$$= \gamma\left(2L + v(t_2^* - t_1^*)\right)$$
$$= 2\gamma L,$$

where the last step uses the fact that $t_2^* = t_1^*$. For E_1 to cause E_2, a signal would have to propagate from E_1 to E_2, and it would have to travel at a speed given by

$$\text{speed} = \frac{\text{distance}}{\text{time}} = \frac{2\gamma L}{2\gamma Lv/c^2} = c\left(\frac{c}{v}\right) > c.$$

Consequently, E_1 cannot be the *cause* of E_2 in any coordinate system.

A nice geometric way to understand the causality problem is based on the concept that an event is a point in a four-dimensional *space-time* coordinate system. An event is specified by three spatial coordinates (x, y, z) and one temporal coordinate (t). I cannot draw (or even visualize) a four-dimensional space, but if, for example, $z = 0$ at all times, then an event can be represented graphically as a point in a three-dimensional projection of the four-dimensional space-time diagram. See Figure 20.7.

Consider two events, call them E_0 and E_1. They are separated by a space-time interval S^2 defined by[5]

$$S^2 = (x_1 - x_0)^2 + (y_1 - y_0)^2 + (z_1 - z_0)^2 - c^2(t_1 - t_0)^2. \tag{20.12}$$

Suppose, for the moment, that event E_0 is the turning on of a light at the origin. The coordinates of E_0 are $(0, 0, 0, 0)$. Further, let E_1 be the arrival of the light beam at some other point (x, y, z, t). The distance traversed by the light is ct and this must be equal to the spatial separation between the points. So for these two events, Equation (20.12) leads to $S = 0$. In the $z = 0$ plane, the light from E_0 spreads out in a circle whose radius is $(x^2 + y^2)^{1/2} = ct$. The spreading of light is represented by a family of circles, each one slightly larger than the previous one and slightly higher along the time axis. These circles generate a cone in the three-dimensional slice of space time, as illustrated in Figure 20.8. Events that are causally related must lie within the same light cone.

20.7 The Twin Paradox

One of the most famous "paradoxes" of special relativity is known as the Twin Paradox. The paradox is this: One twin stays at home (at rest) while his brother travels at a high speed to a distant place, turns around, and comes home. The stay-at-home twin notes that his brother's clocks (heartbeat, etc.) have been "ticking" slowly because he was traveling at a high velocity. The traveling twin will not have aged as much as the stay-at-home twin, so when they meet again, the traveling twin will be younger than his brother!

---

[5] We will consider space-time intervals in more detail in Section 20.9.

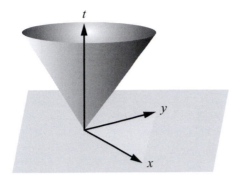

Figure 20.8 The light cone. This is a three-dimensional slice of the four-dimensional space time continuum. The z-axis is not shown. Light emitted at time zero at the origin will lie on the surface of the cone.

At first glance this may appear to be no paradox at all because special relativity is concerned with inertial systems. Two systems moving at constant relative velocity will only meet once, then separate to infinity, never to see each other again. Thus (at first glance) you might be tempted to exclaim, "The problem is poorly posed! The traveling twin turned around and therefore underwent accelerations. He was not in an inertial system. All bets are off!"

However, upon further reflection, you will appreciate that the accelerations of the traveling twin really do not change the nature of the paradox. Suppose the traveling twin (let's call him Flash) accelerates for a short time then travels at a high speed (say $3c/5$) for 10 Earth years. He then turns around and travels back to Earth at $3c/5$ for another 10 years. Flash will have only aged 16 years (work it out!) while his stay-at-home brother (called Homer) has aged 20 years.

Now suppose that instead of traveling for 10 Earth years at $3c/5$, Flash had traveled for 20 Earth years, then returned. In this case, Homer aged 40 years but Flash only aged 32. You can appreciate that the accelerations that Flash underwent were the same in both cases, but the end result was different. Thus, it is not the accelerations that causes the age differences, but rather the long-term, high-speed trip in the inertial frame.

Let us analyze this twin paradox a bit more carefully. Homer and Flash are at home and are the same age when Flash jumps onto a rocket ship that happens to be heading out towards the star Pollux which is 35 light years away. That is, the distance to Pollux from Earth is $35c$ (if you want to express distances in meters, you have to express c in meters/year). The outgoing rocket is an inertial system moving at $3c/5$. Flash rides the rocket out to the star. He then jumps onto an incoming rocket which just happens to be heading back to Earth at $3c/5$. (Ignore the fact that Flash will be flat as a pancake after leaping from one rocket to the other.) Homer says that his twin, Flash, was gone for a total time of 117 years: $(2 \times 35c)/(3c/5)$. Homer is 117 years older now. But Flash notes that once he jumped onto the outgoing rocket, he saw that the distance to Pollux was only $(4/5)(35) = 28$ light years, and therefore it only took him 46.6 years to get there, and another 46.6 years to return on the incoming rocket ship, for a total of 93.2 years. So Flash is only 93 years older, while Homer aged 117 years.[6]

By the way, to get really dramatic results, you need to let Flash go much faster, like at $0.95c$.

Worked Example 20.4 A 30-year-old man has a 10-year-old daughter. The man travels away and back home at (essentially) constant speed. When he returns, both he and his daughter are 60 years old. How fast did the man travel?

---

[6] Some interesting aspects of the twin paradox are discussed in an article by S. Wortel, S. Malin, and M. Semon, "Two Examples of Circular Motion for Introductory Courses in Relativity," *American Journal of Physics*, **75**, 1123 (2007).

Solution In the rest frame of Earth the elapsed time is 50 years ($t^* = 50$). In the moving reference frame the elapsed time is 30 years ($t = 30$). In the rest frame the man traveled a distance l_0 at speed v, so

$$t^* = 50 = \frac{\text{distance}}{\text{velocity}} = \frac{l_0}{v}.$$

Also

$$t = 30 = \frac{\text{distance}}{\text{velocity}} = \frac{l}{v} = \frac{l_0\sqrt{1 - v^2/c^2}}{v}$$

$$30 = \frac{50v\sqrt{1 - v^2/c^2}}{v}.$$

Consequently,

$$\frac{3}{5} = \sqrt{1 - v^2/c^2}$$

and

$$v = \frac{4}{5}c.$$

Exercise 20.11

Assume Flash travels to and back from Pollux at $0.98c$. Determine how much he and Homer have aged. Answer: 71.42 yr and 14.7 yr. ∎

Exercise 20.12

The "triplet paradox" imagines one triplet traveling to the right at speed v, one traveling to the left at speed v, and one staying at home. The traveling triplets turn around and come back home. Will the traveling triplets be the same age or different ages when they meet again? ∎

20.8 Minkowski Space-Time Diagrams

A geometric way to treat relativistic problems in terms of "space-time" diagrams was developed by Hermann Minkowski, who had been one of Einstein's professors. Many complicated relativistic problems, especially those concerning simultaneity, become easier to analyze by using these space-time diagrams because they reduce relativistic situations to simple geometrical constructions.

Suppose observer A is at rest. Draw a two-dimensional rectangular coordinate system with coordinate axes x and ct, as shown in Figure 20.9. Assume the only spatial coordinate of interest is x. Note that the time coordinate has been multiplied by c so that it also represents a distance. During a time interval from 0 to t, a light ray moves a distance $x = ct$ so light rays are always represented by lines at 45° to both the x-axis and t-axis. The vertical line at $x = -4$ is the "world line" for an object at rest in A's frame, so that it is at the same position at all times. The wiggly line starting at $x = -2$ is the world line for an object that moves first in a negative direction, then in a positive direction, and then in a negative direction again. The world line on the right-hand side represents a ray of light propagating in the $+x$ direction.

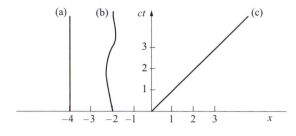

Figure 20.9 The space-time diagram for observer A (assumed at rest). Three "world lines" are shown: (a) a stationary object, (b) an object moving back and forth, and (c) a light ray.

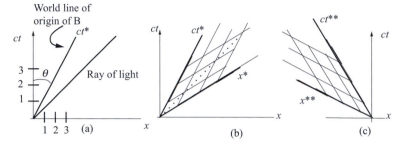

Figure 20.10 Sketch (a) shows the world line for the origin of observer B's coordinate system. Sketch (b) shows the oblique coordinate system for observer B, and sketch (c) shows the coordinate system of observer C who is moving to the left.

Consider a similar space-time coordinate system for observer B who is moving to the right with respect to A at a constant speed v. For simplicity, assume B's origin coincides with A's origin when both of their clocks read zero ($t = t^* = 0$). Because B's origin moves with respect to A, it also has a world line that can be represented on A's space-time diagram. The world line of the origin of B is shown as a sloped line in Figure 20.10(a). Note that this world line has a greater slope than the world line for the ray of light because, of course, B is moving at less than the speed of light. The world line for B's origin makes an angle θ with the ct-axis, and it is easy to appreciate that $\tan\theta = v/c$. You will note that this world line is labeled ct^* because along this line, for any value of time in B's system, $x^* = 0$. In Figure 20.10(b) the space-time diagram for B is completed by adding the x^*-axis at an angle $\theta = \tan^{-1}(v/c)$ with respect to the x-axis. I have also drawn in a set of oblique lines to indicate that B's space-time coordinate system is not rectangular. I have not, however, placed any numerical values along these oblique axes because we do not yet know the *scale* for the axes of the moving coordinate system. Finally, Figure 20.10(c) shows the coordinate system for observer C who is moving to the *left* with respect to A. However, the A and B systems are sufficient for the present analysis.

The space-time diagrams (except for the scales for the B system) are now set up and ready to be used in solving relativistic problems. A good place to begin is with length contraction, because that not only illustrates the use of the Minkowski diagrams, it also yields the correct way to put scales on the B system. In Figure 20.11 I drew a ruler of length l_0 (which you can consider to be one meter long) at rest in the A system. The world lines of the two ends of the ruler in the A system are simply vertical lines. The ruler at rest in A's system is observed by B, the moving observer. To determine the ruler's length, B must see both ends simultaneously. According to B the ruler is an object whose world lines are the inclined lines in the figure. This length is smaller than l_0.

Now if you were very attentive, you would have noticed something strange. The ruler, as *shown* in the figure, is *longer* in B than in A, as you can appreciate from the fact that in B it is the

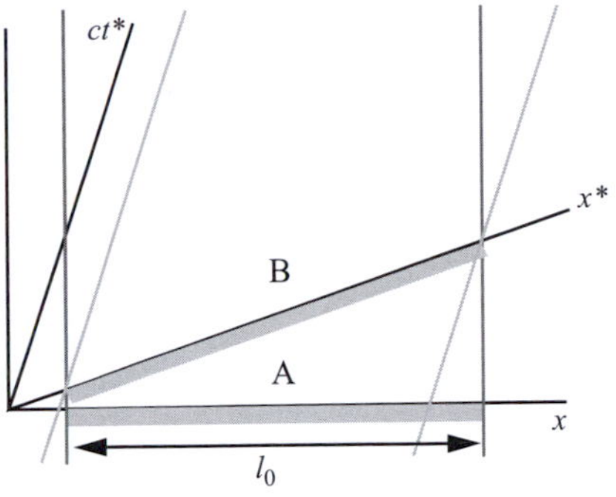

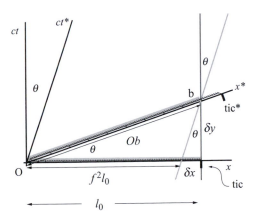

Figure 20.11 A ruler at rest in the A system as seen by A. The world lines of the ends of the ruler are vertical lines. These world lines as seen by B in the moving system are shown as inclined lines.

Figure 20.12 The ruler of length l_0 in the A system appears to have length Ob in the B system. A ruler of length Ob at rest in the B system, appears to have length $f^2 l_0$ in the A system.

hypotenuse of a triangle and in A it is along the adjacent side. You know that from the moving system, the ruler must appear shorter. The problem disappears, however, when you realize that we have not decided on the *scale* of the various axes. (The ruler must be shorter according to B. This means that the "tics" along the x^*-axis must be further apart than the tics on the x-axis. I'll get to this point in a moment.) Just to make things easier to analyze, in Figure 20.12 I moved the origin of coordinates so that one end of the ruler coincides with the origin in both systems.

Figure 20.12 indicates three lengths. The first one is l_0, the length of the ruler in the system A, which you can take to be one meter. The next one is Ob, the length of the ruler as measured by B. It is shortened by a factor f that has not yet been determined. Also we do not (yet) know the scale for the starred axes. With all these limitations, you might well ask how you could possibly determine the length Ob. Well, there is a clever way around the problem, and that involves going back to the A system that does have a scale. Imagine that B happens to have a rod that is exactly the length $Ob = f l_0$. If A observes this rod, it will appear to have shrunk by another factor of f. So in the A system, this rod has a length $f^2 l_0$. All of this fiddling around with rulers and going from one inertial system to another, will allow you to determine the amount f by which moving objects shrink, because now it is a straightforward geometry problem. In the figure the angle $\theta = \tan^{-1}(v/c)$ is indicated in several places. You can see that

$$f^2 l_0 = l_0 - \delta x$$

and $\delta x = \delta y \tan \theta$. But $\delta y = l_0 \tan \theta$, so $\delta x = l_0 \tan^2 \theta$, and hence

$$f^2 = 1 - \tan^2 \theta = 1 - (v^2/c^2).$$

Taking the square root, the length contraction factor is

$$f = \sqrt{1 - \frac{v^2}{c^2}},$$

as you knew all along. This is the graphical derivation of the Lorentz contraction factor. Note that I was careful not to use any lengths in the B system because there was no scale for it. Now, however, you can determine the scale on the starred coordinate system. That is, you can determine where to place the "tic marks" along the B system axes, x^* and t^*. Begin by assuming that according to A the distance l_0 corresponds to one meter. Call it one A-meter. If a distance of one meter (according to B) is marked off along the x^* axis, it must be longer than Ob because according to B the distance Ob is less than one B-meter by a factor of $\sqrt{1 - (v/c)^2}$. So,

$$1\text{B-meter} = \frac{Ob}{\sqrt{1 - (v/c)^2}}.$$

From Figure 20.12,

$$\frac{Ob}{1\text{A-meter}} = \frac{1}{\cos\theta} = \sqrt{1 + (v^2/c^2)}.$$

Therefore, the relation between a B-meter and an A-meter is

$$1\text{B-meter} = \frac{\sqrt{1 + (v^2/c^2)}}{\sqrt{1 - (v^2/c^2)}} (1\text{A-meter}).$$

Exercise 20.13

Show that $\cos\theta = 1/\sqrt{1 + (v^2/c^2)}$.   ■

20.9 4-Vectors

An event is specified by its position and time in a particular coordinate system. It is convenient to introduce a four-dimensional coordinate system in which an event is specified by the four coordinates (x_0, x_1, x_2, x_3) where

$$x_0 = ct,$$
$$x_1 = x,$$
$$x_2 = y,$$
$$x_3 = z.$$

(By multiplying the time coordinate by c, all coordinates are distances, although x_0 is the coordinate that gives the time when the event occurred.)

Three spatial coordinates describe an ordinary three-dimensional vector. Similarly, the four space-time coordinates (x_0, x_1, x_2, x_3) describe a "4-vector" that starts at the origin and ends at some event E. A 4-vector from event E^1 to event E^2 has components $(x_0^2 - x_0^1), (x_1^2 - x_1^1), (x_2^2 - x_2^1), (x_3^2 - x_3^1)$. Note that the superscripts indicate the event. Unfortunately, this makes for a somewhat confusing notation because *powers* are also superscripts.

Just as an ordinary 3-vector dotted into itself gives a scalar, the 4-vector "displacement" (called D_μ) from E^1 to E^2 generates a scalar called the "space-time interval" denoted S^2:

$$S^2 = (x_1^2 - x_1^1)^2 + (x_2^2 - x_2^1)^2 + (x_3^2 - x_3^1)^2 - (x_0^2 - x_0^1)^2 = d^2 - c^2 t^2.$$

Note the change in sign on the temporal term. S^2 was previously introduced in Equation (20.12). Because S^2 is a scalar, its value is unchanged under a Lorentz transformation that expresses S^2 in some other inertial coordinate system. S^2 is called a four-scalar (or sometimes, a "world scalar").[7]

You may recall from Chapter 17 that a three-dimensional vector is defined to be a quantity whose components transform in the same way as the coordinates under the rotation of the coordinate axes. Similarly, a 4-vector is defined to be a quantity whose components transform according to the Lorentz transformation (which is to say, the same as the coordinates). The transformation of the coordinates from the unstarred to the starred system according to the Lorentz transformation was presented in Equations (20.8) which I now repeat using a slightly different notation:

$$x_0^* = \gamma(x_0 - \beta x_1),$$ (20.13)
$$x_1^* = \gamma(x_1 - \beta x_0),$$
$$x_2^* = x_2,$$
$$x_3^* = x_3,$$

where $\beta = v/c$ and $\gamma = 1/\sqrt{1 - \beta^2}$. These equations can be expressed more succinctly by defining an array $a_{\mu\nu}$ as follows

$$a_{\mu\nu} = \begin{pmatrix} \gamma & -\beta\gamma & 0 & 0 \\ -\beta\gamma & \gamma & 0 & 0 \\ 0 & 0 & 1 & 0 \\ 0 & 0 & 0 & 1 \end{pmatrix}.$$

Then

$$x_\mu^* = \sum_\nu a_{\mu\nu} x_\nu.$$ (20.14)

This equation describes the way the four coordinates transform. We define a 4-vector to be a quantity that transforms according to Equation (20.14). That is, the components of the 4-vector A_μ transform thus:

$$A_\mu^* = \sum_\nu a_{\mu\nu} A_\nu.$$

It is traditional in dealing with relativity to use Greek letter subscripts for 4-vectors. Also, you should note that 4-vectors are not written in boldface the way 3-vectors are. We usually represent a 4-vector simply as A_μ.

For example, the 4-vector displacement D_μ is the 4-vector originating at one event and terminating at another. The components of D_μ in the unstarred (rest) system are (D_0, D_1, D_2, D_3) and its components in the starred system are

---

[7] A more elegant way to write S^2 uses the following metric (see Section 5.3.4):

$$g_0 = -1,$$
$$g_1 = g_2 = g_3 = 1,$$

so that

$$S^2 = \sum_{\mu=0}^{3} g_\mu (x_\mu^2 - x_\mu^1)^2.$$

$$D_0^* = \gamma(D_0 - \beta D_1),$$
$$D_1^* = \gamma(D_1 - \beta D_0),$$
$$D_2^* = D_2,$$
$$D_3^* = D_3.$$

The algebra of 4-vectors is straightforward. For example, if s is a 4-scalar, then

$$B_\mu = sA_\mu$$

will be a 4-vector if A_μ is a 4-vector. Similarly, if A_μ and B_μ are both 4-vectors, then the sum $A_\mu + B_\mu$ is also a 4-vector. In many respects, 4-vectors behave like 3-vectors, as well they might, since for $v = 0$ the components A_1, A_2, A_3 of a 4-vector reduce to the three components of an ordinary 3-vector.

The scalar product of two 4-vectors is defined as follows:[8]

$$\left(A_\mu, B_\mu\right) = \sum_\mu g_\mu A_\mu B_\mu = A_1 B_1 + A_2 B_2 + A_3 B_3 - A_0 B_0. \tag{20.15}$$

On the other hand, there is no 4-vector analogue to the cross product.

Exercise 20.14

Prove that if A_μ and B_μ are both 4-vectors, then the sum $A_\mu + B_\mu$ is also a 4-vector.   ■

Exercise 20.15

Show that $\left(A_\mu, B_\mu\right)$ is a 4-scalar, that is, its value is not changed by a Lorentz transformation.   ■

20.9.1 Space-like and Time-like Intervals

Consider two events, E^1 and E^2 and the related scalar S^2. From the definition of S^2 as given by Equation (20.12) we can see that S^2 can be positive or negative. If S^2 is positive, the interval between the two events is called space-like. (In this case each event is within the light cone of the other one.) For such an interval, it is possible to define the real quantity "proper distance" between the two events by

$$\sigma = (S^2)^{1/2}.$$

This quantity is the distance between the two events in a coordinate system in which the events are simultaneous. An example would be when the lights on the tips of the wings of my rocket ship flash on and off at the same time, but they are separated in space.

If S^2 is negative, a real "proper time" can be defined by

$$\tau = \frac{(-S^2)^{1/2}}{c}.$$

Physically, this is the time interval between two events in a coordinate system in which the two events occur at the same place. Then S^2 is negative. A simple example is when a light on the dashboard of my rocket ship flashes twice. In my reference frame the two events occur at the same location but are separated in time.

---

[8] A purist would point out that the scalar product is the product of a contravariant vector with a covariant vector. See Section 9.8.

Exercise 20.16

Show that the proper time is invariant under a Lorentz transformation   ∎

20.9.2 The 4-Vector Velocity

Once you have defined proper time, you can define the 4-vector velocity. This is different than the 3-vector velocity in some important respects. Consider a moving particle that is at position $\mathbf{x}$ at time t and at $\mathbf{x} + d\mathbf{x}$ at time $t + dt$. These two events occur at well-defined space-time coordinates x_μ and $x_\mu + dx_\mu$ where dx_μ is a 4-vector. The proper time between these two events is given by $\tau = (-S^2)^{1/2}/c$, so

$$d\tau = \left[-(dx_\mu, dx_\mu)\right]^{1/2}/c,$$

and the 4-velocity is defined by

$$U_\mu \equiv \frac{dx_\mu}{d\tau}. \tag{20.16}$$

Note that although the spatial part of the 4-velocity (U_μ) looks like an ordinary 3-velocity (u_i), it is really quite different. The 3-velocity of a moving object is determined by measuring the distance it moved and dividing by the elapsed time, where both distance and time are measured in the same coordinate frame (usually assumed to be at rest). However, the 4-velocity is a "distance" measured in some reference frame divided by the proper time, which is a quantity independent of the coordinate system. The difference between them is brought out when one writes the transformation equations for the 3-velocity components and compares them with the transformation equations for the 4-velocity. These are:

3-velocity transformation equations (for relative velocity v along the x-axes)

$$u_x^* = \frac{u_x - v}{(1 - vu_x/c^2)},$$

$$u_y^* = \frac{u_y}{\gamma(1 - vu_x/c^2)},$$

$$u_z^* = \frac{u_y}{\gamma(1 - vu_x/c^2)},$$

4-velocity transformation equations

$$U_0^* = \gamma(U_0 - \beta U_1),$$
$$U_1^* = \gamma(U_1 - \beta U_0),$$
$$U_2^* = U_2,$$
$$U_3^* = U_3.$$

Worked Example 20.5 Express the components of the 4-velocity in terms of the 3-velocity.

Solution Consider a clock at rest in $\mathbf{R}^*$. A time interval read on this clock is the proper time denoted $d\tau$. The time interval according to an observer in $\mathbf{R}$ is dt. These times are related by

$$d\tau = \sqrt{1 - v^2/c^2}dt.$$

The 4-velocity is defined by

$$U_\mu = \frac{dx_\mu}{d\tau},$$

so

$$U_1 = \frac{dx_1}{d\tau} = \frac{v_x dt}{\sqrt{1 - v^2/c^2} dt} = \frac{v_x}{\sqrt{1 - v^2/c^2}} = \gamma v_x.$$

Similarly

$$U_2 = \gamma v_y,$$
$$U_3 = \gamma v_z.$$

Finally,

$$U_0 = \frac{dx_0}{d\tau} = \frac{cdt}{\sqrt{1 - v^2/c^2} dt} = \frac{c}{\sqrt{1 - v^2/c^2}} = \gamma c.$$

Worked Example 20.6   A nuclear physicist has marked off an x-axis and a y-axis on the lab bench and placed a radioactive source at the origin. The source emits an alpha particle that moves at a 30° angle to the x-axis with a speed of $2c/3$. Determine the four components of the 4-velocity.

Solution   The three ordinary velocity components (relative to the laboratory) are

$$u_1 = u_x = v \cos 30° = \frac{2c}{3} \frac{\sqrt{3}}{2} = \frac{\sqrt{3}}{3} c,$$

$$u_2 = u_y = v \sin 30° = \frac{2c}{3} \frac{1}{2} = \frac{1}{3} c,$$

$$u_3 = u_z = 0.$$

The 4-velocity components are

$$U_i = \gamma u_i, \qquad (i = 1, 2, 3)$$
$$U_0 = \gamma c.$$

Note that $\gamma = 1/\sqrt{1 - (2c/3)^2/c^2} = \sqrt{9/5} = 3/\sqrt{5}$. Therefore,

$$U_0 = \gamma c = \frac{3}{\sqrt{5}} c,$$

$$U_1 = \gamma u_1 = \frac{3}{\sqrt{5}} \frac{\sqrt{3}}{2} c = \sqrt{\frac{27}{20}} c,$$

$$U_2 = \gamma u_2 = \frac{3}{\sqrt{5}} \frac{1}{3} c = \frac{c}{\sqrt{5}},$$

$$U_3 = \gamma u_3 = 0.$$

Exercise 20.17

Show explicitly that the components of the 4-velocity transform like a 4-vector. Start with the definition, Equation (20.16). ∎

20.10 Relativistic Dynamics

It is interesting to consider two basic dynamical principles, namely the conservation of momentum and the conservation of energy, using the methods of special relativity.

Having defined the 4-velocity, it is straightforward to define the 4-momentum of a particle by

$$p_\mu = mU_\mu.$$

Here m is the mass of the particle measured in a system in which the particle is at rest (m is called the "rest mass" of the particle). Note that the time and space parts of the momentum are

$$p_0 = \gamma mc,$$
$$p_i = \gamma mu_i \quad (i = 1, 2, 3).$$

For $u \ll c$, the space part of the 4-vector p_μ reduces to the classical momentum 3-vector. For a system of particles, the total 4-vector P_μ has components

$$P_\mu = \sum_j p_{j\mu} = \sum_j m_j U_{j\mu},$$

where m_j is the mass of the jth particle.

Now consider the conservation of momentum. Assume that a group of particles can interact with each other but are isolated from any other particles (no external forces act on the group of particles). Furthermore, the particles all maintain their identity (no particles break apart nor do particles fuse together).

Next, assume that the 4-momentum is conserved. (I will not *prove* that the 4-momentum is conserved, but I assure you that this conservation is confirmed by a vast body of experimental evidence and it can also be derived theoretically.) Then the total 4-momentum is the same at all times, i.e.,

$$\left(P_\mu\right)_{t1} = \left(P_\mu\right)_{t2}. \tag{20.17}$$

What are the consequences of this statement? Equation (20.17) states that the 3-momentum is conserved, and this is as expected, but it also means that P_0 is constant. That is,

$$\sum_j \gamma_j m_j c = \text{constant},$$

where

$$\gamma_j = \left(1 - \frac{u_j^2}{c^2}\right)^{-1/2}.$$

Expanding this expression, using the binomial expansion, yields

$$P_0 = \sum_j \gamma_j m_j c = c \sum_j \left(1 - \frac{u_j^2}{c^2}\right)^{-1/2} m_j,$$

$$= c \sum_j \left\{1 + \frac{1}{2}\frac{u_j^2}{c^2} + \frac{3}{8}\frac{u_j^4}{c^4} + \cdots\right\} m_j,$$

$$= \sum_j m_j c + \frac{1}{c}\sum_j \frac{1}{2}m_j u_j^2 + \cdots$$

or

$$cP_0 = \left(\sum_j m_j\right) c^2 + T + \mathcal{O}(1/c^2). \tag{20.18}$$

The last term is negligible. I wrote T for $\sum \frac{1}{2} m_j u_j^2$ because that is the kinetic energy of the system of particles. Note that, in relativity, energy and momentum have become enmeshed in the single quantity 4-momentum. Einstein identified the quantity $c P_0$ with the total energy E.

$$E = c P_0.$$

An equivalent expression for the total energy is $E = \gamma m c^2$.

Writing Equation (20.18) for a single particle and replacing $c P_0$ with E,

$$E \doteq mc^2 + T, \tag{20.19}$$

which states that the total energy of a particle is given by its kinetic energy plus a quantity mc^2. You might suspect that mc^2 is a constant because it is nearly always acceptable to add a constant to the energy and only consider energy differences. Thus, you might feel that relativity theory has only resulted in adding a constant to the energy. But if the total energy E is constant, this interpretation would mean that T, the kinetic energy, is also constant, and you know that is not necessarily true. In the interaction between two particles (a collision), the kinetic energy is conserved only if the interactions are perfectly elastic. Otherwise, kinetic energy is lost. According to relativity theory, the term mc^2 represents *all other forms of energy* (thermal energy, potential energy, chemical energy, etc.). Equation (20.19) states that the total energy E is equal to the kinetic energy T plus all other forms of energy. Thus, if you heat an object (at rest), the value of mc^2 must increase. Recall that m is the rest mass. This means that the heated object is more massive than the same object when cool.

The total energy of an object at rest (so that $T = 0$) is given by the well-known equation

$$E = mc^2.$$

For example, if you heat a gram of water by one degree centigrade, you supply it with one calorie (4.186 joules) of heat. You will have increased the mass of the water by the unmeasurably small amount $4.18/c^2 = 4.64 \times 10^{-17}$ kg. In like manner, if a nucleus splits apart (fission) and the fragments are less massive than the original nucleus, then energy is emitted. This energy is, as we all know, quite significant.

It is sometimes convenient to write the rest mass of a particle as m_0. Then the momentum of a particle is

$$P_i = \gamma m_0 u_i \qquad (i = 1, 2, 3),$$

and the "relativistic mass" m_{rel} is defined as

$$m_{\text{rel}} = \gamma m_0 = \frac{m_0}{\sqrt{1 - v^2/c^2}},$$

where v is the velocity of the particle. This notation suggests that the mass of a moving particle increases with its velocity and tends toward infinity as the speed approaches c. You might consider the concept of relativistic mass to be a needless variant, and, in fact, modern physicists tend to dismiss it. Nevertheless, it is helpful in solving collision problems involving relativistic particles, particularly because relativistic mass is conserved in collisions.

An interesting and useful relation is obtained by dotting the momentum 4-vector into itself. According to Equation (20.15) we obtain

$$(p_\mu, p_\mu) = -\gamma^2 m^2 c^2 + \gamma^2 m^2 (u_x^2 + u_y^2 + u_z^2) = m^2 \frac{-c^2 + u^2}{1 - u^2/c^2} = -mc^2.$$

But we can also write

$$(p_\mu, p_\mu) = -(p_0)^2 + \mathbf{p} \cdot \mathbf{p} = -E^2/c^2 + p^2,$$

so

$$E^2 - p^2c^2 = m^2c^4.$$

This alternate expression involving E and p is often quite useful because it does not depend explicitly on the velocity.

20.11 Summary

In this brief review of the theory of relativity, I only considered aspects of the theory that relate to mechanics and did not delve at all into electromagnetic phenomena. Nevertheless, you should be aware that it was electromagnetic considerations that led Einstein to develop this theory.

The fundamental concepts of the theory are embodied in the Lorentz transformations, from which time dilation and length contraction follow. These transformations lead to a number of unusual effects, including paradoxes such as the relative aging of moving twins. The Lorentz transformations are

$$x^* = \frac{x - vt}{\sqrt{1 - v^2/c^2}},$$

$$y^* = y,$$

$$z^* = z,$$

$$t^* = \frac{t - \left(xv/c^2\right)}{\sqrt{1 - v^2/c^2}}.$$

Time dilation is described by

$$\tau = \frac{\tau_0}{\sqrt{1 - v^2/c^2}},$$

and length contraction is given by

$$l = l_0\sqrt{1 - v^2/c^2}.$$

The relativistic addition of velocities along the line of motion is given by

$$u = \frac{u^* + v}{1 + \frac{u^*v}{c^2}}.$$

An event is represented by a point in the four-dimensional space time continuum and the interval between two events is

$$S^2 = (x_1 - x_0)^2 + (y_1 - y_0)^2 + (z_1 - z_0)^2 - c^2(t_1 - t_0)^2.$$

Any two events that are causally related must lie within the same light cone.

The Minkowski diagrams introduce no new physics, but they help one to analyze relativistic effects. These lead naturally to the mathematical definition of 4-vectors which can be used to determine the famous expression $E = mc^2$.

In relativity, velocity has four components and these transform according to

$$U_0^* = \gamma(U_0 - \beta U_1),$$
$$U_1^* = \gamma(U_1 - \beta U_0),$$
$$U_2^* = U_2,$$
$$U_3^* = U_3.$$

The fact that the relativistic expression for the space part of the 4-vector momentum is $p_i = \gamma m u_i$ is often interpreted as the mass increase of moving objects and one frequently sees the expression for the mass

$$m = \frac{m_0}{\sqrt{1 - v^2/c^2}}.$$

However, from the point of view of the theory, it is not the mass but the *momentum* that is affected by the motion of the reference frame.

20.12 Problems

Problem 20.1 An observer in **R** measures an event at $x = 3$ m and $t = 7$ ns. The **R*** frame is moving along the common x-axis at $v = 0.6c$. (a) What are the relativistic space-time coordinates measured by the observer in **R***? (b) What space-time coordinates would be measured in **R*** if the Galilean transformation law held?

Problem 20.2 Use the Lorentz transformations to show that an observer moving at speed v toward a clock at rest in some reference frame will find the clock reads a time differing from the time on the clock at rest by $-lv/c^2$. At the instant of the measurements the two observers are at the same location and at a (rest frame) distance l from the clock.

Problem 20.3 John, the observer in **R**, sets out three synchronized clocks at locations $x = -l, 0, +l$. Mary (in frame **R*** traveling at speed v along x) observes the times on the clocks at the instant she passes the origin of John's reference frame. Show that according to Mary the clocks read $+lv/c^2, 0, -lv/c^2$.

Problem 20.4 An electron in a linear accelerator is given an energy of 50 MeV. It is moving through a tube of length 10 meters (according to the laboratory reference frame). From the point of view of the electron, what is the length of the tube?

Problem 20.5 An observer in **R** notes that an event occurred at $x = 3 \times 10^8$ m at time 3 s. Reference frame **R*** has a speed of $0.5c$ in the positive x-direction. (a) Determine the coordinates of the event according to an observer in **R***. (b) Determine the coordinates of the event according to an observer in a reference frame moving at $0.5c$ in the negative x-direction.

Problem 20.6 Consider two events, E_1 and E_2. In **R**, the first event is a light being turned on at some point along the x-axis. The second event occurs one microsecond later when another light is turned on a distance 1000 meters away. (a) There is an inertial reference frame **R*** moving at speed v relative to **R** in which these two events are simultaneous. Explain why this is possible. (b) What is v, the relative speed of **R***? (c) What is the spatial separation of the two events in **R***?

Problem 20.7 A pion is a particle that has a half-life of 1.8×10^{-8} seconds (in its rest frame). Suppose a beam of pions has been produced by collisions between protons and some appropriate material, and it is found that they have a speed of $0.99c$. Measurements a distance 38 meters away show that half of the pions are present in the beam. (a) Show that this result is not consistent according to classical physics. (b) Show that time dilation will account for the measurement.

Problem 20.8 John and Mary both have rocket ships. In their rest frames the ships have a length of 200 m. They pass one another, heading in opposite directions. Mary measures the time for John's ship to pass her and gets a value of 4 μs. (a) What is the relative speed of the two ships? (b) Mary has set up two clocks in her ship, one at either end. Each is turned on when the nose of John's ship is directly opposite. What is the time difference between her clocks?

Problem 20.9 Show that the speed of a particle whose kinetic energy is equal to its rest energy ($m_0 c^2$) is a constant independent of its mass. Determine that speed.

Problem 20.10 An electron is accelerated through a potential difference of 5×10^6 V. Determine its kinetic energy and velocity in the laboratory frame.

Problem 20.11 The clock in Mary's rocket ship is moving along the positive x-axis, which is marked off in meters. As it passes the origin, the clock reads zero and when it passes the 200 m mark, it reads 0.5 μs. How fast is her rocket ship moving?

Problem 20.12 We have seen that if the "clock" illustrated in Figure 20.3 is moving perpendicular to the direction of the light pulse in R^* the rate at which the clock is running in the two reference frames is related by $T = T^*/\sqrt{1 - v^2/c^2}$. (In the notation in the text, T^* is Δt_M and T is Δt_J.) Show that we obtain the same expression for time dilation if the clock is moving in the same direction as the light pulse.

Problem 20.13 Formulate the Lorentz transformation equations in terms of a new variable w defined by $w \equiv ct$. Then show that the Lorentz equations and their inverses for the x- and t-coordinates are given by:

$$x^* = \gamma (x - \beta w), \qquad x = \gamma (x^* + \beta w^*),$$
$$w^* = \gamma (w - \beta x), \qquad w = \gamma (w^* + \beta x^*),$$

where $\beta = v/c$. (Note that these transformation equations are more symmetric than the original set.)

Problem 20.14 The Earth's speed in its orbit around the Sun is 30 km/s. Determine the shortening of the Earth's diameter as observed from a reference frame at rest with respect to the Sun. Is it valid to assume the Earth is moving in a straight line?

Problem 20.15 Some pions are created high in the Earth's atmosphere by collisions between cosmic rays and the nuclei of nitrogen. These pions have very high speeds. Assume $v = 0.998c$. A pion at rest has a mean life of 26 ns. Assuming the pion moves directly downward, determine how far toward the surface such a pion can travel before it decays.

Problem 20.16 Princess Leia is standing at one end of a landing field waiting for Luke Skywalker to fly over. She sees a light flash from the far end of the field (it is 2.7 km away) and 10 μs later she sees a flash from a light that is 200 m away. Luke also observes these flashes but he sees them at the same location. (a) Use this information to determine how fast Luke is traveling. (b) What is the time interval between flashes in Luke's reference frame?

Problem 20.17 Show that the spacetime interval S^2 associated with two events is invariant under a Lorentz transformation. You may assume the events occur on the common x-axis.

Problem 20.18 The theory of relativity correctly predicts the aberration of light. (This phenomenon can be described as follows: If the Earth were at rest and a particular star were directly overhead, you would observe it by pointing your telescope straight up. But because the Earth is moving, you must compensate by pointing your telescope at a small angle away from straight up.) The aberration of light means that a ray of light that makes an angle θ^* with the x-axis in the rest frame will make an angle θ in the moving frame where

$$\tan \theta = \frac{\sin \theta^* \sqrt{1 - v^2/c^2}}{\cos \theta^* + v/c}.$$

Consider a light source at rest in $\mathbf{R}^*$. The fraction of light emitted into a cone of half angle θ^* is

$$f = \frac{1}{2} \left(1 - \cos \theta^* \right).$$

In this problem, you will show that in a moving reference frame, the same amount of light is emitted into a smaller cone of half angle θ. (a) Derive the relationship $f = 0.5 \left(1 - \cos \theta^* \right)$. (b) Assume that $\theta^* = 40°$ and determine f in a coordinate system in which the light source is at rest. (c) Calculate the half angle of a cone in a frame moving at $v = 0.9c$ into which the same amount of light will be emitted. (d) Why do you think this is called the "headlight" effect?

Problem 20.19 Evaluate the work done on an electron when it is accelerated from rest to $0.98c$.

Problem 20.20 Determine the speed and kinetic energy of a particle whose momentum is equal to m_0c.

Problem 20.21 Show that a particle that travels at the speed of light must have zero rest mass.

Problem 20.22 A proton is traveling at $0.995c$ in the laboratory frame. What is its kinetic energy? What is its momentum?

Problem 20.23 Two identical particles of rest mass m_0 approach one another at speeds $\pm 0.5c$, and undergo a completely inelastic collision. What is the momentum and energy of the final (composite) particle? Note that the momentum of a particle is given by $p = \gamma m_0 v$, where $\gamma = \gamma(v)$ depends on the speed of that particular particle.

Problem 20.24 In some particular reference frame (perhaps the laboratory), a particle traveling at $0.8c$ collides with and sticks to an identical particle at rest. (a) Determine the momentum of the resulting combined particle. (b) What is its speed? (c) What does classical physics yield for the speed? (You may assume that when the particles are at rest they have mass m_0.)

Problem 20.25 A particle traveling at $0.8c$ (in the lab frame) collides elastically with a particle at rest. The rest mass of the incoming particle is three times greater than the rest mass of the second particle. Determine the velocities of the two particles after the collision. (Hint Number 1: Transform to a coordinate system in which the total momentum is zero. Hint Number 2: The speed of the center of momentum system is $v_c = P_T c^2 / E$, where P_T and E are the total momentum and energy.)

Problem 20.26 A well-insulated container maintained at $20\,°C$ is placed on an impossibly accurate scale. You put a 50 gram ice cube at $0\,°C$ in the container and wait a long time until the system comes to equilibrium at $20\,°C$. What is change in reading of the scale?

Problem 20.27 A relativistic train has a speed of $(4/5)c$ relative to a worker standing outside, next to the tracks. A passenger inside the train measures its length to be 200 meters. Suddenly, two lightning bolts strike the train, one at the front end and the other at the rear. The passenger (who is at the center of the train) says they struck the train at exactly the same instant. When the lightning bolt hit the rear of the train, the rear of the train was directly opposite the worker. How much time elapses before the worker sees the lightning flash from the bolt that hit the front of the train?

Problem 20.28 A wooden wedge in the shape of a right triangle is on board a rocket traveling at $(4/5)c$. The dimensions of the triangle are $4, 5$, and $\sqrt{41}$ meters in its rest frame. Determine the shape of the triangle (calculate its dimensions) relative to a reference frame at rest. What is the fractional change in its area?

Problem 20.29 Determine $a_{\mu\nu}$ for the general case in which the starred axes are not parallel to the unstarred axes. Show that a subsequent transformation to a third system yields $a'_{\lambda\nu} = \sum a^*_{\lambda\mu} a_{\mu\nu}$.

21 Classical Chaos (Optional)

This book has concentrated on "integrable" problems, that is, problems whose equations of motion can be integrated yielding closed-form solutions.

Some problems do not yield closed-form solutions, nevertheless they can be integrated numerically and are found to have well-behaved solutions. An example of this kind of problem is the motion of the Moon. As you know, the Moon is affected by the gravitational force of the Sun and the planets as well as that of its primary (the Earth). The resulting equation of motion is very complicated and cannot be integrated by analytical methods. However, numerical solutions have been carried out. You will be relieved to know that the Moon will continue orbiting the Earth for many millions of years.

In considering the motion of the Moon, the effects of the Sun and planets are treated by the techniques of perturbation theory in which the dominant motion is integrable and the weak interactions are treated as perturbations. An example of this is found in Section 10.11 where we considered the effect of a small perturbation on a previously circular orbit. (The perturbation was a comet colliding with a planet.)

For most mechanical systems a small perturbation, or a slight change in the initial conditions, will lead to a solution that is not very different from the original solution. Such cases are called "regular" or "normal." However, there are some physical systems for which a very small change in the initial conditions can lead to a dramatically different motion. Such systems and their solutions are referred to as "chaotic." Note, however, that these systems are perfectly *deterministic*. If we start the system again with exactly the same initial conditions, the system will precisely retrace the same steps. The solutions, however, are not *predictable* because they are highly sensitive to the initial conditions which are never known with absolute accuracy.

A completely random system, a system characterized by unpredictable stochastic behavior, is *not* an example of chaos as considered here. Classical chaos is a type of motion that has some degree of regularity. The behavior of a completely random system cannot be predicted, but a chaotic system is deterministic, and if you know the state of the system at any given moment, you can evaluate its future behavior.[1]

If you are wondering why some systems are chaotic and others are not, a glib answer is, "Chaos is a consequence of nonlinear equations of motion."[2] This gives us a reason why chaotic motion occurs, but it does not tell us what chaos *is*. I hope this short chapter will give you a basic understanding of chaos.

---

[1] A nice description of chaotic motion is presented in Chapter 11 of *Classical Mechanics*, 3rd ed., by H. Goldstein, C. Poole and J. Safko, Addison Wesley, San Francisco, 2002.

[2] Nonlinear equations often do not yield analytical solutions, so a study of chaos involves using numerical methods and access to a good computer.

21.1 Configuration Space and Phase Space

The Lagrangian for a physical system is a function of the coordinates and the velocities. You will recall that the Lagrangian for a mass on a spring can be expressed in terms of x and $\dot{x}$ as

$$L = T - V = \frac{1}{2}m\dot{x}^2 - \frac{1}{2}kx^2.$$

At any instant of time, the state of the system is described by the values of x and $\dot{x}$ at that time. On a plot of $\dot{x}$ vs. x, the system would be represented by a single point. As time goes on, the values of x and $\dot{x}$ will change and the point will trace out a path.

The Lagrangian for the double pendulum of Figure 4.2 is a function of θ_1, θ_2 and $\dot{\theta}_1, \dot{\theta}_2$. The system of masses and springs in Figure 11.11 had a Lagrangian depending on the coordinates and velocities, x_1, x_2 and $\dot{x}_1, \dot{x}_2$.

The values of the positions (such as x_1 and x_2) at any instant of time are said to describe the "configuration" of the system, and the plot of x_1 vs. x_2 is said to be a representation of the system in "configuration space."

Another way to describe a physical system is to give the Hamiltonian. You recall that the Hamiltonian is a function of momentum and position. $H = H(p,q)$. For example, the Hamiltonian for a mass on a spring is

$$H = \frac{p^2}{2m} + \frac{k}{2}x^2.$$

At any instant of time, the state of a system is given by the values of p and x. For a one-dimensional system, these can be represented by a point on a p vs. x plot. As time goes on the values of p and x will change and the point will trace out a path in the px-plane. This plane is called "phase space" and the path is called a phase-space trajectory. For most simple systems, the momentum and the velocity are related by $p = m\dot{x}$, so plots of $\dot{x}$ vs. x are also called phase-space plots (or phase-space "portraits").

Although chaotic motion is not periodic, it is convenient to begin our study by considering some aspects of periodic motion. You are familiar with periodic systems such as simple harmonic oscillators and the Kepler (two-body) problem. The Kepler problem is interesting because it involves two different periodicities, namely, the periodicity in the radial motion and the periodicity in the angular motion.

The radial position of a particle in an elliptical orbit is

$$r = \frac{a\left(1 - e^2\right)}{1 + e\cos\theta},$$

and the angular velocity is given by

$$\dot{\theta} = \frac{l}{mr^2}.$$

The unperturbed two-body system, as considered in Chapter 10, is *degenerate*; that is, the period of the radial motion is equal to the period of the angular motion, and the orbit is repeated over and over again. A configuration-space plot of r vs. θ is simply the elliptical orbit of the particle.

The Hamiltonian for a Keplerian particle is a function of p_r, p_θ, r, θ. It is interesting to generate phase-space plots for this motion. One normally plots velocity rather than momentum vs. position, that is, one plots $\dot{r}$ vs. r and $\dot{\theta}$ vs. θ. Figure 21.1 gives these plots for various values of e, using arbitrary units.

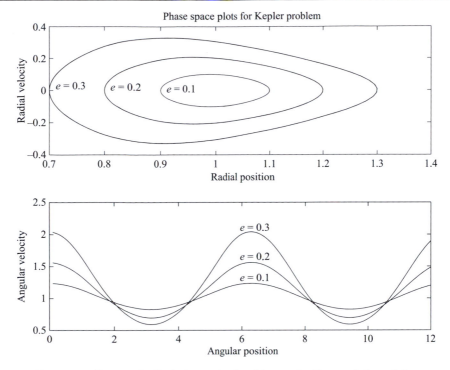

Figure 21.1 Phase-space diagrams for Keplerian motion in arbitrary units. Top panel ($\dot{r}$ vs. r), bottom panel ($\dot{\theta}$ vs. θ).

Exercise 21.1

Letting $k = 1$ and $m = 1$, generate a phase-space plot for a mass on a spring for $H = 1, 2, 3$.

■

Exercise 21.2

What would the plots in Figure 21.1 look like for $e = 0$?

■

21.2 Periodic Motion

Although the Hamiltonian was introduced in Chapter 4 we did not need it to solve the various dynamical problems that have been considered. However, to study chaos, the Hamiltonian approach is indispensable.

Recall that the Hamiltonian is a function of the momenta, the positions, and time, thus:

$$H = H(p_1, p_2, \ldots, p_n; q_1, q_2, \ldots q_n; t)$$

where n is the number of degrees of freedom. We shall only consider Hamiltonians that do not depend on time.

In Section 4.9 we determined that the equations of motion in the Hamiltonian formulation are

$$\dot{p}_i = -\frac{\partial H}{\partial q_i} \quad \text{and} \quad \dot{q}_i = +\frac{\partial H}{\partial p_i}. \tag{21.1}$$

Under most usual conditions, the Hamiltonian reduces to the total energy and one can write

$$H = T + V.$$

Let us consider, as an example of periodic motion, a double oscillator in which the motions in mutually perpendicular directions are independent. The Hamiltonian for such a system can be written as

$$H = \frac{(p_1')^2}{2m_1} + \frac{1}{2}m_1\omega_1^2\left(q_1'\right)^2 + \frac{(p_2')^2}{2m_2} + \frac{1}{2}m_2\omega_2^2\left(q_2'\right)^2 \tag{21.2}$$

which is just $E_1 + E_2 = $ total energy. It is convenient to transform to "normalized" coordinates p_i and q_i defined by

$$p_i = \frac{p_i'}{\sqrt{2m_i}} \quad \text{and} \quad q_i = q_i'\sqrt{\frac{1}{2}m_i\omega_i^2}.$$

In terms of these ps and qs the Hamiltonian is

$$H = p_1^2 + q_1^2 + p_2^2 + q_2^2 = E_1 + E_2.$$

The phase-space plots are simply circles of radius $\sqrt{E}$, as shown in Figure 21.2.

It is interesting to represent both diagrams on a single plot. Assume $\omega_2 \gg \omega_1$. First plot p_1 vs. q_1 as a circle in the horizontal plane. Next plot p_2 vs. q_2 in a vertical plane, selected such that p_2 is perpendicular to the horizontal plane and the q_2 axis points towards the origin. Thus, at a given instant of time, the combined plot will be the two circles shown in Figure 21.3.

At an instant of time, the state of the system is represented by a single point whose position on the "high-frequency" circle gives the values of p_2 and q_2. The instantaneous values of p_1 and q_1 are given by the location of the center of the smaller circle. As time goes on the high-frequency circle moves around on the larger circle, and the point traces out a helix. Eventually all of the locations of this point will generate a torus. The point representing the state of the system moves on the surface of the torus. If the two frequencies are commensurable, that is, if

$$\frac{\omega_2}{\omega_1} = \text{an integer},$$

the trajectory of the system point will be closed and the path will repeat over and over. If the frequencies are not commensurable, the trajectory will never close and the system point will

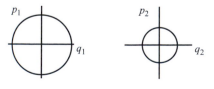

Figure 21.2 Phase-space diagrams for double oscillator for the case $E_2 < E_1$.

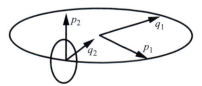

Figure 21.3 A phase-space diagram for the double oscillator.

gradually cover the surface of the torus, eventually coming arbitrarily close to every point on the surface.

For a triple oscillator, the motion takes place on a three-dimensional surface called a 3-torus in six-dimensional space. (The parameters are p_1, q_1, p_2, q_2, and p_3, q_3.)

21.3 Attractors

If a bounded, periodic, dynamical system is subjected to a small perturbation, its behavior will change somewhat, but you would still expect it to be bounded. The Russian mathematician Andrey Kolmogorov in 1954, and later (in the 1960s) Vladimir Arnold and Jurgen Moser, developed and proved a theorem to that effect. It has become known as the KAM theorem and it states that a bounded system subjected to a small perturbation will remain bounded. However, the theorem does allow for chaotic motion to occur in some systems if the initial conditions are just right.

Consider a periodic system acted upon by a small perturbing force. This force will affect the orbit in various ways. For example, the orbit may decay and the phase-space trajectory might end up as a point in phase space and remain there from then on. Or the system might evolve to a stable orbit (called a "limit cycle") and remain in that orbit from then on. For these situations, the point or the stable orbit are referred to as "attractors."

In other words, an attractor is a set of points in phase space to which the solution evolves, usually after a long interval of time. As an example, consider a simple pendulum. The motion of an unperturbed pendulum can be described by a circle in phase space. But if there is a small perturbation, say a weak drag force, then the amplitude of the pendulum swings will grow smaller and smaller and the momentum will gradually decrease, until the pendulum comes to rest. If you were to plot this on a phase-space diagram you would see the trajectory is an inward spiral that ends up at the origin $(p_\phi = 0, \ \phi = 0)$. The origin would be the attractor in this case.

Another example is a planet perturbed by an impact with a comet. As you saw in Section 10.11, the original circular orbit became an elliptical orbit, but the planet thereafter remained in the elliptical orbit. This final orbit is the attractor for the perturbed planet. It is a limit cycle.

If the attractor is a point in phase space (as in the case of the pendulum) it is called an attractor of dimension zero. If the attractor is a limit cycle, as in the case of the planet, it is an attractor of dimension one. If the attractor is the surface of a torus, it has dimension two.

Some attractors associated with chaotic motion have noninteger dimensions and are called "strange attractors." For a strange attractor, the phase-space points visit and revisit a region of phase space, filling it with a series of points. This is often a signature of chaos.

As an example of a system with a one-dimensional attractor, consider the van der Pol equation,

$$m\frac{d^2x}{dt^2} - \epsilon(1 - x^2)\frac{dx}{dt} + m\omega_0^2 x = F\cos\omega_d t. \tag{21.3}$$

This equation actually gives a reasonable description for oscillations of some mechanical, electrical, and biological systems. It has the form of the equation for a nonlinearly damped, driven pendulum. Note that $F = 0$ and $\epsilon = 0$ lead to simple harmonic motion. If $F = 0$ and ϵ is small, the phase-space trajectory tends to a circle (a limit cycle), as indicated in Figure 21.4.

Exercise 21.3

Plot the phase-space trajectory for a damped oscillator. ∎

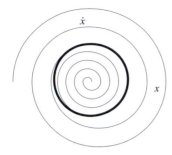

Figure 21.4 Depending on whether x^2 is greater or less than unity, the system spirals in or out to the limit cycle. For stronger damping, the limit cycle is distorted in shape.

21.4 Chaotic Trajectories and Liapunov Exponents

If the phase-space trajectory ends up at a point or on a limit cycle, as described above, the motion is well behaved and the trajectory is confined to a limited region of phase space. Such motion is *not* chaotic.

Sometimes, however, the trajectory will wander about in a seemingly random manner thus filling a *region* of phase space (which is the strange attractor described above). The trajectory never passes through the same phase-space point twice. This is chaotic motion, and it is characterized by three properties.

Property 1: Mixing If I_1 and I_2, are two small regions in the domain of the motion, then an orbit or trajectory that passes through I_1 will eventually pass through I_2.

Property 2: Dense, Quasi-Periodic Orbits Chaotic orbits are called "quasi-periodic" because they repeatedly pass through the whole range of the domain. However, they are not closed and there is no periodicity (no regularity) in the time between visiting and revisiting a particular region. The trajectory visits and revisits every region of available phase space (which is the "mixing" property).

Property 3: Sensitivity to Initial Conditions A very small change in the initial conditions will lead to a very large change in the final state. For example, in the nonchaotic flow of a stream of water, two nearby water particles will remain close to one another, but if the flow is turbulent, they will be separated by an ever-increasing distance.

A measure of the sensitivity to initial conditions is given by the Liapunov exponent. Consider two nearby points on nearly identical orbits. If their separation at time $t = 0$ is s_0, then at a later time their separation is

$$s(t) \sim s_0 e^{\lambda t}, \tag{21.4}$$

where λ is called the Liapunov exponent.

If $\lambda > 0$, the motion is chaotic and s will grow until it has the size of the available coordinate space, after which it will vary randomly with time.

If $\lambda < 0$, the system approaches a regular attractor.

For the planets in our neighborhood, it is believed that $\lambda \approx 3 \times 10^{-10}$/year, so the solar system seems to be chaotic, although barely so. That is, it is a case of "marginal stability."

21.5 Poincaré Maps

Consider again the double oscillator whose Hamiltonian is given by Equation (21.2). The motions of the two oscillators were independent, that is, the motions were uncoupled. The phase-space plots were two circles and showed, as expected, that the two oscillations were independent. Nevertheless, it was possible to represent the motion on a single, albeit somewhat complex, plot.

If, however, the motions are not independent, things get more complicated. For example, perhaps the Hamiltonian contains a term such as $q_1^2 q_2$. In this case the phase-space plots, in general, do not separate and we need to use the entire four-dimensional phase space with axes p_1, p_2, q_1, q_2.

Suppose, however, that the total energy is constant. This gives one constraint. Recall that each constraint reduces the number of independent coordinates by one. Therefore, the motion is confined to a three-dimensional region of phase space. This is a three-dimensional surface called an "energy hypersurface." As an example, consider Keplerian motion. In Cartesian coordinates the position is specified by x and y. The Hamiltonian will be a function of p_x, p_y, x, y. But if the energy is constant, then there is a relation between these coordinates, namely

$$E = T + V = \frac{p_x^2}{2m} + \frac{p_y^2}{2m} - \frac{k}{(x^2 + y^2)^{1/2}}. \tag{21.5}$$

The easiest way to visualize the motion is to take a two-dimensional slice through the energy hypersurface. This two-dimensional slice is called a "Poincaré section" or, more commonly, a "surface of section." The orbit is, of course, just an ellipse, as shown in Figure 21.5.

At points A and A', the x component of momentum is zero. Plotting the values of x, p_x at these two points (i.e., when $y = 0$) gives the extremely simple plot shown in Figure 21.6.

The surface of section contains only two features, the dots at $x = A$ and $x = A'$. I shaded in the left-hand side of the plot because it is usually not shown. Then the surface of section has just one point, at $x = A$. This just means that each time $y = 0$, the value of x is A and the value of p_x is zero. Clearly, the motion is regular.

Now suppose some sort of perturbation causes the elliptical orbit to precess. Then the orbit will have the general appearance of Figure 21.7.

Note that when $y = 0$ the planet is not always at the perihelion and the point going through the x, p_x-plane will gradually move along the trajectory shown in Figure 21.8. (The point representing the system configuration gradually moves from A to B to C and so on, generating the x, p_x curve shown.)

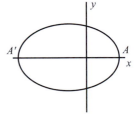

Figure 21.5 A Keplerian orbit in configuration space. Points A and A' are the periapsis and apoapsis.

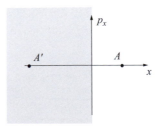

Figure 21.6 The surface of section for a Keplerian orbit.

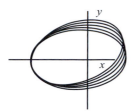

Figure 21.7 A precessing elliptical orbit.

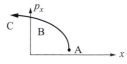

Figure 21.8 The surface of section for a precessing ellipse.

Thus, a Poincaré section is a two-dimensional slice through a three-dimensional energy hypersurface in a four-dimensional phase space. For higher-dimension phase spaces it is possible to draw Poincaré surfaces, but their usefulness is mainly limited to four-dimensional phase space.

21.6 The Henon–Heiles Hamiltonian

Henon and Heiles[3] suggested that the motion of stars in a galaxy could be approximated by the following two-dimensional Hamiltonian:

$$H = \frac{p_x^2}{2m} + \frac{p_y^2}{2m} + \frac{k}{2}(x^2 + y^2) + \lambda\left(x^2 y - \frac{1}{3}y^3\right), \tag{21.6}$$

where k is related to the gravitational constant. To simplify without losing any essential information, let $k = 1$ and $m = 1$. If $\lambda = 1$ the equations of motion are

$$\ddot{x} = -x - 2xy, \tag{21.7}$$
$$\ddot{y} = -y - x^2 + y^2,$$

and the potential energy is

$$V = V(x, y) = \frac{1}{2}(x^2 + y^2) + x^2 y - \frac{1}{3}y^3.$$

This expression has an interesting form, being a nonlinear expression for the potential energy. The subsequent motion in such a potential well depends on the choice for the total energy. For very small values of the total energy, the position variables are small ($x, y \ll 1$) and the cubic terms are negligible, giving $V \approx \frac{1}{2}(x^2 + y^2)$, which is a circle in the xy-plane. Depending on the value of V, the circles are larger or smaller. For larger values of V, but requiring that $V < 1/6$, the equipotential curves have the shapes indicated in Figure 21.9. At $V = 1/6$ the equipotential is the bounding triangle.

Obtaining the surface of section for this Hamiltonian is interesting and instructive. The method is to choose some value for the energy, then integrate the equations of motion (21.1). Whenever $x = 0$, the values of y and $\dot{y}$ are plotted as a point on the $\dot{y}y$-plane.

[3] M. Henon and C. Heiles, "The Applicability of the Third Integral of Motion: Some Numerical Experiments," *Astronomical Journal*, **69**, 73–79 (1964).

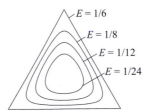

Figure 21.9 Equipotentials for the Henon–Heiles potential.

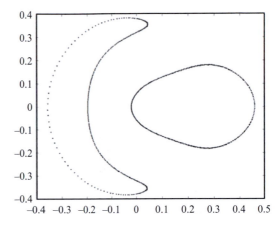

Figure 21.10 Surface of section for the Henon–Heiles potential for initial conditions $y = 0.02$, $\dot{y} = 0.08$, and $E = 1/12$.

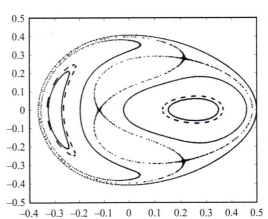

Figure 21.11 Surface of section for the Henon–Heiles potential for $E = 1/12$ and initial conditions $(y, \dot{y}) = (0.02, -0.08)$, $(0.3, 0.246)$, $(0.499, 0)$, $(0.366, 0)$, and $(0.2, 0.05)$.

For example, assume the energy is $1/12$ and the initial conditions are $x = 0, y = 0.02$, and $\dot{y} = 0.08$. The initial value of $\dot{x}$ is determined from the total energy

$$E = \frac{1}{2}\dot{x}^2 + \frac{1}{2}\dot{y}^2 + \frac{1}{2}(x^2 + y^2) + x^2 y - \frac{1}{3}y^3,$$

and is $\dot{x} = 0.4$. New values for x and y can be generated from the equations of motion. Each time the value $x = 0$ appears, the values of y and $\dot{y}$ are plotted. This gives the plot shown in Figure 21.10. If you were to watch the figure being generated, at first you would see a seemingly random set of points being plotted here and there on the $y\dot{y}$-plane. However, after a large number of points have been plotted, you would see the pattern beginning to form.

Using a variety of different initial conditions yields a more complete surface of section, as shown in Figure 21.11.

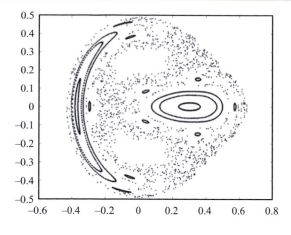

Figure 21.12 The surface of section obtained with the same initial conditions as in Figure 20.11 but for a total energy $E = 1/8$.

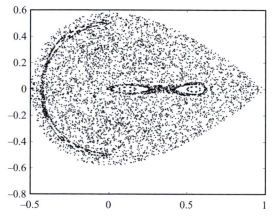

Figure 21.13 The surface of section for $E = 1/6$.

Each of the rather complicated curves represents an orbit that repeats over and over in a regular manner. The four regions inside the looping curve that crosses itself three times contain regular orbits, that is, nested orbits whose circumference depends on the initial conditions. In the limit they shrink to four points called the "elliptic fixed points."

A different value of the energy (but using the same initial conditions as in Figure 21.11) leads to the significantly different surface of section shown in Figure 21.12.

You can see this still has the four regions with stable orbits but outside of them there is a region where the trajectories crossing the $x = 0$ plane show no regularity at all. The curves representing orbits that repeat over and over again now occupy a much smaller region of the plot and the rest of the plot is taken up by points that show no apparent pattern. This region of the plot represents chaotic motion, and is called a strange attractor. Figure 21.13 shows the surface of section obtained for an energy of $E = 1/6$. Now nearly all motion is chaotic, although there are a few small regions with regular orbits.

21.7 Summary

In this very brief introduction to classical chaos, I have attempted to give you some of the vocabulary used in the theory and to give you a general idea of the basic ideas of chaos. Note that nonlinear equations are sometimes very sensitive to initial conditions, in which case they will lead to chaotic motion. Chaos is characterized by the generation of random points on a phase portrait. The effect of initial conditions is quantified by the Liapunov exponent which describes the divergence of two trajectories that start out near each other. As an example, we considered in some detail the Henon–Heiles Hamiltonian. The figures illustrating chaotic motion in the Henon–Heiles potential are instructive and well worth careful study.

21.8 Problem

Problem 21.1 Write the Henon–Heiles Hamiltonian in polar coordinates.

Computational Projects

Computational Project 21.1 Assuming $E = 1/6$, plot the Henon–Heiles potential to show that it is an equilateral triangle. (Hint: Ignore the complex solutions for the value of r vs. θ.)

Computational Project 21.2 Generate a plot like that of Figure 21.4 for the van der Pol equation for $F = 0$ and small ϵ. Explore the consequences of changing ϵ.

Computational Project 21.3 The van der Pol equation can be written in the somewhat simpler form

$$\frac{d^2x}{dt^2} - \mu(1 - x^2)\frac{dx}{dt} + x = 0.$$

Plot $\dot{x}$ vs. x for $\mu < 0, \mu = 0,$ and $\mu > 0$.

Computational Project 21.4 Write computer programs to generate Figures 21.11, 21.12, and 21.13.

Computational Project 21.5 Consider the following equation for a nonlinear, damped, driven pendulum:

$$\frac{d^2\theta}{dt^2} = -\frac{g}{l}\sin\theta - \alpha\frac{d\theta}{dt} + F_0\sin(\beta t).$$

Plot $\theta = \theta(t)$ and $\omega = \frac{d\theta}{dt}$ for $F_0 = 0, 0.4, 1.1$. Assume $\alpha = 0.5$, $g/l = 1$, and $\beta = 0.6$. Let the initial conditions be $\theta(t = 0) = 0.3$ rad and $\omega(t = 0) = 0$ rad/s.

Computational Project 21.6 Consider two identical damped, forced pendulums. Both have the same equation of motion

$$\frac{d^2\theta}{dt^2} = -\frac{g}{l}\sin\theta - \alpha\frac{d\theta}{dt} + F_0\sin(\beta t)$$

but differ slightly in the initial value of θ. You are asked to show that if the F_0 is small (say 0.1) the two pendulums stay in phase, but for larger F_0 (say 1.0), the angular distance between the pendulums grows larger and larger. You may assume $\alpha = 0.5$, $g/l = 1$, $\beta = 0.6$, and $\omega(t = 0) = 0$. The initial difference in displacement $\Delta\theta = \theta_2 - \theta_1$ is small, say 0.001. Plot $\Delta\theta = \Delta\theta(t)$ for the two pendulums on a semilog graph. For the case of diverging $\Delta\theta$, estimate the Liapunov exponent.

Computational Project 21.7 The "logistic map" is a plot of x_n vs. n assuming that

$$x_{n+1} = \mu x_n(1 - x_n) \qquad 0 \le x \le 1.$$

This equation can be considered a model for the population growth of animals. (a) Plot the logistic map for $0 < n < 100$ for $\mu = 2, 3,$ and 4. Connect the dots with straight lines. You may assume $x_0 = 0.6$. (b) If the value of $x_{n+1} = x_n$ so that the value of x does not change, the system has reached a "limiting value." Show that for $\mu = 2.0$ a limiting value is reached. Show that for $\mu = 3.0$ the system oscillates between two limiting values. Show that for $\mu = 4.0$ there is no limiting value (the system is chaotic).

Computational Project 21.8 The Feigenbaum plot is a plot of the limiting value of the logistic equation as a function of μ. (See previous problem.) Let μ range from 1 to 4 and plot x_n vs. μ for $100 < n < 200$.

Computational Project 21.9 The Lorenz equations are

$$\frac{dx}{dt} = \sigma(y - x),$$

$$\frac{dy}{dt} = -xz + rx - y,$$

$$\frac{dz}{dt} = xy - bz.$$

(a) Plot z vs. t for $r = 25, \sigma = 10$, and $b = 8/3$ and initial conditions $x = 1$, and $y = z = 0$. Run the system to a (dimensionless) time $t = 50$. (b) Plot z vs. x. (Note that the time step must be small.)

Computational Project 21.10 The predator–prey system can be modeled by using the equations

$$\dot{x} = 0.6x - 0.02xy,$$

$$\dot{y} = -2y + 0.02xy.$$

Here, x represents the number of prey (say, rabbits) and y represents the number of predators (say, foxes). Plot x and y as functions of time on the same plot and interpret in terms of food supply for the predators and in terms of survival probability for the prey. You may assume that initially the number of prey is 100 and the number of predators is 5.

Computational Project 21.11 In general the predator–prey relation is

$$\dot{x} = ax - bxy,$$

$$\dot{y} = -cy + dxy,$$

where a, b, c, d are positive constants. Assume $a = 2$, $b = 0.2, c = 5$, and $d = 0.2$. Use initial conditions $x = 10, 30, 80$ and $y = 5, 15, 15$. Obtain the configuration plots x vs. y and interpret in terms of the population of predators and prey.

Appendix A
A Formulas and Constants

Universal Constants

Gravitational constant $= G = 6.67 \times 10^{-11}$ Nm^2/kg^2

Speed of light $= c = 3 \times 10^8$ m/s

Astronomical Constants

Mass of Earth	5.97×10^{24} kg
Mass of Sun	1.99×10^{30} kg
Radius of Earth	6.37×10^6 m
Radius of Earth's orbit	1.5×10^{11} m

Definition of Hyperbolic Functions

$\sinh z = \frac{e^z - e^{-z}}{2}$

$\cosh z = \frac{e^z + e^{-z}}{2}$

$\tanh z = \frac{\sinh z}{\cosh z}$

Integrals

Expressions involving terms such as $a^2 \pm x^2$

$$\int \frac{dx}{a^2 + x^2} = \frac{1}{a} \arctan \frac{x}{a}$$

$$\int \frac{dx}{a^2 - x^2} = \frac{1}{a} \tanh^{-1} \frac{x}{a}$$

$$\int \frac{dx}{\sqrt{a^2 - x^2}} = \arcsin \frac{x}{a}$$

$$\int \frac{dx}{\sqrt{x^2 - a^2}} = \ln\left(x + \sqrt{x^2 - a^2}\right)$$

$$\int \frac{dx}{a + bx + cx^2} = \frac{2}{\sqrt{4ac - b^2}} \tan^{-1} \frac{2cx + b}{\sqrt{4ac - b^2}}$$

$$\int \frac{dx}{\sqrt{a + bx + cx^2}} = \begin{cases} \frac{1}{\sqrt{c}} \sinh^{-1}\left(\frac{2cx+b}{\sqrt{4ac-b^2}}\right) & \text{if } c > 0 \\[2ex] \frac{1}{\sqrt{-c}} \sin^{-1}\left(\frac{-2cx-b}{\sqrt{b^2-4ac}}\right) & \text{if } c < 0 \end{cases}$$

Expressions involving trigonometric quantities

$$\int \tan x \, dx = -\ln \cos x$$

$$\int \sec x \, dx = \ln(\sec x + \tan x)$$

$$\int \sin^2 x \, dx = \frac{x}{2} - \frac{1}{4} \sin 2x$$

$$\int \cos^2 x \, dx = \frac{x}{2} + \frac{1}{4} \sin 2x$$

Expressions involving hyperbolic functions

$$\int \sinh x \, dx = \cosh x, \quad \int \cosh x \, dx = \sinh x, \quad \int \tanh x \, dx = \ln \cosh x$$

Expression involving logarithms

$$\int \ln x \, dx = x \ln x - x$$

Error function

$$\mathrm{erf}(x) = \frac{2}{\sqrt{\pi}} \int_0^x e^{-t^2} dt$$

Vector Relations

Divergence theorem (Gauss) $\int_V \nabla \cdot \mathbf{A} \, d\tau = \oint_S \mathbf{A} \cdot \mathbf{da}$

Curl theorem (Stokes) $\int_S (\nabla \times \mathbf{A}) \cdot \mathbf{da} = \oint_c \mathbf{A} \cdot \mathbf{dl}$

Triple product $\mathbf{A} \cdot (\mathbf{B} \times \mathbf{C}) = (\mathbf{A} \times \mathbf{B}) \cdot \mathbf{C} = \mathbf{B} \cdot (\mathbf{C} \times \mathbf{A})$

BAC – CAB rule $\mathbf{A} \times (\mathbf{B} \times \mathbf{C}) = \mathbf{B}(\mathbf{A} \cdot \mathbf{C}) - \mathbf{C}(\mathbf{A} \cdot \mathbf{B})$

Expansions

Binomial expansion $(a + x)^n = 1 + nx + \frac{n(n-1)}{2!} x^2 + \frac{n(n-1)(n-2)}{3!} x^3 + \cdots$

Taylor's series $f(x) = f(a) + (x - a) f'(a) + \frac{1}{2!}(x - a)^2 f''(a) + \frac{1}{3!} f'''(a) + \cdots$

Exponential $e^x = 1 + x + \frac{x^2}{2!} + \frac{x^3}{3!} + \cdots$

Trigonometric $\sin x = x - \frac{x^3}{3!} + \frac{x^5}{5!} - \cdots$

$\cos x = 1 - \frac{x^2}{2!} + \frac{x^4}{4!} - \cdots$

Fourier $\frac{a_0}{2} + \sum_{n=1}^{\infty} \left(a_n \cos \frac{n\pi x}{L} + b_n \sin \frac{n\pi x}{L} \right)$, where

$a_n = \frac{1}{L} \int_{-L}^{L} f(x) \cos \frac{n\pi x}{l} \, dx$

$b_n = \frac{1}{L} \int_{-L}^{L} f(x) \sin \frac{n\pi x}{l} \, dx$

Appendix B
B Answers to Selected Problems

Chapter 1

1.1 35 mph

1.3 10.3 km

1.5 (a) $\frac{v_1+v_2}{2}$, (b) $\frac{2v_1 v_2}{v_1+v_2}$

1.7 (a) 0.69 miles

1.9 (c) $v = v_0[(\sin\theta - \mu\cos\theta)/(\sin\theta + \mu\cos\theta)]^{1/2}$

1.11 29.6 cm

1.13 2.58×10^{16} Nm

1.15 176 watts/person

1.17 (c) -1.10 rad/s^2

1.19 (b) Stable

1.21 (b) 189 s

Chapter 2

2.1 6 m/s

2.5 23.66 m

2.7 23.65° and 66.35°

2.10 77.08°

2.11 5609 m

2.13 15.7 m

2.15 (a) $s = \frac{v_0^2 \cos^2\theta}{2g}\left[\frac{\sin\theta}{\cos^2\theta} + \log(\sec\theta + \tan\theta)\right]$

2.17 $v = [(gx_t^2/2y_t) + 2gy_t]^{1/2}$ at $\theta = \arctan(2y_t/x_t)$

2.19 $R = 455$ km

2.21 (a) $\tau = 2\pi\sqrt{r/a}$

2.31 $\mathbf{v} = (\cos t)\hat{\mathbf{r}} + (1 + \sin t)e^{-t}\hat{\boldsymbol{\theta}}$

2.33 $\theta = \ln(2 + \sin t)$

2.37 1.56 m/s

2.39 $\nabla \cdot \mathbf{F} = 2 + z\cos\phi/\rho + \frac{1}{2}\sqrt{\rho/z}$

2.43 $\rho = r\sin\theta, \phi = \phi, z = r\cos\theta$

2.45 (b) $v = [b^2 + (akt - bkt^2)^2]^{1/2}$

Chapter 3

3.1 $\mathbf{a} = (3\hat{\mathbf{i}} + \hat{\mathbf{j}})$ m/s^2, $|\mathbf{a}| = \sqrt{10}$ m/s^2

3.3 0.51

3.5 3136 N

3.7 (a) $x(t) = (A/m\beta^2)(\cosh \beta t - 1)$

3.9 (b) $F_0/b + (v_0 - F_0/b)e^{-bt/m}$

3.11 (b) $a = g(m_1 - m_2)/(m_1 + m_2 + I/R^2)$

3.19 $x = (5/6)L$

3.21 (a) $x(t) = \frac{Fm_0}{\kappa^2}\left[-\frac{\kappa}{m_0}t - \ln\left(1 - \frac{\kappa}{m_0}t\right)\right]$

3.23 (a) $x(t) = (m/K_x)\ln\left(\frac{v_0 K_x}{m}t + 1\right)$

3.25 9.82 s

3.27 (a) 241 kg

3.29 358 m/s

3.33 $x(t = 4) = 77.33$ m

3.35 $v(t) = -(1/b)\ln(e^{-bv_0} + (Ab/m)t$

3.39 $D = 5.9$ kg-h/km-s

3.41 $t = (1/k)\ln(1 + kv_0/g)$

3.43 $x = 2777$ m

3.45 747 m

3.47 $x = x_0 \cosh\sqrt{b/m}t$

3.49 $\omega = \sqrt{1/m}\sqrt{(k_1 k_2)/(k_1 + k_2)}$ rad/s

3.51 $v^2 = v_0^2 + 2GM_S/d$

3.53 Amplitude = 0.47 m

3.55 (b) $v_0 = -\sqrt{\frac{2Gm}{z} - K}$

3.57 $\simeq 42$ minutes

Chapter 4

4.1 $L = \frac{1}{2}m\dot{r}^2 + \frac{1}{2}mr^2\dot{\theta}^2 + \frac{K}{r}$

4.3 $L = \frac{1}{2}m(R^2\dot{\theta}^2 + R^2\omega^2\sin^2\theta) - mgR\cos\theta$

4.5 $L = \frac{1}{2}m\dot{z}^2 - \frac{3}{2}kz^2$

4.7 $\frac{d}{dt}\left(\frac{m_1 m_2}{m_1 + m_2}r^2\dot{\theta}\right) = 0$

4.9 $(M + m)\dot{x} + ml\dot{\theta}\cos\theta = $ constant

4.11 $\ddot{s} = \frac{g\sin\alpha}{1 - \frac{m}{m+M}\cos^2\alpha}, \ddot{X} = -\frac{m}{m+M}\left(\frac{g\sin\alpha}{1 - \frac{m}{m+M}\cos^2\alpha}\right)\cos\alpha$

4.13 $\ddot{s} = (2/3)g$

4.15 (c) $p_\theta = mR^2\dot{\theta}$

4.17 (b) $x_1 = \frac{1}{2}\left(\frac{M-m}{M+m}g - \frac{F}{Mm}\right)t^2$

4.19 $\ddot{r} = r\dot{\theta}^2 + g\cos\theta - \frac{k}{M}(r - r_0), \ddot{\theta} = -\frac{g}{r}\sin\theta - \frac{2\dot{r}\dot{\theta}}{r}$

4.21 $\omega = \sqrt{\frac{g(a-R)}{(a-R)^2 + \frac{2}{5}a^2}}$

4.23 (a) $\Phi = \sqrt{1 + a^2\theta'^2}$

4.25 $y = \pm 2c\sqrt{x - c^2} + d$

4.27 $\Phi = y\sqrt{1 + y'^2}$

4.29 $\hat{H} = -\frac{\hbar^2}{2m}\left[\frac{\partial^2}{\partial x^2} + \frac{\partial^2}{\partial y^2} + \frac{\partial^2}{\partial z^2}\right] = -\frac{\hbar^2}{2m}\nabla^2$

4.33 $H = \frac{p_\theta^2}{2ma^2} + \frac{p_\phi^2}{2ma^2\sin^2\theta} + mga\cos\theta$

Chapter 5

5.1 49.5 J

5.3 $2kR^2$

5.5 (b) $d\mathbf{s} = \hat{\mathbf{e}}_1 \frac{1}{h_{11}} dq_1 + \hat{\mathbf{e}}_2 \frac{1}{h_{22}} dq_2 = \hat{\mathbf{r}} dr + \hat{\boldsymbol{\Theta}} r d\theta$

5.7 $d\mathbf{s} = a(\sinh^2 u \cos^2 v + \cosh^2 u \sin^2 v)^{1/2} du\hat{\mathbf{e}}_u + a(\cosh^2 u \sin^2 v + \sinh^2 u \cos^2 v)^{1/2} dv\hat{\mathbf{e}}_v + dz\hat{\mathbf{k}}$

5.9

$$\nabla = \hat{\mathbf{e}}_\eta \frac{1}{a\left[\cosh^2 \eta \sin^2 \theta + \sinh^2 \eta \cos^2 \theta\right]^{1/2}} \frac{\partial}{\partial \eta}$$

$$+ \hat{\mathbf{e}}_\theta \frac{1}{a\left[\sinh^2 \eta \cos^2 \theta + \cosh^2 \eta \sin^2 \theta\right]^{1/2}} \frac{\partial}{\partial \theta}$$

$$+ \hat{\mathbf{e}}_\phi \frac{1}{a\left[2\sinh^2 \eta \sin^2 \theta\right]^{1/2}} \frac{\partial}{\partial \phi}.$$

5.13 $\nabla f = \hat{\boldsymbol{\rho}} \frac{\partial f}{\partial \rho} + \hat{\boldsymbol{\phi}} \frac{1}{\rho} \frac{\partial f}{\partial \phi} + \hat{\mathbf{k}} \frac{\partial f}{\partial z} = -\hat{\boldsymbol{\rho}}$

5.15 $|\nabla T| = \sqrt{\nabla T \cdot \nabla T} = 2[1 + 1 + 1]^{1/2} = 2\sqrt{3} = 3.46$

5.17 $v_{\max} = \left[\frac{2}{m} \left\{ \frac{q(Q_1 + Q_2)}{4\pi\epsilon_0} \left(\frac{1}{a} - \frac{1}{\sqrt{b^2 + a^2}} \right) \right\} \right]^{\frac{1}{2}}$

5.19 609 N

5.21 $x = \sqrt{\frac{20}{7} R(d - r)}$

5.23 $x = x_0 + \left[\frac{b}{2m} t^2 + \left(\sqrt{\frac{2}{m}} \sqrt{E + bx_0} \right) t \right]$

5.27 (d) $v = \sqrt{2A/m}$

5.29 (a) $-(2V_0/\delta)(1 - e^{-(s-s_0)/\delta})e^{-(s-s_0)/\delta}$

5.31 $V(x, y, z) = -K[x^2 + y^2 + xy - \text{constant}]$

Chapter 6

6.1 (b) 3.45 m/s

6.3 $v = \frac{m+M}{m}\sqrt{2gh}$

6.7 (a) $v = v_0 m_0 \left[m_0^{4/3} + \frac{4kv_0 m_0}{3K^{2/3}} t \right]^{-3/4}$ $(k = \rho_d \pi$ and $K = m/r^3)$

6.9 $\theta' = \cos^{-1}\left[1 - \left(\frac{1}{2}\right)\left(\frac{m_M}{m_M + m_W}\right)^2 \right]$

6.11 Increased power load $= kmV^2$

6.15 $v = -gt + (2mg/k)\ln\left[(2m)/(2m - kt)\right], m = $ initial mass

6.17 16.2 kg/s

6.19 44 min

6.21 8.70h

6.23 $V_2' = \sqrt{12}$ m/s

6.25 25 km/s

Chapter 7

7.5 $l = -(1/2)v_{0x}F_e t^2 \hat{k}$

7.7 $d = \frac{D\mu R}{h}$, where $h = (b^2 - R^2)^{\frac{1}{2}}$

7.9 1.51×10^4 rad/s

7.13 $I = \frac{1}{2}M(R_2^2 + R_1^2)$

7.15 $I = \frac{5}{4}MR^2$

7.17 $t = \left(1 + \sqrt{2}\right)\frac{\omega_0 I \sqrt{\mu^2 + 1}}{a\mu mg}$

7.19 $\omega \cos\frac{RMg}{I\omega}t\hat{i} + \omega \sin\frac{RMg}{I\omega}t\hat{j} + \frac{RMg}{I\omega}\hat{k}$

Chapter 8

8.1 Energy

8.3 $N = \frac{d\mathbf{L}}{dt}$ is unchanged by reflection.

8.7 Force is directed out of page.

Chapter 9

9.1 $G(0.97\hat{i} - 0.44\hat{j})$

9.3 59.8 kg

9.5 1.03×10^6 m/s

9.7 $T = \sqrt{3\pi/G\rho}$

9.9 $P = Gm_a m_b/4\pi b^4$

9.11 $-\frac{2GM}{R^2 L}\left[L - \sqrt{Z^2 + R^2} - \sqrt{(Z-L)^2 + R^2}\right]$, where z = distance from base of cylinder

9.13 $\frac{2GM}{a^2}[z - \sqrt{a^2 + z^2}]$

9.17 $\Phi = -\frac{GM}{2}\left[\frac{3}{R} - \frac{r^2}{R^3}\right]$, for $r < R$

9.19 $\Phi = -\frac{GM}{b}\left[\frac{\pi}{2} - \tan^{-1}\frac{r}{b}\right]$

9.21 (a) $g = -\frac{GM}{R^3}r$

9.23 $F = \frac{4\pi}{3}R^3\rho Gm(1/z^2)\left[-1 + \frac{1}{8}\left(1 - \frac{R}{4z}\right)^{-2}\right]$

9.25 84 min

9.27 $W = Gmh\left[\frac{M}{R(R+h)} + 2\pi\sigma\right]$

9.29 $\Phi = -\frac{GM}{r} + \frac{1}{4}\frac{GMa^2}{r^2} + \mathcal{O}\frac{a^3}{r^3}$

Chapter 10

10.1 $E = -\frac{1}{n^2}\frac{e^4 m}{2\hbar(4\pi\epsilon_0)^2}$

10.3 (c) $T = \frac{m_1 m_2 \omega^2 b}{m_1 + m_2}$

10.5 $a = 9.59 \times 10^6$ m

10.9 $\Delta\tau = 3v\Delta v(2\pi GM)^{-2/3}\tau^{5/3}$

10.11 (d) $\ddot{r} = \frac{l^2}{m^2 r^3} - G\frac{m_1 + m_2}{r^2}$

10.13 (b) $t = \frac{3\pi}{8}\sqrt{\frac{3ma^5}{2K}}$

10.17 1300 km

10.23 $M_A = 1, M_B = 3$ solar masses

10.25 2.03×10^4 s ≈ 5.6 h

10.27 (c) 2.26×10^4 s

10.29 $r_f = 1.53 r_0$

10.31 (c) $f \propto 1/r^3$

Chapter 11

11.1 $k = b^2/4m$

11.3 $A = \dfrac{\sqrt{\omega_1^2 + \gamma^2}}{\omega_1}$

11.5 $b = 3.1 \times 10^3$ kg/s

11.9 $b = \dfrac{mh}{\delta} + \dfrac{k\delta}{h}$

11.11 $t \le \sqrt{0.6}/\sqrt{\dfrac{k}{m} - \dfrac{b}{2m}}$ for error < 10 percent

11.13 (d) 1.42 s

11.15 $\frac{1}{2} m A^2 e^{-2\gamma t} \omega_0^2$

11.23 $\omega = \sqrt{k \dfrac{m_1 + m_2}{m_1 m_2}}$

11.29 $x = A \cos t + B \sin t + 6 - 2 \cos 2t$

Chapter 12

12.3 $\approx 8.8 \times 10^{-3}$

12.5 (c) $\omega = \sqrt{\dfrac{Af^2}{l} \cos f t + g/l}$

12.7 (a) 2.85 s

12.9 $l_{eq} = 1.73r$

12.11 $T = 2\pi \sqrt{m/2k}$

12.13 $P = \dfrac{4k_0}{\sqrt{gh}} F (\sin(\theta_{max}/2))$

12.15 $\dot{\phi}_{max} = 4.08$ rad/s

12.17 $T = \dfrac{3E}{l} - 3mg \cos \theta$

12.19 $H = \dfrac{p_\theta^2}{2ml^2} + \dfrac{p_\phi^2}{2ml^2 \sin^2 \theta} + mgl \cos \theta$

Chapter 13

13.1 $y(x,0) = \dfrac{8b}{\pi^2} \left(\sin \dfrac{\pi x}{L} - \dfrac{1}{3^2} \sin \dfrac{3\pi x}{L} + \dfrac{1}{5^2} \sin \dfrac{5\pi x}{L} - \cdots \right)$

13.3 (b) $y(x,0) = \frac{1}{6} a L^2 + \sum_{n=0}^{\infty} \left(-\dfrac{aL^2}{\pi^2 n^2} (-1)^n \right) \cos \dfrac{2\pi n x}{L}$

13.5 $y(x,t) = \sum_{m=odd}^{\infty} \dfrac{0.16}{m\pi^2} \cos \sqrt{1000} \dfrac{m\pi}{L} t \sin \dfrac{m\pi x}{L}$

13.9 (b) $\rho \ddot{\phi}_m + b \dot{\phi}_m + [Fm^2 \pi^2 / L^2] \phi_m = 0$

13.13 $y(x,t) = \sum_{n=1,3,5,\ldots} \dfrac{n\pi}{2L} \left(C_n \cos \dfrac{cn\pi}{2L} t + D_n \sin \dfrac{cn\pi}{2L} t \right) \cos \dfrac{n\pi}{2L} x$

13.19 (b) $m\ddot{q}_i = k(2q_i - q_{i+1} - q_{i-1})$ $\qquad i = 1, 2, \ldots, n$

13.21 $\ddot{Z}_n + \dfrac{F\pi^2}{\rho l^2} n^2 Z_n = 0, \quad n = 1, 2, \ldots$

Chapter 14

14.1 The potential matrix is

$$k \begin{pmatrix} 2 & -1 \\ -1 & 2 \end{pmatrix}$$

14.3 $2.62k/m, 0.38k/m$

14.5 frequencies: $k/m, 2k/m, 3k/m$

14.7 (b) $\omega_1^2 = k/m$, $\omega_2^2 = 3k/m$, (c) $\mathcal{A} = \frac{1}{\sqrt{2m}} \begin{pmatrix} 1 & 1 \\ 1 & -1 \end{pmatrix}$

14.9 $\eta_1(t) = C_1 \cos(\omega_1 t + \phi_1) + C_2 \cos(\omega_2 t + \phi_2) + K_1 \sin \omega t$,
$\eta_2(t) = C_1 \cos(\omega_1 t + \phi_1) - C_2 \cos(\omega_2 t + \phi_2) + K_2 \sin \omega t$

14.11 $\mathcal{V} = \begin{pmatrix} k & 0 \\ 0 & k \end{pmatrix}$, $\mathcal{M} = \begin{pmatrix} m+M & -M \\ -M & m+M \end{pmatrix}$

14.13 $\omega^2 = \frac{g}{l} + 4\frac{k}{m} \sin^2 \frac{ka}{2}$

Chapter 15

15.3 $\mathbf{a} = \mathbf{a}''' + 2(\boldsymbol{\Omega} + \boldsymbol{\omega}) \times \mathbf{v}''' + 2\boldsymbol{\omega} \times (\boldsymbol{\Omega} \times \mathbf{r}''') + \boldsymbol{\omega} \times (\boldsymbol{\omega} \times \mathbf{r}_E)$

15.5 $a_{\text{cent}} = (V^2/R)\hat{\mathbf{i}} + (V^2/a)\hat{\mathbf{k}}$

15.7 0.79 cm

15.9 $\approx$4.1 km

15.11 21.3 m/s

15.13 32 m

15.15 3.3×10^{-3} rad

15.17 (a) $2\Omega v \hat{\boldsymbol{\phi}}$

15.19 28.8 m

15.21 (a) $\omega_0 = \sqrt{k/m}$

Chapter 16

16.5 (a) $\mathcal{A} = \begin{pmatrix} \cos\phi & \sin\phi & 0 \\ -\sin\phi\cos\theta & \cos\phi\cos\theta & \sin\theta \\ \sin\phi\sin\theta & -\cos\phi\sin\theta & \cos\theta \end{pmatrix}$

16.8 $(\phi, \theta, \psi) = (\pi/2, 38°29', -\pi/2)$

16.9 Magnitude of 53.27°. Axis in equatorial plane directed at an angle of $119°47' - 90°$ west of Greenwich.

16.11 angle = 98.42°

16.15 (c) $\mathcal{A} = \begin{pmatrix} 1 - 2l_1^2 & -2l_1 l_2 & -2l_1 l_3 \\ -2l_1 l_2 & 1 - 2l_2^2 & -2l_2 l_3 \\ -2l_1 l_3 & -2l_2 l_3 & 1 - 2l_3^2 \end{pmatrix}$, where $\hat{\mathbf{n}} = l_1\hat{\mathbf{x}} + l_2\hat{\mathbf{y}} + l_3\hat{\mathbf{z}}$

16.17
$$\mathbf{x}^{(1)} = \begin{pmatrix} 0 \\ 0 \\ 1 \end{pmatrix}, \mathbf{x}^{(2)} = \begin{pmatrix} i \\ 1 \\ 0 \end{pmatrix}, \mathbf{x}^{(3)} = \begin{pmatrix} 1 \\ i \\ 0 \end{pmatrix}$$

16.21 (a) $d\mathbf{x} = \beta d\Omega \hat{\mathbf{i}} - \alpha d\Omega \hat{\mathbf{j}}$

Chapter 17

17.3 $5Ma^2, 5Ma^2, 8Ma^2$

17.5 $\begin{pmatrix} 2.77 & 0 & 0 \\ 0 & 2.77 & 0 \\ 0 & 0 & 2.94 \end{pmatrix}$

17.15 $\mathbf{L} = I_1\dot{\phi}\sin\theta\sin\psi\hat{\mathbf{e}}_1 + I_1\dot{\phi}\sin\theta\cos\psi\hat{\mathbf{e}}_2 + I_3(\dot{\psi} + \dot{\phi}\cos\theta)\hat{\mathbf{e}}_3$

17.17 (a) $\cos\theta_2 = \frac{1}{2\alpha}[1 - (1 - 4\alpha\cos\theta_1 + 4\alpha^2)^{1/2}] \simeq \cos\theta_1 + \alpha\sin^2\theta_1$

17.19 2.18 rad/s

Chapter 18

18.1 $T_1 = \frac{mg}{\sin\theta + \cos\theta\tan\phi}$

18.3 $\phi = 71.8°$

18.7 Equilibrant $= 1.44\,Mg$ at end of rod.

18.9 A force of 8.66 N towards right at 0.29 above left end.

18.11 The couple is $-F\hat{\mathbf{k}}$ at the origin and $+F\hat{\mathbf{k}}$ acting at $a\hat{\mathbf{i}}$.

18.15 $2\pi^2 Ra^2$

18.17 $h = (3/16)l^2/r$

18.21 $\mu = \frac{m_2\sin\theta_2 - m_1\sin\theta_1}{m_2\cos\theta_2 + m_1\cos\theta_1}$

Chapter 19

19.1 8916 lb = 39 660 N

19.3 (b) 0.5 kg

19.5 $\kappa_T = 1/p$

19.9 $h_2 = 4h_1$

19.19 911 Pa

Chapter 20

20.1 (a) $x^* = 2.175$ m, $t^* = 1.25$ ns

20.5 (b) $x^* = 8.66 \times 10^8$ m, $t^* = 4.04$ s

20.9 $v = \sqrt{3}c/2$

20.11 $v = 0.8c$

20.15 123 m

20.19 3.3×10^{-13} J

20.23 $E = 2.31m_0c^2$

20.25 $v_{1f} = 0.471c, v_{2f} = 0.923c$

20.27 4×10^{-6} s

Chapter 21

21.1 $H = \frac{1}{2}\left(m\dot{r}^2 + mr^2\dot{\theta}^2\right) + \frac{k}{2}(r^2) + \lambda(r^3\cos^2\theta\sin\theta - \frac{1}{3}\sin^3\theta)$

References

Intermediate Level Texts

[1] R. A. Becker, *Introduction to Theoretical Mechanics*, McGraw Hill, New York, 1954.
[2] R. Kolenkow and D. Kleppner, *An Introduction to Mechanics*, Cambridge University Press, Cambridge, 2013.
[3] D. Morin, *Introduction to Classical Mechanics*, 2nd ed., Cambridge University Press, Cambridge, 2013.
[4] K. R. Symon, *Mechanics*, 3rd ed., Addison Wesley, Reading, MA, 1961.
[5] J. R. Taylor, *Classical Mechanics*, University Science Books, Melville, NY, 2005.
[6] S. T. Thornton and J. B. Marion, *Classical Dynamics*, 5th ed., Brooks Cole, Belmont, CA, 2003.
[7] L. N. Hand and J. D. Finch, *Analytical Mechanics*, Cambridge University Press, Cambridge, 1998.

Graduate Level Texts

[8] A. L. Fetter and J. D. Walecka, *Theoretical Mechanics for Particles and Continua*, McGraw Hill, New York, 1980.
[9] H. Goldstein, C. P. Poole, and J. Safko, *Classical Mechanics*, 3rd ed., Pearson, San Francisco, 2011.
[10] P. Hamill, *A Student's Guide to Lagrangians and Hamiltonians*, Cambridge University Press, Cambridge, 2013.
[11] J. Jose and E. Saletan, *Classical Dynamics*, Cambridge University Press, Cambridge, 1998.
[12] L. D. Landau and E. M. Lifshitz, *Mechanics*, 3rd ed., Pergamon Press, Oxford, 1976.

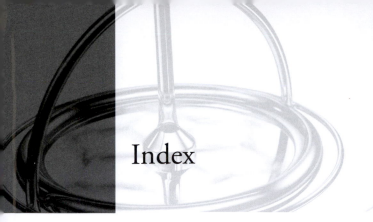

Index